Surface Modification and Mechanisms

Surface Modification and Mechanisms

Friction, Stress, and Reaction Engineering

edited by

George E. Totten

G. E. Totten & Associates, LLC
Seattle, Washington, U.S.A.

Hong Liang

University of Alaska
Fairbanks, Alaska, U.S.A.

CRC Press
Taylor & Francis Group
Boca Raton London New York

CRC Press is an imprint of the
Taylor & Francis Group, an **informa** business

First published 2004 by Marcel Dekker, Inc.

Published 2020 by CRC Press
Taylor & Francis Group
6000 Broken Sound Parkway NW, Suite 300
Boca Raton, FL 33487-2742

First issued in paperback 2020

ISBN 13: 978-0-367-57836-7 (pbk)
ISBN 13: 978-0-8247-4872-2 (hbk)

Visit the Taylor & Francis Web site at
http://www.taylorandfrancis.com

and the CRC Press Web site at
http://www.crcpress.com

Cover: An optical interference image of a thin old film under high pressure rolling contact. Courtesy of L. D. Wedeven, Wedeven Associates, Inc., Edgmont, Pennsylvania.

Library of Congress Cataloging-in-Publication Data
A catalog record for this book is available from the Library of Congress.

Preface

There are many texts and handbooks available describing tribological processes, effects of additives on lubrication, tribochemistry, surface engineering, and heat treating methodologies involved in surface modification. However, few of these texts provide a thorough integration of surface modification reactions and processes to achieve a tribological result and none provides a physical tribochemical (mechanistic) understanding of surface structural changes that occur under various circumstances. This book was written to address these deficiencies and provide a critically important text for this vitally important technology field.

The book was written in four parts:

Part One   Residual Stresses
Part Two   Reaction Processes and Mechanisms
Part Three   Surface Modification by Heat Treatment and Plasma-Based Methods
Part Four   Modeling, Simulation, and Design

An often overlooked aspect of tribological processes is the role of residual stresses. Part One discusses the formation of temperature fields created as a result of friction during heat treatment, welding and cutting, and their role in residual stress formation, with an overview of residual stresses arising during thin-film formation. These are important considerations involved in surface design for tribological applications.

Part Two provides an extensive overview of the role of surface structure and mechanisms as a result of formation of metallic oxides due to wear processes, the role of lubricious oxides formed on various ceramic substrates, structure of plastics, tribochemical surface mechanisms, tribopolymerization, base oil and additive surface interaction, and the role of surface hydrolysis on these processes. The reaction mechanisms of extreme pressure additives with surfaces during wear and mechanisms involved in boundary lubrication are also covered. Chapters 12 and 13 provide an extensive discussion of electrochemical mechanisms and the role of electrochemistry in wear processes.

Part Three continues with the effect of surface modification technologies and related mechanisms. These processes include conventional heat treating such as hardening and tempering, carburizing, nitriding, shot peening, and others. In addition, a comprehensive overview of physical vapor deposition (PVD), chemical vapor deposition (CVD), and ion implantation in addition to various related hybrid processes is provided. Laser impingement and nanometer scale surface modification technologies are also discussed.

Part Four pulls all of these approaches together and provides general guidelines for designing for wear life and frictional performance and a description of the use of simulation methodologies to model interfacial friction on solid surfaces.

This book will be an invaluable resource for material scientists and engineers, designers, mechanical engineers, metallurgists, tribologists and lubrication engineers. It may also be used as a textbook for advanced undergraduate and graduate courses in surface engineering and tribology.

The preparation of this text was an enormous task and we are indebted to the various International experts for their contributions. Special thanks to the staff at Marcel Dekker, Inc. and Richard Johnson for their patience and invaluable assistance.

George E. Totten
Hong Liang

Contents

Contributors

O. O. Ajayi, Ph.D. Department of Tribology, Argonne National Laboratory, Argonne, Illinois, U.S.A.

Asgeir Bardal, Dr.Ing. Hydro Aluminium Technology Center Årdal, Årdal, Norway

Einar Bardal, Dr.Ing. (Emeritus), Department of Machine Design and Materials Technology, Norwegian University of Science and Technology, Trondheim, Norway

Ilia A. Buyanovsky, D.Sc. (Eng) Department of Tribology, Mechanical Engineering Research Institute, Russian Academy of Sciences, Moscow, Russia

Robert C. Cammarata, Ph.D. Department of Materials Science and Engineering, Johns Hopkins University, Baltimore, Maryland, U.S.A.

Lauralice Campos Franceschini Canale, Ph.D. University of São Paulo, San Carlos, Brazil

Paul K. Chu, Ph.D. Department of Physics and Materials Science, City University of Hong Kong, Kowloon, Hong Kong

Ovidio Richard Crnkovic, Ph.D. University of São Paulo, San Carlos, Brazil

Jeff Th. M. De Hosson, Dr.Phys. Department of Applied Physics, University of Groningen, Groningen, The Netherlands

J. Thomas Dickinson, Ph.D. Department of Physics, Washington State University, Pullman, Washington, U.S.A.

Franz Dieter Fischer, Dr. Sc. Montanuniversität Leoben and Austrian Academy of Sciences, Leoben, Austria

Michael J. Furey, Ph.D. Department of Mechanical Engineering, Virginia Polytechnic Institute and State University, Blacksburg, Virginia, U.S.A.

James E. Hammerberg,   Los Alamos National Laboratory, Los Alamos, New Mexico, U.S.A.

Jeffrey A. Hawk, Ph.D.   U.S. Department of Energy, Albany, Oregon, U.S.A.

Brad Lee Holian, Ph.D.   Theoretical Division, Los Alamos National Laboratory, Los Alamos, New Mexico, U.S.A.

Zinaida V. Ignatieva, Ph.D.   Department of Tribology, Mechanical Engineering Research Institute, Russian Academy of Sciences, Moscow, Russia.

Czeslaw Kajdas, Ph.D., D.Sc.   Warsaw University of Technology, Institute of Chemistry at Płock, Płock, Poland

Peter V. Kotvis, Ph.D.   Department of Research and Development, Benz Oil, Inc., Milwaukee, Wisconsin, U.S.A.

Steve C. Langford, Ph.D.   Department of Physics, Washington State University, Pullman, Washington, U.S.A.

Kenneth C. Ludema, Ph.D.   (Emeritus), Department of Mechanical Engineering, University of Michigan, Ann Arbor, Michigan, U.S.A.

Vašek Ocelík, Dr.Phys.   Department of Applied Physics, University of Groningen, Groningen, The Netherlands

Yutao Pei, Dr.Mat.Sc.Eng.   Department of Applied Physics, University of Groningen, Groningen, The Netherlands

Zygmunt Rymuza, Ph.D., D.Sc., MEng.   Department of Mechatronics, Institute of Micromechanics and Photonics, Warsaw University of Technology, Warsaw, Poland

Hon So, Ph.D.   Department of Mechanical Engineering, National Taiwan University, Taipei, Taiwan

Xiubo Tian, Ph.D.   Department of Physics and Materials Science, City University of Hong Kong, Kowloon, Hong Kong

Wilfred T. Tysoe, Ph.D.   Department of Chemistry, University of Wisconsin–Milwaukee, Milwaukee, Wisconsin, U.S.A.

Jože Vižintin, Ph.D.   Faculty of Mechanical Engineering, Centre for Tribology and Technical Diagnostics, University of Ljubljana, Ljubljana, Slovenia

Ewald A. Werner, Dr. Mont.habil.   Department of Materials Science and Mechanics of Materials, Technische Universität München, Munich, Germany

Mathias Woydt, Dr.-Ing.   Department of Component Tribology, Federal Institute for Materials Research and Testing (BAM), Berlin, Germany

Ruvim N. Zaslavsky, Ph.D.   Venchur-N Ltd., Moscow, Russia

Matgorzata E. Ziomek-Moroz, Ph.D.   U.S. Department of Energy, Albany, Oregon, U.S.A.

1

Temperature and Stress Fields Due to Contact with Friction, Surface Heat Treatment, Welding, and Cutting

Franz Dieter Fischer
Montanuniversität Leoben and Austrian Academy of Sciences, Leoben, Austria

Ewald A. Werner
Technische Universität, München, Munich, Germany

I. INTRODUCTION

The main goal of this chapter is to provide the reader with analytic expressions and diagrams for temperature fields due to a moving heat source in contact with the surface of a body. Technological applications of such problem solutions are manifold:

- Frictional contact realized by a pin moving across the surface of a body or by rolling/sliding of a wheel on a rail is a typical case. The temperature field depends on the boundary conditions (see, e.g., Ref. 1 for an insulated surface and Refs. 2 and 3 for a convective surface). The temperature field connected to fretting is reported in Ref. 4. The temperature field in a work roll during strip rolling falls into this category of problems (see, e.g., Refs. 5 and 6). Yet another application is hot spotting in the contact of two ideally parallel planes such as brakes and clutches. The input of heat into the rotating part may be inhomogeneous and frequently will concentrate to local spots because of asperities and/or thermo-elastic instability phenomena (see, e.g., Refs. 7 and 8).
- The welding literature offers many solutions for temperature fields associated with moving heat sources (see, e.g., the books by Radaj [9], chap. 2, and by Grong [10], chap. 1). Both books refer often to the classical solutions by Rosenthal obtained in the 1940s and the Russian literature by, e.g., Rykalin. Especially steady temperature distributions are reported assuming that the source approaches from minus infinity and after passing the point of interest goes to plus infinity. It turned out that the conduction of heat orthogonal to the surface of the body plays the dominant role. However, if thin plates are considered, only heat conduction in the plane of the plate is relevant, and special solutions for thin plates are collected in the two a.m. books. In addition, some specific papers on the cutting and welding of plates dealing also with a moving, mostly circular heat source leaving a cut or

a welding seam behind are cited (see, e.g., Refs. 11 and 12). With respect to cutting of single crystals of silicon, e.g., for wafers, the reader is referred to a recent paper [13].

- Surface hardening by electron or laser beam technology is one of the current methods to produce selectively hardened layers, e.g., on steel substrates (see the overviews by Krauss [14] and Brooks [15]). With respect to the thermal problem a heat source passes several times over the surface of a body giving rise to a temperature field to be controlled (see Ref. 16, the works by Ashby et al. [17–19], the works of Rödel et al. [20–24] and chaps. 5 and 6 of the book by Dowden [25]). Similar applications with respect to the temperature field developing are the deposition of a coating by thermal spraying [23], laser heat-forming processes as described in Ref. 26, or laser nitriding [27].

- Moving heat sources are of importance for turning of metals on a lathe (for details see Ref. 28, parts I–III, or Ref. 29). One must distinguish between the tool, the chip, and the work piece. During turning the shear plane heat source moves along the surface of the work piece. With respect to the chip the same heat source moves as an oblique band through the chip, while it acts as a stationary heat source in the tool. Of course, a proper partition of the total heat generated to the three distinct bodies must be performed. It is also reported in Ref. 28 that the modified solution for an oblique band agrees well with measured temperatures in the chip. There exists also a solution for an inclined heat source band allowing for convection at the surface (see Dawson and Malkin [29]).

- Recently, a moving rectangular and triangular heat source due to grinding was dealt with in Ref. 30. The finite element method was employed to calculate both the temperature field and the residual stresses in an elastic–plastic halfspace subjected to this type of heat input.

There are several reasons why an engineer is interested in the temperature field due to a moving heat source:

- Among others, the temperature field developing in frictional contact decides over the proper functionality of a workpiece or a device.
- In addition to the stress state from a mechanical load, the thermally induced stress state, especially in the case of a repeated heat input, is of major importance to understand the influence of the temperature field on the initiation and propagation of cracks.
- One might be interested in the metallurgical processes such as phase transformations driven by temperature field. In steel, this can be the formation of austenite during heating or of martensite during cooling.

The problem of a moving heat source on the surface of a body has been dealt with several times in the past. The reader is referred here to an early work by Jaeger [31] and the prominent text book by Carslaw and Jaeger [32], Sect. 29, and Refs. 1–3 and 20, and the recent studies by Komanduri and Hou [33–35] with respect to analytical solutions; see also the literature mainly cited in Refs. 1–3. In the last decade, numerical solutions, mainly based on the finite element method, have been applied to this type of problem (see the literature cited in Refs. 2 and 3 and also Refs. 13, 23, 24, and 26 where the numerical verification of the results developed below can be seen). The finite difference method is used in Ref. 36. This chapter intends to provide the reader with some analytical and semianalytical solutions (in the sense of numerical evaluation of analytical expressions) for temperature and stress fields

making possible the evaluation of such fields from easily programmable expressions and diagrams without the necessity to redo the somewhat complicated analysis. Besides already known solutions, this summary also presents new results the surface technology community may not be aware of. This comprehensive treatment of the topic is further motivated by the observation that in technological applications as mentioned above surprisingly often solutions are employed considering a stationary heat source (e.g., from Gecim and Winer [37]) instead of those assuming a moving heat source.

A. The Model

1. The Physical Model

All solutions are restricted to a two-dimensional heat transfer process. Therefore as a physical model a halfplane is assumed with the properties conductivity λ [W m^{-1} K^{-1}], specific heat c_p [W sec kg^{-1} K^{-1}], and density ρ [kg m^{-3}]. In all further analyses the quantities λ, c_p, ρ and hence also the thermal diffusivity $\kappa = \lambda/(\rho c_p)$ [m^2 sec^{-1}] are assumed to be constant in space and time.

A heat source $\dot{q}(x)$ [W m^{-2}] of length $2a$ is moving on the surface with the velocity v from the right to left (see Fig. 1). A local x-z-coordinate system is attached to the leading edge of the heat source. The x-axis is directed to the right, the z-axis (depth direction) into the halfplane. Outside of the interval $[0,2a]$ heatflow to the environment is modeled by $\lambda(\partial T/\partial z) \equiv f(T)|_{z=0}$, $x\notin[0,2a]$, T is the temperature and t is the time. For convenience, two characteristic quantities are introduced, namely, the Péclet number Pe and the thermal penetration depth δ,

$$Pe = va/(2\kappa), \quad \delta = \sqrt{2a\kappa/v} = a/\sqrt{Pe}. \tag{1}$$

These quantities allow to introduce other physically reasonable and nondimensional quantities. For not too small values of Pe, heat conduction in the x-direction can be neglected as it will be done in this chapter (for a detailed discussion on this topic see Ref. 3).

We restrict all further analyses to a constant value $\dot{q}$. This simplification is also acceptable, if frictional heating as a result of contact pressure is considered. This topic is discussed in Ref. 1, chap. I; a comparison between rectangular and triangular distributions of $\dot{q}(x)$ is given in Ref. 30. The authors of Ref. 30 report that the very local distribution of $\dot{q}(x)$

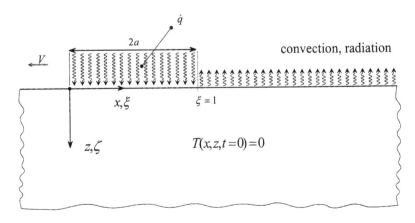

Figure 1 Moving coordinate system for one heat source.

may be relevant for the tool-temperature in turning operations. To this end $\dot{q}$ must be considered as the average value of $\dot{q}(x)$ over the heat input interval $[0,2a]$. With this assumption a proper mathematical model can be established. We formulate an equivalent problem with a stationary heat source and the body moving from the left to right with the velocity v. The energy balance for a body free of heat sources yields $dT/dt = \kappa(\partial^2 T/\partial x^2 + \partial^2 T/\partial z^2)$. The derivative dT/dt is the material derivative of T, $dT/dt = \partial T/\partial t + v(\partial T/\partial x)$.

We apply now two simplifications:

- Heat conduction in the x-direction is ignored. This assumption is fully justified for higher Pe numbers as discussed in Ref. 3. Note that in chapter D of this report a solution is applied which includes heat conduction in the x-direction. In our solutions, the term $\kappa(\partial^2 T/\partial x^2)$ is discarded in the energy balance relation.
- We concentrate on stationary solutions, hence we ignore the term $\partial T/\partial t$. Referring to a discussion by Komanduri and Hou [33] it can be shown that quasi-steady-state conditions are reached very shortly after the start of the motion of the source, approximately after a time period $t_s = 20\kappa/v^2$. Assuming that the length vt_s should be significantly smaller than $2a$, then the above relation for t_s and Eq. (1) require Pe to be a multiple of 5, say $Pe > 25$, for quasi-steady-state conditions.

Both simplifications are therefore reasonable for higher Pe numbers.

2. The Mathematical Model

According to the arguments above, the following initial-boundary value problem must be solved:

$$v\frac{\partial T}{\partial x} = \kappa\frac{\partial^2 T}{\partial z^2}, \qquad \text{with the boundary conditions} \tag{2.1}$$

$$T = 0 \qquad \text{for} \quad x \le 0 \qquad \text{and} \quad 0 \le z < \infty, \tag{2.2}$$

$$-\lambda\frac{\partial T}{\partial z} = \dot{q}(x) \qquad \text{for} \quad 0 \le x \le 2a \qquad \text{and} \quad z = 0, \tag{2.3}$$

$$\lambda\frac{\partial T}{\partial z} = f(T) \qquad \text{for} \quad 2a \le x \le \infty \qquad \text{and} \quad z = 0, \tag{2.4}$$

$$T = 0 \qquad \text{for} \quad -\infty \le x \le \infty \qquad \text{and} \quad z \to \infty. \tag{2.5}$$

The temperature of the surrounding medium as well as the initial temperature of the halfspace are assumed to be zero. A solution for this type of problem is given in Refs. 1–3 for the heating of a halfspace by friction. Introducing a maximum temperature T_{max} as

$$T_{max} = \sqrt{\frac{8}{\pi}}\sqrt{\frac{a\kappa}{v}}\frac{\dot{q}}{\lambda} = \frac{2}{\sqrt{\pi}}\frac{\delta}{\lambda}\dot{q}, \tag{3.1}$$

which is the temperature at the trailing edge of the heat input interval. The surface temperature in the interval $[0,2a]$ is (see, e.g., Ref. 1):

$$T = T_{max}\sqrt{x/2a}, \qquad z = 0, \qquad 0 \le x \le 2a. \tag{3.2}$$

The problem can be reformulated in a dimensionless way. Introducing the reduced coordinates

$$x = 2a\xi, \qquad z = \delta\zeta, \tag{4.1}$$

leads to

$$\frac{\partial T}{\partial \xi} = \frac{\partial^2 T}{\partial \zeta^2}, \tag{4.2}$$

with the initial and boundary conditions:

$$T = 0 \qquad \text{for} \quad \xi \le 0 \qquad \text{and} \quad 0 \le \zeta \le \infty, \tag{4.3}$$

$$\frac{\partial T}{\partial \zeta} = -\dot{q}\frac{\delta}{\lambda} \qquad \text{for} \quad 0 \le \xi \le 1 \qquad \text{and} \quad \zeta = 0, \tag{4.4}$$

$$\frac{\partial T}{\partial \zeta} = \delta\cdot\frac{f(T)}{\lambda} \qquad \text{for} \quad 1 \le \xi \le \infty \qquad \text{and} \quad \zeta = 0, \tag{4.5}$$

$$T = 0 \qquad \text{for} \quad -\infty \le \xi \le \infty \qquad \text{and} \quad \zeta \to \infty. \tag{4.6}$$

In Refs. 1–3, the Laplace transform is used to solve Eqs. (4.2)–(4.6). From this solution the temperature field is obtained:

$$T(\xi, \zeta) = T_{\max}\int_0^\xi \frac{1}{2\sqrt{u}}e^{\left(-\zeta^2/4u\right)}\,\mathrm{d}u \qquad \text{for} \quad 0 \le \xi \le 1, \tag{5.1}$$

$$T(\xi, \zeta) = T_{\max}\int_{\xi-1}^\xi \frac{1}{2\sqrt{u}}e^{\left(-\zeta^2/4u\right)}\,\mathrm{d}u - \frac{\delta}{\lambda}\int_0^{\xi-1} f(T(\xi - u, 0))\frac{1}{\sqrt{\pi u}}e^{\left(-\zeta^2/4u\right)}\,\mathrm{d}u \tag{5.2}$$

for $1 \le \xi \le \infty$.

$T(\xi-u, 0)$ represents the—a priori unknown—surface temperature at $\xi-u$ and therefore in the interval $1 \le \xi \le \infty$.

The integral (5.1) as well as the first integral of Eq. (5.2) can be given in closed form [1–3] leading to

$$\tilde{\theta}(\xi, \zeta) = \int_0^\xi \frac{1}{2\sqrt{u}}e^{\left(-\zeta^2/4u\right)}\,\mathrm{d}u = \sqrt{\xi}e^{\left(-\zeta^2/4\xi\right)} - \frac{\sqrt{\pi}}{2}\zeta\,\mathrm{erfc}\left(\zeta/2\sqrt{\xi}\right). \tag{6}$$

$\mathrm{erf}(\varepsilon) = 2/\sqrt{\pi}\int_0^\varepsilon e^{-t^2}\,\mathrm{d}t$ is the error function, its complement is defined as $\mathrm{erfc}(\varepsilon) = 1 - \mathrm{erfc}(\varepsilon)$.

Inserting Eq. (6) into Eqs. (5.1) and (5.2), introducing the nondimensional temperature $\theta(\xi, \zeta) = T(\xi, \zeta)/T_{\max}$ and substituting $\xi - u$ by $\bar{\xi}$ and therefore $f(T(\xi-u, 0))$ by $T_{\max}\cdot/\tilde{f}(\theta(\bar{\xi}, 0))$ yields

$$\theta(\xi, \zeta) = \tilde{\theta}(\xi, \zeta) \qquad \text{for} \quad 0 \le \xi \le 1, \tag{7.1}$$

$$\theta(\xi, \zeta) = \tilde{\theta}(\xi, \zeta) - \tilde{\theta}(\xi - 1, \zeta) - \frac{\delta}{\lambda}\cdot\frac{1}{\sqrt{\pi}}\int_1^\xi \tilde{f}(\theta(\bar{\xi}, 0))\cdot\frac{e^{\left(-\zeta^2/4(\xi - \bar{\xi})\right)}}{\sqrt{\xi - \bar{\xi}}}\,\mathrm{d}\bar{\xi} \qquad \text{for} \quad 1 < \xi \le \infty. \tag{7.2}$$

The crucial problem is to solve Eq. (7.2) for the surface temperature $\theta(\xi, 0)$. Equation (7.2) resembles a Volterra integral equation of the second kind containing a weak singularity which simplifies for $\zeta = 0$ to

$$\theta(\xi, 0) = \sqrt{\xi} - \sqrt{\xi - 1} - \frac{\delta}{\lambda}\cdot\frac{1}{\sqrt{\pi}}\int_1^\xi \tilde{f}(\theta(\bar{\xi}, 0))\cdot\frac{1}{\sqrt{\xi - \bar{\xi}}}\,\mathrm{d}\bar{\xi}. \tag{8}$$

Such an integral equation can be solved either by following the concept of the successive approximation [38] or by constructing of a resolvent kernel [3].

B. Temperature Fields for a Single Pass

As a cardinal solution a "single-pass solution" is understood. This refers to a heat source moving once along the surface of a halfspace from right to left (see Fig. 1).

1. The Insulated Halfplane

For the sake of completeness only, the dimensionless temperature field, $\theta = T/T_{max}$, is restated for the special case $\tilde{f} = 0$ and reads

$$\tilde{\theta}(\xi, \zeta) = \sqrt{\xi}e^{-\left(\zeta^2/4\xi\right)} - \frac{\sqrt{\pi}}{2}\zeta \, \mathrm{erfc}\left(\zeta/2\sqrt{\xi}\right), \tag{9.1}$$

$$\theta(\xi, \zeta) = \tilde{\theta}(\xi, \zeta) \quad \text{for} \quad 0 \leq \xi \leq 1, \tag{9.2}$$

$$\theta(\xi, \zeta) = \tilde{\theta}(\xi, \zeta) - \tilde{\theta}(\xi - 1, \zeta) \quad \text{for} \quad \xi \geq 1. \tag{9.3}$$

The above solution has been published in the Russian literature already in the early 1980s (see Ryzhkin and Shuchev [39]).

In contrast to solutions (9.1)–(9.3), in Refs. 9 and 10 a steady-state solution for a point source on a halfspace is reported as the "Rosenthal thick plate solution" (see Ref. 10, chap. 1.10.2, and for a line source on a halfplane as the "Rosenthal thin plate solution," see Ref. 10, chap. 1.10.3, and Sect. E of this paper). Grong deals in Ref. 10, chap. 1.10.5.2, with the same problem. As, however, the integration with respect to a line source solution in the sense of a Green's function is performed, a different result was achieved obviously because of the selection of a simplified transient line source solution involving only the depth coordinate z.

2. The Convecting Surface

If the surface may lose heat by convection, $f(T)$ in Eq. (4.5) is substituted by $f(T) = hT$, with h being the heat transfer or film coefficient. Fischer et al. report an analytical solution for this problem in Refs. 2 and 3 in the form of a series expansion. The radius of convergency of this series is extremely large and, because of $x = 2a\xi$, ranges from micrometers to kilometers. For the special case of $\zeta = 0$ the solution of Eq. (8) gives the surface temperature $\theta(\xi, 0)$ which reads with the coefficient of convection $\tilde{h}_c = h\delta/\lambda \cdot 1/\sqrt{\pi}$:

$$\theta\left(\xi, \zeta = 0, \tilde{h}_c\right) = \sqrt{\xi} - \sqrt{\xi - 1} - \tilde{h}_c \tilde{\theta}_1\left(\xi, \zeta = 0, \tilde{h}_c\right), \tag{10.1}$$

$$\begin{aligned}
\tilde{\theta}_1\left(\xi, \zeta = 0, \tilde{h}_c\right) = &\left\{ \frac{\pi}{2} - \xi \arcsin\left(\frac{1}{\sqrt{\xi}}\right) + \sqrt{\xi - 1} - \tilde{h}_c \frac{2\pi}{3} \right. \\
&\times \left[\xi\sqrt{\xi - 1} - (\xi - 1)\sqrt{\xi - 1} \right] \\
&+ \tilde{h}_c^2 \frac{\pi}{4} \left[\pi(2\xi - 1) - 2\xi^2 \arcsin\left(\frac{1}{\sqrt{\xi}}\right) + 2(\xi - 2)\sqrt{\xi - 1} \right] \\
&\left. - \tilde{h}_c^3 \frac{2\pi^2}{15} \left[2\xi^2\sqrt{\xi} - 5\xi + 3 - 2(\xi - 1)^2\sqrt{\xi - 1} \right] \right\},
\end{aligned} \tag{10.2}$$

valid for $1 \leq \xi \leq 1 + \dfrac{1}{10\tilde{h}_c^2}$.

$\tilde{\theta}_1(\xi, \zeta = 0, \tilde{h}_c)$ can be plotted conveniently against a new dimensionless coordinate $\tilde{\xi} = \tilde{h}_c\sqrt{\xi}$. We concentrate here on small values of $\tilde{\xi}$ (i.e., ξ near 1) and show $\tilde{\theta}_1(\tilde{\xi})$ for various values of $\tilde{h}_c$

in Fig. 2. For values of $\tilde{\xi}$ above the sharp bend of the curves in Fig. 2 the function $\tilde{\theta}_1(\tilde{\xi})$ can be approximated very well by

$$\tilde{\theta}_1\left(\tilde{\xi}\right) \approx \frac{\pi}{2}\left(1 - 2\tilde{\xi} + \pi\tilde{\xi}^2 - \frac{4}{3}\pi\tilde{\xi}^3\right) \quad \text{with} \quad \tilde{\xi} = \tilde{h}_c\sqrt{\xi} . \tag{10.3}$$

For $\zeta > 0$ an analytical expression for the temperature field cannot be offered. However, using the approximation $\tilde{f} \approx h(\sqrt{\bar{\xi}} - \sqrt{\bar{\xi}-1})$ in Eq. (7.2) makes possible an analytical evaluation of the integral in Eq. (7.2):

$$\begin{aligned}
\theta_2(\xi,\zeta) = \int_1^\xi \exp\left(-\frac{\zeta^2}{4(\xi-\bar{\xi})}\right)\frac{\sqrt{\bar{\xi}} - \sqrt{\bar{\xi}-1}}{\sqrt{\xi-\bar{\xi}}}\,d\bar{\xi} = &-\frac{\sqrt{\pi}}{4}\operatorname{erfc}\left(\frac{\zeta}{2\sqrt{\xi-1}}\right) \\
&\times \left\{2\zeta\left[\sqrt{\xi}\exp\left(-\frac{\zeta^2}{4\xi}\right) - \sqrt{\xi-1}\exp\left(-\frac{\zeta^2}{4(\xi-1)}\right)\right]\right.\\
&+ \left.\sqrt{\pi}\left[(2(\xi-1)+\zeta^2)\operatorname{erf}\left(\frac{-\zeta}{2\sqrt{\xi-1}}\right) - (2\xi+\zeta^2)\operatorname{erf}\left(\frac{-\zeta}{2\sqrt{\xi}}\right)\right]\right\}\\
&+ \frac{\xi-1}{\sqrt{\pi}}\exp\left(-\frac{\zeta^2}{4(\xi-1)}\right)\cdot\sum_{\ell=1}^\infty(-1)^\ell\frac{2^{2\ell}}{(2\ell+1)!}\Gamma\left(\ell-\frac{1}{2}\right)\\
&\times\left(\frac{\zeta}{2\sqrt{\xi-1}}\right)^{2\ell}\cdot\left[{}_pF_q\left(\left(\frac{1}{2},1,\ell-\frac{1}{2}\right),\left(\ell+1,\ell+\frac{3}{2}\right),1\right)\right.\\
&\left.- \left(\frac{\xi-1}{\xi}\right)^{\ell-\frac{1}{2}}{}_pF_q\left(\left(\frac{1}{2},1,\ell-\frac{1}{2}\right),\left(\ell+1,\ell+\frac{3}{2}\right),\frac{\xi-1}{\xi}\right)\right].
\end{aligned} \tag{11}$$

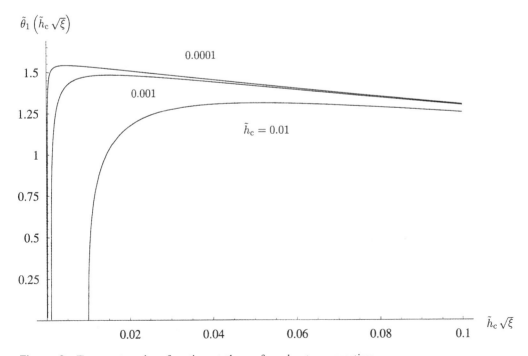

Figure 2 Temperature loss function at the surface due to convection.

Here, $\Gamma(\ )$ and $_pF_q((\ ),(\ ),)$ denote the Euler–Gamma function and the generalized hyper-geometry function, respectively. The infinite series in Eq. (11) converges rapidly, and in practical calculations only the first three terms have to be considered. From Eqs. (7.1), (7.2), (9.1), and (11) the temperature field for a convecting surface is obtained as

$$\theta\left(\xi,\zeta,\tilde{h}_c\right) = \tilde{\theta}(\xi,\zeta) - \tilde{\theta}(\xi-1,\zeta) - \tilde{h}_c\theta_2(\xi,\zeta), \quad \text{valid for} \quad 1 \leq \xi \leq \infty. \tag{12}$$

$\theta\left(\xi,\zeta,\tilde{h}_c\right)$ is plotted in Fig. 3 for the surface ($\zeta=0$) and for the interior of the halfplane ($\zeta=1$ and 2). As a realistic upper value $\tilde{h}_c=0.01$ is chosen. For the sake of comparison the temperature field for an insulated halfplane is replotted in this graph. Although the coefficient of convection is rather large, heat loss due to convection outside of the heat input interval does not alter markedly the temperature field. This small decrement in temperature does ensure, however, a finite temperature even after an infinite number of successive heat input passes (see Sect. C).

Zhai and Chang [36] presented a solution where, instead of the assumption of linear convection, heat extraction due to a viscous lubricant is studied by solving the heat equation for the lubricant, too. They transformed the spatially varying gap geometry by a mapping into a rectangular region and applied finally the finite difference concept.

Finally, it should be mentioned that expressions for the temperature field in a body with a convecting surface are rather rare. Mostly solutions are presented assuming a stationary heat source, see, e.g., the famous approximation by Gecim and Winer [37] or a very recent treatment by Yevtushenko and Ivanyk [4]; these authors present an integral transform solution both for the temperature field and the corresponding stress state. Of course, numerical solutions exist, e.g., by Kou et al. [16] employing the finite difference method or by several authors referenced in Ref. 3 employing the finite element method.

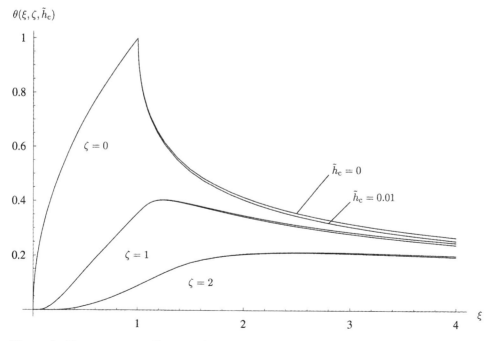

Figure 3   Temperature profiles at various depths due to convection.

3. The Radiating Surface

Very high temperatures can be achieved during electron or laser beam heating of bodies. It is therefore interesting to treat heat loss by radiation, too.

The boundary condition (4.5) now reads:

$$\frac{\partial T}{\partial \zeta} = \frac{\delta}{\lambda}\varepsilon\sigma\left((T+273)^4 - T_e^4\right), \quad f = \varepsilon\sigma\left((T+273)^4 - T_e^4\right). \tag{13}$$

ε is the coefficient of emissivity (≤ 1), σ is the Stefan–Boltzmann constant ($5.67*10^{-8}$ W m^{-2} K^{-4}), and T_e is the temperature of the surrounding medium and is set to 273 K. Furthermore, we introduce the coefficient of radiation $\tilde{h}_r = \varepsilon\sigma T_{max}^3 \delta/\lambda \cdot 1/\sqrt{\pi}$. For steel with $v = 0.01$ m/sec (beam scan velocity), $2a = 0.01$ m, $\kappa = 9.10^{-6}$ m^2/sec (beam diameter), and therefore $\delta = 3.10^{-3}$ m, $\lambda = 45$ W m^{-1}, $T_{max} = 1273$ K, and $\varepsilon = 1$, one obtains $\tilde{h}_r = 0.0045$. Using $\alpha = 273/T_{max}$ the integral equation (8) for the surface temperature reads

$$\theta\left(\xi, \zeta = 0, \tilde{h}_r\right) = \sqrt{\xi} - \sqrt{\xi - 1} - \tilde{h}_r \int_1^\xi \left[\theta^4\left(\bar{\xi}, 0\right) + 4\alpha\theta^3\left(\bar{\xi}, 0\right) + 6\alpha^2\theta^2\left(\bar{\xi}, 0\right) + 4\alpha^3\theta\left(\bar{\xi}, 0\right)\right].$$
$$\cdot\frac{1}{\sqrt{\xi - \bar{\xi}}}d\bar{\xi}, \quad \text{valid for} \quad 1 \leq \xi \leq \infty. \tag{14}$$

Approximate solutions for this nonlinear Volterra integral equation can be found by the method of successive approximation starting from $\theta_o(\bar{\xi},0) = 0$:

$$\theta_{n+1}\left(\xi, \zeta = 0, \tilde{h}_r\right) = \sqrt{\xi} - \sqrt{\xi - 1} - \tilde{h}_r \int_1^\xi \left[\theta_n^4(\bar{\xi}, 0) + 4\alpha\theta_n^3(\bar{\xi}, 0)\right.$$
$$\left. + 6\alpha^2\theta_n^2(\bar{\xi}, 0) + 4\alpha^3\theta_n(\bar{\xi}, 0)\right]\cdot\frac{1}{\sqrt{\xi - 1}}d\bar{\xi}, \quad n = 0, 1, \ldots, \tag{15}$$
$$\text{valid for } 1 \leq \xi \leq \infty.$$

In practice, the calculation of $\theta_2(\xi, \zeta = 0, \tilde{h}_r)$ gives a sufficiently precise approximation for $\theta(\xi, \zeta = 0, \tilde{h}_r)$, as $\tilde{h}_r$ is quite small and appears with high exponents in all further iterations. The same method can be employed to evaluate the temperature field $\theta(\xi, \zeta, \tilde{h}_r)$ of a half-plane losing heat by radiation outside of the heat input interval $0 \leq \xi \leq 1$. As the analytical expression for $\theta_2(\xi, \zeta, \tilde{h}_r)$ is a very lengthy expression [such as Eq. (11)] we provide a numerical evaluation instead for a value of $\tilde{h}_r = 0.045$ (10 times larger than the estimated value) which is plotted in Fig. 4.

C. Temperature Fields for Preheated Bodies (The Multipass Problem)

1. Solution of the Inhomogeneous Boundary/Initial Value Problem

All the above solutions are derived assuming that the whole body initially is at the same temperature level $T_o(x,z) \equiv 0$. One might be interested also in the temperature field $T(x,z)$, if an inhomogeneous initial temperature $T_o(x,z)$ exists in the halfspace. This is of considerable importance because in many technological applications, e.g., the overrolling of a rail by a train, the heat source passes over the surface several times. In this case the initial condition (4.3) for $\xi = 0$ has to be modified to

$$T = T_o(\zeta) \quad \text{for } \xi = 0 \quad \text{and } 0 \leq \zeta \leq \infty. \tag{4.3}$$

Note that $T_o(\zeta) \equiv 0$ is only valid for the very first pass. We will sketch a mathematical concept to deal with $T_o(\zeta) > 0$.

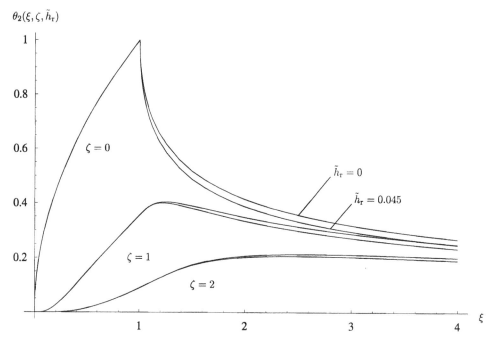

Figure 4 Temperature profiles at various depths due to radiation.

The general solution $T(x,z)$ of the problem can be constructed by superposition of the general "single-pass" solution and a particular solution $T_i(x,z)$ according to the initial temperature $T_o(x,z)$ leading to the following problem formulation:

$$v\frac{\partial T_i(x,z)}{\partial x} = \kappa\frac{\partial^2 T_i(x,z)}{\partial z^2}, \tag{16.1}$$

$$\lambda\frac{\partial T_i(x,z)}{\partial z} = 0 \qquad \text{for} \quad -\infty \le x \le +\infty \quad \text{and} \quad z = 0 \tag{16.2}$$

$$T_i(x,z) = T_o(z) \qquad \text{for} \quad x = 0, \tag{16.3}$$

$$T_i(x,z) = 0 \qquad \text{for} \quad -\infty \le x \le +\infty \quad \text{and} \quad z \to \infty. \tag{16.4}$$

The general solution $T(x,z)$ is the sum

$$T(x,z) = T(x,z)|_{\text{single pass}} + T_i(x,z). \tag{17}$$

To find $T_i(x, z)$ the Laplace transform technique is applied leading to the differential equation

$$\frac{\partial^2 \tau_i}{\partial \zeta^2} - s\tau_i = -T_o(0,\zeta), \quad L(T_i) = \tau_i(s,\zeta). \tag{18}$$

The solution of this differential equation is

$$\tau_i(s,\zeta) = A\sinh(\zeta\sqrt{s}) + B\cosh(\zeta\sqrt{s}) - \frac{1}{\sqrt{s}}\int_0^\zeta T_o(0,\bar{\zeta})\sinh((\zeta-\bar{\zeta})\sqrt{s})\mathrm{d}\bar{\zeta}. \tag{19}$$

A reformulation of Eq. (19) gives

$$\tau_i(s,\zeta) = \sinh(\zeta\sqrt{s})\left(A - \frac{1}{\sqrt{s}}\int_0^\zeta T_o(0,\bar\zeta)\cosh(\bar\zeta\sqrt{s})d\bar\zeta\right)$$
$$+ \cosh(\zeta\sqrt{s})\left(B + \frac{1}{\sqrt{s}}\int_0^\zeta T_o\left(0,\bar\zeta\right)\sinh(\bar\zeta\sqrt{s})d\bar\zeta\right).$$

(20)

From the condition $\lim T_i = \lim \tau_i \equiv 0$ for $\zeta \to \infty$ and the homogeneous boundary condition (16.2) the integration constants can be calculated:

$$A = 0, \quad B = \frac{1}{\sqrt{s}}\int_0^\infty T_o(0,\bar\zeta)e^{-\bar\zeta\sqrt{s}}d\bar\zeta.$$

(21)

Finally, the Laplace transform τ_i of T_i follows as

$$\tau_i(s,\zeta) = \frac{1}{\sqrt{s}}\left\{\int_0^\infty T_o(0,\bar\zeta)e^{-\bar\zeta\sqrt{s}}d\bar\zeta \cdot \cosh(\zeta\sqrt{s}) - \int_0^\zeta T_o(0,\bar\zeta)\sinh((\zeta-\bar\zeta)\sqrt{s})d\bar\zeta\right\}.$$

(22)

Rewriting Eq. (22) and using the Laplace transform of $\left(e^{(-\alpha^2/4\xi)}/\sqrt{\pi\xi}\right)$, i.e., $\frac{e^{-\alpha\sqrt{s}}}{\sqrt{s}}$, one obtains

$$T_i(\xi,\zeta) = \frac{1}{2\sqrt{\pi}}\int_0^\infty \frac{T_o(0,\bar\zeta)}{\sqrt{\xi}}\cdot\left(e^{-(\zeta-\bar\zeta)^2/4\xi} + e^{-(\zeta+\bar\zeta)^2/4\xi}\right)d\bar\zeta.$$

(23)

The substitution $\bar{\bar\zeta}\cdot 2\sqrt{\xi} = \bar\zeta$ and $\tilde\zeta\cdot 2\sqrt{\xi} = \zeta$ allows to eliminate the singularity in Eq. (23) leading to

$$T_i\left(\xi,\tilde\zeta\right) = \frac{1}{\sqrt{\pi}}\int_0^\infty T_o\left(0,2\sqrt{\xi\bar{\bar\zeta}}\right)\cdot\left(e^{-\left(\tilde\zeta-\bar{\bar\zeta}\right)^2} + e^{-\left(\tilde\zeta+\bar{\bar\zeta}\right)^2}\right)d\bar{\bar\zeta}.$$

(24)

Inserting $\xi = 0$, T_o becomes independent of the integration variable $\bar{\bar\zeta}$, which means

$$T_i\left(0,\tilde\zeta\right) = T_o\left(0,\tilde\zeta\right)\cdot\frac{1}{\sqrt{\pi}}\int_0^\infty\left(e^{-\left(\tilde\zeta-\bar{\bar\zeta}\right)^2} + e^{-\left(\tilde\zeta+\bar{\bar\zeta}\right)^2}\right)d\bar{\bar\zeta}.$$

(25)

It can be shown easily that the integral in Eq. (25) takes the value $\sqrt{\pi}$ for all values of $\tilde\zeta$. Hence $T_i(0,\tilde\zeta) = T_o(0,\tilde\zeta)$, which is in accordance with the second boundary condition of Eqs. (16.1)–(16.4).

2. Surface Temperature and Temperature Field After Several Passes

We now consider a second pass following the first one at a distance ξ_a with heat $\dot p$ over the interval $[\xi_a, \xi_e]$ (see Fig. 5). One can use the solution from the single-pass problem (heat input $\dot q$ over the interval $[0,1]$) as initial temperature $T_o(x,z)$ for the second pass (see Sect. 1) or solve the problem directly by modifying the boundary condition (4.5):

$$\frac{\partial T}{\partial \xi} = \delta\cdot\frac{f(T)}{\lambda} \qquad \text{for} \quad 1 \le \xi \le \xi_a,$$

$$\frac{\partial T}{\partial \xi} = -\dot p\frac{\delta}{\lambda} \qquad \text{for} \quad \xi_a \le \xi \le \xi_e,$$

(26)

$$\frac{\partial T}{\partial \xi} = -\delta\cdot\frac{f(T)}{\lambda} \qquad \text{for} \quad \xi_e \le \xi \le \infty.$$

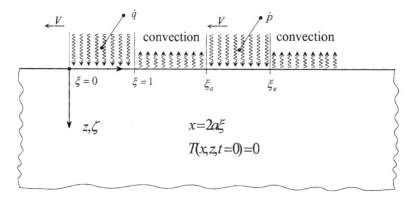

Figure 5   Moving coordinate system for two heat sources.

Using Eq. (9.1) the solution becomes for $\delta = 0$:

$$\theta(\xi, \zeta) = \tilde{\theta}(\xi, \zeta) - \tilde{\theta}(\xi - 1, \zeta) \qquad \text{for} \qquad 1 \le \xi \le \xi_a,$$

$$\theta(\xi, \zeta) = \tilde{\theta}(\xi, \zeta) - \tilde{\theta}(\xi - 1, \zeta) + \frac{\dot{p}}{\dot{q}}\tilde{\theta}(\xi - \xi_a, \zeta) - \frac{\dot{p}}{\dot{q}}\tilde{\theta}(\xi - \xi_e, \zeta).$$

(27)

Assuming for simplicity $\dot{q} = \dot{p}$ and $\xi_e - \xi_a = 1$ and applying the approximation

$$\sqrt{\xi} - \sqrt{\xi - 1} \sim 1/\left(2\sqrt{\xi}\right) \qquad \text{for} \qquad \xi \gg 1$$

(28)

we obtain

$$T(\xi_e, 0) = T_{max}\left(1 + 1/\left(2\sqrt{\xi_e}\right)\right),$$

(29)

which means for an insulated surface that after the first pass the maximum surface temperature increases during the next pass from T_{max} to $T_{max}(1 + 1/(2\sqrt{\xi_e}))$. An estimate for n subsequent passes leads to $T_{max}(1 + (1/2\sqrt{\xi_e}) \cdot \Sigma_{j=1}^{n}(1/\sqrt{j}))$. As $\lim_{n \to \infty} \Sigma_{j=1}^{n} 1/\sqrt{j} \to \infty$, an insulated surface would lead to an infinite heating of the body, at least near the surface. This means that convection or radiation at the surface is a necessary condition for the surface to remain at a finite temperature during an infinite number of subsequent passes.

D. Thermal Stress State

To the best knowledge of the authors no direct analytical expression can be given for the stress state due to the temperature fields presented in Sect. B, not even for the case of an elastic homogenous material with temperature-independent material constants E (Young's modulus) and μ (Poisson's ratio). Therefore the use of the finite element method seems to be the most practical way to solve for the stress field [1,40]. Bryant [41] presented an exponential Fourier transform technique to derive exact solutions for moving thermal distributions over the surface of a halfspace.

   For the sake of completeness we repeat the quasi-steady-state temperature distribution for a line source $\dot{Q} = 2a\dot{q}$ taken from Ref. 24:

$$T(x, z) = \frac{\dot{Q}}{\pi\lambda}e^{-vx/2\kappa} \quad K_0\left(v\sqrt{x^2 + z^2}/2\kappa\right), \quad K_0 \text{ is defined below.}$$

(30)

Note that Eq. (30) takes into account heat conduction also in the x-direction.

This solution is also referenced in the literature as the "Rosenthal thin plate solution" (see, e.g., Grong [10], chap. 1.10.3, for a steady state of the temperature with respect to a moving line source). However, in most practical cases one has to solve several complicated integrals over an infinite interval to achieve a solution $T(x,z)$ for a distributed heat source in $[0,2a]$.

We concentrate here on the solution published by Barber [42]. In the notation of the present elaboration the stress σ_x at the position ξ on the surface of a halfplane under plane strain follows from Barber's single heat source solution as

$$\sigma_x(\xi) = -\frac{\alpha E T_{max}}{1-\mu} \cdot \frac{1}{2\sqrt{\pi Pe}} \int_0^1 \left(-\frac{1}{\xi - \xi'} + 2Pe \cdot e^{2Pe(\xi - \xi')} K_1(2Pe(\xi - \xi')) \right) d\xi', \quad (31)$$

$\xi \geq \xi'$.

Again, Pe is the Péclet number, $K_0(t)$, $K_1(t)$ are the modified Bessel functions of the second kind of orders 0 and 1, respectively. The integral in Eq. (31) can be solved leading to the following expression for the stress:

$$\sigma_x(\xi) = -\frac{\alpha E T_{max}}{1-\mu} \cdot \frac{1}{2\sqrt{\pi Pe}} \cdot [\ln(\xi - 1)/\xi + \exp(2Pe\xi)[K_0(2Pe\xi)$$

$$\times (2Pe\xi - 1) + K_1(2Pe\xi) \cdot 2Pe\xi] - \exp(2Pe(\xi - 1))[K_0(2Pe(\xi - 1)) \quad (32)$$

$$\times (2Pe(\xi - 1) - 1) + K_1(2Pe(\xi - 1)) \cdot 2Pe(\xi - 1)]\}.$$

For $\xi = 1$ the stress becomes

$$\sigma_x(1) = -\frac{\alpha E T_{max}}{1-\mu} \cdot \frac{1}{2\sqrt{\pi Pe}} [-1 - \gamma - \ln Pe + \exp(2Pe) \cdot ((2Pe - 1)$$

$$\times K_0(2Pe) + 2Pe \cdot K_1(2Pe))], \quad (33)$$

$\gamma = 0.577215$.

Approximations for $K_0(a)$, $K_1(a)$ for either small (≤ 0.15) or large arguments a (≥ 2) yield the following estimates for $\sigma_x(\xi)$:

- For small values of $Pe\xi \leq 0.075$

$$\sigma_x(\xi) \sim -\frac{\alpha E T_{max}}{1-\mu} \cdot \sqrt{\frac{Pe}{\pi}} \xi. \quad (34.1)$$

- For any value $\xi \leq 1$, but $Pe\xi \geq 1$

$$\sigma_x(\xi) \sim -\frac{\alpha E T_{max}}{1-\mu} \cdot \sqrt{\xi}. \quad (34.2)$$

- For any value $\xi > 1 + 2/Pe$

$$\sigma_x(\xi) \sim -\frac{\alpha E T_{max}}{1-\mu} \cdot \left(\sqrt{\xi} - \sqrt{\xi - 1} \right). \quad (34.3)$$

The conclusion is that for not too small Péclet numbers $\sigma_x(\xi)$ is in principle proportional to the surface temperature T_{max}. Checking at $\xi = 1$ with the finite element calculations of Ref. 1 shows a very good agreement with Eqs. (34.2) and (34.3).

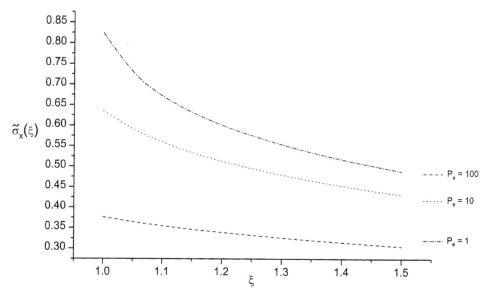

Figure 6 Relative longitudinal stress $\tilde{\sigma}_x(\xi) = -\sigma_x(1 - \mu/\alpha E T_{\max})$ for various Péclet numbers Pe.

Figure 6 demonstrates $\sigma_x(\xi)$ for that range of $Pe\xi$ that cannot be approximated as above.

For the sake of completeness it has to be mentioned that Farris and Chandrasekar [43] presented a relation corresponding to Eq. (32) 10 years ago. However, they did not do a limit analysis with respect to the point $\xi = 1$. In a further publication Moulik et al. [30] gave an analytical expression for a triangular distribution of the heat source; however, they did not arrive at a closed form equivalent to Eq. (32) or (33).

REFERENCES

1. Fischer, F.D.; Werner, E.; Yan, W.-Y. Thermal stresses for frictional contact in wheel-rail systems. Wear 1997, *211*, 156–163.
2. Fischer, F.D.; Werner, E.; Knothe, K. The surface temperature of a halfplane subjected to rolling/sliding contact with convection. ASME J Tribol 2000, *122*, 864–866.
3. Fischer, F.D.; Werner, E.; Knothe, K. An integral equation solution for the surface temperature of a halfplane heated by rolling/sliding contact and cooled by convection. ZAMM 2001, *81*, 75–81.
4. Yevtushenko, A.; Ivanyk, E. Influence of convective cooling on the temperature and thermal stresses in a frictionally heated semispace. J Thermal Stresses 1999, *20*, 635–657.
5. Chang, D.-F. An efficient way of calculating temperatures in the strip rolling process. ASME J Manuf Sci Eng 1998, *120*, 93–100.
6. Chang, D.-F. Thermal stresses in work rolls during the rolling of metal strip. J Mater Proc Technol 1999, *94*, 45–51.
7. Anderson, A.E.; Knapp, R.A. Hot spotting in automotive friction systems. Wear 1990, *135*, 319–337.
8. Kao, T.K.; Richmond, J.W.; Douarre, A. Brake disc hot spotting and thermal judder: an experimental and finite element study. Int J Vehicle Des 2000, *23*, 276–296.

9. Radaj, D. Berlin, et al., Ed.;*Wärmewirkungen des Schweißens*; Springer-Verlag:Berlin, 1988.
10. Grong, Ø. *Metallurgical Modelling of Welding*, 2nd Ed.; The Institute of Materials: London, 1997.
11. Bunting, K.A.; Cornfield, G. Toward a general theory of cutting: a relationship between the incident power density and the cut speed. ASME J Heat Transfer 1975, *116*, 116–122.
12. Kotousov, A. Thermal stresses and fracture of thin plates during cutting and welding operations. Int J Fract 2000, *103*, 361–372.
13. Elperin, T.; Kornilov, A.; Rudin, G. Formation of surface microcrack for separation of nonmetallic wafers into chips. ASME J Electronic Packag 2000, *122*, 317–322.
14. Krauss, G. *Heat Treatment and Processing Principles*; ASM International: Materials Park, Ohio, 1990; 319–350 pp.
15. Brooks, C.R. *Principles of the Surface Treatment of Steels*; Technomic Publishing Co: Lancaster, 1992; 37–60 pp.
16. Kou, S.; Sun, D.K.; Le, Y.P. A fundamental study of laser transformation hardening. Met Trans A 1983, *14A*, 643–653.
17. Ashby, M.F.; Easterling, K.E. The transformation hardening of steel surfaces by laser beams: I. Hypo-eutectoid steels. Acta Metall 1984, *32*, 1935–1948.
18. Li, W.B.; Easterling, K.E.; Ashby, M.F. Laser transformation hardening of steel: II. Hyper-eutectoid steels. Acta Metall 1986, *34*, 1533–1543.
19. Ion, J.C.; Shercliff, H.R.; Ashby, M.F. Diagrams for laser materials processing. Acta Metall Mater 1992, *40*, 1539–1551.
20. Rödel, J. *Beitrag zur Modellierung des Elektronenstrahlhärtens von Stahl*; FLUX Verlag: Chemnitz, 1997.
21. Rödel, J. Werkstoffphysikalische Modelle für das Randschichthärten von Stahl. Härterei Tech Mitteilungen HTM 1999, *54*, 230–240.
22. Rödel, J.; Spies, H.-J. Modelling of austenite formation during rapid heating. Surf Eng 1996, *12*, 313–318.
23. Lugscheider, E.; Barimani, C.; Eritt, U.; Kuzmenkov, A. FE-simulations of temperature and stress field distribution in thermally sprayed coatings due to deposition process. Proceedings of the 15th International Thermal Spray Conference, Nice, 1998; 367–372 pp.
24. Haidemenopoulos, G.N. Coupled thermodynamic/kinetic analysis of diffusional transformations during laser hardening and laser welding. J Alloys Compd 2001, *320*, 302–307.
25. Dowden, J.M. *The mathematics of thermal modeling*; Chapman & Hall/CRC: Boca Raton, London, New York, Washington D.C., 2001.
26. Chiumenti, M.; Agelet de Saracibar, C.; Cervera, M. Computational modeling of laser heat-forming processes. *Plastic and Viscoplastic Response of Materials and Metal Forming*; Neat Press: Maryland, 2000; 213–215 pp.
27. Schaaf, P. Laser nitriding of metals. Prog Mat Sci 2002, *47*, 1–161.
28. Komanduri, R.; Hou, Z.B. Thermal modeling of the metal cutting process: Part I. Temperature rise distribution due to shear plane heat source. Int J Mech Sci 2001, *42*, 1715–1752 - Part II. Temperature rise distribution due to frictional heat source of the tool–chip interface. Int J Mech Sci 2000, *43*, 57–88 - Part III. Temperature rise distribution due to the combined effects of shear plane heat source on the tool–chip interface frictional heat source. Int J Mech Sci 2001, *43*, 89–107.
29. Dawson, P.R.; Malkin, S. Inclined moving heat source model for calculating metal cutting temperatures. ASME J Eng Ind 1984, *106*, 179–186.
30. Moulik, P.N.; Yang, H.T.Y.; Chandrasekar, S. Simulation of thermal stresses due to grinding. Int J Mech Sci 2001, *43*, 831–851.
31. Jaeger, J.C. Moving sources of heat and the temperature at sliding contacts. Proc R Soc New South Wales 1942, *76*, 203–224.
32. Carslaw, H.S.; Jaeger, J.C. *Conduction of Heat in Solids*, 2nd Ed.; Clarendon Press: Oxford, 1959.
33. Komanduri, R.; Hou, Z.B. Thermal analysis of dry sleeve bearings—a comparison between analytical, numerical (finite element) and experimental results. Tribol Int 2001, *34*, 145–160.
34. Komanduri, R.; Hou, Z.B. Analysis of heat partition and temperature distribution in sliding systems. Wear 2001, *251*, 925–938.

35. Hou, Z.B.; Komanduri, R. General solutions for stationary/moving plane heat source problems in manufacturing and tribology. Int J Heat Mass Transfer 2000, *43*, 1679–1698.

36. Zhai, X.; Chang, L. A transient thermal model for mixed-film contacts. Tribol Trans 2000, *43*, 427–434.

37. Gecim, B.; Winer, W.O. Transient temperatures in the vicinity of an asperity contact. ASME J Tribol 1985, *107*, 333–342.

38. Polyanin, A.D.; Manzhirov, A.V. *Handbuch der Integralgleichungen*, 1st Ed.; Spektrum Akademischer Verlag GmbH: Heidelberg, Berlin, 1999. chap. 7.6.2 ff.

39. Ryzhkin, A.A.; Shuchev, K.G. Estimation of contact temperature fluctuations at friction and cutting of metals. J. Friction Wear 1998, *19*, 30–36.

40. Fridrici, V.; Attia, M.H.; Kapsa, P.; Vincent, L. Temperature rise in fretting: a finite element approach to the thermal constriction phenomenon. In: *Advances in Mechanical Behaviour, Plasticity, Damage*; Elsevier: Amsterdam, Lausanne, New York, Oxford, Shannon, Singapore, Tokyo, 2000; 585–590.

41. Bryant, M.D. Thermoelastic solutions for thermal distributions moving over half space surfaces and application to the moving heat source. ASME J Appl Mech 1988, *55*, 87–92.

42. Barber, J.R. Thermoelastic displacements and stresses due to a heat source moving over the surface of a half plane. ASME J Appl Mech 1984, *51*, 636–640.

43. Farris, T.N.; Chandrasekar, S. High-speed sliding indentation of ceramics: thermal effects. J Mat Sci 1990, *25*, 4047–4053.

2

Stresses in Thin Films

Robert C. Cammarata
Johns Hopkins University, Baltimore, Maryland, U.S.A

I. INTRODUCTION

A thin solid film grown on a much thicker solid substrate is generally deposited in a state of stress. This stress can be quite large, often exceeding the yield stress of the material in bulk form, and can lead to deleterious effects such as cracking, spalling, and deadhesion. As a result, attempts are often made to control and to minimize the stress levels during growth, or to relieve the stress after deposition. However, it is sometimes necessary or even desirable for a thin film to be under stress. It is generally required for electronic material applications that a semiconductor film be grown epitaxially, that is, as a single crystal deposited on a single crystal substrate with defect-free lattice matching at the film–substrate interface. If the in-plane equilibrium lattice spacings of the film and the substrate are different, the film will be under stress to achieve this lattice matching. The deposition of a magnetical film with a certain stress state can lead to an enhanced magnetical anisotropy that may be exploited in certain device applications. A material that has a thin film coating in a state of compressive stress can result in enhanced fracture and fatigue resistance compared to an uncoated material.

II. MECHANICS OF THIN FILM STRESS

Before considering the physical origin of stresses in thin films, it is important to consider the mechanics of the problem [1]. This will be illustrated by considering an elastically isotropic film that has been grown on a much thicker substrate. In addition, it will be assumed that the lateral dimensions of the film–substrate system are much greater than the substrate thickness.

Consider a film that is initially deposited stress-free and imagine that it is detached from the substrate (see Fig. 1a). The result is a free-standing film with an in-plane area equal to that of the substrate. Suppose a process that causes the free-standing film to experience a volume change, which results in a strain state relative to the substrate, occurs (Fig. 1b). Let this strain state be described by the principal strains e_{xx}, e_{yy}, and e_{zz} referred to a Cartesian coordinate system, with the x-direction and the y-direction lying in the plane of the film, and the z-direction perpendicular to the plane of the film. For simplicity, the in-plane strain state will be taken as uniform so that $e_{xx} = e_{yy}$. To reattach the film to the substrate so that the film

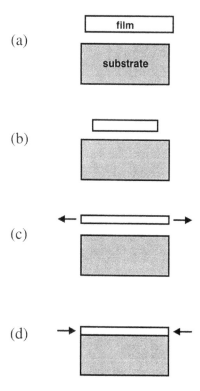

(a)

(b)

(c)

(d)

Figure 1 Schematic diagram illustrating the mechanics of thin film stress formation: (a) film detached from the substrate in a stress-free state; (b) volume change in film resulting in a change in the in-plane area relative to the substrate; (c) tractions around the film edge, imposing a biaxial stress that stretches the film so that the in-plane area matches that of the substrate; (d) superimposed stress on reattached film to remove the edge tractions.

and the substrate have the same in-plane area, it will be necessary to impose a biaxial stress $\sigma = \sigma_{xx} = \sigma_{yy}$ on the film. Consider that this biaxial stress state results from tractions applied around the film edge (Fig. 1c). For the in-plane areas of the film and the substrate to be equal, elastic strain components $\varepsilon_{xx} = \varepsilon_{yy}$ resulting from these tractions must be equal to $-e_{xx}$. Using Hooke's law, the biaxial stress will be given by:

$$\sigma = Y\varepsilon_{xx} = -Ye_{xx},$$ (1)

where $Y = E/(1-v)$ is the biaxial modulus of the film, and E and v are Young's modulus and Poisson's ratio, respectively, of the film. Finally, the film is reattached to the substrate and the normal tractions at the edge of the film are removed by superimposing edge forces of opposite sign (Fig. 1d). If the film is well adhered to the substrate, shear stresses on the film–substrate interface that maintain the in-plane biaxial stress σ will be produced.

Although it is difficult to calculate the detailed nature of the shear stresses, it is possible to infer certain qualitative aspects from solutions obtained for other similar elasticity problems [1–4]. For an infinitely rigid substrate, the shear stresses are expected to be concentrated very near the edge of the film. When the substrate modulus is finite, the shear stresses are lower and extended further into the film, with the maximum interfacial shear stress typically occurring at about one-half the film thickness from the edge [1].

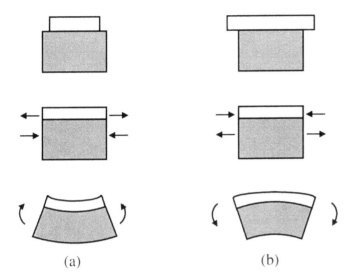

Figure 2 Thin film stress leading to substrate bending: (a) tensile stress; (b) compressive stress.

The above discussion indicates that a film will be deposited in a state of stress if there is good adherence between the film and the substrate, and if the film would spontaneously change its lateral dimensions were it detached from the substrate. This change in the lateral dimensions can be associated with a latent in-plane strain e that leads to a biaxial film stress $\sigma = -Ye$. It is noted that a compressive latent strain (i.e., a latent strain associated with a film that would contract laterally were it detached from the substrate) leads to a film under tension, and vice versa. The mechanisms that have been proposed to explain the generation of latent strains during deposition will be discussed later. One example will be given here. Let α_f and α_s denote the thermal expansion coefficient of the film and the substrate, respectively. Suppose that deposition occurred at a temperature T_1 and the film–substrate is then brought to room temperature T_2. The relative latent strain in the film becomes:

$$e = (\alpha_f - \alpha_s)(T_2 - T_1). \tag{2}$$

The resulting biaxial stress is called the thermal stress.

Suppose a thin film deposited in a state of stress was grown on an initially flat substrate (see Fig. 2). Mechanical equilibrium requires that the net force and the bending moment vanish on the film–substrate cross-section. As a result, the film stress will cause the substrate to bend [5]. It is seen from the figure that a tensile film stress causes the substrate to bend in a concave manner upward, whereas a compressive film stress causes the film to bend in a convex manner downward.

III. EXPERIMENTAL MEASUREMENTS

Thin film stress measurements involve the determination of the strain state of the film, which is then used in an elasticity analysis to obtain the stress state. Most approaches can be classified as either substrate bending or x-ray diffraction methods. Each approach has its advantages and disadvantages, and recent advances that allow both types to conduct real-time in situ measurements as well as postdeposition ex situ measurements have been made.

A. Substrate Bending Methods

As discussed above, a thin film stress will result in the bending of the substrate. Stoney [6] used this effect to measure stress in electrodeposited films nearly a century ago. Over the years, various techniques have developed [1,5,7–12], which employ substrates of different shapes and constraints (e.g., cantilever beam or unclamped circular wafer) and which determine the deflection of the substrate by methods such as optical reflection [10], interferometry [11], and x-ray topography [12]. These techniques can be used to calculate the product of the film stress σ and the film thickness t_f. For a thin film on a much thicker substrate that has the form of a cantilever beam, the product stress–thickness product σt_f can be calculated from the deflection δ of the free end of the beam as [8]:

$$\sigma t_f = Y t_s^2 \delta / 3L^2, \tag{3}$$

where Y is the biaxial modulus of the film, t_s is the substrate thickness, and L is length of the beam length. If the substrate is in the form of an unclamped circular disk (wafer), the stress–thickness product can be determined from the radius of curvature R of the substrate using the Stoney equation:

$$\sigma t_f = Y t_s^2 / 6R. \tag{4}$$

It is important to realize that bare substrates can have a significant curvature $1/R_o$ prior to deposition; if this is so, the curvature $1/R$ in Eq. (4) should be replaced with the change in curvature $(1/R - 1/R_o)$.

Examples of real time wafer curvature measurements during ultrahigh vacuum evaporation on amorphous substrates [13] are shown in Fig. 3. Care should be taken in interpreting the data given on the plots of Fig. 3. The stress σ in the stress–thickness product σt_f represents the mean stress, integrated through the film thickness. The slope of the curve at a particular point on a stress–thickness vs. thickness plot, equal to $d(\sigma t_f)/dt_f$ at that point, represents the stress of the layer of thickness dt_f (assuming that the stress develops instantaneously). It should be recognized that in situ stress measurements can be affected by momentum transfer of the depositing atoms, temperature gradients through the substrate thickness, and temperature changes during deposition [1], all of which can be significant for very thin films [9]. Even in thicker films, there can be a contribution to the measured stress during deposition, which disappears if growth is interrupted [13,14]. This effect can be seen in Fig. 3d, which displays an instantaneous tensile increase in the stress that occurred when the deposition of Al was interrupted at a thickness of 200 nm. A possible mechanism for this behavior is discussed later.

B. X-ray Diffraction Methods

X-ray diffraction methods have proven to be very powerful techniques for characterizing the strain and stress states of thin films [15,16]. The basic principle is that changes in crystallographical interplanar spacings of a stressed film can be measured from the shifts in the angular positions of x-ray diffraction peaks, which can then be used to determine the in-plane strain and stress. For example, consider the simple case of an elastically isotropic film. The interplanar spacing $d_{hkl}{}^*$ for crystal planes (hkl) in the stressed film that are parallel to the plane of the film can be calculated from Bragg's law using the angular Bragg peak positions obtained from a symmetrical reflection x-ray diffractometer. The strain perpendicular to the plane of the film is:

$$\varepsilon_{xx} = (d_{hkl}{}^* - d_{hkl})/d_{hkl}, \tag{5}$$

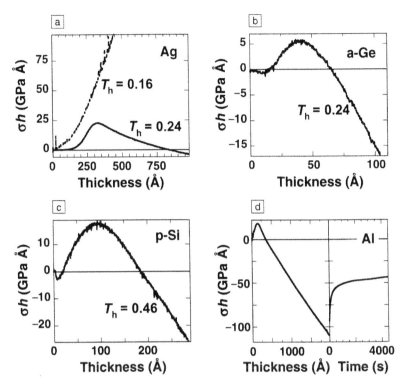

Figure 3 Real-time wafer curvature measurements during ultrahigh vacuum deposition onto amorphous SiO_2: (a) polycrystalline Ag; (b) amorphous Ge; (c) polycrystalline Si; (d) polycrystalline Al. (From Ref. 13.) T_h is the ratio of the deposition temperature to the melting temperature. For Al, the behavior during growth and after growth is interrupted at a film thickness of 2000 Å (200 nm), as shown.

where d_{hkl} is the unstressed interplanar spacing. Using Hooke's law, the in-plane strain is related to the perpendicular strain by:

$$\varepsilon_{xx} = \varepsilon_{yy} = -(1 - \nu)\varepsilon_{zz}/2\nu, \tag{6}$$

and the film stress is given by:

$$\sigma = Y\varepsilon_{xx} = -E\varepsilon_{zz}/2\nu. \tag{7}$$

Although the example given above considered an elastically isotropic film investigated by a symmetrical reflection method, it is possible to use both symmetrical and asymmetrical reflection techniques and a complete elasticity analysis employing the components of the stiffness matrix to study anisotropic films. For this reason, x-ray diffraction methods are particularly useful for investigations of epitaxially grown single crystal films and poly-crystalline films with a strong crystallographical texture [15,16]. One of these methods, grazing angle incidence x-ray scattering (GIXS), is a technique that employs the phenom-enon of total internal reflectance [12,17–19]. The grazing incidence allows for the determi-nation of the complete stress and strain states in films down to very small thicknesses, as well as for high-resolution measurements of the strain variation with film thickness.

It is important to point out that x-ray diffraction measurements of thin film stress can be complicated by issues such as peak broadening, resulting from instrumental effects or from the effects of structural defects. In addition, determining strain using the interplanar spacing of the bulk material as the value for the spacing for an unstressed film (see Eq. (5)) is often not appropriate, so that it is best to directly measure the unstressed interplanar spacing using asymmetrical reflection methods [15,16,20].

IV. INFLUENCE OF SURFACES

One of the principal microstructural features of thin films is the high density of surfaces relative to conventional bulk materials. In addition to the film–substrate interface and the film free surface, there can be grain boundaries in polycrystalline films and interlayer interfaces in multilayered thin films. These surfaces can have a significant effect on the mechanical behavior of thin films, in general, and the internal stress, in particular. For this reason, it is worth briefly discussing certain structural and thermodynamic aspects of surfaces in thin films.

A. Film–Substrate Interface

As mentioned in Sec. I, there can be an epitaxial relationship between a thin film and the substrate, leading to lattice matching at the film–substrate interface. If the lattice matching is perfect, resulting in a defect-free interface, the interface is referred to as coherent. If the film and the substrate have different equilibrium lattice spacings, and the substrate is much thicker than the film, the film will have to be coherently strained to have it in perfect atomic registry with the substrate. Let a_f and a_s denote the bulk equilibrium in-plane lattice spacings of the film and the substrate, respectively. The misfit between the film and the substrate is defined as $m = (a_s - a_f)/a_f$. For a film perfectly lattice-matched to the substrate, the in-plane coherency strain is equal to the misfit.

As long as the misfit is not too large, it is generally possible to grow a coherently strained epitaxial film, at least at smaller thicknesses [21]. As will be discussed later, there is a critical thickness above which it is thermodynamically favorable for the film to elastically relax, resulting in a loss of the perfect lattice matching at the film–substrate interface. One way in which this can occur is by the formation of an array of edge dislocations at the film–substrate interface that can accommodate some or all of the misfits. As long as the spacing of these misfit dislocations is not too small, so that there is still a significant amount of residual lattice matching at the film–substrate interface, the interface is said to be semicoherent. If the misfit dislocation spacing is less than a few lattice spacings, or if there is no epitaxial relationship between the film and the substrate, the interface is called incoherent.

B. Surface Thermodynamics

Consider a solid–vapor interface (i.e., the free surface of a solid). There are two thermodynamic quantities associated with the reversible work to change the area of the surface [22]. One of these is the surface free energy γ, which can be defined by setting the reversible work to create a new surface of area A by a process such as cleavage equal to γA. The other surface thermodynamic quantity is the surface stress tensor f_{ij}, which can be defined by setting the reversible work to introduce a surface elastic strain $d\varepsilon_{ij}$ on a surface of area A equal to $f_{ij} A d\varepsilon_{ij}$. For simplicity, it will be assumed that the surface stress is isotropic and can be taken as a scalar f (this is valid for a surface that displays a threefold or higher rotational symmetry). It can be shown that the surface stress and the surface free energy are related by the expression [22]:

$$f = \gamma + \partial\gamma/\partial\varepsilon_s, \tag{8}$$

where ε_s is an in-plane linear surface strain. Unlike the surface free energy, which must be positive (otherwise, solids would spontaneously cleave), the surface stress can be positive or negative. Experimental measurements and theoretical calculations for the low index surfaces of many metals, semiconductors, and ionic solids give positive values for the surface free energy and the surface stress of order 1 N/m.

For finite-size solids in mechanical equilibrium, the surface stress will induce a volume elastic strain [22]. For a spherical solid of radius r, this will result in a pressure difference ΔP (called the Laplace pressure) between the solid and a surrounding vapor given by:

$$\Delta P = P_s - P_v = 2f/r, \tag{9}$$

where P_s and P_v denote the pressure of the solid and the vapor, respectively. Similarly, for a thin disk of thickness t, a surface stress f acting on the top and the bottom surfaces will result in a radial Laplace pressure of $2f/t$. Because of this Laplace pressure, the lattice spacing in the interior of the solid at equilibrium will be different from the equilibrium bulk spacing. Using Hooke's law, this difference in lattice spacing can be described in terms of an in-plane elastic strain:

$$\varepsilon = -2f/Yt. \tag{10}$$

As with the free solid surface, a solid–solid interface has an associated surface free energy that will be referred to as the interface free energy. Because the phases on either side of solid–solid interface can be independently strained, resulting in different strain states at the interface, there are two surface stresses that can be associated with this interface [22]. These will be referred to as interface stresses. For a thin film–substrate interface, it is convenient to define these interface stresses in the following manner. Let ε' represent an interface strain associated with an in-plane deformation of the film keeping the substrate fixed. Such a strain would lead to a change in the misfit dislocation density at a semicoherent interface. An interface stress g associated with this type of deformation can be defined by taking the reversible work to strain an interface of area A by an amount $d\varepsilon'$ equal to $gAd\varepsilon'$. Now consider an interface strain resulting from deformation of the film and the substrate by the same amount in the plane of the interface, and let it be denoted as e. An interface stress h associated with this type of deformation can be defined by taking the reversible work to strain an interface of area A by an amount de equal to $hAde$.

Suppose a film–substrate system with a semicoherent film–substrate interface has a misfit m. Let $a_f{}^*$ be the in-plane lattice spacing of the strained film. The coherency strain can be defined as $\varepsilon_c = (a_f{}^* - a_f)/a_f$. If the film is fully relaxed, so that it has its bulk equilibrium lattice spacing a_f, $\varepsilon_c = 0$; if the interface is completely coherent, $\varepsilon_c = m$. A simple model [21] for the interface free energy Γ of a semicoherent interface of misfit m and coherency strain ε_c leads to the expression:

$$\Gamma = \Gamma_0[1 - \varepsilon_c/m], \tag{11}$$

where Γ_0 is the interface free energy when the film is completely relaxed. Based on this model, the following approximate expression for the interface stress g has been given [22]:

$$g \approx -\Gamma_0/2m. \tag{12}$$

It is noted that for $|m| \ll$, 1, $|g| \gg \Gamma_o$. A similar analysis for the interface stress h indicates that for a semicoherent interface, h is on the order of $-10\Gamma_o$, which is consistent with experimental measurements for metal–metal interfaces [22]. On the other hand, the interface stress h for an interface between a crystalline film and an amorphous substrate can be positive.

C. Growth Modes

Three basic thin film growth modes [5] (see Fig. 4) that can be associated with relationships involving the values of surface free energies have been identified. Volmer–Weber growth involves three-dimensional growth of islands that eventually coalesce to form a continuous film. This growth mode is favored when $\gamma_s < \gamma_f + \Gamma$, where γ_s and γ_f are the surface free energy of the free surface of the film and the substrate, respectively. The Frank–van der Merwe mode is a two-dimensional layer-by-layer growth mode and represents "ideal" epitaxial growth. The Stranski–Krastanov mode involves two-dimensional growth for one or two monolayers, followed by three-dimensional islandlike growth. The switchover from two-dimensional to three-dimensional is not completely understood, but is apparently related to effects of stress relaxation. Frank–van der Merwe and Stranski–Krastanov growth modes are favored when $\gamma_s \geq \gamma_f + \Gamma$.

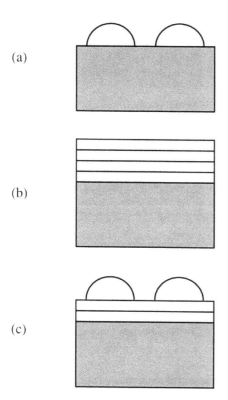

(a)

(b)

(c)

Figure 4   Thin film growth modes: (a) Volmer–Weber (island growth); (b) Frank–van der Merwe (layer-by-layer growth); (c) Stranski–Krastanov (layer-by-layer followed by islandlike growth).

V. PHYSICAL ORIGIN OF THIN FILM STRESS

A. Epitaxial Growth

For layer-by-layer epitaxial growth, it is expected that the film stress is principally a result of lattice matching at the film–substrate interface leading to coherency stresses. As long as the misfit is not too large, the film will initially grow as a completely coherent overlayer. From a thermodynamic point of view, this is because the work to form misfit dislocations at the interface is greater than the work to elastically strain the film to accommodate the misfit. Because the volume strain energy of the coherent film is proportional to the film thickness, there will eventually be a critical thickness above which it is thermodynamically favorable to introduce misfit dislocation at the interface to relieve some of the elastic strain energy.

An expression for the critical thickness t_c can be given in terms of relevant thermodynamic quantities as [22]:

$$t_c = (f_s + g - \gamma_s - \Gamma)/Ym, \tag{13}$$

where f_s and γ_s are the surface stress and the surface free energy of the film, respectively. The contribution of the free film surface has been ignored, effectively taking $f_s = \gamma_s$. In many (but not all) cases, this is an acceptable approximation. Assuming this to be true, it is possible to calculate the critical thickness by substituting Eqs. (11) and (12) into Eq. (13) and by employing an expression for Γ_o that involves the self-energies for an array of interface dislocations that completely accommodates the misfit. If the elastic moduli for the film and the substrate can be taken as approximately the same, such a model leads to the following expression for the critical thickness [5,21]:

$$t_c = [b \ln(t_c/b + 1)]/8\pi(1 + v)m, \tag{14}$$

where b is the Burgers vector of the misfit dislocations.

The process of stress relaxation can involve the nucleation of misfit dislocations at the interface or slip of preexisting dislocations. It is often found experimentally that it is possible to grow a completely lattice-matched film to thicknesses greater than the critical thickness. This is presumably a result of kinetic limitations associated with the stress relaxation process.

B. Nonepitaxial Island Growth

Results from recent experiments investigating the development of thin film stress during Volmer–Weber (island) growth by ultrahigh vacuum evaporation onto an amorphous substrate for a variety of film materials are shown in Fig. 3. It is seen that the stress behavior, which occurs for both crystalline and amorphous films, generally involves an initial compressive stress, followed by a tensile increase, and then back to a compressive stress. This has been termed CTC behavior. (Some previous studies [23,24] have reported a similar behavior for the deposition of "high-mobility" materials, but during deposition of some "low-mobility" materials, the final compressive stage is absent.) The initial compressive regime occurs during the formation and growth of islands before coalescence. The rapid tensile rise initiates around the onset of coalescence and reaches a maximum when the film becomes continuous. The final compressive stage occurs during further growth of the continuous film. Different mechanisms that have been proposed for each regime will be reviewed below.

Consideration is first given to the early stage of island growth [25,26]. An isolated island can be modeled as a disk of diameter d and thickness t. As discussed earlier, surface stresses acting on this disk exert a size-dependent Laplace pressure that results in an equilibrium lattice spacing different from the bulk equilibrium lattice spacing. Let d_o and t_o represent the size of an island when it first becomes firmly attached to the substrate so that a film stress can be generated. As the island grows, the equilibrium spacing will change, but the island is constrained by the substrate not to deform laterally. Thus, the change in equilibrium spacing leads to a latent strain that is manifested as a film stress. A simple elasticity analysis gives the following approximate expression for the island stress [25]:

$$\sigma = (f_s + h)(1/t - 1t_o) + \beta f_z (1/d - 1/d_o), \tag{15}$$

where f_z is the free surface stress for the curved surface of the island and β depends on the elastic constants of the disk. For an elastically isotropic island, $\beta = (1-3v)/(1-v)$. It is generally expected that the first term on the right-hand side of Eq. (15) will dominate and that $(f_s + h)$ will be positive. As a result, the firmly attached islands will produce a compressive film as they grow, consistent with experiment.

It is well established that the rise in tensile stress observed in Fig. 3 is correlated with island coalescence. This is consistent with a model that has been popular over the years, which involves tensile stress generation resulting from relaxation of grain boundaries formed during coalescence [7,9,27,28] and which has been the subject of recent reinterpretations and reformulations [29–31]. When two islands impinge, a grain boundary is formed and two free surfaces disappear. This results in a lowering of the surface free energy by an amount $\Delta\gamma = (2\gamma_s - \gamma_{gb})$, where γ_{gb} is the interface free energy for the grain boundary. As the distance between neighboring growing islands is reduced, there will be a critical interaction distance where it will be thermodynamically favorable for the islands to elastically deform and "snap" together to form a grain boundary. The elastic strain energy created that causes the islands to impinge is compensated by the surface energy reduction $\Delta\gamma$. If the islands are modeled as having impinging faces that are flat, the contribution to the film stress can be approximated as [28,29]:

$$\sigma \approx (E\Delta\gamma/d_i)^{1/2}, \tag{16}$$

where d_i is the size of the grain boundary formed at impingement. Using $E = 100$ GPa, $\Delta\gamma = 1$ N/m, and $d_i = 100$ nm, Eq. (16) gives a value for the stress contribution of about 1 GPa. A more sophisticated analysis employing contact mechanics [30] gives the following expression for the average stress contribution resulting from the coalescence of a square array of hemispherical islands of diameter d_i:

$$\sigma \approx 4\Delta\gamma/d_i. \tag{17}$$

Substituting the values given above into Eq. (17) gives $\sigma = 40$ MPa, consistent with results from a finite element model [29] and is on the order of what is often observed experimentally. The values that can be obtained from Eqs. (16) and (17) can be viewed as approximate upper and lower limits for the stress contribution from the grain boundary relaxation mechanism.

Significant grain growth is sometimes observed during and after island coalescence [1]. This process is driven by the reduction in grain boundary area per unit volume of the film because the surface free energy of a grain boundary is positive, and thus reduction in the grain boundary area reduces the free energy of the film. Because a region near a grain boundary is expected to display a lower atomic density than the interior of the grain, the elimination of grain boundaries leads to a negative latent strain and therefore a tensile film stress. An approximate expression for the stress generated by this mechanism in a film as a function of the average grain size d is [1,30]:

$$\sigma \approx Ew(1/d_i - 1/d) \tag{18}$$

where w is the excess volume per unit area associated with a grain. When dd_i, this stress approaches the value of Ew/d_i. Using values of $E = 100$ GPa, $d_i = 100$ nm, and $w = 0.1$ nm leads to a value of $Ew/d_i = 100$ MPa, suggesting that grain growth can be a significant contributor to the tensile component of the film stress. It should be noted that an increased grain size often leads to a reduced flow stress, so that increasing the grain size may allow for stress relaxation by plastic flow.

Unlike the initial compressive stress regime during island growth and the tensile stress stage during coalescence, where plausible models have been proposed to explain these behaviors, the origin of the compressive stress generated after coalescence is less clear. As noted previously, this regime is sometimes not observed. When it does occur, it may be a continuation of the compressive stress that was being generated during island growth that was temporarily masked by the tensile jump during coalescence. It is noted from Eq. (15) (considering only the first term on the right-hand side) that this mechanism leads to an asymptotic stress value of $-(f_s + h)/t_o$. If this asymptotic value is larger in magnitude than the tensile stress generated by the grain boundary relaxation mechanism, a superposition of the general compressive behavior resulting from surface stress effects (Eq. (15)) with a step-function-like tensile jump at coalescence can qualitatively explain the CTC behavior [13].

Recently, another mechanism that can lead to a contribution to the final stage compressive stress during deposition has been proposed [33], which involves the incorporation of excess atoms at the grain boundaries. The basic idea is that depositing atoms creates a large concentration of adatoms on the surface, much larger than the equilibrium concentration when there is no deposition. Consider a film–substrate system in the absence of deposition. At equilibrium, the chemical potentials for atoms on the film surface and in the grain boundary are equal. If the film is then exposed to a flux of depositing atoms, there will be an increase in the concentration of surface adatoms, resulting in an increase in their chemical potential. This chemical potential increase will create a driving force for the diffusion of atoms from the surface into the grain boundaries, resulting in a compressive film stress. If the deposition process is stopped, the chemical potential of surface adatoms will be reduced, inducing a flow of atoms out of the grain boundaries to the surface, relaxing the compressive stress. This type of relaxation after may explain the behavior shown in Fig. 3d.

C. Other Mechanisms

The stress-generating mechanisms discussed would appear to be generic for island growth and therefore are expected to occur during a variety of deposition processes. It has been suggested that other (generally tensile stress-generating) mechanisms may also be occurring during island growth. Examples include annihilation of an excess concentration of crystalline defects and phase transformations such as crystallization of a growing film that was deposited with an amorphous structure during the initial stage of growth [1,5]. In some cases, mechanisms that are specifically associated with particular deposition conditions or methods have been identified. The presence of a large partial pressure of residual gases during deposition may result in the incorporation of these gases and result in a compressive stress. Sputtered films are often found to display a large compressive stress. It is believed that this is principally a result of atoms of the sputtering gas (often argon) colliding with the film surface and creating an "atomic peening" effect. Incorporation of the atoms may also occur, although it is thought to be a less important effect.

D. Stress Relaxation in Nonepitaxial Films

If a film stress becomes large enough, certain stress relaxation processes may operate, sometimes leading to disastrous consequences. A large tensile stress can result in cracking, or the formation of voids. Stresses of either sign can induce flow by slip, or by diffusion (creep). Stress relaxation processes often encounter effects in thin films that impede or modify their operation compared to the case of conventional bulk materials. The presence of the substrate and, in some cases, a surface oxide, can strongly influence deformation processes [1]. For example, differences in the elastic properties of the film, substrate, and oxide may introduce dislocation image forces that can inhibit slip. Diffusional processes may be limited by a lack of vacancy sources and sinks in thin film microstructures. The constraint of the substrate and a surface oxide will also limit the amount of relaxation that can occur by grain boundary diffusion. As a result, diffusion often proceeds inhomogenously in a thin film. Because of this inhomogeneity, it has been observed that the relaxation of a compressive stress in polycrystalline metal films at elevated temperatures occurs by the formation of whiskerlike protrusions called hillocks [1,34]. Hillocks generally have a diameter similar to the grain size and can extend in height several times the film thickness. The formation of hillocks in metal film devices that undergo repeated cycling between a large temperature difference, resulting in a thermal stress, is a significant cause of device failure.

ACKNOWLEDGMENTS

The author gratefully acknowledges support from the National Science Foundation as administered through the Materials Science and Engineering Center at the Johns Hopkins University.

REFERENCES

1. Doerner, M.F.; Nix, W.D. CRC Crit. Rev. Solid State Mater. Sci. 1988, *14*, 225.
2. Tranter, C.J.; Craggs, J.W. Philos. Mag. 1947, *38*, 214.
3. Aleck, B.J. J. Appl. Mech. 1949, *16*, 118.
4. Blech, I.A.; Levi, A.A. J. Appl. Mech. 1981, *48*, 442.
5. Ohring, M. *The Materials Science of Thin Films*; Academic Press: Boston, 1992; 413 pp.
6. Stoney, G.G. Proc. R. Soc. Lond., Ser. A 1909, *82*, 172.
7. Hoffman, R.W. Phys. Thin Films 1966, *3*, 211.
8. Campbell, D.S. *Handbook of Thin Film Technology*; Maisel, L.I., Glang, R., Eds.; McGraw-Hill: New York, 1970; 123 pp.
9. Hoffman, R.W. *Physics of Non-Metallic Thin Films, NATO Advanced Study Institute, Series B*; Dupuy, C.H.S., Cachard, A., Eds.; Plenum: New York, 1976; Vol. 14, 273.
10. Flinn, P.A.; Gardner, D.S.; Nix, W.D. IEEE Trans. Electron Devices 1987, *34*, 689.
11. Floro, J.A.; Chason, E. *In Situ Real-Time Characterization of Thin Films*; Auciello, O., Krause, A.R., Eds.; Wiley: New York, 2001; 191 pp.
12. Cammarata, R.C.; Bilello, J.C.; Greer, A.L.; Sieradzki, K.; Yalisove, S.M. MRS Bull. 1999, *24* (2), 34.
13. Floro, J.A.; Chason, E.; Cammarata, R.C.; Srolovitz, D.J. MRS Bull. 2002, *27* (1), 29.

14. Shull, A.L.; Spaepen, F. J. Appl. Phys. 1996, *80*, 6243.
15. Murakami, M. CRC Crit. Rev. Solid State Mater. Sci. 1984, *11*, 317.
16. Murakami, M.; Segmüller, A.; Tu, K.N. *Analytical Techniques for Thin Films, Treatise on Materials Science and Technology*; Tu, K.N., Rosenberg, R., Eds.; Academic Press: Boston, 1988; Vol. 27, 201.
17. Doerner, M.F.; Brennan, S. J. Appl. Phys. 1988, *63*, 126.
18. Noyan, I.C.; Huang, T.O.; York, B.R. CRC Crit. Rev. Solid State Mater. Sci. 1995, *20*, 125.
19. Malhotra, S.G.; Rek, Z.U.; Yalisove, S.M.; Bilello, J.C. J. Appl. Phys. 1996, *79*, 6872.
20. Fewster, P.F.; Andrew, N.L. Thin Solid Films 1998, *319*, 1.
21. Matthews, J.W. *Epitaxial Growth, Part B*; Matthews, J.W., Ed.; Academic Press: New York, 1975; 559 pp.
22. Cammarata, R.C. Prog. Surf. Sci. 1994, *46*, 1.
23. Koch, R.; Abermann, R. Thin Solid Films 1985, *129*, 71.
24. Abermann, R. Mater. Res. Soc. Symp. Proc. 1992, *308*, 25.
25. Laugier, M. Vacuum 1981, *31*, 155.
26. Cammarata, R.C.; Trimble, T.M.; Srolovitz, D.J. J. Mater. Res. 2000, *15*, 2468.
27. Dojack, F.A.; Hoffman, R.W. Thin Solid Films 1972, *12*, 71.
28. Pulker, H.K. Thin Solid Films 1982, *89*, 191.
29. Nix, W.D.; Clemens, B.M. J. Mater. Res. 1999, *14*, 3471.
30. Freund, L.B.; Chason, E. J. Appl. Phys. 2001, *89*, 4866.
31. Sheldon, B.W.; Lau, A.; Rajamani, A. J. Appl. Phys. 2001, *90*, 5097.
32. Seel, S.C.; Thompson, C.V.; Hearne, S.J.; Floro, J.A. J. Appl. Phys. 2000, *88*, 7079.
33. Chason, E.; Sheldon, B.W.; Freund, L.B.; Floro, J.A.; Hearne, S.J. Phys. Rev. Lett. 2002, *88*, 1561.
34. Tu, K.N.; Mayer, J.W.; Feldman, L.C. *Electronic Thin Film Science for Electrical Engineers and Materials Scientists*; Macmillan: New York, 1992; 373 pp.

3
Metallic Tribo-oxides: Formation and Wear Behavior

Hon So
National Taiwan University, Taipei, Taiwan

I. INTRODUCTION

When two metallic surfaces are loaded together and one surface slides over the other, considerably thick oxide films can be formed on the sliding surfaces if the normal pressure and sliding speed are appropriate. Such oxide films will strongly affect the behavior of friction and the wear of contacting metals. This kind of oxidation can be defined as tribo-oxidation. Many investigators have carried out studies relevant to tribo-oxidation over many decades. The topics of research into this area can roughly be divided into three groups: (a) the formation and effects of oxides on friction characteristics of metals; (b) the thermal aspect of sliding contact; and (c) the wear rate and wear mechanism of tribo-oxidation. The last one is commonly known as oxidational wear. Because the formation of metallic oxides on sliding surfaces depends on frictional heating and the wear rate is affected by the type of oxides as well as the combination of sliding velocity and normal load, many studies take the three groups as a whole to obtain a complete view on the problem. A brief review on such topics is given below, whereas a more detailed discussion is provided later on.

It has been known for many decades that a layer of oxide about 10^{-8} m thick may form on the free surfaces of many metals and their alloys [1,2], when they are exposed to normal atmosphere. It has been pointed out that oxide layers have profound effects on the friction and the wear of parent metals. For instance, the frictional force caused by rubbing between metals can be reduced in the presence of oxide films. On the other hand, due to frictional heating, considerably thick oxide films can be formed on the rubbing surfaces of metals, when the normal load and sliding speed are high enough. Quinn [3] found that the thickness of an oxide film could be as large as 3 μm. However, So [4] and So et al. [5] obtained the oxide films on rubbing surfaces with a thickness even larger than 30 μm. The mechanical strengths of the oxide films produced in such a situation greatly affect the tribological properties of the metals in a normal atmospherical condition. If the oxide is very hard, such as Al_2O_3, SnO_2, or Mn_2O_3, it will be embedded in the underlying metals and may cause heavy wear to the intimate surface. If the oxide is softer than the underlying metal, such as Fe_3O_4, Cu_2O, or ZnO, and so on, it will function as lubricant on the rubbing surfaces and will reduce the frictional force as well as wear of the parent metals. Relevant results presented by Johnson, Dies, Finch, and Rabinowicz were summarized in the books written by Bowden and Tabor

[1] and Rabinowicz [2]. Quinn [3,6], Sullivan et al. [7], and Hong et al. [8] found that the latter case of wear fell in the category of mild wear defined by Archard and Hirst [9]. The so-called mild wear is defined such that the wear rate (i.e., volume loss per unit distance slid) is on the order of 10^{-13}–10^{-12} m^3/m. Then, Quinn proposed that as the oxide reached a critical thickness, usually 1–3 μm, the oxide broke up and appeared as wear debris. Therefore, it was suggested that the wear rate was proportional to the rate of oxidation, and the wear loss in such a situation was called oxidational wear. However, there is no convincing evidence to confirm such a proposal that the oxide films have to reach a critical thickness of 1–3 μm before they break up. On the other hand, So [4] and So et al. [5] found that the thicknesses of oxide films have magnitudes from 2 to 30 μm and are usually between 10 and 20 μm before they break up, and that the thickness of wear debris is only a part of the whole thickness of any oxide film. In addition, oxides break up due to the propagation of fatigue cracks. Quinn [10] also realized such a mechanism.

Another important influence on tribo-oxidation is the temperature of real contact area. Many investigators tried to obtain such a temperature as accurate as possible. Bowden and Tabor [1] used a metallic sliding pair to form a thermocouple and obtained the contact temperature. However, what they obtained was an average temperature at the nominal contact surface. Blok [11], Jaeger [12], and Barber [13] presented analytical results for temperature at sliding contact. Because of neglecting the exact value of the real contact area, their results did not match the real situation of sliding contact completely. Quinn and Winer [14] obtained the "hot spots" on the sapphire and steel interface during sliding and estimated the temperatures of hot spots by their color. However, the color of a hot spot cannot provide the exact temperature of that spot. In addition, by using a guessed number of contact spots, Quinn and Winer [15] computed the flash temperature at the real contact area. Because there is no rule for selecting the correct number of contact spots, the accuracy of their results was questionable.

By modifying Quinn's method [7,8] of measuring the temperatures at two or three points on the surface of the pin in a pin-on-disc wear test, So [16] used the basic equations of heat transfer instead to obtain the temperature distribution in the pin. Then, the average temperature at the apparent contact surface, which is defined as the nominal contact temperature or bulk temperature T_b in this article, can be obtained. By using this nominal contact temperature together with the hardness of the pin material at elevated temperatures, the temperature of real contact area, which is defined as the real contact temperature (or simply contact temperature) T_c, can be calculated with a trial-and-error method. A detailed description will be provided later on.

Bowden and Tabor [1] pointed out that the temperature at the contact junctions of a sliding pair was very high even when a liquid lubricant was present. Massive oxide is usually not found on such contact surfaces. However, if some antiwear or extreme pressure additives are added to the liquid lubricant, chemical films will be produced on the rubbing surfaces. If there are some kinds of chemical films formed on the rubbing surfaces, these films will have a profound influence on the tribological behavior of the metallic bodies under boundary lubrication conditions [17–19]. However, such relevant research is out of the scope of this article.

II. FRICTIONAL HEATING AND SURFACE TEMPERATURE

A. Frictional Heating

According to Bowden and Tabor [1], when two solid surfaces are loaded together, they can only touch at the tips of some asperities and make adhesion over such small contact areas. Because the real contact area is very small, the true contact pressure is sufficiently high to cause the contacting asperities to deform plastically. There are three main mechanisms liable to contribute to the loss of energy at the contacting surfaces, when relative motion takes place between these contacting surfaces. The material beneath the adhesive junctions and the loose particles retained between the contact surfaces can be deformed. The deformation can be elastic and plastic, or viscous. Both plastic and viscous deformation can cause a loss in energy. The other mechanisms of losses of energy are due to shearing of adhesive junctions, breaking up of asperities and loose particles into smaller debris, and cutting or ploughing of the softer surface by the hard surface or particles. Shearing, cutting, and ploughing always accompany plastic deformation. All such losses in energy require external work.

If the external work required to overcome frictional resistance can transform to heat, the frictional heat can be estimated with an equation, as follows:

$$Q = \mu L v, \tag{1}$$

where Q is the rate of heat generated in the contacting bodies (i.e., the total quantity of heat generated per unit time), μ is the coefficient of kinetic friction, L is the normal force on the contacting surfaces, and v is the relative velocity of the sliding surfaces. Part of the heat flows into one member, whereas the remainder flows into the other. However, not all the energy mentioned above will be liberated as heat. In fact, only the work bringing about plastic or viscous deformation may be liberated as substantial heat. Therefore, only the plastic or viscous zones may be the most possible heat sources. The use of Eq. (1) will obviously overestimate the quantity of frictional heat.

As soon as frictional heat is generated, the plastic zones and the material near the contact interface encounter a temperature rise. Such a temperature rise must depend on the rate of deformation (i.e., the rate of plastic strain or viscous strain). When the rate of plastic (viscous) strain is low, or the rubbing bodies have excellent properties of heat conduction, the heat has sufficient time to conduct away. Consequently, the increase in temperature during heat generation will not be remarkable. When the rate of plastic (viscous) strain is high, or the rubbing bodies have poor heat conduction properties, such conditions may be considered an adiabatic case and the increase in temperature in the plastic (viscous) zones will be substantial. A high rate of strain always results from a high sliding velocity between the rubbing bodies.

Another factor that affects the temperature rise in plastic zones is the temperature gradient in the subsurface. The rate of heat flowing into one member of a sliding pair is dependent on temperature gradient ∇T, such that $Q_1 = -kA\nabla T$, where k is the thermal conductivity and A is the cross-sectional area. Consequently, the greater the temperature gradient is, the less is the heat quantity remaining in the plastic (viscous) zones. In other words, the temperature rise dies down quickly after only one pass of rubbing. However, if the rubbing action is repeatedly or continuously carried out, frictional heating will accumulate to cause a gradual temperature rise in both of the rubbing bodies. Eventually, a steady state in heat transfer can be reached, in which the rate of generation of frictional heat is equal to the rate of heat transferred away. In such a steady-state condition, both the temperature distributions in the pin and the disc become constant. Figure 1 indicates such a condition. The nominal contact temperature (bulk temperature) shown in Fig. 1a slightly fluctuates due to the variation in friction (Fig. 1b). On the whole, the temperature may be considered to be constant once the sliding distance is over 4 km.

B. Measurement of Frictional Temperature

In general, the total real area of contact is only a small part in the whole apparent area of contact. The contact temperature at one junction may not be the same as that at the others.

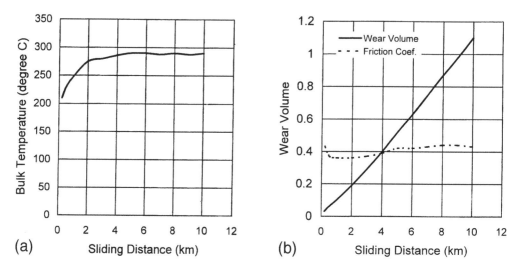

Figure 1 Variations in nominal contact temperature (bulk temperature). (a) Coefficient of friction and wear volume (in 10 mm^3) (b) with sliding distance for an AISI 4140 steel pin rubbing on an AISI 4340 steel disc at a normal pressure of 2.54 MPa and a sliding speed of 2 m/sec.

The observation of "hot spots" at contact interfaces carried out by Quinn and Winer [14] confirmed such a condition. In their experiment, Quinn and Winer used tool steel pins sliding against a rotating sapphire disc without lubrication, so that "hot spots" between a pin and the disc surfaces could be seen and photographed. The temperature of the hot spots could be estimated by infrared scanning, or by comparing photographs of heated coupons of the same material. Bowden and Tabor [1] conducted another kind of experiment to find the contact temperature. They used the rubbing surfaces as a pair of thermocouple to measure the contact potential. However, the measured electrical potential was only an integrated value of the potentials at all the contact junctions; such an electrical potential only indicated an average temperature at the apparent contact surface. Furthermore, if there are some oxides formed on the surfaces, or if there are some loose particles from the soft surface sticking on the hard surface, it will be a great error in using the measured potential to indicate the exact temperature at the real contact surface. On the other hand, although Quinn and Winer [14] could identify the areas as well as the color of the hot spots, the temperatures of the hot spots were not directly measured with some devices. Hence, the temperatures they obtained were only estimated values. Therefore, it can be concluded that the measurement of exact temperature at the real contact surface of two rubbing bodies is almost impossible.

C. Calculation of Temperature at Contact Surfaces

Instead of measuring the exact temperatures at real contact surfaces, some investigators [11–13,20] used analytical methods, or partly analytical partly experimental methods to calculate the temperature of a real contact area. In most analyses, two assumptions—(a) that the temperature at the subsurface of the contact, which is equivalent to the bulk temperature, was the same as the ambient temperature, and (b) that all the frictional work was dissipated into heat—were made. These assumptions are almost true for rubbing surfaces if the moving surface slides over the stationary surface in only one pass. However, if the rubbing action is continuous and repeated for a long time and the rubbing bodies encounter wear loss, the prediction of contact temperature by using those assumptions will result in a big error. Realizing such a situation, Allen et al. [21] arranged for a pin-on-disc wear test rig to measure the temperatures at the stationary pin. Using these measured temperatures, Rowson and Quinn [22] could obtain the heat flowing to the pin in a steady-state condition. Then, Quinn and Winer [15] could compute the contact temperature by guessing a number of contact plateaus. In their experiments, Allen et al. [21] employed a pin-on-disc configuration similar to Fig. 2, except that the pin was mounted in a thermal insulator mounted within a copper calorimeter up to the upper thermocouple. If the positions of the two thermocouples are fixed, two temperatures of the pin at two known positions and the temperature of the calorimeter could be measured during rubbing. The portion of the total frictional heat flow rate that flows to the pin could be computed after a laborious derivation of equations has been done [3,15,21,22].

Instead of following Quinn's method of measuring temperatures on an insulated stationary pin and making tedious calculations, So [16] allowed the pin to be exposed to air in a pin-on-disc test rig (Fig. 2) and measured the temperatures at three fixed positions on the surface of the pin (Fig. 3). Then, by employing the basic theory of heat transfer, So can obtain the rate of heat flowing to the pin as well as the nominal contact temperature (bulk temperature) at the apparent contact area in a quasi-steady-state condition. If a natural convection condition is assumed for the stationary pin, the heat conducted away from the clamped end of the pin dominates over that by convection through air. Furthermore, if the diameter of the pin is much smaller than its length, one-dimensional heat conduction can be assumed. Consider the energy balance between any two cross-sections of area A_n separated by a small axial distance dx of the cylindrical pin (Fig. 3). The rate of heat flowing into this axial element q_x at x should be equal to the rate of heat flowing out this element q_{x+dx} at

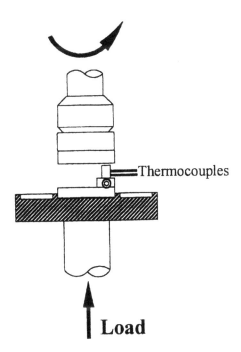

Figure 2 Schematic diagram of the pin-on-disc arrangement.

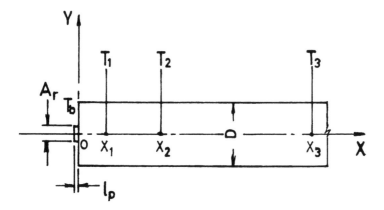

Figure 3 Coordinate system for temperature measurement and calculation.

$x + dx$ plus the rate of heat $q_{\Delta A}$ dissipating to the air from the cylindrical surface of the element $C_o dx$ by convection. Then:

$$q_x = q_{x+dx} + q_{\Delta A}. \tag{2}$$

The conduction terms become:

$$q_x - q_{x+dx} = -\frac{dq_x}{dx}dx = -\frac{d}{dx}\left(-k_p A_n \frac{dT}{dx}\right)dx \tag{3}$$

and the convection term is:

$$q_{\Delta A} = C_o dx (T - T_a). \tag{4}$$

Substituting Eqs. (3) and (4) in Eq. (2) yields:

$$\frac{d}{dx}\left(k_p A_n \frac{dT}{dx}\right) - hC_o(T - T_a) = 0, \tag{5}$$

where k_p is the thermal conductivity of the pin material, h is the convection heat transfer coefficient, C_o is the circular perimeter of the pin, and T_a is the ambient temperature. If the thermal conductivity and the convection coefficient are assumed to be constants and equal to the average values, respectively, the general solution for Eq. (5) is:

$$T(x) = T_a + C_1 \exp(mx) + C_2 \exp(-mx), \tag{6}$$

where:

$$m = \left(\frac{C_o h}{k_p A_n}\right)^{\frac{1}{2}}. \tag{7}$$

If the temperatures T_1 and T_2 at any two points x_1 and x_2 on the pin are measured, the constants C_1 and C_2 can be obtained:

$$C_1 = \frac{(T_1 - T_a)\exp(-mx_2) - (T_2 - T_a)\exp(-mx_1)}{\exp(m(x_1 - x_2)) - \exp(-m(x_1 - x_2))} \tag{8}$$

$$C_2 = \frac{(T_1 - T_a)\exp(mx_2) - (T_2 - T_a)\exp(mx_1)}{\exp(-m(x_1 - x_2)) - \exp(m(x_1 - x_2))}. \tag{9}$$

The temperature at any position x_3 of the pin can be shown as:

$$T_3 = T_a + C_1 \exp\left(x_3\sqrt{\frac{C_o h}{k_p A_n}}\right) + C_2 \exp\left(-x_3\sqrt{\frac{C_o h}{k_p A_n}}\right). \tag{10}$$

If this temperature is measured, the constants C_1, C_2, and h can be obtained by a trial-and-error method.

Alternatively, the convection heat transfer coefficient h can be simply approximated as:

$$h = 1.42\left(\frac{\Delta T}{x_2 - x_1}\right)^{\frac{1}{4}}, \tag{11}$$

where:

$$\Delta T = \frac{1}{2}(T_1 + T_2) - T_a.$$

The average temperature, the nominal contact temperature T_b at the apparent contact area, is found to be:

$$T_b = T_a + C_1 + C_2, \tag{12}$$

when the origin is located at the center of the apparent contact surface. The portion of heat flowing to the pin per unit time is found as follows:

$$Q_p = -k_p A_n \frac{dT(0)}{dx}. \tag{13}$$

From Eq. (6), we arrive at:

$$Q_p = k_p A_n m(C_2 - C_1), \tag{14}$$

provided that this rate of heat can induce an average temperature T_b at the apparent contact area A_n although the heat is assumed to be generated in the integrated plastic zone. In a real situation, the real contact area is made from many small contact plateaus. The area of each plateau is different from the most of the others. Under each plateau, there is a plastic zone, the depth of which is also different from that of most others. If the frictional heat is generated from such plastic zones, these zones will be oxidized with first priority. Therefore, the thickness of oxide films will not be uniform. Figure 4 indicates such a situation. The taper sections shown in the figures were chosen arbitrarily. Therefore, if the heat generation as well as the temperature at individual contact plateaus are taken into consideration alone, the problem will be very complicated. To make the problem simpler, an integrated area A_r of all small contact plateaus is considered as a real contact area, and an average temperature T_c at such an integrated contact area is assumed to be a representative temperature. Such a temperature can then be obtained easily by Eq. (13) with suitable variables as follows:

$$T_c = T_b + \frac{l_p Q_p}{k_p A_r}, \tag{15}$$

where l_p is the average depth of all small plastic zones below the contact interface. This value can be approximated by the average thickness of oxide films that cover the apparent contact surface of the pin. Under such a condition, the thermal conductivity of the pin k_p should be replaced with the thermal conductivity k_{ox} of the oxides.

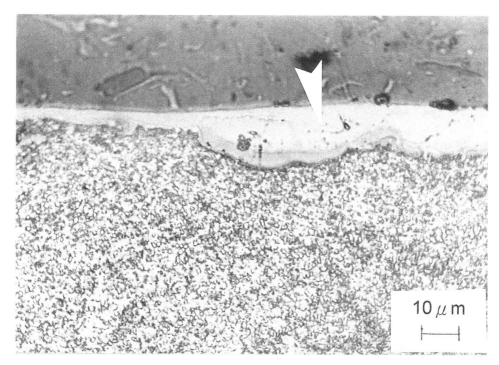

Figure 4   Taper section showing the thickness distribution of an oxide film formed on the surface of a pin made of the tool steel W1-100C rubbing on the same steel disc at a sliding speed of 1 m/sec and at a normal pressure of 2.75 MPa, and on the surface of a bearing steel pin rubbing on a die steel disc at 3 m/sec and 4.4 MPa for 3 min.

The real contact area can be obtained with the well-known method such that:

$$A_r = \frac{L}{H_a},$$ (16)

where L is the normal load, and H_a is the smaller value of the material hardness of the pin and the disc at room temperature, respectively. Quinn [3], Quinn and Winer [15], Allen et al. [21], Rowson and Quinn [22] suggest that $A_r = L/H_b$, where H_b is the smaller value of the two material hardnesses of the pin and the disc at the nominal contact temperatures of the pin and the disc, respectively. In a continuously rubbing process, the temperatures at the real contact surface and at the nominal contact surfaces will gradually rise to higher values, when the rubbing process is maintained for a while. According to Bowden and Tabor [23], the contact junctions will grow plastically when a tangential force is applied to the contact pair in addition to the normal load. On this behalf, the growth of plastic junctions as well as the real contact area must be determined by the yield strengths of the contacting materials at elevated temperatures that may be the real contact temperatures. Therefore, it is more reasonable to use the smaller value of the two metal hardnesses of the pin and the disc at the contact temperature H_c than that at the nominal contact temperature. Thus:

$$A_r = \frac{L}{H_c},$$ (17)

Substituting Eq. (17) into Eq. (15) and assuming that the hardness of the pin is smaller, one can obtain:

$$T_c = T_b + \frac{H_c l_p Q_p}{L k_p}.$$ (18)

Because the contact temperature T_c and the hardness H_c in Eq. (18) are dependent on each other, they ought to be computed with a trial-and-error method. To find T_c, one should try a value for T_c first, and then use the hardness of the pin material H_c at T_c in Eq. (18). If the assumed T_c is equal to the sum of the right-hand side, T_c is the required temperature. As usual, the hardness H_c can be replaced by three times the value of the yield stress in the

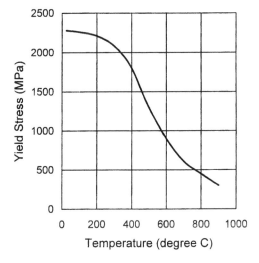

Figure 5 The plot of yield stress in compression against the temperature for W1-90C tool steel.

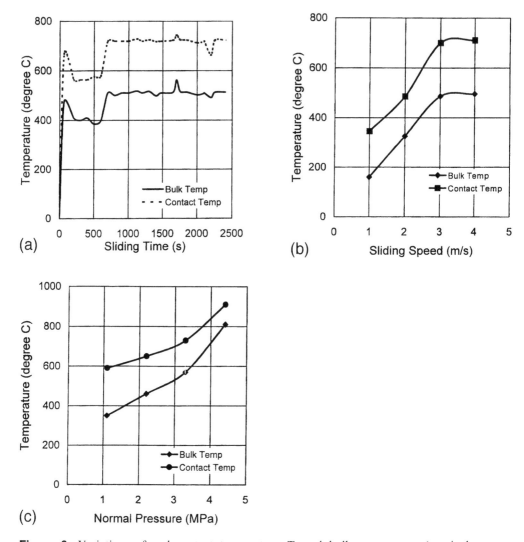

Figure 6 Variations of real contact temperature T_c and bulk temperature (nominal contact temperature) T_b (a) with sliding time at a normal pressure of 2.2 MPa and a sliding speed of 4 m/sec; (b) with sliding speed at a normal pressure of 2.2 MPa; and (c) with normal pressure at a speed of 4 m/sec. The data in (b) and (c) were computed in steady-state conditions.

compression of the material at T_c. In accordance with the yield stresses in compression at elevated temperatures for W1-90C tool steel shown in Fig. 5, the contact temperatures of W1-90C tool steel pins rubbing on die steel discs in various conditions can be obtained. Some typical results are indicated in Fig. 6.

The average temperature at the contact surface of the disc can be estimated with the heat partition Q_d generated on the disc surface:

$$Q_d = \mu L v - Q_p - Q' \tag{19}$$

where Q' is the rate of energy dissipation for creating new surfaces and for accelerating the debris. However, Q' is difficult to determine and is therefore neglected. Neglecting Q' will cause the computed temperature T_{bd} at the apparent contact surface of the disc to be underestimated. Then:

$$T_{bd} = T_c - \frac{l_d Q_d}{k_p A_r}, \tag{20}$$

where l_d is the average depth of all small plastic zones in the disc below the contact interface. Again, this value can be approximated by the average thickness of oxide films on the disc, where k_d is the thermal conductivity of the disc.

III. EXPERIMENTS

A. Test Rig

To study the formation and the wear mechanism of metallic tribo-oxides, it is sensible to use a pin-on-disc arrangement as the wear test rig. Here, the pin is kept stationary while the circular disc is rotated. A schematic diagram is shown in Figs. 2 and 3. The load is applied by means of dead weights through a stationary vertical rod. The pin is mounted at an adapter mounted on the top end of this rod and is forced against the bottom surface of the rotating disc. The temperature of the pin is measured with two K-type thermocouples welded on the pin at distances of 2 and 6 mm from the top of the pin, respectively. Besides, the temperature near the bottom end of the pin that is mounted on an adapter is measured with the third thermocouple. The three thermocouples give three readings for temperatures, T_1, T_2, and T_3 of the pin at any moment. In such an arrangement, the distances from the thermocouples to the contact surface might change at every moment as the pin is worn by the rubbing action. The amount of wear is measured with a linear variable differential transformer (LVDT) and a data acquisition system.

Table 1 Chemical Compositions of the Metals Used for the Specimens

	Fe	C	Mn	Ni	Cr	Mo	Si	V	Co	W
Medium carbon steel, AISI7 1045	Bal.	0.42~0.48	0.60~0.90							
Tool steel, W1-90C	Bal.	0.90~1.00	<0.50				<0.35			
Tool steel, W1-100C	Bal.	1.00~1.10	<0.50c				<0.35			
Die steel	Bal.	1.40~1.60	<0.60	<0.4	11~13	0.8~1.20	<0.4	0.20~0.50		
Cr–Mo steel, AISI 4140	Bal.	0.38~0.43	0.60~0.85		0.90~1.20	0.15~0.30				
Ni–Cr–Mo steel, AISI 4340	Bal.	0.36~0.43	0.60~0.90	1.60~2.0	0.60~1.0	0.15~0.30				
Bearing steel, AISI 52100	Bal.	0.95~1.10	<0.50		1.30~1.60					
Stellite alloy 6	2.5	1.0	1.0	2.5	27.0		1.0		60	5

Table 2 Vickers Hardness Numbers of Alloys (Heat-Treated) Used in Present Tests

Medium carbon steel	Tool steel, W1-90C	Tool steel, W1-100C	Die steel	Cr–Mo steel, AISI 4140	Ni–Cr–Mo steel, AISI 4340	Bearing steel	Stellite alloy 6
671	765	800	812	970	750	826	580 ~ 650

B. Specimens

The pin specimens were 4.75 mm in diameter and 15 mm in length, and those of the discs were 32 mm (or 55.5 mm) in diameter and 7 mm (or 10.88 mm) thick. Several kinds of steel were used for pin specimens, which included a medium carbon steel (AISI 1045), tool steels (AISI W1-90C and AISI W1-100C), die steel, bearing steel (AISI 52100), Cr–Mo steel (AISI 4140), Ni–Cr–Mo steel (AISI 4340), laser-clad stellite alloy 6, and so on. The disc specimens were made from the medium carbon steel and tool steel, die steel, Ni–Cr–Mo steel, and laser-clad stellite alloy 6. The chemical compositions of the alloys are listed in Table 1, whereas the average hardness number of each alloy after being heat-treated is listed in Table 2.

C. Test Procedures

The experiments were carried out at nominal sliding speeds from 0.6 to 8 m/sec. The load was set at values from 9.8 to 392 N to yield apparent contact pressures from 0.55 to 22 MPa. The duration of a test depended upon the rubbing materials, the speed, and the load. Except those tests for special purposes (e.g., for studying the growth of oxide films), most tests were conducted until the thermal condition or the wear rate had reached a steady-state condition.

IV. FORMATION OF METALLIC OXIDES

In a continuous rubbing process, the sliding speed not only governs the rate of heat generation but also governs the rate of deformation. If the rate of heat generation is higher than that of heat conduction, the temperature at the real contact surface will rise until the thermal condition arrives at a steady state. On the other hand, the applied normal load governs the sizes and the number of plastic zones. The greater the applied load is, the larger is the plastic zone size and the more are the plastic zones. The type and the thickness of oxides formed on sliding surfaces are governed by the real contact temperature and the sizes of plastic zones. Experimental evidence indicating such phenomena will be shown later on.

A. Growth of Oxide Film on Rubbing Surfaces

It is known that oxides with thickness on the order of 10^{-8} m on almost all the free metallic surfaces are easily found when metals are exposed to the normal atmosphere [1,2]. On the other hand, it is observed that the thickness of oxide films formed on metallic surfaces through rubbing action is two orders of magnitude larger than that on free metallic surfaces.

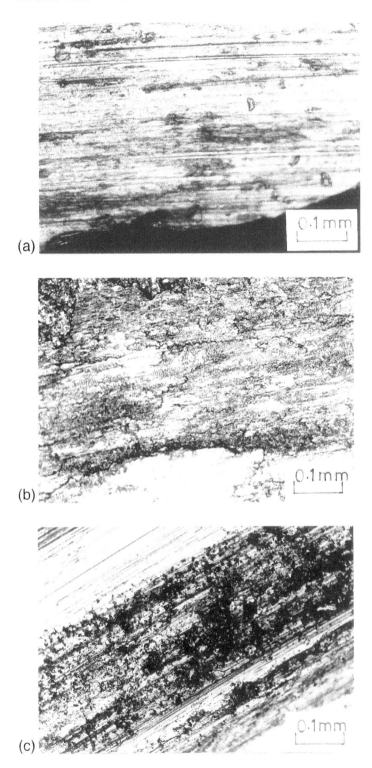

Figure 7 Photographs showing the increase in the density of oxide on the surfaces of pins made of W1-90C steel with sliding time at a normal pressure of 4.4 MPa and at a sliding speed of 3 m/sec for (a) 10 sec, (b) 30 sec, and (c) 60 sec of sliding.

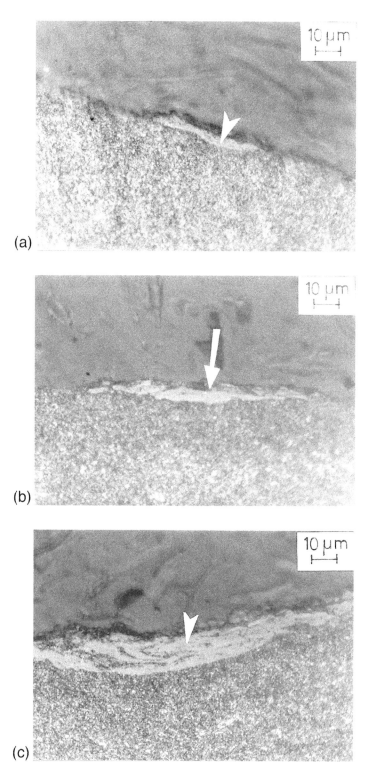

Figure 8 Photographs showing the taper sections of oxide films under the conditions depicted in Fig. 7 for (a) 10 sec, (b) 30 sec, and (c) 60 sec of sliding.

Quinn [3,6] developed a theory for the growth of oxide films, in which it was assumed that parabolical oxidation occurred at the contact surfaces. This means that the increase in thickness of an oxide film obeyed a parabolical equation in terms of sliding time. Later, as Quinn found that the parabolical oxidation did not satisfy most experimental results, he proposed a linear oxidation theory instead. However, it is still difficult to confirm linear oxidation from experiments [4,5,16].

Recent experiments on the oxidational wear of some steels show that oxidation immediately occurred at some relatively high asperities of the two rubbing surfaces at the beginning of a rubbing process. Such relatively high asperities were squashed down as soon as they were in contact with the intimate asperities and were oxidized simultaneously. The next higher asperities on intimate surfaces were then in contact with one another and were oxidized, and so on. Under such a situation, the area of oxide films quickly spread over the apparent contact surfaces in some 10 sec as the rubbing action proceeded. In the mean time, some oxide films cracked due to intermittent rubbing and then spalled off. In general, the thicknesses of the cracked oxide films were not the same, neither was that of the debris. Moreover, the increase in the thickness of oxides is not significant during rubbing. Figure 7 indicates the increase in density of oxides formed on the rubbing surface with time. Figure 8 depicts that the increase in the thickness of oxide films was not caused by the growth of the initial oxides, but was directly caused by the increase of plastic zone sizes, which, on the other hand, were governed by the rise of the nominal and real contact temperatures when the normal load and sliding speed were kept constant. These temperatures could lower the hardnesses of the pin and the disc so as to cause larger plastic zone sizes. Generally, the thickness of an oxide film ranges from as small as 2 μm to as large as 30 μm. Obviously, the oxides come from the squashed asperities. The depth of a plastic zone must be the most important factor that governs the thickness of an oxide film formed on the rubbing surface.

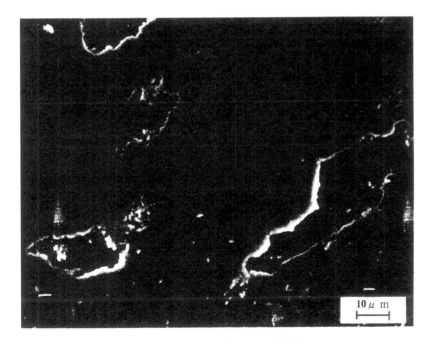

Figure 9   Photograph showing the spalling of oxide film.

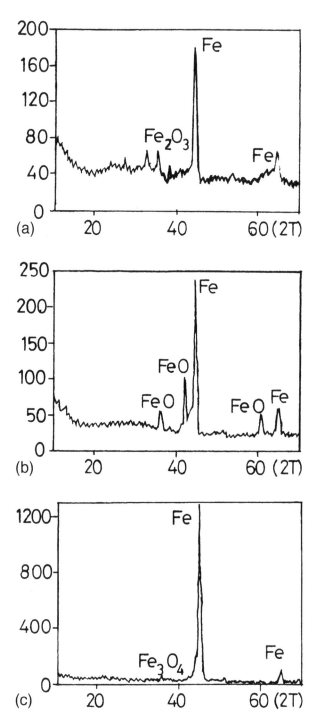

Figure 10 X-ray diffraction analysis showing the types of oxides formed on the rubbing surfaces of the pins made of (a) die steel sliding at 2 m/sec and at 1.1 MPa; (b) die steel sliding at 4 m/sec and at 2.2 MPa; and (c) W1-100C tool steel sliding at 3 m/sec and at 1.1 MPa.

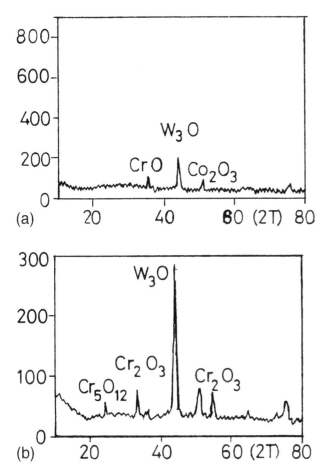

Figure 11 X-ray diffraction analysis showing the types of oxides formed on the rubbing surfaces of pins made of stellite alloy 6 rubbing with the die steel at a sliding speed of 4 m/sec and a normal pressure of (a) 4.4 MPa and (b) 8.8 MPa.

In other words, the thickness of an oxide film depends mainly on the applied load and the height of the asperity. The formation and the spalling of oxides occur simultaneously. In addition, both the thick and thin oxides may be spalled off. Therefore, the growth of oxide films is not an important phenomenon that controls the wear rate of the rubbing bodies. One can observe that oxide films in various thicknesses occur at the same time on the same rubbing surface of the specimen, and that cracks have been created in oxides liable to spalling (Figs. 4a and 8b). Figure 9 depicts the wear surface of an AISI 4340 steel disc rubbing on an AISI 4140 steel pin. Some parts of the oxide film have already spalled off the worn surface.

B. Types of Oxides

The compositions of oxides produced on the rubbing surfaces not only depend on the materials but also rely on the sliding speed and the normal pressure on the rubbing surfaces. Suitable normal pressures can enhance the effect of sliding speed on the formation of oxides.

A half century ago, Dies [1] found that both Fe_2O_3 and FeO could be formed on the rubbing surfaces made from mild steel in the atmosphere. Johnson, Godfrey, and Bisson [1] pointed out that the oxide αFe_3O_4 was more effective as a lubricant than Fe_2O_3. Quinn [24] confirmed that when the temperature at the contact spots on a steel surface was lower than 450°C, the dominant oxide was rhombohedral oxide (αFe_2O_3); when the temperature ranged from 450°C to 600°C, the dominant oxide was spinel oxide (Fe_3O_4); and when the temperature was over 600°C, the oxide structure changed to ferrous oxide (FeO). Examples are given in Fig. 10, wherein the three types of oxides were detected in three different conditions. Similarly, different oxides can be found on the surfaces of laser-clad stellite 6, when they are rubbing with AISI 4340 steel discs in different conditions (Fig. 11). Obviously, the types of oxide formed on stellite 6 mainly depend upon the contact temperature [5]. If one takes the nominal contact temperatures as the reference, Co_2O_3 will be the dominant oxide at a temperature lower than 500°C. When the nominal contact temperature is increased to 800°C, CrO appears on the rubbing surface. When the temperature is increased further to above 1000°C, Cr_2O_3 and Cr_5O_{12} are the dominant oxides.

V. EFFECT OF OXIDES ON FRICTION AND WEAR

A. Wear Mechanism

It is common knowledge for tribologists that different combinations of sliding speed and normal loads applied to two metallic surfaces in contact may result in different wear mechanisms. In this regard, Lin and Ashby [25] constructed a wear mechanism diagram for steel in accordance with experimental data. In the diagram, they used normalized velocity (sliding velocity times radius of pin/thermal diffusivity) as well as sliding velocity as abscissas, and normalized pressure (normal force/apparent contact area/hardness) as ordinate. The contours of constant wear rate were also plotted. They divided the diagram into six regions to cover a wide range of wear rates. When the normalized pressure is over unity at any velocity, seizure is a dominant mechanism. When both the pressure and the sliding velocity are low, it is an ultra mild wear region at the bottom left corner of the map. Between these two extreme regions, there are four other fields of different wear mechanisms. At the low-speed range where the normalized velocity is smaller than 50, delamination wear dominates over other wear mechanisms. When the normalized velocity ranging from 50 to 1000 (it is roughly equivalent to a sliding velocity ranging from 0.6 to 10 m/sec) at any normalized pressure is smaller than unity, it falls in a region of mild oxidational wear. When the normalized velocity is higher than this range, melt wear occurs at higher normalized pressures, whereas severe oxidational wear occurs at lower pressures. When rubbing pairs made of different steels are subjected to normal pressures from 1.1 to 4.4 MPa at sliding speeds from 1 to 4 m/sec, the wear mechanism of the rubbing pairs falls in the region of mild oxidational wear. In such a region, oxide films of different thicknesses are formed on the rubbing surfaces as the rubbing action continues. In the mean time, it can be observed that some cracks start to initiate in the oxide and then propagate. Some parts of oxide films spall off and a typical wear pattern occurs on the rubbing surfaces. Figure 12 shows the wear surface of a die steel pin rubbed on an AISI 4340 steel disc at a speed of 4 m/sec under a normal pressure of 4.4 MPa. The nominal contact temperature in such a condition can be as high as 815°C. The real contact temperature may rise to 910°C. At such a high temperature, the flow stress of the die steel decreases substantially. One sees that many cracks occur in the oxide and that the surface of the oxide film is quite smooth, except the spots where the oxide has been spalled off. When normal pressure is smaller than 0.55 MPa and the sliding speed is lower than 1 m/sec, the wear of the rubbing surfaces is not caused by spalling of oxide films

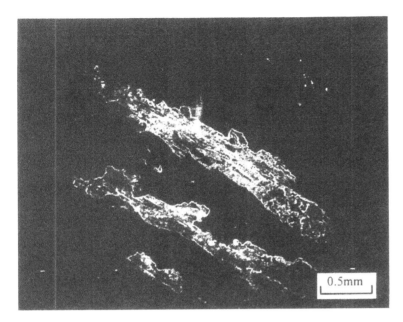

Figure 12   Wear appearance of a die steel pin rubbing on an AISI 4340 steel disc at a sliding speed of 4 m/sec and a normal pressure of 4.4 MPa.

because there is no sufficiently thick oxide film formed on the rubbing surfaces subjected to spalling. It returns to some kind of mechanical wear.

Under a normal pressure greater than 2.75 MPa with a sliding speed from 2 to 4 m/sec, one may observe that layers of the pin material with thickness larger than 10 µm were intermittently extruded from the contact interface, sometimes with sparkles. The experimental observation indicates that the higher the pressure and sliding speed are, the thicker

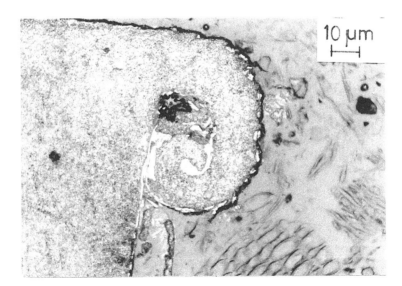

Figure 13   Photograph showing a layer of the pin made of W1-90C tool steel extruded from the rubbing surface at 4 m/sec and 3.3 MPa sliding for 5 min.

are the extruded layers and the shorter is the time interval for a layer to be extruded off the contact interface. Figure 13 indicates an extruded layer that makes the pin become a mushroom shape [16]. Generally, under a normal pressure greater than 8.8 MPa (equivalent to a normalized pressure of 1×10^{-3} for tool and die steels) with a sliding speed higher than 3 m/sec, the surface of a steel pin will be continuously extruded from contact with intense sparkles. The wear of the pin becomes severe.

B. Wear Debris

The wear debris collected from wear tests under different conditions can be examined with a scanning electron microscope. In general, the average sizes of the debris under higher normal pressures are larger than that under lower pressures. Figure 14 [4] indicates such a situation. The debris shown in the photographs was collected from the wear tests of die steel pins rubbing on AISI 4340 steel discs under different normal pressures but at the same sliding speed. Obviously, the debris comes from spalling of oxide films on the rubbing surfaces. Figure 15 shows the debris from the wear of stellite pins rubbing on AISI 4340 steel discs. Similarly, the greater the normal pressure is, the larger is the debris size. The thickness of most debris ranges from 1 to 6 μm. Such sizes are smaller than the thickness of oxide films formed on the rubbing surfaces. The sizes of most debris observed under a scanning electron microscope are also smaller than that immediately spalled off. Each wear particle had to encounter many times of rubbing and squeezing between the two contacting surfaces before they appear as wear debris. Subjected to such squeezing and rubbing actions, almost all the particles break into smaller fragments. However, comparably large debris may be observed by chance.

C. The Definition and the Regime of Oxidational Wear

From the experimental evidence presented above, it is shown that oxide films of thickness from 2 μm to over 30 μm can be formed on the rubbing surfaces of steels under suitable conditions. After sliding for some distance and being subjected to a continuously rubbing action, some parts of oxide films are spalled off in turn. According to Quinn [6], such a wear mechanism can be defined as oxidational wear. In such a category of wear, the wear is mild and the wear rate is on the order of 10^{-13} m^3/m. When two elements made of plain carbon steel or alloy steel are loaded together with a normal pressure from 1.1 to 4.4 MPa and have a sliding speed from 1 to 4 m/sec, the wear mechanism of the rubbing surfaces is expected to fall in such a category of wear.

When two rubbing surfaces have a sliding speed lower than 0.6 m/sec and are subjected to a normal pressure smaller than 4.4 MPa, no thick-enough oxide film (i.e., larger than 1 μm) can be found on the rubbing surfaces. Although one can obtain rhombohedral oxide (Fe_2O_3) from the debris, the wear cannot be classified as oxidational wear. On the other hand, if the rubbing bodies are subjected to a normal pressure greater than 6.6 MPa at a sliding speed over 2 m/sec, mass loss by extrusion from the pin will dominate over the wear loss from other mechanisms of the rubbing pair. Under such a condition, the wear loss increases substantially as normal pressure and sliding speed are increased. Such a wear mechanism can neither be classified as oxidational wear, although oxide films of considerable thick are formed on the outermost surfaces of the extruded layers.

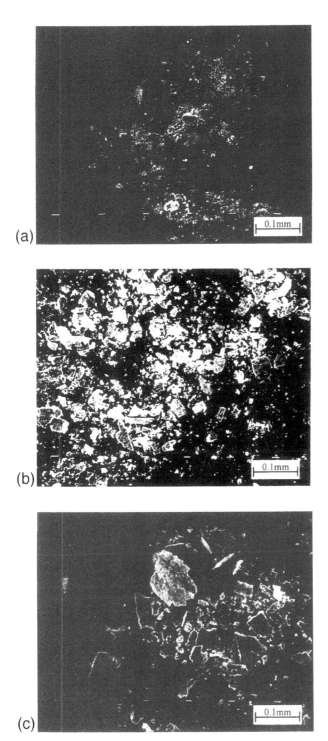

Figure 14 Micrographs of the debris from the die steel pins rubbing on AISI 4340 steel discs at a sliding speed of 2 m/sec and a normal pressure of (a) 1.1 MPa, (b) 2.2 MPa, and (c) 4.4 MPa. (Reprinted from Ref. 4; Elsevier Science.)

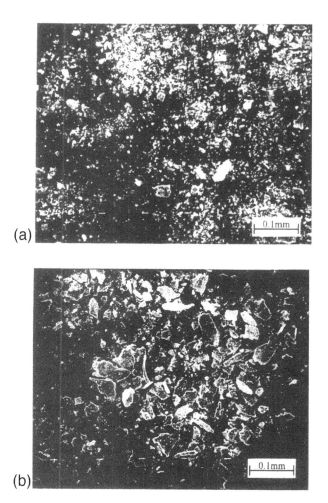

Figure 15 Micrographs showing the debris from the rubbing of laser-clad stellite pins and AISI 4340 steel discs at (a) 2 m/sec and 1.1 MPa; and (b) 4 m/sec and 4.4 MPa.

D. Wear Rate

It has been found that the trend of wear rate for die steel and tool steel does not match the contours plotted in the wear mechanism map presented by Lim and Ashby [25] in the regime of oxidational wear. In addition, there is no evidence to confirm that the wear rate of steel is directly proportional to the normal load under the condition of oxidational wear, although it is almost true that an increase in normal load can cause an increase in wear rate. Such experimental evidence is shown in Figs. 16 and 17. Figures 16 and 17 indicate the plots of wear rate against normal pressure for the medium carbon steel and W1-90C tool steel, respectively. The wear rates indicated in these figures were obtained under a steady-state condition, after the nominal contact temperature had already arrived at a constant.

The effect of sliding speed on the trend of wear rate is quite random. For a tool steel at low normal pressure, an increase in sliding speed causes a decrease in wear rate, although at a high normal pressure greater than 3.3 MPa, an increase in sliding speed causes an increase in wear rate. Figure 18 indicates the relevant results. On the other hand, a different trend obtained for die steel is shown in Fig. 19.

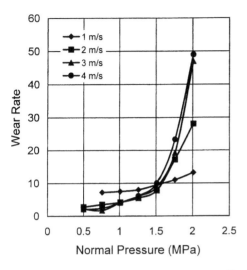

Figure 16 Variation in wear rate (in 10^{-13} m³/m) with normal pressure at various sliding speeds for medium carbon steel rubbing on medium carbon steel.

E. Effect of Oxides on Friction

As mentioned previously, two loaded surfaces can adhere over some part of the apparent contact area and make adhesive junctions deform plastically. If a tangential force is applied in addition to the normal load to the contact pair, the adhesive junctions will grow further. When the tangential stresses at the junctions reach the smaller value of the shear strength of the contacting materials, the junctions are sheared and relative sliding occurs at the junctions. The frictional force is therefore dependent upon the shear strength of the softer body of the contact pair. If there are oxide films on the two contacting surfaces and the shear strengths of the oxide films are smaller than that of the contact bodies, the frictional force as

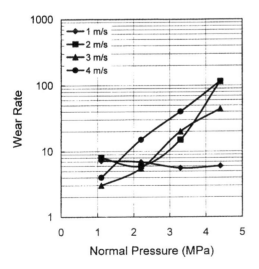

Figure 17 Variation in wear rate (in 10^{-13} m³/m) with normal pressure at various speeds for W1-90C tool steel pins rubbing on die steel discs.

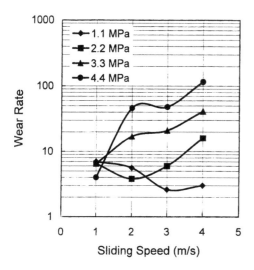

Figure 18 Variation in wear rate (in 10^{-13} m^3/m) with sliding speed at different normal pressures for W1-90C tool steel pins rubbing on die steel discs.

well as the coefficient of friction will obviously be smaller. Therefore, the type of oxides may play an important role as solid lubricant in the sliding contact of two metallic bodies. Especially, the contact bodies are made from some kinds of steel. It is known that the rhombohedral oxide Fe_2O_3 is less effective as a lubricant than the spinel oxide Fe_3O_4. However, Fe_2O_3 is formed on the rubbing surfaces at a contact temperature lower than 450°C. This means that the rubbing pair must be subjected to a lower normal pressure and sliding speed. In such a condition, oxide films can cover only a small part of the contact surfaces, as adhesive and abrasive mechanisms are simultaneously taking place on the metallic surfaces. Figure 20 indicates such a result for the rubbing of medium carbon steel on

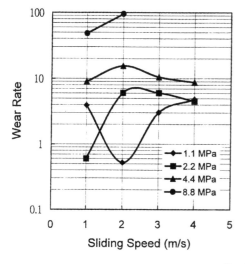

Figure 19 Variation in wear rate (in 10^{-13} m^3/m) with sliding speed at various normal pressures for die steel pins rubbing on 4340 steel discs.

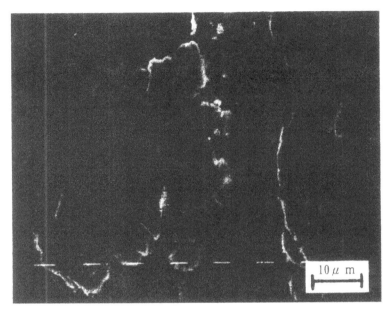

Figure 20 Wear appearance of a medium carbon steel pin sliding on a medium carbon steel disc at a sliding speed of 3 m/sec and at a normal pressure of 0.75 MPa.

medium carbon steel at a sliding speed of 3 m/sec and a normal pressure of 0.75 MPa. The temperatures at the real and apparent contact surfaces were about 340°C and 160°C, respectively. The frictional force caused by metallic adhesion cannot be neglected in computing the coefficient of friction in such a condition. On the other hand, when Fe_3O_4 and FeO are formed, the rubbing bodies are usually subjected to a relatively high pressure (i.e., usually over 2.2 MPa) and a high sliding speed (i.e., usually over 2 m/sec). Under such a condition, the oxide films are thick and cover a larger part of the rubbing surfaces (Fig. 12). Therefore, the two metallic surfaces are almost separated by oxide films. The coefficient of friction in such a condition is therefore comparably lower for shearing oxide junctions. Figure 21 shows some plots of the coefficient of friction against normal pressure at different sliding speeds. In general, under the conditions of the present experiments, the higher the sliding speed and normal pressure are, the smaller is the coefficient of friction for most steels under an oxidational wear condition. The largest value of friction coefficient is around 0.8 at low sliding speed and low normal pressure. Although the smallest value of friction coefficient can be as low as 0.3 at a sliding speed over 4 m/sec and a normal pressure greater than 4.4 MPa, it must be pointed out that at a high speed over 3 m/sec and a high normal pressure larger than 4.4 MPa, the frictional force is mainly contributed by plastic shearing of materials from the contact surface of the pin and not by shearing of adhesive junctions of oxide films (Fig. 13). The real contact temperature in such a condition is usually over 900°C; the shear strength of the material of the pin at such a temperature is as small as one eighth that at room temperature (Fig. 5).

VI. CONCLUDING REMARKS

The article presents recent experimental evidence to the formation of oxides and the wear mechanisms of some steels in the presence of oxide films. The results show that the formation

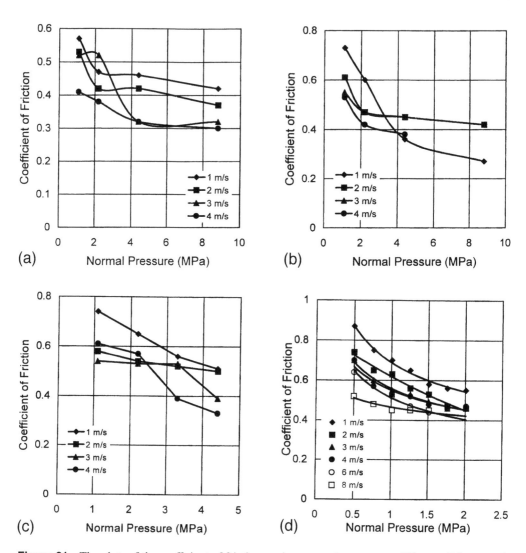

Figure 21 The plots of the coefficient of friction against normal pressure at different sliding speeds for (a) laser-clad stellite pins sliding on 4340 steel discs; (b) die steel pins sliding on 4340 steel discs; (c) W1-90C tool steel pins sliding on die steel discs; and (d) medium carbon steel pins sliding on medium carbon steel discs.

of thick oxide films on rubbing surfaces needs an appropriate condition of suitable normal pressures and sufficiently high sliding speeds. In many tests, the condition that a normal pressure range from 1.1 to 4.4 MPa and a sliding speed range from 1 to 4 m/sec was able to provide an advantage environment for the formation of thick oxide films on the rubbing surfaces. A normal pressure in this range could cause a number of surface asperities to deform plastically. A sufficiently high sliding speed could cause the asperities to deform at a high rate of strain. Consequently, the temperature of the real contact area rose to a level that could provide a suitable condition for the iron to be oxidized. Therefore, the plastic zones became the prior zones to be oxidized. Because the sizes of the plastic zones were not the same, the thicknesses of oxide films were also different. After an oxide film had been formed, the growth of this film was not significant. In a condition of high normal pressure and high sliding speed, oxide films could be formed over larger parts of rubbing surfaces. Consequently, the friction coefficient decreased as normal pressure and sliding speed increased. The coefficient of friction for the steels used in the present studies could be as large as 0.8 at a low sliding speed of 1 m/sec and a low normal pressure of 1.1 MPa. On the other hand, it might be as small as 0.3 at a high sliding speed of 4 m/sec and a high normal pressure of 4.4 MPa.

As soon as oxide films were formed on the rubbing surfaces, they were immediately subjected to intermittent rubbing and shearing actions. Cracks arose in the oxides after the rubbing pair had slid over 10 m. Then, some parts of the oxide films spalled off due to rubbing. The definite sliding distance for crack to initiate and for oxides to spall off depends upon the normal pressure as well as sliding speed. In fact, the spalling of oxide films is a fatigue behavior. Such a wear mechanism is termed oxidational wear. However, if the normal pressure is greater than 2.75 MPa at a sliding speed higher than 2 m/sec, surface layers of the pin will be extruded from the contact interface intermittently. The higher the normal pressure and sliding speed are, the more is the mass of the pin extruded from the interface. Such a condition is no longer considered to be oxidational wear.

The article also presents a method for calculating the temperatures at the real contact area (the real contact temperature) and at the apparent contact surface (the nominal contact temperature or bulk temperature) of the stationary pin. In the calculation, the temperatures at two or three different positions on the surface of the pin were measured beforehand. In addition, the hardnesses of the pin and the disc materials at elevated temperatures were used instead of the hardnesses at room temperature or bulk temperature. The use of such hardnesses is to obtain a more reasonable real contact area at an elevated temperature, so as to predict a more reasonable contact temperature.

REFERENCES

1. Bowden, F.P.; Tabor, D. *The Friction and Lubrication of Solids*; Oxford University Press: Oxford, 1954.
2. Rabinowicz, E. *Friction and Wear of Material,* 2nd Ed.; John Wiley and Sons, Inc.: New York, 1995.
3. Quinn, T.F.J. Review of oxidational wear: Part I. The origins of oxidational wear. Tribol. Int. 1983, *16* (5), 257–271.
4. So, H. The mechanism of oxidational wear. Wear 1995, *184*, 161–167.
5. So, H.; Chen, C.T.; Chen, Y.A. Wear behaviours of laser-clad stellite alloy 6. Wear 1996, *192*, 78–84.
6. Quinn, T.F.J. Oxidational wear. Wear 1971, *18*, 413–419.
7. Sullivan, J.L.; Quinn, T.F.J.; Rowson, D.M. Developments in the oxidational theory of mild wear. Tribol. Int. 1980, *12*, 153–158.
8. Hong, H.; Hochman, R.F.; Quinn, T.F.J. A new approach to the oxidational theory of mild wear. STLE Trans. 1988, *31*, 71–75.
9. Archard, J.F.; Hirst, W. The wear of metals under unlubricated conditions. Proc. R. Soc., A 1956, *236*, 397–405.
10. Quinn, T.F.J. The thermal aspects of wear in tribotesting. Proc. Inst. Mech. Eng. C 1988, *183*, 253–259.
11. Blok, H. Surface temperature under extreme pressure lubricating conditions. Proc. Second World Pet. Congr. 1937, *3*, 471–486.
12. Jaeger, J.C. Moving sources of heat and the temperatures at sliding contacts. Proc. R. Soc. N.S.W. 1942, *26*, 203–224.

13. Barber, J.R. Distribution of heat between sliding surfaces. J. Mech. Eng. Sci. 1967, *9*, 351–354.
14. Quinn, T.F.J.; Winer, W.O. An experimental study of the "hot spots" occurring during the oxidational wear of tool steel on sapphire. ASME Trans. J. Tribol. 1987, *109*, 315–320.
15. Quinn, T.F.J.; Winer, W.O. The thermal aspects of oxidational wear. Wear 1985, *102*, 67–80.
16. So, H. Characteristics of wear results tested by pin-on-disc at moderate to high speeds. Tribol. Int. 1996, *29* (5), 415–423.
17. So, H.; Lin, Y.C.; Huang, G.G.S.; Chang, T.S.T. Antiwear mechanism of zinc dialkyl dithiophosphates added to a paraffinic oil in the boundary lubrication condition. Wear 1993, *166*, 17–26.
18. So, H.; Lin, Y.C. The theory of antiwear for ZDDP at elevated temperature in boundary lubrication condition. Wear 1994, *177*, 105–115.
19. So, H.; Hu, C.C. Effects of friction modifiers on wear mechanism of some steels under boundary lubrication conditions. In *Bench Testing of Industrial Fluid Lubrication and Wear Properties Used in Machinery Applications, ASTM STP 1404*; Totten, G.E., Wedeven, L.D., Dickey, J.R., Anderson, M., Eds.; American Society for Testing and Materials: West Conshohocken, PA, 2001; 125–139 pp.
20. Archard, J.F. The temperature of rubbing surfaces. Wear 1958/1959, *2*, 438–455.
21. Allen, C.B.; Quinn, T.F.J.; Sullivan, J.L. The oxidational wear of high-chromium ferritic steel on austenitic stainless steel. ASME Trans. J. Tribol. 1986, *108*, 172–179.
22. Rowson, D.M.; Quinn, T.F.J. Frictional heating and the oxidational wear theory. J. Phys., D Appl. Phys. 1980, *13*, 209–219.
23. Bowden, F.P.; Tabor, D. *The Friction and Lubrication of Solids, Part II*; Oxford University Press: Oxford, 1964.
24. Quinn, T.F.J. Review of oxidational wear: Part II. Recent developments and future trends in oxidational wear research. Tribol. Int. 1983, *16* (6), 305–315.
25. Lim, S.C.; Ashby, M.F. Wear-mechanism maps. Acta Metall 1987, *35* (1), 1–24.

4

Review on Lubricious Oxides and Their Practical Importance

Mathias Woydt
Federal Institute for Materials Research and Testing (BAM), Berlin, Germany

I. INTRODUCTION

Designers preferably seek to reduce the amount of lubricants delivered to the tribosystems as well as the film thicknesses (smoother surfaces already in manufacturing), which increase the portion of mixed/boundary lubrication and dry friction and, as a consequence, the wear rate.

In the future, tribologically stressed machine elements in the automotive industry [1], civil and mechanical engineering, and metallurgy and apparatuses will operate

1. with liquid film thicknesses lower than 1 μm or
2. with biologically fast, degradable, low additivated, and nontoxic fluids that request new low toxic additive chemistry or
3. used oils in long drain or maintenance intervals or
4. with wear-resistant, triboactive materials [2]

   –under deficient lubrication or
   –under dry conditions in air.

Besides these environmental market drivers, there is an additional trend to reduce maintenance costs by using wear-resistant materials or to increase the specific solicitation in conjunction with equipment downsizing.

A continuous increase in performance has characterized the development of lubricants in the past (naturally within certain cost limits). Liquid lubrication offers above or inside this approximate frame the safest and most reliable long-term operation associated with low friction [3]. In contrast, the oil-free operation cannot be avoided above a sump temperature of 200 °C or peak temperatures of 350–400 °C, where mineral or synthetic oils are thermally unstable or decomposed.

If traditional liquid lubricants with their known lubricity cannot be used, their tribological functions must be taken over by other lubricating mechanisms or materials. As a consequence, the solid contacts between surface asperities will grow and demand more wear-resistant materials.

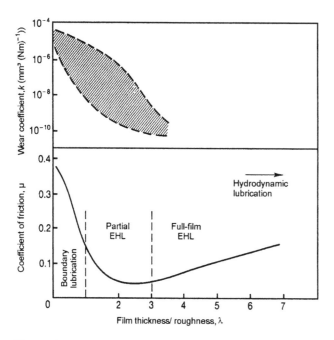

Figure 1 Regions of coefficient of friction and wear rates for the different lubrication regimes of metals (EHL = elastohydrodynamic lubrication) [7].

Even with the use of the latest knowledge from research, there is no dry friction mechanism to substitute liquid lubrication [3–5] at $P \times V$ values greater than 10 MPa m/sec in a wide range of speed and temperature [6], except for gas film lubrication. Because of the physical characteristics, there are boundaries for oils and greases with regard to high-temperature stability and rheology. Above these limits ($T > 350°C$), where liquid lubricants will be thermally unstable, dry lubricants can take over the tasks of lubrication if they have similarly low friction and wear rates as well as low $P \times V$ values.

The three friction regimes of the well-known Stribeck curve (Fig. 1) can be related to the worldwide-used wear rate or wear coefficient in mm³/N m, which describes the wear expectation of a couple under certain operating conditions, but is not a material [7] property. It is defined as the quotient of the wear volume divided by the acting normal force and effective sliding distance.

Under mixed/boundary lubrication, wear rates between 10^{-7} and 10^{-9} mm³/N m are typical; below 10^{-10} mm³/N m, elastohydrodynamic lubrication (EHL) occurs. Under dry friction, wear rates [5] of 10^{-8} mm³/N m for ceramics and below of 10^{-6} mm³/N m for metals and polymers are rare. A wear rate of 10^{-8} mm³/N m stands for a material loss in the microcontact of only one atomic layer at each surpassing, which is difficult to imagine. This aspect illuminates the borderline between dry friction and mixed/boundary lubrication.

In the following, it will be presented how innovative tribomaterials using tribo-oxidation can exhibit a dry wear rate and/or friction coefficient as under liquid lubrication.

II. TRIBOLOGICAL EXPERIMENTS

The high-temperature test rig [8] in Fig. 2 and the materials [9,10] used conducting to the tribological results presented in this paper were intensively described and published else-

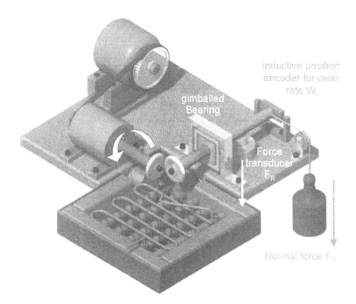

Figure 2 Animated view of the BAM high-temperature tribometer.

where. Important points are here summarized and brought into mind for better under-standing.

A. Test Rig

For unlubricated sliding friction and wear tests, a special high-temperature tribometer [8,11] was built up according to DIN 50324 or ASTM G-99 (Fig. 2). A spherical, stationary specimen ($R_1 = 6$ mm or $R_1 = 21$ mm and $R_2 = 21$ mm) is pressed against the plane surface of the rotating specimen by a dead weight of 10 N. With increasing wear, the contact pressure decreases according to the test geometry with a defined point contact at the beginning, so that statements concerning the influence of the mean contact pressure to wear rate are possible. The sliding distance was 1000 m. All materials were investigated in the same test rig with identical geometries. Friction force, total linear wear, and the sample temperature were continuously measured. The wear volume was calculated from the wear scar dimensions and the profilometry across the wear tacks. The total wear coefficient or rate in Fig. 10 is the sum of the wear rate of the stationary and the rotating specimen and the average from two tests per parameter.

B. Materials

Most of the materials presented in the figures for comparative purposes are briefly described here. The SiC–TiC [12] specimens that contained 48.50 wt.% TiC, 1 wt.% Al_2O_3, and 2 wt.% carbon were made by hot pressing. Si_3N_4–TiN [10] specimens with 32 wt.% TiN were prepared by gas pressure sintering. The microstructure of both polycrystalline materials is a distinct two-phase microstructure with a homogeneous distribution of the minor phase in the matrix.

The sintered SSiC [13] is a commercially available material for sliding bearings and seals with a closed porosity of <3.5% in volume.

(Ti,Mo)(C,N) is a commercially available binary titanium carbonitride composite [TM10] with a homogeneous microstructure of a grain size around <3 μm and is cemented with 13 wt.% nickel and 2 wt.% molybdenum as binder. The Ti/Mo ratio in weight percent is 2:1.

EK3245 is a commercially available, antimony-impregnated, fine-grained carbon [14].

The partially stabilized zirconia (ZN40, 3.3 wt.% MgO) has a uniform distribution of 40–50% of tetragonal ZrO_2 in a cubic ZrO_2 matrix and of 5% of the monoclinic ZrO_2 mainly on the grain boundaries.

The alumina [A1999.7] contains 99.7 wt.% Al_2O_3 with a grain size of 2–8 μm.

III. STATE OF THE ART OF UNLUBRICATED LOW FRICTION AND WEAR COUPLES

A. Solid Lubricants

The use of dry lubricants under boundary conditions has been established for some decades. In accordance to different applications, dry lubricants exist as pure solid lubricant, as solid bonded films, as pure coating, as additives inserted in high-temperature materials, or as composite up to 1000 °C. Fine ceramics or ceramic coatings are thereby increasingly enforced.

According to the common approach, the tribological behavior depends on the formation of a transfer film of polymer materials [15–19], of extrinsic or intrinsic solid lubricants [20–22], as well as soft metals [20] (In, Pb, Au), under unlubricated (dry) friction [23].

This well-known wear mechanism is also valid for molybdenum disulfide [24].

Graphite usually forms various pallets of carbonaceous reaction products in the transfer layer [14].

Load, sliding speed, environment, contact geometry, roughness, adhesion tendency, etc. influence the formation of the layer [21,25].

The characteristics of the materials forming a tribological layer determine the friction and wear behavior of polymers and solid lubricants. The surface roughness of the disk (counter body) determines the thickness of the transfer film. For each combination, "solid lubricant/disk" or "polymer/disk," an optimal thickness of the layer is well known. A thickness optimum of this transfer layer causes low friction and low wear rates. After the run-in and the formation of the transfer layer, the solid lubricant runs against itself with more or less no contact with the counter body. While the run-in occurs, the wear rate will be high because of required materials transfer.

The aim to present sliding couples under solid conditions that are comparable to liquid lubricants can be fulfilled only by avoidance of the transfer film.

B. Unlubricated Friction in the Millirange

The most surprising low friction and wear rates were achieved and confirmed in the last few years with dry sliding couples of DLC and MoS_2, which showed up at special coatings that reach friction coefficients up to 10^{-3} (millirange) [26–28] under dry conditions.

Thin diamond-like carbon (DLC) films of ~1-μm thickness derived from hydrogenated methane plasmas with high hydrogen-to-carbon content [29], achieved in dry nitrogen friction (hertzian contact pressure ~1.04 GPa and $v = 0.5$ m/sec) as self-mated couples [27], have coefficient of friction as low as 0.003. Their wear rates can reach ~4×10^{-10} mm^3/N m, but the tribological performance of DLC films are very sensitive to operating conditions (Fig. 3). When deposited on smooth and rigid sapphire, the DLC films can provide friction coefficients low as 0.001 in dry nitrogen.

The steady state friction [26] coefficients in the millirange (down to ~0.002) of pure (less than 4 at.% O) MoS_2 coatings of less than 100-nm thickness on silicon wafer against AISI 52100 balls were demonstrated at 10^{-7} Pa at low sliding speeds of <0.002 m/sec. The crystallite orientation of MoS_2 was carefully controlled.

At low sliding speeds of v <0.015 m/sec at room temperature, coefficients of friction [30,31] as low as 0.05 were demonstrated for sapphire balls sliding mirror-like finished (C.L.A. ~0.005 μm) disks of borided steels subjected to a short-duration annealing (or preoxidation) at 750°C. Boric acid (H_3BO_3) with triclinic structure formed on surfaces owns an easily shearable, layered structure. Unfortunately, boric acid decomposes at about 165°C and is critical regarding ecotoxicological properties.

Friction [28] in the millirange of 0.009 was also observed with a silicon nitride ball ($\oslash$ = 8 mm) sliding (F_N = 0.1 N; v = 4.2 mm/sec) in nitrogen atmosphere (7.4×10^4 Pa) on a silicon disk coated with CN_x film (IBAD). In repeated tests, a coefficient of friction of 0.007 was observed (Fig. 4).

Finding the answer to the question "What is the origin of the frictional behavior of solid contacts in the millirange?" is of particular interest in tribology. From a theoretical point of view, frictionless sliding between two microcontacts can occur under the following conditions:

1. atomistically smooth surfaces (C.L.A. <2 nm),
2. elastic microcontacts,
3. no surface contamination and wear debris,
4. weak interaction forces between sliding atoms.

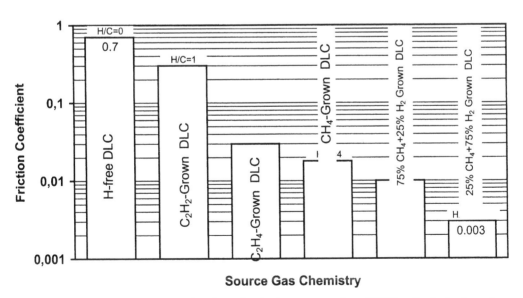

Figure 3 Effect of source gas chemistry [or hydrogen-to-carbon (H/C) ratio in hydrocarbon plasma] on friction coefficients of resultant DLC films (test conditions: load = 10 N; speed = 0.5 m/sec; temperature = 22°C; environment = dry N_2, 9.5-mm diameter M50 balls).

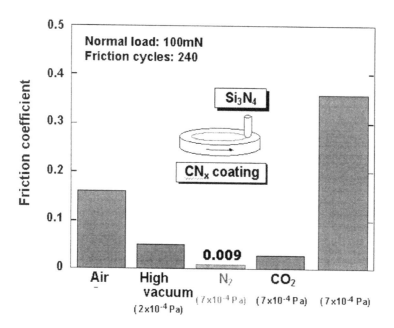

Figure 4 Influence of the surrounding ambient conditions on the coefficient of friction of the couple CN_x/Si_3N_4. [28].

Conditions relevant to the frictionless sliding observed on thin coatings of DLC and MoS_2, such as

1. frictional power loss of 10 mW/mm^2,
2. defined atmospheres (UHV, H_2, N_2),
3. constant operating conditions, and
4. oriented shear layers,

cannot be realized or are not existing in any mechanical constructions and automotive industry on earth. Nowadays, such requirements and operating conditions are only relevant in the field of micromechanics and astronautics.

The shearing of the upper surface areas of the microcontacts under dry friction is, in any case, the only possible mechanism to accommodate the velocity between two surfaces; under liquid lubrication, this is performed through the shearing of the liquid film.

C. Fine Ceramics

Investigations in the last years with different fine ceramics under dry frictional conditions at high temperatures have shown that coefficients of friction less than $\mu \leq 0.1$ are not even possible [32,33]. In these investigations, the wear rates are also not lower than $K_v > 5 \times 10^{-7}$ mm^3/N m. An overview of dry friction and wear of ceramics is given in Refs. 34 and 35. Investigations have been carried out mainly under atmospheric conditions at sliding speed less than 1 m/sec and loads of 10 N. Wear coefficients of $k_v \leq 1 \times 10^{-7}$ mm^3/N m, which were partly reached, always caused friction coefficients of $\mu \geq 0.3$.

Typical lubricated tribological systems achieve coefficients of friction between 0.15 and 0.05 under mixed or boundary conditions. Furthermore, the coefficient of friction will decrease up to 10^{-3} under hydrodynamic conditions. Dry running tribological systems should have to show friction coefficients between 0.001 and 0.015 to be competitive against lubricated systems.

IV. TRIBOCHEMICAL REACTIONS

There have been numerous studies on the oxidational wear of metals in the past 40 years because it is relevant in many practical situations. Most metals are thermodynamically unstable in air and react with oxygen to form oxides (Fe_2O_3, Fe_3O_4, NiO, Cr_2O_3, $FeCr_2O_4$, and $NiCr_2O_4$) as well as with humidity to form hydroxides [40,36] [α-, β-, and γ-FeOOH, $Fe(OH)_2$ or limonite $FeO(OH)_nH_2O$]. Sliding wear can be significantly influenced by heat, either frictionally or externally applied, because it can facilitate oxidation of the contacting metal or alloy surfaces.

The model for mild oxidational wear was developed by Quinn [37] and the principles of tribo-oxidation were summarized in Refs. 38–40.

This chapter is focused on the beneficial effect of tribo-oxidation in ceramics and ceramic composites to reduce friction and/or wear rate.

The most often used sacrificial nonoxide solid lubricants have lamellae crystal structures associated with a low shear strength, but these properties degrade in air or in humid air by oxidation or tribo-oxidation. Oxidative stable substances would be ideal high-temperature lubricants for any application in moving mechanical assemblies in air. Unfortunately, the common oxides of these nonoxide solid lubricants are more "abrasive" than "lubricious" with poor shear properties.

In all cases except ultrahigh vacuum, the nonoxide surfaces of the friction pair are inevitably covered with oxides, hydroxides, and adsorbed gases that fundamentally changes the frictional properties of the tribocontacts [40].

Tribological reactions [1] between monolithic materials and oxygen and/or water vapor represents on earth a promising strategy for reduction of friction and wear under unlubricated (or also lubricated) conditions. The following reactions offer a high potential for reducing friction and wear:

1. adsorption of water [8,41],
2. formation of hydroxide [42,43],
3. formation of oxides [32,44], and
4. vapor phase lubrication [45,46].

In this work, the tribological relevance of hydroxides and understochiometric oxides will be outlined. The tribological effects that were obtained so far through the tribochemical formation of oxides on nonoxide ceramics is summarized in detail in Refs. 41–44.

Tribochemical reactions between ceramic surfaces and water vapor provide a natural opportunity for lubrication of dry running ceramics. However, the adsorption tendency of water is reduced as the temperature is increased to about >100°C [47,48], and the formation of oxides is favored over hydroxides.

A. Formation of Stable Hydroxides

Hydroxides or hydrates are the two possible reaction layers that can be formed on oxides. The two main common materials are alumina and zirconia; mixtures of them also occurs.

This section first reports about a sliding couple made of monolithic materials, which exhibit, under dry friction in a wide range of operating conditions, coefficients of friction in the millirange or close to 0.01. Such a triboelement can be formed, for example, by using as a stationary sample an antimony-impregnated graphite sliding against polished (R_{pK} ~0.011 μm) rotating sample of zirconia partially stabilized with 3.3 wt.% MgO. The best ambient conditions for this couple is air with steam.

By means of laser Raman spectroscopy (LRS) and small spot ESCA (XPS), the presence of carbon could only be detected in the pores of MgO–ZrO$_2$ and not on the surface plateaus of the wear track [49] (Fig. 5). In contrast, the tribochemical formation of Zr(OH)$_4$ [CAS: 14475-63-9] was detected on the wear track by means of LRS and XPS.

The coefficient of friction presented in Figs. 6 and 7 reaches the millirange only at 3 m/sec, and in one individual test as an average value of 0.005. However, at other sliding speeds under dry friction, it exhibit highly interesting values between 0.01 and 0.02. The surface roughness of MgO–ZrO$_2$ (R_{pK} ~0.011 μm) conduct to the lowest friction and wear rate of the carbon.

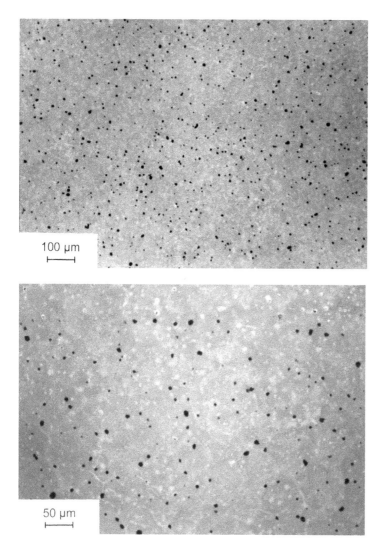

Figure 5 Morphology of the wear track of MgO–ZrO$_2$ sliding against antimony-impregnated graphite (EK3245) at 400°C and $v = 3$ m/sec in air with H$_2$O steam.

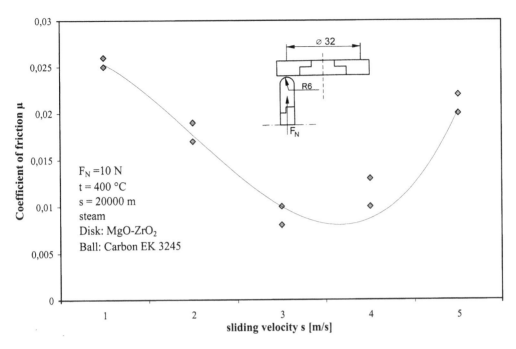

Figure 6 Solid state coefficient of friction as a function of sliding speed of antimony-impregnated graphite against MgO–ZrO$_2$ ($R_{pK} \approx 0.01$ μm).

Figure 5 illustrates that this couple fulfils one of the above-mentioned conditions for ultralow friction because it forms no graphite transfer film, which is contradictory to what is given in the literature on tribology of carbon materials.

In the temperature range between 200 and 450°C, the wear rate for the antimony-impregnated graphite is lower than 10^{-7} mm^3/N m (see Fig. 3). The lowest wear rate measured in a single test at 3 m/sec was 3×10^{-8} mm^3/N m. No wear on MgO–ZrO$_2$ can be detected at all tests by means of stylus profilometry with an atomic force microscope.

The validation of the actual tribological results must take into consideration that the tribochemical effective layer composed of Zr(OH)$_4$ was formed during sliding and was not present from the beginning. In the future, the thermal stability in air of zirconium hydroxides up to 650°C opens the possibility to directly coat MgO–ZrO$_2$ samples by means of PVD or reactive sputtering.

Up to now, engineers are looking for a dry sliding couple with wear rates on the level of mixed or boundary lubrication associated with coefficients of friction near or below of 0.01 capable to run at higher temperatures and in a technical important range of sliding speeds. The couple antimony-impregnated graphite sliding dry against MgO–ZrO$_2$ merits scientific and technical attention. These materials are monolithic.

The effect of friction under unlubricated conditions is not limited to MgO–ZrO$_2$. A similar frictional behavior (Fig. 8) was observed with alumina (Al$_2$O$_3$; 99.9%, A1999.7). The coefficient of friction of the couple EK3245/Al$_2$O$_3$ is slightly lower, even if the wear rate is by a factor of 6 higher, compared to EK3245/MgO–ZrO$_2$. The elevated wear rate is certainly due to the selection of a surface topography for Al$_2$O$_3$, which is favorable for MgO–ZrO$_2$, and the optimum for the harder Al$_2$O$_3$ has to be determined. In contrast, as shown in Fig. 8,

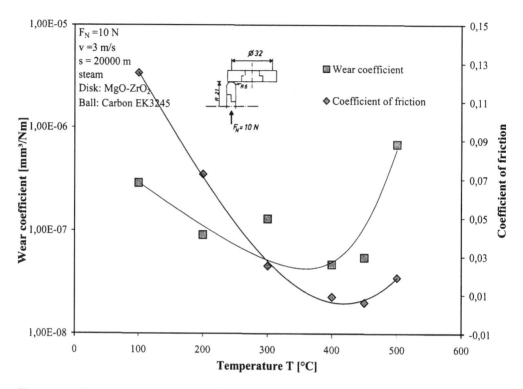

Figure 7 Wear rates carbon and coefficients of friction as averages of two tests of antimony-impregnated graphite sliding at 3 m/sec against MgO–ZrO$_2$ ($R_{pK} \approx 0.01$ µm) as a function of ambient temperature.

the couple EK3245/Al$_2$O$_3$ exhibits after running-in a much lower coefficient of friction as MgO–ZrO$_2$.

Gates et al. [50,51] gave evidence on decreased wear rates and coefficient of friction in H$_2$O caused by the formation of a transfer layer as gibbsite [Al(OH)$_3$] or boehmite [AlO(OH), $H_V \sim 8.000$ MPa] on Al$_2$O$_3$. Relatively weak bonds reflect the structure of both hydroxides and hydrates. The only difference is the kind of layer stacking. Boehmite will be formed above 194°C and gibbsite above 100°C.

At room temperature under oscillating motion, the wear rates of self-mated MgO–ZrO$_2$ couples are sensitive to increasing relative humidity, but surprisingly not to the coefficient of friction (see Fig. 9), which remains high. The wear rate is reduced by increasing relative humidity. The arrow shows increasing humidity. Unfortunately, the origin of this contradictory frictional behavior (Fig. 8) is not understood. The main reason represents the thin and probably amorphous upper surface regions on the MgO–ZrO$_2$ wear scars and tracks.

B. Formation of "Lubricious Oxides"

Thermo- and tribo-oxidatively formed, low shear strength oxides would be ideal for high-temperature solid lubricants for unlubricated and critical applications in air or humidity on earth as well as for contacts under mixed or boundary lubrication [52], and thus fulfilling the industrial demand of automotive [53–55] and aerospace industries [56].

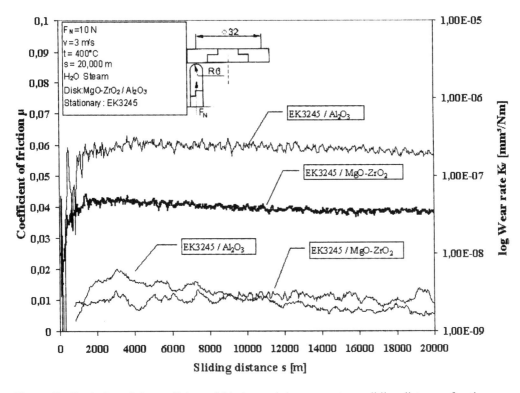

Figure 8 Evolution of the coefficient of friction and the wear rate vs. sliding distance of antimony-impregnated carbon sliding against oxides at 400°C.

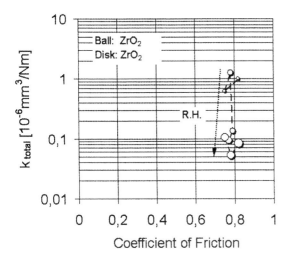

Figure 9 Wear rate vs. coefficient of friction of self-mated MgO–ZrO$_2$ couples at different relative humidities under dry oscillating friction (number of cycles: 1 million; $T = 22°C$; $\Delta x = 200$ μm; $v = 20$ Hz; $F_N = 10$ N; relative humidities = 3–100%; air).

In the past, some papers [52,57] were published on materials with extremely low friction and/or low wear rates, which were attributed to thin reaction layers of unknown structure [32,58]. In this context, the term "lubricious oxides" (LO) was created [59,60] and used.

Historically, at the author's organization, the first step was taken at the beginning of the 1980s. The experimental evidence in tribological tests with different humidities and as function of the oxygen partial pressure, as well as by means of surface analysis, shows that oxide layers exercise a beneficial effect on the tribology of TiN and TiC coatings [61].

To improve the tribological behavior of monolithic ceramics even at high temperatures, the development of ceramic–ceramic composites containing TiN or TiC [62], or finally to use solid solutions of Ti and Mo cations, such as (Ti,Mo)(C,N), was the natural exploitation of the previous lessons learned [32].

In these intensive works [32,44,58], it was hypothesized that the exceptional friction and wear behavior is attributed to special oxides formed by tribo-oxidation on self-mated couples of (Ti,Mo)(C,N); but it was only at the end of the nineteenth century that the structural and crystallographic origin was experimentally proven [63] by testing hot-pressed, dual-phase $60\%Ti_4O_7/40\%Ti_5O_9$ composite, both Magnéli phases of TiO_2.

Magnéli [64] first recognized that oxides of titanium [65] and vanadium, as well as of molybdenum and tungsten [66], form homologous series with planar faults according to the common principles $(Ti,V)_nO_{2n-1}$ or $(W,Mo)_nO_{3n-1}$.

As a consequence, the origin of the well-known and proven wear resistance of hard metals, especially of tungsten carbide (WC), is not only related to the high hardness (H_v =

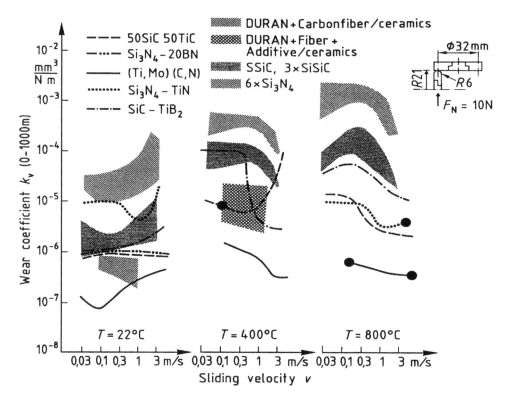

Figure 10 Total volumetric wear rates of self-mated tribocouples subjected to dry sliding according to DIN 50324 or ASTM G-99.

~ 17–18 GPa) and elastic modulus (E = ~ 700 GPa; very small microcontact area!), but is today explained by the tribo-oxidative formation of Magnéli-type oxides of tungsten, such as $(W,Mo)_nO_{3n-1}$, on a hard substrate, which define the shear strength of the microcontact area. The microcontact area is defined by the elastic modulus and hardness as well as by the compression strength [67].

The experimental results with Magnéli-type phases indicate more a strong wear-resistant behavior than low friction [63].

Figure 10 summarizes tribological results with unlubricated self-mated, monolithic couples of commercial ceramics, ceramic–ceramic composites, as well as with fiber-reinforced glasses against ceramics, and illustrates the progress made on reducing the total volumetric wear rates (sum of pin and disk wear volume!). Under dry friction, the self-mated couples of (Ti,Mo)(C,N) displayed in a temperature range up to 800°C and sliding speeds up to 5 m/sec wear rates lower than 10^{-6} mm^3/N m, which lie on a level comparable to those known for mixed/boundary lubrication.

Compared to ceramic–ceramic composites, the tribosystem composed of self-mated (Ti,Mo)(C,N) + Ni couples achieved a further wear reduction at 22°C and up to 800°C. At 800°C, the wear rate of the stationary (pin) specimen decreased from 3.82×10^{-7} mm^3/N m at 0.12 m/sec down to 9.5×10^{-8} mm^3/N m at 3.68 m/sec, whereas the wear rate of the rotating (disc) specimen was only detectable at 3.68 m/sec with a wear rate of 3.5×10^{-7} mm^3/N m. The lowest wear rate was achieved at 6.17 m/sec and 800°C with 3×10^{-8} mm^3/N m for the stationary specimen and 2×10^{-7} mm^3/N m for the rotating specimen.

The values of the coefficient of friction of self-mated TM10 couples were between 0.2 and 0.6, but the wear scars and tracks are free of adhesive wear mechanisms (see Fig. 11). At 800°C, the coefficient of friction drops down from 0.48 at 0.1 m/sec to 0.23 at 3.7 m/sec. The

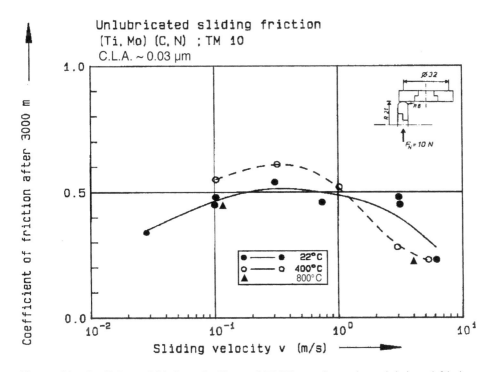

Figure 11 Coefficient of friction of self-mated TM10 couples under unlubricated friction.

tribo-oxidative formation of reaction [32,58] layers composed of [CS] structures of γ-Ti_3O_5, Ti_5O_9, Ti_9O_{17}, and $Mo_{0.975}Ti_{0.025}O_2$, as well as double oxides $NiTiO_3$ and β-$NiMoO_4$, determine the decrease of the coefficient of friction (see Fig. 11) and wear rates (see Fig. 10) at higher sliding speeds because of the hot spot temperatures ΔT in the microasperities. This effect can be enhanced by means of preoxidation [56] of the surfaces.

The differential wear rate in Fig. 12 was calculated at 1.000-m sliding distance, when after running-in, the wear rate was stationary. The differential wear rate represents the wear rate without running-in wear or the stationary wear rate in the low-wear regime.

In the case of dry running sliding couples of self-mated Si_3N_4, the wear rates were above 10^{-5} mm^3/N m and increased with increasing temperature. This example shows that tribo-oxidation is not always beneficial. Also, it can be seen in Fig. 12 that the wear resistance of the composite Si_3N_4-BN suffers under the oxidative degradation of hex BN.

The wear rate of Si_3N_4 is distinctly reduced (up to orders of magnitude) when TiN as secondary phase was added to the silicon nitride matrix (compare Fig. 10 with Fig. 12). This wear-reducing effect is attributed to the tribochemical formation [44,58] of stable layers composed of Magnéli-type phases (TiO_2, Ti_9O_{17}). This stable reaction layer is wear resistant in the low-wear regime as the geometric contact pressure is lower than 15 MPa. Unfortunately, the coefficients of friction of Si_3N_4–TiN remain greater than 0.45 under all test conditions in Fig. 12.

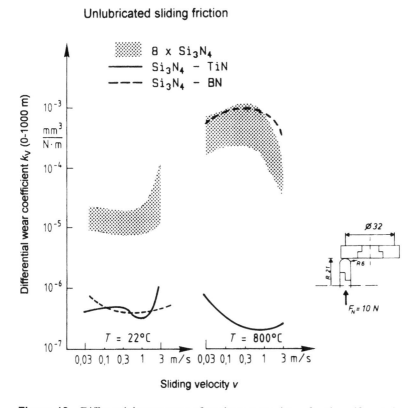

Figure 12 Differential wear rates of stationary specimen for the self-mated composites Si_3N_4–TiN and Si_3N_4–BN in comparison with conventional Si_3N_4 materials at 22 and 800°C.

Kustas and Buchholtz [68] also reported from unlubricated sliding tests up to 800°C with TiC–Si$_3$N$_4$ and TiN–Si$_3$N$_4$ composites that the friction was high, but the wear rate for composites with 30 wt.% titanium compound was as low as $1.8 \times 10^{-7} \text{mm}^3/\text{N m}$ and independent of temperature.

Tribo-oxidation is per se an uncontrolled but useful mechanism to especially reduce wear, but is not limited to (Ti,Mo)(C,N) + Ni. It can be applied to all nonoxide ceramics, such as SiC [69], Si$_3$N$_4$–TiN, SiC–TiC, or coatings of TiN and TiC. These surfaces will form by static oxidation as

1. oxides, such as SiO$_2$, MoO$_3$, TiO$_2$, Ti$_n$O$_{2n-1}$,
2. oxonitrides [10,70], such as SiO$_x$N$_y$, Si$_2$N$_2$O$_2$, YSiO$_2$N, or TiN$_x$O$_y$, and
3. oxocarbides [32], SiC$_x$O$_y$.

By means of preoxidation in water vapor or at higher temperatures, or just after weeks of exposure in "humid" air, the running-in wear is reduced, because most of the later reaction products are already present ab initio.

V. INDUSTRIAL APPLICATIONS

In the 1980s up to the beginning of the 1990s, adiabatic or thermally insulated engines of ISUZU and ADIABATICS, solid lubricated and/or lubricated with alternative fluids, acquired many experiences with alternative engine oils [71,72] (e.g., polyolester), because the oil sump temperature was 100K higher than today. This temperature level increases the friction modes of mixed/boundary lubrication and dry friction, thus requiring additional solid lubricants and ceramics.

An outcome of the work on "adiabatic" or "ceramic" engines was the conclusion that components made of monolithic ceramic, such as liners, are difficult to implant in a cost-efficient way in mass production.

The topic of long drain intervals [73] or lifetime lubricated passenger car engines came up in the mid-1990s followed in Europe by the requirement of an added value as fast biodegradable engine oils and now additionally as oils also non-toxic to aquatic species. In the frame of this efforts, coatings attracted the attention of engine developers.

The following gives a brief summary of the industrial interest on lubricious oxides. The automotive industry investigate them in view of their wear resistance in combination with liquid lubrication and the aerospace industry more as wear-resistant coatings under dry high-temperature friction.

A. FORD

In contrast to previous work on ceramic engines, FORD has published since 1997 some papers and patents related to understoichiometric and/or triboactive oxides [74] of the type Mey0 (Me = Ni, Fe, Cu). The defect or understoichiometry is more on the cation side for FeO$_{0.95-1.05}$, NiO$_{0.75-1.25}$, CuO$_{0.40-0.60}$, and MoO$_{2.5-3.2}$.

The benefits of these oxides, preferably deposited by means of thermal spraying, is the substitution of extreme pressure additives [75] as well as the reduction of sulfur- and phosphorus-containing compounds. Additionally, they reduce the friction on cylinder liners [76]. The tribological properties of the systems Fe–Fe$_x$O or Ni-hex. BN are published in Ref. 77.

More suited are FeO and Fe$_3$O$_4$ in a matrix of a low-alloyed steel [78] on AlSi7–10 or cast iron.

B. DaimlerChrysler AG

For lifetime lubrication, DaimlerChrysler investigates lubricious oxides (LO) as coatings on tribosystems of internal combustion engines to avoid antiwear (AW) and extreme pressure (EP) additives as well as viscosity index improvers. The LOs [53] consist of Magnéli phases of titanium dioxide (γ-$Ti_3O_5 \Rightarrow Ti_{10}O_{19}$), of vanadium oxide ($\gamma$-$V_3O_5 \Rightarrow V_8O_{15}$), as well as of molybdenum and tungsten oxides ($W_{18}O_{49} \Rightarrow W_{40}O_{118}$), in general $(Ti,V)_nO_{2n-1}$ or $(Mo,W)_nO_{3n-1}$, with $3.4 \leq n \leq 10$, which can also be formed by means of tribo-oxidation.

The work [79] is focused on the phases $Ti_4O_7 \Rightarrow Ti_6O_{11}$ and TiO_2 deposited by atmospheric plasma spraying (Rotaplasma®). The sprayed TiO_x phase exhibit for a ceramic relatively low Vickers hardness between 8500 MPa $< H_V <$ 11,000 MPa. Wear debris are in consequence less abrasive and the coatings do not require a too smooth honing.

Hot-pressed and thermally sprayed Magnéli phases of TiO_2 exhibit under dry friction at room temperature $P \times V$ values up to 121 W/mm^2 or geometric contact pressures up to 160 MPa for the low-wear regime [80].

C. Renault SA

The metallurgy of Renault is complementary to this initiated by FORD and DaimlerChrysler and shares the strategy of "triboactive oxides" with the patent applications [54,81]:

1. Oxides with cation defects "Revêtements triboactifs à base d'oxydes métalliques présentant des défauts en cations métalliques," $Ni_{1-y}O$, $Co_{1-y}O$, CrO_{2-x}, $FeO_{1.06-1.20}$, $Fe_{3-y}O_4$, and Mn_yO with $0.001 < y < 0.15$.
2. Oxides with planar defects "Pièce mécanique de friction recouverte d'oxydes triboactifs stabilisés par des oligoéléments," $(Ti,V)_nO_{2n-1}$ or $(Mo,W)_nO_{3n-1}$ with the "oligo elements" Ni, Al, Zr, Ni, Cr, Fe, Co, Ti, K, Mn, Nb <3000 ppm and the mixture $Ti_{n-2}O_{2n-1} + 15\%$ Fe_2O_3.

Especially the potential high resistance against oxidation up to 1500°C of $Ti_{n-2}Cr_2O_{2n-1}$, with $6 \leq n \leq 9$, and the phase stability of the planar defects makes this system attractive for atmospheric spraying.

D. IAV GmbH

The Ingenieurgesellschaft Auto Verkehr (IAV), a subsidiary of Volkswagen AG, has presented one and three cylinder zero emission engines [82] based on the Rankine cycle, which operate with steam up to 100 bar and 800°C. The engine fulfills SULEV without any exhaust gas treatment system. The tribosystems run dry or unlubricated under steam atmospheres with component temperatures up to 450 °C.

The most promising couples [49,83] are carbons impregnated with antimony sliding against MgO–ZrO_2 or Al_2O_3 (see Sect. IV.A).

E. MAN Technologie AG

The Magnéli phases were investigated also as triboactive coating on C/SiC (carbon-reinforced SiC) for movable hot structures (e.g., control surfaces) for future space reentry vehicles [84], such as the body flap of the X38, which can operate up to 1600°C.

The tribosystem [56] can, as a function of the temperature and the mission profile, be coated with oxides or mixed oxides in monoclinic, triclinic, or tetragonal crystal structure formed from MeO_6 octahedrons with planar oxygen defects from the group consisting of TiO_{2-x}, Ti_5O_9, γ-Ti_3O_5, Ti_9O_{17}, $Ti_{10}O_{19}$, Mo_8O_{23}, high-V_3O_5, WO_3, $W_{20}O_{58}$, β-$NiMoO_4$, $Ti_{n-2}Cr_2O_{2n-1}$, $V_{0.985}Al_{0.015}O_2$, tetragonal tungsten oxide bronzes, and vanadium oxide bronzes ($M_xV_2O_5$), x being a variable and n being a natural integer. In view of the stability against oxidation and stability of the planar defects, $Ti_{n-2}Cr_2O_{2n-1}$ is the most promising phase system.

F. National Aeronautics and Space Administration "Oil-Free Turbomachinery Program"

The NASA seeks to transfer their technology and solid lubricants or lubricious oxides for advanced foil air bearings and seals. The components are oil-free turbochargers, introduced into market by Caterpillar [85] and a small oil-free aeropropulsion engines (Williams International V-Jet-II aircraft [86]).

The substitution of chromium carbide (US 5,034,187) in the NASA PS200-series [87] self-lubricating coating 87 based on 30–70% chromium carbide, 5–20% soft and nobles (Ag) metals, 5–20% fluorides, and 20–60% metallic binder by chromium oxides leading to the NASA PS300 series (US P 5866518) generic composed of 60–80% Cr_2O_3, fluoride of gr. I and II, Ag, Au, Pt, Pd, Rh, or Cu, and NiCr binder seems to be motivated by cost and process aspects.

Disks coated with PS300-series [88,89] exhibited wear rates greater than 6.6×10^{-5} mm^3/N m, when pin of INCX750 slid on them at 1 m/sec up to 800°C.

REFERENCES

1. Woydt, M. Materials-based concepts for an oil-free engine. In: Hutchings, I.M., Ed.; *New Directions in Tribology, Plenary and Invited Papers from the First World Tribology Congress, London*; MEP Mechanical Engineering Publications Ltd.: London, 459–468 pp. ISBN 1-86058-099-8.
2. Woydt, M. Werkstoffkonzepte für den Trockenlauf (Materials-based concepts for unlubricated tribosystems). Tribol. Schmierungstechnik 1997, *44* (1), 14–19.
3. Woydt, M. Werkstoffbedingte Trends in der Schmierungstechnik (Plenarvortrag). Proc. 11. Int. Colloquium Tribology, 13–15 January 1998, Esslingen; Technische Akademie Esslingen: 73740 Ostfildern, Germany, 1998, Vol. I, 23–33. ISBN 3-924813-39-6.
4. Klamann, D. *Schmierstoffe und verwandte Produkte*; Herstellung-Eigenschaften-Anwendungen,Verlag Chemie: Weinheim, 1982.
5. Möller, U.J.; Boor, U. *Schmierstoffe im Betrieb*; VDI Verlag: Düsseldorf, 1987.
6. Woydt, M.; Habig, K.-H. Tribological criteria and assessments for the life of unlubricated engines. Lubr. Eng. 1994, *50* (7), 519–522.
7. Czichos, H.; Habig, K.-H. Mixed lubrication and lubricated wear. In: Dowson, D., Taylor, C.M., Godet, M., Berthier, D. Eds.; *Proc. 11th Leeds-Lyon Symposium on Tribology*; Butterworths: London, 1985; 135–147 pp. ISBN 408 221 909.
8. Woydt, M.; Habig, K.-H. High temperature tribology of ceramics. Tribol. Int. 1989, *22* (22), 75–88.
9. Woydt, M.; Kadoori, J.; Hausner, H.; Habig, K.-H. Development of ceramic materials from a tribological point of view. J. Ger. Ceram. Soc. (cfi) 1990, *67* (4), 123–130.
10. Skopp, A.; Woydt, M.; Habig, K.-H. Tribological behaviour of silicon nitride materials under unlubricated sliding between 22C and 1000C. Wear 1995, *181/183*, 571–580.
11. Woydt, M.; Gienau, M.; Habig, K.-H. Hochtemperaturtribometer für Reibungs-und Verschleißprüfung bis 1000C. Materialprüfung 1987, *29* (7/8), 197–199.

12. Rabe, T.; Wäsche, R. *Gasdrucksintern von SiC/TiC-Keramiken, Tagungsband (Proceeding) der Jahrestagung 1993*; der Deutschen Keramischen Gesellschaft: Weimar, Germany, *06.-08.10*; 266–268 pp.

13. Technical Data Sheet, Engineered ceramics—general characteristics, Elektroschmelzwerk Kempten GmbH (ESK, now Wacker Ceramics), 87405 Kempten, Germany.

14. Thiele, W. Tribologisches Verhalten von Kohlegraphit (Tribological behavior of carbons). *Tribologisches Verhalten keramischer Werkstoffe*; Kontakt&Studium Band 431, Expert Verlag: Renningen, Germany, 1993.

15. Czichos, H.; Habig, K.-H. *Tribologie-Handbuch*; Vieweg Verlag: Braunschweig/Wiesbaden, 1992. ISBN 3-528-06354-8.

16. Gardos, M.N. Self-lubricating composites for extreme environment applications. Tribol. Int., 273–283.

17. Czichos, H.; Feinle, P. *Tribologisches Verhalten von thermoplastischen, gefüllten und glasfaser-verstärkten Kunststoffen*, BAM-Forschungsbericht Nr. 83, Juli 1982. ISSN 0172-7613.

18. Uetz, H.; Wiedemeyer, J. *Tribologie der Polymere*; Carl Hanser Verlag: München, 1985. ISBN 3-446-14050-6.

19. Friedrich, K. *Friction and Wear of Polymer Composites*; Elsevier Science Publishers BV: Amsterdam, 1986. ISBN 0-444-42524-1.

20. Wäsche, R.; Habig, K.-H. *Physikalisch-chemische Grundlagen der Feststoffschmierung-Litera-turübersicht*, BAM-Forschungsbericht Nr. 158, April 1989. ISBN 3-88314-889-x. Berlin.

21. Kanakia, M.D.; Peterson, M.B. *Literature Review of Solid Lubrication Mechanisms*, Interim Report BFLRF No. 213, November 1986, contract No. DAAK70-85-C-007.

22. Zechel, Z., et al. *Molykote*; Dow Corning GmbH: Wiesbaden, Germany, 1990.

23. Zambelli, G.; Vincent, L. *Matériaux et contacts—Une approche tribologique*; Presse polytechniques et universitaires romandes, EPFL, Centre Midi: Lausanne, CH, 1998. ISBN 2-88074-338-9.

24. Lansdown, A.R. Molybdenum disulphide lubrication. Tribology Series 1999, *35*. Elsevier Science BV, Amsterdam, 0-444-50032-4.

25. Kragelski, I.W.; Dobycin; Kombalov. *Grundlagen und Berechnung von Reibung und Verschleiß*; Carl Hanser Verlag, 1983.

26. Donnet, C.; Belin, M.; Le Mogne, T.; Martin, J.M. Tribological behaviour of solid lubricated contacts in air and high-vacuum environments; The third body concept. In: Dowson, D., Ed.; Tribology Series *31*, Elsevier, 1996; pp. 389–400.

27. Erdemir, A.; Eryilmaz, O.L.; Nilufer, I.B.; Fenske, G.R. Synthesis of super low-friction carbon films from highly hydrogenated methane plasmas. Surf. Coat. Technol. 2000, *133–134*, 448–454.

28. Tatsuno, M.; Umehara, N.; Kato, K. Nitrogen lubricated sliding between CNx coatings and ceramic balls. Proc. Int. Tribology Conference, Nagoya 29.10-02.11.2000, Vol. II, 1007–1012. ISBN 4-9900139-4-8. Japanese Society of Tribology, Tokyo 105-0011.

29. Erdemir, A. Surf. Coat. Technol. 2001, *146–147*, 292.

30. Erdemir, A. Lubrication from Mixture of Boris Acid with Oils and Greases, US P 5,431,830.

31. Erdemir, A.; Bindal, C. Formation and self-lubricating mechanisms of boric acid on borided steel surfaces. Surf. Coat. Technol. 1995, *76–77*, 443–449.

32. Skopp, A.; Woydt, M. Ceramic and ceramic composite materials with improved friction and wear properties. Tribol. Trans. 1995, *38* (2), 233–242.

33. Buckley, H.; Miyoshi, K. *Friction and Wear of Ceramics*, Wear 1984, Elsevier Science, 1984; Vol. 100, 333–353 pp. ISSN 0043-1648.

34. Booser, E.R. CRC Handbook of Lubrication and Tribology 1994, vol. III. CRC Press Inc.: Boca Raton FL, 1994. ISBN 0-8493-3903-0.

35. Woydt, M., Ed.; *Tribologie keramischer Werkstoffe (Tribology of ceramics), Grundlagen-Werkstoffneuentwicklungen-Industrielle Anwendungsbeispiele, Kontakt & Studium, Band 605*; Expert Verlag: Renningen, Germany, 2000. ISBN 3-8169-1744-5.

36. Mehner, H.; Klaffke, D.; Schierhorn, E. Mössbauer analysis of wear particles produced by oscillatory sliding in steel/ceramics couples. Hyperfine Interact. 1994, *93*, 1823–1829.

37. Quinn, T.F.J. Oxidational wear. Wear 1971, *18*, 413–419.
38. Scott, F.H. The influence of oxidation on the wear of metals and alloys. In: Hutchings, I.M., Ed.; *New Directions in Tribology: Plenary and Invited Papers Presented at the First World Tribology Congress, London, U.K., 08.–12. September*; MEP Mechanical Engineering Publications Ltd.: London, 391 pp. ISBN 1-86058-099-8.
39. Quinn, T.F.J. Review on oxidational wear, Part I: The origins of oxidational wear. Tribol. Int., 257–271.
40. Iliuc, I. Tribology of thin layers. Tribology Series No. 4,. Elsevier, 1980. 225 pages, ISBN 0.444-99768-7.
41. Sasaki, S. Effects on environment on friction and wear of ceramics. Bull. Mech. Eng. Lab. 1992, *58*, Ibaraki-ken 305 Japan, ISSN 0374-2725. *report.*
42. Sasaki, S. The effects of surrounding atmosphere on the friction and wear of aluminia, zirconia, silicon carbide and silicon nitride. Int. Conf. Wear Mater. 1989, *II*, 409–417.
43. Erdemir, A. Tribological properties of boric acid and boric acid forming substrates. Part I: Crystal chemistry and mechanism of self-formation of boric acid. STLE Preprint No. 90-AM-3E-1, presented during. 45th Annual STLE-Meeting.
44. Woydt, M.; Skopp, A.; Dörfel, I.; Wittke, K. Wear engineering oxides/Antiwear oxides. Tribol. Trans. 1999, *42* (1), 21–31.
45. Sawyer, W.G.; Blanchet, T.A. High temperature lubrication of combined rolling/sliding contacts via directed hydrocarbon gas streams. Wear 1997, *211*, 247–253.
46. Klaus, E.E.; Jeng, G.S.; Duda, J.L. A study of tricresylphosphate as vapor delivered lubricant. Lubr. Eng. 1989, *45* (11), 717–723.
47. Westbrok, J.H. The temperature dependence of hardness of sone common oxides. Rev. Int. Hautes Temp. Refract. 1966, *3*, 45–57.
48. Lindan, P.J.D.; Harrison, N.M.; Holender, J.M.; Gillan, M.J. First-principles molecular dynamics simulation of water dissociation on TiO_2. Chem. Phys. Lett. 1996, *261*, 246–252.
49. Kleemann, J.; Woydt, M. Friction in the millirange at 400C without intrinsic solid lubricants. Proc. 13th Int. Colloquium Tribology "Lubricants, Materials and Lubrication Engineering," Ostfildern, Supplement. Technische Akademie Esslingen, 73740 Ostfildern, Germany, 37–51 pp. ISBN 2-924813-48-5.
50. Gates, R.S.; Hsu, S.M.; Klaus, E.E. Tribochemical mechanism of alumina with water. Tribol. Trans. 1989, *32* (3), 357–363.
51. Gates, R.S.; Hsu S.M.; Klaus, E.E. *Ceramic Tribology: Methodology and Mechanisms of Alumina Wear*, NIST Special Publication 758, September 1988, Library of Congress Catalog Card Number 88-600582.
52. Gardos, M.N.; Hong, H.-S.; Winer, W.O. The effect of anion vacancies on the tribological properties of rutile (TiO_{2-x}). Part II: Experimental evidence. Tribol. Trans. 1990, *22* (2), 209–220.
53. Patent application DE 195 48 718 C1, Reibungsbelastetes Bauteil eines Verbrennungsmotors (Tribological stressed components of internal combustion engines).
54. Patent application FR 2 793 812 (EP 00401232.4), Pièce mécanique de friction recouverte d'oxydes triboactifs stabilisés par des oligoéléments.
55. Chavanes, A.; Pauty, W. *Fuel Pump and Injector Components Made from Wear Resistant Cermets, Powder Metal Applications and Components*, SAE 2002-01-0610.
56. Patent applications DE 196 51 094 A, FR 2 756 887, US 6,020,072, US 6,017,592, Tribosysteme und Verfahren zu seiner Herstellung.
57. Fischer, T.E.; Tomizawa, H. Interaction of tribochemistry and microfracture in the friction and wear of silicon nitride. Wear 1985, *105*, 29–45.
58. Woydt, M.; Skopp, A.; Hantsche, H. Ceramic wear track characterization by advanced X-ray diffraction. Rigaku J. 1992, *9* (2), 9–16.
59. Gardos, M.N. Tribo-oxidative behaviour of rutile forming substrates. *New Materials Approaches to tribology: Theory and Applications*Materials Research Society Symposium Proc. 1989, Vol. 140, 325–338.

60. Gardos, M.N. The effect of Magnéli phases on the tribological properties of polycrystalline rutile. Proc. 6th Int. Congress on Tribology 1993 (EUROTRIB '93), Budapest, Vol. 3, 201–206.

61. Habig, K.-H. Chemical vapor deposition and physical vapor deposition coatings: Properties, tribological behaviour and applications. J. Vac. Sci. Technol., A Nov./Dec. 1986, *4* (6), 2832–2843.

62. Woydt, M.; Kadoori, J.; Hausner, H.; Habig, K.-H. Development of engineering ceramics according to tribological considerations (bilingual). Cfi/Ber. DKG. J. Ger. Ceram. Soc. 1990, *67* (4), 123–130.

63. Woydt, M. Tribological characteristics of polycrystalline titanium dioxides with planar defects. Tribol. Lett. 2000, *8* (2–3), 117–130. Special issue "Lubricious Oxides".

64. Magnéli, A. Structures of the ReO_3-type with recurrent dislocations of atoms: "Homologous series" of molybdenum and tungsten oxides. Acta Crystallogr. 1953, *6*, 495–500.

65. Anderson, S.; Magnéli, A. Diskrete titanoxydphasen im Zusammensetzungsbereich $TiO_{1.75}$–$TiO_{1.90}$. Die Naturwiss. 1956, *43* (21), 195–196.

66. Magnéli, A.; Anderson, S.; Westmann, S.; Kihlborg, L.; Holmberg, B.; Åsbrink, S.; Nordmark, C. *Studies on the Crystal Chemistry of Titanium, Vanadium and Molybdenum Oxides at Elevated Temperatures*, Final technical report 1, DA-91-591-EUC-935, 1959; Stockholms Universitet, Dept. of Inorganic Chemistry, S-10691 Stockholm, Sweden.

67. Woydt, M.; Willmann, G.; Klo, H. Temperaturberechnungen an den Artikulationsflächen beim künstlichen Hüftgelenk, (Calculation of the temperature at the articulating surfaces in artificial hip joints). Mat.-wiss. Werkstofftechnik. 2001, *32*, S.200–S.210.

68. Kustas, F.M.; Buchholtz, B.W. Lubricious surface silicon nitride rings for high-temperature tribological applications. (STLE preprint No. 95-AM-6A-1, presented during). 50th Annual STLE-Meeting, May 14–19, 1995, Chicago.

69. Habig, K.-H.; Woydt, M. Sliding friction and wear Al_2O_3, ZrO_2, SiC and Si_3N_4. Proc. 5th Eurotrib Helsinki 1989; The Finnish Society of Tribology: 02150 Espoo, Finland, Vol. 3, 106–113.

70. zu Köcker, G.M.; Santher, E.; Rabe, T.; Wäsche, R. Tribological behaviour of nanocrystalline ceramics and coatings. Proc. Int. Symp. on Advanced Ceramics for Structural and Tribological Applications, 20–24 August 1995, Vancouver; Canadian Institute of Mining, Metallurgy and Petroleum: Montréal, Québec, 285–295 pp.

71. Kita, H.; Kawamura, H.; Unno, Y.; Sekiyama, S. *Low Frictional Ceramic Material*, SAE paper 950981.

72. Bryzik, W.; Kamo, R. High temperature tribology for future diesel engines. Conference Proceedings 589, 82nd NATO-AGARD-Meeting, 06.–08. Mai 1996, Sessimbra, Portugal. paper 18.

73. Korcek, S.; Jensen, R.K.; Johnson, M.D. Assessment of useful life of current long drain and future low phosphorus engine oils. Proc. Int. Tribology Conference, September 2001, Vienna.

74. Rao, V.D.N. Method of depositing composite metal coatings, PCT WO 97/13884.

75. Korcek, S.; Sorab, J.; Johnson, M.D.; Jensen, R.K. Automotive lubricants for the next millenium. 12th Int. Kolloquium "Tribology 2000-plus", 11.–13. January 2000; Technische Akademie Esslingen: 73740 Ostfildern, Germany, Vol. I, 243–253. ISBN 3-924813-44-2.

76. Korcek, S.; Sorab, J.; Jensen, R.K.; Johnson, M.D. Automotive lubricants for the next millenium. Ind. Lubr. Tribol. 2000, *52* (5), 209–220.

77. Rao, V.D.N.; Kabatt, D.M.; Cikanek, H.A.; Fucinari C.A.; Wuest, G. *Material Systems for Cylinder Bore Applications—Plasma Spray Technology*, SAE paper 970023.

78. Barbezat, G.; Schmid, J. Plasmabeschichtungen von Zylinderkurbelgehäusen und ihre Bearbeitung durch Honen MTZ. Mot. Tech. Z. 2001, *62* (4), 2–8.

79. Storz, O.; Gasthuber, H.; Woydt, M. Tribological properties of thermal-sprayed Magnéli-type coatings with different stoichiometries (Ti_nO_{2n-1}). Surf. Coat. Technol. 2000, *140*, 76–81.

80. Storz, O. Herstellung und Charakterisierung plasmagespritzter Titansuboxide (Ti_nO_{2n-1}) als Beschichtungen für Zylinderlaufbahnen, Dissertation (Ph.D. thesis), Technical University of Berlin (Germany), 2001.

81. FR 2 795 095, Pièce mécanique de friction recouverte d'oxydes triboactifs présentant un défaut de cations métalliques.

82. Buschmann, G.; Clemens, H.; Hoetger, M.; Mayr, B. Zero Emission Engine—Der Dampf-motor mit isothermer Expansion. MTZ, Mot. Tech. Z. *61* (5), 2–10.

83. Buschmann, G.; Clemens, H.; Hötger, M.; Mayr, B. The steam engine—Status of development and market potential. MTZ, Mot. Tech. Z. *62* (5), 2–10.

84. Dogigli, M.; Wildenrotter, K.; Lange, H.; Woydt, M.; Gienau, M. CMC hinge joints for hot structures. . Proc. 3rd European Workshop on Thermal Protection Systems. 457–467. ESA Publications Division: Noordwijk, NL, 25–27 March 1998. publication number: WPP-141.

85. Shelley, T. Oil-free bearings turn at highest speeds. Eur. Automot. Des., 56–57.

86. Buss, W. Running on air…bearings. Lubr. Eng., 15–19.

87. Bogdanski, M.S.; Sliney, H.E. The effect of processing and compositional changes on the tribology of PM 212, Lubr. Eng. August 1995, *51* (8), 675–683.

88. Dellacorte, C.; Fellenstein, J.A. The effect of compositional tailoring on the thermal expansion and tribological properties of PS300: A solid lubricant composite coating. Tribol. Trans. *40* (4), 639–642.

89. DellaCorte, C. The evaluation of a modified chrome oxide based high temperature solid lubricant coating for foil gas bearings. 54th Annual Meeting Society of Tribologists and Lubrication Engineers, Las Vegas. Tribol. Trans. *23* (2), 257–262.

5
Plastics

Zygmunt Rymuza
Warsaw University of Technology, Warsaw, Poland

I. INTRODUCTION

Plastics are very interesting materials in the construction of machines, in particular miniature mechanisms. They are used in manufacturing elements of tribosystems, and in the case of miniature mechanisms often the whole mechanism is produced by injection molding, which is very cheap. Two factors have contributed to the very rapid growth in plastic usage: facility of being able to shape plastics under mild temperature conditions and high resistance of these materials to chemical attack. Plastics are useful in mechanical applications as the elements can reduce vibrational effects and the tribological behavior is easy to control by selecting and combining different plastic materials between them or with other materials such as metals (steel) and ceramics.

The number of plastics is almost limitless, but in practical use considering the costs of manufacturing and the properties they possess the number of commercially significant materials is restricted to about 50. As they are built from relatively simple substances of low molecular weight, they are sometimes called "mers," or more usually, monomers, and the reaction by which they are joined together is called "polymerization." The product of such a reaction is a polymer. Such material can be very pure and may not give optimum performance, so a wide range of additives has been developed to match the properties of a plastic compound to its requirements.

II. POLYMERIZATION MECHANISMS

A polymer molecule contains a large number of similar repeated units linked to each other by covalent bonds. If all repeated units are identical, the polymer is termed a homopolymer, and the number of repeated units is the degree of polymerization, n. Polymer molecules can be formed by a sequence of classical chemical reactions, starting with small molecules called monomers.

Polymerization can produce a linear polymer, so in this case the functionality (linking capability) of the monomer units is two. Functional groups are sites for reaction. Polymerization can also give nonlinear chains and network structures that both require branch points, which are units with a functionality greater than two. Such structure can also be formed by interconnecting of the branches of a linear polymer resulting in a network structure. This can also be obtained by deliberately cross-linking existing polymer chains.

Two distinct reaction mechanisms are found in polymerization: addition and condensation [1–5]. In the first case the polymer chain grows by successive addition of monomer units. The sequences of the polymerization process can be presented as follows: $nA \rightarrow [A]_n$ $NCH_2 = CHR \rightarrow \sim [CH_2 - CHR]_n \sim$, where $-R-$ is atoms of H, Cl, or groups of atoms $-COOH$, $-COOCH_3$, $-CN$, $-C_6H_5$. For the initiation of addition polymerization an initiator or catalyst is needed. Depending on the initiator used the chain radical, ion (anion or cation), and coordination polymerization can be distinguished [5].

During the process of condensation we can obtain a polymer with the formation of simple chemical compounds such as water, ammonia, etc. The sequences of the polymerization process are as follows:

$$nX - R - X + nY - R` - Y \rightarrow X - [R - Z - R`]_n - Y + (n-1)XY$$

$$HOOC - R - COOH + H_2N - R` - NH_2$$

$$\rightarrow HO[OC - R - CO - NH - R` - NH]H + H_2O$$

The functional groups X and Y except the carboxyl group $-COOH$ and amide group $-NH_2$, as it was presented in the above reaction, can also be hydroxyl groups $-OH$, sulfone groups $-SO_3H$, etc.

Condensation polymerization occurs without any initiator; during the process the monomer disappears in the initial stages of the process and the molecular weight of the polymer slowly increases. The initiation and growth of molecular chain are the same reactions; after the end of the reaction the polymer has reactive groups at the ends of the chain.

Polymer composed of only one type of monomer is called a homopolymer and that built of different monomers is called a copolymer. The monomers in a copolymer can be distributed irregularly in random copolymers (e.g., ethylene/vinyl/acetate) while certain pairs of monomers, which have particular reactivities toward each other, may form alternating copolymers. The molecules of a block copolymer contain long uninterrupted sequences of each type of repeated unit: examples are the thermoplastic rubbers. Graft copolymers consist of branches of one species attached to the main chain of a different repeated unit.

III. CHAIN LENGTH AND MOLECULAR WEIGHT

The length of the molecular chain is generally different as its formation is the result of random events, and the chain length reflects the growth history. Under very carefully controlled conditions during anionic polymerization it may be possible to produce a polymer that is virtually monodispersed, i.e., the macromolecules are all of the same size. In contrast during free-radical polymerization, because of the randomness of both chain initiation and termination, the polymers yielded have a wide molecular size distribution (i.e., they are polydispersed) [1,6]. In contrast to typical chemical compounds, polymers do not have a well-defined molecular weight. As polymers are constructed from macromolecules with different degrees of polymerization n, their molecular weight therefore is an averaged value. Average molecular weight reflects the ratio of the weight of polymers to the total number of molecules. The number average molecular weight M_n is defined below:

$$M_n = \frac{\sum n_i M_i}{\sum n_i}$$

The number average molecular weight is the summation of the product of each molecular weight value, M_i, and the number of such molecules present, n_i, is divided by the total number of all molecules present.

All macromolecules have similar effect on the value of average molecular weight. By the weight the average value of molecular weight is calculated by the assumption of the weight part of the macromolecules with a given molecular weight in relation to the total weight of the polymer. The effect of macromolecules with a higher molecular weight is higher. The weight average molecular weight M_w is defined as follows:

$$M_w = \frac{\sum w_i M_i}{\sum w_i}$$

In contrast to the number average molecular weight, the weight average molecular weight sums the weight fraction of each species (w_i is the weight of each component). The ratio of M_w/M_n describes the degree of nonuniformity of molecular weight; for mono-dispersed polymers $M_w/M_n = 1$, and for polydispersed polymers $M_w/M_n > 1$. The molecular weight determines the many physicochemical properties and mechanical behavior of polymers, e.g., melting and plasticity temperatures, solubility, crystallization ability, ability to form fibers and membranes, elongation strength resistance to multiple deformation, elasticity modulus, chemical resistance, thermal resistance, and viscosity in liquid state.

IV. STATES OF AGGREGATION

Typical polymers are in amorphous or crystalline state [6–8]. Amorphous polymers are glassy solids at low temperatures, as molecular motion is severely restricted. The temperature at which a polymer softens is known as the glass transition temperature, T_g. As the temperature increases further, all types of molecular chain-segment motion become feasible; the molecules then assume a state of continuous, random motion, which defines the melt state of glassy polymers. A specific linear amorphous polymer, such as poly(methyl methacrylate) or polystyrene, can exist in a number of states according to the temperature and the average molecular weight of a polymer.

In amorphous polymers the macromolecules form disordered structures (Fig. 1) with weak energy interactions. In such state only ordered structures are formed on the short distances of interactions of about 1 nm.

If polymer chains possess sufficient chemical and geometrical regularity, small regions of local order are developed on cooling the melt; these are known as crystallites (small crystals). The process of crystallization never reaches completion, as chain entanglements preclude the unlimited growth of a crystallite, so the term semicrystalline is often used for such polymers. A single chain may participate in several different crystallites at different parts of its length. Well-known examples of crystalline polymers are polyethylene, acetal resins, and polytetrafluoroethylene.

The nucleation rate and the subsequent crystallite growth rate are strongly dependent on temperature; it is possible therefore to change the degree of crystallinity and the texture or morphology by thermal treatment. Crystallites are anisotropic. Crystallinity is characterized

Figure 1   Amorphous structure of polymers—random array of chains of macromolecules.

at the crystalline melting point (T_m), which is usually independent of molecular weight, but is modified by the chemical structure of the chains.

There have been, over the years, profound changes in the theories of crystallization in polymers. For many years it was believed that the crystallinity present was based on small crystallites of the order of a few nanometer units in length. The crystallites consisted of a bundle of segments from separate molecules that had packed together in a highly regular order. The general picture is such that the crystallites are embedded in an amorphous matrix (Fig. 2a). This theory, known as the fringed micelle theory or fringed crystallite theory, helped to explain the many properties of crystalline polymers, but it was difficult to explain the formation of certain structures such as spherulites, which could possess a diameter as large as 0.1 mm. In many circumstances the polymer molecules can fold upon themselves at intervals of about 10 nm to form lamellae (Fig. 2b), which appear to be fundamental units in a mass of crystalline polymer. The study of crystals grown from polymer solution, notably by Keller [9], indicated a laminar structure with the polymer chain folding back and forth, perpendicular to the plane of the lamina (Fig. 2b) with regular and irregular forms (Fig. 2c and d). There is strong evidence that the folded chain structure is present in polymers crystallized from the melt, but the structure is not universal, for crystals along the length of a polymer chain can be found, as can mixed structures, the so-called shish-kebab (Fig. 2e).

Polymers that are crystallized from the melt under quiescent conditions frequently form spherically symmetrical structures known as spherulites [6–8]. The birefringence of the crystallites, when aggregated into a spherulite, leads to a Maltese Cross extinction pattern when viewed through crossed polars. Spherulites are similarly important with respect to

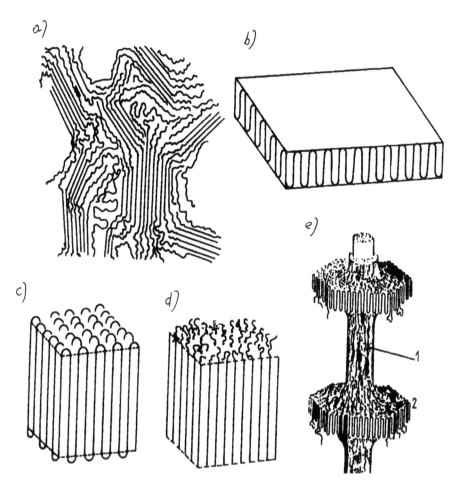

Figure 2 Models of crystalline structures of polymers: (a) fringed micelle, (b) folded chain (lamellae), (c) folded chain regular structure, (d) folded chain irregular structure, (e) shish-kebab structures. 1—core, 2—buildup.

properties such as transparency and toughness. Crystallization process spreads by the growth of individual lamellae as polymer molecules align themselves into position and start to fold. A variety of reasons such as point of branching or some other irregularity in the structure of molecule growth would then tend to proceed in many directions. In effect this would mean an outward growth from the nucleus and the development of spherulites; so in this concept a spherulite is simply caused by growth of the initial crystal structure, whereas in the fringe micelle theory it is generally postulated that formation of a spherulite required considerable reorganization of the disposition of the crystallites. Closer packing of the molecules results in increased density. The decreased intermolecular distances will increase the secondary forces holding the chain together and increase the value of properties such as tensile strength, stiffness, and softening point.

Some polymers (e.g., homopolymers of *p*-hydroxybenzoic acid) demonstrate liquid crystal structure [6–8]. Such polymers flow like liquids but retain some long-range order because of restricted rotational mobility of the molecules. Liquid crystal polymers consist of rod-like molecules that, during shear, tend to orient in the direction of shear. Because of the molecular order the molecules flow past each other with comparative ease and melts have a low viscosity. When the melt is cooled the molecules retain their orientation, giving self-reinforcing materials that are extremely strong in the direction of orientation.

With increasing departure from linearity, the precise description of a polymer chain becomes more difficult; a branched polymer chain, for example, requires a description of the distribution of chain lengths with, additionally, the distribution of branch lengths and the number of branches per molecule. For network structures, it is even more difficult to determine the characterizing parameters, as such materials are insoluble, although it is possible to assess the molecular weight between cross-links. Both branching and the formation of network structures interfere with the capability of a polymer to crystallize; indeed, the presence of a dense array of cross-links effectively eliminates crystallinity. A cross-linked polymer can generally be placed into one of two groups: lightly cross-linked materials and highly cross-linked materials. In the lightly cross-linked polymers (e.g., the vulcanized rubbers) the main purpose of cross-linking is to prevent the material deforming indefinitely under load. The chains can no longer slide past each other, and flow, in the usual sense of the word, is not possible without rupture of covalent bonds. If the degree of cross-linking is increased the distance between cross-links decreases and a tighter, less flexible network will be formed. Segmental motion will become more restricted as the degree of cross-linking increases so that the transition temperature will eventually reach the decomposition temperature. In polymers of such degree of cross-linking only amorphous rigid state will exist. This is the state commonly encountered with, for example, the technically important phenolic, aminoplastic, and epoxide resins. Many modern materials effectively cross-link on cooling to room temperature after processing but which on reheating appear to lose their cross-links. Ionic cross-linking, hydrogen bonding, triblock, and multiblock copolymerizing approaches to such fugitive cross-linking have been made in recent years and have come to commercial fruition [1–3].

With a glassy or near-glassy resin, in conjunction with a rubbery polymer, when suitably paired, it is possible to produce rigid compounds with a high degree of toughness, particularly on impact; a combination of properties that tend to be lacking in most polymers. Such compositions, frequently referred to as polyblends, may be exemplified by such well-known products as high-impact polystyrene, ABS, and impact-modified PVC [1–3,10].

Polymers can exist in a number of states: amorphous, rubbers or fluids, or crystalline structures. Molecular and crystal structures can be monoaxially or biaxially oriented. Heterogeneous blends of polymers in different states of aggregation enable materials to be produced with a combination of properties not shown by single polymers. The form that a polymer composition will take at any given temperature will depend on the glass transition temperature, the ability of the polymer to crystallize, the crystalline melting point, any orientation of more crystal structures that may have been induced, and the type and extent of cross-linking, whether or not polymers in different states of aggregation have been heterogeneously blended [10].

V. RELATION OF STRUCTURE TO CHEMICAL AND PHYSICAL PROPERTIES

The properties of polymers depend very much on the structure, in particular on the chemical structure. The properties of polymers, e.g., mechanical behavior, result from the bonds that hold atoms and molecules together [3,6–8]. The atoms of a molecule are held together by primary bonds. The attractive forces that act between molecules are usually referred to as secondary bonds, secondary valence bonds, intermolecular forces, or van der Waals forces. The primary bond formation takes place by various interactions between electrons in the outermost shell of two atoms resulting in the production of a more stable state. The ionic, covalent, and coordinate bonds can be met in polymers. The most important interatomic bond in polymers is the covalent bond, which is formed by the sharing of one or more pairs of electrons between two atoms. The secondary bonds are caused by intermolecular forces, which can be of four types: dipole forces, induction forces, dispersion forces, and the hydrogen bond. In polymers, in the absence of hydrogen bonding, the intermolecular force is primarily due to dispersion effects.

The chemical properties of polymers relate to the solubility characteristics, the effect of specific chemicals on their molecular structure (degradation or cross-linking reactions), the effect of specific chemicals or environments on polymer properties at elevated temperatures, the effect of high energy of irradiation, the aging and weathering of the material, permeability and diffusion characteristics, and toxicity.

Solubility (i.e., coexisting on the molecular scale and no tendency to separate) depends on the compatibility of the molecules of the solvent chemical and the polymer. Molecules of two different species will be able to coexist if the force of attraction between two different molecules is not less than the forces of attraction between two like molecules of either species. The suitable measure of the forces of attraction holding molecules together can be the energy of vaporization per unit weight and the energy of vaporization per molar volume. The latter term is known as the cohesive energy density. More commonly encountered is the square root of the cohesive energy density, which is known as the solubility parameter. The solubility parameter depends on the functional groups of a polymer and is high when the polymer contains $-CO-$ or $-CH(CN)-$, $-S-$ groups, and low for polymers with $-CF_2-$ or $-C(CH_3)_2-$ groups [3].

Polymers are affected, to varying extent, by exposure to thermal, photochemical, and high-energy radiation [1,3,6–8,11]. These forms of energy may cause such effects as cross-linking, chain scission, modifications to chain structure, and modifications to the side group of the polymer, and they may also involve chemical changes in the other ingredients present. In the absence of other active substances, e.g., oxygen, heat stability is related to the bond energy of chemical linkages present. Heat stability is higher in aromatic than in aliphatic bonds. The high stability of polytetrafluoroethylene (PTFE) is due to the fact that only $C-C$ and $C-F$ bonds are present, both of which are very stable. Poly-p-xylene contains only the benzene ring structure (very stable thermally) and $C-C$ and $C-H$ bonds, and these are also very stable. Built-in to the main chain, some groups (such as polar, high volume, double bonds, cyclic structures in particular aromatic) prevent free rotation effects, resulting in increase of stiffness, and increase the glass and melting temperatures of polymers. Similar effects result from cross-linking. Some side groups prevent packaging of macromolecules and formation of crystalline structures—so they have influence on T_g and T_m. The linear correlation function $T_g = (0.50-0.66)$ and T_m can be observed in the crystalline phase of polymers. Generally, the high heat stability of polymer can be reached when macromolecules are rigid, symmetrical, and the bonding energy is high. Summarizing, the groups attached to the backbone, effect on the increase of glass temperature, rigid structures (e.g., phenylene groups incorporated in the backbone of the molecule), the packing of substituents around the main chain, secondary bonding between chains (e.g., hydrogen bonding), primary bonding between chains (e.g., cross-linking), length of side chains, molecular weight, copolymerization, and plasticization affect strongly the glass transition. Most polymers are affected by exposure to light, particularly sunlight. This is the result of the absorption of radiant light energy by chemical structures. The lower the wavelength the higher the energy. Whether or not damage is done to a polymer depends also on the absorption frequency of a bond. Under normal terrestrial conditions the carbonyl bond causes much trouble. Polytetrafluoroethylene and other fluorocarbon polymers have good light stability because the linkages present normally have bond energies exceeding the light energy. Sometimes carbonyl or other groups may prove to be a site for photochemical action and, e.g., PE and PVC have only limited light stability. To some extent, the light stability of a polymer may be improved by incorporating an additive that preferentially absorbs energy, at wavelengths that damage the polymer linkage.

In analogy with thermal and light radiations, high-energy radiation may also lead to scission and cross-linking. Most polymers of monosubstituted ethylene cross-link, while most polymers of disubstituted ethylenes degrade. Exceptions are polypropylene, which degrades, and PVC, which either degrades or cross-links depending on the conditions.

Aging and weathering depend on many factors. The following agencies may cause a change in the properties of a polymer: chemical environments, heat, ultraviolet light, and high-energy radiation. As different polymers and additives respond in different ways to the influence of chemicals and radiant energy, weathering behavior can be very specific.

Diffusion of small molecules into, out of, and through a polymer is of importance in the processing and usage of the latter. High diffusion and permeability is sometimes desirable but at other times undesirable. Diffusion through a polymer occurs by the small molecules passing through voids and other gaps between the polymer molecules. The diffusion rate will therefore depend, to a large extent, on the size of the small molecules and the size of the gaps. The size of the gaps in the polymer will depend, to a large extent, on the physical state of the polymer, i.e., whether glassy, rubbery, or crystalline. In the case of amorphous polymers above the glass transition temperature (i.e., in the rubbery state) molecular segments have considerable mobility so there is a high likelihood that a molecular segment will, at the same stage, move out of the way of a diffusing small molecule and so diffusion rates are higher in rubbers than in other types of polymer.

Toxicity problems of polymers associated with the finished product arise from the nature of the additives and seldom from the polymer. Such polymers as poly(vinyl carbazole) and polychloroacrylates can cause unpleasant effects when monomer is present.

Fire performance is important in many applications. By this it is meant that plastic materials should resist burning and, in addition that, levels of smoke and toxic gases emitted should be negligible. The development of new polymers of intrinsically better performance and the development of flame retardants are two approaches to fire-resistant polymers. Generally, the higher the hydrogen-to-carbon ratio in the polymer the greater is the tendency to burning (other factors being equal). Some polymers, on burning, emit blanketing gases that suppress burning.

Most plastic materials may be considered as electrical insulators [6–8,12]. These materials may be divided roughly into two groups: polymers with outstandingly high resistivity, low dielectric constant, and negligible power factor; and moderate insulators with lower resistivity and higher dielectric constant and power factor affected by the conditions of the test (temperature, frequency, humidity). The latter group is often referred to as polar polymers. The properties of these two groups of materials relate to the molecular structure. The dielectric properties of polar materials depend on whether or not the dipoles are attached to the main chain. If the dipoles are attached to the main chain, dipole polarization will depend on segmental mobility, so at low temperature below the glass transition temperature such polymers are therefore better insulators below the glass temperature than above it. The influence of humidity on the electrical insulating properties of plastics is higher for the absorbing water materials than for nonabsorbent materials.

Some special polymers such as poly-acetylenes, poly-*p*-phenylene, poly(*p*-phenylene sulphide), polypyrrole, and poly-1,6-heptadiyne are conductive materials [1,8,12]. Their conductivity is comparable with some of the moderately conductive metals. Most of such polymers have important disadvantages such as improcessability, poor mechanical strength, instability of the doped materials, sensitivity to oxygen, poor storage stability leading to loss in conductivity, and poor stability in the presence of electrolytes.

Optical properties such as refractive index, clarity, haze and birefringence, color, transmittance, and reflectance are closely linked with molecular structure. Amorphous polymers free of fillers or other impurities are transparent unless the chemical groups present absorb visible light radiation. Crystalline polymers may or may not be transparent. If the crystalline structures such as spherullites are smaller than the wavelength of light, then they do not interfere with the passage of light and the polymer is transparent.

Density (the mass per unit volume) is a function of the weight of individual molecules and the way they pack. Hydrocarbons do not possess "heavy" atoms and therefore the mass of the molecule per unit volume is rather low [3]. Where large atoms are present, e.g., chlorine atoms, the mass per unit volume is higher (so PVC, a substantially amorphous polymer, has a specific gravity of about 1.4). If a polymer can crystallize then the molecular packing is much more efficient and higher densities can be achieved. The conformation adopted by a molecule in the crystalline structure will also affect the density.

By knowledge of the glass transition, the ability to crystallize, and, where relevant, the crystalline melting point, general statements may be made regarding the properties of a given polymer at a specified temperature. In the case of a crystalline polymer the maximum service temperature is largely dependent on the crystalline melting point. When the polymer possesses a low degree of crystallinity, the glass transition temperature will remain of paramount importance.

The melt viscosity of a polymer at a given temperature is a measure of the rate at which chains can move relative to each other. This is controlled by the ease of rotation about the backbone bonds, i.e., the chain flexibility, and on the degree of entanglement. The higher the chain flexibility of polymers the lower is the viscosity in the melting state. For a specific polymer, melt viscosity is considerably dependent on the (weight average) molecular weight. The higher the molecular weight the greater the entanglements and the greater the melt viscosity. Chain branching also has an effect.

In the comparison of the yield and elastic moduli of amorphous polymers well below their glass temperature it is observed that the differences between polymers are quite small. The effect of molecular weight is small. In the case of engineering crystalline polymers wider differences are to be noted. Many PE have a yield strength below 14 MPa, while PA may have a value of 83 MPa [1,3,13,14]. In these polymers the intermolecular attraction, the molecular weight, and the type and amount of crystalline structure all influence the mechanical properties. The position of the glass transition temperature and the facility with which crystallization can take place are fundamental to the impact strength of a polymer. Well below the glass transition temperature amorphous polymers break with a brittle fracture but they become tougher as the glass transition temperature is approached. A rubbery state will develop above the glass transition and the term impact strength will cease to have significance. In the case of crystalline materials the toughness will depend on the degree of crystallinity; large degrees of crystallinity will lead to inflexible masses with only moderate impact strengths. The size of the crystalline structure formed will also be a significant factor, large spherulitic structures leading to masses with low impact strength.

Surface properties of polymers such as adhesive properties expressed by the free energy depend mainly on the chemical structure of polymers [3,15–19]. The surface free energy is the effect of intermolecular interactions unbalanced on the surface. The surface free energy of polymers is composed of two components: polar and dispersion. As typical low energy groups such as $-CH_2-$, $-CF_2-$, $CH{=}CH-$, $-COO-$ can be found in the chemical structure of macromolecules of engineering polymers the surface free energy of these materials is relatively low, 20–50 mJ/m^2 [3].

VI. PROPERTIES OF POLYMERS

The properties of polymers depend on the chemical composition and physical state [1,3,6–8,14]. Ambient conditions such as temperature and humidity also have very great influence. For mechanical or tribological applications, the mechanical and adhesive properties of polymers play a very important role. Table 1 shows typical (averaged) mechanical and adhesive properties (surface free energy) of commercial thermoplastic engineering polymers used in the construction of machine or small mechanisms. The glass and softening (loss of stiffness) temperatures are also indicated.

The thermal properties of some engineering plastics are listed in Table 2. Thermal conductivity, thermal expansion, and specific heat are most important in engineering applications. Table 3 shows the electrical properties of several thermoplastic engineering polymers. Typical engineering polymers demonstrate electrical properties characteristic of materials used in isolators and semiconductors. Electrical conductivity depends on the dynamics of dissociation and ionization of molecules and atoms, recombination of carriers, and the diffusion processes of charged particles in electrical field. The conductivity of polymers has both ionic and electron character.

Cohesive energy density, surface free energy (with dispersion and polar components), solubility parameter, and absorptivity of water for some engineering polymers are listed in Table 4. Cohesive molar energy E_c is the energy needed in the total decomposition of intermolecular bonds in 1 mol of polymer, so the cohesive energy is strongly connected with the heat of evaporation ΔH, so $E_c = \Delta H - RT$ (where R is gas constant, T is absolute temperature). Table 5 (according to Ref. 3) shows the effect of functional groups on the density, glass transition temperature, melting temperature, surface free energy, solubility parameter, and dielectric constant of polymers.

In many technical applications (e.g., in tribosystems on rubbing elements) the surface properties of polymers play a very important role. Several methods are used to modify the surface properties of plastics: chemical treatment in oxidizing liquid, flame treatment, corona-discharge at atmospheric pressure, plasma treatments, or activation by radiation (UV, γ radiation, laser-, ion-, or electron-beam treatment) [15–18]. Chemical treatment has a rather historical meaning. Plasma treatments of polymers are nowadays an effective way to modify the properties of the surface layer of a polymer [15,16]. Low-pressure plasma processing of polymers can be divided into three categories: etching (removal of material), deposition (addition of new material to the surface), and modification (morphological, structural, and physicochemical change of the surface or near-surface region).

Low-pressure plasmas (so-called cold plasmas or glow discharges) are the basis of a large variety of nonequilibrium processes for leading-edge surface modification processes, usually included in the category of thin film technology, that allow the design of substrate surfaces by means of the deposition or modification of thin surface films. In

Table 1 Mechanical Properties of Some Engineering Plastics

Polymer	Tensile strength, MPa	Compression strength, MPa	Shear strength, MPa	Elasticity modulus, MPa	Elasticity modulus at compression, MPa	Glass temperature, K	Softening temperature, K
PA	66	61	40	1500	–	320	360
PETP	70	90	–	2700	–	350	410
HDPE	26	–	25	1800	2000	220	360
PP Isotactic	30	40	25	1300	1400	250	370
POM	80	130	–	3000	–	220	410
PTFE	20	10	–	600	700	–	–
PCTFE	40	60	–	1400	1500	315	400
PC	50	60	30	2500	–	420	430
PPSU	70	–	–	2500	2600	430	450
PS	25	100	40	3500	–	370	375
PMMA	60	80	40	3300	3300	370	385

this class, plasma-enhanced chemical vapor deposition (PE-CVD) of well adherent deposited thin films (10–1000 nm) is a major technological product. Another important and dynamic field in this category is the technology for the modification of the very first surface layers of solid materials by grafting chemical functionalities. The principal feature of thin films designed by glow discharge technology is the possibility of continuous variation of the chemical composition of treated surfaces and, obviously, their physical and chemical properties (such as wettability, adhesivity and dyebility, refractive index, hardness and tribological properties, barrier characteristics, chemical inertness, biocompatibility, etc.) in a very broad range. A plasma reactor can be simply made with a vacuum (or even can work at atmospheric pressure [ ]) including two metal plates as electrodes, connected to a radio-frequency generator through an impedance matching unit. Reactive species are atoms, radicals and charged particles. Fluoropolymer (also said Teflon-like), metal-containing, silicon-like or silicon oxide/nitride and

Table 2 Thermal Conductivity, Thermal Expansion, and Specific Heat of Some Thermoplastic Polymers

Polymer	Thermal conductivity, W/mK	Thermal expansion, 10^{-4} 1/K	Specific heat, kJ/kg K
PA 6	0.29	0.9	1.68
PA 66	0.23	0.9	1.68
PETP	0.15	0.6	1.20
HDPE	0.48	1.5	1.55
PP isotactic	0.34	1.7	1.20
POM	0.40	0.8	1.38
PTFE	0.25	0.9	1.05
PC	0.23	0.7	1.25
PPOX	0.20	0.3	1.35
PPSU	0.19	0.54	1.30
PS	0.16	0.7	1.25
PMMA	0.19	0.6	1.47

Table 3 Electrical Properties of Some Engineering Polymers

Polymer	Volume resistivity, Ωm	Dielectric constant (800 Hz)	Dielectric loss factor (800 Hz)
PA 6 (0.2% H_2O)	10^{12}	5.0	0.02
PA 6 (3% H_2O)	10^{8}	8.0	0.15
PA 66 (0.2% H_2O)	10^{12}	3.8	0.02
PA 11	6×10^{12}	3.5	0.03
PETP	3×10^{14}	3.3	0.002
HDPE	10^{14}	2.3	0.0003
POM	10^{12}	3.7	0.04
PTFE	10^{16}	2.0	0.0003
PP	10^{14}	2.3	0.0015
PS	10^{14}	2.5	0.0005
PC	10^{12}	3.0	0.001
PPSU	3×10^{14}	3.9	0.003
PMMA	10^{13}	3.2	0.004
PPOX	10^{15}	2.5	0.0004

diamond-like carbon films are typical products of PE-CVD technology used to modify the surface of polymers.

The most important feature of plasma techniques is that the surface properties of the treated material can be modified without changing their intrinsic bulk properties. The plasma treatments can simultaneously bring about many changes on the surface of polymers, such as degradation of polymer chain, formation of free radicals, chemical function-

Table 4 Cohesive Energy Density, Surface Free Energy (with Dispersion and Polar Components), Solubility Parameter, and Absorptivity of Water for Some Thermoplastic Engineering Polymers

Polymer	Density of cohesive energy, kJ/mol	Surface free energy, mJ/m^2	Dispersion component of surface free energy, mJ/m^2	Polar component of surface free energy, mJ/m^2	Solubility parameter, 10^3 J$^{1/2}$ m$^{-3/2}$	Absorptivity of water, %
PA 6	163.8	47	40.8	6.2	30.4	1.3–1.9
PA 66		45			27.7	0.9–1.3
PA 11	33					
PA 12						0.25
PETP	60.5	47.3	43.2	4.1	19.8–21.8	0.1
PBTP						0.1
LDPE	8.4	33.2	33.2			
HDPE					15.7–17.0	0.01
POM	10.5	39			20.8–22.4	0.22
PTFE	6.7	19.1	18.6	0.5	12.6	0
PCTFE	14.7	31			14.7–16.1	0
PP	14.5	29			16.7–18.8	0.01–0.03
PS	33.1	42	41.4	0.6	17.3–19.0	0.03–0.10
PC		42			20.2	0.1–0.2
PMMA	30.3	40.2	35.9	4.3	18.6–26.1	0.2–0.4

Table 5 The Effect of Functional Groups on Density (ρ), Glass Transition Temperature (T_g), Melting Temperature (T_t), Surface Free Energy (γ_s), Solubility Parameter (δ), and Dielectric Constant (ε) of Polymers

$\rho, \dfrac{kg}{dm^3}$	$\gamma_s, \dfrac{mJ}{m^2}$	T_g, K	T_t, K	δ $2.04 \cdot 10^3 \dfrac{J^{1/3}}{m^{3/2}}$	ε

alization, etching, and cross-linking. Usually for applications such as improvement of the adhesion of polymers to metals for example, polar functional groups are introduced at the surface. It was found [16] that not only ions and radicals but also UV (ultraviolet radiation) in a polymer plasma treatment can initiate reactions on polymer surfaces. The specific contribution of each depends largely on the treatment conditions. In a downstream remote process in which no ions are present near the sample and the radical density is rather low, UV radiation can be very important, in particular when gases with intense emission lines in this region as hydrogen or noble gases are used in the feed. The penetration depth of the radiation is considerably higher than with ion bombardment, where the heaviest ion bombardment in a plasma affects only the outermost nanometers of a substrate. The UV-induced reactions in the subsurface are often governed by photolysis reactions such as cross-linking or double-bond formation.

Plasma surface modifications have been widely used in the synthesis and fabrication of biomaterials. Plasma treatments and depositions are particularly useful for permeation barriers, enhancing biostability, improving lubricity, permitting adhesion in the assembly of medical devices, and altering material wettability. Plasma treatments can be used for modifying blood interactions in vivo and in vitro. The exceptionally smooth nature of many plasma-deposited films provides a good foundation onto which one can engineer surfaces with chemistries and features of molecular and supramolecular size.

Cold-plasma modification of polymers and deposition of thin polymer films is a branch of science characterized by increasing popularity in the last years for a large number of new industrial processes that have been realized by its applications. The unique common need of a wide variety of applications (e.g., electronics, automotive components, optics, food and pharmaceutical packaging, biomedical and surgical equipment, etc.) is that the products feature "surfaces" with tailored and unusual properties, which enable their use where otherwise would be impossible with conventional materials.

VII. CHARACTERIZATION OF POLYMERS

The determination of chain structure is simple as the development of physical methods of analysis, especially infrared (IR) spectroscopy and nuclear magnetic resonance (NMR) spectroscopy, has enabled the structures of small molecules to be determined precisely [20–23].

Traditional techniques based on pyrolysis, followed by examination of degradation products by gas chromatography, mass spectrometry, or infrared spectroscopy, are still much used in the analysis of polymers. Elemental analysis is limited in the information it yields, as many polymers can have the same composition, e.g., high-density polyethylene, low-density polyethylene, linear low-density polyethylene, polypropylene, polybutene, and ethylene–propylene rubber, and blends and copolymers all have the empirical formula $H-(CH_2)_n-H$.

The chain length (or molecular weight) can be determined by the use of many methods [1,20]. Some methods are direct measurements, others reply on the measurement of properties that are considerably affected by molecular weight. Molecular weights may be averaged with other kinds of weighting, which are of value in some applications: the viscosity average molecular weight is that derived from solution viscosity measurements. It is usually closer to weight average than to number average molecular weight. The following methods give the number average molecular weight: end group analysis, increase in boiling point of solvent by solution of polymer, decrease in melting point of solvent by solution of polymer, and osmometry. In the case of the last three methods the polymer must be in true solution, a state that is difficult to achieve with crystalline polymers, except at high temperatures. The weight average molecular weight is generally determined by measurement of the light scattering of polymer solutions; less often, by sedimentation in an ultracentrifuge. A useful practical monitor of chain length and especially of changes in chain length is afforded by the measurement of solution or melt viscosity [1,20,24].

Characterizing the molecular weight distribution can be achieved by fractionating the polymer, making use of the fact that some properties of polymers depend on molecular weight. Solubility is one such property and its dependence on molecular weight can be used to effect a separation of a polydispersed polymer into a number of fractions, differing in molecular weight. Fractionation can be carried out by changing either the temperature or the solvent power of the liquid by varying the solvent/nonsolvent ratio. Another new method of determining the molecular weight distribution is by gel permeation chromatography, the principle of which is that chains of differing length behave differently when a solution passes through a column packed with a gel with uniform and appropriate pore size. While the smaller molecules can diffuse freely into the pores of the gel and are thereby retarded in their passage through the column, the larger chains are not so delayed and emerge first from the column.

The glass transition temperature can be determined easily using dilatometric techniques as specific volume of a polymer increases with temperature above T_g. Other techniques that can be used to characterize glass transition temperature are based upon physical properties that change markedly at the point of the transition; these include thermal analysis (heat content), thermomechanical analysis (thermal expansion), optical methods (refractive index), nuclear magnetic resonance, and temperature scans of mechanical (dynamic modulus) or electrical (permittivity) properties [20].

Determination of crystallinity can be achieved by a variety of methods. The most fundamental technique is x-ray diffraction, the crystal planes leading discrete diffraction angles in the polycrystalline material [1,20]. By making assumptions about the level of x-ray scattering from the amorphous polymer, the fraction of crystalline polymer can be determined. The density of a crystalline polymer is usually higher than that of the amorphous material so the density of a sample is directly related to its fractional crystallinity. While the density of the crystalline polymer can be calculated from the unit cell by x-ray diffraction, the density of the amorphous phase can only be estimated generally, and further assumption made that the crystalline in amorphous phases does not interfere with one another. Differential scanning calorimetry (DSC) was adapted to the assessment of crystallinity; the latent heat of fusion is measured to monitor the crystalline content of a sample. The method depends on knowing the heat of fusion of pure crystalline polymer, a state that cannot be achieved; therefore the method requires calibration against a sample of the same polymer of known crystallinity. Furthermore, there is always the danger that the crystallinity level is increased by heating the polymer during the measurement.

The texture of amorphous polymers is simple but crystalline polymers, on the other hand, may have very complex texture that is strongly dependent on processing conditions, in particular, the thermal history. Conversely, examination of the texture allows comment to be made on the likely thermal treatment imposed on a product. Polarized light microscopy is a technique much used in the examination of microtomed thin sections of crystalline polymers. Other textural features investigated by this method include nucleated crystallites where nuclei are provided by a foreign surface or by a shear plane. For the examination of texture beyond the resolving power of the light microscope, the scanning electron microscope (SEM) or atomic force microscope (AFM) is a valuable tool [19,20]. Orientation (uniaxial, biaxial, or planar orientation) occurs in both amorphous and crystalline polymers and is a consequence of the ordering of polymer chains during flow with freezing of the morphology before relaxation occurs. Anisotropy in mechanical or tribological properties results from the directional character of the polymer chains, which in their length are sustained by covalent bonds (strong), while an assembly of chains is kept together only by much weaker forces. The usual method of assessing orientation is by measuring the birefringence, which is the difference in refractive index in the orientation and transverse directions. Shrinkage is also a satisfactory monitor of orientation, especially for amorphous polymers, when they are heated to a temperature marginally higher than the glass transition temperature.

The characterization of the surface properties is important in the study of tribological and adhesive properties. Atomic force microscopic assessment of surface topography is fruitful together with the determination of surface free energy (both dispersion and polar components) by the measurement of the contact angle during wetting by the test liquid [19,25]. Atomic force microscopy can also be used to determine the nanomechanical properties of the thin surface layer, which plays a very important role in the tribological behavior of polymers [19].

The characterization of the final polymer product, which often contains additives, is a rather difficult problem. Frequently, the additive is present only in small concentrations, hopefully well dispersed in the matrix polymer; almost without exception the analysis depends on isolating the additive, e.g., by extracting the additive with a solvent, or for inorganic additives the polymer and any organic additives can be removed by pyrolysis and the residue analyzed [6]. Sometimes the problem devolves into identifying a small amount of one polymer in the bulk of a different one. In such a case infrared spectroscopy may help, while for crystalline polymers, differential thermal analysis might identify the polymer. The distribution of the additives in a polymer matrix can be examined on a very small scale by an optical or light microscope. Refractive index differences in the constituents of a plastic compound are the basis for analysis by differential interference contrast microscopy. There are many other variants of microscopy, including ultraviolet and fluorescence methods, which contribute to an understanding of the structure of plastics, the latter are particularly useful in monitoring polymer degradation.

REFERENCES

1. Brydson, J.A. *Plastics Materials*; 6th Ed.; Butterworth Heinemann: Oxford, 1996.
2. Tonell, A.E. *Polymers From the Inside Out: An Introduction to Macromolecules*; Wiley: Chichester, 2001.
3. van Krevelen, D.W. *Properties of Polymers*; Elsevier: Amsterdam, 1990.
4. Elias, H.G. *An Introduction to Polymer Science*; Wiley: Chichester, 1997.
5. Kuran, W. *Principles of Coordination Polymerisation*; Wiley: Chichester, 2001.
6. Birley, A.W.; Haworth, B.; Batchelor, J. *Physics of Plastics*; Hanser: Munich, 1991.
7. Sperling, L.H. *Introduction to Physical Polymer Science*; 3rd Ed.; Wiley: Chichester, 2001.
8. Strobl, G.R. *The Physics of Polymers*; 2nd Ed.; Springer: Berlin, 1997.
9. Keller, A. Polyethylene as a paradigm of polymer crystal morphology. Plast. Rubber Process. Appl. 1984, *4*, 85–92.
10. Donald, R.P.; Bucknall, C.B. *Polymer Blends*; Wiley: Chichester, 1999.
11. Rabek, J.F. *Photodegradation of Polymers*; Springer: Berlin, 1996.
12. Hadziioannou, G.; van Hutter, P.F. *Semiconducting Polymers*; Wiley: Chichester, 1999.
13. Aharoni, S.M. *n-Nylons—Their Synthesis, Structure and Properties*; Wiley: Chchester, 1997.
14. Mark, J.F. *Physical Properties of Polymers Handbook*; Springer: Berlin, 1996.
15. Garbassi, F.; Morra, M.; Ochiello, E. *Polymer Surfaces: From Physics to Technology*; Wiley: Chichester, 1997.
16. d'Agostino, R., Favia, P., Fracassi, F., Eds.; *Plasma Processing of Polymers*; Kluwer Academic Publishers: Dordrecht, 1997.

17. Feast, W.J., Munro, H.S., Eds.; *Polymer Surfaces and Interfaces*; Wiley: Chichester, 1987.
18. Zenkiewicz, M. *Adhesion and Modification of Surface Layer of High-Molecular Materials*; WNT: Warszawa, 2000. *in Polish*.
19. Tsukruk, V.V., et al. *Advances in Scanning Probe Microscopy for Polymer Characterization*; Wiley: Chichester, 1999.
20. Pethrick, R.A.; Dawkins, J.V. *Modern Techniques for Polymer Characterization*; Wiley: Chichester, 1999.
21. Chalmers, J.M. *Polymer Spectroscopy*; Wiley: Chichester, 1999.
22. Zerbi, G.; Siesler, H.W.; Noda, I.; Tasumi, M.; Krimm, S. *Modern Polymer Spectroscopy*; Wiley: Chichester, 1999.
23. Stuart, B. *Polymer Analysis*; Wiley: Chichester, 2002.
24. Teraoka, I. *Polymer Solutions: An Introduction to Physical Properties*; Wiley: Chichester, 2002.
25. Neumann, A.W., Spelt, J.K., Eds.; *Applied Surface Thermodynamics*; Marcel Dekker: New York, 1996.

6
Tribochemistry

Czeslaw Kajdas
Warsaw University of Technology, Płock, Poland

I. INTRODUCTION

A. Chemical Reactions

Reactions initiated and proceeding under boundary and/or mixed lubrication processes concern tribochemistry, which deals with the chemical changes of both solids and lubricant molecules due to the influence of friction operating conditions. Generally, it is accepted that chemical reactions are initiated either by temperature or by a kind of irradiation [for instance, ultraviolet (UV) irradiation and nuclear irradiation], or just by mechanical treatment and/or impact. Thus, under boundary lubrication conditions, chemical reactions are also initiated by mechanical forces in solids, which support shear strains. Shear changes the symmetry of a solid and/or molecule, and is therefore more effective in stimulating reactions than in simple isotropic compression. Consequently, the mechanical action at solid surfaces tends to promote chemical reactions and produce surface chemistry that may be entirely different from those observed in static conditions. Such mechanical activity is caused by symmetry breaking, which destabilizes the electronic structure of bonding and makes the solid prone to chemical reactions.

Chemical reactions are characterized by chemical kinetics, concerned with their velocity and also the mechanism by which they occur. The reaction mechanism is applied to demonstrate the step-by-step sequence of the events that are postulated to proceed at the molecular level as reactants are changed to given organic, organometallic, or inorganic products. It is of note that a clear mechanism should also consider intermediates and/or transition states. Bearing in mind chemical reactions, it is necessary to emphasize that the summation of all the changes that occur, expressed as net reaction, is not the whole situation because the net reaction change usually consists of several consecutive reactions. This is of particular importance for better understanding tribochemical processes. Chemical reactions of tribological additives proceeding during the boundary lubrication process involve the formation of a film on the contact surface and its protection during friction. They are beneficial if they generate tribochemical films that reduce adhesive wear and/or minimize friction coefficient.

Tribochemical reactions initiated by mechanical actions in the direct contact zone relate to chemical changes of both solid mating elements and lubricant molecules due to the influence of boundary and/or mixed lubrication operating conditions. Surface interaction between a solid (particularly strained or being strained) and a medium (which features adsorption) can give rise to failure and rearrangement of interatomic bonds in the solid (dissolution, adsorption-induced strength reduction, and various kinds of corrosion) or in the adsorbed molecules (heterogeneous catalysis or certain mechanochemically evolved phenomena) [1,2]. Wear particles are minute debris generated during friction as a result of fracture by the forces transmitted through the microscopical spots, which form the area of real contact. The physics and chemistry of wear mostly relate to specific phenomena generated (e.g., triboemission, fractoemission, and reactions initiated in these microscopical contacts). Of importance are also dangling bonds and issues combined with the formation of "virgin" (nascent) spots. Thus, the application of mechanochemical energy associated with friction releases a greater number of physical processes, which, in turn, can be the cause of tribochemical reactions of solids with molecules of the environment. Pondering solids in terms of the energy gap difference, those with the lowest energy gap (metals) differ significantly from insulators (ceramics) demonstrating the highest energy gap. Because the energy gap relates to the chemical hardness of solids, it might provide some connections between the physics and the chemistry of solids from the viewpoint of the tribochemical reaction process. This is a very important issue because tribochemical reactions are distinct from those of thermochemical ones. The same is due to catalysis and tribocatalysis because the latter is enhanced by mechanical factors. High-energy nascent (fresh) surfaces being produced during boundary friction link the physics and the chemistry of wearing processes, leading to microscopical aspects of wear and tribochemical mechanisms.

    Actually, chemical reactions under friction conditions include several types of reaction processes: (a) genuine tribochemical reactions (i.e., specific chemical reaction process initiated only by the mechanical action of the system; mechanochemistry); (b) reactions of thermochemical processes (e.g., oxidation, polymerization, and degradation); (c) bulk heat-enhanced tribochemical reactions; (d) flash temperature-enhanced tribochemical reactions; (e) some heterogeneous catalytic processes controlled by thermionic emission; and (f) tribocatalytic reactions supposed to be enhanced by triboemission [2]. Typical physical phenomena evolved by wearing processes comprise increased contact temperature and

Action of emitted electrons with lubricant molecules

$$AB + e \rightarrow A^- + B^{\cdot}$$
$$H_2O + e \rightarrow HO^- + O^{\cdot}$$

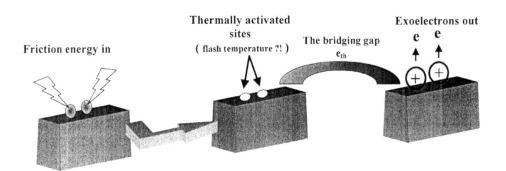

Figure 1   Tribochemistry and tribocatalysis.

triboemission. Emitted low-energy electrons can initiate tribochemical reactions governing tribology processes by forming wear-reducing or corrosive wear films [3]. Figure 1 attempts to look for a common denominator of above listed specific tribochemical/catalytic reactions. This denominator is mostly related to low-energy electrons either being a part of tribo/fractoemission processes or thermionic emission. Nakayama [4] emphasizes that (a) during tribological processes, unexpected physical phenomena appear and the unanticipated chemical reaction products are generated; and (b) many unsolved tribological problems are connected with different tribochemical reactions that cannot be explained only by frictional temperature. Accordingly, the main tribophysical processes clearly govern tribochemical reactions.

The present chapter aims at better understanding some tribochemical findings that relate to both metal and ceramic substrates by reviewing and by discussing selected tribological experimental results. It also aims at answering several specific questions: How far can the triboemission process influence the tribochemistry of the tribological system? What is the part of the negative ion radical action mechanism (NIRAM) in initiating tribochemical reactions of metals and ceramics? What are their specific reactive intermediates? Can tribochemistry be interconnected with selected heterogeneous catalytic reaction processes? What is the role of flash temperature in triggering tribochemical reactions? What is the major difference between catalysis and tribocatalysis? How can present knowledge of tribochemistry be applied in the elucidation of detailed mechanisms by which boundary lubricants and/or additives (e.g., fatty acids, fatty alcohols, esters) reduce wear? How can the boundary friction process affect base oils, particularly model base oils (e.g., hexadecane)? What is the major reason causing the degradation of very thin lubricant films used for magnetical storage apparatus? How far can the hard and soft acids and bases (HSAB) principle be applied to tribochemistry? An additional objective is to attempt to prove the hypothesis saying that specific intermediate reactive species for both tribochemical reactions and heterogeneous catalytic reactions are formed by the same mechanism mostly governed by the NIRAM approach. Another important question ("What is the effect of fresh solid surface sites on sorption processes of lubricant components?") is considered in Chap. 8. A detailed tribochemistry of boundary lubrication processes is included in Chap. 11.

B. General Approach to Tribochemistry

Lubricant components either adsorbed or chemisorbed on friction solid surfaces form a lubricant film that, under boundary lubrication conditions, can react with the surface or on the surface. Chemical reactions of antiwear (AW) and extreme pressure additives, proceeding under boundary lubrication conditions, generate a film reducing the wear and damage of contact solids. Mechanical energy associated with friction releases physical processes resulting in triggering tribochemical reactions of solids with lubricant molecules. Figure 2 demonstrates a general approach to physical and chemical events under boundary lubrication conditions. From the viewpoint of chemical reactions at surfaces, a very comprehensive review of mechanically initiated reactions is described in Ref. 5. It is stressed that many tribological problems are concerned with various tribochemical reactions that cannot be accounted for by frictional heat alone. Another review [6] emphasizes the technological importance of the mechanical initiation of chemical reactions in such processes as grinding, drilling, crushing, and cutting, which are often facilitated by chemical compound formation at the worked interface. Mori [7] considers factors that activate surface reactions during boundary lubrication under very high vacuum conditions. It was shown that reactions under boundary lubrication conditions result in the activity of solid surfaces,

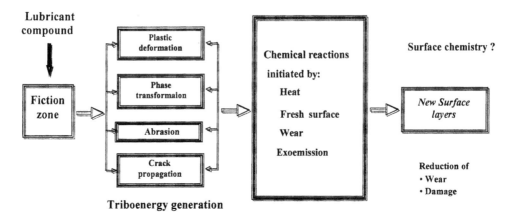

Figure 2   Major physical and chemical events in the boundary lubrication contact.

particularly of fresh metal surfaces generated by friction. Tribochemical reactions are distinct from those of thermochemical ones because the activation energy (E_a) for the latter is much higher. For example, E_a values for iron oxide formation from iron and oxygen differ very significantly if thermochemical reactions ($E_a = 54$ kJ/mol) are compared with tribochemical reactions ($E_a = 0.7$ kJ/mol) [5]. A similar dependence may be observed for other reactions. In comparison with the thermally activated reaction, an altered temperature dependence has been found in tribochemical reactions. Numerous reactions are independent of temperature in a wide measuring range. A good example of that is depicted in Fig. 3 by the steady-state effect of mechanical energy on the general course of tribochemical reaction.

The occurrence of reaction velocity independent of temperature shows clearly that the energy for releasing the reactions is applied to the solid through the mechanically activated regions only. Switching to catalysis at this point, it is necessary to note that chemically stimulated exoelectron emission (EEE) occurs continuously from silver catalysts during

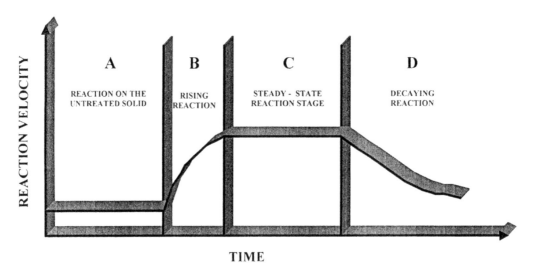

Figure 3   Digramatical presentation of the tribochemical reaction course. (From Ref. 5.)

partial oxidation of ethylene and the emission rate is proportional to the rate of ethylene oxide formation [8]. This chapter is an interdisciplinary review of an extremely fast-developing research on tribochemistry and also leads to a pathway combining the HSAB concept with the NIRAM approach.

C. Boundary Lubrication and Tribochemistry

1. Brief Information on Boundary Lubrication

Boundary lubrication can be defined as the lubrication process that reduces either friction or wear (or both) between surfaces of mating elements in relative motion when neither hydrodynamic (HD) nor elastohydrodynamic (EHD) film is formed. Accordingly, the lubricating boundary film between the two surfaces is no longer a liquid/semiliquid layer but only a very thin film of molecular dimensions. Figure 4, demonstrating the Stribeck curve in the log scale, shows that a compromise must be made with respect to lubricant viscosity between the friction losses in the region of HD lubrication and bearing wear when passing through the regime of mixed lubrication. The viscosity increase enables the lubricant to withstand the high contact stresses and to inhibit contacts between the surfaces of mating elements. The other factor to be considered is that the surfaces deform elastically, thereby enlarging the area of support. This leads to more effective lubrication. The friction coefficient for this lubrication region is the smallest (Fig. 4). Therefore, with EHD lubrication, the material properties of both the lubricant and the solids are of importance. On the other hand, from the viewpoint of the boundary lubrication process, generally, it is assumed that viscosity does not relate to these process controlling factors. Thus, boundary lubrication is a specific condition of lubrication in which friction and wear between two solids in relative

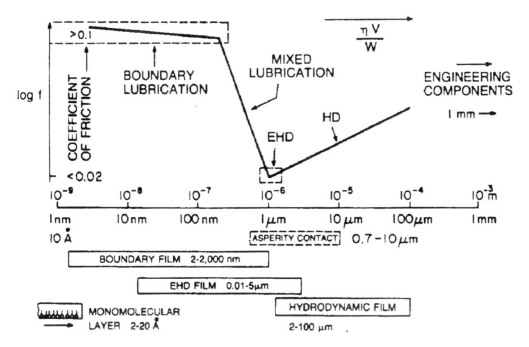

Figure 4 Comparison of lubrication film thickness with average size of lubricant molecules. (From Ref. 9.)

motion are mostly controlled by such properties of the lubricant as antifriction (AF) and load carrying, and by the nature of the mating surfaces.

Boundary lubrication has also been described as being determined by the properties of the solid surfaces and the lubricating properties of the lubricant, usually other than viscosity. On the other hand, recent work demonstrates that the viscosity of the base oil can control the wear process under boundary lubrication when it includes some low concentrations of palmitic acid [10]. Figure 5 demonstrates such finding.

An excellent and detailed review of the boundary lubrication history is described by Dowson [11]. Boundary lubrication is quite complex and is even difficult to clearly define. However, it is an extremely exciting research area due to still existing unknowns, even in respect of the antiwear behavior of fatty acids or tribochemistry of fatty alcohols, and also due to its very interdisciplinary character. Boundary lubrication is often described as a condition of lubrication, in which the friction and wear between two surfaces in relative motion are determined by the properties of the surfaces and by the properties of the lubricant other than viscosity. This definition is closely related to the first approach to the boundary lubrication process by Hardy [12]. Taking into account what Cambell [13] wrote over 30 years ago, one can note that not too much understanding of the boundary lubrication complexity was gained in this period. Cambell stressed that boundary lubrication is perhaps the most confusing and complex aspect of the subject of friction and wear prevention.

Mechanisms of film formation under boundary lubrication conditions include [14]:

Layers of molecules adsorbed by van der Waals forces
High-viscosity layers by the reaction of oil components in the presence of rubbed
 metal surfaces
Thin reacted layers plus smoothing
Thick reacted inorganic layers.

A most recent review [15] concerns boundary lubricating films and their mechanical properties, the nature of surfaces, lubricants, and their reactions along with organometallic

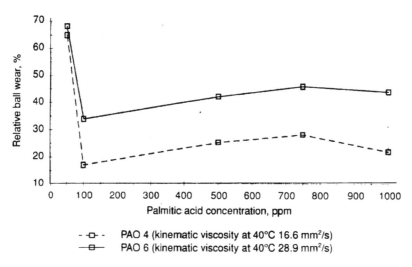

Figure 5 Ball relative wear of the system lubricated with 50–1000 ppm solutions of palmitic acid in PAO 4 and PAO 6. (From Ref. 10.)

chemistry and tribochemistry. It concludes that now we can identify the effects of tribo-chemistry, lubrication film formation, and the failure mechanism of boundary lubrication much better; however, our ability to predict lubrication effectiveness in any given instance for a particular system still needs development. Specific information on boundary lubrication comprising the effect of surface morphology, surface chemistry, lubricating film deformation, and other issues is described and discussed in Chap. 11.

Considering the influence of solid mating element surfaces along with the above listed film formation mechanisms, the following two questions are steadily of interest: (a) How does the chemical and structural nature of the metal surface influence the behavior of additives in the control of friction and wear? (b) Why are the chemical reactivities of rubbing surfaces higher than those of nonrubbing surfaces?

Both questions should also be addressed to other friction solids, particularly to ceramics, because mechanisms of boundary film formation involve similar physical and chemical processes that are due to metals. Because most surfaces generated under friction conditions are more reactive toward lubricant molecules or the environment than are surfaces under static conditions, it is necessary to look at specific physical phenomena, including triboemission and flash temperature (Fig. 1). These specific tribological phenomena can trigger tribochemical reactions and thereby are of particular importance for tribochemistry. Major physical and chemical events in the boundary lubrication contact are illustrated in Fig. 2.

2. General Approach to Triboemission

Disruption of the surface oxide layer generally results in a wide variety of physical processes and usually in higher friction and severe wear. The wear debris produced is entrapped at the interface, and plowing and microcutting of the surfaces occur. During these interactions, (a) forces are transmitted; (b) energy is consumed; (c) physical and chemical natures of these materials are changed; and (d) surface topography is altered. Under these conditions, the friction force arises predominantly from plowing and the wear is controlled primarily by an abrasive-type wear mechanism. Understanding the nature of these interactions is addressed to tribophysics [16], which describes the basic mechanisms that govern interfacial behavior and illustrates how the basic theories can be applied to provide practical solutions to important friction and wear problems. Wear mechanisms can be classified into two groups: (a) those dominated primarily by the mechanical behavior of solids, and (b) those dominated primarily by the chemical behavior of materials. The latter mechanism is clearly related to tribochemistry. Thus, what determine the dominant wear behavior are mechanical properties, chemical stability of materials, operating conditions and tribochemistry of lubricant, and/or environment molecules toward the mating element surfaces.

When oxides of the contact area are thick and porous, they are ruptured easily. That results in both plastic deformation of the sliding surfaces and physical phenomena, particularly those related to the triboemission process. Triboemission is defined as emission of electrons, charged particles, lattice components, photons, etc., under conditions of boundary friction conditions, and/or surface damage caused by fracture processes. Figure 6 illustrates the triboemission process connected with the surface enlargement during friction [17]. One of the most important components of triboemitted particles encompasses electrons, particularly low-energy ones (exoelectrons). Exoelectrons are related to chemical and structural changes in the surface films on the friction solids. Hence, the triboemission process is of particular significance for both the boundary friction process as such and the tribochemistry of the boundary lubrication process. According to Nakayama [4], the major

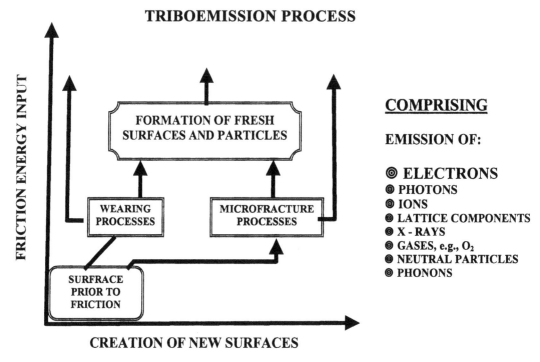

Figure 6 Physical processes evolved by friction. (From Ref. 17.)

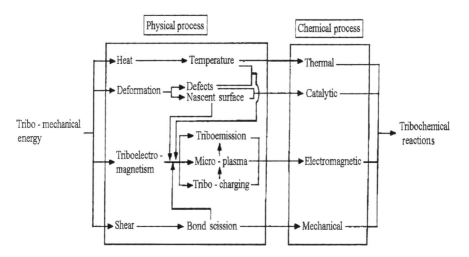

Figure 7 Physical processes and chemical processes in friction. (From Ref. 4.)

factors relating to tribochemical reactions accompany triboelectromagnetics as depicted in Fig. 7. Generally, exoelectron emission can be considered as the emission process occurring after the solid surfaces are specifically activated, for instance, by ultraviolet photons or heat.

II. NEGATIVE ION RADICAL ACTION MECHANISM

A. General Description

1. Importance of the Exoelectron Emission Process

Under rubbing conditions, two types of activated sites on friction surfaces can be created (i.e., thermally activated sites and sites activated by the exoelectron emission process). Exoelectrons are electrons of low energy spontaneously emitted from most fresh surfaces, or they are electrons that are emitted from surface atoms under certain conditions that provide enough energy to provoke their evaporation. Apart from low-energy electrons, other particle types, mostly ions and lattice components, are also emitted from solids during rubbing process, as depicted in Fig. 6. The formation of negative ions and their decomposition process are simulated by electron attachment mass spectrography [18]. This type of mass spectrography uses electrons of an energy range similar to those of exoelectrons (1–4 eV). Accordingly, the energy of triboelectrons (exoelectrons) is sufficient to form negative ions or negative ion radicals of a wide variety of compounds, which, in turn, as reactive intermediates, may easily react with the positively charged surface sites of the friction solid [metal, ceramic, diamond-like carbon (DLC), composite] counterparts, yielding given chemical compounds. Figure 8 presents schematically the triboelectron emission process. A most recent research [19] provided clear evidence that the major part of negatively charged particles emitted from diamond, alumina, and sapphire is of low energy, from 0 to 5 eV.

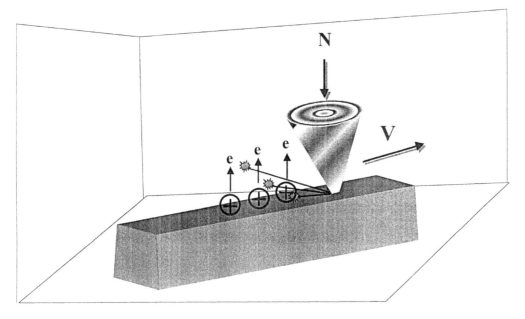

Figure 8 Schematic of the electron emission process during rubbing.

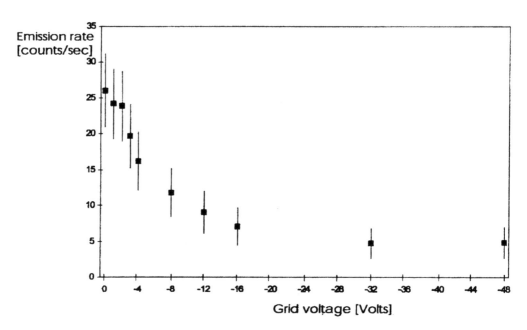

Figure 9 Retarded energy spectrum for negatively charged triboemission from diamond-on-sapphire. Background: 0.7 cps. (From Ref. 19.)

This finding (Fig. 9) allows us to consider the NIRAM approach as an important factor initiating typical tribochemical reactions of the environment compounds with mechanically treated solids or, more specifically, under boundary lubrication conditions. The typical activation energy of thermochemical reactions expressed in bond dissociation energy for (C–C) in –CH$_2$–CH$_2$– and (C–O) in –CH$_2$–O– bonds equals about 340–360 kJ/mol. Thus, when one triboemitted electron of the kinetic energy less than 4 eV is attached to an organic molecule, it shall immediately initiate the tribochemical reaction process. The same is due to thermionic electrons. Consequently, formation of tribochemical reaction products as a result of chemical interactions between the elements of a tribosystem could be initiated by low-energy electrons. Specific tribochemical reactions proceed according to both the free radical mechanism and the ionic mechanism. The shearing process generates free radicals, and other tribochemical reactions comprise ionic and free radical reaction types. Negative ion radical intermediate species are of particular interest. Therefore, some 20 years ago, the negative ion radical action mechanism approach was proposed [20–22]. Recently, this concept has been reviewed and its importance for lubricant compound reactions with the friction surfaces has been presented [3].

2. Major Phases of the NIRAM Approach

The general anionic radical lubrication model is based upon the ionization mechanism of lubricating oil components caused by exoelectrons. Electrons with an average energy of 3 eV are used in the mass spectrography of negative ions [18]. The energy of these electrons seems to be close to the energy of exoelectrons emitted from freshly formed surfaces during the friction of solids, including diamond-like carbon. Therefore, the simulation of ionization and fragmentation processes of lubricating oil components can be based on negative ion mass spectroscopy. The most important chemical processes are degradation of a lubricant molecule and chemisorption of active components (reactive intermediates) on a solid friction contact surface (metal, ceramic, DLC, and composite). Two basic factors arise in assessing the role of reactive intermediates of lubricant components in the formation of load-carrying layers by tribochemical reactions: (a) mechanism of reactive intermediates formation, and (b) mechanism of their degradation. The general NIRAM model assumes the creation of two types of activated sites on friction surfaces [i.e., thermally activated sites and sites activated by the exoelectron emission process (surface activation process)]. The situation is presented in Fig. 10. In comparison with a thermally stressed solid, mechanically treated solids show a reactivity often increased by several orders of magnitude, particularly in the low-temperature range. For tribochemical reactions, only limited use can be made of the relationship of classical thermodynamics; thus, reactions with a positive free enthalpy also appear during mechanical treatment [5,6].

The NIRAM approach considers the following stages.

♦ Low-energy electron emission process (exoemission) and creation of positively charged spots on the friction contact surface (Fig. 8). The process depends on friction conditions (e.g., type of material, load, speed, and atmosphere). Two molecule types inherently combined with the tribological system environment (oxygen and water) are of particular importance for the boundary lubrication process. This relates to their promptness to interact with triboemitted electrons generating very active reactive intermediates HO^- and O^-. HO^- is easily generated in the argon plasma negative ion source of the electron attachment mass spectrograph [18]. The same is due to oxygen negative ion O^-. Experimental appearance potential (AP) for O^- generation from oxygen molecules (O_2) is in the range of 3.6–5.4 eV and the theoretical one is estimated as 3.2–3.7 eV [23]; the production of O^- from oxygen radicals proceeds at lower energy not exceeding 1 eV. On the other hand, from carbon monoxide, this negative ion is formed at a much higher electron energy of about 10 eV. Interestingly, the O^- species can be much easily obtained from either NO_2 or NO than from oxygen molecules. Negative ions are known to be generally formed by resonance

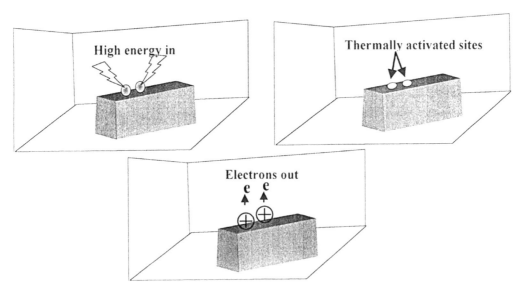

Figure 10   Surface activation during friction.

capture (i.e., $AB + e \Rightarrow AB^-$). The minimum electron energy of AP necessary to produce a given ion from its precursor neutral of AB^- formed by resonance capture is given by

$$AP(AB^-) = \sum E + EA(AB)$$

where ΣE is the excitation energy of AB^- and $EB(AB)$ is the electron affinity of AB. Negative ions are usually induced via resonance capture mechanism of very-low-energy electrons (<1 eV) [23]. To produce negative ions from radicals (compare the O^- formation from oxygen radicals and O_2), the electron energy needed is still lower. Sufficient energy for such processes can be provided by secondary and or degraded primary electrons. The dissociative resonance capture reactions yielding appropriate anions and radicals is observed at energies of over 2 eV [18,23]. The number of ions being formed and their kind will depend on the chemical character of the compound and on the type of solid along with rubbing conditions.

♦ Action of the emitted electrons with lubricating oil component molecules—when very close to the contact area—causing the formation of negative ions and radicals. It is widely known from the mass spectrometry analytical technique that when the energy of electrons is sufficiently high, generally greater than 10 eV, the impact of electrons with a molecule yields the positive parent ion that may or may not dissociate to give rise to fragment ions. As the energy of triboelectrons is much smaller, usually it is not sufficient to produce positive ions. However, in the conditions of the electron attachment mass spectrography [18], negative ion spectra can be obtained. Electrons of an average energy of about 3 eV are attached to the molecules and negative ions are formed according to the following three mechanisms: resonance capture, dissociative resonance capture, and ion attachment. The latter one is of particular significance in the formation of reactive intermediates from paraffin hydrocarbons due to the attachment of HO^- to any higher-molecular-weight paraffin (see Sec. III.A.3).

Because the energy of triboelectrons is sufficient to cause the ionization of lubricating oil component molecules, and the conditions in friction contacts approximate those existing in the plasmatic electron source of the electron attachment mass spectrography, it is possible to compare processes taking place in both the friction contact zone and the ion source. Reactions of various compounds that take place in the ion source of this mass spectrograph may be roughly compared with reactions taking place in the friction contact area. The number of ions being formed and their kind depend on the chemical character of the compound. The difference in the negative ion formation of various compounds can be explained in terms of the difference in their electron affinity. Simulation of tribochemical reactions in the plasmatic ions source of the electron attachment (EA) mass spectrograph might be supported by the magma–plasma model [5,24] and the most recently discovered tribomicroplasma [25]. Figure 11 depicts the magma–plasma model. Another paper [26], considering physical and chemical processes under friction conditions, also discusses and combines the triboemission process with the triboplasma. The short life of triboplasma causes no Maxwell–Boltzmann distribution so that the equilibrium temperature cannot be given and the chemical process taking place in this excitation phase cannot be described by the laws of thermodynamics [18,27]. The conversions in triboplasmas are of stochastic nature. The discovery of plasma generated in the microscopical gap around a sliding contact, having an elliptical shape with a horseshoe pattern and with a size beyond a hundred micrometers, provides further possibility to enhance and to improve the NIRAM approach. It was demonstrated [25] that the microplasma generated under a sliding velocity of as low as 2 cm/s and a load of as low as 30 mN emits mostly invisible ultra-

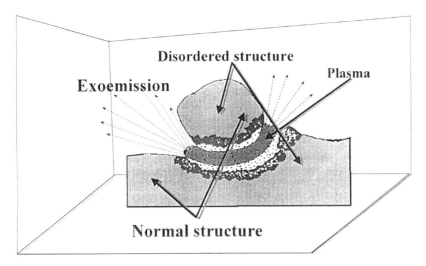

Figure 11 Illustration of the magmaplasma model. (From Ref. 24.)

violet photons and, to a lesser extent, infrared (IR) photons. The plasma is almost invisible and completely different from the well-known visible and IR photons emission due to frictional heating. Nakayama and Nevshupa [25] also emphasized that this tribomicroplasma (a) is different from the well-studied electrostatic discharge phenomena, which occur after complete separation of the tribocharged bodies; and (b) explains the mechanisms of the abnormal tribochemical reactions and introduces a new mechanism of energy and tribocharge dissipation under friction.

♦ Reactions of negative ions with solid (metal, ceramic, and DLC) surface microspots, reflected mostly as chemisorption process and other reactions (e.g., free radical reactions, generation of bidendate compounds, inorganic compound formation, and/or formation and depositing of a tribopolymeric film) (see Chap. 7). Figure 12 shows the chemisorption process of the negative ion. The kinetics of such reactions depends on the susceptibility of

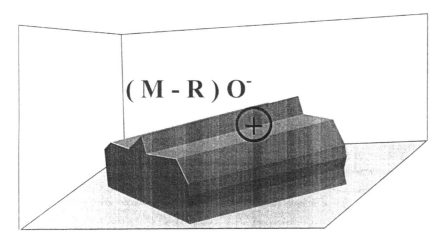

Figure 12 Chemisorption of negative ions $(M–R)O^-$.

the lubricant molecule to negative ion formation. The free radicals, generated in the previous step, may undergo oxidation and/or form friction resins.

♦ If shear stress is high (extreme pressure friction conditions), it can cause cleavage of chemical bonds producing inorganic load-carrying films and reaction by-products (Fig. 13). This process is controlled by the bond strength of a given lubricant molecule. Generally, the load-carrying ability of the film formed increases as the ease of scission of the closest bond to the element chemisorbed on the surface decreases.

♦ Destruction of an inorganic soft layer, protecting surfaces from seizures, connected with the creation of high temperature spots, electron emission and high energetic wear debris formation processes (Fig. 14). The activated surface spots, presented in Fig. 14, form a new protective film.

Figure 15 summarizes the essential features of this lubrication model, which presents a specific reaction cycle of lubricant components with friction solid surface spots. It is important to note that the created positively charged spots (Fig. 8) can also be considered as Lewis acids. Bearing in mind that the emitted triboelectrons are involved in the negative ion formation process (e.g., HO^- or O^-, which are hard bases), it is easy to realize that the HSAB concept should be very helpful in understanding specific features of tribochemical reactions, including reactive intermediates.

B. Specific Reactive Intermediates

1. The NIRAM–HSAB Concept

To account for the behavior of lubricant molecules during friction, it is essential to have clear information on their reactive intermediates. Based on the NIRAM approach, it is possible to predict types of reactive intermediates formed from lubricant components. Formation of the protective layers proceeds according to the anionic reaction process and

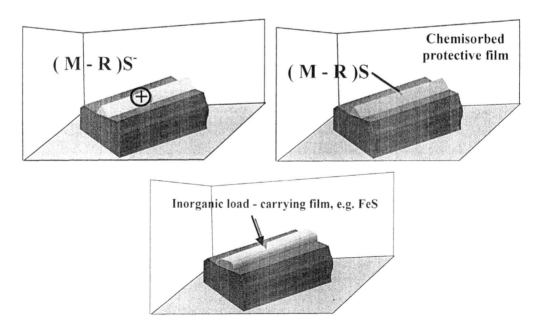

Figure 13   Schematic of load-carrying film formation under EP condition.

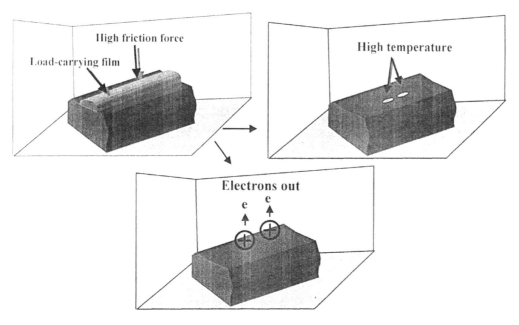

Figure 14   Destruction of the load-carrying film and creation of activated surface spots.

the destruction of the inorganic layer, which generates new Lewis acid sites. The latter process is associated with the exoelectron emission phenomenon. The emitted low-energy electrons produce Lewis bases by reducing lubricants and/or environment gas molecules. Accordingly, the reaction cycle of lubricant components on solid contacts during rubbing can be expressed in terms of the reduction–oxidation process. Reactions of the formed anions and positively charged spots may also be considered in terms of the acid–base concept, which can provide additional information on lubrication mechanisms. This is due

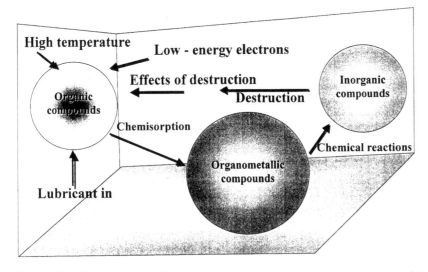

Figure 15   Reaction cycle of lubricant components on solid contacts during friction.

to the fact that all cations are Lewis acids and all anions are Lewis bases; hence, all salts (e.g., soaps) are automatically acid–base complexes. The typical acid (A)–base (B) reaction is presented as follows:

$$A+ : B \rightarrow A : B.$$

The species A:B is called a coordination compound or an acid–base complex. Acid A is the electron acceptor and base B is the donor (:B). The application of the HSAB theory [28] to the NIRAM concept is described in Ref. 29. Figure 31 in Chap. 8 ("Hydrolysis") shows the general approach to the NIRAM–HSAB action mechanism.

The hard acids prefer to associate (react) with hard bases, and soft acids prefer to associate (react) with soft bases. Neutral acids and bases have strengths proportional to the local dipoles at the acceptor or donor sites. The features that bring out hard behavior are small size and high positive oxidation state. Soft behavior is associated with low or zero oxidation state and/or with large size. For instance, iron as bulk metal is soft, Fe^{2+} is borderline, and Fe^{3+} is a hard acid. Bulk metals and metal atoms are classified as soft acids, which react much more easily with soft bases than others. Examples of hard bases include H_2O, HO^-, ROH, F^-, and Cl^-. However, Br^- relates already to the borderline bases. Soft bases encompass sulfides (R_2S), thiols (RSH), and negative ions such as RS^- or R^-. These base species employ a doubly occupied orbital in initiating a reaction.

From the tribochemical point of view, the hardness of an acid is a function of the oxidation state of the acceptor atom, which relates to the work function of solids being rubbed. Additionally, bear in mind that the friction process can change the work function of the solid being rubbed. In other words, a hard acid can be defined as positively charged and with no valence electrons that are easily distorted or removed. Thus, any electron emission process leads the adequately treated solid to a higher acid state of the given element (e.g., iron). Al^{3+}, Ti^{4+}, and Mg^{2+}, along with Fe^{3+}, are hard acids. Ni^{2+}, Cu^{2+}, and Bi^{3+}, along with Fe^{2+}, belong to borderline acids. Generally, for cations, increased charge and decreased radius make for strong acids. These acid species employ an empty orbital in initiating a reaction.

For anions, increased charge and decreased radius also increase base strength. Accordingly, O_2^- is a stronger base than HO^-. On metal oxide surfaces, the oxygen anions are Lewis base sites and the electron-deficient metal atoms are Lewis acid sites. Under boundary friction conditions, due to the exoelectron emission process, mostly hard and borderline acids are generated in the form of M^{n+} cations. The emitted electrons acting with lubricant components generate bases, which in turn react with adequate acid sites. Based on the NIRAM–HSAB approach, it is possible to explain tribochemical reactions proceeding under specific cutting conditions. For example, alcohols under aluminum cutting as described in Ref. 30 are more reactive than propionic acid because alcohols are hard bases that prefer to react with hard acids. Although the electronic structure of typical NIRAM species seems to be a very important factor in initiating the reaction process, individual tribochemical reactions may depend not only on this factor but also on the specific matching of that structure with electronic structures of other species composing the system. This fact could be applied in accounting for some very complex chemistries of triboreaction products.

The hard and soft acids and bases principle complements well the NIRAM concept. Soft acid–soft base interactions refer mostly to the adsorption process of lubricant components on the rubbing surface. The hard acid–hard base interactions are associated with typical tribochemical processes. This novel approach combines the reactivity of lubricant components and tribological solids according to reactions induced by low-energy electrons and reactions affected by the acid–base interaction. As the HSAB principle [28] is only semiquantitative, it tells us nothing about the absolute reaction rate; however, in combination with the NIRAM concept, it predicts well which of two competing reactions will have the most favorable equilibrium. The approach adds to our knowledge of tribochemical reactions proceeding under boundary lubrication conditions.

2. Example of Specific Reactive Intermediates

This example considers reactive intermediates generated by low-energy electrons and UV irradiation. In Ref. 31, it has been hypothesized that low-energy electron-mediated degradation of perfluoropolyether (PFPE) lubricants under sliding conditions of computer head on hard disk in ultra-high vacuum chamber (UHVC) is the dominant mechanism. The hypothesis was very helpful in better understanding and accounting for all the major experimental findings. Another work [32] demonstrated that the negative ions are emitted from the Demnum S200 lubricant during irradiation with low-energy electrons (less than 14 eV). They found that the most intense fragment ion relates to F^- at $m/e = 19$. This is not unexpected because F^- is a very stable ion (hard base). Usually, fluorinated compounds have very low values of APs (e.g., the AP value for F^- generation from fluorine is less than 2 eV) [23].

Figure 16 presents the discussed mass spectrum. Other ions of strong intensity are $C_3F_5O_2^-$ and $C_3F_7O^-$ at $m/e = 163$ and $m/e = 185$, respectively. The negative ion $(m/e)^- = 185$ is generated via C–O bonding cleavage at one end of the Demnum S200 molecule. However, the mechanism by which ion $(m/e)^- = 163$ is formed was unknown. A paper [33] aimed at accounting for this specific ion formation mechanism. Accordingly, the formation mechanism of the negative species $(m/e)^- = 163$ proceeds as follows: (a) splitting off the $(m/e)^- = 185$ anion $(CF_3CF_2CF_2O^-)$ from the Demnum S200 molecule; (b) the left free

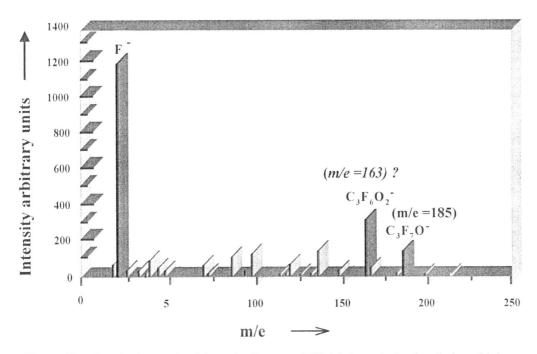

Figure 16 Negative ions emitted from the Demnum S200 lubricant during irradiation with low-energy electrons. (From Ref. 32.)

radical $\cdot$O–CF$_2$–CF$_2$–CF$_2$–O–CF$_2$–CF$_2$–CF$_2$ interacting with a low-energy electron forms the negative ion radical reactive species $\cdot$O–CF$_2$–CF$_2$–CF$_2$–O$^-$ with a m/e ratio of 182. This reactive intermediate splits off one free radical (F$\cdot$) generating the negative ion $(m/e)^- = 163$ (O=CF–CF$_2$–CF$_2$–O$^-$), which corresponds to the strong signal found in the negative ion mass spectrum of the Demnum S200 lubricant (Fig. 16). Interestingly, the NIRAM approach can also be applied in accounting for experimental results concerning the chemical bonding of PFPE lubricant films with DLC under sliding conditions. A review paper [34] discusses and analyzes the literature concerning the interaction and degradation mechanisms of perfluoropolyether lubricants with carbon-protective overcoats used for magnetical media.

III. TRIBOCHEMISTRY OF SELECTED CH AND CHO COMPOUNDS

A. Hydrocarbons (CH)

1. General Remarks

Important compounds of mineral base oils and some synthetic ones (e.g., poly-α-olefin oils) relate to aliphatic hydrocarbons. Hexadecane (n-C$_{16}$H$_{34}$), also known as cetane, is widely used as the low-viscosity model base oil to investigate both tribological effectiveness and tribochemical reaction mechanisms of various antiwear and extreme pressure additives, as well as friction modifiers. It is a nonpolar compound and thus does not compete with tribological additives on the surface adsorption process. Generally, it is accepted that freshly purified hexadecane is a far poorer lubricant than a material that has stood for some time, particularly in a transparent glass bottle. Early works on the chemistry of boundary lubrication of steel by hydrocarbons demonstrated that the sliding behavior of steel lubricated by hydrocarbons under boundary lubrication conditions could be related to chemical reactions at the sliding surfaces involving metal, hydrocarbon, and oxygen. The results suggested that the reactions occur at sites where fresh metal surface was exposed by rubbing. A most recent investigation [35] clearly demonstrates that the oxidation process under friction conditions is very specific and the major oxidation compounds from hexadecane under boundary friction conditions relate to oxygenates other than carboxylic acids.

2. Reaction Mechanism

It has been found that hexadecane (cetane) under boundary friction conditions forms very specific complex compounds with aluminum [36]. Fourier transform infrared (FTIR) spectra taken from the wear track clearly demonstrated two new significant absorption bands at 1547 and 1657 cm^{-1}. The most specific situation arises from the fact that similar peaks are observed in hydroxyl group-containing compounds [37]. The similarity of spectra taken from hexadecane-lubricated wear tracks with those lubricated with alcohols clearly shows that hexadecane during the friction process is oxidized either to alcohols or just transformed to alkoxide anions. The latter might be generated from $(C_nH_{2n+2}OH)^-$ ions. Figure 17 depicts the general fragmentation mechanism of negative ions produced from aliphatic hydrocarbons.

These reactions are based on previous experimental data concerning the negative ionization and fragmentation of straight paraffin hydrocarbons [38] in the electron EA mass spectrography [18]. To obtain negative ions of paraffins, HO$^-$ ions have to be attached. The stability of (M + HO$^-$) ions decreased with a decrease in the molecular weight of the

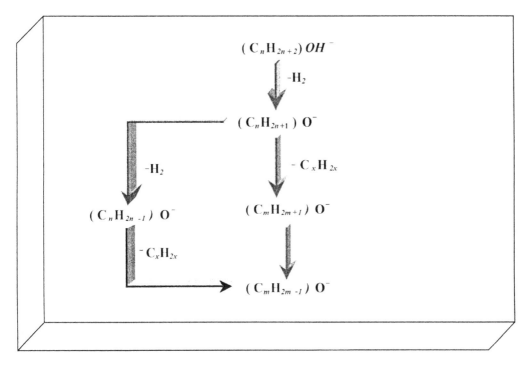

Figure 17 Fragmentation of $(C_nH_{2n+2}OH)^-$ ions.

paraffins. The intensity of $M + 17$ peaks increase from n-heptane to n-nonane in the ratio 1:2.5:6. It is important to note that major ions formed closely relate to those that are generated from alcohols, as described in Ref. 37. This clearly evidences the similarity of FTIR spectra taken from wear scars lubricated with alcohols and those taken from wear scars generated under boundary friction conditions lubricated with hexadecane.

3. Tribochemistry of Hexadecane

An earlier work [36] demonstrated that from an aluminum disk, it is possible to obtain very simple FTIR spectra from the wear track lubricated with hexadecane. Friction tests described in Ref. 35 were carried out on the steel-on-steel and steel-on-aluminum tribological systems. The first important observation was that, in each case, the wear product spectrum differed drastically from that of hexadecane taken either from the steel disk or the aluminum disk before the friction test. Figure 18 compares an FTIR spectrum obtained from the steel disk wear scar region lubricated with hexadecane and the hexadecane reference spectrum taken from the same disk.

Thus, there is no question that tribochemical reactions occurred on the steel surface. This was also confirmed by the total absorbance map of the aluminum microregion surface presented in Fig. 19. It is worthy to note that the surface region analyzed is very small (around 1 mm^2). The map also includes three spectra taken from different spots of the wear region. The two upper spectra relate to the wear track edges and the third one is generated from the wear contact zone. Although these spectra are neither corrected by zapping of spurious bands and normalizing of baseline nor smoothed, one can note that both edge spectra are very similar and include new absorption bands in the same regions, as demonstrated in Fig. 18. The spectrum taken from the wear contact zone does not have

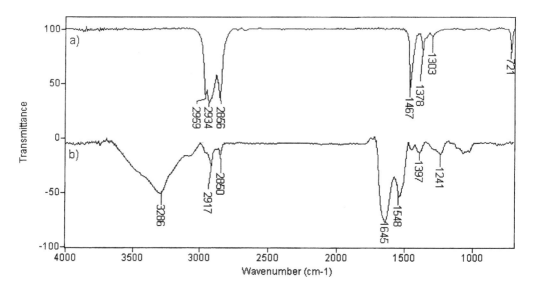

Figure 18 Comparison of FTIR spectra of *n*-hexadecane: (a) before friction test on steel surface, and (b) after *n*-hexane-lubricated friction test on a 3000-m sliding distance. (From Ref. 35.)

any clear peaks relating to absorption bands around 1550, 1650, and 3290 cm^{-1}. Based on this finding, it is plausible to suggest that most of the reaction products are moved from the friction contact zone to the wear scar edges. It is worthy of note at this point that the FTIR spectrum taken from the four-ball machine wear scar lubricated with SN 400 mineral base oil (Fig. 20) is quite similar to the spectrum taken from the disk wear track of the ball-on-disk test lubricated with hexadecane (see Fig. 18).

Systematical tribological tests were also performed on disks made from steel, aluminum, brass, and bronze. The obtained test results described in Ref. 35 clearly demonstrated that distinct and very similar spectra were obtained from the wear scar regions for each of the materials tested. Figure 21 presents FTIR spectra taken from various metal disks lubricated with hexadecane after a 500-m sliding distance. This finding allows to state that in all these cases, similar tribochemical reactions proceed on different material surfaces. All the major absorption bands present in Fig. 21 are also very close to the spectrum shown in Fig. 20.

Figure 22 shows two spectra taken from steel and aluminum disks lubricated with hexadecane after a 500-m sliding distance. These spectra are most distinct and have identical absorption bands at 1548 and 1649 cm^{-1}. The third intensive peak appears at 3280 cm^{-1} for steel and at 3290 cm^{-1} for aluminum. The first two peaks can be compared with absorption bands found in deposits formed under sliding of steel balls on the aluminum substrate lubricated with alcohols [37]. These absorption bands encompass mostly the region of 1530–1550 cm^{-1}, which usually is combined with absorption peaks around 1600–1640 cm^{-1}. They are assigned to chelate compounds including double bonding. The results present evidence that the tribochemistry of hexadecane is very complex and appears to involve its reactions with different metals to form chelate-type compounds, probably via prior oxidation to alcohols. To better understand the specific tribochemistry of one of very simple CH compounds, a more detailed research has been performed. For example, Makowska et al. [40], using a pin-on-disk apparatus, proved that under hexadecane-lubricated boundary friction processes, the following compounds are formed: aldehydes,

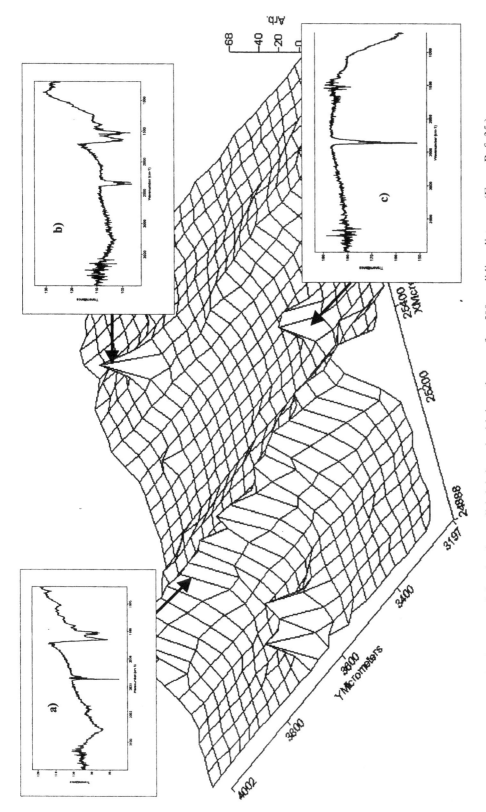

Figure 19 Surface projection map of the aluminum disk lubricated with hexadecane after 500 m sliding distance. (From Ref. 35.)

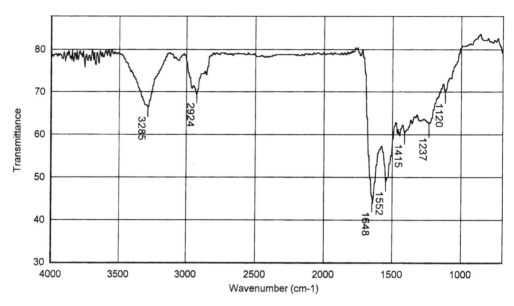

Figure 20 FTIR spectrum taken from wear scar of a four-ball machine test lubricated with a mineral base oil (SN 400). (From Ref. 39.)

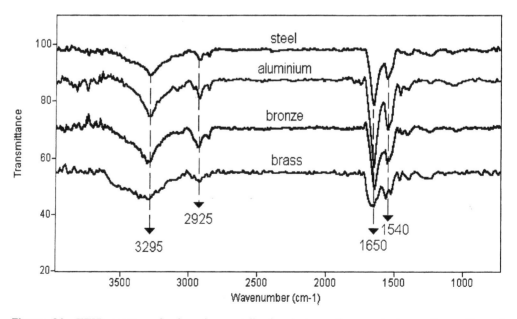

Figure 21 FTIR spectra of *n*-hexadecane tribochemical reaction products on the surface of different materials. (From Ref. 35.)

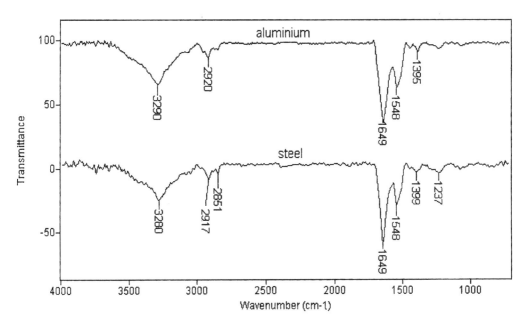

Figure 22 FTIR spectra of hexadecane tribochemical reaction products on the surface of steel and aluminum after 500 m sliding distance. (From Ref. 35.)

alcohols, and carboxylic acids. This finding clearly indicates that tribochemical reactions cause significant changes of the apparently nonreactive paraffin hydrocarbon. These tribochemical processes lead to the formation of a homologous series of the found compounds having smaller molecular weights than hexadecane. The reaction process is initiated by friction energy. Another work [41] aimed at investigating tribochemical reactions of hexadecane proceeding in a tribosystem lubricated by hexadecane at ambient and elevated temperatures. It was hypothesized that at ambient temperature, reactions are mostly initiated by the mechanical action of the system (friction energy), and at elevated temperature (200°C), thermochemical reactions should be dominant. An experimental study was performed on a ball-on-disk machine with a steel-on-steel mating elements. To analyze wear tracks, Fourier transform infrared microspectrophotometry (FTIRM) and electron spectroscopy for chemical analysis (ESCA)/x-ray photoelectron spectroscopy (XPS) were applied. To investigate the chemical changes of the bulk lubricant, gas chromatography mass spectrometry (GC/MS) was applied. The obtained results provided clear evidence for the hypothesis that two types of oxygenation processes of hexadecane under boundary lubrication processes must be considered: the first one is at ambient temperature, which is controlled by mechanical action, and the second one is clearly controlled by temperature. The formation of some reaction product types including (a) compounds having Fe–O bonding (salts and chelates), (b) carbonyl compounds, and (c) iron carbide was evidenced. Both the FTIRM and ESCA results suggest that at ambient temperature, reactions between iron and carboxylic acids lead to the formation of salts and/or chelates. Specific IR absorption bands appeared at around 1550, 1650, and 3300 cm^{-1}. However, characteristic infrared absorption bands around 3500–3000, 1740, and 1600 cm^{-1} occurring in the spectrum presented in Fig. 23 are different from those obtained at ambient temperature (Fig. 24). The observed signals are most probably related to carbonyl compounds (e.g.,

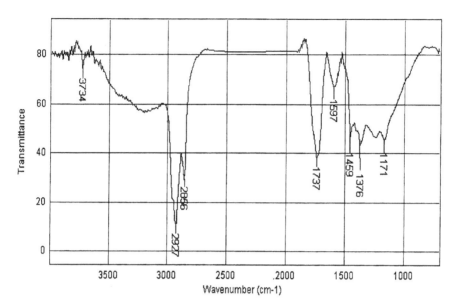

Figure 23 FTIR spectrum of triboreaction products formed during friction process carried out at elevated temperature (200°C). (From Ref. 41.)

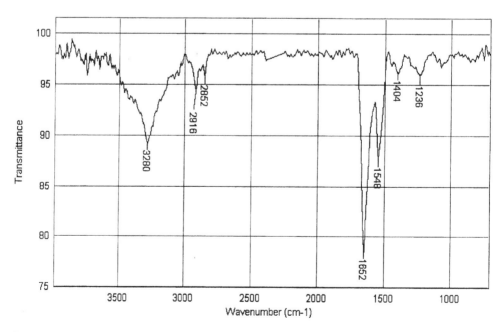

Figure 24 FTIR spectrum of triboreaction products formed during friction process carried out at ambient temperature (20°C). (From Ref. 35.)

esters, aldehydes, and ketones). It appears that the thermal oxidation process proceeding in the bulk lubricant at elevated temperatures is more enhanced and leads to the formation of different oxygen-containing compounds. It seems that the absence of bands assigned to –COOFe groups is caused by a relatively thick organic layer.

The XPS spectra of the iron, oxygen, and carbon regions revealed shifts in binding energy values due to chemical bonding. Figure 25 demonstrates that in spectra corresponding to external layers formed at elevated temperatures, there are no signals coming from iron. Furthermore, peaks appear at about 708 eV (Fe_{2p}), 532 eV (O_{1s}), and 284 eV (C_{1s}). In the carbon photoelectrons region, there are some overlapped signals at 288,28, 286,40, and 284,58 eV (Fig. 26). Obtained results show that carbon might be at least in two forms: one in Fe_3C, and another in oxygen-containing organic compounds with C=O bonding. Iron is the best candidate for electron donation, especially with increasing depth, indicating the formation of iron carbide Fe_3C near steel. The existence of large amounts of carbon— determined using ESCA—and the detection of organic compounds with FTIRM analysis clearly evidence that a very specific tribochemical reaction layer in the disk wear track was formed. The layer consists of compounds including organic ligands and iron–oxygen bonding along with Fe–C bonding in the formed iron carbide (Fe_3C). Figure 27 shows graphically the high-resolution spectra of iron, oxygen, and carbon photoelectrons corresponding to products of hexadecane tribochemical changes layered on steel surface at ambient temperature. The peak at about 711 eV is assigned to the iron compound form (for elemental iron, this is 706,4 eV) and is typical for Fe–O bond. The binding energy of ~530 eV, somewhat around ~288,5 eV (C=O) and ~285 eV (C–O), corresponds to oxygen in organic compounds (e.g., iron carboxylate).

The above results demonstrate that the reaction between iron and carboxylic acids leads to the formation of salts or chelates. It is worthy to note again at this point that their characteristic infrared absorption bands are different from those obtained at elevated temperatures (see Figs. 23 and 24). These results also provide an evidence for the hypothesis that two types of the oxygenation processes of hexadecane under boundary lubrication processes should be considered. The first one is at ambient temperature, which is controlled by mechanical action, and the second one is clearly controlled by temperature.

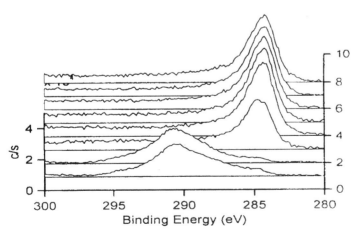

Figure 25 Exemplary XPS spectrum of Fe_{2p}, O_{1s}, and C_{1s} photoelectrons recorded on triboreaction products layer formed at elevated temperature (200°C). (From Ref. 41.)

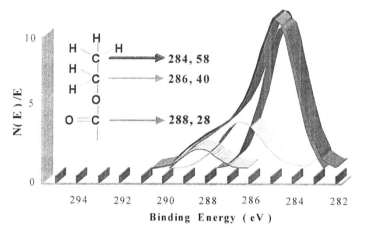

Figure 26 Results of overlaying peaks separation in C_{1s} photoelectrons spectrum. (From Ref. 41.)

Makowska et al. [40] demonstrated that characteristic infrared adsorption peaks, taken from the wear scar of a steel disk lubricated with hexadecane, appear at around 3300, 1650, and 1550 cm^{-1}. Figure 28 presents the full spectrum and result of a 1548-cm^{-1} band deconvolution demonstrating that surface reactions lead to salt and/or complex compounds formation. To check if nitrogen is incorporated into tribochemical reaction products, elemental composition of the wear track surface was determined by energy dispersive

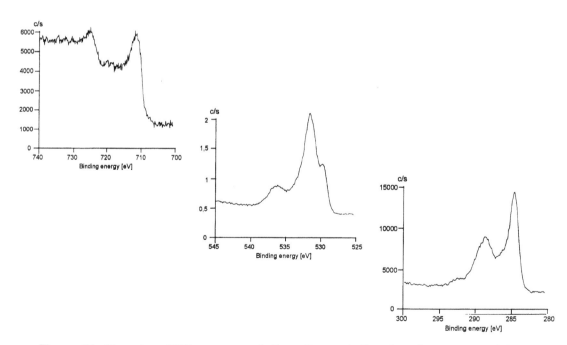

Figure 27 Exemplary XPS spectrum of Fe_{2p}, O_{1s}, and C_{1s} photoelectrons recorded on triboreaction products layer formed at ambient temperature (20°C). (From Ref. 41.)

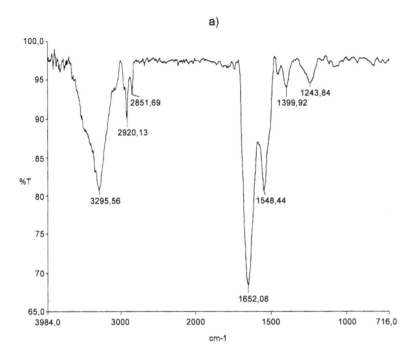

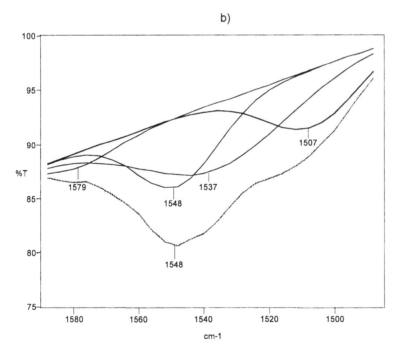

Figure 28   FTIRM spectrum of tribochemical changes products: (a) whole range spectrum, and (b) result of 1548 cm^{-1} IR adsorption band deconvolution for tribochemical products of hexadecane. (From Ref. 40.)

spectroscopy (EDS) analysis. The spectra recorded on steel surface before and after hexadecane-lubricated friction tests under air (78% N) are presented in Fig. 29.

B. Oxygenated Compounds

1. Tribochemistry of Aliphatic Alcohols

Higher-molecular-weight alcohols, similar to fatty acids, are effective tribological additives. The effect of the chain length of fatty alcohols on friction coefficient has been reviewed in Refs. 13 and 42. Physical properties of an alcohol are best understood if we recognize the following fact: Structurally, an alcohol is a composite of an alkane and water. It contains an alkanelike alkyl group and waterlike hydroxyl group. The hydroxyl (OH) group provides the alcohol its characteristic physical properties, and the alkyl group that, depending upon its size and shape, modifies these properties. The hydroxyl group is quite polar and, most importantly, contains hydrogen bonded to the highly electronegative element oxygen. Through the hydroxyl group, an alcohol is capable of hydrogen bonding to (a) its fellow alcohol molecules, (b) other neutral molecules, and (c) anions. Of the varied chemical properties of alcohols, there is one pair that underlines all the others: their acidity and basicity [43]. These properties reside, of course, in the functional group of alcohols: the hydroxyl group, OH. Alcohols are weak acids and weak bases—roughly about as acidic and as basic as water.

Recently, Jahanmir [44] investigated the chain length influence of alcanoic alcohols C_{12}–C_{18} on the friction coefficient of n-hexadecane containing 1 wt.% of the tested alcohol. He found that the coefficients of friction for both the alcohols and acids with the same chain length are very close and their carbon number increase causes only a slight decrease in the friction coefficient. Alcohols have long been recognized to be effective antifriction and antiwear lubricant additives, particularly in the boundary lubrication region with aluminum steel systems [45–47]. The practical and theoretical aspects of lubrication with lubricants containing fatty alcohols have been studied for many decades. Nevertheless, not everything is known about both their detailed antiwear behavior as a function of alcohol chain length, and the action mechanism of these compounds. The latter has been discussed in detail in Ref. 22, where the model of the anionic radical lubrication mechanism of alcohols was derived. Generally, during sliding under boundary lubrication conditions, friction effectiveness is almost the same for long-chain alcohols and acids having equal numbers of carbon atoms in the molecule. Gates and Hsu [48] provide direct evidence of chemical reactions between silicon nitride and alcohols during wear tests; the proposed chemical reaction sequence is the adsorption and reaction of the alcohol group on the oxide/hydroxide surface of the silicon nitride, followed by subsequent tribochemical reaction in the contact to form free silicon (poly) alkoxides.

The effect of straight-chain alcohols on the wear of steel-on-steel tribological elements was investigated recently [37,49]. Pure compounds and their 50 wt.% mixtures with hexadecane have been tested in a pin-on-disk tribometer. The aim of that work was to establish the alcohol chain length influence on the steel-on-steel wear process. The test conditions were designed to result in boundary lubrication at the sliding interface. Nine pure primary alcohols C_4–C_{14} and nine 50 wt.% mixtures of these alcohols with hexadecane were selected as lubricants for the research. Hexadecane was used as a reference lubricant. Table 1 summarizes the obtained results. Figure 30 illustrates the relationship between the wear scar diameter and the chain length of investigated alcohols. Two major issues concerning these results are to be noticed are: (a) the wear decreases with the alcohol chain length increase,

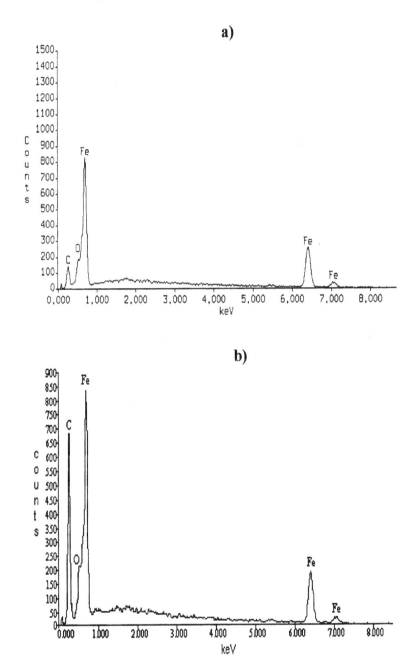

Figure 29 Results of disk surface EDS analysis: (a) before and (b) after the friction test lubricated with *n*-hexadecane. (From Ref. 40.)

Table 1 Ball Wear Scar Diameter Lubricated with (a) Pure Aliphatic Alcohols C_4–C_{14} and (b) Their 50% (m/m) Mixtures with Hexadecane

Alcohol name	Alcohol carbon number	Wear scar diameter $(mm \times 10^{-1})$	
		Pure compound	50% (m/m) mixture with n-hexadecane
1-Butanol	4	5.16	4.13
1-Pentanol	5	4.63	3.68
1-Hexanol	6	4.13	3.23
1-Heptanol	7	3.99	3.05
1-Octanol	8	3.69	2.90
1-Nonanol	9	3.51	2.89
1-Decanol	10	3.52	3.55
1-Dodecanol	12	2.92	2.55
1-Tetradecanol	14	2.40	2.06

Source: Ref. 49.

and (b) low-molecular-weight compounds, 1-butanol and 1-pentanol, show higher wear than that of pure straight-chain hydrocarbon (hexadecane).

Alcohols are known to form a physically adsorbed film on both base and noble metals [13,42], and the greater part of the literature assumes that alcohols do not react with metallic surfaces. Furthermore, it has been emphasized that liquid alcohols, which are highly polar and are presumably well adsorbed at the metal surface, are poor boundary lubricants and scarcely better than liquid paraffins, and that, on unreactive surfaces, liquid fatty acids are almost as ineffective as liquid paraffins and alcohols. Figure 30 clearly demonstrates that 1-butanol and 1-pentanol show higher wear than hexadecane. Thus, direct evidence is found of a corrosive action of these two alcohols. This effect is probably caused by tribochemical reactions between steel surface and alcohols. Accordingly, it is permissible to assume that (a)

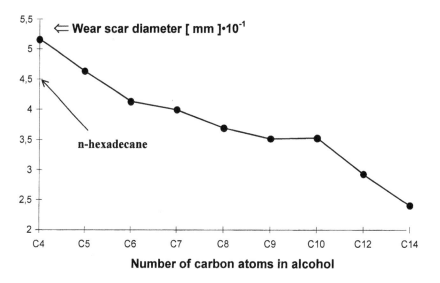

Figure 30 Influence of aliphatic alcohol chain length on the steel bar wear. (From Ref. 49.)

alcohols do react with the steel surface under boundary lubrication conditions, and (b) low-molecular-weight compounds 1-butanol and 1-pentanol, and lower alcohols are even too reactive and therefore produce corrosive wear. To test this hypothesis, more two sets of experiments using 1-propanol and ethanol as lubricants were performed. The average wear scar diameters found were 5.64×10^{-1} and 6.24×10^{-1} mm for 1-propanol and ethanol, respectively [37].

The primary chemical reaction considered is alkoxide formation. However, at this point, we need to bear in mind that metal alkoxides exhibit great differences in physical properties, mostly depending on the position of the metal in the periodical table. Some alkoxides are stable and some are not. For example, alkoxides, including those of sodium, potassium, magnesium, aluminum, zirconium, and titanium, are stable and commercially important. Iron alkoxides are not stable. On the other hand, alkoxide of nonmetals (e.g., silicon alkoxides) are stable. This fact enabled to find direct evidence of chemical reactions between silicon nitride and alcohols as presented in Ref. 48. There is also work showing the importance of the reactivity of alcohols toward aluminum during sliding. For instance, it was found [45] that 1-octanol, 1-dodecanol, 1-hexadecanol, and 1-octadecanol react with the aluminum surface when sliding occurs and the wear mechanism is largely chemical. Pentaerythritol partial ester reacts with surface aluminum to form amorphous substances, such as aluminum complex or salt, during lubrication [47]. Another work [50] evidenced that 2% 1-hexadecanol in hexadecane produced an adherent coating on the sliding steel surface, which was composed of many compounds, including alkynes, dialkynes, alkenes, and dialkenes, with a maximal molecular weight rise of about 100%. By using 1% hexadecan-1,16-diol in hexadecane as the lubricant, many compounds, similar to those formed from 1-hexadecanol, were found in the coating formed on the sliding surface. In this case, the molecular weight rise was about 50%. The above listed findings were then accounted for in terms of the NIRAM approach as described in Ref. 22. The thesis of the model takes into account the formation of anions and radical anions of alcohol molecules and the chemisorption of these ions on the positively charged spots of rubbing metal surfaces, connected with a recombination of radical anions. Accordingly, it is supposed that the same mechanism controls the formation of iron alkoxides under boundary lubrications of the steel-on-steel system. As these alkoxides are unstable, they decompose to form other compounds. The corrosivity of low-molecular-weight alcohols has been accounted account for in terms of their acidity; the acidity of ethanol is about the same as that of water, and higher-molecular-weight alcohols tend to be less acidic [51]. Thus, it was assumed that the dissociation constant of alcohols (pK_a) might be a factor controlling wear behavior. The effect of pK_a value on salt and chelate-type film formation by carboxylate anion and alkoxide anion is shown in Chap. 8 (Fig. 28). Data presented in Table 1 also illustrate the second set of experiments concerning alcohol mixtures with a nonpolar base fluid (hexadecane). The data demonstrate two major findings: (a) all alcohols tested in 50 wt.% mixtures with hexadecane generate less wear than hexadecane, and (b) 1-decanol shows a clear wear maximum. The found maximum is related to the fact that 1-decanol presents the boundary solubility of alcohols in water. However, to better understand this phenomenon, further detailed research, including friction surface analysis along with chemical changes of the bulk lubricants, should be carried out.

Overall, these results, although still in their initial stage, show that the tribological behavior of alcohols under boundary lubrication is complex and appears to involve their chemical reaction with the surface (i.e., formation of alkoxides). There is also another specific issue concerning the tribochemistry of alcohols. Using FTIR wear track analysis, new absorption bands were found in deposits generated on the aluminum substrate. As already mentioned in Sec. III.A, these absorption bands [37] encompass mostly the region of $1530-1550$ cm^{-1}, which, in many cases, is strictly combined with absorption peaks around $1600-1650$ cm^{-1}. They are assigned to complex (chelate) compounds comprising Al–O–C and C=C bondings. Figure 29 in Chap. 8 illustrates the formation mechanism of tribofilm elucidated for the ethanol-lubricated steel-on-aluminum friction system. The formation of compounds of this type can be well explained using the NIRAM approach and confirms experimentally the anion radical triboreaction mechanism of alcohols as hypothesized in Ref. 22.

2. Tribochemistry of Fluorinated Alcohols

Fluorinated lubricants containing hydroxyl groups are used for such specific purposes as lubrication of head–disk interface in computer hard disk drives. Substitution of the majority of hydrogen atoms with fluorine provides necessary thermal and chemical stability, whereas a polar hydroxyl group strengthens bonding between the lubricant and the diamond-like carbon layer covering the magnetical disk. Lubricants such as Z-DOL and Demnum SA contain a –CF$_2$–CH$_2$–OH group at both ends (Z-DOL) or at one end (Demnum SA). The tribochemistry of these compounds is of particular importance from the viewpoint of the head–disk interface durability. These lubricants undergo decomposition under boundary lubrication conditions, leading to the formation of volatile products and thus to lubricant loss and interface failure. Several approaches have been considered and reviewed [34]; the NIRAM model of triboreaction initiated by exoelectrons was used with success to explain the specific degradation mechanism. Another recent research [52,53] was focused on the tribochemistry of fluorinated alcohols in steel/steel and steel/aluminum contacts, studied by Fourier transfer infrared micro analysis (FTIRMA). That work was undertaken to find the correlations between the action mechanism of Z-DOL and fluorinated alcohols containing the same Z-DOL –CF$_2$–CH$_2$–OH functional group. It was found that under friction conditions of either aluminum or steel, fluorinated alcohols produce organometallic compounds similar to those proposed for alcohols, as presented in Ref. 37. Among the triboreaction products of Z-DOL on steel surface, compounds having the same structural characteristics were also found. A reaction path, based on the NIRAM model, leading to the formation of these organometallic compounds was proposed. Another interesting finding concerned an extremely big difference in antiwear properties between fluorinated alcohols having different numbers of hydrogen atoms in the proximity of hydroxyl group. Alcohols with two methylene groups nonsubstituted with fluorine ($1H,1H,2H,2H$-perfluorooctanol) exhibited much better performance than alcohols with only one –CH$_2$– group ($1H,1H,5H$-perfluoropentanol). However, the difference in carbon chain length (eight and five carbon atoms) could also have an influence on the tribological performance of these alcohols. The most recent work by Przedlacki et al. [53] presents data concerning the structure of organometallic triboreaction products of alcohols on steel and aluminum surfaces. In order to enhance information on their structures earlier determined by FTIRM, XPS and scanning electron microscopy (SEM)/EDS techniques were applied for analysis of the wear tracks. It was found that the highest wear was produced by $1H,1H,5H$-perfluoropentanol; $1H,1H,2H,2H$-perfluorooctanol caused lower wear, but still higher than for dry sliding test. On the basis of FTIR results, a probable structure of organometallic compounds formed by fluorinated alcohol on steel is presented in Fig. 31.

Fluorinated alcohols produce organometallic reacted films on metal surfaces subjected to friction. The antiwear and antifriction effects of these reacted films are clearly seen

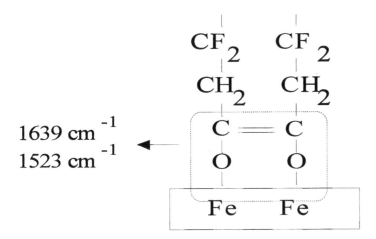

Figure 31   The chemical structure formed by 1*H*,1*H*,2*H*,2*H*-perflourooctanol on steel surface under boundary friction conditions. (From Ref. 53.)

in the case of steel; however, in aluminum substrates, fluorinated alcohols cause increased wear (corrosive wear). The stability of these layers depends strongly on the number of hydrogen atoms in THE vicinity of the hydroxyl group. From the two straight-chain alcohols having the same number of carbon atoms, better antiwear and antifriction performances on steel surface were exhibited by 1*H*,1*H*,2*H*,2*H*-perfluorooctanol (four hydrogen atoms near the hydroxyl group) than 1*H*,1*H*-perfluorooctanol (two hydrogen atoms). Figure 32 presents the antiwear effectiveness of fluorinated alcohols in the steel/steel contact. Surface analysis showed that a strong organometallic reacted film was formed under boundary friction conditions. Figure 33 shows XPS spectra gathered from steel disk surface lubricated with 1*H*,1*H*,2*H*,2*H*-perfluorooctanol inside and outside the wear scar. It was the same disk from which the XPS spectrum presented in Figs. 33 and 34 was taken. There are distinct differences in the chemical characteristics of the surface inside and outside the wear scar. Oxygen and carbon spectra taken from the worn surface contain strong new peaks not seen in the spectrum taken from the area not subjected to wear. These new signals are located at 293.5 and 288.0 eV (binding energy scale) for C_{1s} and can be attributed to C–F and C–O combinations, respectively. In the O_{1s} spectrum, there is a new signal at 535.8 eV, which can be attributed to oxygen atoms in proximity to strongly electronegative fluorine atoms, although in reference to the literature, there are no data for oxygen signals situated at such high binding energy. Other O_{1s} signals located at the binding energies of 531.0 and 530.0 eV present in both spectra relate to hydroxyl groups and iron oxides, respectively. Fluorine signal, not shown in Fig. 33, was located at 689.1 eV and can be attributed only to the C–F combination in perfluorinated molecules and there is no evidence for iron fluoride formation in the worn area. The above observations confirm that in the area subjected to friction, 1*H*,1*H*,2*H*,2*H*-perfluorooctanol undergoes tribochemical reactions with the metal surface, leading to the formation of a protective layer containing carbon, oxygen, and fluorine. Similar XPS spectra taken from the steel disk lubricated with 1*H*,1*H*,5*H*-perfluoropentanol did not give any evidence that a protective layer was formed. It seems that the ability to form a stable reacted film on steel surface is much smaller for 1*H*,1*H*,5*H*-perfluoropentanol that for 1*H*,1*H*,2*H*,2*H*-perfluorooctanol. This correlates well with friction and wear measurements showing much higher wear and friction coefficient values in the case of 1*H*,1*H*,5*H*-perfluoropentanol [53].

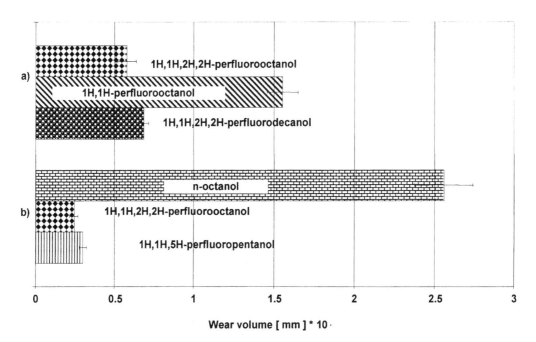

Figure 32 Wear volume of steel disk in steel disk–steel ball contact lubricated with compounds containing hydroxyl groups. Optimol SRV tester: load, 400 N; frequency, 25 Hz; amplitude, 1 mm; test time, 0.5 hr: (a) at 60°C and (b) at room temperature. Error bars represent standard deviation from three tests. (From Ref. 53.)

Figure 34 presents the F_{1s} signals in XPS spectra gathered from the wear scar areas on Al 2024 disks lubricated with $1H,1H,2H,2H$-perfluorooctanol and $1H,1H,5H$-perfluoropentanol. In contrast to the spectra taken from steel surfaces, F_{1s} signals for all those investigated consist of two main peaks: one located in the binding energy range of 688.9–688.3 eV and another at 685.2 eV. The first signal relates to carbonfluorine bonding in an organic fluorinated molecule, whereas the second can be attributed to metal fluoride, in this case AlF_3. In the case of $1H,1H,5H$-perfluoropentanol, the AlF_3 signal is relatively much stronger than for $1H,1H,2H,2H$-perfluorooctanol and has strength similar to the signal at 688.9 eV, whereas for $1H,1H,2H,2H$-perfluorooctanol, the majority of fluorine atoms exist in organic compounds.

Octanol provided a much better tribological performance in aluminum steel contact than $1H,1H,2H,2H$-perfluorooctanol (Fig. 35). The XPS spectra showed that under boundary friction conditions in the presence of fluorinated alcohols, AlF_3 was formed. The wear rate of aluminum depends on a tendency of the fluorinated compound to react with the aluminum surface. It is taken for granted that the AlF_3 formation accelerated the wear process. The volumetrical wear data of Al 2024 lubricated with C_5- and C_8-fluorinated alcohols (Fig. 35) exhibited corrosive effects on aluminum, causing much larger wear than that observed under dry friction conditions.

In the case of steel/steel contact, fluorinated alcohol proved to be much more effective in wear and friction reduction than a nonfluorinated one, having the same chain length. There was no clear evidence found for iron fluoride formation. Taking into account the above experimental results, we have proposed a reaction path leading to the formation of

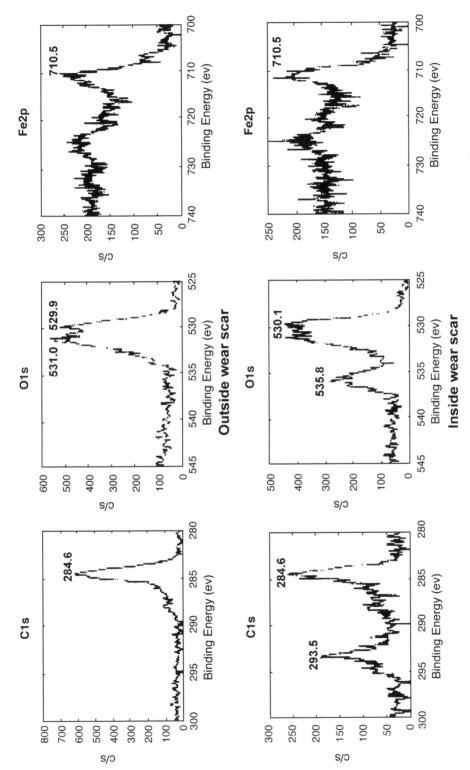

Figure 33 XPS spectra taken from steel disk surface lubricated with $1H,1H,2H,2H$-perflourooctanol. (From Ref. 53.)

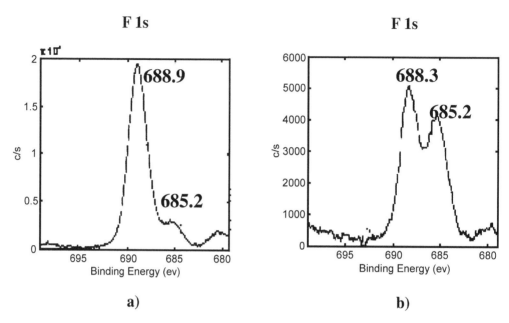

Figure 34 Fluorine signal in XPS spectra taken from Al 2024 disk surface lubricated with (a) 1*H*,1*H*,2*H*,2*H*-perflourooctanol, and (b) 1*H*,1*H*,5*H*-perflouropentanol. (From Ref. 53.)

these reacted films according to the NIRAM hypothesis. Alcohols form anions or anion radicals, which in turn react with positively charged metal surface spots. The reaction of anions and anion radicals with a positively charged surface, followed by cross-linking between unsaturated anions and anion radicals reacting with the surface, leads to an oligomeric organometallic surface film. The first suggestion of this specific type of reaction process is schematically presented in Fig. 36. The ability of fluorinated alcohols to produce stable and effective antiwear films is governed by the number of hydrogen atoms in the vicinity of a hydroxyl group. The difference between tribological properties of 1*H*,1*H*,2*H*,2*H*-perfluorooctanol and 1*H*,1*H*-perfluorooctanol is based upon the fact that 1*H*,1*H*,2*H*,2*H*-perfluorooctanol has more hydrogen atoms that can be removed from the molecule during interactions with exoelectrons than 1*H*,1*H*-perfluorooctanol.

The energy of C–F bonding is 473 kJ/mol, and that bonding is much stronger than C–H (414 kJ/mol). Thus, it is easier to remove a hydrogen atom from the molecule than a fluorine atom. The removal of H atoms from the alcohol molecule results in the formation of unsaturated or radical species, which can undergo cross-linking. In 1*H*,1*H*,2*H*,2*H*-per-fluorooctanol molecule, two carbon atoms near the hydroxyl group can form chemical bondings with carbon atoms from other molecules, whereas for 1*H*,1*H*-perfluorooctanol, only one carbon atom may take part in that interaction. Accordingly, fluorinated alcohols with more hydrogen atoms near the hydroxyl group can form more stable and durable protective layers on the metal surface due to their better ability to form polymeric-type structures. This results in their better antiwear and friction reduction performance. Considering only the neighborhood of the hydroxyl group number, the higher that number is, the better is the tribological performance. In the steel/steel contact, however, fully hydrogenated alcohol is much less effective than a partially fluorinated one (Fig. 32). The cause of that phenomenon still remains unclear.

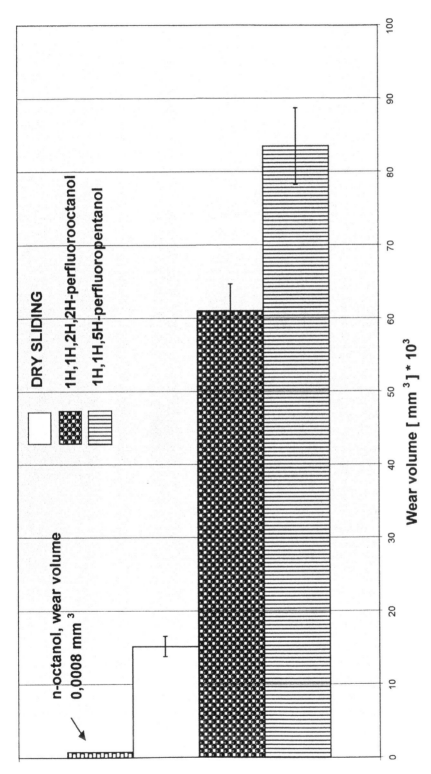

Figure 35 Wear volume of aluminum in Al 2024 disk–steel ball contact without lubrication and lubricated with compounds containing hydroxyl groups. Optimol SRV tester: load, 15 N; frequency, 10 Hz; amplitude, 0.5 mm; test time, 3 min; room temperature. (From Ref. 53.)

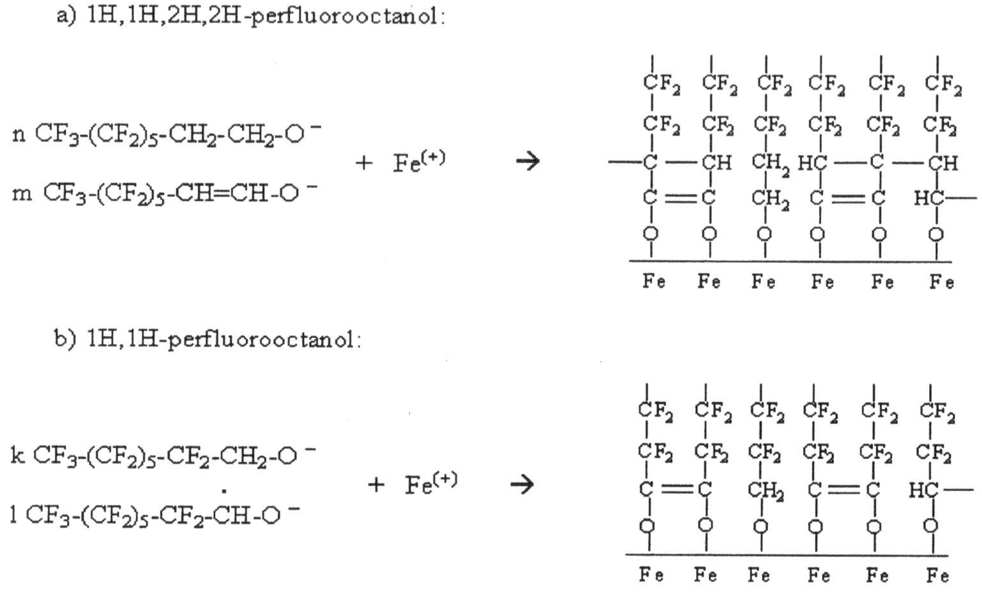

a) 1H,1H,2H,2H-perfluorooctanol:

$$n \ CF_3\text{-}(CF_2)_5\text{-}CH_2\text{-}CH_2\text{-}O^- \\ \qquad\qquad\qquad\qquad\qquad + \ Fe^{(+)} \ \rightarrow \\ m \ CF_3\text{-}(CF_2)_5\text{-}CH=CH\text{-}O^-$$

b) 1H,1H-perfluorooctanol:

$$k \ CF_3\text{-}(CF_2)_5\text{-}CF_2\text{-}CH_2\text{-}O^- \\ \qquad\qquad\qquad\qquad\qquad + \ Fe^{(+)} \ \rightarrow \\ l \ CF_3\text{-}(CF_2)_5\text{-}CF_2\text{-}CH\text{-}O^-$$

Figure 36 The reaction between anions and anion radicals formed from: (a) 1*H*,1*H*,2*H*,2*H*-perfluorooctanol and (b) 1*H*,1*H*-perfluorooctanol. (From Ref. 53.)

In summary, the tribochemical reaction process of fluorinated alcohols proceeds in three phases including:

(a)   The interaction between lubricant molecules and low-energy electrons generates anion and radical anion species.

(b)   The reaction of anions and anion radicals with a positively charged surface, followed by cross-linking between unsaturated anions and anion radicals reacting with the surface, leads to oligomeric organometallic layers that protect the surface (Fig. 36).

(c)   If shear strength is high, it can cause the breaking of chemical bonds within the organometallic layer, leading to decomposition products (mainly fluorine containing radicals) that react with the metal surface and produce metal fluorides. The formation of AlF_3 is connected with corrosive wear of aluminum; on the other hand, iron fluorides seem to have good load-carrying capabilities.

3.   Tribochemistry of Fatty Acids

An early review of the fatty acids' importance in boundary lubrication was presented by Cambell [13], in which all the details related to the friction coefficient tests under various operating conditions are described. These include the effect of load and speed on friction, the effect of atmosphere, as well as the effect of temperature. For example, it had been emphasized that at low loads, the adsorbed film is oriented approximately perpendicular to each surface, with its active carboxylic group attached to the metal surface. The influence of fatty acids on the wear process under boundary lubrication conditions has not been investigated so well as the friction process. Moreover, some pieces of evidence exist in the literature [54,55] to indicate that antiwear properties of these additives mostly concern higher fatty acid concentrations, usually 0.4–2.0%. Additionally, the major tested fatty acids relate to higher-molecular-weight compounds, mostly from lauric acid (C_{12}) to stearic acid (C_{18}). A most recent paper [56] discussed the effect of fatty acids (C_6–C_{18}) at low concentrations (0.1–0.5%) and very low concentrations (50–750 ppm) in hexadecane on the wear of steel under boundary lubrication conditions. Further research aiming at providing more information on and a better understanding of the wear process in the presence of lubricants containing fatty acids under boundary lubrication conditions of steel-on-steel mating elements is presented in Ref. 57. The primary objective of that work was twofold: (a) to present a specific antiwear behavior of fatty acids in the low additive concentration range of 0.005–0.1% in hexadecane, and (b) to demonstrate how the different base fluids influence the antiwear behavior of fatty acids. The ball wear results generated in a ball-on-disk apparatus for steel lubricated with hexadecane plus 50, 100, 500, 750, and 1000 ppm each of caproic acid (C_6) and myristic acid (C_{14}) are illustrated in Fig. 37, which clearly demonstrates the existence of two different kinds of curves. The first one shows a maximum at the 750-ppm concentration of myristic acid; the second curve kind shows a minimum at the 500-ppm concentration of caproic acid. Consequently, in this specific behavior of fatty acids, two kinds of events can be distinguished: (a) a clear corrosive effect at 750 ppm of C_{14}, C_{16}, and C_{18} fatty acids, and (b) a distinct antiwear effect at a 500-ppm concentration of caproic acid (C_6). To find out if the corrosive effect also appears in other base fluids, palmitic acid was selected for further investigation in saturated hydrocarbons of typical synthetic base oils (PAO 4 and PAO 6) and paraffinic–naphthenic fraction isolated from SN 400 base oil. The obtained results are very interesting and different from those of hexadecane base lubricants. Figure 5 demonstrates that in the case of synthetic base fluids, a clear minimum wear value exists at 100 ppm concentration of palmitic acid; however, at 750 ppm concentration of the acid, only a slight maximum is observed. In comparison with the wear data for hexadecane-based lubricants, both PAO 4– and PAO 6–based lubricants have these maxima at the relative wear reduction below 50%. Thus, no corrosive effect appears for the tested synthetic saturated base oils. Surprisingly, it was found that the addition of 50 and 100 ppm of palmitic acid to SN 400–saturated hydrocarbons increases wear by around 50% [57]. The same is due to the mineral base oil SN 400. However, in this case, the corrosive effect is even

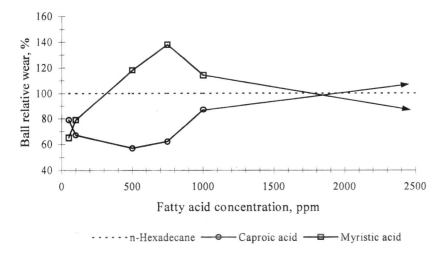

Figure 37 Ball relative wear of the systems lubricated with 50–1000 ppm solutions of caproic acid and myristic acid in hexadecane. (From Ref. 57.)

bigger; the wear was increased by a factor of two. When the concentration of palmitic acid was increased to 1000 ppm, the antiwear behavior of both oils was improved.

The results obtained evidently confirm the complexity of the boundary lubrication approach. Nevertheless, it is possible to suggest that three points seem to be of importance: (a) a kind of molecular interaction exists between the fatty acid additive and the chemistry of base fluids; (b) viscosity of the base fluids seems to be a factor in the boundary lubrication process; and (c) 100 ppm concentration of longer-chain fatty acids in a mixture of lower viscosity hydrocarbons can reduce the steel-on-steel wear significantly under boundary lubrication conditions. For example, it was found [58] that fatty acids C_6–C_{18} reduce dramatically the wear of steel-on-steel tribological systems lubricated with low-viscosity petroleum fraction containing 100 ppm of any of these acids (Table 2). Overall, these results demonstrate that the tribochemical behavior of alcanoic fatty acids at very low concentrations in different base fluids is complex and more research is needed to clarify the new findings. Kajdas and Majzner [58] also show that the tribological wear does not decrease with growth of any fatty acid concentration in a diesel fuel component; the dependence of fatty acid molecular weight on tribological wear is not a characteristic neither of the pin-on-disk apparatus nor the high-frequency reciprocating rig (HFR2 apparatus). Figure 38 shows

Table 2 Tribological Wear in Systems Lubricated with 50–1000 ppm C_6–C_{18} Fatty Acid Solutions in a Low-Sulfur Diesel Fuel Component (Pin-on-Disk Apparatus)

Fatty acid	Fatty acid concentration (ppm)	Wear scar diameter (μm)		Ball relative wear (%)
		Average value	SD	
Caproic acid	50	299	27	15.4
	100	240	24	6.4
	500	272	26	10.6
	750	237	35	6.0
	1000	299	37	15.5
Capric acid	50	305	17	16.8
	100	183	25	2.2
	500	300	23	15.6
	750	234	39	5.8
	1000	232	36	5.6
Lauric acid	50	203	3	3.3
	100	173	9	1.7
	500	210	34	3.8
	750	238	35	6.2
	1000	197	32	2.9
Palmitic acid	50	216	13	4.2
	100	217	33	4.3
	500	197	15	2.9
	750	200	21	3.1
	1000	207	28	3.6
Stearic acid	50	191	15	2.6
	100	206	21	3.5
	500	221	37	4.6
	750	196	22	2.8
	1000	187	19	2.3

Source: Ref. 58.

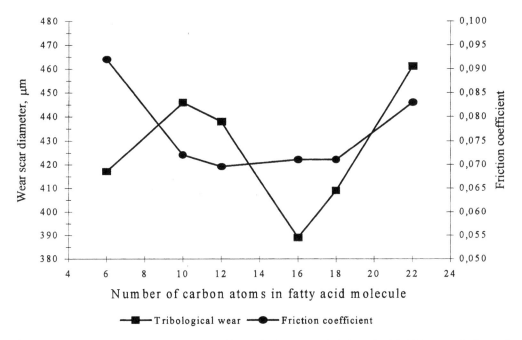

Figure 38 Tribological wear friction coefficient in the system lubricated with 500 ppm solutions of fatty acids C_6–C_{22} in a low-sulfur diesel fuel components (HFR2 apparatus). (From Ref. 58.)

that there is no correlation between wear and friction coefficient for C_6–C_{22} fatty acids. The antiwear behavior of fatty acids under boundary lubrication conditions is influenced by the rate of acid molecule diffusion to the friction surface; this conclusion is drawn from results of experiments obtained for systems lubricated with palmitic acid solutions in various viscosity synthetic oils (poly-α-olefins); it has also been found that in some cases, the structure of solvent molecules has a predominant influence on wear (e.g., interaction between molecules of hexadecane and fatty acid).

Extensive research [42,58–67] emphasized the importance of the chemical attack of carboxylic acids and the formation of salts (soaps) in boundary lubrication. Results of experiments carried out using radioactive foils [59] and electron diffraction studies [60] correlate with the lubricating properties of acids and support the view that tribochemical reactions may occur during friction. "Covalent soap" species, different from typical salts, have been identified in IR spectra of nickel–acid system [61]. According to Ref. 62 during friction process performed in the presence of fatty acids dissolved in an aromatic base oil, abraded metal surfaces donate electrons to aromatic compounds and adsorbed radical anions are created. The tribochemistry of stearic acid on copper has been studied in detail by applying FTIRM and surface-enhanced Raman spectroscopy [63,64]. Results of these experiments reveal that during friction process, different structures of carboxylic acid salts are formed and bidentate complex is responsible for the protection of tribological pair surfaces. Other specific investigations [65,66] revealed that stearic acid chemisorbs initially in both monodentate and bidentate sites, but after tribological reaction, the monodentate chemisorbed stearic acid is transformed to a bidentate surface configuration. Tribochemical changes of a nanometer-thick stearic acid film on a smooth copper surface were detected during the mechanical disruption of surface bonds by a multidiamond tip scratching using FTIR [67].

Kajdas [68] proposed a new approach concerning the tribochemistry of carboxylic acids. The approach takes into account the formation of anions and radical anions of acid molecules and the chemisorption of these species on positively charged spots of a rubbing solid surface connected with a recombination of radical anions:

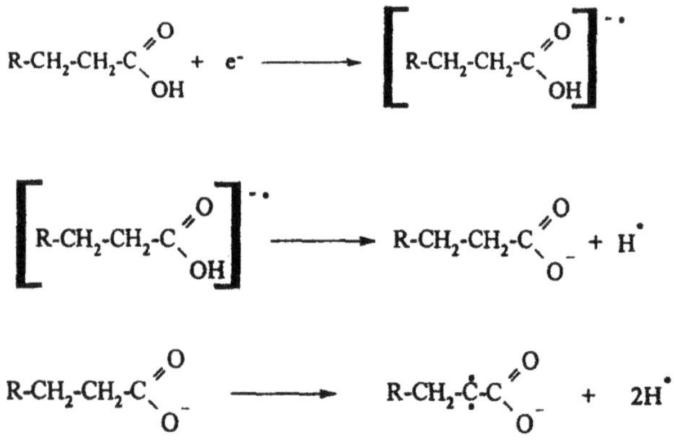

$$R\text{-}CH_2\text{-}CH_2\text{-}C \overset{O}{\underset{OH}{\diagup}} \; + \; e^- \;\longrightarrow\; \left[R\text{-}CH_2\text{-}CH_2\text{-}C \overset{O}{\underset{OH}{\diagup}} \right]^{-\cdot}$$

$$\left[R\text{-}CH_2\text{-}CH_2\text{-}C \overset{O}{\underset{OH}{\diagup}} \right]^{-\cdot} \;\longrightarrow\; R\text{-}CH_2\text{-}CH_2\text{-}C \overset{O}{\underset{O^-}{\diagup}} \; + \; H^\bullet$$

$$R\text{-}CH_2\text{-}CH_2\text{-}C \overset{O}{\underset{O^-}{\diagup}} \;\longrightarrow\; R\text{-}CH_2\text{-}\overset{\bullet}{C}\text{-}C \overset{O}{\underset{O^-}{\diagup}} \; + \; 2H^\bullet$$

(1)

(1) $\longrightarrow$ (2)

$$R\text{-}\overset{\bullet}{C}H\text{-}\overset{\bullet}{C}H\text{-}C \overset{O}{\underset{O^-}{\diagup}} \;\longrightarrow\; R\text{-}CH\text{=}CH\text{-}C \overset{O}{\underset{O^-}{\diagup}}$$

(2)

A most recent research by Majzner et al. [69] aimed at: (a) identifying products of tribochemical reactions taking place in the presence of carboxylic acids; (b) considering the possibility of formation of these products according to the NIRAM approach; and (c) comparing products of triboreactions to products of reactions performed for carboxylic acids without mechanical action (i.e., under static thermal reaction conditions). Three compounds were used: caprylic acid, 3-cyclopentylpropionic acid, and 4-phenylbutyric acid. Tribological tests were performed using a pin-on-disk machine. After the friction process, elements of tribological pairs were rinsed with hexane and dried to prepare the surface for analyses. For examination of wear surfaces, FTIRMA and SEM/EDS techniques were used. Each obtained IR spectrum was compared with spectra of reference standards in order to identify new absorption bands. To identify products of reactions taking place under static conditions, the FTIR method was applied. A drop of caprylic acid was put on steel disks. After 33 min, 66 min, and 24 hr, FTIR reflection spectra were recorded for the drop and for the surface of disks after washing them with hexane. Transmission mode FTIR spectra of reference standards were taken. The reference compounds included: sodium caprylate ($CH_3(CH_2)_6COONa$), sodium acrylate ($H_2C{=}CHCOONa$), sodium oleate ($CH_3(CH_2)_7CH{=}CH(CH_2)_7COONa$), and hydroxyaluminum diacetate (($CH_3COO)_2AlOH$).

Caprylic acid was used in pure state as a lubricant in three material systems: steel ball–steel disk, steel pin–aluminum disk, and steel ball–copper disk [69]. Deposit generated during friction of the steel ball-on-steel disk was analyzed using the EDS technique. Figure 39 shows that the concentration of carbon and oxygen is higher within than beyond the area of the wear track. In addition, SEM/EDS analysis of the aluminum disk surface after friction process carried out in the steel ball-on-aluminum disk system confirmed that within the wear track, deposits of an organic substance exist [69]. A SEM image obtained for the ball surface after tribological tests carried out in the steel ball–copper disk system (Fig. 40) as well as EDS dot map (Fig. 41) revealed that copper was transferred from the copper disk onto the ball surface. On transferred copper, material compounds containing oxygen and carbon were detected.

The FTIR spectra taken from wear scars of the tested metal systems confirmed the presence of organic compounds within the area of wear scars and/or wear tracks. Figure 42 demonstrates an exemplary spectrum for caprylic acid in the steel ball–copper disk system. It is important to note that in the FTIR spectrum taken from the wear track on the aluminum disk, additional peaks located at 3697 and 982 cm^{-1} were found. These bands were also observed in the spectrum of hydroxyaluminum diacetate (Fig. 43) [69]. Peri [70] demonstrated that the infrared absorption band 3700 cm^{-1} relates to the C-site isolated hydroxyl group ions on the surface of γ-alumina. Four other hydroxyl ions (A, B, D, and E) show characteristic infrared bands between 3733 and 3800 cm^{-1}. Accordingly, it is permissible to suggest that during the friction process, a kind of structure film, as presented in Fig. 44, is formed.

Another interesting finding in that research relates to the spectrum recorded from the steel disk surface after the friction process performed in the steel ball–steel disk system. The FTIRM spectrum (Fig. 45) includes: a "triangle" signal with maximum in the region 3288 cm^{-1}; a weak band placed at 3064 cm^{-1}; peaks located from 3000 to 2800 cm^{-1}; a strong band situated at 1646 cm^{-1}; a signal positioned at 1546 cm^{-1}; peaks placed in the area 1500–1400 cm^{-1}; a band situated at 1241 cm^{-1}; and a signal located at 1115 and at 1011 cm^{-1}.

This kind of spectra was also taken from the wear track on the disk surface after tribological tests carried out for a 1000-ppm solution of caprylic acid in n-hexadecane in the steel ball–steel disk system. It proves that the formation of this compound during friction process in the presence of an acid dissolved in hydrocarbon base oil is also possible. It is also important to remember that a very similar spectrum was obtained from the wear track of the same system lubricated with hexadecane (see Fig. 24). Identical spectra were obtained from the wear track on steel disk after the friction process performed in the presence of an acid containing a straight saturated chain and a saturated ring attached to the end of the chain (3-cyclopentylpropionic acid) (Fig. 46). Characteristic spectra peaks in Fig. 43 extended from 3000 to 2800 cm^{-1} are attributed to symmetrical and asymmetrical stretching vibrations of CH bonds in CH_2 and CH_3 groups; however, they are much more weaker than those in spectra of typical monodentate structures. Therefore, it seems that chains of acids are broken during the friction process. To verify this hypothesis, the acid containing a straight chain and an aromatic ring attached to the end of the chain was applied (4-phenylbutyric acid). Because 4-phenylbutyric acid is solid at ambient temperature, it was used as a 1000-ppm solution in 1-methylnaphthalene. Figure 47 presents the spectrum of a tribological reaction product obtained from the wear track on the steel disk lubricated with the solution of 4-phenylbutyric acid in 1-methylnaphthalene. The spectrum does not include any characteristic absorption bands of monosubstituted aromatic ring and it is very similar to that presented in Fig. 46. Thus, the hypothesis seems to be plausible. However, the origin

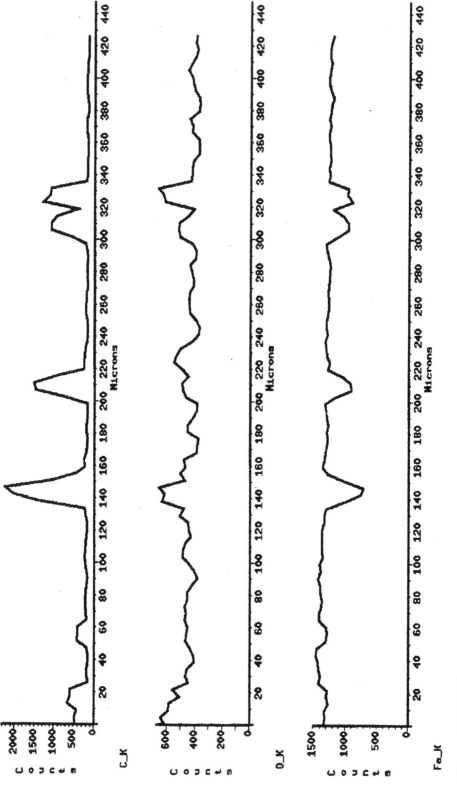

Figure 39 EDS line scan of the wear track on the steel disk after tribological tests carried out in the presence of caprylic acid in the steel ball–steel disk system: (a) carbon, (b) oxygen, and (c) iron. (From Ref. 69.)

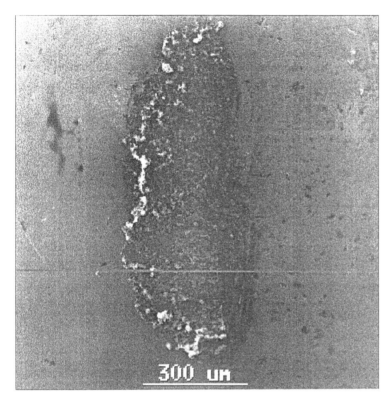

Figure 40 SEM image of the steel ball surface after tribological tests carried out in the presence of caprylic acid in the steel ball–copper disk system. (From Ref. 69.)

of other signals appearing in these spectra taken from steel disk surface after the friction process is unknown. Peaks located at 1646 cm^{-1} (Fig. 45) can be assigned to the stretching vibration of C=C. To verify this assumption, two salts with different locations of double bond in the molecule (a) sodium oleate and (b) sodium acrylate were analyzed [69]. Based on very detailed analytical results, it can be concluded that the formed compounds during the friction process are chelating bidentate complexes having double bonds in the molecule in the position depicted in Fig. 48. The finding emphasizes the importance of the NIRAM approach in accounting for the formation of these untypical compounds. Anions react with the positive charged spots of the rubbed surface forming typical salts (monodentate structure). On the other hand, due to the recombination of radicals, a compound having a double bond in the molecule can be formed.

It is worthy note that under static conditions, the reaction of caprylic acid and iron does not take place during the period of time equal to the time of the tribological tests. Spectrum recorded for the drop of caprylic acid after 24 hr confirmed the reaction of caprylic acid and iron under static conditions. Similarly, in the case of the static test performed for 66 min, no organic species was found on the steel surface after washing it with hexane. A comparison of the characteristic spectrum taken from the wear track on the steel disk surface after tribological tests performed for caprylic acid in the steel ball–steel disk system and the spectrum of caprylic acid drop put on the steel disk surface after 24 hr confirmed that products of tribochemical reactions and products of reactions performed under static conditions differ very significantly.

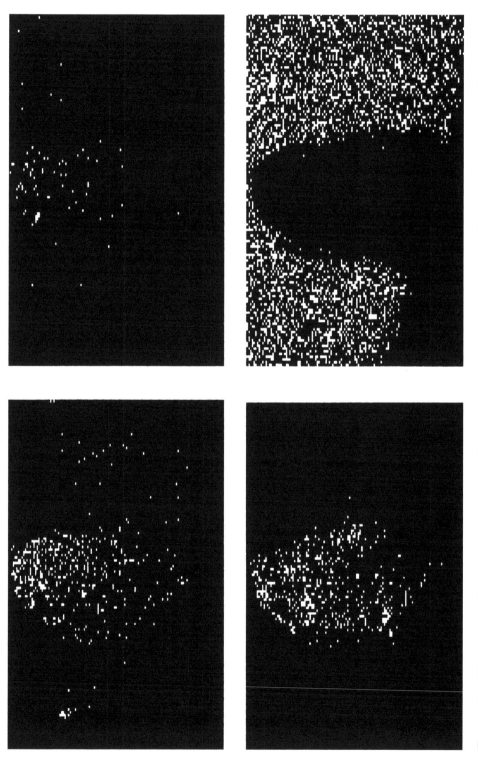

Figure 41 EDS dot map of the steel ball surface after tribological tests carried out in the presence of caprylic acid in the steel ball–copper disk system. (From Ref. 69.)

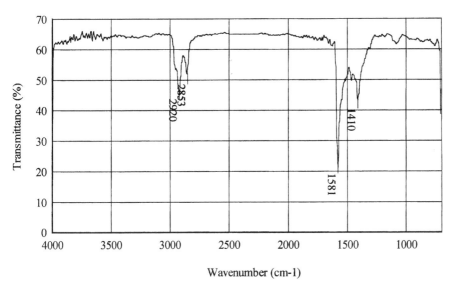

Figure 42 FTIR spectrum recorded from the copper material transferred on the steel ball after tribological tests carried out in the presence of caprylic acid in the steel ball–copper disk system. (From Ref. 69.)

The SEM/EDS analysis performed for wear areas after tribological tests carried out in the steel ball–steel disk, steel pin–aluminum disk, and steel ball–copper disk systems in the presence of carboxylic acids showed that within the wear scars and wear tracks, deposits of organic compounds are present. FTIR spectra recorded for those areas confirmed that during rubbing on the steel surface and on the copper surface, typical salts are created (monodentate structures). In the case of the steel pin–aluminum disk system, formation of

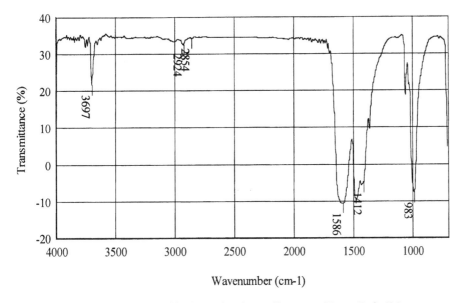

Figure 43 FTIR spectrum of hydroxyaluminum diacetate. (From Ref. 69.)

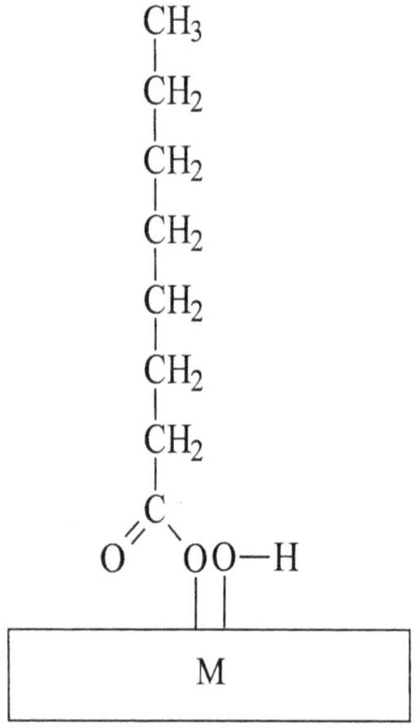

CH₃
|
CH₂
|
CH₂
|
CH₂
|
CH₂
|
CH₂
|
CH₂
|
C
O⫽ `OO—H

```
+-----------------------------+
|              M              |
+-----------------------------+
```

Figure 44 "Hydroxysalt" of caprylic acid and aluminum (monodentate structure) formed during friction process. (From Ref. 69.)

the hydroxysalt on aluminum surface was observed. On the other hand, spectra of chelating symmetrical bidentate complexes having double bonds in molecules were obtained from the steel disk surface after tribological tests conducted in the steel ball–steel disk system. These compounds were not detected after reactions taking place under static conditions. All these results enhance our knowledge on the tribochemistry of carboxylic acids, which was contributed mostly in the last decade [63–66]. Hsu et al. [67] presented tribochemistry induced by nanomechanical scratches and demonstrated that mechanical disruption of surface bonds can be achieved by a multidiamond tip scratching a nanometer-thick stearic film on a smooth copper surface. Tribochemical changes were observed by FTIR showing the link between mechanical bond disruption and chemical reaction at room temperatures. On the other hand, bearing in mind the triboemission process, it is suggested that the tribochemistry of carboxylic acids is also controlled by the NIRAM concept or its modified version, the HSAB–NIRAM approach [29].

4. Tribochemistry of Unsaturated/Ether Compounds

This section aims at determining the chemical structure of friction products generated from compounds containing unsaturated bonding or ether functional group, or both. A better understanding of tribochemical reactions controlling the steel-on-steel wear process by organic additives containing double bonding and/or an ether group is the important boundary lubrication issue. Effects on the wear and tribochemical reactions of selected unsaturated additives in dissolved hexadecane were investigated under boundary lubrication conditions in the ball-on-disk contact. Several modern microanalytical (FTIRM, SEM/EDS, and XPS) techniques were used to get more detailed information concerning the tribochemistry of the tested compounds. Although the tribochemical reactions seem to be very complex, the presence of carboxylate structures combined with the steel substrate

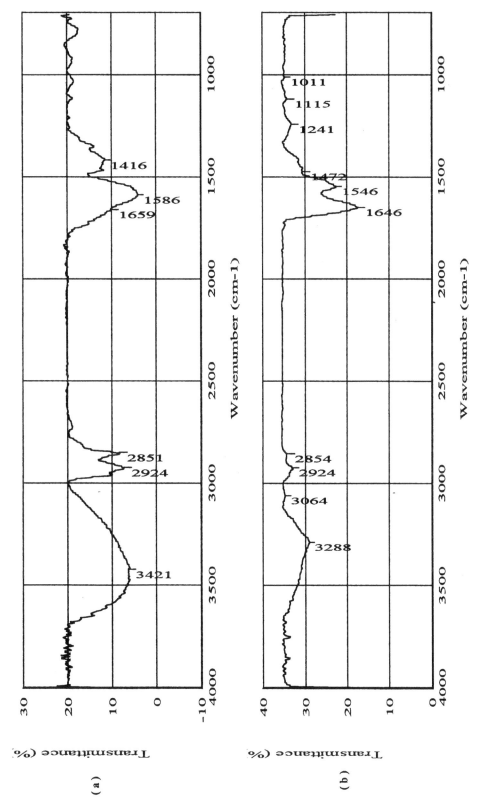

Figure 45 FTIR spectra recorded from the wear track on the steel disk after tribological tests carried out in the presence of caprylic acid in the steel ball–steel disk system. (From Ref. 69.)

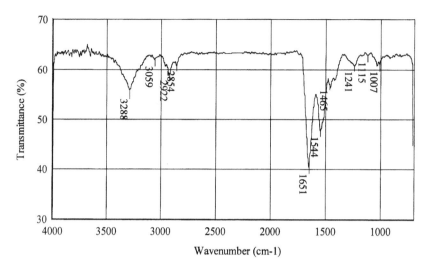

Figure 46 FTIR spectrum recorded from the wear track on the steel disk after tribological test carried out in the presence of 3-cyclopentylpropionic acid in the steel ball–steel disk system. (From Ref. 69.)

provides a good evidence for the oxidation process of the tested compound [71–75]. Thus, it can be concluded that the double bond undergoes the oxidation reaction. It is also of importance that the hydroxyl group of that molecule takes part in triboreactions.

Solutions of unsaturated compounds (1 wt.%) containing various functional groups (Table 3) dissolved in *n*-hexadecane were tested as antiwear additives. A diameter of wear scar on the ball (arithmetic average of three parallel results) was used to determine the

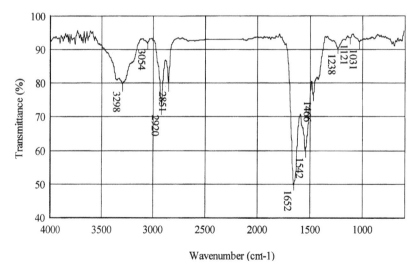

Figure 47 FTIR spectrum recorded from the wear track on the steel disk after tribological test carried out in the presence of a 1000-ppm solution of 4-phenylbutyric acid in 1-methylnaphthalene in the steel ball–steel disk system. (From Ref. 69.)

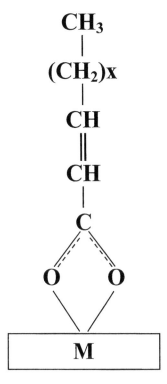

Figure 48 Hypothetical structure of a compound formed during friction process carried out in the presence of carboxylic acid (caprylic acid; chelating bidentate structure). (From Ref. 69.)

Table 3 Chemical Structures of Investigated Additives

H_2C—CH $\overset{O—CH_2}{\underset{}{}}$ CH=CH_2 $\underset{O}{}$	Allyl-2,3-epoxypropyl ether
CH_3—$(CH_2)_{13}CH$=CH_2	*n*-hexadecene
	4-Allyl-2-methoxyphenol

antiwear efficiency of the tested lubricants as compared to hexadecane. Figure 49 presents wear scar diameter on the balls and shows that the composition containing 1 wt.% of 4-allyl-methoxyphenol provides the best antiwear properties (36% wear reduction) among tested lubricants. FTIRMA results of the formed triboreaction products formed on the disk wear track are collected in Fig. 50. All the obtained spectra confirm the formation of organic products as a result of friction. Absorption bands in the range of 1540–1650 cm^{-1} are characteristic for vibrations of carboxylic anions. Besides, in all the abovementioned spectra, there are peaks assigned to symmetrical vibrations of carboxylic anion (at 1417, 1432, and 1426 cm^{-1}, respectively). The absence of bands at 3080 cm^{-1} corresponding to double bonds shows that they are active centers for the oxidation to carboxylic acids. The last ones can react with iron atoms from the steel surface to form carboxylates and/or chelate compounds. Moreover, it was found that the remaining functional groups (hydroxy- and epoxy-) present in molecules also take place in triboreactions [72]. The SEM/EDS results (Fig. 51) indicate the deposition of triboreaction products containing oxygen and carbon within the wear scar.

The work described in Refs. 71–75 demonstrated that antiwear efficiency of unsaturated compounds is controlled by tribochemical reactions leading to the formation of specific deposits on the metal surface. Although the tribochemical reactions seem to be very complex, the presence of carboxylate structures combined with the steel substrate provides a good evidence for the oxidation process of the tested compound. Thus, it can be concluded that the double bond undergoes the oxidation reaction to carboxylate structure. The carboxylate structure formation was additionally confirmed by XPS analysis [72].

5. Tribochemistry of Esters

It is well known that esters under boundary lubrication conditions form soaps with the friction metal surface. The lubrication related literature suggests that the soap formation by

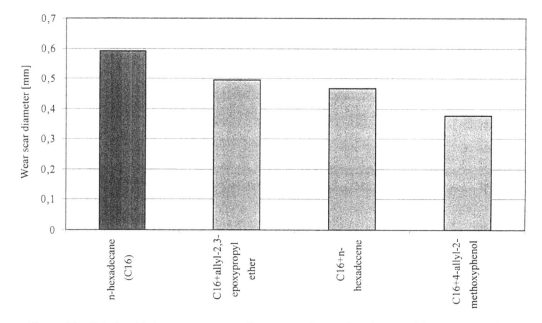

Figure 49 Relationship between wear scar diameters and type of used composition. (From Ref. 72.)

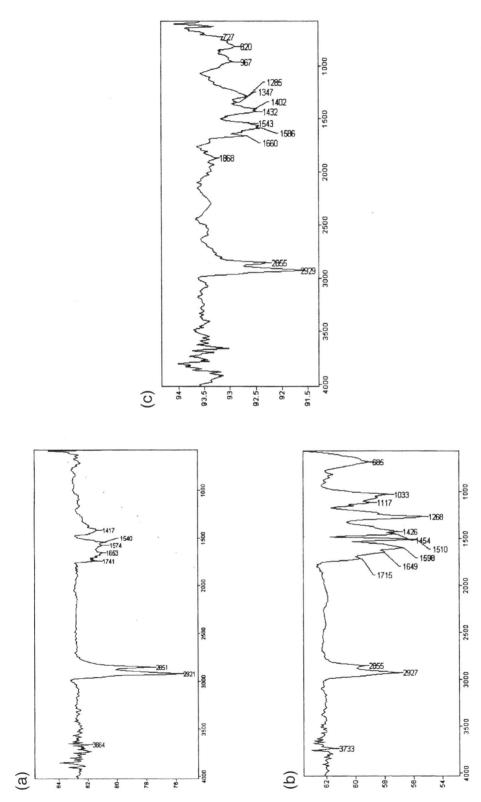

Figure 50 FTIRMA spectra of tribochemical conversion products: (a) *n*-hexadecane with 1 wt.% of allyl-2,3-epoxypropyl ether; (b) *n*-hexadecane with 1 wt.% of *n*-hexadecene; and (c) *n*-hexadecane with 1 wt.% of 4-allyl-2-metoxyphenol. (From Ref. 72.)

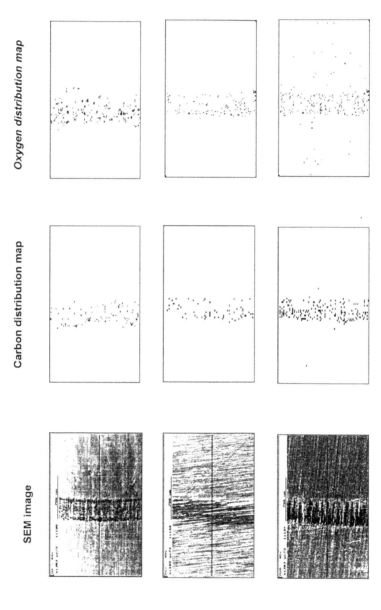

Figure 51 Results of SEM/EDS analysis after the friction process with the participation of: (Left) C_{16} + allyl-2,3-epoxypropyl ether; (Center) C_{16} + n-hexadecene; and (Right) C_{16} + 4-allyl-2-methoxyphenol. (From Ref. 72.)

the ester under boundary lubrication is due to the hydrolysis process of the ester. This leads to the formation of a very small amount of the fatty acid, which, when adsorbed, attaches to the metal to form the corresponding soap film. On the other hand, it is supposed that soap formation proceeds according to the NIRAM approach. This topic is described in detail in Chap. 8 ("Hydrolysis"). It was assumed that soap formation is due to the ester C–O bond cleavage leading to the formation of the $RCOO^-$ species, which reacts with the positively charged surface spot to form the soap. In hydrolysis, the nucleophile is a water molecule and the leaving group is an alcohol; in esterification, the roles are exactly reversed. Based on the above information, it is difficult to realize that under typical boundary lubrication conditions, mineral acid might be present to catalyze the ester hydrolysis process. The most striking finding concerning the tribochemistry of esters, described in Chap. 8, relates to the fact that the acid/alcohol mixtures have the same effect on the steel-on-steel wear process. Their antiwear property does not change with the concentration in n-hexadecane. Additionally, the mixtures affect dramatically the wear reduction of steel under boundary lubrication conditions. The results enabled to state that the hypothesis saying "During the friction process under boundary lubrication conditions, lubricated by aliphatic esters, the ester hydrolysis process cannot proceed without an adequate catalyst" was proven for aliphatic esters dissolved in hexadecane. Another interesting finding relates to the best wear reduction ability of n-hexadecyl palmitate and the equimolar mixture of palmitic acid and 1-hexadecanol. This finding was accounted for in terms of the chain-matching effect with hexadecane. More research is needed to clarify these interactions.

IV. TRIBOCHEMISTRY OF CERAMICS

A. General Information Focused on Tribochemistry

Ceramics, particularly oxides and nitrides, have a very low free energy of formation and therefore are very stable chemically in a nontribological sense. However, under boundary lubrication conditions, they become quite reactive due to the fact that solids support shear strains. As ceramics are relatively inert, hard, and more resistant to abrasive and erosive wear than typical steels, they offer some advantages over conventional tribological mating elements. The aim of the present chapter section is to discuss only selected issues relating to ceramic tribochemistry because under boundary friction conditions, the lubrication of ceramics is widely dominated by tribochemical reactions. It is even possible to say that the friction and wear behavior of ceramics might be more sensitive to the environment than friction and wear behavior of metals. Accordingly, the importance of a comprehensive understanding of the ceramic surface physics and chemistry cannot be overestimated. A review paper [76] discussed the initiation process of ceramic tribochemical reactions and mostly aimed at (a) presenting some tribochemical reactions that play an important role in the wear reduction of ceramics, and (b) proposing a model of such reactions. The proposed model is based on the NIRAM approach of a lubricant/environment component with the rubbing ceramic surface. Tribochemical reactions of ceramics are very complex because several physical processes may be involved in the formation of wear-reducing films. However, the application of the NIRAM–HSAB approach [77] to account for tribochemical reactions of silicon nitride with water or alcohols allows to better understand all the major findings related to the silicon nitride tribochemistry. Many studies of ceramic lubrication involve alcohols, gases, water, and other chemicals [78–86].

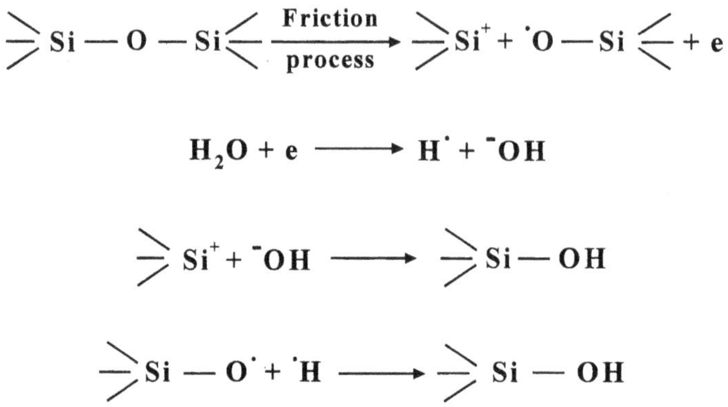

$$\text{>Si}-\text{O}-\text{Si<} \xrightarrow[\text{process}]{\text{Friction}} \text{>Si}^+ + \,{}^\cdot\text{O}-\text{Si<} + e$$

$$H_2O + e \longrightarrow H^{\cdot} + {}^-OH$$

$$\text{>Si}^+ + {}^-OH \longrightarrow \text{>Si}-OH$$

$$\text{>Si}-O^{\cdot} + {}^\cdot H \longrightarrow \text{>Si}-OH$$

Figure 52 NIRAM–HSAB model for reaction of water with silica under friction conditions.

B. Tribochemistry of Silicon Nitride, Silicon Carbide, Alumina, and Silica

Earlier extensive work concerning the tribochemistry of silicon nitride was presented in detail in Refs. 87 and 88 and then summarized in two papers [48,86]. A similar one is due to silicon carbide [89]. Present understanding of fundamental friction and wear properties of ceramics in water is discussed in Ref. 79 with emphasis on practical usefulness of water lubrication of ceramics. A most recent work [89] provides an overview of fundamental friction and wear properties of carbon nitride (CN_x) coatings sliding against silicon nitride balls. An approach to the tribochemistry of silicon nitride in terms of hydrolysis is discussed in Chap. 8 ("Hydrolysis") in which also the NIRAM–HSAB model for the tribochemical reaction of water with silicon nitride was presented. The same NIRAM–HSAB approach can also be applied to the tribochemistry of silica. The suggested tribochemical reaction process of silica with water under rubbing conditions is depicted in Fig. 52 [2].

   The well-known fact of hydroxide formation from silica under friction conditions in the presence of water molecules can also be explained in terms of the NIRAM–HSAB approach. Under rubbing conditions, the emitted electrons produce OH^- and $H^{\cdot}$ species

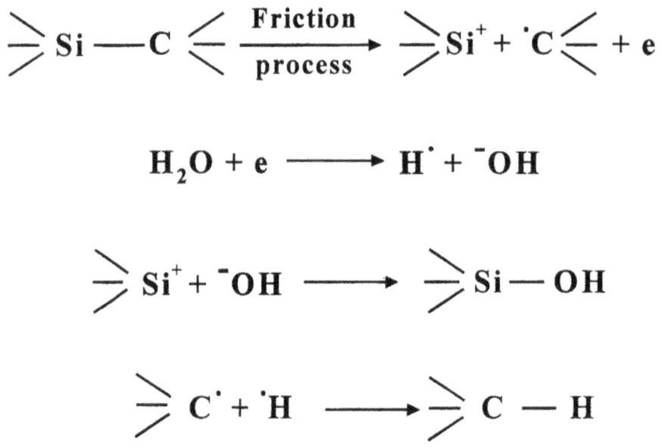

$$\text{>Si}-\text{C<} \xrightarrow[\text{process}]{\text{Friction}} \text{>Si}^+ + \,{}^\cdot\text{C<} + e$$

$$H_2O + e \longrightarrow H^{\cdot} + {}^-OH$$

$$\text{>Si}^+ + {}^-OH \longrightarrow \text{>Si}-OH$$

$$\text{>C}^{\cdot} + {}^\cdot H \longrightarrow \text{>C}-H$$

Figure 53 NIRAM–HSAB model for reaction of water with silicon carbide under friction conditions.

the ester under boundary lubrication is due to the hydrolysis process of the ester. This leads to the formation of a very small amount of the fatty acid, which, when adsorbed, attaches to the metal to form the corresponding soap film. On the other hand, it is supposed that soap formation proceeds according to the NIRAM approach. This topic is described in detail in Chap. 8 ("Hydrolysis"). It was assumed that soap formation is due to the ester C–O bond cleavage leading to the formation of the $RCOO^-$ species, which reacts with the positively charged surface spot to form the soap. In hydrolysis, the nucleophile is a water molecule and the leaving group is an alcohol; in esterification, the roles are exactly reversed. Based on the above information, it is difficult to realize that under typical boundary lubrication conditions, mineral acid might be present to catalyze the ester hydrolysis process. The most striking finding concerning the tribochemistry of esters, described in Chap. 8, relates to the fact that the acid/alcohol mixtures have the same effect on the steel-on-steel wear process. Their antiwear property does not change with the concentration in n-hexadecane. Additionally, the mixtures affect dramatically the wear reduction of steel under boundary lubrication conditions. The results enabled to state that the hypothesis saying "During the friction process under boundary lubrication conditions, lubricated by aliphatic esters, the ester hydrolysis process cannot proceed without an adequate catalyst" was proven for aliphatic esters dissolved in hexadecane. Another interesting finding relates to the best wear reduction ability of n-hexadecyl palmitate and the equimolar mixture of palmitic acid and 1-hexadecanol. This finding was accounted for in terms of the chain-matching effect with hexadecane. More research is needed to clarify these interactions.

IV. TRIBOCHEMISTRY OF CERAMICS

A. General Information Focused on Tribochemistry

Ceramics, particularly oxides and nitrides, have a very low free energy of formation and therefore are very stable chemically in a nontribological sense. However, under boundary lubrication conditions, they become quite reactive due to the fact that solids support shear strains. As ceramics are relatively inert, hard, and more resistant to abrasive and erosive wear than typical steels, they offer some advantages over conventional tribological mating elements. The aim of the present chapter section is to discuss only selected issues relating to ceramic tribochemistry because under boundary friction conditions, the lubrication of ceramics is widely dominated by tribochemical reactions. It is even possible to say that the friction and wear behavior of ceramics might be more sensitive to the environment than friction and wear behavior of metals. Accordingly, the importance of a comprehensive understanding of the ceramic surface physics and chemistry cannot be overestimated. A review paper [76] discussed the initiation process of ceramic tribochemical reactions and mostly aimed at (a) presenting some tribochemical reactions that play an important role in the wear reduction of ceramics, and (b) proposing a model of such reactions. The proposed model is based on the NIRAM approach of a lubricant/environment component with the rubbing ceramic surface. Tribochemical reactions of ceramics are very complex because several physical processes may be involved in the formation of wear-reducing films. However, the application of the NIRAM–HSAB approach [77] to account for tribochemical reactions of silicon nitride with water or alcohols allows to better understand all the major findings related to the silicon nitride tribochemistry. Many studies of ceramic lubrication involve alcohols, gases, water, and other chemicals [78–86].

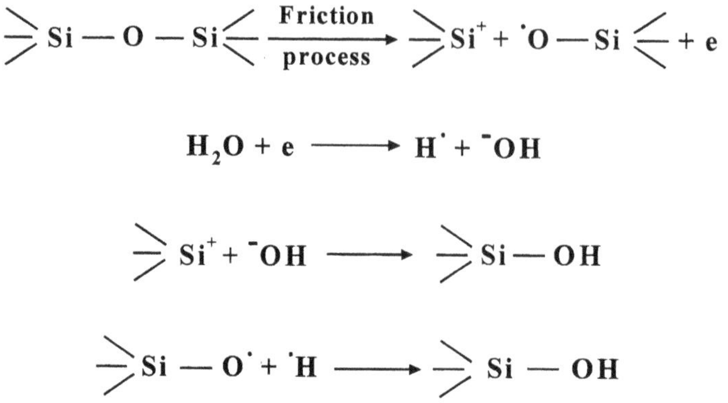

$$\ce{>Si-O-Si<} \xrightarrow[\text{process}]{\text{Friction}} \ce{>Si+ + \cdot O-Si< + e}$$

$$\ce{H2O + e -> H\cdot + {}^-OH}$$

$$\ce{>Si+ + {}^-OH -> >Si-OH}$$

$$\ce{>Si-O\cdot + \cdot H -> >Si-OH}$$

Figure 52 NIRAM–HSAB model for reaction of water with silica under friction conditions.

B. Tribochemistry of Silicon Nitride, Silicon Carbide, Alumina, and Silica

Earlier extensive work concerning the tribochemistry of silicon nitride was presented in detail in Refs. 87 and 88 and then summarized in two papers [48,86]. A similar one is due to silicon carbide [89]. Present understanding of fundamental friction and wear properties of ceramics in water is discussed in Ref. 79 with emphasis on practical usefulness of water lubrication of ceramics. A most recent work [89] provides an overview of fundamental friction and wear properties of carbon nitride (CN_x) coatings sliding against silicon nitride balls. An approach to the tribochemistry of silicon nitride in terms of hydrolysis is discussed in Chap. 8 ("Hydrolysis") in which also the NIRAM–HSAB model for the tribochemical reaction of water with silicon nitride was presented. The same NIRAM–HSAB approach can also be applied to the tribochemistry of silica. The suggested tribochemical reaction process of silica with water under rubbing conditions is depicted in Fig. 52 [2].

The well-known fact of hydroxide formation from silica under friction conditions in the presence of water molecules can also be explained in terms of the NIRAM–HSAB approach. Under rubbing conditions, the emitted electrons produce OH^- and $H\cdot$ species

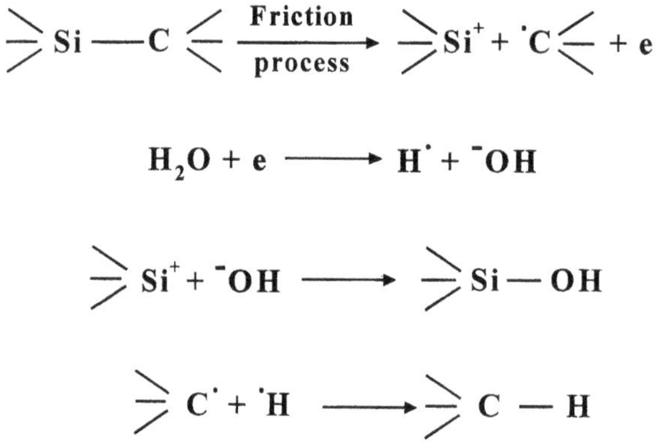

$$\ce{>Si-C< } \xrightarrow[\text{process}]{\text{Friction}} \ce{>Si+ + \cdot C< + e}$$

$$\ce{H2O + e -> H\cdot + {}^-OH}$$

$$\ce{>Si+ + {}^-OH -> >Si-OH}$$

$$\ce{>C\cdot + \cdot H -> >C-H}$$

Figure 53 NIRAM–HSAB model for reaction of water with silicon carbide under friction conditions.

from water and positively charged sites (Si^+) and free radicals ($Si–O^•$) from the silica substrate. HSAB interaction and radical recombination produce the hydroxyl groups ($Si–OH$). Similar tribochemical processes can be considered for silicon carbide (Fig. 53).

V. TWO MAJOR SPECIFIC ISSUES RELATED TO TRIBOCHEMISTRY

A. Links Between Tribochemistry and Catalysis

Numerous chemical reactions relating to catalytic processes can proceed with the same velocity at significantly reduced temperatures in comparison to the thermal reference procedure [5]. This situation is very similar to the noncatalytic thermochemical reactions and their tribochemical analogues. Therefore, from the viewpoint of the reaction mechanism, both catalytic processes and typical tribochemical reactions might relate to the same driving force—governed by the affect of exoelectrons. Heterogeneous catalytic processes can be initiated by thermally emitted electrons. Tribocatalytic process is the typical catalytic reaction enhanced by the action of triboelectrons produced in the test system. This hypothesis can be evidenced by earlier experiments described in Ref. 8. That work studied electron emission from silver catalyst during partial oxidation of ethylene by measuring simultaneously the emission rate and the rate of the ethylene oxide formation. Figure 54 demonstrates the effect of temperature on the emission rate of exoelectrons and on the formation rate of ethylene oxide, for which the temperature of the silver catalyst was raised stepwise from 25°C to 210°C followed by the descent to 25°C. Simultaneous measurements of the exoelectron emission rate and the formation of ethylene oxide at each level of

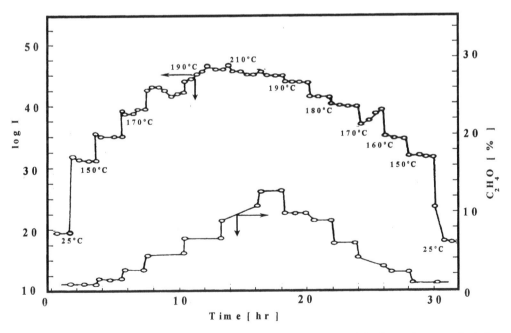

Figure 54   Effect of temperature on the exoelectron emission rate and the formation rate of ethylene oxide. (From Ref. 8.)

temperature were made as a function of time. In Ref. 8, the exoelectron emission from silver catalyst during partial oxidation of ethylene was accounted for as thermoelectron emission in a thin semiconducting oxide on silver.

Actually, it is well known and acknowledged that heterogenous catalytic reactions take place at centers where the electron exchange in the catalyst–substrate system is favored, and that electron emission arises from centers at the surface where electrons are mostly weakly bound [90]. At this point, it is important to emphasize that thermally stimulated electron emission (TSEE; thermionic emission) usually occurs in the same range of temperatures typical for heterogeneous catalysis (Fig. 54). Because thermionic emission has been observed during catalytic reactions, catalysis and thermionic emission seem to be well correlated. Another reproducible relation was found between catalytic activity and thermionic emission capacity in different groups of material such as (a) Pt/SiO_2 with different Pt contents, and (b) ZnS/Cu with different copper contents [91]. The catalytic oxidation of hydrogen and carbon monoxide at NiO, ZnO, and platinum black was found with the occurrence of electron emission. Interestingly, emission of both electrons and ions was observed. It was also demonstrated that highly hydrated oxides (e.g., alumina and magnesia) show a strong gas evolution accompanied by electron emission [91]. TSEE intensity exhibits a maximum positioned at the primary energy of the electrons of about 3 eV in NiO, ZnO, and CuO [92].

Considering the discussed results, it is hypothesized that some heterogeneous catalytic processes proceed according to the NIRAM–HSAB approach, similar to tribochemical reactions. However, typical pure tribochemical reactions are initiated by triboelectrons and heterogeneous catalytic processes can be initiated by thermally emitted electrons. On the other hand, a tribocatalytic process can be considered as the catalytic process enhanced by the action of low-energy electrons emitted from the friction contact (Fig. 55).

Actually, two factor types of the tribochemical reaction initiation process, other than thermal, have been taken into account. They include (a) the mechanically triggered chemistry (mechanochemistry), and (b) the "thermally" triggered chemistry at the contact of asperities caused by the flash temperature effect. Bearing in mind that the duration of flash temperatures is very short, it is necessary to ask the question: "Can this reaction initiation process be considered in terms of overcoming the activation energy by heat or by another form of energy?" This author's answer to that question is comprised in the following hypothesis: The common denominator of tribochemical reactions is that they are mostly initiated by low-energy electrons enhanced by ultraviolet photons. This is partly consistent with the negative ion radical action mechanism approach assuming that tribochemical reactions are initiated by low-energy electrons in the energy range of 1–4 eV [20]. Based on measurements of triboemitted charged particles from ceramics, a first evidence for such energy range was recently demonstrated [19].

Generally, chemical reactions are initiated either by temperature or by a kind of irradiation (e.g., UV photons, nuclear irradiation), or just by mechanical treatment and/or impact. New sophisticated techniques, which allow to measure friction, wear, and lubricant thickness along with surface topography and adhesion, all on a microscale to nanoscale, has led to the development of molecular tribology or atomic scale tribology. It is important to stress that: (a) the nature and origin of tribochemistry have always been subjected to many conjectures, and (b) there are three sources of tribochemistry [67]: (i) normal conventional chemistry, which occurs under the interphase temperatures of the contact zone; (ii) mechanically induced chemistry (fresh surface, electron emission), which will not occur under normal circumstances; and (iii) thermally induced chemistry, which occurs at the asperity tips due to flash temperatures.

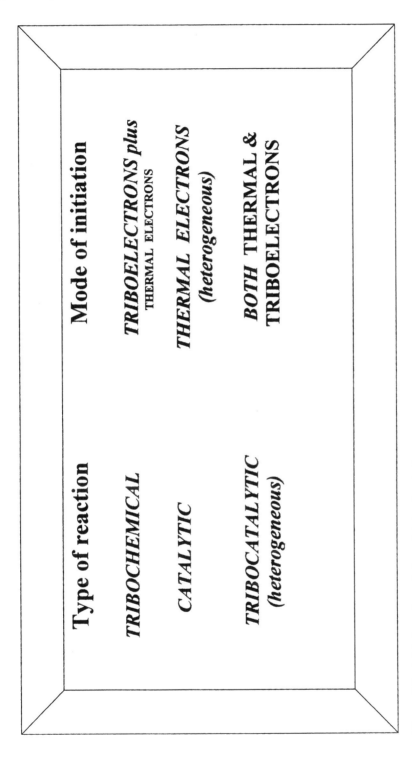

Figure 55 Initiation of catalytic and tribochemical processes.

B. Flash Temperature and Thermionic Emission

Under friction process conditions, the contacting asperities of each surface produce heat and the transient high temperatures at the tips are evolved. These high temperatures are generally known as flash temperatures. They are a microsecond in duration and highly localized [93]. The flash temperature occurs at areas of real contact due to frictional heat dissipation. To initiate thermochemical reactions, an adequate heat amount has to be supplied to overcome the activation energy. Even very high calculated flash temperatures as short-lived rather cannot initiate chemical reactions by heat. This author, linking tribochemistry with catalysis and tribocatalysis [2], has assumed that flash temperature, expressed as the maximum computed friction temperature, can also be expressed in the form of the thermionic emission. Most recently, that assumption was clearly confirmed by an examination of thermionic emission due to frictionally generated temperatures [94]. The emission of electrons from a surface due to heating was investigated theoretically for sliding contact. A thermal model previously developed by Vick and Furey [95,96] for sliding contact was used to predict the temperature rise over the surface and the Richardson–Dushman equation for thermionic emission was then used to estimate the corresponding current density from the surface. The obtained computed results demonstrate that high local temperatures generated by friction of the contacts between rubbing surfaces can activate the emission of electrons. Thus, similar to the typical TSEE, thermionic emission due to frictionally generated temperatures in sliding contact could have a number of important consequences, including activation of tribochemical reactions according to the NIRAM approach and enhancement of surface activity as in the case of tribocatalysis (Fig. 55).

Looking at other evidences for the importance of low-energy electrons in triggering tribochemical reactions, the most recent discovery of microplasma generation in a gap around a sliding contact [25] should be also taken into account. A specific issue of this discovery is that the plasma emits mostly invisible ultraviolet photons and, to a lesser extent, infrared photons. From the viewpoint of the microplasma components and the general triboemission and/or fractoemission processes, it is necessary to consider an interaction of negatively charged particles with photons. It is known that the absorption of a single photon with an energy $h\nu$ larger than the electron affinity results either in direct photoemission from negatively charged clusters, or in a fast thermalization of energy and a subsequent delayed emission [97].

Bearing in mind all the discussed physical facts, it is proposed to understand the flash temperature effect as the factor initiating tribochemical reactions by both (a) thermal electrons and (b) the result of UV photons interaction with negatively charged particles (clusters), leading to the generation of additional low-energy electrons that, according to the NIRAM approach, are responsible for trigging tribochemical reactions. Thus, it is possible to conclude that low-energy electron stream (exoelectrons plus thermal electrons) and UV photons might be considered as the governing factor of tribochemical reaction initiation processes. Accordingly, flash temperatures might also be expressed in the form of electronic energy. This is consistent with the NIRAM approach assuming that tribochemical reactions are initiated by low-energy electrons in the energy range of 1–4 eV. It seems to be justifiable to conclude that the most recent specific and sophisticated research on tribochemistry induced by nanomechanical scratches [67] provides a good deal of evidence for the importance of the NIRAM concept in triggering tribochemical reactions.

Another evidence relates to the most recent work [98] clearly showing that the PFPE lubricants in computer head–hard disk contacts are mainly decomposed by low-energy electrons emitted from the rubbing surface and by mechanical shear. The results demonstrate that the effect of catalytic reaction on the degradation of the PFPE lubricants is negligible in sliding conditions because of the short contact time, small contact area, and low interface temperature. On the other hand, it was found that illumination with ultraviolet light accelerates the decomposition of the lubricant, reducing the head–disk interface durability and causing more gaseous fragments because low-energy electrons (photoelectrons) created by the illumination interact with the lubricant molecules, activating and breaking up the molecules [98]. Another specific work [99] studied the decomposition mechanism of PFPE lubricants by means of mass spectrometry and FTIR; by comparing the products generated during mechanical decomposition of PFPE lubricants, it was concluded that the decomposition process in sliding is an electron-mediated reaction because most of the product fragment distribution was very similar to the electron decomposition fragment distribution. It is also important to note that the bonding of PFPE thin films under illumination of 185 nm UV light takes place through interaction of the PFPE molecule, with a low-energy photoelectron created by excitation of the substrate by the UV photons [100]. The activated molecule in turn undergoes dissociative electron attachment, resulting in the formation of a radical and a negative ion. All the above information shows the existence of specific physical and chemical phenomena that are responsible for the initiation process of tribochemical reactions that differ from thermochemical ones.

In summary, the above discussed major specific issues related to tribochemistry, alongside with other factors (e.g., chemical hardness), plus the most recently discovered tribomicroplasma [25] evidently demonstrate the importance of a typical entanglement of physical and chemical tribological phenomena for a better understanding of tribochemistry. Considering now only the triboplasma phenomenon, it is necessary to clearly emphasize that the tribomicroplasma is totally different from the well-known regular plasma. The latter is defined as ionized gas, containing approximately equal concentrations of positive ions and electrons, that is electrically conductive and relatively field-free [101]. It is also important to bear in mind that electrons having higher kinetic energies can produce positive ions along with low-energy electrons. On the other hand, it is easier to realize the "flash temperature" phenomenon in terms of the thermionic energy (a flash of thermal electrons) because a thermionic energy converter transforms heat into electricity by "evaporating electrons from a hot emitter and condensing them on a cooler collector." The converter is just a heat engine utilizing electron gas as the working fluid [101]. Therefore, it seems plausible to compare a tribological surface micro "hot spot" with the electron emitter.

Bearing in mind the extension of the HSAB principle and the frontier orbital concept to solid interactions as demonstrated in Ref. 102 and the NIRAM–HSAB concept application to both (a) tribochemistry of PFPE lubricants (see Secs. II.B.2 and III.B.2) and (b) tribochemistry of ceramics (Sec. IV), it is good to come back to the issue of low-energy electron attachment to perfluoropolyethers. Matsunuma and Hosoe [103] present semiempirical molecular orbital simulation on electron-induced degradation of PFPEs with different types of segment. The calculated diagram demonstrated that the anion radical states of these species had lower energy levels compared to the neutral states. The anion radical was supposed to decompose to an anion and a radical. These findings are in line with the NIRAM approach.

VI. CONCLUSIONS

This chapter presents a specific review of extremely fast-developing research on tribochemistry. Its specificity is in that it mostly proposes a new approach to the initiation of tribochemical reactions and thereby deals only with selected lubricant components and tribological solids. This review demonstrates the importance of a pathway combining the NIRAM approach with the HSAB concept in initiating tribochemical reactions. The most important conclusion is that the NIRAM approach is associated with both tribochemistry and heterogeneous catalysis. Other typical tribochemical processes are presented and discussed in Chap. 11 ("Tribochemistry of Boundary Lubrication Processes").

The present author believes that there is already a good deal of evidence demonstrating that low-energy electrons are involved in the initiation process of tribochemical reactions. To emphasize the importance of the hypothesis that the common denominator of tribochemical reactions is that they are mostly initiated by low-energy electrons enhanced by ultraviolet photons, it seems adequate to add some features related to photochemical reactions. According to Ref. 105: (a) they proceed through repulsive excited states, thus effectively competing with quenching by the substrate; (b) photodissociation thresholds are red-shifted from the gas values by nearly 1 eV; and (c) both dissociation and desorption channels are observed with some products remaining chemisorbed on the surface; results of that work provide strong evidence that photochemistry on metal surfaces involves substrate excitation. The likely mechanism is the production of hot electrons, which then propagate through the lattice and tunnel into the affinity level of the absorbate *inducing dissociative electron capture* [105].

Characteristic IR common features of hexadecane tribochemistry are reflected in very specific absorption bands of wear products appearing at around 1550, 1650, and 3300 cm^{-1} for different metals. Similar IR spectra relate also to tribochemical products of alcohols and carboxylic acids.

It seems that new findings, for instance, related to the microtriboplasma provide a good deal of additional evidence for the importance of NIRAM approach in initiating tribochemical reactions.

More evidence for the significance and quite broad application of the NIRAM concept can be found in Chap. 7 ("Tribopolymerization as a Mechanism of Boundary Lubrication") and Chap. 8 ("Hydrolysis"). The latter demonstrates that aliphatic esters under boundary lubrication conditions do not hydrolyze and their tribochemistry is controlled by the NIRAM approach. However, one of the most recent detailed research described in Ref. 104 concludes that for ester systems, using time-of-flight secondary mass spectrometry (ToF-SIMS) analysis of model compounds, carboxylic anions and Fe-alcoholate cations were detected as reaction products between the *hydroxylates* of esters and the ferrous material. In the present author's view, exoelectron interaction with an ester is responsible for producing RCOO$^-$ and RO$^-$ negative ions. Figure 27 in Chap. 8 shows the whole process and indicates two types of C–O bond dissociation. The first bond cleavage type produces carboxylate anion and the second generates the alkoxide anion. The exoelectron dissociative attachment to an ester molecule yielding two types of negative ions is distinctly evidenced by the electron attachment mass spectrographic results [18]. Both anions interact with positively charged surface sites to form the iron carboxylate and iron alkoxide, respectively.

The limited space of this chapter excluded a possibility to describe in detail major issues of triboemission and related phenomena/processes that are also of importance for tribochemistry. Such information can be found in another recent review work [17].

REFERENCES

1. Shchukin, E.D. Interactions between adsorbed species and strained crystals. In *Atomistic of Fracture*; Latanisation, R.M., Pickens, J.R., Eds.; Plenum Press: New York, 1983; 421–423.

2. Kajdas, C. Tribochemistry. In *Tribology 2001: Scientific Achievements, Industrial Applications, Future Challenges*; Franek, F., Bartz, W.J., Pauschitz, A., Eds.; Austrian Tribology Society: Vienna, 2001; 39–46.
3. Kajdas, C. Importance of anionic reactive intermediates for lubricant component reactions with friction surfaces. Lubr. Sci. 1994, *6*, 203–228.
4. Nakayama, K. Tribophysical phenomena and tribochemical reaction. Jpn. J. Tribol. 1997, *42*, 1077–1084.
5. Heinicke, G. *Tribochemistry*; Akademie-Verlag: Berlin, 1984; 97–180.
6. Fox, P.G. Review. Mechanically initiated reactions in solids. J. Mater. Sci. 1975, *10*, 340–360.
7. Mori, S. Surface reactions during boundary lubrication under ultra-high vacuum conditions. Jpn. J. Tribol. 1991, *36*, 223–232.
8. Sato, N.; Seo, M. Chemically stimulated exoelectron emission from silver catalyst during partial oxidation of ethylene. J. Catal. 1972, *24*, 224–232.
9. Kajdas, C. Industrial lubricants. In *Chemistry and Technology of Lubricants*; Mortier, R.M., Orszulik, S.T., Eds.; Blackie Academic and Professional: London, 1997; 228–263.
10. Kajdas, C.; Majzner, M. Boundary lubrication of low-sulphur diesel fuel in the presence of fatty acids. Lubr. Sci. 2001, *14*, 83–108.
11. Dowson, D. *History of Tribology*; Longman: London, 1979.
12. Hardy, W.B. *Collected Works*; Cambridge University Press: Cambridge, 1936.
13. Cambell, W.F. Boundary lubrication. In *Boundary Lubrication: An Appraisal of World Literature*; Ling, F.F., Klaus, E.E., Fein, R.S., Eds.; ASME: New York, 1969; 87–117.
14. Bovington, C.H. Friction, wear and the role of additives in their control. In *Chemistry and Technology of Lubricants*; Mortier, R.M., Orszulik, S.T. Blackie Academic and Professional: London, 1997; 320–348.
15. Hsu, S.M.; Gates, R.S. Boundary lubrication and boundary lubricating films. *Modern Tribology Handbook*; Bhushan, B., Ed.; CRS Press: Boca Raton, 2001; Vol. 1, 455–492 pp.
16. Suh, N.P. *Tribophysics*; Prentice-Hall: Englewood Cliffs, NJ, 1986.
17. Kajdas, C.; Furey, M.J.; Ritter, A.L.; Molina, G.J. Triboemission as a basic part of the boundary lubrication regime. A review. Lubr. Sci. 2002, *14*,223–254.
18. von Ardenne, M.; Steinfelder, K.; Tuemmler, R. *Elektronenanlagerungs-Massenspektrographie Organischer Substanzen*; Springer-Verlag: Berlin, 1971; 235–237.
19. Molina, G.J.; Furey, M.J.; Ritter, A.L.; Kajdas, C. Triboemission from alumina, single crystal sapphire, and aluminum. Wear 2001, *249*, 214–219.
20. Kajdas, C. On a negative-ion concept of EP action of organo-sulfur compounds. ASLE Trans. 1983, *28*, 21–30.
21. Kajdas, C. About a negative-ion concept of the antiwear and antiseizure action of hydrocarbons during friction. Wear 1985, *101*, 1–12.
22. Kajdas, C. About an ionic-radical concept of the lubrication mechanism of alcohols. Wear 1987, *116*, 167–180.
23. Melton, C.E. Negative ion mass spectra. In *Mass Spectrometry of Organic Ions*; Mc Lafferty, F.W., Ed.; Academic Press: New York, 1963; 163–202.
24. Thiessen, P.A.; Meyer, K.; Heinicke, G. *Grundlagen der Tribochemie*; Akademie-Verlag: Berlin, 1966.
25. Nakayama, K.; Nevshupa, R.A. Plasma generation in a gap around a sliding contact. J. Phys., D. Appl. Phys. 2002, *35*, 1–4.
26. Nakayama, K. Triboemission from wearing solid surfaces. In *ITC Proceedings, Yokohama Satellite Forum on Tribochemistry;* Tokyo, October 28, 1995; 25–30.
27. Hu, L.; Chen, J.; Liu, W.; Xue, Q.; Kajdas, C. Investigation of tribochemical behavior of Al–Si alloy against itself lubricated by amines. Wear 2000, *242*, 60–67.
28. Pearson, R.G. Acids and bases. Science 1966, *151*, 172–177.
29. Kajdas, C. A novel approach to tribochemical reactions: generalized NIRAM–HSAB action mechanism. In ITC Proceedings, Yokohama Satellite Forum on Tribochemistry, Tokyo, October 28, 1995; 31–35.

30. Mori, S.; Suginoya, M.; Tamai, Y. Chemisorption of organic compounds on a clean aluminium surface prepared by cutting and high vacuum. ASLE Trans. 1982, *25*, 261–266.

31. Zhao, X.; Bhushan, B.; Kajdas, C. Lubrication studies of head–disk interfaces in a controlled environment: Part 2. Degradation mechanisms of perfluoropolyether lubricants. Proc. Inst. Mech. Eng. Part J 2000, *214*, 547–559.

32. Vurens, G.H.; Gudeman, C.S.; Lin, L.J.; Foster, J.S. Mechanism of ultraviolet and electron bonding of PFPEs. Langmuir 1992, *8*, 1165–1169.

33. Kajdas, C. Major tribochemical reaction pathways governing the nanotribochemistry processes. Proceedings of the ITC Conference, Nagasaki 2000, Satellite Forum Tribochemistry, Tsukuba; 57–62.

34. Kajdas, C.; Bhushan, B. Mechanism of interaction and degradation of PFPEs with a DLC coating in thin-film magnetic rigid disks: a critical review. J. Info. Storage Proc. Syst. 1999, *1*, 303–320.

35. Kajdas, C.; Makowska, M.; Gradkowski, M. Tribochemistry of hexadecane under boundary lubrication conditions. Proceedings of the Nordtrib 2000 Conference, Porvoo, Finland. VTT: Espoo, 2000; Vol. 2, 594–603.

36. Kajdas, C.; Al-Nozili, M. Wear behaviour and tribochemistry of oxygenates in 1-methylnaphthalene under boundary lubrication of the steel-on-steel system. Tribologia 1999, *30*, 301–314.

37. Kajdas, C.; Obadi, M. Wear behaviour and tribochemical reactions of low molecular alcohols under boundary lubrication of the steel-on-aluminium system. Tribologia 1999, *30*, 273–287.

38. Kajdas, C. Negative ionisation of n-paraffins C_6–C_9, C_{20}, C_{22} and C_{24} in EA-Mass Spectrograph. Rocz. Chem. 1971, *45*, 1771–1775. *in Polish.*

39. Kajdas, C. Tribological aspects of the lubricated process. Tribologia 2000, *31*, 981–998.

40. Makowska, M.; Kajdas, C.; Gradkowski, M. Interactions of *n*-hexadecane with 52100 steel surface under friction conditions. Tribol. Lett. 2002, *13*, 65–70.

41. Kajdas, C.; Makowska, M.; Gradkowski, M. Influence of temperature on tribochemical reactions of hexadecane. Lubr. Sci. 2003, *15*, 329–340.

42. Bowden, F.P.; Tabor, D. *Friction and Lubrication of Solids*; Clarendon Press: Oxford, 1950.

43. Morrison, R.T.; Boyd, R.N. *Organic Chemistry*; Prentice-Hall: Englewood Cliffs, NJ, 1992.

44. Jahanmir, S. Chain length effects in boundary lubrication. Wear 1985, *102*, 331–349.

45. Montgomery, R.S. Chemical effects on wear in the lubrication of aluminium. Wear 1965, *8*, 289–302.

46. Montgomery, R.S. The effect of alcohols and ethers on the wear behaviour of aluminum. Wear 1965, *8*, 467–473.

47. Hironaka, S.; Sakurai, T. The effect of pentaerythritol partial ester on the wear of aluminium. Wear 1978, *50*, 105–114.

48. Gates, R.S.; Hsu, S.M. Silicon nitride boundary lubrication: lubrication mechanism of alcohols. Tribol. Trans. 1995, *38*, 645–653.

49. Kajdas, C.; Obadi, M. Influence of C_4–C_{18} alcohols on the wear process of a sliding steel-on-steel tribological system. Tribologia 1998, *29*, 367–380. *in Polish.*

50. Stinton, H.C.; Spikes, H.A.; Cameron, A. A study of friction polymer formation. ASLE Trans. 1982, *25*, 355–360.

51. Bonder, G.M.; Pardue, H.L. *Chemistry—An Experimental Science*; Wiley: New York, 1989.

52. Kajdas, C.; Przedlacki, M. Wear behaviour and degradation mechanism of some perfluorinated compounds. Pr. Nauk.-Politech. Radom, Poland 2000, *17*, 62–70. *in Polish.*

53. Przedlacki, M.; Kajdas, C.; Liu, W. On lubrication mechanism of fluorinated alcohols in steel/steel and steel/aluminium contacts. Tribologia. 2003, *34*, 123–143.

54. Wachal. The influence of chemical structures of additives on antiseizure and anti-wear properties of oils. Tech. Smarownicza Trybol. 1977, *8*, 161–165. *in Polish.*

55. Kajdas, C.; Luczkiewicz, J.; Ozimina, D.; Wawak-Pardyka, E. The influence of concentration of alcohols and acids on tribological properties of paraffin oil. Tribologia 1980, *11* (6), 22–24. *in Polish.*

56. Kajdas, C.; Majzner, M.; Konopka, M. The effect of fatty acids at low concentrations in

n-hexadecane on wear of steel in boundary lubrication. Tribologia 1997, *28*, 221–237. *in Polish.*

57. Kajdas, C.; Majzner, M. Influence of fatty acids solutions in hydrocarbons and petroleum fractions on anti-wear properties of the steel-on-steel system. Tribologia 1998, *29*, 285–317. *in Polish.*

58. Kajdas, C.; Majzner, M. Boundary lubrication of low-sulphur diesel fuel in the presence of fatty acids. Lubr. Sci. 2001, *14*, 83–108.

59. Bowden, F.P.; Moore, A.C. Physical and chemical adsorption of long chain compounds on metals. Research 1949, *2*, 585–586.

60. Menter, J.W.; Tabor, D. Orientation of fatty acid sop films on metal surfaces. Proc. R. Soc. A 1951, *204*, 514–524.

61. Eischens, R.P. Catalysis studies related to boundary lubrication. In *Boundary Lubrication: An Appraisal of World Literature*; Ling, F.F. Klaus, E.E., Fein, R.S., Eds.; ASME: New York, 1969; 61–86.

62. Goldblatt, I.L. Surface fatigue initiated by fatty acids. ASLE Trans. 1973, *16*, 150–159.

63. Hu, Z.S.; Hsu, S.M.; Wang, P.S. Tribochemical and thermochemical reactions of stearic acid on copper surface studied by infrared microspectroscopy. Tribol. Trans. 1992, *35*, 189–193.

64. Hu, Z.S.; Hsu, S.M.; Wang, P.S. Tribochemical reactions of stearic acid on copper surface studied by surface enhancement Raman spectroscopy. Tribol. Trans. 1992, *35*, 417–442.

65. Fischer, D.A.; Hu, Z.S.; Hsu, S.M. Tribochemical and thermochemical reactions of stearic acid on copper surfaces in air as measured by ultra-soft x-ray absorption spectroscopy. Tribol. Lett. 1997, *3*, 35–40.

66. Fischer, D.A.; Hu, Z.S.; Hsu, S.M. Molecular orientation and bonding of monolayer stearic acid on copper surface prepared in air. Tribol. Lett. 1997, *3*, 41–45.

67. Hsu, S.M.; Zhang, J.; Yin, Z.; Shen, M.C.; Zhan, X. Tribochemistry induced by nano-mechanical scratches. Proceedings of the International Tribology Conference, Nagasaki 2000, Satellite Forum Tribochemistry, Tsukuba. JST: Tokyo, 2000; 25–30.

68. Kajdas, C. On a negative-ion-radical lubrication mechanism of fatty acids and fatty alcohols. Zesz. Nauk. WSI. Radom. Pol. Mater. Chem. 1985, *7*, 157–194.

69. Majzner, M.; Kajdas, C. Tribochemical reactions of carboxylic acids. Tribologia. *in press.*

70. Peri, J.B. A model for surface of γ-alumina. J. Phys. Chem. 1965, *69*, 220–230.

71. Molenda, J.; Gradkowski, M.; Makowska, M.; Kajdas, C. Tribochemical characteristic of some vinyl-type compounds in aspect of anti-wear interaction. Tribologia 1998, *29*, 318–329.

72. Molenda, J.; Gradkowski, M.; Kajdas, C. Study of chemical nature of organic products forming during friction on steel surface lubricated by unsaturated compounds. Tribologia. 2003, *34*, 93–102.

73. Kajdas, C.; Molenda, J. Some aspects of the friction polymer formation. In *Tribologia Sprzyja (Wy)trwalosci. Wroclaw, Oficyna Wyd. Polit, Wroc.*; 20–31.

74. Kajdas, C.; Molenda, J.; Makowska, M.; Gradkowski, M. Investigation of tribochemical behaviour of some unsaturated organic additives in steel–steel contact. Proceedings of the Symposium on Lubricating Materials and Tribochemistry; CAS, 1998; 83–94.

75. Molenda, J.; Kajdas, C.; Makowska, M.; Gradkowski, M. Unsaturated oxygen compounds as antiwear additives for lubricants. Tribologia: Lanzhou, China 1999, *3*, 323–331.

76. Kajdas, C. Tribochemistry of ceramics. Tribologia 1998, *29*, 148–168.

77. Kajdas, C. A novel approach to tribochemical reactions: Generalized NIRAM–HSAB action mechanism. Proceedings of ITC, Yokohama 1995, Satellite Forum on Tribochemistry, Tokyo, October 28, 1995; 31–35.

78. Smith, J.C.; Furey, M.J.; Kajdas, C. An exploratory study of vapor-phase lubrication of ceramics by monomers. Wear 1995, *181–183*, 581–593.

79. Kato, K. Water lubrication of ceramics. In *Tribology 2001: Scientific Achievements, Industrial Applications, Future Challenges*; Franek, F., Bartz, W.J., Pauschitz, A., Eds.; Austrian Tribology Society: Vienna, 2001; 51–58.

80. Fischer, T.E.; Tomizawa, H. Interaction of tribochemistry and microstructure in the friction and wear of silicon nitride. Wear 1985, *105*, 29–45.

81. Hibi, Y.; Enomoto, Y. Tribochemical wear of silicon nitride in water, alcohols and their mixture. Wear 1989, *13*, 133–245.

82. Ishigaki, H.; Kawaguchi, I.; Iwasa, M.; Toibana, Y. Friction and wear of hot pressed silicon nitride, and other ceramics. In *Wear of Materials*; Ludema, K.C., Ed.; ASME: New York, 1985; 13–21.

83. Sugita, T.; Ueda, K.; Kanemura, Y. Material removal mechanism of silicon nitride during rubbing in water. Wear 1984, *97*, 1–8.

84. Tomizawa, H.; Fischer, T.E. Friction and wear of silicon nitride and silicon carbide in water: hydrodynamic lubrication at low sliding speed obtained by tribochemical wear. ASLE Trans. 1987, *30*, 41–46.

85. Hibi, Y.; Enomoto, Y. *Tribochemistry of alcohol/silicon-based ceramics*, Proceedings of the JAST Tribology Conference, Marioka, October; 353–356.

86. Gates, R.S.; Hsu, S.M. Silicon nitride boundary lubrication: effect of oxygenates. Tribol. Trans. 1995, *38*, 607–617.

87. Gates, R.S.; Hsu, S.M. Boundary Lubrication of Silicon Nitride. NIST Special Publication 876; U.S. Government Printing Office: Washington, DC, 1995.

88. Gates, R.S.; Hsu, S.M.; Klaus, E.E. Ceramic Tribology: Methodology and Mechanisms of Alumina Wear. NIST Special Publication 758; U.S. Government Printing Office: Washington, DC, 1988.

89. Kato, K. Friction and wear of carbon nitride coatings. Keynotes and Abstracts of Nordtrib; Royal Institute of Technology: Stockholm, 2002; 35–46.

90. Glaefeke, H. Exoemission. In *Thermally stimulated relaxation in solids*; Brauelich, P., Ed.; Topics in Applied Physics, Springer-Verlag: Berlin, 1979; 225–273.

91. Krylova, I.V. Usp. Biol. Him. 1976, *45*, 2138. According to Ref. 90.

92. Hiernaut, J.P.; Forier, R.P.; van Lakengerghe, J. Influence of oxygen on electron-trapping by the surfaces of metal oxides. Vacuum 1972, *22*, 471–473.

93. Furey, M.J. Surface temperature in sliding contact. ASLE Trans. 1964, *7*, 133–146.

94. Vick, B.; Furey, M.J.; Kajdas, C. An examination of thermionic emission due to frictionally generated temperatures. Wear 2002, *13*, 147–153.

95. Vick, B.; Furey, M.J. A basic theoretical study of the temperature rise in sliding contact with multiple contacts. Nordtrib 2000 Proceedings, VTT: Porvoo, Finland, 2000; Vol. 2, 389–398.

96. Vick, B.; Furey, M.J.; Fao, S.J. Thermal analysis of sliding contact using a boundary integral equation method. Numer. Heat Transf. A. Appl. 1991, *20*, 19–40.

97. Weidele, H.; Kreisle, D.; Recknagel, E.; Schulze Icking-Konert, G.; Handschuh, H.; Gabtefoer, G.; Eberhardt, W. Thermionic emission from small clusters: direct observation of the kinetic energy distribution of the electrons. Chem. Phys. Lett. 1995, *237*, 425–431.

98. Zhao, X.; Bhushan, B. Studies on degradation mechanisms of lubricants for magnetic thin-film disks. Proc. Inst. Mech. Eng. Part J 2001, *215*, 173–188.

99. Vurens, G.; Zehringer, R.; Saperstein, D. The decomposition mechanism in perfluoropoly-ether lubricants during wear. In *Surface Science Investigation in Tribology*; Chung, Y.W. Homola, A.M., Street, G.B., Eds.; ACS Symposium Series, Vol. 485, 169–180.

100. Vurens, G.H.; Gudeman, C.S.; Lin, L.J.; Foster, J.S. The mechanism of ultraviolet bonding of perfluoropolyether lubricants following far-UV irradiation. Langmuir 1990, *6*, 1522–1524.

101. Huffman, F. Thermionic energy conversion. Encyclopedia of Physical Science and Technology. Academic Press: New York, 1987, Vol. 16, 621–633.

102. Lee, L.H. Physicochemical interactions between metals and polymers: HSAB principle for surface tribointeraction. In *ITC Proceedings, Nagoya*; 1990; 1153–1158.

103. Matsunuma, S.; Hosoe, Y. Molecular orbital simulations on electron-induced degradation of perfluoropolyethers with several types of segments. Tribol. Int. 1997, *30*, 121–128.

104. Murase, A.; Ohmori, T. ToF-SIMS analysis of model compounds of friction modifier adsorbed onto friction surface of ferrous materials. Surf. Interface Anal. 2001, *31*, 191–199.

105. Hatch, S.R.; Zhou, X.Y.; White, J.M.; Campion, A. Surface photochemistry 15: on the role of substrate excitation. J. Chem. Phys. 1990, *92*, 2681–2682.

7
Tribopolymerization as a Mechanism of Boundary Lubrication

Michael J. Furey
Virginia Polytechnic Institute and State University, Blacksburg, Virginia, U.S.A.

Czeslaw Kajdas
Warsaw University of Technology, Płock, Poland

I. INTRODUCTION

Two major classes of compounds have been developed from the concept of tribopolymerization proposed by Furey and Kajdas. The first, condensation-type monomers, formed the core of Furey's early work. It was suggested that high surface temperatures generated in tribological contact initiated the "in situ" polymerization of adsorbed monomers [1]. The second class, addition-type monomers, was also investigated, spurred on by the ideas proposed by Kajdas, who postulated that the polymerization of these compounds (e.g., vinyl-type monomers) was initiated by the emission of low-energy electrons from the surfaces in contact [2]. Experimental evidence supporting the action of each monomer class was obtained. We believe that we now have a better understanding of the mechanism(s) of tribopolymerization in reducing wear. However, several questions concerning detailed action remain. It is a complex and challenging problem of tribochemistry.

Limited space does not permit a detailed discussion of the factors believed to be important in the initiation of surface polymerization of condensation and addition-type monomers. However, appropriate references will be given for those seeking further information.

A. What Is Tribopolymerization?

Tribology—derived from the Greek "τρίβω" (to rub)—is the study of friction, wear, and lubrication [3]. By tribopolymerization, we mean the *planned, intentional, and continuous* formation of protective polymeric films directly on tribological surfaces by the use of minor concentrations of selected monomers capable of forming polymer films "in situ" either by polycondensation or addition polymerization [4,5]. The approach involves the *design* of molecules which will form thin, deposited polymeric surface films in critical regions of lubrication—thus reducing contact and wear.

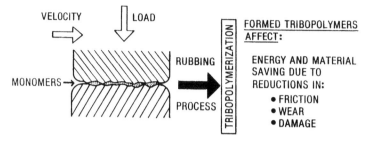

Figure 1 Tribopolymerization as a mechanism of boundary lubrication.

An oversimplified view of the process is shown in Fig. 1. Although the detailed mechanisms of surface film formation are not yet fully understood, the central and unique feature of tribopolymerization is that it is primarily a process of controlled *deposition on* a solid surface rather than *reaction with* the surface (e.g., as is the case with most conventional antiwear additives). Thus, the mechanism should work with a variety of solids, including metals, various alloys, ceramics, and composites. It is believed that tribopolymerization can be initiated by high surface temperatures generated by friction, high contact pressures, and exoelectron emission from rubbing surfaces. The protective polymeric films formed are extremely thin and generally invisible. The films formed reduce adhesion and wear in a continuous formation/removal/replenishment process. It is important to note that we do not coat mechanical elements with polymers prior to rubbing contact; the coatings would wear off and would no longer be effective. Neither do we add polymers to a fluid carrier. By the concept of tribopolymerization, we need to use the embryonic monomeric molecules in a carrier fluid to form the protective polymeric surface films continuously only where needed—in critical regions of boundary lubrication—and not on all parts of an engine or system. It is a technique of molecular design in which three main aspects are important: (1) the structure and surface orientation of monomeric molecules prior to polymerization, (2) the process of tribopolymerization, and (3) the properties of the films formed (e.g., adhesion, thermal and mechanical stability, durability).

II. THE ACTION OF CONDENSATION TYPE MONOMERS

A. Initial Research on "In Situ" Surface Polymerization as a Mechanism of Boundary Lubrication

Early work by Furey while at the Esso Research and Engineering Company, carried out 30 years ago, clearly demonstrated that a mechanism of "in situ" surface polymerization led to dramatic increases in load-carrying/antiscuff properties of fuels as well as wear reductions with lubricants. Results of these initial studies were presented at the Conference on Physico-Chemical Mechanics of Friction and Wear in Kiev, USSR, and published in 1973 in the article in Wear—"The Formation of Polymeric Films Directly in Rubbing Surfaces to Reduce Wear" [6]. A summary of this early work was included and presented at the 2nd International Symposium on Tribochemistry in Janowice, Poland in 1997, and published in the proceedings [1]. Therefore, it will not be repeated here. However, some of the more important findings relating to proof of the surface polymerization concept will be restated in this section.

In an initial test of the concept of surface polymerization, an equimolar mixture of C_{36} dimer acid and a specially synthesized C_{16} glycol was used in jet fuel to test the possible reaction below.

Rubbing
surfaces

$$HOOC—R—COOH + HO—R'—OH \longrightarrow HO—[OC—R—COOR'—O—]_nH + n\,H_2O \uparrow$$

diacid $\qquad\qquad$ glycol $\qquad\qquad$ polymer film

The results, summarized in Table 1, supported the idea—showing a dramatic increase in Ryder Gear Scuff Rating. Dimer acid alone had little effect.

In another approach, the primary compounds synthesized in this early work were of the A–R–B type derived from C_{36} dimer acids—notably the C_{36} dimer acid/glycol mono-esters; in particular, the ethylene glycol derivative. The postulated action is shown below.

Rubbing
surfaces

$$n\,HOOC—R—COO—C_2H_4OH \longrightarrow HO—[OC—R—COO—R'—O—]_nH + n\,H_2O \uparrow$$

monoester $\qquad\qquad$ polyester

The branched structure of the monoester can be seen in Fig. 2.

In Ryder Gear tests using various fuel-type carrier fluids, the addition of 0.1% of the dimer acid/ethylene glycol monoester resulted in dramatic increases in antiscuffing loads, as illustrated by the data in Table 2. Dimer acid alone was ineffective, thus showing that adsorption and/or metal soap formation by a long-chain organic acid is not responsible for the antiscuff action of the monoester [6].

As shown in Table 3, other glycol monoesters of the dimer acid were also effective in increasing the Ryder Gear Scuff Rating. However, the original ethylene glycol derivative is the best in terms of effectiveness and cost.

In ball-in-cylinder tests with AISI 52100 steel-on-steel, two versions of radioactively (C-14)-tagged methyl esters of the above compound—shown in Table 4 —were used as 1% solutions in a solvent 100 Neutral paraffinic mineral oil under severe conditions in which the additives caused significant decreases in metallic contact and wear.

Conclusive proof of film formation in the cylinder wear track was found for Case I, but not for Case II, thus supporting the idea of surface polymerization. An example of an autoradiograph obtained in Case I is shown in Fig. 3.

Table 1 Effect of C_{36} Dimer Acid/C_{16} Glycol Mixture on Ryder Gear Performance of a Jet Fuel

Additive in jet fuel	Ryder gear scuff rating (N/cm)
None	ca. 700
0.1% Equimolar mixture of C_{36} dimer acid and C_{16} glycol	2500
0.1% C_{36} dimer acid	840

Figure 2 Structure of the dimer acid/ethylene glycol monoester.

The effectiveness of the C_{36} dimer acid/ethylene glycol monoester in reducing valve train wear was also demonstrated in engine tests using radioactive valve lifters. Table 5 shows that the compound reduced valve lifter wear by about 90% and was at least as effective as the additive zinc dialkyldithiophosphate.

It is important that the embryonic, shorter-chain monomeric units be used rather than oligomers, as shown by the data in Table 6. Increasing the prepolymerization of the monomer in glassware prior to engine testing significantly reduces the antiwear effectiveness. This is consistent with the behavior of high-molecular-weight polymers (VI improvers), which show little benefit on valve lifter wear under the high shear rates existing between the cam and lifter.

Table 2 Effect of C_{36} Dimer Acid/Ethylene Glycol Monoester on the Antiscuff Properties of Fuels

	Ryder gear scuff rating (N/cm)	
Base fluid	Base fluid	Base + 0.1% monoester
Jet Fuel	700	2600
Turbo Fuel	900	3700
JP-4	350	5200
Xylene	900	6100

Table 3 Effect of Glycol Type on the Antiscuff Properties of C_{36} Dimer Acid Monoesters

Additive in isoparaffinic jet fuel[a]	Ryder gear scuff rating (N/cm)
None	700
C_{36} Dimer Acid Monoester of	
Ethylene Glycol	2600
Triethylene Glycol	1900
Neopentyl Glycol	1400
1,6-Hexane Diol	3200

[a] At 0.1% concentration.

Other valve train wear engine tests using (A–R–A)/(B–R′–B) condensation–monomer combinations demonstrated significant wear reduction as well as beneficial carry-over effects—evidence of the formation of durable surface films.

Selected results of more recent research with the monoester at Virginia Tech—in collaboration with Polish researchers Kajdas and Kempinski with graduate students—will be summarized in the next section.

Table 4 Radioactively Tagged Monoesters

CASE I: ETHYLENE GLYCOL PORTION LABELED

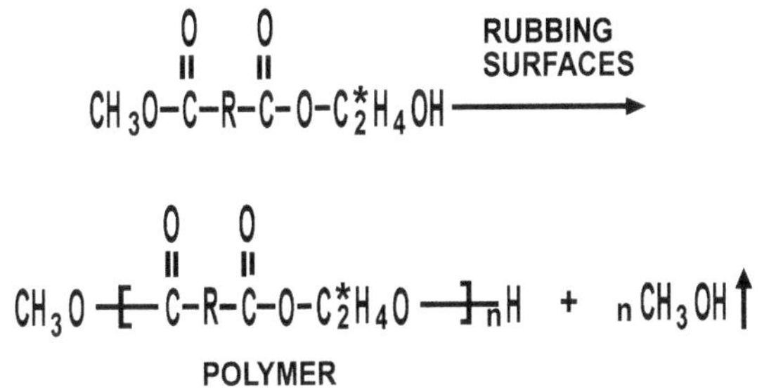

$$CH_3O-\overset{\overset{O}{\|}}{C}-R-\overset{\overset{O}{\|}}{C}-O-C_2^*H_4OH \xrightarrow{\text{RUBBING SURFACES}}$$

$$CH_3O-\left[\overset{\overset{O}{\|}}{C}-R-\overset{\overset{O}{\|}}{C}-O-C_2^*H_4O\right]_n H \ + \ n\,CH_3OH\uparrow$$

POLYMER

CASE II: METHANOL PORTION LABELED

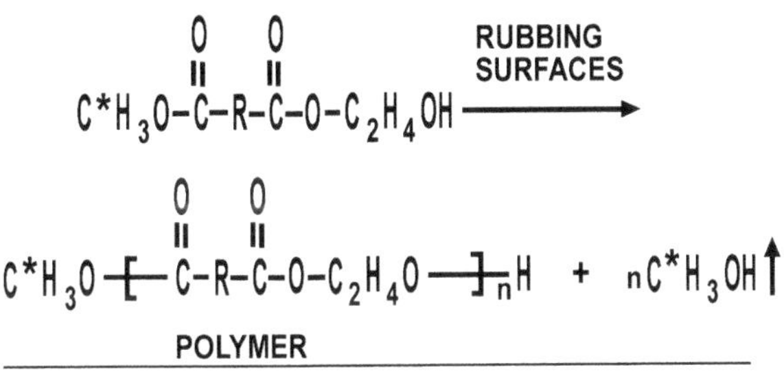

$$C^*H_3O-\overset{\overset{O}{\|}}{C}-R-\overset{\overset{O}{\|}}{C}-O-C_2H_4OH \xrightarrow{\text{RUBBING SURFACES}}$$

$$C^*H_3O-\left[\overset{\overset{O}{\|}}{C}-R-\overset{\overset{O}{\|}}{C}-O-C_2H_4O\right]_n H \ + \ nC^*H_3OH\uparrow$$

POLYMER

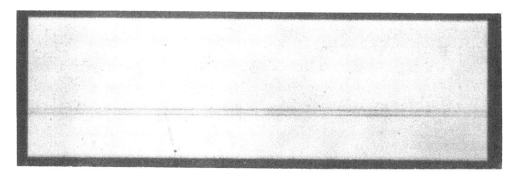

Figure 3 Autoradiograph of cylinder surface lubricated with C-14 labeled monoester (Case I).

B. Continued Research on Dimer Acid/Glycol Monoesters

Continued research on the mechanism of antiwear action of C_{36} dimer acid/ethylene glycol monoesters was carried out during the past several years in the Tribology Laboratory at Virginia Tech. Results were presented at the 1997 Symposium on Tribochemistry at Janowice, Poland [1], as well as in various conferences and papers [5,7–11]. Thus, this information will not be repeated here even as review. However, a few key findings will be presented, focusing on what we believe to be the mechanism of action of these compounds. Our research has shown that the dimer acid/ethylene glycol monoester is effective in reducing wear in pin-on-disk tests with steel-on-steel as well as ceramic-on-ceramic systems (e.g., alumina, zirconia).

An example of the effect of the monoester on wear with steel is shown in Fig. 4. In this study, highly colored interference films were observed in wear tracks in which the monoester was used as an additive at 1 wt.% concentration in hexadecane.

Fourier-transform infrared (FTIR) surface analysis was used to study film formation in the wear tracks. An example of FTIR spectra for various compounds as well as the wear track is shown in Fig. 5.

We have found that, in both steel and ceramic sliding contact, there is strong evidence of a surface reaction to form a metallic soap as well as chain growth (oligomer/polymer formation).

Evidence of material formed from the monoester in the region of contact in an Al_2O_3-on-Al_2O_3 (alumina) pin-on-disk test is shown by the scanning electron micrograph in Fig. 6.

The original mechanism proposed—physical adsorption followed by surface polymerization—is apparently incorrect, at least based on the conditions existing in the pin-on-disk experiments. Therefore, instead of Model I (as shown in Fig. 7), we suggest that a surface reaction occurs to form a soap (e.g., an Al soap in the case of the ceramic alumina), which is

Table 5 Effect of the C_{36} Dimer Acid/Ethylene Glycol Monoester on Valve Train Wear

Compound in paraffinic mineral oil	Relative valve lifter wear rate[a]
None	100
1% c_{36} Dimer acid/Ethylene glycol monoester	8
1% Zinc di(C_6) alkyl dithiophosphate	12

[a] 28-hr valve train wear tests in V-8 engine equipped with radioactive valve lifters.

Table 6 Effects of Degree of Pre-polymerization on the Antiwear Properties of C_{36} Dimer Acid/Ethylene Glycol Esters

Additive in paraffinic mineral oil	Actual molecular weight (and theoretical)	Relative valve lifter wear rate[a]
None	—	100
1% Monoester	716 (609)	8
1% Diester	1264 (1200)	12
1% Tetraester	1975 (2364)	85

[a] 28-hr valve train wear tests in V-8 engine equipped with radioactive valve lifters.

then followed by chain growth illustrated by Model II. This could lead to a strongly bonded higher molecular oligomeric or polymeric surface film—a mechanism that could be much more effective in reducing wear.

C. Tribological Behavior of Condensation-Type Monomers Under More Severe Conditions

It has been demonstrated that the principle of tribopolymerization can be used as a model for the design and selection of specific molecular structures effective in reducing wear of ceramics in both liquid- and vapor-phase lubrication [12,13].

In an experimental study by Furey et al. [14] to extend the range of frictional energy generation in this research, effects of load and speed on ceramic wear were determined for each of six selected monomers in pin-on-disk tests using alumina-on-alumina. This was carried out in a two-factor, two-level designed experiment for each of six monomers at

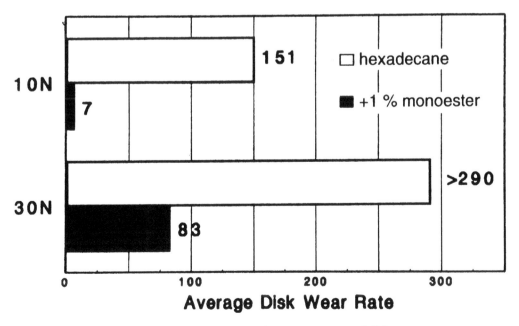

Figure 4 Effect of the dimer acid/ethylene glycol monoester on steel disk wear.

1% concentration in the carrier fluid, hexadecane. The six monomers chosen consisted of one condensation type—a dimer acid/glycol monoester—and five addition-type, chiefly vinyl, compounds. The loads and speeds were each varied by a factor of four, thus resulting in a 16-fold variation in frictional energy. Although the focus of the present section is on polycondensation, the results obtained in the designed experiments are included because of the similarity in the behavior of condensation- and addition-type monomers at higher speeds.

Statistically significant main effects and interactions were found among the three variables—applied load, velocity, and monomer structure. For example, at the low sliding velocity (0.25 m/sec), all monomers were very effective in reducing alumina wear, with reductions ranging from 40% to 98%; five of the six monomers selected significantly reduced wear at both loads (40 and 160 N). The most surprising and significant first-order interaction involved the effect of velocity. Monomers effective in reducing alumina wear at the lower speed had no effect at the higher speed (1.0 m/sec); in fact, in some cases, monomers actually increased wear at the higher sliding velocity.

Fourier-transform infrared microscopy (FTIRM) examination of worn alumina surfaces provided useful information on tribochemical reactions in the contact region.

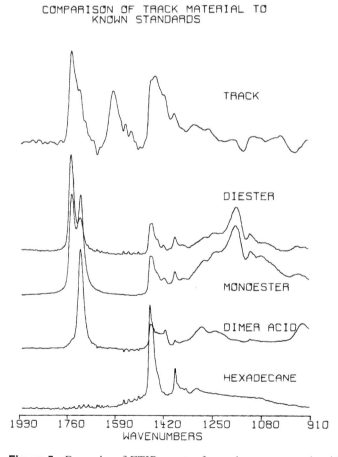

Figure 5 Examples of FTIR spectra for various compounds within wear tracks on steel disks.

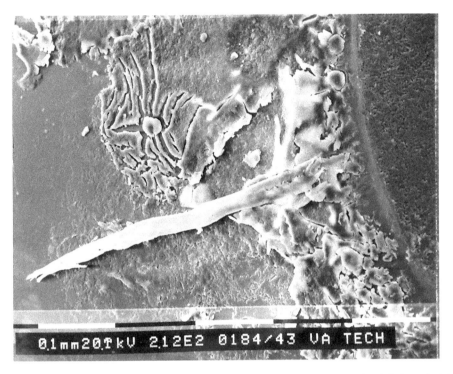

Figure 6 Scanning electron micrograph of an alumina ball contact region lubricated with monoester in hexadecane.

There was evidence of tribopolymerization on the ceramic wear regions, but not in all cases of antiwear action. Other tribochemical reactions also occur.

It was suggested that the loss of antiwear action of the compounds selected at higher velocities is a result of the higher surface temperatures generated by friction—thus leading to possible thermal degradation of surface polymers formed in the contact region.

In an attempt to model and understand the behavior of monomers used in our research on ceramic lubrication, a second phase of this study was conducted in three areas: (1) the use of a chemical modeling computer program to examine the possible orientation of monomers on ceramic surfaces prior to tribochemical reactions with and on the surfaces, e.g., as in Ref. 15; (2) the calculation of surface temperatures produced by friction using a more sophisticated approach developed by Vick; and (3) the testing of hypotheses designed to explain the results of the first phase of this study. Indeed, there appears to be a relationship between wear volume and calculated surface temperatures assuming contact area governed by either elastic or plastic deformation. There is an effective intermediate temperature range for the monomers; above this (e.g., above ca 400–500°C), the compounds used in this first study do not reduce ceramic wear.

To test this hypothesis, more recent studies involving new classes of additives based on tribopolymerization but designed to function at higher surface temperatures were carried out in pin-on-disk tests, using low concentrations of compounds in a hydrocarbon carrier, hexadecane. Previous work involved aliphatic straight or branched-chain monomers of two main types—condensation monomers and vinyl-type addition monomers. In the new studies, many of the compounds investigated contain aromatic rings. The reasons

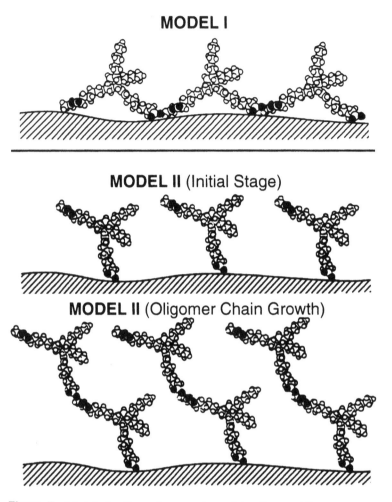

MODEL I

MODEL II (Initial Stage)

MODEL II (Oligomer Chain Growth)

Figure 7 Models for the antiwear action of the dimer acid/ethylene glycol monoester.

are twofold: (1) to use monomers that will polymerize at higher surface temperatures and (2) to produce polymeric films with greater thermal stability. Suggested effects of surface temperature on tribopolymerization with condensation-type monomers are illustrated in Fig. 8.

As an example, Fig. 9 summarizes the effects of two aromatic hydroxy acids on wear in the steel-on-steel system. The addition of 1% of 4-hydroxycinnamic acid to hexadecane reduced wear by an average of 96%. Significant dark and adherent deposits formed in the wear tracks from the hydroxy acids tested.

As another example, Fig. 10 shows the effects of two aromatic amino acids on wear with steel under the same conditions. The compound 4-aminophenylacetic acid is particularly effective—reducing wear by 93%.

These and other condensation-type monomers of the structure A–R–B, as well as combinations of A–R–A and B–R′–B, have been evaluated under more severe conditions in the pin-on-disk machine and found to be effective in reducing wear. One special class of condensation-type monomers, i.e., lactams, will be discussed briefly in the next section.

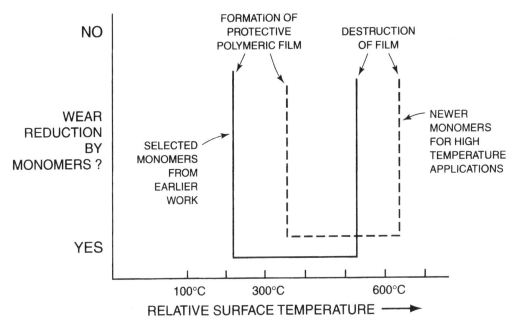

Figure 8 Possible relationship between surface temperature and antiwear effectiveness of selected condensation-type monomers.

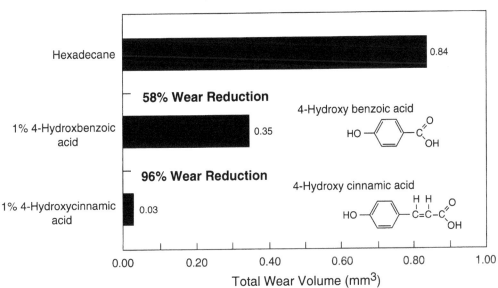

Figure 9 Effects of two aromatic hydroxy acids on wear with steel-on-steel.

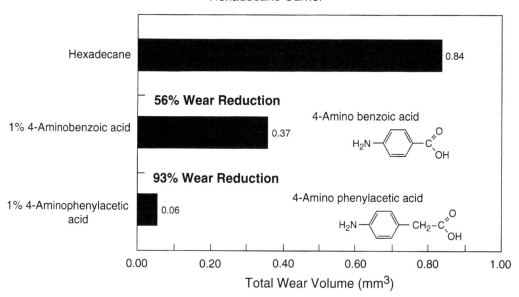

Figure 10 Effects of two aromatic amino acids on wear with steel-on-steel.

The results of this study show that, indeed, it is possible to develop antiwear additives capable of reducing ceramic or steel wear by as much as 99% under high load/high speed conditions and at low concentrations. The FTIRM examination of wear tracks shows complex tribochemistry and evidence of oligomer/polymer formation. Unusual debris found in and near the contact region may also play a role in antiwear action. More work is needed on the identification and characterization of the surface films formed with these compounds. This information is used to develop new models of ceramic lubrication by tribopolymerization—a technique shown to be effective for metals as well.

D. Lactams as Condensation-Type Monomers for Reducing Wear

Our research on lactams as antiwear additives for steel and ceramic systems—including lactam/monoester combinations for "no-oil" hot run tests with small engines during manufacture—has been described elsewhere [16–18]. It is not our intention to repeat those findings here, but only to summarize key results as part of the tribochemical behavior of condensation-type monomers.

The general structure of lactams and some examples are illustrated in Table 7.

Most of our work was carried out with caprolactam—the cheapest and most readily available lactam, which is used to synthesize Nylon-6. Pin-on-disk tests with caprolactam added to hexadecane show striking reductions in alumina wear, as seen by the data in Tables 8 and 9. Significant antiwear effects were observed even at concentrations of 200 ppm (0.02 wt.%) in hexadecane.

Table 7 Lactam Structure and Examples

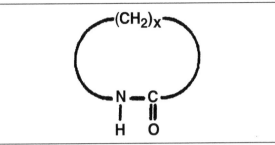

X	Compound name(s)
2	2-azetidinone
5	ε-caprolactam (2-oxohexamethyleneimine)
6	2-azacyclooctane
7	2-azacyclooctane
11	2-azacyclotridecanone (laurolactam)

The lactams are also effective as antiwear compounds in the steel-on-steel system, as illustrated by the data in Table 10. The addition of 1% caprolactam to hexadecane reduced average wear by 93%; the antiwear effect persists at much lower caprolactam concentrations.

To determine the effects of caprolactam on wear in the ceramic system alumina-on-alumina at two speeds (0.25 and 0.75 m/sec) and two loads (40 and 80 N), a designed experiment was carried out in the pin-on-disk machine. The results demonstrated that caprolactam is effective in reducing wear at all conditions.

In a separate study relating to the production of small engines—one that involved pretreatment with minor amounts of compounds rather than using a conventional charge of oil—it was found that combinations of two condensation-type monomers, caprolactam and the dimer acid/glycol monoester, were unusually effective in reducing wear, as illustrated by Fig. 11 [16,18].

In plant production tests, this and related compositions used to pretreat engine components (Table 11) were found to be extremely effective in hot-run, "no oil" tests as shown by the results summarized in Table 12.

Although the surface films formed by this combination in the steel-on-aluminum system used have not been chemically identified, it is obvious that some interesting and probably complex tribochemistry is involved in producing such dramatic antiwear effects. This is being studied further.

Table 8 Effect of Caprolactam on Alumina Wear. Alumina-on-Alumina: Structure Type A

Test fluid	Average total wear volume [mm³]	Average wear reduction
Hexadecane	27.10	—
Hexadecane + 1 wt.% Caprolactam	0.09	99.8%
Above caprolactam mixture filtered	0.10	99.6%

40 N load, 0.25 m/sec sliding velocity, 250 m total sliding distance.

Table 9 Effect of Caprolactam Concentration in Hexadecane on Alumina Wear. Alumina-on-Alumina: Structure Type B

Caprolactam concentration in hexadecane [wt.%]	Wear [mm^3]			Average wear reduction
	Ball	Disk	Total	
0.00	0.208	1.045	1.253	—
1.00	0.066	0.136	0.201	84%
0.10	0.095	0.142	0.237	81%
0.02 (200 ppm)	0.165	0.296	0.461	63%

40 N load, 0.25 m/sec sliding velocity, 250 m total sliding distance.

III. THE ACTION OF VINYL ADDITION-TYPE MONOMERS

A. Introduction

Earlier, we have demonstrated that some vinyl-type monomers are effective tribological additives in both the liquid-phase [19] and vapor-phase lubrication [13]. This section is a brief review of major experimental results of the research pertinent to the addition-type tribopolymerization. It concerns the wear and/or friction reduction of metals and ceramics by the use of specifically selected unsaturated monomers capable of polymerizing under boundary friction conditions. The formation of specific protective polymeric films on friction surfaces results in effective antiwear action of many monomers; however, their action mechanism is complex. Actually, the polymerization can be caused by many factors, e.g., high temperature, high pressure, and catalytic action of the freshly exposed friction solid surface. The latter process is very special because it involves a wide variety of physical phenomena. For example, the surface enlargement during the frictional energy input clearly creates new surfaces along with different types of emission, such as electrons, photons, ions, etc. Most recently, the importance of these phenomena for the boundary friction regime was discussed in a review work [20]. It is important to note at this point that, for more than the past 10 years, the authors of this chapter demonstrated the importance of triboemitted electrons (exoelectrons) in models of tribopolymerization antiwear mechanisms [4]. The approach was based on the negative-ion-radical action mechanism (NIRAM) proposed in 1983 [21]. The NIRAM approach assumes that electrons of an average energy of 3 eV are attached to the molecules and negative ions are formed mostly according to resonance capture and dissociative resonance capture mechanisms. Molina et al. [22] proved that the major part of emitted exoelectrons is in the energy range 0–5 eV. Figure 12 [20] diagrammatically shows the known and unknown factors that relate to the complexity of the tribo-emission/tribopolymerization issue. Creation of fresh surfaces is well reflected in schematic

Table 10 Effect of Caprolactam on Wear with Steel. AISI52100 Steel Balls on 1045 Steel Disks

Additive in hexadecane	Wear [mm^3]			Average wear reduction
	Ball	Disk	Total	
None	0.014	0.821	0.835	—
1 wt.% Caprolactum	0.003	0.054	0.057	93%

40 N load, 0.25 m/sec sliding velocity, 250 m total sliding distance.

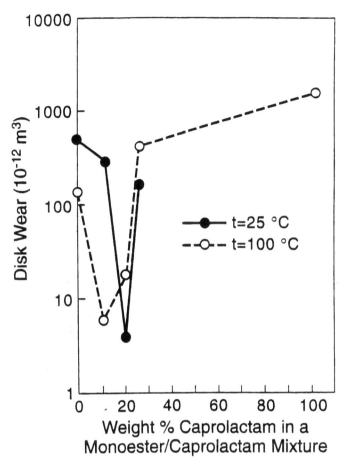

Figure 11 Effect of caprolactam/monoester mixtures on aluminum wear.

illustration of energy dissipation as presented in Ref. 23. Figure 13 depicts the most likely primary energy-absorbing mechanisms during abrasive wear and some resulting dissipation processes. The tribological literature gives much information concerning the possible polymerization process during friction of unsaturated monomers evolved either from base oil components, additives, or model compounds. Below is a review of papers that are most relevant to the tribopolymerization concept, as proposed by Furey [6].

Table 11 Small Engine Pre-treatment for "Hot-Run" Production Tests

---

4-Cycle, 5-HP Tecumseh Engines
Engine Parts Pretreated
- Piston rings
- Cylinder walls
- Main bearing assembly
- Connecting rod bearings
- Cam/lifter contact region

Total quantity of pre-treatment composition used for the entire engine: 5 g

---

Table 12 Results of In-Plant Engine Pretreatment Tests with 4-Cycle, 5-HP Tecumseh Engines

Plant (location)	Pre-treatment composition[a]	No. of engines	Results
Tecumseh (New Holstein, WI)	TC-420	16	All excellent; no failures
Tecumseh (New Holstein, WI)	TC-429	12[b]	9 excellent; 3 with C.R. damage
Tecumseh (Dunlap, TN)	TC-419	24	All excellent; no failures

[a] For critical connecting rod bearing assembly; variations used for other engine parts.
[b] Selected group of minimum clearance engines.

B. Brief Overview of Vinyl-Type Tribopolymerization-Related Experimental Results

Most of the early experiments that might be related to mechanically initiated addition-type polymerization were performed in the 1970s [24–28], but at that time, these works were not addressed to tribopolymerization as an antiwear action mechanism. Experiments involving polymerization of addition-type additives on freshly formed surfaces were investigated in Ref. 24. Both polymerization and grafting processes were studied by the interaction of various oxides, metals, and salts dispersed by vibration in a monomer solution. Campbell and Lee [25] studied polymer formation on sliding metals in the vapor phase. St. Pierre et al. [26,27] reported on the action of olefins as highly reactive toward aluminum under boundary

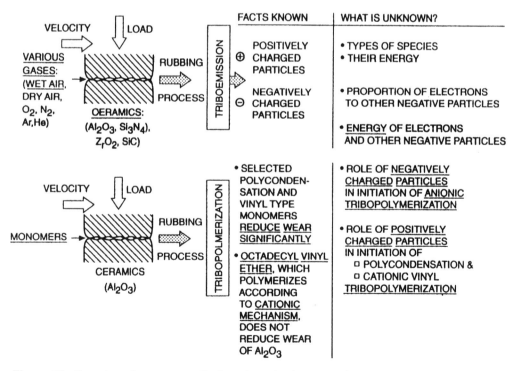

Figure 12 Overview of one aspect of tribopolymerization research at Virginia Tech.

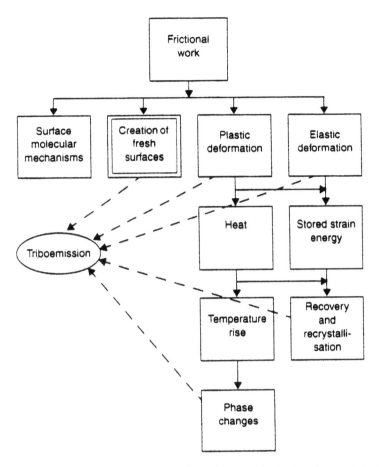

Figure 13 Schematic representation of energy dissipation in a tribological process.

lubrication. In an extension of this work, Owens and Barnes [28] suggested the use of unsaturated hydrocarbons as boundary lubricants for stainless steels. An earlier paper [29] studied the slow formation of polymer by contacting abraded aluminum with an aqueous solution of acrylonitrile. Buckley [30] found evidence for polymerization of vinyl chloride on an iron disk rubbed with an aluminum oxide slider. The polymer film reduced the friction but no mention was made about its effect on wear. Tamai [31] investigated the addition-type polymerization of several vinyl monomers on mechanically activated powders of silica, aluminum, and iron. He clearly demonstrated that methyl methacrylate (MMA) yields considerably more polymer when contacting with vibromilled aluminum power then with iron powder. The mechanochemical polymerization was measured by immersing the powders in MMA and obtaining their increase in weight. Smolin [32] found that styrene (ST) polymerized during the friction of aluminum-on-aluminum.

Tamai's work [31] precisely demonstrated the role of mechanochemical activity in the boundary lubrication process. Thus it may be treated as a bridge between generally accepted known polymerization mechanisms and the NIRAM based, i.e., mechanically activated, tribopolymerization mechanism. Further investigations on addition-type tribopolymeriza-tion, clearly designed toward the antiwear action mechanism, were presented and described in Refs. 33–35. In one of these studies, an interesting feature of vinyl-type tribopolymerization was found [34]. This is depicted in Fig. 14, in which wear scar diameter is plotted against load for steel-on-steel in a four-ball test. It is seen that the addition of 0.5% diallyl phthalate to a mineral base oil decreased the wear scar diameter by more than 50% within the load ranging from 500 to 650 N. At the lowest load and the highest load, the effect of diallyl phthalate on wear was minimal. Obviously, there is an optimum load within which this approach is effective in reducing wear. Several works [36–38] investigated the lubricating properties of olefins, e.g., 1-dodecene for aluminum [37] and other olefins for aluminum alloys and different metals [36]. Igari and Mori [38] discussed the lubrication of aluminum alloys with tribochemical reaction products from unsaturated compounds. Another work [39] tested different unsaturated compounds on a wide variety of metals.

C. Role of Exoelectrons in the Tribopolymerization Process

1. General Information on the NIRAM Approach

The term exoectron emission process [40] (Kramer effect) originates from investigation of the emission in freshly treated metals, which were accounted for as a consequence of exothermal transformation process of the surface. Presently, the triboemission process includes exoelectrons in the negative charge particles (Fig. 12). This is because friction leads not only to exoectron emission (EEE), but also to other particle emission including negatively charged particles and positively charged particles. Triboemission, along with triboemission-related terminology, can be found in Ref. 20. Work presented in Refs. 21, 41 and 42 has shown that exoelectrons form negative ions or negative-ion radicals in a wide

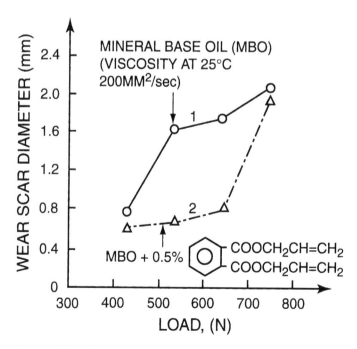

Figure 14   Effect of diallyl phthalate on wear in a four-ball test.

variety of compounds, which may then, as reactive intermediates, easily react with the positively charged surface of the friction solid yielding surface films. This forms the basis for the negative-ion-radical lubrication model. The model assumes creation of two types of activated sites on friction surfaces: thermally activated sites and sites activated by the EEE process [43]. The situation is illustrated in Fig. 15. The principal thesis of this model is that lubricant components form negative ions, which are then chemisorbed on the positively charged areas of rubbing metal surfaces. If shear strength is high, the chemical bonds may break, producing further radicals. Destruction of the generated boundary film leads to formation of new activated spots and the reaction cycle repeats.

2. Early Application of the NIRAM Approach to Vinyl-Type Tribopolymerization

Research described in Ref. 33 was focused on the tribopolymerization mechanism of unsaturated compounds in the steel-on-steel system. The combination of styrene (ST) and methyl methacrylate (MMA) as a model system for elucidation of the polymerization mechanism were selected. Composition of the ST–MMA copolymers depends on the chemical mechanism of the polymerization process. The differences in copolymer composition are caused by various influences of the substituents of carbon–carbon double bond on the ability of this bond to combine to a radical, cation, or anion. The phenyl group has stronger electron-donor properties than the ester group of the MMA molecule. The active center in cationic copolymerization willingly reacts with ST molecules. This fact manifests itself in an increase of the ST content in the copolymers. MMA much more easily combines with an anionic polymerization center because of smaller double bond electron density. In radical polymerization, ST and MMA form alternating copolymers (ca. 50% ST in copolymer). By comparing IR spectra of frictional ST–MMA copolymers with that of standard mixtures, it was found that the content of ST in the copolymers was about 50%. At that time, these data pointed to a pure radical mechanism of tribopolymerization [33].

The formation of free radicals is associated with a high temperature. For example, creation of free radicals from hydrocarbons usually takes place if the temperature is higher than 300°C. Evidently, the temperature for radical formation can be lowered because of the mechanochemical effects associated with the friction process. On the other hand, friction of solids causes low-energy electron emission. Thus these electrons may initiate the tribopolymerization at a lower temperature as well. Tamai [31] demonstrated that MMA yielded much more polymer when contacted with vibromilled aluminum powder than with an iron powder. The "mechanochemical polymerization" decreased with the electron emission decay. Without vibromilling, no polymerization took place. Exoelectron emis-

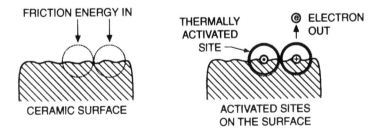

Figure 15 Surface activation during friction.

sion was probably a determinant factor here. Although the results described in Ref. 33 pointed to the free radical mechanism of tribopolymerization, it seemed reasonable to suggest another mechanism, i.e., the formation of radical–anion reactive intermediate $CH_2 = CHX + e \rightarrow [CH_2 = CHX]^{\bullet-}$ where X is a substitute, e.g., phenyl, and e is a low-energy electron. This approach connected with the NIRAM concept was discussed in Ref. 4. The addition of a monomer to the radical–anion gives a species that contains one radical end and one anion end. Such species can add monomer from the two ends by different mechanisms. However, two radical ends may dimerize, leaving a divalent anion to propagate.

The approach explains the results described in Ref. 33, and is coherent with Tamai's results [31] obtained at the same time, i.e., in 1980. The considered mechanism seems also to be consistent with other experimental results concerning the addition-type tribopolymerization presented in Ref. 44; however, tribochemical reactions should be taken into account as well. The work [45] investigating 4-allyl-2-methoxyphenol and 3-allyloxy-1,2-propandiol demonstrated that antiwear efficiency of these unsaturated and oxygen containing compounds is controlled by tribochemical reactions leading to formation of oxygen-rich deposits on the metal surface. Although these tribochemical reactions are very complex, the presence of carboxylate structures combined with the steel substrate provided a good evidence for the oxidation process of the tested compound. Thus it can be concluded that, in some unsaturated compounds, the double bond undergoes the oxidation reaction to carboxylate structure.

The developed NIRAM approach for the vinyl-type tribopolymerization, called Model III in Ref. 4, has been tested in collaborative research conducted at Virginia Tech. According to this model, it was predicted that the highest tendency for tribopolymerization will occur with vinyl monomers that polymerize according to both anionic and free radical mechanisms. Vinyl monomers that polymerize only according to the anionic mechanism should provide tribopolymers. However, it is not expected that vinyl monomers that polymerize only according to the cationic mechanism will form any tribopolymers. On the other hand, they may react with surfaces under boundary friction conditions. It was also assumed that other monomers, aside from tribopolymerization reactions, may react with solid surfaces. Tribochemical reactions mostly relate to vinyl esters, e.g., MMA, because low-energy electrons can produce carboxylate ions from ester molecules [46]. Taking into account the mechanism by which the monomers polymerize, the general Model III discussed encompasses four submodels—a, b, c, and d (Table 13).

Table 13 Predicted Ability of Selected Vinyl Monomers to Form Tribopolymers Under Sliding Conditions

Submodel	Mechanism by which the monomers polymerize	Expected tribopolymer formation
IIIa	Both anionic and free radical	Large amounts of tribopolymers should be formed
IIIb	Anionic	Significant amounts of tribopolymers should be formed
IIIc	Free radical	Smaller amounts of tribopolymers are expected
IIId	Cationic	No tribopolymer formation

D. Experimental Results of the Collaborative Research

1. Vinyl-Type Tribopolymerization on Metal/Metal Systems

The objective of the research was to evaluate the validity of the NIRAM approach concerning the potential addition-type polymer formers. Two metal systems were investigated: (1) a hard steel ball (AISI 52100 steel) on a softer steel plate (1045 steel), and (2) a hard steel ball on an aluminum plate (6061 aluminum). Both metals are exoelectron emitters under boundary tribological conditions. However, aluminum was found to give off more exoelectrons than steel [47]. To carry out tribological tests in the main and most extended study pertinent to vinyl tribopolymerization on metal-on-metal systems [48], a fretting machine was used [49]. Operating conditions were kept constant for all the tests and are summarized in Table 14. The contact geometry involved in each test consisted of loaded ball fretting against a fixed disk, as presented in Fig. 16. Tested monomers (MMA and ST) were used in hexadecane solutions at the 1 wt.% treat rate.

A key part of the research was the attempt to identify and characterize the potential films on the tribological surfaces using Fourier-transform infrared microscopy (FTIRM). A summary of the main results obtained under fretting conditions for MMA and ST tested on steel-on-steel and steel-on-aluminum system is presented in Table 15 [48]. The additive effectiveness in reducing friction and wear ranges from poor to very good; for example, ST reduced aluminum disk wear volume by 66%. Evidence of polystyrene (PS) was found via FTIRM at the edge of the scars and within the brown deposit around the scars. It is emphasized that this polymer formation was observed for the case in which the best wear reduction has been recorded. This finding clearly supports the vinyl-type tribopolymerization concept. The fact that the PS film was detected outside the scar suggests that it was formed in a significant quantity and that it was not broken up in small molecules when removed from the contact area. Interestingly, this observation recalls the studies carried out by Day [50] and Raciti [51]. Investigating polymeric coating films in fretting systems, they found an unexpectedly high film life of the very brittle PS.

Furthermore, in addition to a PS film, the FTIRM analysis detected a material containing carbonyl groups and carboxylate anions on the aluminum disk. This product was slightly different from the possible oxidation products derived from the base oil. It may result from the opening of the vinyl bond followed by reaction with the metal substrate that may also yield a polymerized form. As seen in the literature review, St. Pierre et al. [26,27] found that vinyl-type compounds reacted with an aluminum surface by opening of their double bond, thus reducing friction and wear in a sliding system. Another hypothesis would be that styrene underwent oxidation reactions.

Table 14 Test Operating
Conditions in Fretting
Experiments

Parameter	Value
Load	90 N
Frequency	65 Hz
Amplitude	200 μm
Run time	1 hr
Temperature	20–23 °C

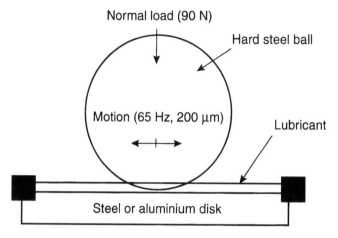

Figure 16 Contact geometry of the experimental set-up for the fretting study.

Additionally, almost no black frothy material believed to correspond to products derived from the base oil was observed on the aluminum disks. It can be suggested that the reactions involving the ST, i.e., polymerization, formation of oxidation products, or opening of the vinyl bond and reaction with aluminum, competed with the base oil oxidation process. Thus it is speculated that ST may have acted as an antioxidant by minimizing the base oil oxidation. This would be consistent with Weatherford et al.'s work [52], and indicates a strong correlation between protection from wear and protection from oil oxidation in fretting systems. In the steel-on-steel systems, the ST solution did not reduce wear significantly. No evidence of PS was found here. This observation is consistent with the

Table 15 Summary of Friction, Wear, and FTIRM Results from Fretting Contact Tests

Tribological systems/results[a]	Hexadecane	Methyl methacrylate[b]	Styrene[b]
Steel-on-steel			
Reductions in friction	—	48%	15%
Reductions in wear area	—	32%	4%
FTIRM analysis	Carboxylate products	Carbonyl and carboxylate products	Carbonyl and carboxylate products
Steel-on-aluminum			
Reductions in friction		9%	15%
Reductions in wear area	—	25%	36%
Reductions in wear volume	—	36%	66%
FTIRM analysis	Carboxylate products	Carbonyl and carboxylate products	Polystyrene, carbonyl, and carboxylate products

[a] Compared to hexadecane base fluid.
[b] At 1% concentration in hexadecane.

model of tribopolymerization for the addition-type monomers: the steel-on-aluminum system emits more exoelectrons than the steel-on-steel system.

Moreover, the concept based on the NIRAM approach would have predicted that MMA would polymerize more easily than ST, because of its better electron receptivity. However, the reductions in wear in the steel-on-aluminum system were not as striking with MMA, although they were significant. On the other hand, the most significant additive effectiveness in reducing wear in the steel-on-steel system was found with MMA. This may be explained by the fact that both the polymeric film formed and the monomer react with the surfaces. Because aluminum is a better electron emitter than steel, chemical reactions may proceed faster, thus involving some corrosion. This would be consistent with the work of Day [50], who showed that polymethyl methacrylate coating films had a very short life although they had physical properties very similar to those of the PS.

These results are consistent with the finding that MMA, under fretting conditions used, (1) polymerizes and (2) reacts with the metal surface. Because aluminum emits more exoelectrons than steel under tribological conditions, MMA reacts chemically with aluminum, leading to some corrosive wear, which decreases the wear protective function of the polymeric film. This could account for the better performance of ST vs. MMA on aluminum. It is also mentioned that the maximum calculated temperature rise reached 33°C for the steel-on-steel system, and was only equal to 2°C for the steel-on-aluminum system. Thus, low-energy electrons emitted during the friction process can initiate surface vinyl-type tribopolymerization even at relatively low surface temperatures.

2. Vinyl-Type Tribopolymerization in Ceramic-on-Ceramic Systems

Most of the works relating to the lubrication of ceramics involve classical lubricants and additives. The benefits are not particularly striking, at least compared to what is normally observed with such additives in steel systems. By and large, these conventional approaches to the lubrication of ceramics include: (1) incorporation of various materials into the ceramic, (2) surface treatment and coatings, (3) dispersed solid lubricants in a fluid, and (4) conventional lubricants containing soluble antiwear additives in a carrier fluid. Recently, Furey and Kajdas [53] demonstrated that the concept of tribopolymerization can be used as an effective approach to designing specific molecular structures for boundary lubrication of ceramic materials. The approach encompassed both the liquid-phase and vapor-phase lubrication under boundary friction conditions.

Liquid-Phase Lubrication. Our recent work [19] has shown that the NIRAM concept may also be applied to explain the behavior of vinyl monomers under boundary lubrication of ceramic, e.g., alumina. The approach is somewhat supported by an earlier work of Nakayama et al. [54], demonstrating that aluminum oxide film emits negatively charged particles under friction conditions even in atmosphere. Importantly, it is a well-known fact [55] that exoelectrons make the major part of the negatively charged particles emitted from oxide-covered aluminum during tensile deformation under ultrahigh vacuum conditions. It is also known that alumina, under scratching conditions, emit charged particles in ambient atmosphere [56,57]. Most recently, Molina et al. [22] measured emission from diamond-on-alumina, diamond-on-sapphire and diamond-on-aluminum, and from alumina-on-alumina. They found no significant difference between emission from polycrystalline alumina and sapphire. Decaying emission was always observed after the contact ceased for these three ceramics. Energy spectra of negatively charged triboemission from these ceramic systems extended for higher energy than the test maximum level (e.g., 48 eV) and they demonstrated that important fractions of these particles were emitted in the 0–5 eV range, with only decaying fractions extending to higher energy. These experimental results are consistent with the NIRAM hypothesis.

The major objective of the most recent research by Furey et al. [58], discussed in this section, was to test the validity of the NIRAM approach to initiate anionic/free radical tribopolymerization in the ceramic alumina-on-alumina system via FTIRM. Four vinyl monomers were used as the additives in this study. They are listed in Table 16, along with their polymerization mechanisms. All the monomers were used at 1 wt.% concentration in hexadecane.

The tribological tests were carried out using the pin-on-disk type contact geometry consisting of a spherical ball loaded against a rotating disk. The device is shown schematically in Fig. 17. The test conditions (see Table 17) were designed to result in boundary lubrication at the sliding interface. After the wear tests, the worn ceramic surfaces were examined using FTIRM to determine the chemical nature of the worn surfaces and wear debris. Alumina disks and balls were used as the mating elements of the tribological system. One of the tested monomers, vinyl octadecyl ether (VOE) (see Table 16) polymerizes only according to the cationic mechanism. Thus, based on the NIRAM approach, its polymerization cannot be initiated under friction conditions in the liquid phase. FTIRM analysis of wear scars was carried out using a Spectra Tech Infrared Microscope coupled to a Nicolet FTIR spectrometer. A detailed description of this technique may be found in Ref. 59.

The ball wear results for alumina lubricated with pure hexadecane and hexadecane + 1% of each of the four vinyl monomer additives are summarized in Fig. 18. Results are shown for eight separate tests on each lubricant carried out under the same conditions. Interestingly, three of the four monomers—vinyl acetate, lauryl methacrylate and diallyl phthalate—were effective in substantially reducing alumina wear. The average wear reductions ranged from 57% to 80%, with diallyl phthalate being the most effective. However, vinyl octadecyl ether was only marginally effective in reducing wear under these conditions. Its effect was statistically significant but small (9% average wear reduction). The disk wear results also showed similar trends. The average disk wear reductions by vinyl acetate, lauryl methacrylate, and diallyl phthalate ranged from 55% to 70%. Vinyl octadecyl ether reduced disk wear only by 5% [58].

The first important observation related to FTIRM spectra is that, in each case, the wear product spectra differs drastically from that of the monomer used. Figure 19 shows an exemplary FTIRM spectra for diallyl phthalate. The standard spectrum for diallyl phthalate shows peaks in the region of 2880–3080 cm^{-1}, which correspond to CH stretching vibrations of the molecules. The saturated bonds appear at frequencies below 3000 cm^{-1}, whereas the unsaturated bonds appear above 3000 cm^{-1}. However, for the spectrum taken from the wear products, all the peaks corresponding to CH stretching vibrations appeared below 3000 cm^{-1}. This clearly suggests the disappearance of or a decrease in the number of double bonds present at the surface, most probably due to polymerization. Additionally, the relative

Table 16   List of Vinyl Monomers Tested

Monomer	Polymerization mechanism
Vinyl Acetate	Radical
Lauryl Methacrylate	Anionic and Radical
Diallyl Phthalate	Radical
Vinyl Octadecyl Ether	Cationic

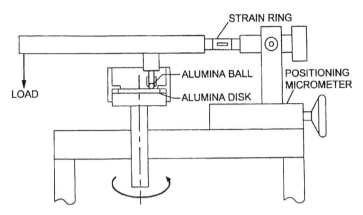

Figure 17 Schematic diagram of the pin-on-disk apparatus.

intensity of the CH stretching peak due to saturated bonds is greater in the wear product spectrum when compared to the standard spectrum. This difference relates to an increase in the concentration of the CH_2 saturated groups, and can only be explained by a polymerization reaction of the diallyl phthalate molecules in which the double bonds were broken for oligomer/polymer products.

The spectrum from the diallyl phthalate wear products shows much reduced peaks in the characteristic diallyl phthalate absorption regions at 1725 and 1280 cm^{-1}, relating to the CO and COC stretching of the ester groups, respectively. Instead, intense duplex peaks are observed in the regions of 1530–1600 and 1370–1410 cm^{-1}. This suggests that the bonds that formed ester groups were broken and some new products were formed. The first region may be associated with the CO asymmetric stretching vibration of the carboxylates, which usually appears in the region of 1610–1560 cm^{-1}. The second region could also be attributable to carboxylates as the CO symmetric stretching appears in the region of 1400–1300 cm^{-1}. Thus these peaks suggest a chemical reaction between diallyl phthalate and alumina to form alumina soaps. Another peak (2350 cm^{-1}) was identified because of the CO_2 present in the air. The infrared spectra for diallyl phthalate, depicted in Fig. 19, show that, in this case—similarly for methyl methacrylate tested under fretting conditions in the metal-on-metal systems—there is a dual reaction mechanism. The infrared spectra for vinyl acetate also support a dual reaction mechanism—a chemical reaction of the monomer with the alumina surface and a polymerization reaction [58]. Thus the FTIRM analysis provides clear evidence of a chemical reaction between each of three of the vinyl monomers examined and the alumina surface; for two of the monomers, vinyl acetate and diallyl phthalate, there

Table 17 Experimental Set-up and Test Conditions for Pin-on-Disk Experiments

Geometry	Sphere-on-flat (fixed alumina ball on rotating alumina disk)
Specimen size	3.2-mm diameter (1/8 in.) ball on 25-mm diameter (1 in.) disk
Lubricants	Hexadecane carrier, and 1 wt.% solution of vinyl monomer
Applied load	20 N
Sliding velocity	0.25 m/sec
Sliding distance	500 m
Ambient temperature	23°C

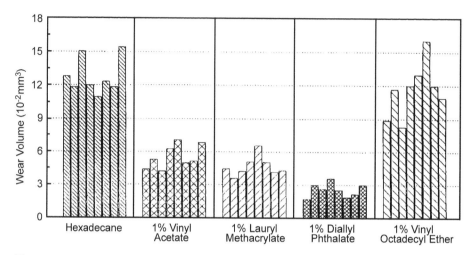

Figure 18 Effects of four vinyl monomers on alumina ball wear in the liquid phase.

is also evidence of a polymerization reaction forming oligomers/polymer. The spectra obtained with lauryl methacrylate were very noisy; however, peaks corresponding to the carboxylates are clearly seen in the wear track, supporting the hypothesis of lauryl methacrylate-alumina reactions.

Based on the FTIRM analysis, we believe that the tribochemical reactions with this monomer involve a breakdown of the ester groups to form carboxylates with alumina, in addition to a tribopolymerization reaction. The results presented here, and discussed in detail in Ref. 58, suggest that the NIRAM approach can also be applied to the tribopolymerization mechanism of vinyl monomers on alumina. We believe that this polymerization process is responsible for the reduction in alumina wear. Based on molecular mechanics energy calculations, it is possible to determine anion-radical species derived from the above-discussed monomers. Such reactive intermediates are responsible for both tribochemical reactions and tribopolymerization processes. This topic is discussed by Kempinski et al. [15].

As vinyl octadecyl ether can polymerize only by a cationic mechanism, it is not expected to reduce wear of alumina under the frictional conditions used in these tests. The small wear reduction (9%) may be a result of the formation of some surface reaction products. If so, then these products are not particularly beneficial, as far as wear reduction is concerned. These results show that the tribochemical behavior of the selected polar—oxygen containing—vinyl monomers on alumina is complex and that further research is needed to clarify these surface reactions, which are not initiated by low-energy electrons, for example, vinyl ethers. Interestingly, vinyl octadecyl ether is as effective as other tested vinyl monomers in the vapor-phase lubrication.

Vapor-Phase Lubrication. At present, there is an increased interest in advanced structural ceramics for use in tribological applications, particularly for engine parts. However, such applications need lubrication at higher temperatures. This led us to develop the vapor-phase lubrication of an alumina-on-alumina tribological system by vinyl monomers selected on the basis of the concept of tribopolymerization as a mechanism of boundary lubrication. The compounds selected for this study consisted of three oxygen-containing addition-type monomers. They included diallyl phthalate, lauryl methacrylate, and vinyl octadecyl ether. Experimental tests were performed using a modified pin-on-disk

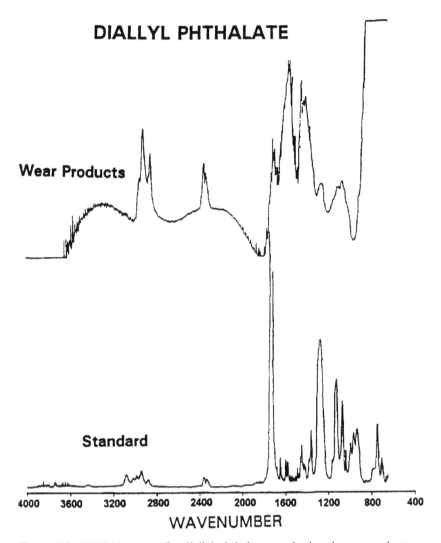

Figure 19 FTIRM spectra for diallyl phthalate standard and wear products.

machine [13]. Dry nitrogen gas was used as the carrier fluid for the tested compounds. The same types of balls and disks as for the liquid-phase lubrication—described in the preceding subsection—were used for tribological contacts in the discussed vapor-phase lubrication study. Experimental set-up and test conditions were similar as those used for the liquid-phase lubrication (Table 17); however, the applied load was 5 N. Selected surfaces were examined by FTIRM in order to determine the chemical composition of surface films left on the wear scar.

An example of ball wear reduction results for the alumina-on-alumina elevated bulk temperature (145°C) treatments is summarized in Fig. 20. Monomer temperatures were raised to the temperatures indicated in Fig. 20 for the duration of the experiments. This example clearly demonstrates large reduction in wear for various treatments when compared to the nitrogen standard. Dramatic reductions in wear occur, particularly for those mono-mers delivered at the highest temperatures (Fig. 21). Percentage wear reductions range from 35% to as high as 99%. Furthermore, all reductions are statistically significant at the 95%

confidence level when compared with the nitrogen standard. The effects of the monomer vapors at the higher temperatures are extraordinary—with wear reductions ranging from 94% to 99% for the three compounds studied. Each increase in monomer delivery temperature leads to a reduction in wear. The trend is apparent in Figs. 20 and 21. However, for vinyl octadecyl ether, the result at 130°C is better than for diallyl phthalate at 135°C. But the downward trend in wear with increasing monomer delivery temperature was observed in all cases. It is emphasized that very low concentrations of monomer vapor provide substantial reductions in wear; the concentrations ranged from 0.3 to 7.3×10^{-5} mol/L.

At ambient bulk temperature, the coefficient of friction for dry nitrogen alone is approximately 0.4. At 145°C bulk temperature, the coefficient of friction for the nitrogen is about 0.8. Under these conditions, both lauryl methacrylate and diallyl phthalate vapor reduced friction coefficients to 0.4, while vinyl octadecyl ether reduced the coefficient to 0.5. This demonstrates that the monomer vapors reduced friction of alumina-on-alumina system by approximately 37–50% at higher temperatures.

Visual inspection of wear scars resulting from monomer vapor treatments showed clear, obvious, and unusual deposits. The deposits, along with FTIRM spectra obtained from the surface deposits, are described in detail elsewhere [13]. These analytical findings provide evidence for both the (1) polymerization reaction of the monomer and (2) its reaction with the substrate to form a soap. For example, the results confirm the diallyl phthalate tribochemistry of the same alumina/alumina system lubricated by this monomer tested at 1% concentration in hexadecane described in the previous subsection. However, in the vapor-phase lubrication wear product, spectrum intensities of the peaks around 1725 and 1280 cm^{-1} are much stronger than in the liquid-phase lubrication. This difference may suggest that, under vapor-phase lubrication conditions, the surface polymerization process is much more advanced than under the liquid-phase lubrication. This could be due to higher temperatures contributing to the free-radical polymerization process.

Spectra for the wear products of lauryl methacrylate and vinyl octadecyl ether do not provide clear evidence for tribopolymerization. However, it is believed that polymerization

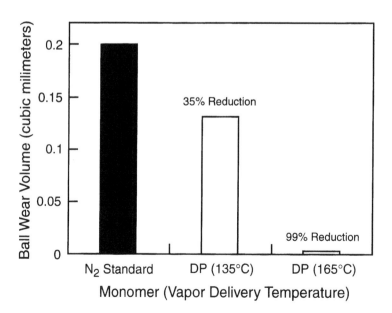

Figure 20 Effect of diallyl phthalate on alumina wear in the vapor phase.

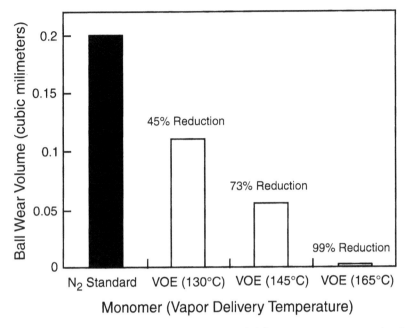

Figure 21 Effect of vinyl octadecyl ether and delivery temperature on alumina wear in the vapor phase.

of either or both of these monomers cannot be conclusively ruled out. This assumption is supported by the fact that the saturated CH stretching peaks, seen in the standards below $3000 \, \text{cm}^{-1}$, reappear in the wear products spectra for both monomers, but the concentration of CH_2 groups cannot be determined. On the other hand, both of these spectra do include the characteristic peaks near 1600 and 1400 cm^{-1}, clearly supporting the occurrence of monomer/alumina reactions. Although strong evidence for soap formation was found, the reaction mechanism is not completely understood, particularly for vinyl octadecyl ether. The question is how the ether has been oxidized in the nitrogen atmosphere? Was it due to trace oxygen present in the nitrogen gas? Is there another possible oxygen source under that friction condition? The first question cannot be answered because we did not analyze the nitrogen gas. It was just described as "pure dry nitrogen for laboratory applications." Bearing in mind the fact that oxygen has been found as a neutral component of the tribo-emitted particles from alumina [47], it is also possible to suggest that the emitted oxygen contributed to the ether oxygen process.

E. Summary of Experimental Results Relating to Vinyl-Type Tribopolymerization

The most significant result of the overviewed research was the confirmation of the hypothesis that vinyl-type tribopolymerization can be initiated by low-energy electrons emitted under the boundary lubrication conditions. The study of that type tribopolymerization under fretting contact conditions resulted in obtaining a clear evidence for the polystyrene formation. The evidence was based on the detailed FTIR microanalysis data [48]. The mean temperature rises estimated by the Archard or Vick models [60–63] for the steel-on-aluminum system were equal to 10–17°C; therefore, the free radical initiation of the tribopolymerization process has been excluded. Accordingly, the interpretation of the results obtained by Kajdas et al. [33] in 1980 has been modified.

Methyl methacrylate (MMA), under fretting conditions, was found to polymerize and react with the metal surface. For that compound and others (e.g., diallyl phthalate), clear evidence was obtained for their similar behavior under friction conditions of alumina-on-alumina. Interestingly, this is due for both the liquid-phase and vapor-phase lubrication. However, the most striking result of the vapor-phase lubrication research was the discovery that three quite different monomers—selected as models for the tribopolymerization concept of boundary lubrication—can cause extraordinary reductions in alumina-on-alumina wear when used at low concentration in the vapor phase. Actually, the model monomers were selected on the basis of the tribopolymerization concept; however, the results obtained clearly show that the formation of protective films on the rubbing alumina surfaces in our study is complex. The complexity increases even more for the liquid-phase lubrication under higher sliding velocity (1.0 m/sec). For example, at that sliding speed and 40 N load, all monomers tested (vinyl acetate, lauryl methacrylate, vinyl octadecyl ether, diallyl phthalate) either had no effect on wear or increased alumina wear. This particular finding and other results concerning lubrication of ceramics by tribopolymerization at high speeds and high loads are described elsewhere [14]. Interestingly, even more clearly pronounced tribopolymerization is observed under a higher load (160 N) at the sliding velocity 0.25 m/sec.

Vinyl octadecyl ether was found ineffective under mild operating conditions; this finding was in line with what the NIRAM approach predicted. The situation dramatically changed under vapor-phase lubrication conditions and under higher loads (40 and 160 N) in liquid-phase lubrication at the sliding velocity 0.25 m/sec. Why is this monomer ineffective in reducing alumina wear in hexadecane under mild operating conditions, and yet is so (1) very effective (wear reduction up to 99%) in the vapor-phase lubrication and (2) effective in hexadecane at the higher loads? Close examination and comparison of selected FTIRM spectra show similarities as well striking differences. This should not be too surprising, because the conditions—particularly the expected surface temperatures generated under the vapor phase conditions (e.g., vapor delivery temperature 165°C and bulk temperature 145°C) and mild liquid-phase lubrication conditions—are vastly different. But how does vinyl octadecyl ether react with alumina or aluminum hydroxide to form a carboxylate soap under vapor phase conditions? One speculation may consider the oxidation of ether to carboxylic acids, which, in turn, could reduce the wear. Higher temperatures generated by higher load in the liquid-phase experiments can also explain a higher oxidation rate of the ether.

The observed effectiveness of vinyl octadecyl ether (VOE) might additionally relate to vinyl-type tribopolymerization initiated by higher energy exoelectrons able to generate cations. Accordingly, cationic polymerization of VOE could proceed. At present, however, no data are available regarding the energy of emitted electrons from alumina surfaces rubbed under high loads, high speeds, and elevated temperatures. Assuming that higher loads and temperatures contribute to emission of higher energy electrons, it is conceivable to expect the higher wear reduction by VOE, as higher energy electrons can produce cations from that monomer.

F. Conclusions on Vinyl Tribopolymerization

It was found that the proposed NIRAM approach to the vinyl tribopolymerization mechanism predicts well the behavior of vinyl monomers on metals and alumina in the liquid-phase lubrication under mild operating conditions. This approach also accounts for the results related to vinyl-type tribopolymerization obtained by other researchers. It explains both the polymerization mechanism and reaction mechanism with the surface. At present, the NIRAM concept cannot be clearly applied to the vapor-phase lubrication by vinyl monomers and liquid-phase lubrication under some more severe operating conditions. The NIRAM approach is based on the action of low-energy exoelectrons, and thus does not concern the tribopolymerization process proceeding according to the cationic mechanism. However, it is expected that under severe friction conditions (e.g., high speed, high temperature), more electrons of higher energy might be emitted, which then could initiate the cationic tribopolymerization.

IV. APPLICATIONS OF TRIBOPOLYMERIZATION TECHNOLOGY

Several applications of tribopolymerization have been described elsewhere. Space does not permit detailed discussion here, but key results are summarized in a number of papers, including "Tribopolymerization: An Advanced Lubrication Concept for Automotive Engines and Systems of the Future" [12], "Tribopolymerization as a Novel Approach to Ceramic Lubrication" [7], "Applications of the Concept of Tribopolymerization in Fuels, Lubricants, Metalworking, and 'Minimalist' Lubrication" [64], and "Recent Developments in Environmentally Friendly Antiwear Additives and Lubricants from Tribopolymerization" [65].

Important areas of application are summarized in Fig. 22 and include:

- Ashless antiwear or "lubricity" additives for fuels, including jet fuel, (e.g., to minimize fuel pump wear), diesel (e.g., to reduce fuel injector wear occurring with low sulfur fuels), and gasoline (two-stroke engines).
- Vapor-phase applications of this technology for high-temperature gaseous systems.

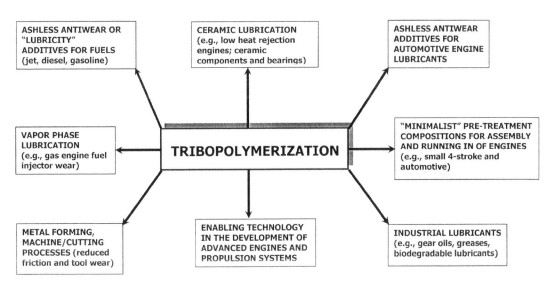

Figure 22 Applications of tribopolymerization technology to reducing wear, friction, environmental effects, and costs.

- New ashless lubricants for existing and future automotive engines to reduce exhaust catalyst poisoning and environmental emissions.
- Machining and cutting applications using minimal thin films to reduce friction and ceramic tool wear, and minimize lubricant waste and disposal.
- Special "run-in" pretreatment films in production (e.g., of engines).
- Lubrication of ceramic engines (e.g., low heat rejection diesel engines) or ceramic components (e.g., turbochargers, cam followers, valve guides, and bearings).

For tribological applications, ceramics offer several advantages over conventional materials. But conventional approaches to the lubrication of ceramics are limited (e.g., surface coatings or treatments) or are often ineffective [53]; and commonly used lubricants and additives designed for steel systems generally do not work. Our discovery that the principle of tribopolymerization is a useful and often strikingly effective approach to the lubrication of ceramic as well as metals was selected as 1 of 12 out of 600 proposals in the Department of Energy's (DOE) Energy-Related Inventions Program. An extract from the Executive Summary of the NIST Evaluation of this invention states:

> ...The invention holds promise of being a better way to lubricate adiabatic engines than current techniques. Thus it could become an enabling technology for ceramic internal combustion engines.

What are the environmental benefits? For one, significant savings in fuel consumption (and therefore, fuel costs) as well as reduced pollution and CO_2 emission.

- Other environmentally friendly applications, e.g., industrial "Green Lubricants," refrigerant oils, and food-grade lubricants.

Several patents have been issued on applications of this technology [17,18,66–69] and more are underway. Through an agreement with Virginia Tech Intellectual Properties, the newly formed company, Tribochem International, has been granted exclusive rights for the synthesis manufacture use, and marketing of several classes of the compounds, additives, and lubricants discussed in this paper.

This activity has been further enhanced by recent agreements of collaboration between our research groups, including Tribochem International, and two Polish organizations—the Institute of Terotechnology at Radom and the Central Petroleum Laboratory in Warsaw.

V. CONCLUSIONS

Tribopolymerization, a process of molecular design developed by Furey and Kajdas, is defined as the planned and continuous formation of protective polymeric films directly on rubbing surfaces by the use of selected compounds designed to form films, "in situ." These very thin and generally invisible films are particularly effective in reducing wear and are continuously replenished during sliding contact.

Two classes of tribomonomeric compounds have been discussed—(1) condensation-type and (2) addition-type. Although mechanism studies have shown that the process of "in situ" surface polymerization is more complex than originally hypothesized, we believe that the initiation of surface polymerization of condensation monomers is chiefly controlled by high surface temperatures generated by friction while the action of addition-type monomers is triggered by the triboemission of low-energy electrons. More research is continuing along both lines.

It has been demonstrated in a variety of high contact stress tribological systems (e.g., Ryder Gear, automotive engine cam/valve lifter, ball-on-cylinder, pin-on-disk machines) that the principle of tribopolymerization is a useful and often strikingly effective approach to the boundary lubrication of metals (e.g., alloy steels), as well as ceramics. For example, in a liquid hydrocarbon carrier fluid—hexadecane or cetane—wear reductions of 80% and greater of ceramic materials (e.g., alumina, zirconia) were achieved via the use of minor concentrations of selected or synthesized compounds based on the tribopolymerization concept. This discovery was selected as 1 of 12 out of 600 proposals, in the DOE's Energy-Related Inventions Program. The evaluation stated that "...The invention holds promise of being a better way to lubricate adiabatic engines than current techniques. Thus it could become an enabling technology for ceramic internal combustion engines." For tribological applications, ceramics offer several advantages over conventional materials. For example, they can be used at much higher temperatures, while some ceramics are lighter in weight than the alloy steels commonly used in engines.

It has also been found that compounds selected on the basis of the tribopolymerization concept are also effective in reducing wear and friction in the vapor phase. For example, in an alumina-on-alumina system at high contact stress, low vapor concentrations of compounds in nitrogen gas were extremely effective in reducing wear and friction—the beneficial effects being more pronounced at higher temperatures, where wear reductions of 99% were observed. But the concept has broader significance and potential.

The antiwear compounds developed from this technology are effective with metals as well as ceramics, and in the liquid and vapor phases. Furthermore, they are ashless and contain no harmful phosphorus or sulfur; and many are biodegradable. Thus potential applications of this technology are diverse and include a variety of cost/performance/ energy/environmental advantages.

Examples include the following: (1) machining and cutting applications using minimal thin films to reduce friction and ceramic tool wear and which can be used with difficult metals; (2) the lubrication of ceramic engines (e.g., low heat rejection diesel engines) or ceramic components (e.g., turbochargers, cam followers, valve guides, cylinder liners, seals, and bearings); (3) the development of ashless lubricants for existing and future automotive engines to reduce exhaust catalyst poisoning and environmental emissions; (4) ashless antiwear or "lubricity" additives for fuels, including gasoline (two-stroke engines), diesel (e.g., to reduce fuel injector wear occurring with low sulfur fuels), and jet fuel (e.g., to minimize fuel pump wear); (5) vapor-phase applications of this technology to high-temperature gaseous systems or to fuel injector wear problems associated with the use of natural gas engines; and (6) the use of the concept of tribopolymerization as an enabling technology in the development of new engines and new automotive propulsion systems.

Several patents based on this technology have been issued and several others are pending. Arrangements for licensing, field testing and evaluation, and marketing can be made by contacting Tribochem International and its Polish collaborating organizations.

ACKNOWLEDGMENTS

The authors express their appreciation to two agencies for their support of this research, the Energy-Related Inventions Program of the Department of Energy, and especially the Surface Engineering and Tribology Program of the National Science Foundation. In addition, we wish to acknowledge the help of Dr. Brian Vick on surface temperatures, Dr. Roman Kempinski, former Fulbright Scholar, and several former graduate students who worked with us in research on tribopolymerization, including Dr. Bhawani Tripathy, Dr. Gustavo Molina, Mr. Benjamin Tritt, Mr. J.C. Smith, Ms. Pascal Lefleche, and Mr. Jeffrey Valentino.

REFERENCES

1. Furey, M.J.; Kajdas, C.; Kempinski, R. "Tribopolymerization: I. Surface Temperatures and the Antiwear Action of Condensation-Type Monomers." Proceedings of the 2nd Symposium on Tribochemistry, Janowice, Poland. Polish Tribology Society and participating universities: Lodz, Cracow, and Plock, Sept. 15–17, 1997; 115–125.
2. Kajdas, C.; Furey, M.J.; Kempinski, R. "Tribopolymerization: II. NIRAM Applications to the Antiwear Action for Addition-Type Monomers." Proceedings of the 2nd Symposium on Tribochemistry, Janowice, Poland. Polish Tribology Society and participating universities: Lodz, Cracow, and Plock, Sept. 15–17, 1997; 127–139.
3. Furey, M.J. Tribology. *Encyclopedia of Materials Science and Engineering*; Pergamon Press: Oxford, 1986; 5145–5158.
4. Furey, M.J.; Kajdas, C. "Models of tribopolymerization as an anti-wear mechanism." Proceedings Japan International Tribology Conference, Nagoya 1990, Vol. II, 1089–1094.
5. Furey, M.J.; Kajdas, C. "Tribopolymerization as a lubrication mechanism for high-energetic contacts of solids." 6th International Tribology Colloq. Technische Akademie Esslingen: Esslingen, Germany, Jan. 12–14, 1998.
6. Furey, M.J. "The formation of polymeric films directly on rubbing surfaces to reduce wear." Wear 1973, *26*, 369–392.
7. Furey, M.J.; Kajdas, C. "Tribopolymerization as a novel approach to ceramic Lubrication." Proceedings 4th International Symposium on Ceramic Materials and Components for Engines Elsevier: Goteborg, Sweden, June 10–12, 1991.
8. Tripathy, B.S.; Furey, M.J.; Kajdas, C. Mechanism of Wear Reduction of Alumina by Tribopolymerization. Wear 1995, *181–183*, 138–147.
9. Furey, M.J.; Tritt, B.R.; Kajdas, C.; Kempinski, R. "Models for Ceramic Lubrication by Tribo-polymerization at High Loads and Speeds," presentation at the World Tribology Congress, London, September 8–12, 1997.
10. Furey, M.J.; Ghasesmi, H.; Tripathy, B.S.; Kajdas, C.; Kempinski, R.; Hellgeth, J.W. Tribochemistry of the antiwear action of a dimer acid/glycol monoester on alumina. Tribol. Trans. 1994, *37*, 67–74.
11. Furey, M.J.; Kajdas, C.; Kempinski, R.; Tripathy, B.S. Tribopolymerization and the Behavior of Oxygen-Containing Monomers in Reducing Ceramic Wear. Proceedings of the EUROTRIB '93 Congress, Budapest, Hungary Aug. 30–Sept. 2, 1993, Vol. 2.
12. Furey, M.J.; Kajdas, C. "Tribopolymerizaton: an advanced lubrication concept for automotive engines and systems of the future." Proceedings, June 16–19, 30th International Symposium on Automotive Technology and Automation, Florence, Italy, Session on The Motor Vehicle and The Environment–Entering a New Century, 241–249.
13. Smith, J.C.; Furey, M.J.; Kajdas, C. An exploratory study of vapor phase lubrication of ceramics by monomers. Wear 1995, *181–183*, 581–593.
14. Furey M.J.; Tritt, B.R.; Kajdas, C.; Kempinski, R. "Lubrication of ceramics by tribopolymeri-zation: a designed experiment to determine effects of monomer structure, load, and speed on wear," Preprint No. 99-AM-9, 54th Annual Meeting of STLE, Las Vegas, Nevada, May 23–27, 1999; 1–9
15. Kempinski, R.; Furey, M.J.; Kajdas, C.; Tripathy, B.S. "Tribopolymerization: III. Computer modelling of monomer/surface interactions." 2nd International Symposium on Tribochemistry, Janowice, Poland, Sept. 15–17, 1997; 141–147.
16. Kajdas, C.; Furey, M.J.; Kempinski, R. "New condensation-type monomer combinations as

ashless antiwear compositions." June 11–12, 2000 paper, NORDTRIB 2000, Vol. 1, 299–310.

17. Wear Reduction Using Cyclic Amide Compounds, Inventors, Michael J. Furey and Czeslaw Kajdas, U.S. Patent No. 5,851,964 issued December 22, 1998.

18. Wear Reducing Compositions and Methods for Their Use, Inventors, Michael J. Furey and Czeslaw Kajdas, U.S. Patent No. 5,880,072 issued March 9, 1999.

19. Furey, M.J.; Kajdas, C.; Kempinski, R.; Tripathy, B.S. Action mechanism of selected vinyl monomers under boundary lubrication of an alumina-on-alumina system. Lubr. Sci. 1997, *10*, 3–25.

20. Kajdas, C.; Furey, M.J.; Ritter, A.L.; Molina, G.J. Triboemission as a basic part of the boundary friction regime: a review. Lubr. Sci. 2002, *14*, 223–254.

21. Kajdas, C. On a negative-ion-concept of EP action of organo-sulfur compounds. ASLE Trans. 1985, *28*(1), 21–30.

22. Molina, G.J.; Furey, M.J.; Ritter, A.L.; Kajdas, C. Triboemission from alumina, single crystal sapphire, and aluminum. Wear 2001, *249*, 214–219.

23. Moore, M.A. Energy dissipation in abrasive wear. In *Wear Materials*; ASME: New York, 1979; 636–638.

24. Kargin, V.A.; Plate, N.A. Polymerization and grafting process on freshly formed surfaces. J Polym. Sci. 1961, *52*, 155–158.

25. Campbell, W.E.; Lee, R.E. Polymer formation on sliding metals in air saturated with organic vapors. ASLE Trans. 1962, *5*, 91–103.

26. St. Pierre, L.E.; Ownes, R.S.; Klint, R.B. Chemical effects in the boundary lubrication of aluminum. Nature 1964, *202*, 4939.

27. St. Pierre, L.E.; Owens, R.S.; Klint, R.B. Chemical effects in the boundary lubrication of aluminum. Wear 1966, *9*, 160–168.

28. Owens, R.S.; Barnes, W.J. The use of unsaturated hydrocarbons as boundary lubricants for stainless steels. ASLE Trans. 1967, *10*, 77–86.

29. Ferroni, E.; Baistrocchi, R. The kinetics of the polymerization of acrylonitrile in aqueous solution. Ann. Chem. (Rome) 1953, *43*, 555–558.

30. Buckley, D.H. Friction-induced surface activity of some simple organic chlorides and hydrocarbons with iron. ASLE Trans. 1974, *17*, 36–43.

31. Tamai, Y. "Role of mechanochemical activity in boundary lubrication". In *Fundamentals of Tribology*; Suh, N.P., Saka, N. Eds.; MIT Press: Cambridge, MA, 1980; 975–980.

32. Smolin, A.O. Polymerization of styrene during friction of similar metals. Trenie i Iznos. 1981, *2*(5), 925–927.

33. Kajdas, C.; Hombek, R.; Adamski, Z. Polymerization mechanism of unsaturated compounds in friction zone. Zesz Nauk WSI Radom 1980, *1*, 25–39. in Polish.

34. Kajdas, C.; Hombek, R.; Adamski, Z. "Tribopolymerization Initiators and Inhibitors." In *Lubricants and Their Applications, III*; Hebda, M., Kajdas, C., Hamilton, G.M., Eds.; Wydawniciwa Komunikacji I Lacznosci: Warszawa, Poland, 1982; 136–144.

35. Hombek, R.; Kajdas, C. Über die Möglichkeit der Verschleiss-schutzschichten. Bildungdurch Tribopolymerization 1983, *14*, 13–16.

36. Igari, S.; Mori, S. Lubricating properties of olefins for aluminum alloy and other Metals. J. Jpn. Soc. Tribol. 1993, *38*, 644–648. in Japanese.

37. Igari, S.; Takikawa, Y.; Mori, S.; Masahiro, Y. Tribochemical reaction of 1-dodecene with aluminum. J. Jpn. Soc. Tribol. 1993, *38*, 1083–1088. in Japanese.

38. Igari, S.; Mori, S. Lubrication of aluminum alloy with tribochemical reaction products from olefins. J. Jpn. Soc. Tribol 1993, *37*, 605–608. in Japanese.

39. Nakano, T.; Hiratsuka, K.; Sasada, T. Conditions of polymer formation during friction. Jpn. J. Tribol. 1993, *38*, 345–356.

40. Kramer, J. *Der Metallische Zustand*; Vandenhoek and Ruprecht: Goettingen, Germany, 1950.

41. Kajdas, C. About a negative-ion concept for antiwear and antiseizure action of hydrocarbons during friction. Wear 1985, *101*, 1–12.

42. Kajdas, C. About an anionic-radical concept of the lubrication mechanism of alcohols. Wear 1987, *116*, 167–180.

43. Kajdas, C. Importance of anionic reactive intermediates for lubricant component reaction with friction surfaces. Lubr. Sci. 1944, *6*, 203–228.

44. Wan, Y.; Xue, Q. Boundary lubrication of an aluminum alloy by tribopolymerization of a monomer under a fretting contact condition. Wear 1997, *205*, 15–19.

45. Molenda, J.; Kajdas, C.; Makowska, M.; Gradkowski, M. Unsaturated oxygen compounds as antiwear additives for lubricants. Tribologia 1999, *3*, 323–331.

46. vonArdenne, M.; Steinfelder, K.; Tummler, R. *Elektronenanlagerungs—Massenspektrographie Organischer Substanzen*; Springer-Verlag: Berlin, 1971.

47. Connelly, M.; Rabinowica, E. Detecting wear and migration of solid-film lubricants using simultaneous exoelectron emission. ASLE Trans. 1983, *20*, 139–143.

48. Kajdas, C.; Lafleche, P.M.; Furey, M.J.; Hellgeth, J.W.; Ward, T.C. A Study of tribopolymerization under fretting contact conditions. Lubr. Sci. 1993, *6*, 51–89.

49. Furey, M.J.; Eiss, N.S.; Mabie, H.H.; Sweitzer, K.A. "Effects of thin polymeric surface films on fretting corrosion and wear." Proc. Int. Conf. On Organic Coatings Science and Technology, Athens, Greece, 8–12 July 1983; 29–45.

50. Day, K.A. "The use of thin polymeric coatings to prevent fretting corrosion and metallic contact in steel-on-steel systems," March 1986, Masters Thesis, Dept. Mech. Eng., VPI&SU: Blacksburg, VA.

51. Raciti, R. "The effects of film thickness on the behavior of polystryene-coated steel discs under fretting conditions," 1987, Masters Thesis, Dept. Mech. Eng., VPI&SU: Blacksburg, VA.

52. Weatherford, W.D.; Valtierra, M.L.; Ku, P.M. Mechanisms of wear in misaligned splines. J Lubr. Technol. 1968, *90*, 42–48.

53. Furey, M.J.; Kajdas, C. "Tribopolymerization as a novel approach to ceramic lubrication." In *Proc. of the 4th int'l Symp. On Ceramic Materials and Components for Engines*; Carlsson, R. Johnansson, T., Kahlman, L., Eds.; Elsevier Applied Science: London, 1992; 1211–1218.

54. Nakayama, K.; Hashimoto, H.; Fakuda, Y. "Triboemission from aluminum oxide film in atmosphere." Proc. Japan International Tribology Conference, Nagoya, 1990; 1141–1146.

55. Rosenblum, B.; Braunlich, P.; Himmel, L. Spontaneous emission of charged particles and photons during tensile deformation of oxide-covered metals under ultrahigh-vacuum conditions. J. Appl. Phys. 1977, *48*, 5262–5273.

56. Nakayama, K.; Hashimoto, H. Triboemission from various materials in atmosphere. Wear 1991, *147*, 335–343.

57. Nakayama, K.; Hashimoto, H. Triboemission of charged particles and photons from wearing ceramic surfaces in various gases. Tribol. Trans. 1992, *35*, 643–650.

58. Furey, M.J.; Kajdas, C.; Kempinski, R.; Tripathy, B.S. Action mechanism of selected vinyl monomers under boundary lubrication of an alumina-on-alumina system. Lubr. Sci. 1997, *10*, 3025.

59. Furey, M.J.; Ghasemi, H.; Tripathy, B.S.; Kajdas, C.; Kempinski, R.; Hellgeth, J.W. "Tribochemistry of the Antiwear Action of a Dimer Acid/glycol Monoester on Alumina," May 17–20 1993, Preprint No. 93-AM-6E-1, 48th Annual Meeting of STLE, Calgary, Alberta, Canada.

60. Archard, J.F. The temperature of rubbing surfaces. Wear 1959, *2*, 438–455.

61. Vick, B.; Furey, M.J.; Foo, S.J. Thermal analysis of sliding contact using a boundary integral equation method. Numer. Heat. Transf. A Appl. 1991, *20* (1), 19–40.

62. Furey, M.J.; Vick, B.; Foo, S.J.; Weick, B.L. "A theoretical and experimental study of surface temperatures generated during fretting." Proceedings of the Japan International Tribology Conference, Nagoya, Oct. 29–Nov. 1, 1990.

63. Vick, B.; Furey, M.J.; Iskandar, K. "Surface temperatures generated by friction with ceramic materials." STLE 54th Annual Meeting, Las Vegas, Nevada, May 23–27, 1999.

64. Furey, M.J.; Kajdas, C.; Kempinski, R. "Applications of the concept of tribopolymerization in fuels, lubricants, metalworking, and 'minimalist' lubrication." June 11–14, 2000; 9th Nordic Symposium on Tribology, NORDTRIB, Porvo, Finland, also published in Lubr Sci, 2000, Vol. 2, 583–592.

65. Furey, M.J.; Kajdas, C. "Recent developments in environmentally friendly antiwear additives and lubricants from tribopolymerization." June 9–12, 2002. Paper on CD-ROM, 10th Nordic Symposium on Tribology, NORDTRIB 2002, Stockholm, Sweden.
66. Compositions for Reducing Wear on Ceramic Surfaces, Inventors, Michael J. Furey and Czeslaw Kajdas, U.S. Patent 5,407,601, issued 18 April 1995.
67. Compositions for Reducing Wear on Ceramic Surfaces, Inventors, Michael J. Furey and Czeslaw Kajdas, U.S. Patent No. 5,637,558, issued June 10, 1997.
68. Method for Reducing Ceramic Tool Wear and Friction in Machining/Cutting Applications, Inventors, Michael J. Furey and Czeslaw Kajdas, U.S. Patent No. 5,651,648 issued July 29, 1997.
69. Method for Reducing Friction and Wear of Rubbing Surfaces Using Anti-Wear Compounds in Gaseous Phase, Inventors, Michael J. Furey and Czeslaw Kajdas, U.S. Patent No. 5,716,911 issued Feb. 10, 1998.

8
Hydrolysis

Czeslaw Kajdas
Warsaw University of Technology, Płock, Poland

I. INTRODUCTION

A. *Hydro* Derived from Hydrogen

The prefix *hydro* is closely associated with either hydrogen or water. Well-known terms concerning the manufacture of modern lubricant base oils include hydrogenation, hydro-treating, and hydrocraking. These terms also refer to fuel production processes. The term hydrogenation is applied mostly to reactions of the hydrogen molecule (H_2) with organic compound multiple bonds to saturate them. Accordingly, one can say it is just the addition of hydrogen to an unsaturated bond in the molecule that can include only carbon (C) and hydrogen (H)—for example, olefins—or unsaturated molecules encompassing more elements. The latter may include sulfur (S), oxygen (O), and/or nitrogen (N). A typical hydrogenation process is based on the reaction in which hydrogen adds to a double bond of CH, CHS, CHO, CHOS, CHON, or even CHOSN compounds to produce the most environmentally desirable saturated CH or CHO components of base oils and/or fuels. It is of note that the hydrogenation process is usually accomplished via heterogeneous catalytic reactions.

Widely applied hydrofinishing process is effective for removing mostly nitrogen and partly sulfur/oxygen to improve poor color and chemical stability of base oils. Some sulfur-containing compounds might be retained because they tend to act as natural oxidation inhibitors to the base oil. This is attributable to the antiseizure and antiwear properties of the oil. The types of reactions that occur in catalytic hydrogenation processes are [1]:

- Hydrogenation of aromatics and other unsaturated molecules
- Ring opening, particularly of multiring molecules
- Cracking to lower molecular weight products
- Isomerization of alkanes and alkyl side-chains
- Desulfurization
- Denitrogenetion.

Hydrotreatment is a broader term. A mild version of this process—hydrofinishing—relates to the conventional base oil production stage. If more severe operating conditions are used, hydrogenation of aromatic hydrocarbons and ring opening reactions become important. Hydrocracking is the dominant reaction mechanism when hydrotreatment is very severe. The hydrocracking process causes a significant chemical change of the treated feed. Hydrocracking is a highly exothermic reaction.

Although all the above-discussed terms include the prefix *hydro*, they have nothing specific in common with *hydrolysis*, a specific chemical reaction in which splitting of a molecule by water occurs. In the terms hydrogenation, hydrotreatment, and hydrocracking, the prefix *hydro*—as already mentioned above—comes from the hydrogen element.

B. *Hydro* Derived from Water

An intermediate prefix *hydra*, combining the terms *water* with *hydro*, encompasses such words as *hydrant*, *hydrate* (chemical compound of water with another substance; this term also means to combine with water to make a hydrate), *hydraulic*, and even *hydra*. Other water-derivative terms begin with the prefix *hydro*, for example, *hydro*foil (*hydro*plane), *hydro*electric, *hydro*power, *hydro*mechanics, *hydro*pathy (use of water in the treatment of disease), *hydro*metallurgy, *hydro*phobia (disease marked by strong contractions of the muscles of the throat and consequent inability to drink water), *hydro*philic, etc. Another term, *hydro*xylation, is defined as the addition of hydroxyl groups (OH) to each carbon or atom of a double bond. Hydrophilic group is a polar group that likes water and seeks an aqueous environment. *Hydrogen bond* is defined as a strong dipole–dipole interaction that occurs between a hydrogen atom acting on another strongly electronegative atom, particularly oxygen and nitrogen.

In the case when two atoms of different electronegativities form a covalent bond as in the water molecule, the electrons are not shared equally between them. Oxygen—being the atom of greater electronegativity—draws the electron pair closer to it, and a polar covalent bond results. The oxygen atom pulls the bonding electrons to it and thereby makes the hydrogen atom somewhat electron-deficient and gives it a partial positive charge. Accordingly, an electron excess in the oxygen atom provides it with a partial negative charge. Therefore, water is prone to form hydrogen bonding easily. Similarly, in the case of H_2 addition to an unsaturated bonding, water can also be added to such hydrocarbons. A good example of the water molecule (H_2O) addition is its reaction with alkynes to produce ketones (Fig. 1). The reaction is mostly catalyzed by sulfuric acid and mercuric ions.

The unstable product (vinylic alcohol) is initially produced that rearranges instantly to a ketone. This kind of rearrangement involves the loss of a proton (H^+) from the hydroxyl

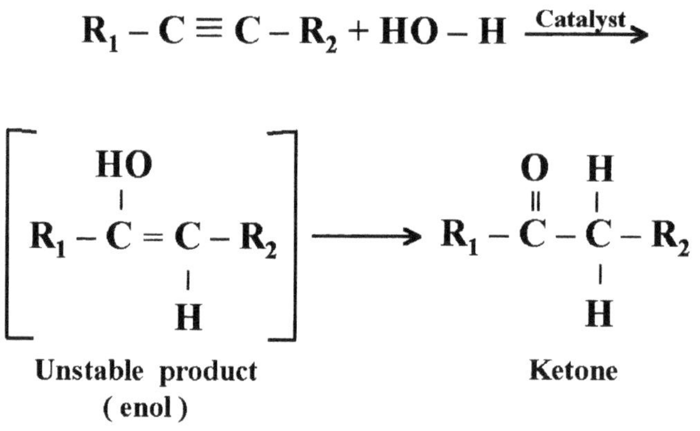

$$R_1 - C \equiv C - R_2 + HO - H \xrightarrow{\text{Catalyst}}$$

$$\left[\begin{array}{c} HO \\ | \\ R_1 - C = C - R_2 \\ | \\ H \end{array} \right] \longrightarrow R_1 - \overset{O}{\overset{\|}{C}} - \overset{H}{\overset{|}{C}} - R_2 \\ \overset{|}{H}$$

Unstable product (enol)   Ketone

Figure 1 Water addition to alkynes.

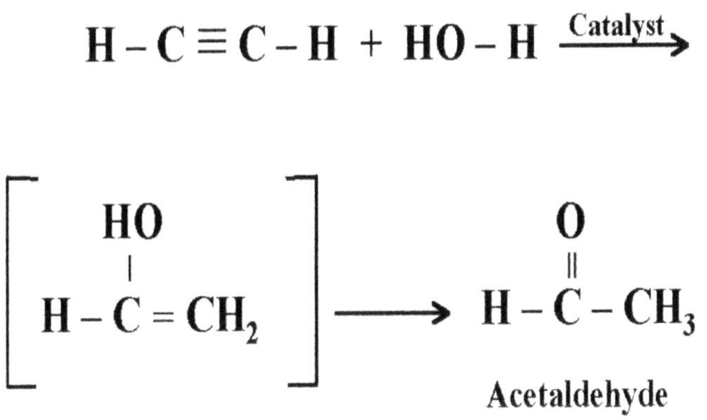

$$H - C \equiv C - H \; + \; HO - H \xrightarrow{\text{Catalyst}}$$

$$\left[\begin{array}{c} \overset{\displaystyle HO}{\underset{\displaystyle |}{}} \\ H - C = CH_2 \end{array} \right] \longrightarrow \; H - \overset{\displaystyle O}{\overset{\displaystyle \|}{C}} - CH_3$$

Acetaldehyde

Figure 2 Reaction of acetylene with water.

group, the addition of a proton to the vicinal carbon atom, and the relocation of the double bond. This rearrangement is known as *tautomerization*. Tautomers are constitutional isomers that are easily interconverted. Keto and enol tautomers, for example, are rapidly interconverted in the presence of acids and bases [2]. The enol accepts a proton at one carbon atom of the double bond to yield a cationic intermediate. The formed intermediate loses a proton from the oxygen atom to produce a keton. It is of note that, when acetylene (HCCH) itself undergoes addition of water molecule, acetaldehyde is the reaction product. Figure 2 demonstrates this reaction.

C. General Information on the Esterification Reaction

Previous reactions related to water addition to hydrocarbons having unsaturated bonding, a reaction in which water is eliminated, are known as the condensation reaction, resulting in joining two chemically different molecules. The condensation reaction of carboxylic acids with alcohols produces esters, and the process is known as esterification (Fig. 3). The esterification reaction process is acid catalyzed.

Reaction of acyl chlorides [RC(O)Cl] with alcohols also leads to synthesis of esters. Another way to synthesize esters is the transesterification process. Figure 3 demonstrates that the esterification reaction is reversible. Adding of water to the ester molecule leads to the formation of the ester substrates carboxylic acid and alcohol. The reaction process relates to hydrolysis. However, esters not only undergo acid hydrolysis (as shown in Fig. 3), they also

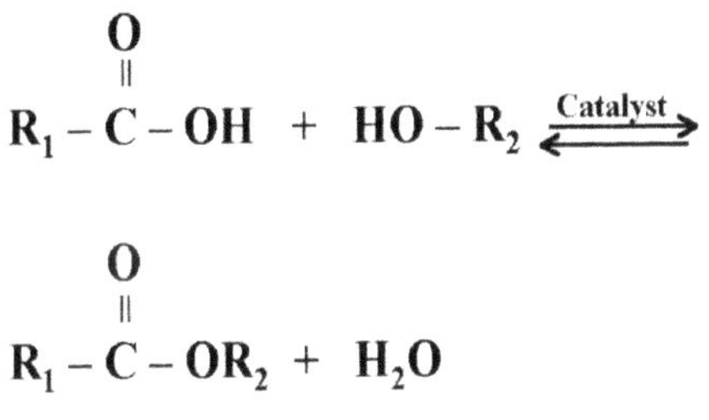

$$R_1 - \overset{\displaystyle O}{\overset{\displaystyle \|}{C}} - OH \; + \; HO - R_2 \underset{\xleftarrow{\hspace{1.2em}}}{\overset{\text{Catalyst}}{\xrightarrow{\hspace{1.2em}}}}$$

$$R_1 - \overset{\displaystyle O}{\overset{\displaystyle \|}{C}} - OR_2 \; + \; H_2O$$

Figure 3 Esterification reaction.

undergo base-promoted hydrolysis. Actually, base-promoted hydrolysis is often called saponification (soap formation). For example, refluxing an ester with aqueous sodium hydroxide produces the sodium salt (soap) of the carboxylic acid and an alcohol. The hydrolysis process is described in Section II.

D. Major Synthetic Ester Base Oils and Their Physicochemical, Tribological, and Biological Properties

Early main lubricants were natural esters contained in animal or vegetable oils. Then, a range of synthetic oils was developed. These oils included esters of long-chain alcohols and acids. Further development of esters—particularly neopentyl esters—was focused on aviation gas turbine lubricants.

Because of their specific properties, esters are presently used in many applications (engine oils, gear oils, hydraulic fluids, compressor oils). The inherent biodegradability of ester molecules provides added advantages to such applications. Typical synthetic esters used as base stocks include aliphatic monoesters, C_{36} dimer acid esters, diesters, penta-erythritol esters, trimethylolpropane esters, neopentylglycol esters, and aromatic compounds (which include phthalate esters, trimellitate esters, and complex esters). Figures 4 and 5 represent the general chemistry of aliphatic and aromatic esters, respectively.

Typical diester oils include adipates, azelates, sebacates, dodecanoates, and C_{36} dimer acid esters. Dimer acid/glycol monoester was shown [3] to be very effective as additive reducing wear. The formation of polymeric films directly on rubbing surfaces to reduce wear was suggested. Detailed description of that issue is presented in the chapter on tribo-polymerization.

Polyol esters are produced by reacting a polyhydric alcohol, such as pentaerythritol (PE), trimethylol propane (TMP), or neopentyl glycol (NPG), with a monobasic acid to give the desired ester, as shown in Fig. 4. Ester base stocks are produced from compounds having uniform molecular structures, which provide properties that can be tailored to specific tribological applications. A wide variety of raw materials, along with some fatty acid synthesis waste products [4], can be applied for the preparation of ester-type base fluids, and this can affect a number of typical properties such as hydrolytic stability, biodegradability, antifriction, and antiwear characteristics.

Important physicochemical properties of ester base stocks or lubricants include rheological properties, pour point, solvency, and thermal and hydrolytic stability. The viscosity and viscosity index of an ester are mostly controlled by both the carbon chain length of the acid, the carbon chain length of the alcohol, and the number of ester groups. Other factors relate to the size and degree of the molecule branching and the presence of cyclic groups in the molecular backbone. The pour point of the ester fluid can be decreased by [1]:

* Decreasing the acid chain length
* Decreasing the internal symmetry of the molecule
* Increasing the amount of branching
* The positioning of the branch; branching in the center of the molecule provides better pour points than branches near the end.

Viscosity and viscosity index of an ester fluid is controlled by both the chain length of the acid and the alcohol. It is noteworthy at this point that one disadvantage of very long chain molecules is their tendency to shear into smaller fragments under stress. On the other hand, increasing the linearity of the molecule increases the viscosity index. However, the presence of cyclic groups in the backbone lowers the viscosity index even more than aliphatic

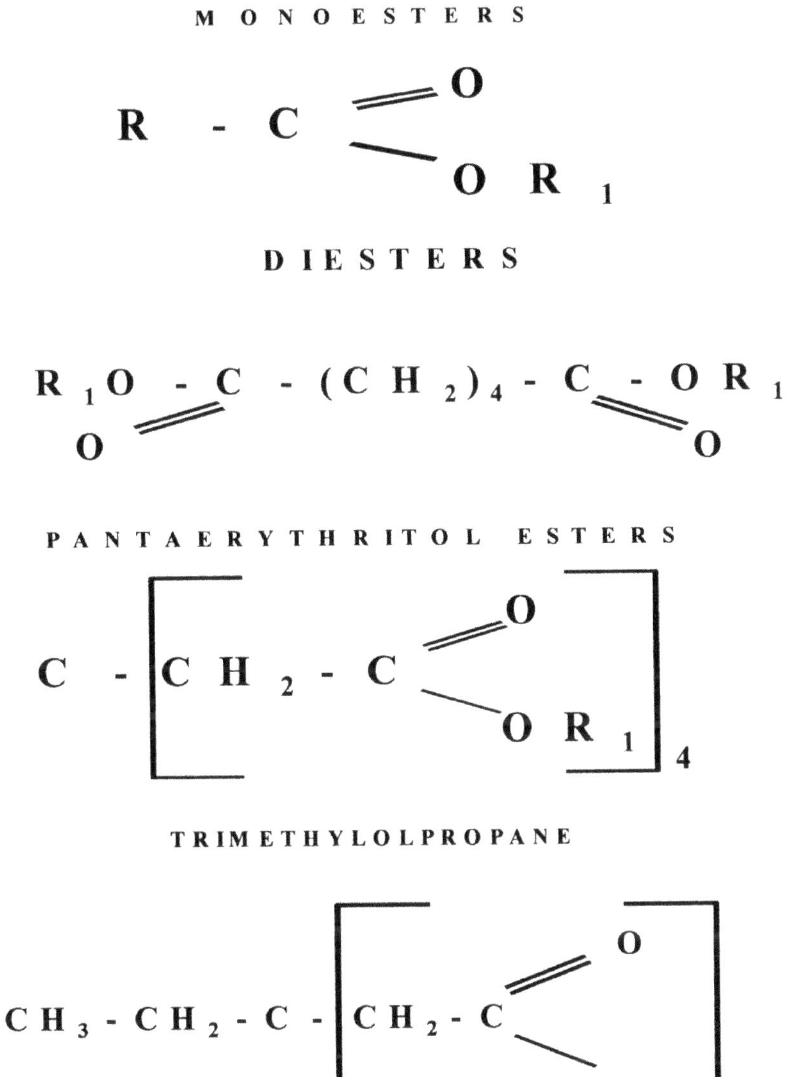

MONOESTERS

$$R - C \begin{matrix} \diagup\!\!\diagup O \\ \diagdown O R_1 \end{matrix}$$

DIESTERS

$$R_1 O - \underset{\underset{O}{\diagdown\!\!\diagdown}}{C} - (CH_2)_4 - \underset{\underset{O}{\diagdown\!\!\diagdown}}{C} - O R_1$$

PANTAERYTHRITOL ESTERS

$$C - \left[CH_2 - C \begin{matrix} \diagup\!\!\diagup O \\ \diagdown O R_1 \end{matrix} \right]_4$$

TRIMETHYLOLPROPANE

$$CH_3 - CH_2 - C - \left[CH_2 - C \begin{matrix} \diagup\!\!\diagup O \\ \diagdown O R_1 \end{matrix} \right]_3$$

Figure 4 Chemistry of aliphatic synthetic esters.

branches. Fluids prepared from mixtures of branched and straight acid of the same molecular weight have viscosity indices between those of the normal and branched acid esters; however, they demonstrate lower pour points than esters made from either straight or normal acids. Viscosity indices of polyol esters have tendency to be somewhat lower than their diester analogs and this is attributable to their molecular configuration; polyol molecule has a more compact configuration (see Fig. 4).

Another important property of esters relates to hydrolytic stability, which is controlled by both (1) molecular geometry and (2) processing parameters. If the final processing parameters of esters are not tightly controlled, they can have a major effect on the hydrolytic stability of the esters.

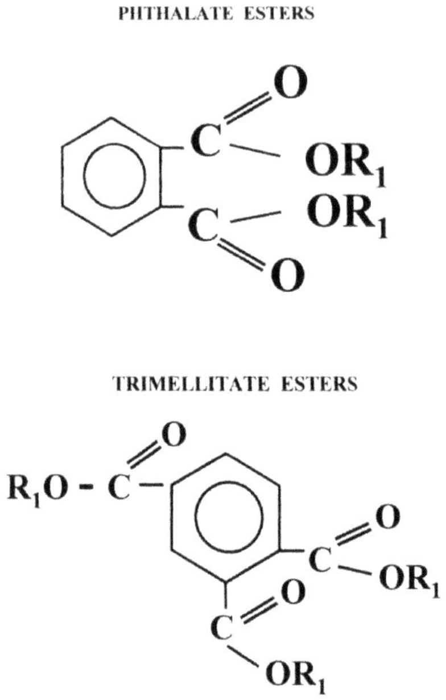

PHTHALATE ESTERS

TRIMELLITATE ESTERS

Figure 5 Chemistry of aromatic synthetic esters.

Table 1 lists the viscosity of the most important ester types at two temperatures (40 and 100°C), along with viscosity index data and other physical and biological (biodegradability) properties. More detailed discussion concerning biodegradability is presented in Section IV.

Solvency of esters is considered as compatibility with other lubricants, additives, and elastomers. Usually, they are wholly compatible with mineral base oils. This property provides them with a number of major advantages [1]:

1. There are no contamination problems, thus esters can be used in machinery that previously used mineral oil; additionally, they can be blended with mineral oil (semisynthetics) to boost their performance.
2. Most additive technology is based on mineral oil and this technology is usually directly applicable to esters.

Table 1 Selected Properties of Esters [1]. Summary of Ester Properties

	Diesters	Phthalates	Trimellitates	C_{36} dimer esters	Polyols	Polyoleates
Viscosity at 40°C	6 to 46	29 to 94	47 to 366	90 to 185	14 to 35	8 to 95
Viscosity at 100°C	2 to 8	4 to 9	7 to 22	13 to 20	3 to 6	10 to 15
Viscosity index	90 to 170	40 to 90	60 to 120	120 to 150	120 to 130	130 to 180
Pour point (°C)	−70 to −40	−50 to −30	−55 to −25	−50 to −15	−60 to −9	−40 to −5
Flash points	200 to 260	200 to 270	270 to 300	240 to 310	250 to 310	220 to 280
Thermal stability	Good	Very good	Very good	Very good	Excellent	Fair
Conradson carbon	0.01 to 0.06	0.01 to 0.03	0.01 to 0.40	0.20 to 0.70	0.01 to 0.10	?
% Biodegradability	75 to 100	46 to 88	0 to 69	18 to 78	90 to 100	80 to 100
Costs (PAO = 1)	0.9 to 2.5	0.5 to 1.0	1.5 to 2.0	1.2 to 2.8	2.0 to 2.5	0.6 to 1.5

3. Esters can be blended with other synthetics such as polyalphaolefins (PAOs). This gives esters great flexibility, while blending with other oils provides unrivalled opportunities to balance the cost of a lubricant blend against its performance.

Two possible types of ester interaction concerning elastomer compatibility comprise mostly of (1) physical interaction and, to some extent, (2) chemical interaction. The physical interaction relates to the ester diffusion through the elastomer network causing absorption of the ester lubricant by the elastomer (swelling process) and/or extraction of soluble components out of the elastomer (shrinkage process).

Because ester groups are polar, they clearly contribute to the antiwear property (lubricity). According to van der Waal [5], esters can be classified in terms of their polarity or nonpolarity by using the following formula

$$\mathrm{Non-polarity\ index} = \frac{\mathrm{Total\ number\ of\ C\ atoms} \times \mathrm{molecular\ weight}}{\mathrm{Number\ of\ carboxylic\ groups} \times 100}$$

The general rule concerning this relationship is that increasing molecular weight improves overall lubricity. On the other hand, it must be emphasized that physical and chemical interactions of esters with both metal and ceramic surfaces are extremely complex. Generally, esters have superior antiwear and friction-reducing properties than hydrocarbon oils. When added to mineral oil, esters usually improve the lubricant antiwear property, but their antiwear effectiveness is clearly controlled by the chemical structure that reflects in their polarity. Ester groups are polar, and thereby they affect the efficiency of antiwear additives as follows. Very polar ester base fluid can cover the solid tribological component surfaces instead of antiwear additives, which may result in higher wear characteristics. Accordingly, although esters provide superior antiwear properties than mineral base oil, they are generally less efficient than classical antiwear additives. However, there are specific ester chemistries that demonstrate both very effective wear reduction and friction reduction. From the viewpoint of nonpolarity index, it is possible to state that the higher the index, the lower the affinity for the metal surface.

Increasing the molecular weight improves the overall antiwear property of the ester. It is noteworthy, however, that the scatter is large because the formula takes no account of the structural aspects of the ester lubricant. Even the simplest structural difference corresponding to chain linearity or branching is not taken into account. Esters terminated by linear acids or alcohols have better lubricating properties than those made from branched acids/alcohols, and esters made from mixed acids/alcohols have intermediate lubricating properties between esters of linear acids/alcohols and esters of branched acids/alcohols [6].

Fatty acid methyl esters, commonly known as biodiesel, and other esters have also been recently proposed as diesel fuel lubricity improvers [7–9]. In the case of dicarboxylic acid esters, concentration levels of esters ranging from 500 to 750 ppm were necessary to bring the high frequency reciprocating rig (HFRR) wear scar diameter (WSD) value within the required limit of 460 µm, and any extra addition of dicarboxylic acid esters did not give any significant improvement in the lubricity of the fuels. Interestingly, among the diesters from the same dicarboxylic acid [9], an increase in the chain length of the alcohol involved in the esterification reaction leads to a higher lubrication performance. On the other hand, if the chain length of the alcohol is kept constant, an increase in dicarboxylic acid chain length does not cause significant improvement in lubricity. In the case of ethyl esters [10], the effective concentration for lowering the lubricity value to within the acceptable range of less than 460 µm is between 0.25% and 0.5%, while further addition of fatty acid esters did not lead to any spectacular increase in lubricity.

It is well known that esters under boundary lubrication conditions form carboxylic acid salts (soap) with the friction metal surface. The lubrication-related references suggest that the soap formation by ester under boundary lubrication is attributable to the hydrolysis process of the ester. This leads to the formation of a very small amount of fatty acid which, when adsorbed, attach the metal to form the corresponding soap film. Another approach assumes that the soap formation proceeds according to the negative-ion-radical action mechanism (NIRAM). This issue is discussed in detail in Section V.

Natural esters comprising of fats and oils derived from animal and vegetable sources have a long history of use in lubrication. It is also notable that vegetable oils are used as feedstock for the oleochemical industry to produce a wide variety of chemical products. These chemicals, specifically described in Ref. 11, are used in many applications, including lubricants, either as CHO additives or for the production of synthetic materials such as ester base fluids. Typical vegetable oils have the triglyceride structure, in which the fatty acids are mostly straight chains ranging from 8 to 22 carbon atoms, and may be saturated, mono-unsaturated, or polyunsaturated. Each vegetable oil is characterized by its fatty acid composition, which may nevertheless vary somewhat, depending on such factors as weather, soil conditions, and variety [12,13]. The typical acid composition of the most common vegetable oils, which together account for over 90% of world production, are presented in Table 2 [12].

Fatty acid components of vegetable esters usually contain an even number of carbon atoms. The fatty acids in natural triglycerides display positional isomerism and they are not randomly distributed among the three possible positions of glycerol. For example, palmitic acid (C_{16}) in soya and olive esters (oils) is present predominantly on the 1 and 3 positions rather than the central position of glycerol. Typical vegetable oils are composed of tri-esters and are therefore prone to hydrolysis, and this may pose problems for some applications. On the other hand, a small degree of hydrolysis may be acceptable, or even beneficial, because free acids, produced as a result of hydrolysis, are generally active as effective surface modifiers from the viewpoint of the friction reduction process [13]. Of particular importance is the application of castor oil in lubricants. Castor oil is adequate for very effective friction surface modification because it is a triglyceride with a very high content (over 80%) of ricinoleic acid (C_{18}), including also one hydroxyl group. A combination of ester functional group with hydroxyl one ensures very high adsorptivity on solid tribological components. This is because the free hydroxyl group is able to interact with polar sites on solid surfaces, thereby imparting superior lubricity. Soaps of 12-hydroxystearic acid, derived from castor oil, have superb thickening and lubricity properties and are widely used in greases. Another very specific vegetable ester oil that is of interest in lubricants is jojoba oil as a replacement of sperm whale oil, which is now banned [13]. It is emphasized that jojoba oil is not a tri-glyceride but is a wax ester composed mainly of straight-chain esters of C_{20}–C_{22} monoun-saturated acids and alcohols.

E. Lubricating Oil Additive Hydrolysis

The hydrolysis process is of particular interest for all types of ester lubricating oil additives. These additives include antiwear agents such as trialkyl phosphate esters, alkyl acid phosphate esters, triaryl phosphate esters, metal dialkyldithiophosphates (MDTPs), etc. These types of compounds can control friction and/or wear by forming films at tribological surfaces. Mechanistic studies of organo-phosphorus esters have identified two different types of reaction films: (1) those that include tricresyl phosphate (TCP), which forms thin films (0.1–2 nm) of low shear strength ($FePO_4$ and $FePO_4 2H_2O$), and (2) those which give rise to thick films (100–300 nm) of a polymeric nature. These consists of, in the main,

Table 2 Typical Fatty Acid Composition of Some Common Vegetable Oils; Values in Parentheses Are World Annual Production [12,13]

Nonsystematic name and numerical representation		Typical % fatty acid composition								
		Soya (15.3)	Sunflower (7.3)	Olive (1.9)	Peanut (3.7)	Cotton (3.6)	Palm (9.1)	Coconut (3.6)	Rape (high erucic) (7.5)	Rape (low erucic)
Caprylic	8:0							8.0		
Capric	10:0							6.0		
Lauric	12:0							47		
Myristic	14:0						1	17.5		
Palmitic	16:0	11	6.5	14	10	22	45	9	2	4
Palmitoleic	16:1					1				
Stearic	18:0	3.5	4	2	3	3	4.5	3	1	1
Oleic	18:1	22	21.5	64	42	19	38	7	15	60
Linoleic	18:2	54	66	16	38	53	10	1.8	15	20
Linolenic	18:3	8							7	9
Arachidic	20:0				1.5					
Gadoleic	20:1				1				7	2
Behenic	22:0				3					
Erucic	22:1								50	2

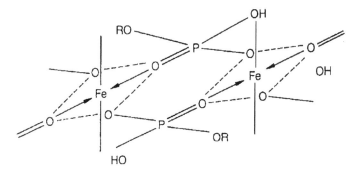

Figure 6   The iron(III) mono-alkyl/aryl phosphate oligomer. (From Ref. 14.)

iron(III) mono-alkyl/aryl phosphate oligomer, as depicted in Fig. 6 [14]. The formation mechanism of these films is believed to involve the loss of an alkyl group by hydrolysis, which generates two PO ligands for coordination.

II.   CONDITIONS LEADING TO THE ESTER HYDROLYSIS PROCESS AND ITS MECHANISMS

A.   Hydrolysis and Terms Related to the Hydrolysis Process

The hydrolysis process leads to a cleavage of a compound to generate another compound(s) via the action of water. In the case of monoesters, hydrolysis results in the formation of two compounds: alcohol and carboxylic acid. Triesters of glycerol with carboxylic acids (fat, triglyceride) under hydrolysis yields glycerol and three molecules of one or two carboxylic acid or different acids. Most of natural acids have saturated straight chain and even number of carbon atoms. Unsaturated fatty acids have the *cis* structure of double bonds. Hydrolysis of a trimellitate ester (1,2,4-benzene tricarboxylate) generates three alcohol molecules and 1,2,4-benzene tricarboxylic acid.

A general term solvolysis (*cleavage by solvent*) is defined as a nucleophilic substitution reaction in which the nucleophile is a molecule of a solvent. Considering that the solvent is water, solvolysis relates to hydrolysis. Carrying our similar reaction in methanol, the reaction is called a methanolysis. Such reactions involve the initial formation of carbocation (species in which trivalent carbon atom carries a positive charge) and the subsequent reaction of that cation with a molecule of the solvent. Nucleophile is an electron pair donor (*Lewis base*) that seeks a positively charged site in a molecule. Therefore, a nucleophilic substitution reaction is the reaction initiated by a nucleophile, in which the nucleophile reacts with the substrate to replace a substituent, which departs with an unshared electron pair. The simplest substitution is the unimolecular S_N1 reaction. The S_N1 reaction of an alkyl halide with water is an example of solvolysis—cleavage by the solvent (Fig. 7). As the solvent in this case is water, the reaction is called a hydrolysis. Should we consider another solvent, for instance methanol, the reaction will be called a methanolysis. Both these reactions involve the initial generation of a carbocation (see Fig. 7) and the subsequent reaction of that cation with a molecule of the solvent. Final solvolysis products are *tert*-butyl alcohol and methyl *tert*-butyl ether, respectively, and hydrochloric acid (HCl).

A negatively charged nucleophile is always a more reactive nucleophile than the conjugated acid (*the molecule or ion that forms when a base accepts a proton*). Accordingly,

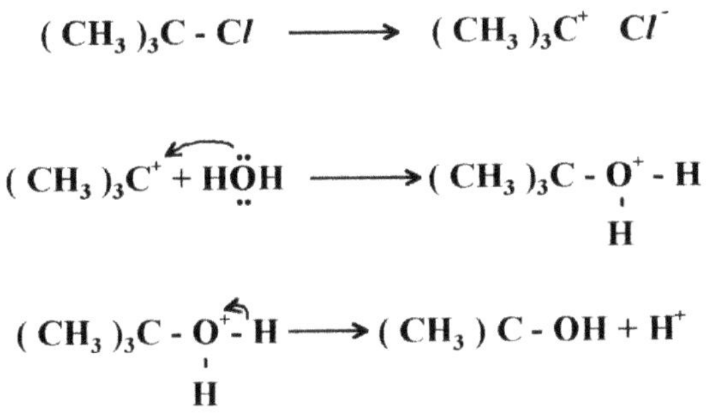

$$(CH_3)_3C \text{-} Cl \longrightarrow (CH_3)_3C^+ \quad Cl^-$$

$$(CH_3)_3C^+ + H\ddot{O}H \longrightarrow (CH_3)_3C \text{-} \underset{\overset{|}{H}}{O^+} \text{-} H$$

$$(CH_3)_3C \text{-} \underset{\overset{|}{H}}{O^+} \text{-} H \longrightarrow (CH_3) C \text{-} OH + H^+$$

Figure 7 Hydrolysis of tert-butyl chloride as an example of the solvolysis process.

HO^- is a better nucleophile than H_2O. It is also of note that in the alcoholysis reaction, RO^- is better than alcohol (ROH). In a group of nucleophiles in which the nucleophilic atom is the same, nucleophilicities parallel basicities. For example, oxygen compounds show the following order of reactivity [2]:

$$RO^- > HO^- >> RCO_2^- > ROH > H_2O$$

This is also their order of basicity. An alkoxide ion (RO^-) is a slightly stronger base than a hydroxide ion (HO^-), and a hydroxide ion is a much stronger base than a carboxylate (RCO_2^-).

Hydrolysis is a part of the hydrolytic processes that encompasses the reactions of both inorganic and organic chemistry, in which the water action effects a double decomposition with another compound. Figure 8 depicts exemplary reactions in which the hydroxyl group is attached to one component and hydrogen to another.

Although the term hydrolysis denotes decomposition by water, reactions in which water brings about effective hydrolysis unaided is rare. Typical hydrolytic processes of esters relate to (1) acid-catalyzed reactions, (2) base-promoted reactions, and (3) the enzyme lipase-catalyzed reactions. High temperatures and pressures are usually required. Accordingly, in the field of organic chemistry, the term hydrolysis is also applied to cover the reactions in which alkali or acid is added to water. A typical example of a base (alkaline)-promoted hydrolytic process is the hydrolysis of esters to produce alcohols and salts (soap).

B. Reverse Reactions of the Hydrolysis Process—Esterification

In ester production, a carboxylic acid is reacted with an alcohol. Figure 3 clearly demonstrates that this reaction is combined with the elimination of water. Esterification

$$AB + HO \text{-} H \longrightarrow A \text{-} OH + B \text{-} H$$

$$NaCN + HO \text{-} H \longrightarrow Na \text{-} OH + HCN$$

$$RCl + HO \text{-} H \longrightarrow R \text{-} OH + HCl$$

Figure 8 Examples of hydrolytic processes.

reactions are acid catalyzed. They proceed very slowly in the absence of strong acids. However, they reach equilibrium in a matter of few hours when an acid and alcohol are refluxed with a small amount of concentrated sulfuric acid (H_2SO_4) or hydrogen chloride (HCl). When benzoic acid reacts with methanol (CH_3OH) that has been labeled with ^{18}O, the labeled oxygen appears in the ester [2]. This means that the formed water molecule is produced from the acid hydroxyl group (OH) and the alcohol hydrogen atom (H), as shown in Fig. 9.

The result of the labeling experiment and the fact that esterifications are acid-catalyzed are both consistent with the detailed mechanism for acid-catalyzed esterification as described in Ref. 2. Following the forward reaction in Fig. 9, we have the acid-catalyzed esterification of an acid.

On the other hand, following the reverse reactions, we have the acid-catalyzed hydrolysis of the ester. The same is due to the reaction presented in Fig. 3. The final reaction result depends on the reaction conditions. To hydrolyze an ester R_1COOR_2, it is necessary to apply a large excess of water. To esterify an acid R_1COOH, we have to use an excess of the alcohol R_2OH or remove the formed water. Usually, either the alcohol or acid substrate, depending on whichever is more volatile, is added in excess. After the reaction is completed, the excess of the more volatile substrate is removed by distillation.

C. Acid-Catalyzed Ester Hydrolysis Process

The reaction between pure ester and water is very slow. Ester lubricants containing some 0.1% water can be stored for several years at ambient temperature and undergo essentially no reaction [15]. Therefore, to promote the hydrolysis reaction at a significant rate, a catalyst is required. The hydrolysis process of esters via the use of water and a mineral acid conducts to an equilibrium mixture of alcohol, carboxylic acid, and the remaining ester. As mentioned previously, complete reaction can only be achieved by the removal of alcohol or acid from the equilibrium. Because esters have poor solubility in water, the reaction rate in dilute acids is fairly low. To facilitate the reaction, some emulsifiers such as sulfonated aromatic compounds or sulfonated oleic acid are added. In the acid-catalyzed hydrolysis, a hydrogen ion (H^+) adds to the carboxyl oxygen of the ester linkage, transiently converting it to a carbonium ion, which rapidly adds water to form a positively charged tetrahedral intermediate. This tetrahedral intermediate then separates into carboxylic acid and alcohol,

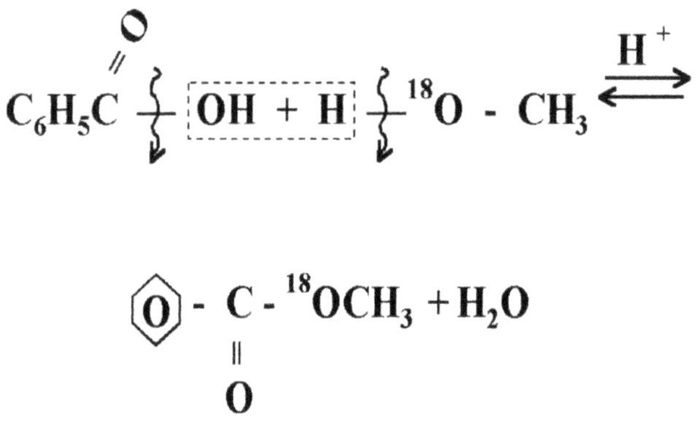

$$C_6H_5C \overset{O}{\|} \{ \overline{OH + H} \} \overset{18}{} O - CH_3 \overset{H^+}{\rightleftharpoons}$$

$$\underset{\|}{\overset{}{(O)} - C - {}^{18}OCH_3 + H_2O}$$

Figure 9 Water splitting mechanism in the esterification reaction of benzoic acid with ^{18}O-labeled methanol.

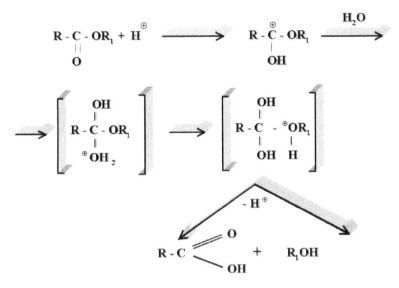

Figure 10   Acid-catalyzed ester hydrolysis.

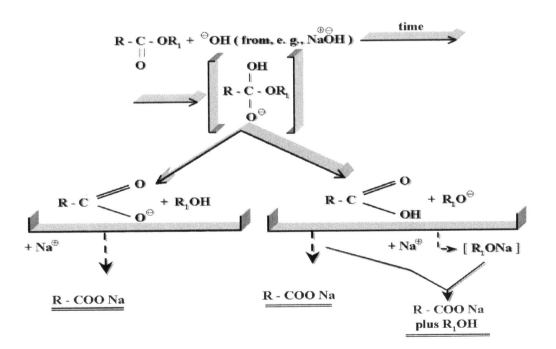

Figure 11   Based-promoted ester hydrolysis process.

regenerating a proton that can then catalyze further reaction. The mechanism of acid-catalyzed hydrolysis is depicted in Fig. 10. It is of note that reactions involving tetrahedral intermediates are subject to steric and electronic effects. Electron-withdrawing substituents facilitate basic hydrolysis, while electron-donating and bulky substituents exert the opposite effect [16]. Steric effects in acid-catalyzed hydrolysis are similar to those in base-catalyzed hydrolysis; however, electronic effects are much less important in acid-catalyzed reactions.

D. Base-Promoted Hydrolysis Process

In base-promoted hydrolysis, also known as saponification, the ester is hydrolyzed with a stoichiometric amount of alkali. The irreversible formation of carboxylate anion leads the reaction to completion. Figure 11 demonstrates that hydroxide ion attacks the carbonyl group to form—similar to the case of acid-catalyzed hydrolysis—a tetrahedral intermediate. Loss of alkoxide generates the acid, which is rapidly deprotonated to the carboxylate anion in basic solution. The hydroxide ion used up in this reaction is not regenerated at the end. Thus, it is not the typical catalytic process. Actually, it is the process converting a strong base (HO^-) into carboxylate anion, being a weak base. Accordingly, it is to emphasize that the net effect is that the base behaves just as a stoichiometric reagent.

III. HYDROLYSIS OF DIFFERENT ESTER TYPES

A. General Information on Hydrolytic Stability

As already shown, hydrolysis applied to organic molecules can be considered a reversal of such reactions as esterification. The same is due to amide formation. Examples of other hydrolytic processes are depicted in Fig. 12.

In inorganic chemistry, the hydrolysis process is also called *aquation*. This type of process might also be referred to "hydrolysis of ceramics." That specific approach is presented and discussed in Section V.C.

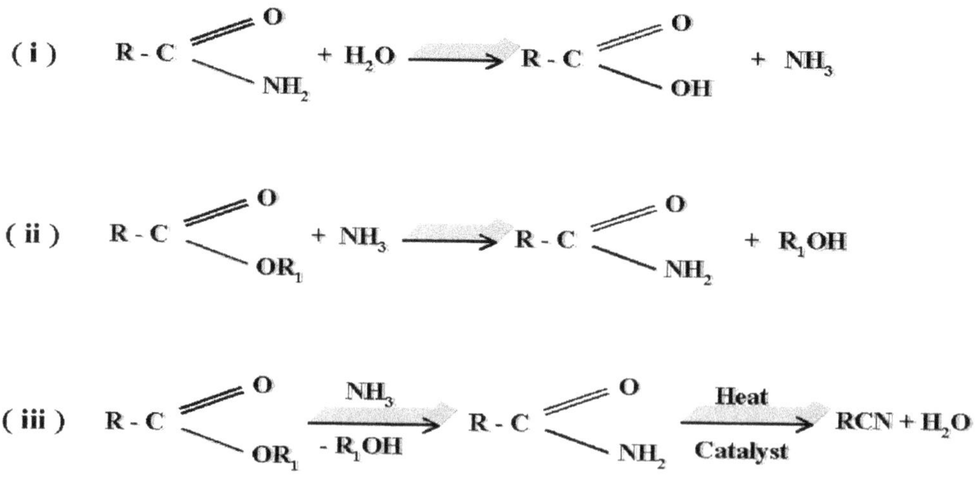

Figure 12 Examples of other than aliphatic ester's hydrolysis processes: (i) amide hydrolysis, (ii) ester aminolyis, (iii) ester aminolysis.

In contrast to conventional mineral oils and synthetic hydrocarbons, e.g., polyalphaoleffins, esters are unique in that they include the polar ester group that provides a strong affinity for tribological contact solid surfaces, particularly those made of metals. Considering the wide variety of ester choices presently available, and bearing in mind any type of contact with water, it is necessary to know the ester hydrolytic stability because hydrolysis results in poor lubrication. Usually, ester structure and processing parameters determine water solubility. This is of particular importance for esters used in metalworking fluid formulations. The type and number of polar moieties in the ester structure determines its behavior with water. Hydrophilic esters will readily absorb moisture-promoting hydrolysis; hydrophobic esters do not combine with water and are more resistant to hydrolysis. However, the rate of hydrolysis is mostly controlled by the presence of other materials and impurities, such as acids and bases, because the hydrolysis process proceeds either according to the mechanism of acid-catalyzed or base-promoted reactions. Esters containing higher amounts of residual acid from unreacted raw materials, higher initial total acid numbers, are more prone to hydrolysis. As previously noted, strong mineral acids are used to catalyze the esterification process. Because acid-catalyzed esterification and hydrolysis proceed by essentially the same mechanism, the carboxylic acid content must be considered when testing an ester for hydrolysis. Another important factor considers the additive effect on rate of hydrolysis. According to Boyde [15], the selection of additive chemistry for ester base lubricants in applications where hydrolytic stability is an issue must take into account the effect of initial acid value on the rate of ester hydrolysis. Usually, commercial ester base fluids have very low acid values, so acidic additives, even at a very low treatment rate, may make a large proportionate change in the acid value. This is clearly demonstrated by the data in Table 3.

It is of note that evident care is required with, for instance, fatty acid anticorrosion agents or friction modifiers. A less obvious factor, perhaps, is that phosphate ester antiwear additives may have significant acid values, and may themselves be prone to hydrolysis reactions, which generate strong acids, often more rapidly than organic esters [15]. In practice, ester lubricants are required to be hydrolytically stable because they are often exposed to humid atmospheres during use. Actually, hydrolysis has proven to be a less serious problem than theoretical predictions would indicate. The problem becomes more serious when ester-type compounds, particularly antiwear additives, e.g., phosphate esters, MDTPs, are coming into contact with appreciable quantities of water. That occurs in water-

Table 3 Effect of Acidic Additives on the Hydrolytic Stability of Polyol Esters (PE) as Determined Using 2-Week Beverage-Bottle Test [15]

Ester	Additives	Initial AV	Final AV	Viscosity change (cSt)
PE linear		0.05	1.98	−0.7
PE linear	AW	0.33	56.45	−2.5
PE linear	AW	0.06	19.54	−1.2
PE linear	AW + AC	0.28	43.45	−1.9
PE branched		0.06	0.25	−0.06
PE branched	AW	0.34	0.89	−0.4

AV = acid value (mg KOH/g).
AW = acidic antiwear additive package.
AC = epoxide acid catcher.

based metalworking fluids, engine oils, etc. Generally, the hydrolytic stability of esters depends on four major factors: (1) the molecular geometry of the ester used, (2) the processing parameters of the ester, (3) the additives used in the ester, and (4) the system in which it is used [1,6]. The final processing parameters of the ester need to be well controlled because they have a major impact on the hydrolytic stability of the product. The molecular geometry of esters can affect the hydrolytic stability in several ways. By sterically hindering the ester portion of the molecule (see Fig. 4), it is possible to retard hydrolysis. This is best achieved by sterically hindering the ester linkage of the acid portion of the ester. Conversely, hindrance of the alcohol has relatively little effect. To achieve adequate steric hindrance, geminal (*substituents that are at the same atom*) dibranched acids (such as neoheptanoic acid) and 2-ethyl hexanoic acid have been used [6]. The hydrolytic stability of the polyols is generally considered as superior to that of the diesters. The first step in hydrolysis is cleavage at the ester linkage. Thus, if the ester linkage can be hindered, hydrolysis will occur at much smaller rate. One clear way for causing hindrance is to use a branched acid, especially those branched near the ester linkage, for example, 2-hydroxyethylheksyl acid. Acidic additives can have a major negative effect on the stability of the ester. High levels of acidity can autocatalyze the breakdown of the ester because acids act as a catalyst.

B.   Hydrolysis of CHO Esters and Their Hydrolytic Stability

The CHO esters encompass a wide variety of products starting from low-molecular-weight compounds, e.g., methyl formate, ethyl acetate, vinyl acetate, via glyceryl triproprionate and di(2-ethylhexyl)phthalate, up to glycerol esters, trimethylolpropane (TMP) and neopentyl-glycol (NPG) esters. Properties, production, and sales of very many carboxylic esters along with their specific uses are listed in Ref. 16. Monohydric alcohol esters of dibasic acids and polyol esters of monobasic acids are typical synthetic lubricants. Other works [1,6] describe typical ester lubricant products. Usually, they are prepared from the following carboxylic acids and alcohols: (1) C_6–C_{10} monobasic acids such as heptanoic, nonanoic, lauric, stearic, isostearic acids, and C_6–C_{10} dibasic acids, mostly including adipic, azelaic, and sebacic acids; and (2) C_8–C_{13} monohydric alcohols such as 2-ethylhexyl, isooctyl, isodecyl, isotridecyl alcohols, and polymethylol chemicals such as trimethylolpropane, pentaerythritol, and dipentaerythritol. They are used as base oils in high-performance lubricants. Weller and Perez [17] studied the effect of structure of different base stocks by using a standard four-ball wear test machine to measure friction and wear. Several esters based on various chemical structures of the alcohol and acid moieties used to synthesize the ester were evaluated. Carboxylic acids varied in length of straight chain, and comparisons included both branched and linear compounds. The alcohol substrates included neopentyl glycol (NPG), trimethyl-olpropane (TMP), and pentaerythritol (PE). The results presented in Ref. 17 demonstrate that alcohol structure, acid chain length, and acid chain branching affect the delta wear with ester base stocks. Actually, only acid chain length influences the friction coefficient when working with these base fluids. Difficulty exists in predicting additive responses in the fluids.

As performance requirements become more demanding, the importance of ester products is growing in both lubricant and fuels applications. One generic area of substantial need is that of enhanced oxidative, thermal, and hydrolytic stability. This manifests itself in a number of ways: (1) the desire for longer drain intervals for engine oils, (2) the need to slow down viscosity growth of lubricants, (3) control of inlet valve and combustion chamber deposits, etc. Among the esters presently available to produce high-quality lubricating oils, further enhancements are possible. Detailed information concerning both hydrolysis and formation of esters of organic acids (CHO esters) along with hydrolysis mechanisms is desribed [18]. It is noteworthy at this point that neopentyl glycol esters encompass saturated esters, e.g., aviation polyols, synthesized from short-chain carboxylic acids, and oleochemical esters, e.g., TMP oleate, made from longer-chain caboxylic acids.

For aromatic esters, the carbon–oxygen double bond (CO) of the ester group is conjugated with the π-bonding of the benzene ring and this conjugation stabilizes the CO bond relative to the two CO single bonds, which are present in the tetrahedral intermediate during hydrolysis [15]. Accordingly, the activation energy for hydrolysis of aromatic esters is higher than that for simple polyol esters and diesters. In polyol esters, branching at the 2 or 3 position relative to the ester carbon very significantly reduces the rate of hydrolysis compared with analogous linear esters, as shown in Table 4 [15].

Ester compounds demonstrate much better friction surface affinity than aliphatic hydrocarbons. Compounds containing mainly linear acid or alcohol groups can provide a more coherent, tightly packed surface film, and have therefore superior lubricity compared to esters containing mostly branched groups. According to Boyde [15], higher-molecular-weight esters form a thicker surface film, so lubricity generally increases with increasing molecular weight.

By and large, it is possible to say that, from the viewpoint of friction and wear characteristics, many ester-type lubricants, which are also biodegradable, are similar to or even better than common mineral lubricating oils. Some esters can also be used as antiwear and/or friction modifiers. Presently, the enhancement of the hydraulic fluid fire resistance and environmental characteristics are of particular significance. With respect to these types of demands, generally esters and particularly, TMP and PE esters, have attracted considerable attention in the industry. Major physical properties, tribological characteristics, and performance of fatty acid esters used as hydraulic fluids are described in Ref. 19. Most recently, several cholesterol ester liquid crystals (e.g., cholesteryl acetate, cholesteryl pelargonate) were investigated as film formation additives in nano scale [20]. The technique of relative optical intensity was applied to investigate the effect of molecular-order degree of cholesterol esters in a model base oil (hexadecane) on lubricating properties in thin film lubrication regime. It was found that film thickness is closely related to the number of carbon atoms in the liquid crystal alkyl chain, its polarity, the concentration, and the applied external direct current (DC) voltage. An external electrical field makes lubricant film thicker. When the film is thicker than 30 nm for hexadecane with various liquid crystal additives, further increase of the voltage has little influence on the thickness. Shen et al. [20] emphasize that film-forming properties in thin film lubrication regime are possibly related to the molecular order degree of lubricant.

Table 4 Comparison of the Hydrolytic Stability of Different Ester Structures (2-Week Beverage-Bottle Test, ASTM D-2619) [15]

Ester type	ISO grade	Viscosity change (cSt)	Initial AV (mg KOH/g)	Final AV (mg KOH/g)	Cu weight change (%)
Diester	46	−1.6	0.09	2.08	−0.01
Phthalate	46	+0.3	0.02	0.25	−0.03
Trimellitate	46	−1.2	0.04	0.43	−0.06
Linear polyol	32	−0.7	0.05	1.93	−0.06
Branched polyol	46	−0.6	0.06	0.19	0.00
TMP oleate	46	−1.7	2.00	14.02	−0.05

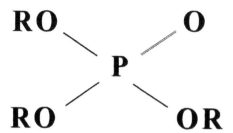

Where R represents an alkyl or aryl group or a mixture
of alkyl and / or aryl substitutnets

Figure 13   General structure of phosphate esters.

C. Hydrolysis of Phosphate Esters

Phosphate esters have gained importance as both synthetic base fluids (for hydraulic and compressor oils) and antiwear additives. However, their initial use in lubrication was as antiwear additives. Being esters of orthophosphoric acid, they have the structure as depicted in Fig. 13.

These ester types can be broadly divided into trialkyl-type phosphates and triaryl-type phosphate esters. Properties of phosphate esters vary depending on the choice of substituents, and these are usually selected to give optimum performance for a specific application. Accordingly, these products can range from low-viscosity and water-soluble liquids, to insoluble solids. They are fire-resistant and provide superior lubricity. However, use of these products is limited mostly because of their low hydrolytic thermal stability. From the viewpoint of phosphate ester hydrolytic stability, aryl esters are superior to the alkyl esters. Increasing chain length and degree of branching of the alkyl group lead to significant improvement in hydrolytic stabilities. Alkylaryl phosphates tend to be more susceptible to hydrolysis than the triaryl or trialkyl esters [1]. It is emphasized that hydrolysis can lead to serious problems, because acid esters are produced that may involve corrosion and also catalyze further degradation. Both hydrolysis (Fig. 14) and thermal degradation of phosphate esters (Fig. 15) lead to acid ester and alcohol or olefin, respectively.

The thermal stability of triaryl phosphates is significantly superior to that of the trialkyl esters, which degrade thermally via a mechanism analogous to the carboxylic esters. The use of neopentyl alcohol or its homologs yields a β-hindered ester, in which this

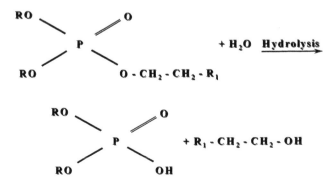

Figure 14   Hydrolysis process of phosphate esters.

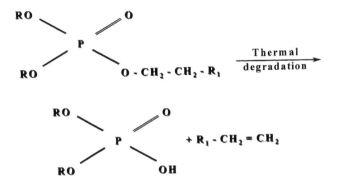

Figure 15 Thermal degradation of phosphate esters.

decomposition mechanism is blocked, leading to much improved thermal stability. Phosphate esters are widely applied as fire-resistant hydraulic fluids. Fire resistance, tribological characteristics, and performance of different phosphate ester chemistries are presented and discussed in Ref. 21. Thus, considering the friction surface modification process under sliding conditions lubricated either with phosphate esters or any lubricant including a phosphate ester antiwear agent, hydrolysis generates acid ester that forms the phosphate ester antiwear film. Certain types of Lewis acids, e.g., metallic halides, or chlorine compounds can have a catalytic action on the hydrolysis process.

According to Bovington [14], the process of wear control by reactive film formation involves the use of a corrosive wear process to limit the rate of an adhesive wear process. Then, the protective film formed is removed during the sliding process and has to be replaced in the time between successive contacts. Detailed knowledge of the role of hydrolysis in that wear-corrosion process remains rather unclear at this stage. Figure 16 depicts the influence of phosphorus-containing antiwear additives of different chemical reactivity.

D. Hydrolysis of Metal Dialkyldithiophosphates

Metal dialkyldithiophosphates (MDTPs), particularly ZnDTPs, are still the most commonly used load-carrying additives, imparting excellent antiwear properties to lubricating

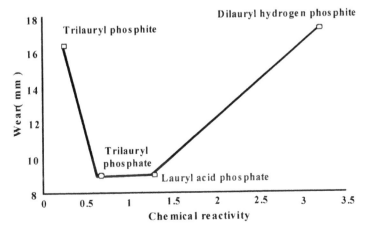

Figure 16 The influence of anti-wear agents of differing chemical activities. (From Ref. 14.)

oils. They have found an extensive application as multifunctional additives in various lubricants. Both the dialkyl and diaryl ZnDTPs are used to control cam and tappet wear. These materials are unique, because in addition to providing valve train wear protection, they also act as copper–lead bearing corrosion inhibitors. As very effective oxidant inhibitors, they also control the oil-thickening tendency in automotive crankcase application. Hydrolysis of MDTPs, especially ZnDTPs, is of significant importance. The hydrolytic stability of these compounds is controlled by the chemistry of thiophosphates. Excellent hydrolytic stability of nDTPs is provided by primary dithiophosphates, followed by secondary ones. Aryl ZnDTPs have bad hydrolytic stability. ZnDTF's effectiveness as wear protection agents is in line with their hydrolytic stability. The higher hydrolytic stability, the better friction surface wear protection. Thermal stability of these compounds is just opposite to the hydrolytic stability data. The most thermally stable are aryl ZnDTPs, which show the worst hydrolytic stability.

IV. HYDROLYTIC STABILITY OF ESTERS AND THEIR BIODEGRADABILITY

Biodegradation is the oxidation conversion process of organic chemicals or materials into carbon dioxide (CO_2), H_2O, and bacterial body via action of adequate bacteria. Accordingly, biodegradability is the proneness of a substance to be decomposed by microorganisms. Fatty acids, amino acids, etc. are usually readily biodegradable. The biochemistry of microbial attack on natural and synthetic esters is quite well known. Biodegradation of lubricants is described in Ref. 22. Pitter and Chudoba [23] present the importance of ester hydrolysis in the biodegradation process. The hydrolysis process is the first step in biodegradation of esters, followed by β-oxidation of long chain hydrocarbons and, in the case of aromatic esters, oxygenase attack on aromatic nuclei. Versino and Novaria [24]

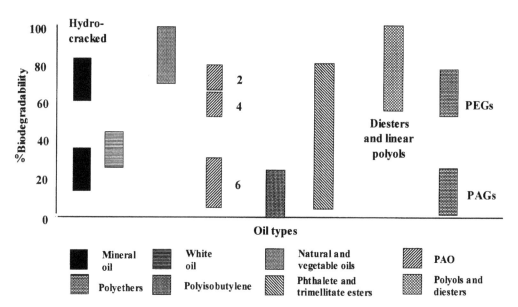

Figure 17 Biodegradability of different lubricants as measured by the CEC-L-33-A-94 test. (From Ref. 1.)

compared the biogradability of several polyol esters and diesters with rape-seed oil, and found them to be very similar. On the other hand, diethyl phthalate was much slower to degrade. Other works [25,26] presented the esterase-catalyzed steps of catabolic pathways for phthalate esters. Table 1 provides the typical ranges of biodegradability of major synthetic carboxylic acid esters used as lubricants. The best biodegradability show: polyols (90–100%), polyoleates (80–100%), and diesters (75–100%). Trimellitates are the least prone to biodegradability. The main feature that slow or reduce microbial breakdown encompass [1] the following: (1) position and degree of branching reducing the β-oxidation process, (2) degree to which ester hydrolysis is inhibited, (3) degree of saturation in the molecule, and (4) increase in molecular weight of the ester. Figure 17 depicts the biodegradabilities of a wide range of lubricants.

V.   ACID-CATALYZED AND BASE-PROMOTED HYDROLYSIS VS. TRIBOCHEMICAL REACTIONS OF ESTERS

A.   Are Typical Boundary Lubrication Operating Conditions Able to Initiate Hydrolysis of Esters?

Esters as polar compounds affect the efficiency of antiwear additives. Using a very polar base fluid, it may cover the metal surfaces instead of the antiwear additives. Esters having superior tribological properties than hydrocarbon oils when added to mineral oil usually improve the lubricant antiwear property. Esters under boundary lubrication conditions produce salt (soap) with the friction metal surface. The lubrication-related literature suggests that the soap formation by ester under boundary lubrication is the result of the hydrolysis process of the ester. This leads to the formation of a very small amount of fatty acid which, when adsorbed, attach the metal to form the corresponding soap film. On the other hand, it is supposed [27,28] that the soap formation proceeds according to the negative-ion-radical action mechanism (NIRAM) approach. The NIRAM concept assumes that the soap formation is a result of the ester C–O bond cleavage leading to the formation of the $RCOO^-$ species that reacts with the positively charged surface spot to form the soap.

It has been assumed [29] that alcohols are physically adsorbed on the reactive metal surfaces; esters with reactive metals such as zinc and cadmium, however, cause chemical attack. It has been emphasized that the reaction with ester is a result of hydrolysis [29,30]. As emphasized earlier, the hydrolysis reaction is known to be catalyzed by acids or promoted by bases. An additional example of base-promoted hydrolysis is that of the saponification of monolayers of α-monostearin [31]; the resulting glycerine dissolved, while the stearic acid anion remained in a mixed film with the reactant. Bruice [32] studied the reverse type of process, the lactonization of γ-hydroxystearic acid (on acid substrates). Alkaline and acidic hydrolysis of esters can be summarized as follows [2,6,33,34]. A carboxylic ester is hydrolyzed to a carboxylic acid and on alcohol with *aqueous acid or aqueous base*. Base promotes hydrolysis of esters by providing the strongly nucleophilic reagent HO^-. It is self-evident that under alkaline conditions, the carboxylic acid is obtained as its salt, e.g., the sodium salt, from which it can be liberated via the addition of mineral acid. The alkali-promoted reaction is essentially irreversible because of a resonance-stabilized carboxylate anion, which shows little tendency to react with an alcohol.

Based on information mentioned above, it is difficult to realize that under typical boundary lubrication conditions, mineral acid might be present to catalyze the ester hydrolysis process. To check this, we have carried out a set of experiments aiming at finding if typical conditions existing under boundary lubrication conditions could cause the hydrolysis process. Under these conditions, but without a catalyst, we have not found any evidence for the acid formation from aliphatic esters. Accordingly, it was hypothesized that, also during the friction process under boundary lubrication conditions, lubricated by aliphatic esters, the ester hydrolysis process cannot proceed without an adequate catalyst. Research described in Ref. 35 aimed at testing the validity of the hypothesis presented above and to discuss if the NIRAM approach is applicable to account for the soap formation mechanism from esters under boundary lubrication conditions. It was the first phase of a broader study and related only to one base fluid *n*-hexadecane. The results obtained, presented below, allowed us to answer the question asked in this subsection—"Are typical boundary lubrication operating conditions able to initiate the classical CHO ester hydrolysis without an adequate catalyst?" The answer is: No, they cannot! However, they can initiate tribochemical reactions with of the ester with the contact solids.

B. Experiments and Results

1. Apparatus and Experimental Details

The wear tests were carried out by using a pin-on-disc tribometer made in the Terotechnology Institute in Radom, Poland. Both tribological elements, balls and disks, were made from bearing steel LH15 (AISI 52100). The specimens were clamped in place with stainless steel holders that also contained the lubricant fully flooding the contact region. A dead-weight loading system applied the normal force. Aliphatic esters and equimolar mixtures of their substrates (acid plus alcohol) were used as the additives in the performed experiments. All these additives were used at six concentrations, ranging from 0.03 to 1.5% wt., in hexadecane. Tests were carried out under the same operating conditions. Table 5 summarizes the test conditions.

The test conditions were designed to result in boundary lubrication at the sliding interface. Hexadecane was used as the model base oil for the all tested fluids. The additives included: (i) five aliphatic esters of palmitic acid (C_{16}) and the following alcohols: 1-octanol (C_8), 1-decanol (C_{10}), 1-dodecanol (C_{12}), 1-hexadecanol (C_{16}), and 1-ocotdecanol (C_{18}); and (ii) five equimolar mixtures of these ester substrates. All the additives were used at six concentrations—0.03, 0.1, 0.5, 1, 1.2, and 1.5% wt. All the esters were synthesized in our laboratory. The synthesis was carried out in toluene in the presence of *p*-toluenesulfonic acid as catalyst. The ester structure was confirmed by Fourier-transform infrared (FTIR) spectrophotometry. Palmitic acid and the alcohols used in this study were commercially obtained (of the highest-grade purity available). Their purity was confirmed by measuring their refractive index.

Table 5 Experimental Set-up and Operating Conditions

Material system	Steel-on-steel
Geometry	Sphere-on-flat
Specimens	
Ball	3.18 mm diameter; 63 HRC bearing steel; $R_a = 0.3$–0.35 μm
Disk	25.4 mm diameter; 7 mm thickness; $R_a = 0.5$–0.55 μm
Applied load	9.81 N
Sliding velocity	0.25 m/sec
Sliding distance	500 m
Wear track radius	8 mm
Ambient temperature	25°C

Prior to use, the steel test specimens were ultrasonically cleaned in acetone for 20 min. The same cleaning procedure was used for the seal elements of the test rig. A minimum of three tests was performed for each lubricant. The seals were cleaned before testing the next lubricant. The volume of material worn from the ball was calculated from the diameter of the spherical segment removed (wear scar diameter); this was measured after unloading the specimens, using a photomacroscope. Wear results obtained for each lubricant are presented as the relative wear. This is 100 minus the ratio of the ball wear volume obtained for a lubricant including an additive and the ball volume wear for the base fluid, hexadecane, multiplied by 100.

2. Test Results

The relative ball wear reduction results for steel lubricated with the tested lubricants are summarized in Table 6 [35]. PAL stands for palmitate and C_n denotes the alcohol used to make the ester. Thus, PAL-C_8 is octyl palmitate and PAL-C_{16} is hexadecyl palmitate. The mixtures of palmitic acid with different alcohols are denoted by $C_{16} + C_n$, where C_{16} stands for palmitic acid and C_n relates to the alcohol. Accordingly, $C_{16} + C_8$ is the equimolar mixture of palmitic acid and 1-octanol; $C_{16} + C_{16}$ denotes the mixture of palmitic acid and 1-hexadecanol. The wear reduction for the system lubricated by plain n-hexadecane equals zero.

3. Discussion of the Results

Experiments performed in the present work were designed in a way that aimed at checking the effect of a possible breaking of the ester bond by hydrolysis on the lubricant antiwear property. Thus, two series of lubricants have been selected. The first lubricant series includes n-hexadecane with ester additives at the concentrations ranging from 0.03 to 1.5% wt. The second series of lubricants relates to mixtures of these ester substrates (acid plus alcohol), added to the base fluid (hexadecane) at the same concentrations as for the esters. The equimolar ratio of the acid to alcohol was used for all the mixtures applied. If, under the testing conditions, the ester hydrolysis actually involves the breaking of the ester bond to

Table 6 Relative Ball Wear Reduction for Tested Additives at Various Concentrations

	Relative ball wear reduction [%]					
	Additive concentration in n-hexadecane [% wt.]					
Additive	1.5	1.2	1	0.5	0.1	0.03
PAL – C_8	78.0	75.5	69.2	48.4	21.5	6.4
$C_{16} + C_8$	77.2	77.2	77.3	77.2	77.1	77.2
PAL – C_{10}	77.0	74.0	67.5	46.0	20.8	6.7
$C_{16} + C_{10}$	81.1	81.2	81.1	80.9	81.2	80.9
PAL – C_{12}	60.1	57.3	51.4	37.0	18.0	8.7
$C_{16} + C_{12}$	90.1	90.2	90.1	90.2	89.9	90.1
PAL – C_{16}	79.8	79.3	78.7	55.0	27.8	18.8
$C_{16} + C_{16}$	96.3	96.3	96.4	96.3	96.4	96.5
PAL – C_{18}	63.0	59.9	53.8	38.9	19.6	10.0
$C_{16} + C_{18}$	91.6	91.8	91.6	91.7	91.7	91.8
$C_{16} + C_{14}$	92.4	92.5	92.6	92.4	92.3	92.5

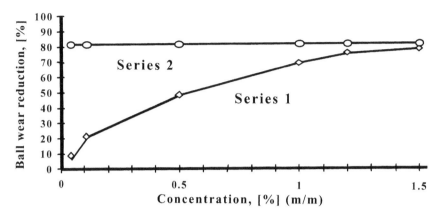

Figure 18 Influence of the solution concentration of octyl palmitate (Series 1) and equimolar mixture of palmitic acid and 1-octanol (Series 2) in hexadecane on the ball wear.

produce an acid and an alcohol, the antiwear behavior of the esters should correspond to the antiwear behavior of their substrates. In other words, the equimolar acid/alcohol mixture at very low concentration in hexadecane should not provide higher relative wear reduction of the test ball than the ester at a much higher concentration.

The wear reduction data collected in Table 6 clearly demonstrate that the lowest wear reduction is observed for the lowest concentration of the esters; however, the mixtures of palmitic acid and different alcohols reduce wear very dramatically at this concentration. This finding is of particular importance for the salt (soap) formation mechanism under boundary lubrication conditions. Figures 18–22 illustrate the ball wear reduction vs. the additive concentration in n-hexadecane. The most striking finding, clearly demonstrated in all these figures, relates to the fact that all the acid/alcohol mixtures investigated have almost the same effect on the steel-on-steel wear process. Their antiwear property does not change with the concentration in hexadecane. The wear reduction range is from around 77% for the

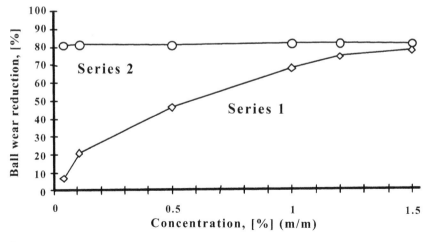

Figure 19 Influence of the solution concentration of decyl palmitate (Series 1) and equimolar mixture of palmitic acid and 1-decanol (Series 2) in n-hexadecane on the ball wear we account for in terms of the chain matching effect with hexadecane molecules. (From Ref. 36.)

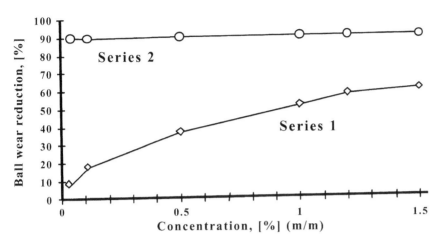

Figure 20 Influence of the solution concentration of dodecyl palmitate (Series 1) and equimolar mixture of palmitic acid and 1-dodecanol (Series 2) in n-hexadecane on the ball wear.

mixture of palmitic acid and 1-octanol (Fig. 18) to 96% for palmitic acid and 1-hexadecanol (see Fig. 22).

The ball wear results for steel lubricated with *n*-hexadecane plus each of the five esters, at concentrations ranging from 300 ppm up to 1.5% wt., summarized in Figs. 18–22 (bottom curves), clearly show that all the five tested esters provide very similar increasing trend in the steel wear reduction. The major difference in the behavior of both additive types, acid/alcohol mixtures and esters is that the distance between the two curves (Δ wear reduction) in the figures becomes smaller as the additives' concentration increase. However, only in the case of *n*-octyl palmitate and the mixture of palmitic acid and 1-octanol (Fig. 18), does the antiwear effect of both additives almost reach the same level for concentrations approaching 1% wt. The difference for *n*-decyl palmitate (Fig. 19) is slightly greater and approaches the wear reduction curve for the ester equimolar substrates at 1.5% wt. The other three esters at all the tested concentrations show lower wear reduction than the appropriate mixtures of the ester substrates. The best wear reduction ability of *n*-hexadecyl palmitate and the equimolar

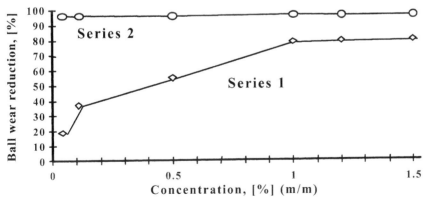

Figure 21 Influence of the solution concentration of hexadecyl palmitate (Series 1) and equimolar mixture of palmitic acid and 1-hexadecanol (Series 2) in n-hexadecane on the ball wear.

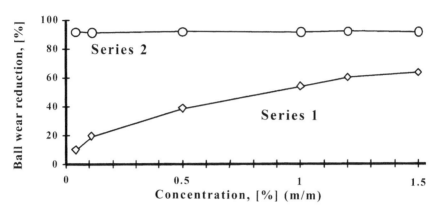

Figure 22 Influence of the solution concentration of octadecyl palmitate (Series 1) and equimolar mixture of palmitic acid and 1-octadecanol (Series 2) in n-hexadecane on the ball wear.

mixture of palimitic acid and 1-hexadecanol we account for in terms of the chain matching effect with hexadecane molecules [36].

The results obtained in Ref. 35 enable us to state that the hypothesis stating "during the friction process under boundary lubrication conditions, lubricated by aliphatic esters, the ester hydrolysis process cannot proceed without an adequate catalyst" was proved for aliphatic esters dissolved in the paraffinic model base fluid, i.e., hexadecane.

C. Investigation of Selected Esters in an Aromatic Base Fluid

Further research [37] was carried out with another base fluid, 1-methylnaphthalene, representing mineral base oil aromatic hydrocarbons. Three palmitates (PAL-C_8, PAL-C_{12}, and PAL-C_{16}) selected from the five palmitates investigated earlier in hexadecane were tested in the aromatic model fluid [methylnaphthalene (MN)]. Apparatus, test conditions, and the test procedure were the same as described in Section V.B.1. As the ester additives were not effective in MN at the lowest concentration (300 ppm) tested in hexadecane, further tests in the aromatic model fluid were performed in higher concentrations, starting from 0.1% wt.

Wear results for the steel system lubricated with the tested three esters and their substrates in methylnaphthalene (MN) are presented in Table 7. The tested additives concentration in MN ranged from 0.1 to 1.5% wt. Wear scar diameter and the ball wear volume obtained for pure MN equals 3.6×10^{-1} mm and 20.8×10^{-13} m^3, respectively. The determined WSD is an average of three tests. The wear reduction value for the system lubricated by plain MN equals zero. The obtained wear reduction data, collected in Table 7, clearly demonstrate that, resembling tests performed in hexadecane, the lowest wear reduction is observed for the lowest concentration of the esters.

Figures 23–25 illustrate the ball wear reduction vs. the additive concentration in MN. All three esters show an increase in the ball wear reduction value with the increasing ester concentration. Interestingly, all these esters demonstrate similar increasing trend in the wear reduction of the steel ball, as that observed previously for these and two other esters in hexadecane (see Figs. 18–22). Again, the most striking finding relates to the fact that all the acid/alcohol mixtures tested in MN have the same rough effect on the steel ball wear as in the paraffinic base fluid (hexadecane). Their antiwear property does not change with the concentration in both MN and hexadecane.

Table 7 Relative Ball Wear Reduction for Tested Additives in Methylnaphthalene

| | Wear reduction [%] | | | | |
| | Additive concentration in methylnaphthalene [% wt.] | | | | |
Additive	1.5	1.2	1	0.5	0.1
PAL – C_8	51.3	49.9	48.2	28.8	15.0
C_{16} + C_8	56.5	56.3	56.5	56.2	56.4
PAL – C_{12}	51.2	49.7	48.0	28.3	14.8
C_{16} + C_{12}	56.2	56.1	56.3	56.1	56.0
PAL – C_{16}	52.0	51.5	51.2	29.2	15.0
C_{16} + C_{16}	56.8	56.7	56.5	56.8	56.6

All these mixtures significantly affect the wear reduction of steel under boundary lubrication conditions in comparison with the wear reduction observed for esters at concentrations below 1% wt. Table 7 shows that the wear reduction values for all the esters tested at different concentrations are very close, ranging from 56.1% to 56.8%. The steel ball wear reduction of the same steel-on-steel system lubricated with equimolar acid/alcohol mixtures at various concentrations in hexadecane is much more significant (Figs. 18–22). The best wear reduction ability of 1-hexadecyl palmitate and the equimolar mixture of palmitic acid and 1-hexadecanol can be accounted for in terms of the chain matching effect with hexadecane molecules. However, it is noteworthy at this point that the ball wear scars determined for both pure base oils are quite dissimilar. The WSD for hexadecane and MN is 450 and 360 μm, respectively.

Figure 26 summarizes the effect of the tested esters at various concentrations in MN on steel ball wear reduction. Comparison of the results obtained for the lowest concentration of

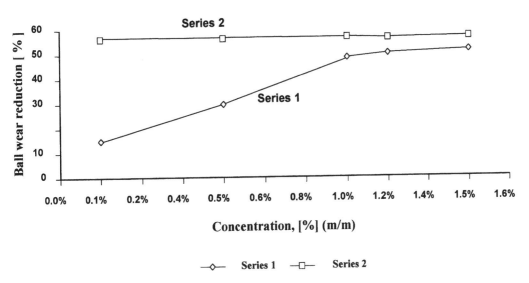

Figure 23 Influence of the solution concentration of octyl palmitate (Series 1) and equimolar mixture of palmitic acid and 1-octanol (Series 2) in methylnaphthalene.

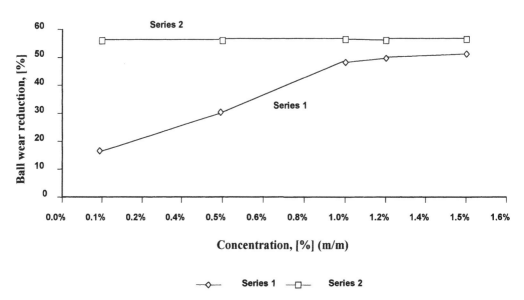

Figure 24 Influence of the solution concentration of dodecyl palmitate (Series 1) and equimolar mixture of palmitic acid and 1-dodecanol (Series 2) in MN on the ball wear.

these esters with results for mixtures of their substrates, as depicted in Figs. 23–25, enables us to say that the previously hypothesized approach suggesting another action mechanism of aliphatic ester in hexadecane is also due to the tested aromatic base fluid (MN). Accordingly, the results generated in Ref. 37 allow us to state that the hypothesis stating "During the friction process under boundary friction conditions, lubricated with solutions of 1-methyl-naphthalene containing aliphatic esters as antiwear additives, the ester hydrolysis mechanism does not control the wear process" is also proved for the aromatic solvent.

We propose that, also in this case, the dominant formation process of reactive species forming the antiwear salt (soap) film on the steel friction tribological contact area proceeds according to the NIRAM approach. Some difference in the antiwear effect of aliphatic esters

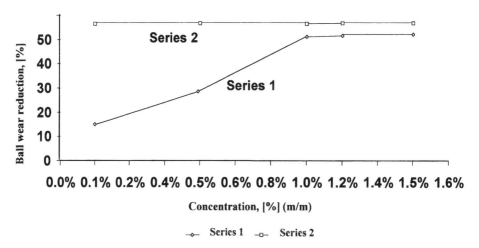

Figure 25 Influence of the solution concentration of hexadectyl palmitate (Series 1) and equimolar mixture of palmitic acid and 1-hexadecanol (Series 2) in MN on the ball wear.

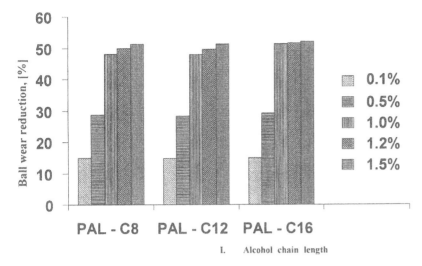

Figure 26 Influence of various palmitic esters concentrations on the ball wear reduction.

in hexadecane and methylnaphthalene relates to specific molecular interactions between the additive molecules and base model fluids. More research is needed to clarity these interactions. The first further research approach should be focused on investigating the ester-type additive chemistry in mixtures of both aliphatic and aromatic model base fluids as well as other real lubricant base oils and/or their components.

D. Is NIRAM Responsible for Initiating the Tribochemical Reactions of Esters?

The ester base-promoted hydrolysis process presented in the simplest way can be expressed by the following reactions:

$$RCOOR_1 + {}^-OH(\text{strong base}) \rightarrow RCOOH + [R_1O^-] \rightarrow [RCOO^-] + R_1OH$$

that lead to the ester substrates: the carboxylic acid in the form of anion $[RCOO^-]$ and alcohol (R_1OH), along with two reactive intermediate $[R_1O^-]$. For more details, see Fig. 10. Actually, the strong base $({}^-OH)$ is the conjugate base of the water molecule; pK_a of conjugate acid (H_2O) equals 15.7. $[R_1O^-]$ and $[RCOO^-]$ are also conjugate bases of their acids. In the first case, the acid relates to an alcohol, and in the second one, the considered acid is a carboxylic acid. The molecule or ion that forms when an acid—as already mentioned, either water or alcohol molecules are also considered as the acid—loses its proton (H^+) is called the *conjugate base* of that acid. The stronger the acid, the weaker will be its conjugate base. Thus, we can relate the strength of a base to the pK_a of its conjugate acid. The larger the pK_a value of the conjugate acid, the stronger is the base. For example, the conjugate base of ethyl alcohol $(pK_a = 16)$ is stronger than the conjugate base derived from acetic acid $(pK_a = 4.75)$ [2].

Now, taking into account the water molecule interaction with a low-energy electron (exolectron) emitted under boundary friction conditions, the ${}^-OH$ strong conjugate base along with hydrogen free radical $(H^{\cdot})$ are produced:

$$HOH + e \rightarrow HO^- + H^{\cdot}$$

It can be realized that, in this case, the hydrolysis process is initiated by the HO⁻ strong conjugate base, generated during the water molecule splitting process by the attached exo-electron. Thus, it is conceivable to consider the exoelectron-promoted hydrolysis without any catalyst. On the other hand, it is necessary to bear in mind another, and even more important, process—the exoelectron interaction with an ester molecule, producing directly two types of negative ions RCOO⁻ and RO⁻. Figure 27 depicts in detail the whole process and shows two types of C–O bond cleavage. The first bond cleavage type produces carboxylate anion (RCOO⁻) and the free radical R¡. The second bond cleavage generates the alkoxide anion (RO⁻) and the R–'C=O free radical that undergoes further reactions. To produce free radicals and thereby initiate the free radical chain reaction process, either heat or catalyst is needed. Therefore, an electron attachment in this case acts as a catalyst or heat pulse (flash temperature). The exoelectron dissociative attachment to an ester molecule yielding two types of negative ions is clearly evidenced by the electron attachment mass spectrographic results [38].

The carboxylate anion RCOO⁻ interacts with a positively charged friction solid surface site to produce a salt (soap). This fact supports the mechanism presented in Fig. 28. Please note that the caboxylate anion's pK_a value is lower than that of the alkoxide value. This means that the carboxylate ion is a more acidic species than that of the alkoxide anion or, in other words, the carboxylate anion is a more stable species than the alkoxide one. Alkoxide ions (R–CH$_2$–O⁻) are known to split off two hydrogen atoms [39]. Recently, it was found that such ions might form very special film on wear tracks, as elucidated by Fourier-transform infrared microscopy (FTIRM) analysis [40]. As Fig. 28 suggests, the ester reaction tribofilm formed under boundary lubrication conditions should be very complicated. A possible complex film produced in the presence of aliphatic alcohols was postulated in an earlier work [41]. Figure 29 depicts the formation mechanism of the tribofilm elucidated for the ethanol lubricated steel-on-aluminum friction system. Most recently, some evidence for such specific film was also presented for fluorinated alcohols [42].

Therefore, it is conceivable to give the positive answer to the question asked in this subsection: "Is the NIRAM approach responsible for initiating tribochemical reactions of

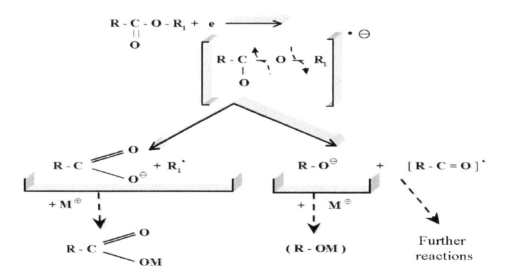

Figure 27   Ester reactive intermediates produced via the dissociative electron attachment.

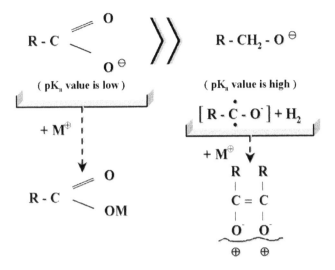

Figure 28 Salt and chelate type film formation by carboxylate anion and alkoxide anion, respectively.

esters?" One conclusion of an early review on mechanically initiated chemical reactions in solids [43] was as follows: "Although not new, the study of mechanically induced chemical reactions in solids has received scarce attention probably because the subject lies uncomfortably between chemistry and materials science." That review examined a number of possible ways in which mechanical energy can initiate chemical reactions. Fox [43] clearly emphasized that in many cases, this appears to take place through the generation of heat although examples exist in which electrical charging and direct bond excitation play an important role. The new mechanically initiated hydrolysis mechanism presented in this chapter is precisely associated with the above approach to tribochemistry. More detailed and very sophisticated research is needed to find unanimous evidence for this mechanism and to elucidate the complexity of the surface tribofilms produced by ester lubricants or ester antiwear and/or friction reducing additives.

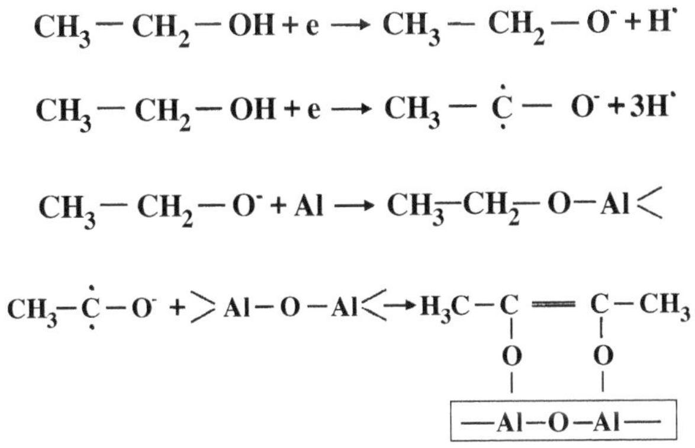

$$CH_3 - CH_2 - OH + e \longrightarrow CH_3 - CH_2 - O^{\cdot} + H^{\cdot}$$

$$CH_3 - CH_2 - OH + e \longrightarrow CH_3 - \overset{\cdot}{C} - O^{\cdot} + 3H^{\cdot}$$

$$CH_3 - CH_2 - O^{\cdot} + Al \longrightarrow CH_3{-}CH_2{-}O{-}Al\!<$$

$$CH_3{-}\overset{\cdot}{\underset{\cdot}{C}}{-}O^{\cdot} + {>}Al{-}O{-}Al\!<{\longrightarrow} H_3C{-}\underset{\underset{\boxed{{-}Al{-}O{-}Al{-}}}{O}}{C} = \underset{O}{C}{-}CH_3$$

Figure 29 Chemical reactions of the tribofilm formation process under boundary friction conditions of steel-on-aluminum lubricated with ethanol; Al in the third equation is considered as $>Al^+$. (From Ref. 40.)

E. Mechanical Hydrolysis

1. General Remarks

Ceramics, particularly oxides and nitrides, have a very low free energy of formation and are therefore very stable chemically in a nontribological sense. However, under boundary lubrication conditions, they become quite reactive because solids support shear strains. As ceramics are relatively inert, hard, and more resistant to abrasive and erosive wear than typical steels, they offer some advantages over conventional tribological mating elements. Under boundary lubrication conditions, lubrication is widely dominated by tribochemical reactions. It is even possible to say that the friction and wear behavior of ceramics might be more sensitive to environment than friction and wear behavior of metals. Thus, the importance of a comprehensive understanding of the ceramic surface physics and chemistry cannot be overestimated. A work [44] discusses the initiation process of ceramic tribochemical reactions. It mostly aimed at (1) presenting some tribochemical reactions that play an important role in the wear reduction of ceramics and (2) proposing a model of such reactions. The proposed model is based on the NIRAM approach of a lubricant/environment component with the rubbing ceramic surface.

   Two reaction steps are considered. The first one includes the reactions of a ceramic material, e.g., silicon nitride, alumina with water/oxygen molecules, and the second step involves the reactions of the modified ceramic surface with additive or lubricant molecules. This mechanism clearly explains (1) the formation of hydroxide groups on silicon nitride and alumina surfaces under boundary lubrication conditions and (2) further reactions with selected lubricant components. Tribochemical reactions of ceramics are very complex because several physical processes may be involved in the formation of wear reducing films. However, the application of the NIRAM–Hard and Soft Acids and Bases (HSAB) approach to account for the tribochemical reaction of silicon nitride with water or alcohols allows us to better understand all the major findings related to the silicon nitride tribochemistry.

2. How Can We Understand the Mechanical Hydrolysis Process?

Lubrication, particularly under boundary friction conditions, is extremely important for ceramic materials in tribological systems. There are many papers considering various lubricants/additives for that purpose. Only few works will be considered here, particularly those relating to esters and the "mechanical" hydrolysis process combined with tribochemistry of silicon nitride (Si_3N_4). For example, in Ref. 45, in the vapor-phase lubrication of alumina-on-alumina tribological system by selected esters, dramatic reduction in wear occurred, particularly for those vapors delivered at higher temperatures. Under these conditions, lauryl methacrylate and diallyl phthalate reduced alumina ball wear by 94–99%. At elevated bulk temperatures (145°C), the monomers in the vapor phase reduced friction coefficients by as much as 50%. Figure 30 presents the wear results obtained for diallyl phthalate at two temperatures.

   Many studies of ceramic lubrication involve gases and or/water vapor [46,47]. In fact, the wear rate of silicon nitride decreased in the presence of water vapor or liquid water [46,48]. Along with mechanical modes of wear such as plastic deformation, material transfer, and microstructure, tribochemical reactions were also identified to be important in the mechanisms of material removal. Grinding of silicon nitride in water produces ammonia [49]. On the other hand, it has been found [50] that in air, this material begins to oxidize at 700–750°C; a distinct amorphous silica surface layer results after 24 h at 750°C.

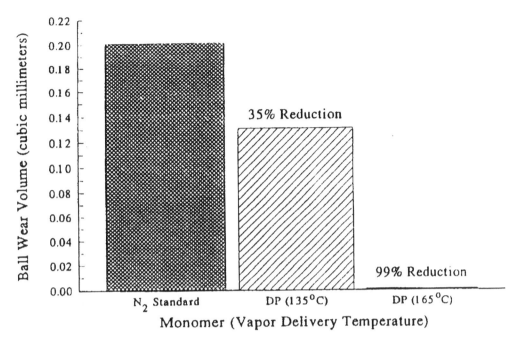

Figure 30   Diallyl phthalate vapor phase ball wear; 145°C bulk temperature. (From Ref. 45.)

3.   NIRAM–HSAB Concept of Ceramic Tribochemistry

The concept is based on the ionization mechanism of ceramic lubricant compound molecules caused by the action of exoelectrons. As already mentioned earlier, the principal thesis of the model is that lubricant components form anions, which are then chemisorbed on the positively charged areas of rubbing surfaces. This approach has also been applied to account for the vinyl-type tribopolymeriyation mechanism [51]. Most recently, another work [52] described a generalized NIRAM–Hard and Soft Acids and Bases (HSAB) theory and provided its possible application to account for some tribochemical processes under boundary conditions. The approach combines the reactivity of lubricant components and tribological solids according to reactions induced by low-energy electrons and reactions affected by acid–base interaction. Major HSAB interactions are related to tribochemical reactions, initiated by low-energy electrons. Figure 31 depicts a general approach to the NIRAM–HSAB action mechanism.

4.   Tribochemistry of Silicon Nitride Discussed in Terms of Hydrolysis

Most recently, the possible application of the NIRAM–HSAB approach to tribochemistry of ceramics was considered in a general way [53]. This subsection aims at detailed accounting for tribochemical wear reactions of silicon nitride with water from the viewpoint of mechanical hydrolysis. Sugita et al. [54] considered the effect of water on the lubrication of silicon nitride in terms of the hydration process. Conducting wear tests on silicon nitride using water as lubricant, it reported a high wear rate of around 10^{-3} mm^3/N m, but very smooth surfaces have been found. These results were attributed as chemical reaction between silicon nitride and water to form an amorphous hydrate on the surface that was removed during the friction process. It was assumed that some form of water-soluble silicon

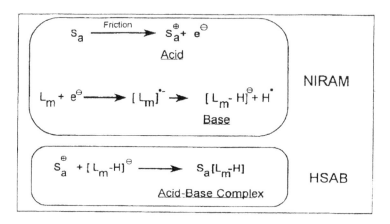

Figure 31 NIRAM-HSAB lubrication mechanism approach: S_a—tribological microsurface area; L_m—lubricant molecule; e—electron emitted during sliding. (From Ref. 52.)

containing compound had been formed. Two years later, it was demonstrated that after an initial high wear rate, water generated a strong lubrication effect with silicon nitride [55]. The wear rates were lowered by several orders of magnitude, and a friction coefficient of 0.02 was observed. All these findings led Tomizawa and Fischer [55] to suggest a tribochemical reaction between silicon nitride and water to form silica (SiO_2). The following chemical reactions had been postulated:

$$Si_3N_4 + 6H_2O \rightarrow 3SiO_2 + 4NH_3 \tag{1}$$

$$SiO_2 + 2H_2O \rightarrow Si(OH)_4 \tag{2}$$

The postulated mechanism was also in line with the work of Kanno et al. [49]. Straight chain alcohols effectively reduce the friction and wear of silicon nitride even at lower sliding velocity; as the carbon number in the paraffin chain increases, the wear decreases. Research presented in Ref. 56 allowed us to show that silica, amines, and polymeric compounds are formed as a result of the tribochemical interaction between silicon nitride and an alcohol, according to the following reaction scheme:

$$Si_3N_4 + ROH \rightarrow SiO_2 + RNH_2 + Polymer \tag{3}$$

Very specific investigation presented in Ref. 57 tested a wide variety of oxygen-containing compounds neat and/or at 1% wt. in a paraffin oil. Compounds containing hydroxyl functional groups were more effective compared to a base case of neat paraffin oil. In most cases, films were observed in and around the wear scar, suggesting that chemical reactions had taken place in the contact. A study of several C_8 compounds with specific oxygen-containing functional groups demonstrated that the primary alcohol had the strongest lubricating effect. Under the conditions of the tests, aldehyde, ketone, ether, ester, and dialkylcarbonate chemical structures were not effective. Wear tests conducted with controlled water concentrations demonstrated that the lubricating effect of these alcohols is not merely due to the effect of dissolved water. A direct surface chemical reaction involving the hydroxyl functional group of these organic compounds was suggested. Another work [58], performed by the same authors, found that thermally induced reactions produced distinctively different products than those obtained from the rubbing contacts. For instance, in the reaction products from an alcohol-lubricated silicon nitride wear test, a range of high molecular weight polymeric materials were found that contained silicon. The presence of silicon in the product clearly demonstrated the role of lubricant—surface tribochemical reactions in the overall chemical mechanism. It was speculated [58] that bond strain introduced by rubbing may account for an enhanced reactivity on the surface of certain materials, resulting in chemical reactions, or higher reaction rates, that would not be observed otherwise.

Results of another sophisticated investigation [59] clearly showed that ammonia originates from the mechanical grinding of any kind of silicon nitride, irrespective of the preparation method adopted. The likely reaction is *mechanically activated hydrolysis* by water vapor, i.e., a reaction caused by direct contact during the milling procedure between disturbed Si–N bonds and water, yielding silica and ammonia. The same reaction was suggested as that proposed in Ref. 55 [Eq. (1)]. On the other hand, a mechanistic model of ammonia formation from silicon nitride in the presence of water has been proposed [60]. Figure 32 presents models of how water reacts with both silica (SiO_2) and silicon nitride (Si_3N_4) surface. A water molecule reacts to form a SiO–H and Si–OH, and a SiN–H and Si–OH, respectively. For silicon nitride, ammonia (NH_3) should be formed.

To find out the relationship between adsorption and/or reaction of water with silicon nitride and generation of uncondensable gases, the moles of evolved gas have been reported as a function of adsorbed amounts of water [59]. Figure 33 shows a linear relationship between the two quantities, which indicates that a definite fraction of uptaken water is

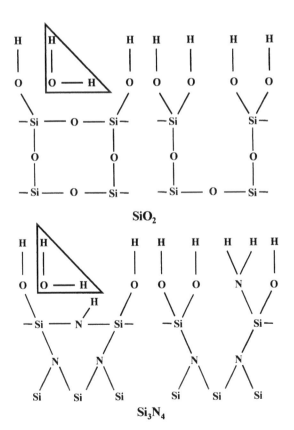

Figure 32 Models for tribochemical water reaction with silica and silicon nitride. (From Ref. 60.)

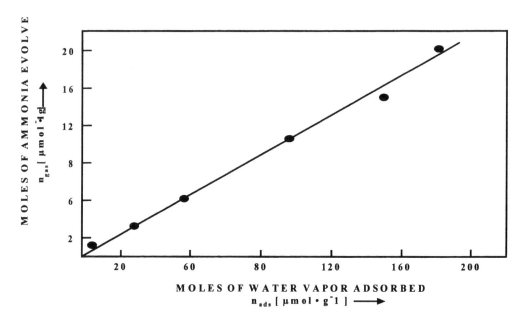

Figure 33 Moles of gases evoled (n_{gas}) plotted against moles of H_2O adsorbed (n_{ads}). (From Ref. 59.)

involved in a chemical reaction. The ratio n_g/n_a is around 0.11. On the other hand, the average ratio n_g/n_a for Eq. (1) is 0.666, which is several times higher that the ratio determined experimentally.

The difference can be explained in terms of the NIRAM–HSAB approach, as depicted in Fig. 34. This mechanism distinctly demonstrates that the consumed molecule of water does not produce any gas. Of course, further steps of the reaction chain will produce ammonia. Assuming that not all hydrogen radicals recombine with only one nitrogen atom, it is easy to realize that even 10 water molecules might be adsorbed to produce only one ammonia molecule. Further reaction steps relate to generation of $–NH_2$ species and finally ammonia (NH_3) is produced.

The well-known fact of hydroxide formation from alumina under friction conditions in the presence of water molecules can also be explained in terms of the NIRAM–HSAB approach. Chemical reactions in ceramics under rubbing conditions can be triggered by mechanical forces because solids support shear strains. Taking into account the well-known

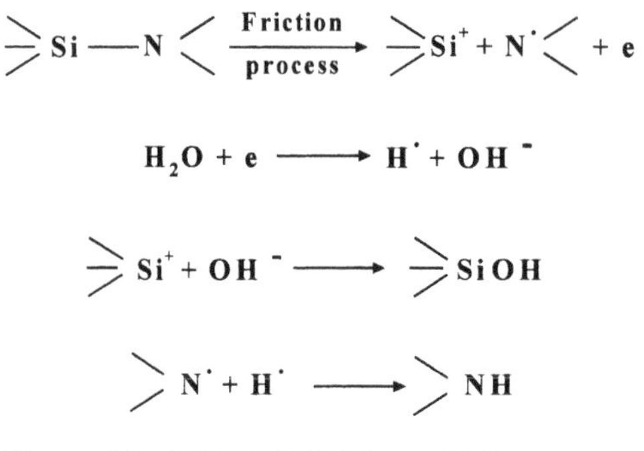

$$\geq Si-N \leq \xrightarrow[\text{process}]{\text{Friction}} \geq Si^+ + N^{\cdot} \leq + e$$

$$H_2O + e \longrightarrow H^{\cdot} + OH^-$$

$$\geq Si^+ + OH^- \longrightarrow \geq SiOH$$

$$\geq N^{\cdot} + H^{\cdot} \longrightarrow \geq NH$$

Figure 34 NIRAM-HSAB model for reaction of water with silicon nitride under friction.

triboemission process, particularly related to ceramics, it is possible to consider the tribochemical reactions of ceramics with water as a kind of a very specific, mechanically initiated hydrolysis proceeding according to the NIRAM–HSAB approach. Thus, it is feasible to conclude that the tribochemistry of ceramics mostly relates to tribological phenomena generated, e.g., emission of electrons and chemical reactions initiated in microscopic solid/solid contacts, unless the tribological process is carried out under ultra-high vacuum or in the presence of totally inert environment.

REFERENCES

1. Brown, M.; Fotheringham, J.D.; Hoyes, T.J.; Mortier, R.M.; Orszulik, S.T.; Randles, S.J.; Stroud, P.M. Synthetic base fluids. In *Chemistry and Technology of Lubricants*; Mortier, R.M., Orszulik, S.T., Eds.; Blackie Academic & Professional: London 1997; 34–74.
2. Solomons, T.W.G. *Fundamentals of Organic Chemistry*, 5th Ed.; Wiley & Sons: New York 1997; 760–780.
3. Furey, M.J. The formation of polymeric films directly on rubbing surfaces to reduce wear. Wear 1973, *26*, 369–392.
4. Beran, E. Application of waste carboxylic acids to the manufacture of biodegradable polyuolester base oils. J. Synth. Lubr. 2001, *18*, 39–50.
5. van der Waal, G. The relationship between chemical structure of ester base fluids and their influence on elastomer seals and wear characteristics. J. Synth. Lubr. 1985, *1*, 281–294.
6. Randles, S.J. Esters. In *Synthetic Lubricants and High-Performance Functional Fluids*; Shubkin, R.L., Ed.; Marcel Dekker: New York 1993; 41–65.
7. Kardasz, K.; Kedzierska, E.; Kajdas, C.; Okulicz, W. Effectiveness of methyl esters of rapeseed oil as low-sulfur diesel fuel anti-wear additive in comparison to other plant and synthetic origin substances. Tribologia 2001, *32*, 139–149. *in Polish*.
8. Karonis, D.; Anastopoulos, G.; Lois, E.; Stournas, S.; Zannikos, F.; Serdari, A. Assessment of the lubricity of Greek road diesel and effect of the addition of specific types of biodiesel. SAE Paper 1999-01-1471.
9. Anastopoulos, G.; Lois, E.; Zannikos, F.; Kaligeros, S.; Teas, C. Influence of aceto acetic esters and di-carboxylic acid esters on diesel fuel lubricity. Tribol. Int. 2001, *34*, 749–755.
10. Anastopoulos, G.; Lois, E.; Karonis, D.; Zannikos, F.; Stournas, S.; Kaligeros, S. The impact of aliphatic monoamines and ethyl esters of fatty acids on the lubricant properties of ultra-low sulphur diesel fuels. Fuels Int. 2000, *1*, 31–44.
11. Zinkel, D.F., Russel, J., Eds.; *Naval Stores: Production—Chemistry—Utilization*; Pulp Chemicals Association: New York, 1989.
12. Gunstone, F.D., Harwood, J.L., Padley, F.B., Eds.; *The Lipid Handbook*; Chapman and Hall: London, 1994.
13. Brawford, J.; Psaila, A.; Orszulik, S.T. Misscellaneous additives and vegetable oils. In *Chemistry and Technology of Lubricants*; Mortier, R.M., Orszulik, S.T., Eds.; Blackie Academic & Professional: London 1997; 196–202.
14. Bovington, C.H. Friction, wear and the role of additives in their control. In *Chemistry and Technology of Lubricants*; Mortier, R.M., Orszulik, S.T., Eds.; Blackie Academic & Professional: London 1997; 320–348.
15. Boyde, S. Hydrolytic stability of synthetic ester lubricants. J. Synth. Lubr. 2000, *16*, 297–312.
16. Tau, K.D.; Elango, V.; McDonough, J.A. Organic esters. In *Kirk–Othmer Encyclopedia of Chemical Technology*, 4th Ed.; John Wiley & Sons: New York, Vol. 9, 781–812.
17. Weller, D.E. Jr.; Perez, J.M. A study of the effect of chemical structure on friction and wear: Part I—Synthetic ester base fluids. Lubr. Eng. 2000, *56* (10), 39–44.
18. Bamford, C.H.; Tipper, C.F.H. Chemical Kinetics. In *Ester Formation and Hydrolysis and Related Reactions*; Elsevier: Amsterdam, 1972; Vol. 10, 57–207.

19. Hiromichi, S. Fatty acid ester type hydraulic fluids. Jpn. J. Tribol. 1997, *42*, 817–825.
20. Shen, M.; Luo, J.; Wen, S.; Yao, J. Investigation of the liquid crystal additive's influence on film formation in nano scale. Lubr. Eng. 2001, *58* (3), 18–23.
21. Atarashi, Y. Phosphate ester-type hydraulic fluids. Jpn. J Tribol. 1997, *42*, 807–815.
22. Cain, R.B. Biodegradation of lubricants. In *Biodeterioration and Biodegradation*; Rossmore, H.W., Ed.; Elsevier Applied Science: Oxford, 1990; 249–275.
23. Pitter, P.; Chudoba, J. *Biodegradability of Organic Substances in the Aquatic Environment*; CRC Press: Boca Raton, Florida, 1990.
24. Versino, C.; Novaria, M. Biodegradability test for synthetic esters. J. Synth. Lubr. 1987, *4*, 3–23.
25. Eaton, R.W.; Ribbon, D.W. Metabolism of dibuthyphtalate and phthalate by *Micrococcus* sp. Strain 12B. J. Bacteriol. 1982, *151*, 48–57.
26. Ribbon, D.W.; Kayser, P.; Kunz, D.A.; Taylor, B.F.; Eaton, R.W.; Anderson, B.N. Microbial degradation of phthalates. In *Microbial Degradation of Organic Compounds*; Gibson, D.T., Ed.; Marcel Dekker: New York, 1984; 371–379.
27. Kajdas, C. Importance of anionic reactive intermediates for lubricant component reactions with friction surfaces. Lubr. Sci. 1994, *6*, 203–228.
28. Kajdas, C. A novel approach to lubrication mechanism under metalworking conditions. Euro-metallworking 1994, *2*, 2–8.
29. Bowden, F.P.; Moore, A.C. Physical and chemical adsorption of long compounds on metals. Research 1949, *29*, 585–586.
30. Bowden, F.P.; Tabor, D. *The Friction and Lubrication of Solids*; Clarendon Press Oxford, 1950.
31. Lower, E.S. Oleochemical metal processing lubricants. Ind. Lubr. Tribol. 1992, *43* (3), 4–10.
32. Bruice, P.Y. *Organic Chemistry*, 2nd Ed. Prentice-Hall: Upper Saddle River, New Jersey, 1998; 690–705.
33. Morrison, R.T.; Boyd, R.N. *Organic Chemistry*; Prentice-Hal: Englewood Cliffs, New Jersey, 1992; 690–705.
34. Ouellette, R.J. *Introduction to General, Organic, and Biological Chemistry*; Maxwell Macmillan, Inc., Toronto, 1992.
35. Kajdas, C.; Shuga'a, A.K. Investigation of AW properties and tribochemical reactions of esters of palmitic acid and aliphatic alcohols in the steel-on-steel system. Tribologia 1998, *29*, 398–402.
36. Askwith, T.C.; Cameron, A.; Crouch, R.F. Chain length of additives in relation to lubricants in the thin film and boundary lubrication. Proc. R. Soc. Ser. A 1966, *291*, 500–519.
37. Kajdas, C.; Shuga'a, A.K. Anti-wear properties and lubrication mechanism of aliphatic esters in a model aromatic fluid. Tribologia 1999, *30*, 289–299.
38. von Ardenne, M.; Steinfelder, K.; Tuemmler, R. *Elektronenanlagerungs-Massenspektrographie organischer Substanzen*; Springer Verlag: Berlin, 1971; 235–237.
39. Melton, C.E.; Rudolf, P.S. Negative-ion mass spectra of hydrocarbons and alcohols. J. Chem. Phys. 1959, *31*, 1485–1488.
40. Kajdas, C.; Obadi, M. Wear behavior and tribochemical reactions of low molecular alcohols under boundary lubrication of the steel-on-aluminum system. Tribologia 1999, *30*, 273–287.
41. Kajdas, C. About an anionic-radical concept of the lubrication mechanism of alcohols. Wear 1987, *116*, 167–180.
42. Kajdas, C.; Przedlacki, M. Tribochemistry of fluorinated compounds containing hydroxyl group. Tribologia 2000, *31*, 155–175. *in Polish*.
43. Fox, P.G. Review. Mechanically initiated chemical reactions in solids. J. Mater. Sci. 1975, *10*, 340–360.
44. Kajdas, C. Tribochemistry of ceramics. Tribologia 1998, *29*, 148–168.
45. Smith, J.C.; Furey, M.J.; Kajdas, C. An exploratory study of vapor-phase lubrication of ceramics by monomers. Wear 1995, *181–183*, 581–593.
46. Fischer, T.E.; Tomizawa, H. Interaction of tribochemistry and microstructure in the friction and wear of silicon nitride. Wear 1985, *105*, 29–45.
47. Hibi, Y.; Enomoto, Y. Tribochemical wear of silicon nitride in water, alcohols and their mixture. Wear 1989, *13*, 133–245.

48. Ishigaki, H.; Kawaguchi, I.; Iwasa, M.; Toibana, Y. Friction and wear of hot pressed silicon nitride, and other ceramics. In *Wear of Materials*; Ludema, K.C., Ed.; ASME: New York, 1985; 13–21.

49. Kanno, Y.; Suzuki, K.; Kuwahara, Y. Ammonia formation caused by presence of water in the grinding of silicon nitride powder. Yag Yo-k Yokaishi 1983, *91*, 386–391.

50. Kiehle, A.J.; Heung, L.K.; Gielisse, P.J.; Rockett, T.J. Oxidation behavior of hot-pressed silicon nitride. J. Am. Ceram. Soc. 1975, *58*, 17–20.

51. Furey, M.J.; Kajdas, C.; Kempinski, R.; Tripathy, B.S. Action mechanism of selected vinyl monomers under boundary lubrication of alumina-on-alumina system. Proc. 10th International Colloquium; Technische Akademie Esslingen, 1966; Vol.1, 1847–1864.

52. Kajdas, C. A novel approach to tribochemical reactions: generalized NIRAM–HSAB action mechanism. Proc. of ITC, Yokohama 1995, Satellite Forum on Tribochemistry, Tokyo 28 Oct 1995, 31–35.

53. Kajdas, C. Tribochemistry. In *Tribology 2001: Scientific Achievements, Industrial Applications, Future Challenges*; Franek, F., Bartz, W.J., Pauschitz, A., Eds.; Austrian Tribology Society Vienna, 2001; 39–46.

54. Sugita, T.; Ueda, K.; Kanemura, Y. Material removal mechanism of silicon nitride during rubbing in water. Wear 1984, *97*, 1–8.

55. Tomizawa, H.; Fischer, T.E. Friction and wear of silicon nitride and silicon carbide in water: hydrodynamic lubrication at low sliding speed obtained by tribochemical wear. ASLE Trans. 1987, *30*, 41–46.

56. Hibi, Y.; Yenomoto. Tribochemistry of alcohol/silicon-based ceramics. Proceed. of JAST Tribology Conf., Marioka, October 1992; 353–356.

57. Gates, R.S.; Hsu, S.M. Silicon nitride boundary lubrication: effect of oxygenates. Tribol. Trans. 1995, *38*, 607–617.

58. Gates, R.S.; Hsu, S.M. Silicon nitride boundary lubrication: lubrication mechanism of alcohols. Tribol. Trans. 1995, *38*, 645–653.

59. Volante, M.; Fubini, B.; Giamello, E.; Bolis, V. Reactivity induced by grinding silicon nitride. J. Mater. Sci. Lett. 1989, *8*, 1076–1078.

60. Mizuhara, K.; Hsu, S.M. Tribochemical reaction of oxygen and water on silicon surfaces. In *Wear Particles;* Dowson, D., et al., Ed.; Elsevier Science Publishers: Amsterdam, 1992; 323–328.

9
Oil Surface: Additive Reaction Mechanisms

Jože Vižintin
University of Ljubljana, Ljubljana, Slovenia

I. THE TRIBOSYSTEM AND LUBRICATION REGIMES

The structure of a tribosystem consists of four elements: the base body (1), the faced body (2), the interacting environment (3), and the surrounding environment (4), as shown schematically in Fig. 1.

The interactions between the elements of a tribosystem depend on many conditions, as shown in Fig. 2 [1–4].

When the contact geometry of interacting surfaces and the operating conditions are such that the load is fully supported by a fluid film, the contact surfaces are completely separated. This condition is referred to as the hydrodynamic lubrication (HD) regime. In this case, the Reynolds equations and the continuum mechanics for fluid-film design can be used [5].

When the load is high or the speed is low, the hydrodynamic pressure may not be sufficient to fully support the load; under such conditions, the surfaces come into contact. At the asperities of the interacting surfaces, the extent of this asperity contact depends on many factors, which include surface roughness, fluid-film pressure, normal load, hardness, and elasticity of the asperities. During the triboprocess, many of the asperities undergo an elastic deformation as a result of the contacting conditions, and the normal load is supported by the asperities and the thin fluid film. Dowson and Higginson [6] termed this lubrication as elastohydrodynamic lubrication (EHL). The theory of EHL assumes continuum mechanics and does not take into account the effect of wear, the chemical interactions between asperities, and the presence of a third body.

A further increase in the contact pressure, so as to exceed the EHL conditions, causes the contacting asperities to plastically deform and the number of asperity contacts to increase as thickness of the fluid film decreases. When the average fluid-film thickness falls below the average relative surface roughness, the surface contact becomes a major part of the load-supporting system. At this stage of the triboprocess, the mechanical interactions of the asperity contacts produce friction, wear, and frictional heat, all of which affect the chemical reactions between the lubricant molecules and the asperity contact surface. The combination of the load sharing by asperities and the occurrence of the chemical reactions constitutes a boundary lubrication (BL) regime [7]. Figure 3 shows a schematic relationship for these regimes as they relate to the coefficient of friction and the contact severity on the interaction surfaces of the tribosystem [8]. The operating principle of the boundary

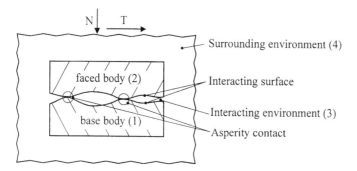

Figure 1   Structure of a tribosystem.

lubrication regime can perhaps be best illustrated by considering the coefficient of friction. In very simple terms, the coefficient of friction (f) is defined as the ratio of the frictional force (F) and the load (W) applied normal to the surface:

$$f = F/W \qquad\qquad (1)$$

Under boundary lubrication conditions, interactions between the two surfaces take place in the form of asperities colliding with each other. The number of asperities that come into contact and the manner in which they act depend on the applied load, the properties of asperity material, and the small-scale topography of the surfaces. Thus contact is limited to a relatively small area, and the rest of the surface is held apart. The sum of the parts of the real contact areas between the asperities is referred to as the real contact area (Fig. 4) [15,16]. Assuming that the major part of the frictional force (F) is a result of adhesion between

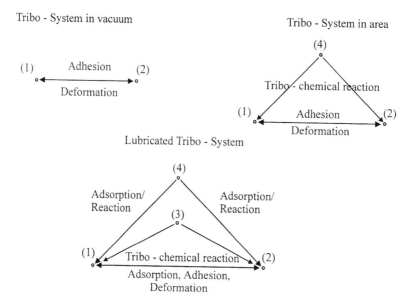

Figure 2   Interaction between the elements of a tribosystem: (1) base body; (2) faced body; (3) interacting environment (vacuum, area, lubricant, etc.); (4) surrounding environment (load, speed, temperature, humidity, etc.). (From Ref. 1)

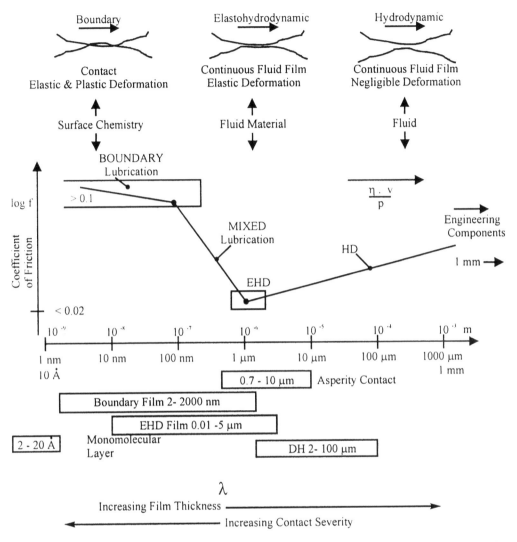

Figure 3 Comparison of lubrication film thickness with average size of lubricant molecules under different contact severities. (From Ref. 8.)

asperities, the frictional force can be determined as the product of the real contact area (A_r) and the effective shear stress (T) of the material in contact. If the applied load (W) is determined as the product of the plastic flow stress (P_y) of the material and the real contact area (A_r), and both terms (F, W) in Eq. (1) are substituted, the coefficient of friction can be rewritten as [16]:

$$f = T/P_y \tag{2}$$

From Eq. (2), we can see that to obtain a low coefficient of friction, a material of low shear strength and high hardness is required. Thus, in general terms, the fundamental principles behind boundary lubrication (BL) and extreme pressure (EP) lubrication involve the formation of low-shear-strength lubricating films on hard interacting surfaces.

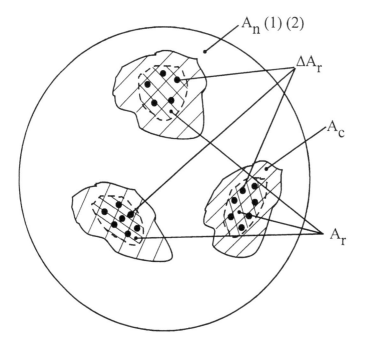

Figure 4 General scheme of the contact of solid bodies. A_n = nominal contact area; A_c = contour contact area; A_r = real contact area; ΔA_r = the part of the real contact.

As it relates to friction, wear, and lubrication, the contact can be examined on one hand as the topography of the contact and on another hand as the interaction between the elements of a tribosystem [17–20].

II. THE TOPOGRAPHY AND PROPERTIES OF THE INTERACTING SURFACES

A. Surface Topography

A solid interacting surface has a complex structure and complex properties depending on the nature of the solid, the method of surface processing, and the interaction between the surface and the environment. The properties of solid surfaces are crucial to surface interactions because surface properties affect the real area of contact, the friction, the wear, and the lubrication. Surface topographies formed in modern engineering considerably vary. Figure 5, for example, shows the topographies of surfaces that have been ground (a), mechanically polished (b), ground and nitrided (c), and mechanically polished and coated with a diamond-like carbon (DLC) thin layer (d) [9–11].

The electron micrographs show that although such surfaces may feel smooth, they are in fact covered with asperities, machining marks, flaws, and scratches, depending on the method used to produce the surface.

The scale of the world of the tribologist is essentially determined by the size of the individual contact areas between surfaces. The principal instruments used to study surface topography are the scanning electron microscope (SEM) and the stylus profilometer with a

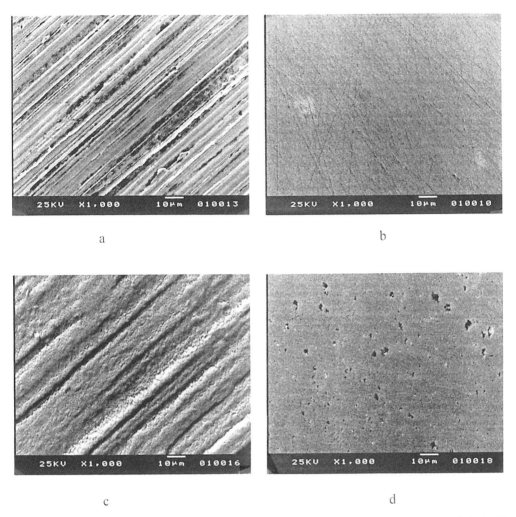

Figure 5 Electron micrographs of surfaces that are (a) ground, (b) mechanically polished, (c) ground and nitrided, or (d) mechanically polished and coated with a DLC thin layer.

three-dimensional analyzer. In practical tribology, the SEM has two disadvantages: the specimen size is limited and the SEM cannot quantify roughness. Consequently, the stylus profile analyzer is the most widely used instrument for surface analysis. When the stylus travels over the surface, it rises and falls. The graph drawn by the pen represents the vertical displacement of the stylus as a function of the distance traveled across the surface. The graphical representation of the surface profile generated by a stylus profilometer contains three major components: roughness, waviness, and error of form. The distinction between roughness and form error is arbitrary, although it clearly involves the horizontal scale of the irregularity [12,13].

Roughness is formed by fluctuations on the surface with short wavelengths; it is characterized by asperities (local maxima) and valleys (local minima) of varying amplitudes and spacings. Asperities are referred to as peaks in a profile (two dimensions) and summits in a surface map (three dimensions).

Waviness is the surface irregularity with longer wavelengths and is referred to as macroroughness. Waviness may result from such factors as machine or workpiece deflections, vibration, heat treatment, or warping strains.

Errors of form are the result of machine marks and flaws; they have very long wavelengths. Flaws are unintentional, unexpected, and unwanted interruptions in the topography.

Surface roughness most commonly refers to variations in the height of the surface relative to a reference plane. It is measured either along a single-line profile or along a set of parallel-line profiles (surface maps). It is usually characterized by a center-line average of roughness (R_a), standard deviation or variance (σ), or root mean square (RMS) with symbol R_q and two other statistical height descriptors, which are known as skewness (S_k) and kurtosis (K) [13,14].

A surface with a high negative skewness has a larger number of local maxima above the mean line, compared to a Gaussian distribution; for positive skewness, the opposite is true. Similarly, a surface with a low kurtosis has a larger number of local maxima above the mean line compared to that of a Gaussian distribution. For a high kurtosis, the opposite is true, as shown in Figs. 6 and 7 [14].

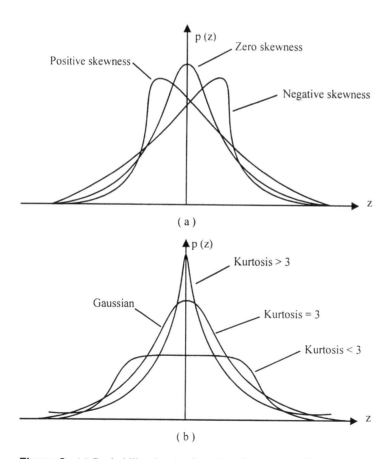

Figure 6 (a) Probability density functions for random distributions with different skewness, and (b) symmetrical distribution (zero skewness) with different kurtosis. (From Ref. 14.)

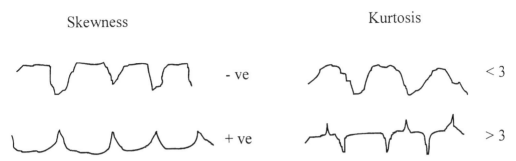

Skewness

Kurtosis

- ve

+ ve

< 3

> 3

Figure 7 Schematic illustration for random functions with various skewness and kurtosis values. (From Ref. 14.)

The graph of a surface profile generated by stylus contains most of the information needed to describe the topography of the surface along the single direction, but gives no information about the distribution of surface features in a direction parallel to the machining marks. Using a profilometer, it is possible to generate a three-dimensional picture of the surface. The procedure involves making a large number of profilometer traverses across a specimen and slightly displacing the specimen between each traverse. Three-dimensional pictures of ground and mechanically polished surfaces generated by automatic stylus profilometers are shown in Fig. 8.

The topography of a freshly machined surface depends on the machining process used to generate it as well as on the nature of the material. Much useful information is contained in the amplitude density function and the bearing area curve. Examples of amplitude density functions (ADFS), in the form of experimentally determined histograms and bearing curves for four surfaces, are shown in Figs. 9 and 10. Figure 9(a) shows a ground surface with a negative skewness ($S_k = -0.00829$) and a positive kurtosis ($K = 2.82$). This surface has a very symmetrical distribution compared to Gaussian distribution ($S_k = 0$; $K = 3$). The mechanical polishing process produces a surface with a symmetrical distribution and a positive skewness ($S_k = +0.0852$) and a lower kurtosis ($K = 2.52$) compared to a Gaussian distribution. For practical application in a tribosystem, it is very important to know that a thermochemical treatment process such as nitriding and a process such as DLC coating have a great effect on the value of the skewness and the kurtosis parameters [Fig. 10(a,b)].

The bearing area is known as the real area of contact. When the plane first touches the surface at a point, the bearing ratio is zero. As the line is further moved downward, the length over which it intersects the surface profile increases, and therefore the bearing ratio increases. Finally, as the line reaches the bottom of the deepest valley in the surface profile, the bearing ratio rises to 100%.

B. Typical Surface Layers

A solid surface has a complex structure and complex properties depending on the nature of the solids, the method of surface preparation, and the interaction between the surface and the environment. Properties of solid surfaces are crucial to surface interactions because surface properties affect the real area of contact, the friction, the wear, and the lubrication. A solid surface, depending on the method of formation, contains irregularities or deviations from the prescribed geometrical form [8,9]. Metals and alloys are by far the most widely used materials in practical tribological systems and the surface properties of such materials

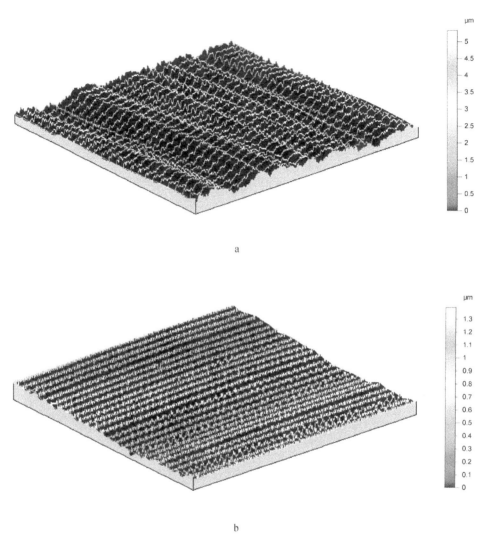

Figure 8 Examples of three-dimensional profile maps: (a) ground and (b) mechanically polished surface.

change in different environments. If a metal is taken from its ordinary environment, placed in a vacuum and mildly heated, the surface liberates water. Desorption with mild heating indicates that bonding to the surface is weak and of a physical nature. Beneath this outer layer of adsorbed water and gases, metals have a metallic oxide layer and several zones with physicochemical properties particular to the bulk material, as indicated in Fig. 11 [21,22].

Surface oxides of iron can be Fe_2O_3, Fe_3O_4, or FeO. The formation of a particular oxide depends on the environment, the amount of oxygen available to the surface, and the oxidation mechanism. The oxides present on an alloy surface depend on the concentration of alloying elements, the affinity of the alloying elements for oxygen, the ability of oxygen to diffuse into surface layers, and the segregation of alloy constituents to the surface. For

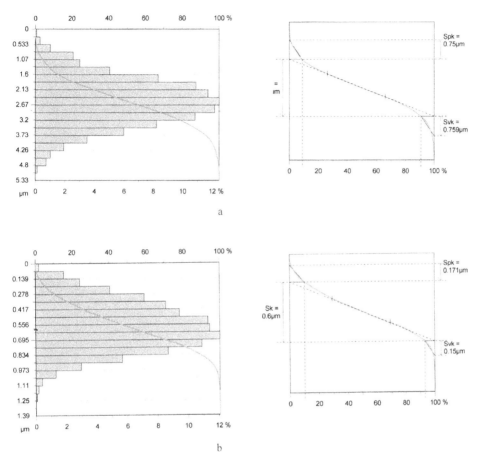

Figure 9 Amplitude density functions in the form of histograms and bearing curves; (a) ground surface ($S_K = -0.00829$; $K = 2.82$); (b) mechanically polished surface ($S_K = +0.0852$; $K = 2.52$).

example, a certain concentration of nickel and chromium is necessary to ensure the formation of nickel and chromium oxides that passivate the surface. The chemical affinity of these elements also has a bearing on the surface oxide that is formed: gold has no affinity for oxygen, copper has a limited affinity, and titanium readily reacts with oxygen.

At the base of the surface layer, there is a zone of work-hardened or deformed material, on top of which is a region with a microcrystalline or amorphous structure that is called the Beibly layer (Fig. 11). A finished mechanical component may have been machined, ground, cast, forged, extruded, or prepared by some other forming process. The energy that goes into the near-surface region can result in strain hardening, recrystallization, and texturing [23,24].

C. Properties of Real Surfaces

We can distinguish two types of solids: crystalline and amorphous. Amorphous solids, in contrast to crystalline solids, lack any sort of regular pattern or array of atoms. Crystalline

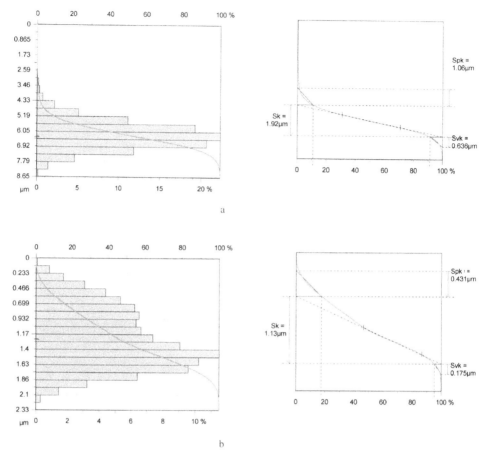

Figure 10 Amplitude density function in the form of histograms and bearing curves; (a) ground and nitrided surface (S_K = 0.666; K = 5.3); (b) mechanically polished and coated with a DLC thin layer (S_K = +0.366; K = 2.22).

solids include most metals and alloys as well as inorganic solids, and even some solid lubricants [25]. In this part, we shall limit the discussion to crystalline solids.

1. Crystalline Structure

An important aspect of any tribological material is its structure. The properties of any tribological material are closely related to this feature. The crystal structure of an ideal surface is defined as an orderly array of atoms in space.

Metals, alloys, ceramics, and solid lubricants are crystalline in nature with their atoms or molecules arranged in accordance with a particular structure. Metals can have a variety of structures: copper (Cu), nickel (Ni), silver (Ag), gold (Au), platinum (Pt), and aluminum (Al) have a face-centered-cubic structure (fcc); iron (Fe), niobium (Nb), vanadium (V), chromium (Cr), molybdenum (Mo), and tungsten (W) are body-centered cubic (bcc); zinc (Zn), cadmium (Cd), cobalt (Co), zirconium (Zr), magnesium (Mg), and titanium (Ti) have a hexagonal structure (hcp).

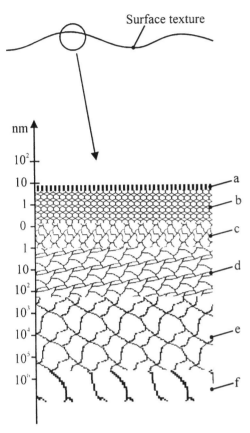

Figure 11 Surface texture and typical surface layers: (a) adsorbed gases and water; (b) oxide layer; (c) Beilby layer; (d) heavily deformed layer; (e) lightly deformed layer; (f) base material.

In both the fcc and bcc systems, there are three principal planes, i.e., (100), (110), and (111), which are referred to when discussing the crystalline nature of the solids. The (111) planes have the closest atomic packing and the lowest surface energy in the face-centered-cubic system; therefore they are the least likely to chemically interact with environmental constituents. The (110) planes in the face-centered-cubic system are the least densely packed; they have higher surface energies than the (111) planes, and are therefore much more reactive [23,25].

All engineering surfaces vary from this ideal situation. Grain boundaries, which develop during solidification as large defects, exist in the solid and extend to the surface. Other defects include subboundaries, twins, dislocations, interstitials, and vacancies. During the wear process, the surface will generally have undergone a high degree of strain and may contain large amounts of lattice distortion and defects such as dislocations. As a result, the microhardness is generally higher in grain boundaries than in grains. With plastic deformation, the strain generally produces a reduction in the recrystallization temperatures of material at the surface. But the combination of strain and temperature can result in surface recrystallization, a process that relieves lattice strain and stored energy and reduces the concentration of surface defects. During sliding, rolling, or rubbing, surface layers may be strained many times.

2. Solid State Bonding

As shown in Fig. 12, the bonding in crystalline solids can be of four types: van der Waals, ionic, metallic, and covalent [25].

The van der Waals bond is the weakest bond holding solids together. van der Waals forces are attributed to fluctuations in the charge distribution within the atoms or molecules. As shown in Fig. 12(a), these forces are responsible for bonding the atoms of an inert gas, e.g., solid argon. van der Waals bonding is frequently encountered in tribological systems with the physisorption of liquid lubricants and gases to a solid surface.

Ionic bonding occurs in solid compounds of two or more different elements, e.g., NaCl, as shown in Fig. 12(b). Ionic bonds form between two oppositely charged (cations–anions) ions, which are produced by the transfer of electrons from one atom to the other. Ionic bonding can be very strong because of this electron charge transfer. A classic example of ionic bonding is aluminum oxide, a tribological material that has an inherently high strength.

In the case of metallic bonds [Fig. 12(c)], the sharing of electrons between neighboring atoms becomes delocalized. The metallic state can be visualized as an array of positive ions with a common pool of electrons, the so-called free-electron cloud. The electrons freely move between the cores of ions, allowing for, and contributing to, many of the properties of metallic materials. This freedom of electrons is responsible for the good thermal and electrical conduction properties of metals.

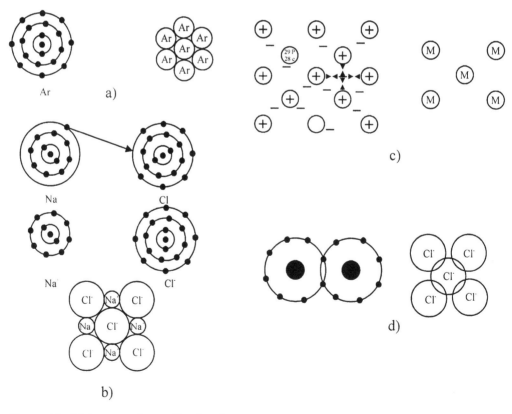

Figure 12 Main types of crystalline bonding: (a) van der Waals (crystalline Ar); (b) ionic (NaCl); (c) metallic; and (d) covalent (C diamond). (From Ref. 25.)

Covalent bonding is shown in Fig. 12(d). This type of bond involves the sharing of electrons between neighboring atoms. The neutral atoms appear to be bound together by the overlapping of their electron distribution. Covalent bonding produces an extremely strong bond, as evidenced by the high melting points and the high hardness of the material. At the same time, the covalent bond found in organic molecules, in polymers, and in lubricants, is relatively weak.

3. Surface Energy

The surface energy of most engineering surfaces is difficult to determine. Rabinowicz [26] has suggested that the surface energy of a solid can be approximated using the surface energy of the liquid at the melting point of the material. The agreement between this approximation and the experimentally determined value for simple pure materials is surprisingly good. However, the real interacting surfaces of most materials are covered with oxides, other impurities, and mechanically altered layers. When the surface roughness is superimposed on this, the energy of an engineering surface is very difficult to estimate [7]. Thermodynamically,

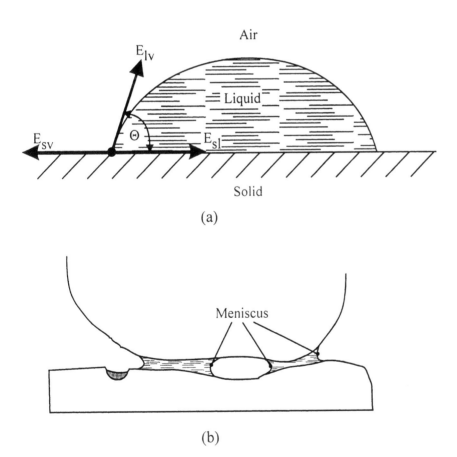

(a)

(b)

Figure 13 (a) Definition of contact angle; (b) formation of meniscus bridge as a result of liquid present at an interface ($0 \leq \theta < 90°$—liquid wets surface; $\theta > 90°$—liquid does not wet surface; $\theta = 90°$—liquid completely wets surface). (From Ref. 28.)

the higher the surface energy, the more reactive the surface will be. Therefore knowing the surface energy will give an indication of the potential chemical reactivity toward oxygen, water, and lubricants. Experimentally, contact-angle (Θ) measurements with a series of well-known liquids have been used to estimate the surface energy of a smooth solid surface. The results are useful for understanding wettability and the spreading out of liquids on a solid surface, but caution must be used in generalizing to other surfaces [7,27].

A liquid wets a surface if there is a suitable contact angle between the liquid–vapor surface and the liquid–solid surface, as shown in Fig. 13 [28]. The contact angle is determined by Young's equation:

$$\cos \Theta = (E_{sv} - E_{sl})/E_{lv} \tag{3}$$

where E_{lv} is the surface tension between the liquid and the vapor interface, E_{sv} is the surface tension between the solid and vapor interface, and E_{sl} is the surface tension between the solid and liquid interface.

Usually, the contact angle is measured by aligning a tangent with the drop profile at the point of contact with the solid surface [27].

If the liquid wets the surface with $0° \leq \theta < 90°$, the liquid surface is constrained to lie parallel with the surface and the complete liquid surface must therefore be concave [Fig. 13(b)]. If the contact angle is $\Theta = 0$, then the liquid completely wets the interacting surface. The liquid does not wet the interacting surface if the contact angle is $\Theta > 90°$ [27, 29,30]. Surface tension results in a pressure difference across any meniscus surface, referred to as capillary pressure, and is negative for a concave meniscus. The negative capillary pressure results in an intrinsically attractive (adhesive) force that depends on the interface roughness, the surface tension, and the contact angle. During sliding, frictional effects due not only to external load but also to intrinsic adhesive forcesneed to be overcome [28].

III. INTERACTIONS BETWEEN THE SOLID SURFACE AND THE ENVIRONMENT

The mechanism of lubrication partly depends on the nature of interactions between the lubricant and the solid surface. These interactions, which include reconstruction, segregation, physisorption, chemisorption, and chemical reactions, are shown in Fig. 14.

A. Surface Reconstruction

Surface reconstruction takes place when the outermost layers of atoms at the solid surface undergo a structural change [Fig. 14(a)]. This surface reconstruction results in a remarkable change in the coefficient of friction, while the segregated surface changes the properties of the surface compared to the nonsegregated surfaces [25].

B. Segregation

Segregation takes place in solids containing more than a single element; atoms from the bulk can diffuse to the surface and segregate there. The segregation of alloy species to grain boundaries in alloys and its effect on mechanical properties have been known to metallurgists for some time [Fig. 14(b)]. The same segregation occurs at a solid surface and segregation has been observed to exert considerable influence on adhesion, friction, and wear.

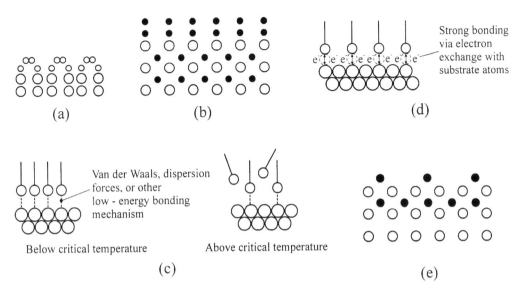

Figure 14 Schematic diagrams of various surface interactions: (a) reconstruction, (b) segregation, (c) physisorption, (d) chemisorption, and (e) compound formation.

The segregation of alloying elements toward the grain boundaries influences the surface energy of an interface, which has a direct effect on the energy of adhesion. This process was found to be irreversible. When the alloying elements segregate to the surface, they remain there for some alloys, e.g., a carbon-to-iron surface, a sulfur-to-iron surface, aluminum to the surfaces of both iron and copper, and indium and tin to a copper surface [25].

C. Physisorption

Physisorption, or physical adsorption, is the classical form of the adsorption of species on a solid surface. The physical adsorption process is a relatively weak process. Molecules of the adsorbate may be attached from a surface without any irreversible changes to the surfaces or adsorbate. A very small amount of energy is required to remove the physisorbed atoms. In physisorption, van der Walls or dispersion forces provide the bonding between the substrate and adsorbate, as illustrated in Fig. 14(c). It is observed that, if the interaction involves less than 10 Cal/mol, the process is one of physisorption [25]. The strength of physisorption depends on the electronic structure. Polar molecules tend to adsorb with their molecules perpendicular to the surface. A useful byproduct of physisorption is the so-called Rehbinder effect—the reduction of the modulus and yield stress of metals and nonmetals in the presence of an adsorbed film [32]. Physisorption is effective in reducing friction during the running in of a bearing surface provided that temperatures do not rise much above the ambient temperature. When the critical temperature (ambient temperature) is exceeded, the adsorbed film disorders and ceases to function as a lubricating layer, as shown schematically in Fig. 14(c).

D. Chemisorption

Chemisorption or chemical adsorption is an irreversible or partially irreversible monolayer process that involves some degree of chemical bonding between adsorbate and substrate, as illustrated in Fig. 14(d). Chemisorption is a much stronger form of bonding than that associated with physisorption, with the bond strengths a function of the chemical activity of the solid surface, i.e., surface energy, degree of surface coverage of the adsorbate, reactivity of the adsorbing species and its structure. The higher the surface energy of the solid surface, the stronger the tendency to chemisorb. It was observed that oxygen will chemisorb relatively strongly to copper, weakly to silver, and not at all to gold. The reactivity of the adsorbent is also very important; for instance, fluorine will more strongly adsorb than chlorine [25]. However, not all adsorbates are sufficiently reactive to initiate chemisorption. A long-chain alcohol did not show any retention, even with base metals, while stearic acid showed permanent retention with base metals such as zinc and cadmium, but not on the noble metals such a gold [31]. The strength of the chemical bonding between the adsorbate and substrate, which affects the friction transition temperature, depends on the reactivity of the substrate material.

E. Compound Formation

Compound formation on solid surfaces plays an important role in tribological systems [Fig. 14(e)]. Physisorbed and chemisorbed films are very effective in reducing friction and mild wear during a friction process. They fail fairly easily under severe rubbing conditions and therefore are not very effective in preventing wear and seizure. The naturally occurring oxides present on metals prevent their destruction during rubbing, but once they are removed the reoxidation of the surface may be too slow to be effective. With metals in contact, and with metals and nonmetals, compound formation by chemical reaction has been observed to occur on the solid surface; an example of this is phosphorus on iron [33–35]. The compound formation produces strong interfacial bonds at the contacting surfaces and influences the adhesion behavior. Similarly, more active chemicals can be added to the lubricants to react with the rubbing surfaces and produce protective films. Suitable films include sulfides, phosphides, and phosphates, and any other chemical that will react with the rubbing surface to produce such a substance will be effective in producing protective films [25].

F. Organic Compounds and Polarity

All organic compounds contain carbon (C) atoms. Carbon, in combination with hydrogen, oxygen, nitrogen, and sulfur, results in a large number of organic compounds. With four electrons in its outer shell, carbon forms four covalent bonds, with each bond resulting from two atoms sharing a pair of electrons. The number of electron pairs that two atoms share determines whether the bond is single or multiple. In a single bond, only one pair of electrons is shared by the atoms. Carbon can also form multiple bonds by sharing two or three pairs of electrons between atoms. An organic compound is classified as saturated if it contains only a single bond, and unsaturated if the molecules possess one or more multiple carbon–carbon bonds [28,36,37].

The polarity of a bond is determined by the difference in the electronegativity values of the atoms forming the bond. If the electronegativities are the same, the bond is nonpolar and the electrons are equally shared. In this type of bond, there is no separation of positive and negative charge between the atoms. If the atoms have very different electronegativities, then the bond is very polar. A dipole is a molecule that is electrically asymmetrical, causing it to be oppositely charged at two points [28]. Table 1 summarizes the polar and nonpolar groups commonly used in the construction of hydrophobic and hydrophilic molecules [28]. Table 2 lists organic groups in increasing order of polarity.

Table 1 Some Examples of Polar (Hydrophilic) and Nonpolar (Hydrophobic) Groups

Name	Formula
Polar	
Alcohol (hydroxyl)	-OH
Carboxyl	-COOH
Aldehyde	-COH
Ketone	$\begin{matrix} O \\ \parallel \\ R\text{-}C\text{-}R \end{matrix}$
Ester	-COO-
Carbonyl	$>C=6;O$
Ether	R-O-R
Amine	-NH$_2$
Amide	$\begin{matrix} O \\ \parallel \\ \text{-C-NH}_2 \end{matrix}$
Phenol	OH on benzene ring
Thiol	-SH
Trichlorosilane	-SiCl$_3$
Nonpolar	
Methyl	-CH$_3$
Trifluoromethyl	-CF$_3$
Aryl (benzene ring)	benzene ring

Source: Ref. 28.

Table 2 Organic Groups Listed in Increasing Order of Polarity

Alkanes
Alkenes
Aromatic hydrocarbons
Ethers
Trichlorosilanes
Aldehydes, ketones, esters, carbonyls
Thiols
Amines
Alcohols, phenols
Amides
Carboxylic acids

Source: Ref. 28.

IV. LUBRICANTS

Modern lubricants are formulated from a range of base fluids and chemical additives. The base fluid has several functions; primarily, it is the lubricant. It also provides a fluid layer separating the moving interacting surfaces and removes heat and wear particles from the contact. Many of the properties of the lubricant are created by the addition of special chemical additives to the base fluid. The base fluid also functions as the carrier for these additives and must therefore be able to keep the additives in solution under all normal working conditions.

A. Base Oils

1. Base Oils Made from Mineral Oils: Composition and Structure

Base oils are made from crude oil, an extremely complex mixture of organic chemicals. The main constituents of crude oil are hydrocarbons, organic oxygen, sulfur, and nitrogen compounds as well as compounds with trace elements. Table 3 shows the typical components of crude oil. Each oil field contains a different type of crude oil, varying in chemical composition and physical properties.

Mineral base oils are mixtures of hydrocarbons of various molecular sizes and molecular structures [39]. Figure 15 shows the general structure of basic hydrocarbons in schematic form. There are certain basic types of hydrocarbon mixtures, which allows them to be classified as chain and ring-shaped hydrocarbons. Both of these types can occur as saturated or unsaturated compounds, characterized by one or more double bonds. Saturated hydrocarbons are chemically unreactive, whereas unsaturated hydrocarbons are chemically reactive. It is usual to find combinations of chain and ring-shaped saturated and unsaturated hydrocarbons. These combinations are then called paraffin-based, naphthene-based or aromatic, depending on which type of hydrocarbon predominates in the compound [38,39,47–50].

The molecular structure and the molecular weight of the hydrocarbons determine their chemical and physical behavior. Three main characteristics, which are important for base lubricant oils, are dependent on these molecular properties: viscosity and viscosity–temperature behavior; state of aggregation, particularly the liquid range; and oxidative and thermal stability [39].

The viscosity of the base oil increases with increasing molecular size, chain branching, and increasing intermolecular cohesion forces. On the other hand, molecular flexibility reduces viscosity [39]. Figure 16 schematically shows this relationship. Increasing temperatures tend to increase molecular flexibility and reduce the influence of chain branching and the forces of cohesion. For example, naphthenes and aromatics demonstrate increasing

Table 3 Typical Component of Crude Oil

Component	Concentration (wt.%)
Carbon	80–85
Hydrogen	10–17
Sulfur	<7
Trace elements	<1
(e.g., N, Cl, P, Na, Mg, V, O_2, etc.)	

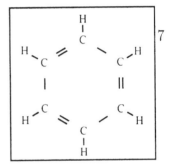

Basic components of
hydrocarbon compound.

Alkane- n - paraffins (normal
paraffins) $C_n H_{2n-2}$. The C atoms
are arranged in a chain; free
valencies are combined with H
atom. The chains are all unbranched.

i - paraffins (iso - paraffins)
$C_n H_{2n-2}$. More or less branched
C chains are saturated with
hydrogen.

Alkene - olefin $C_n H_{2n}$.
Chain - like or branched
arrangement of C atoms
with one or more double
bonds; the remaining
valencies are saturated with
H atoms.

Cycloalkane - naphthenes
$C_n H_{2n}$. Cyclic arrangement
of 5 - 7 atoms. The C atoms
are saturated with H atoms.

Benzene, basic element of
aromatic. Circular arrangement
of 6 C atoms with 3
double bonds. The other 6
free valencies are saturated
with H atoms.

Figure 15   General structure of hydrocarbon compounds. (From Ref. 39.)

molecular flexibility with temperature, which results in a marked viscosity–temperature
dependence, whereas the paraffins, which are already fairly flexible, are less affected.

The dependence of the state of aggregation, primarily of the liquid range, vs. the
molecular weight and the molecular structure are shown in Fig. 17 [39]. We can see that as
the molecular weight increases, the melting point and the boiling point of paraffins,
naphthenes, and aromatics, for the same molecular weight, are shifted toward higher
temperatures in this sequence. A very important characteristic of base oil made from crude
oil is its oxidative and thermal stability. While oxidative stability is the resistance to a
reaction with oxygen, thermal stability is the resistance of a molecule to decomposition
because of the effects of heat [39]. Figure 18 schematically shows the effects of structural
changes in the hydrocarbon on the liquid range, the viscosity, the viscosity–temperature
behavior (v–t), the oxidative and the thermal stability. As we can see in Fig. 18, paraffins
and napthenes, which have single bonds, are more resistant to oxidation than olefins and
aromatics, which have double bonds in their structure. On the other hand, chain-like
compounds are made to vibrate more by the effect of heat than ring-shaped compounds,

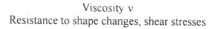

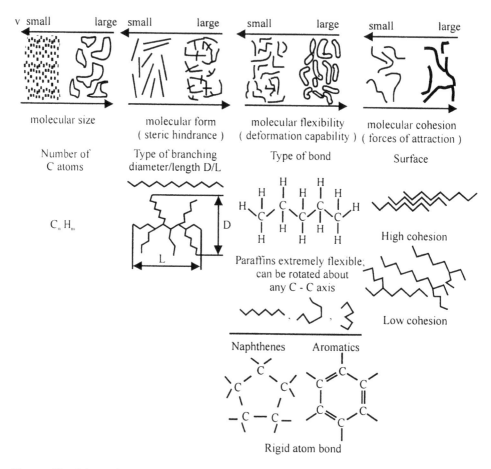

Figure 16 Schematic representation of the effect of the structure on viscosity. (From Ref. 39.)

which can store more thermal energy before the molecule decomposes. Therefore naphthenes and aromatics have a better thermal stability than paraffins (Fig. 18). In the
application of mineral oils, it is always necessary to consider which characteristics are
the most important.

2. Synthetic Base Oils

Synthetic lubricants have been used for many years. However, lubricants manufactured
from synthetic base oils do not represent an all-encompassing substitute for mineral-oil-
based lubricants, rather they are to be regarded as a supplement for particular applications. The requirements for lubricants, particularly military and aeroengine lubricants, to
perform over an increasingly wide temperature range has stimulated the continuing development of synthetic lubricant technology. Synthetic lubricants are now found in all areas
of lubrication—automobiles, trucks, marine diesels, transmissions, and industrial lubricants
as well as aviation and aerospace lubricants. However, synthetic base oils have advantages
and disadvantages, and these are listed in Tables 4 and 5. There are no synthetic base oils

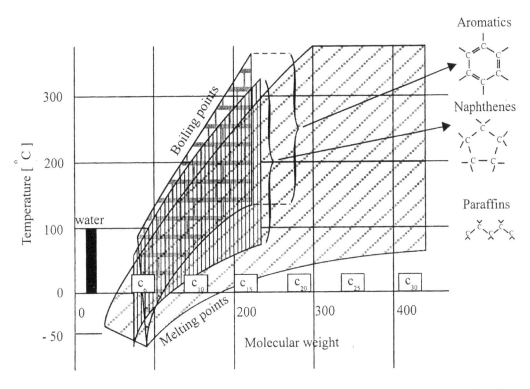

Figure 17 Relationship between the liquid range of hydrocarbon and molecular weight and molecular structure. (From Ref. 39.)

that combine all the listed properties as either advantages or disadvantages. When selecting a synthetic lubricant, it is necessary to rank the values of the characteristics required or expected [40,51].

Synthetic liquids are mostly long-chain molecules that are produced by chemical reactions to obtain specific characteristics. They are made from petrochemicals, animal and vegetable oils, and coal-derived feed stocks. Synthetic liquids can be assigned to certain types or groups on the basis of their chemical composition (Table 6).

Synthetic hydrocarbons contain paraffin-like liquid substances that consist only of carbon and hydrogen. Because of the chain length, the degree of branching and the position of the branches, these compounds possess a certain viscosity, a high viscosity index, and a low pour point [41,42]. Polyalphaolefins have found increasing acceptance because of their compatibility with petroleum base oils, thermal stability, and excellent viscosity–temperature relationship. Alkylbenzenes are mainly used in low-temperature applications as hydraulic oils and greases [7].

Polyglycols of ethylene oxide, propylene oxide, or higher alkyl ethers are called polyalkylene glycols. As with the synthesis of polyalphaolefine (PAO), the initial material is ethylene or an ethylene derivative, which is then oxidized to a cyclic ether in a reaction with oxygen [39]. Polyglycol ethers (polyethers) provide a variety of viscosity and molecular-weight grades. They are used mostly in water-based, fire-resistant hydraulic fluids, brake fluids, and rubber molding lubricants. The volatility and oxidation stability of the polyethers are generally similar to those of mineral base oils. They have low pour points, good

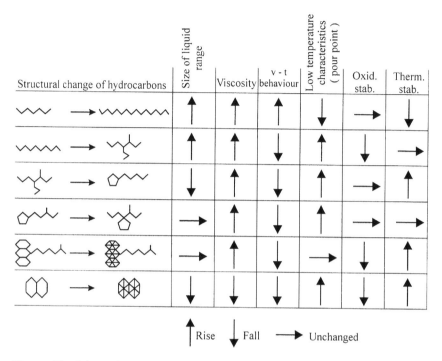

Figure 18 Schematic representation of the effect of structural change on various characteristics of hydrocarbons. (From Ref. 39.)

compatibility with rubber, and good sludge and varnish resistance. They are available in water-soluble and oil-soluble forms [7].

Synthetic esters are organic, oxygen-containing compounds, which can be obtained from the reaction of alcohol with an organic acid [41,43–45]. They consist of fatty acid esters, polyol esters, polyesters, phosphate esters, and silicate esters. Most of these compounds exhibit high boiling points, excellent viscosity–volatility characteristics, and high temperature stability. Esters generally exhibit good solvency and have good friction and wear characteristics [7].

Silicon oils are semiorganic polymers and copolymers that contain a repeated lattice of silicon oxygen units and organic side chains bonded to silicon [41,46]. Silicones have excellent fluid properties and have been used as lubricants in systems designed to run under

Table 4 Advantages of Synthetic Liquids

Thermal stability
Oxidative stability
Viscosity-temperature characteristic
Flow characteristics at low temperatures
Volatility at high temperatures
Temperature application range
Resistance to radiation
Low flammability

Table 5 Disadvantages of
Synthetic Liquids

Hydrolytic behavior
Corrosion behavior
Toxic behavior
Compatibility with other additives
Solubility for additives
Availability:
  in principle
  in specific viscosity situations
Cost

hydrodynamic and elastohydrodynamic conditions where the bearing systems are non-ferrous. They have the best viscosity–temperature characteristics of any synthetics, with very low volatility and excellent oxidative and thermal stability. They are not miscible with mineral base oils [7].

Polyphenyl ethers can be used as high-temperature lubricants with high oxidative stress or high radiation loading [41,46].

Table 6 Classification of Synthetic Liquids on the Basis
of Their Chemical Composition

Synthetic liquid	Composition
Synthetic hydrocarbons	
Poly-alpha-olefins (PAO)	C, H
= synth. Hydrocarbon fluids (SHC)	
Alkylbenzenes	C, H
Monoalkylbenzenes	
Dialkylbenzene	
Polyglycols (polyethers)	
Copolymers of ethylene oxide,	C, H, O
propylene oxide, etc.	
Carbonic acid esters	
Monocarbonic acid esters	C, H, O
Dicarbonic acid ester	C, H, O
Polyolester	C, H, O
Phosphoric acid esters	
Triaryl- , trialkyl-, and alkylarylesters	C, H, O, P
of phosphoric acid	
Silicone oil	
Dimethylsilicone oil	C, H, O, Si
Methylphenylsilicone oil	C, H, O, Si
Phenylsilicone oil	C, H, O, Si
Polyphenylether	
Meta- and *para*-polyphenylether	C, H, O
Polyfluoroalkylether	C, F, O

The general structures of some synthetic lubricants are shown in Fig. 19. The physical and chemical characteristics of synthetic base oils are the viscosity, the temperature behavior, the low-temperature flow properties, the compatibility with seals, the miscibility with mineral oil, the lubricity, the thermal stability, the hydrolytic stability, the solvency, and the biodegradability.

3. Fatty Base Oils

The early liquid lubricants were water-insoluble fatty substances extracted from plant, animal, and marine sources. Fatty oils decompose to fatty acids at slightly elevated temperatures and as a result of certain chemical reactions. These days, such oils are rarely used alone but are usually employed as an additive to increase the lubricity of petroleum-based oils. Figure 20 shows the general structure of fatty base oils.

B. Lubricant Additives—Function and Properties

The properties and the quality of lubricants depend on the provenance and viscosity of the base oil and the process parameters used in its production, as well as on the type and quantity of the additives used. The additive range extends from a few ppm to concentrations of about 30%. The following additives, categorized according to their functions, are used: friction modifiers, antiwear additives, extreme-pressure additives, antioxidants, corrosion inhibitors, detergents, dispersants, pour-point dispersants, deferments, and viscosity-index im-

Polyalphaolefines e.g.	$(- CH_2 - CH_2 - CH_2 - CH_2 -)_n \ldots CH_2 - CH_2 - CH = CH_2$
ESTERS E.G.	
• Diesters e.g.	$C_8H_{17} - O - CO - C_8H_{16} - CO - O - C_8H_{17}$
• Phosphate esters e.g.	$(CH_3 - C_6H_4 - O_3)P{=}O$
• Silicate esters e.g.	$Si(O - C_8H_{17})_4$
• Polyglycol esters e.g.	$CH_2 - (- CH_2 - O - CH_2)_n - CH_2$ $\quad\quad\quad\ \|\quad\quad\quad\quad\quad\quad\quad \|$ $\quad\quad\quad OH \quad\quad\quad\quad\quad\quad OH$
Polyglycols e.g.	$OH - CH_3 - CH_3 - O - CH_3 - CH_3 - O \ldots CH_3 - CH_3 - OH$
Silicones e.g.	$\begin{array}{ccc} CH_3 & & CH_3 \\ \| & & \| \\ CH_3 - Si - \left[O - Si \right] O - Si - CH_3 \\ \| & & \| \\ CH_3 & & CH_3 \end{array}$
Perfluoropolyalkylethers e.g.	$F - \left[\begin{array}{cc} F & Cl \\ \| & \| \\ C & - C - \\ \| & \| \\ CF_3 & F \end{array} \right]_n \begin{array}{c} F \\ \| \\ O - C - CF_3 \\ \| \\ F \end{array}$

Figure 19 General structure of synthetic lubricating oils.

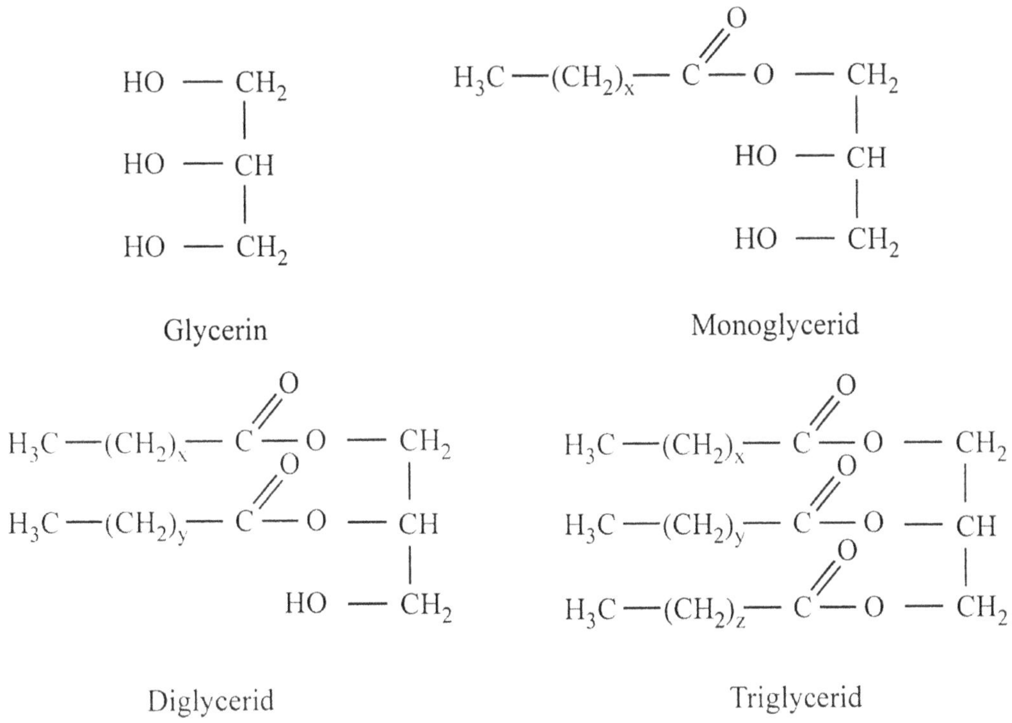

Figure 20 General structure of fatty base oils.

provers. It should be remembered that the required lubricant properties can rarely be achieved with a single type of additive; therefore mixtures of different additives are normally used. The function and properties of additives can generally be divided into three groups: physical property improvers for the lubricant, chemical properties improvers for the lubricant, and physical and chemical property improvers for interacting surfaces [54].

1. Physical Property Improvers for Lubricants

a. Viscosity-Index Improvers. The viscosity index (VI) accurately describes the viscosity–temperature characteristic of base oils. Different base oils may have different viscosity–temperature characteristics. The viscosity–temperature relationship of base lubricating oils is such that, for some applications, the viscosity is too high in the low-temperature range and too low at the operating temperature. This mainly applies to engine oils, but can also be of significance for gear oils and hydraulic oils. To improve the viscosity–temperature behavior, i.e., to raise the viscosity index, additives that have a better solubility at high temperatures than at low temperatures are used [39,52]. This means that at a given concentration, they increase the viscosity of base oil relatively less at low temperatures than at high temperatures. This function is performed by oil-soluble polymers with high molecular weights, of the order of 100,000 or more [7].

From the chemical point of view, the following types of polymers are used as viscosity-index improvers: polyisobutylenes, olefin copolymers, styrene copolymers, polymethacrylates, and polyesters. By incorporating polar groups, the viscosity-index improvers can also produce dispersant additives [55,56].

b. Pour-Point Improvers. The lowest temperature at which a lubricant flows is called the pour point. The fluidity of a lubricant in an engine is very important under all circumstances. When starting up an engine from cold, it is important that the mechanical parts freely move with sufficient lubrication. When cooled to low temperatures, lubricating oil can undergo a number of changes. The circumstances under which this occurs are dependent on the thermal history, the cooling rate, and on the composition of the lubricating oil. Naphthenic oils solidify, forming an opaque solid phase and having a low pour point. Paraffinic oils of high normal paraffins tend to produce relatively strong structures made up of large crystals. Isoparaffins give smaller crystals of a weaker structure in a solidified oil. If the cooling rate is rapid, then crystal growth may be rapid and the oil will not remain fluid. If a microcrystalline wax is formed, then the crystals can swell, behaving like a sponge, with the subsequent absorption of free oil. Rapid cooling of an oil can produce a hard gel that is easily sheared [57,58].

Pour-point improvers are used for improving the low-temperature characteristics of oils. These pour-point improvers act through adsorption onto the wax crystals. The resulting surface layer of a pour-point improver inhibits the growth of the wax crystals and their capacity to adsorb oil and form gels. The effectiveness of pour-point improvers depends on the nature of the lubricating oil and the concentration of the pour-point improvers. Pour-point improvers can be classified into the following groups: naphthalene and phenol condensation products, polymethacrylates, fumaric ester copolymers, and olefine ester copolymers [28].

c. Dispersants. Dispersant additives are intended to suspend impurities, stop the formation of sludge-like products from solid and liquid impurities, and neutralize acid substances. They are also intended to redissolve existing deposits. There are two types of dispersants: ash-producing and ash-free additives.

The following ash-free organic dispersant additives groups can be used: polyisobutenyl succinic acid derivatives, methacrylate copolymers, and fumarates [39,57]. These dispersant additives stop sludge formation from solid and liquid impurities at temperatures below the operating temperatures.

The following products can be used as ash-producing organometallic dispersant additives: naphthenates, sulfonates, phenolates, and phosphates [39,57]. These dispersant additives stop sludge formation at high temperatures.

d. Foam Inhibitors. With lubricating oils, foam formation can impair the oil flow if the quantity and the viscosity of the surface foam become too great. We distinguish between surface foam, which can be suppressed with defoaming agents, and "aeroemulsions," which may be influenced by additives. To disperse liquid droplets or gas bubbles in the oil, the surface of the dispersed phase must be increased. The greater the surface tension of the oil, the greater is the work required, and the lower is the stability of the foam. As detergents reduce tension, they can be the cause of foam formation. The most familiar foam additives are the polyaklyl siloxanes or silicones [39,57].

2. Chemical Property Improvers for Lubricants

a. Oxidation Inhibitors. Oxidation is the most common form of lubricant deterioration. It is impossible to prevent oxygen being present in a lubricating oil. During the oxidation process, acids form, and at an advanced stage, oil-insoluble polymer form, which leads to varnish and sludge-like deposits. The oxidation process proceeds through free radical reactions, which are accelerated by temperatures and catalyzed by metal parts, metal abrasion, combustion products, O_2 content, and aging products. The oil oxidation process proceeds relatively slowly at temperatures below 100°C; however, it accelerates at higher temperatures. The following main groups of oxidation inhibitors are important [39]:

–Sterically hindered phenols, bisphenols, and thiobisphenols.
–Metal dialkyldithiophosphates and phosphonates.
–Overbased metal sulfonates, alkylphenolates, and alkylphenol sulfides.
–Aromatized amines.

Of course, inhibitors cannot completely prevent the oxidation process. Depending on their molecular structure, different inhibitors reveal their full effectiveness over a limited temperature range. The selection and dosing of oxidation inhibitors depends on the type of lubricant to be formulated.

b. Corrosion Inhibitors. Corrosion is an electrochemical process in which the attacked metal oxidizes and the attacking medium is reduced. A prerequisite for this process is the presence of oxygen or air. Corrosion inhibitors are those substances that quickly form dense protective films on metallic surfaces and thus stop direct contact between the base material and the aggressive media or atmospheres. Corrosion inhibitors can usefully be grouped into those for ferrous metals and those for nonferrous metals [59].

When iron and steel rust, the corrosion takes place in the presence of corrosive substances on the metallic surface. These corrosive substances include water, metal chlorides, and bromides, oxides of sulfur and nitrogen, as well as organic and inorganic acids. The best rust inhibitors are those compounds that carry a strongly polar group on a long-chain alkyl residue. Widely used rust inhibitors include metal sulfonates (barium, calcium, magnesium) based on sulfonic acids with average molecular weights of 400–450 [55].

The corrosion of parts made of nonferrous metals is primarily caused by acids and occurs on bronze and white metal parts. Protection against corrosion is provided by those substances that form an adsorption layer, a chemisorption layer, or a chemical reaction layer on the surface of the metal. The reaction products of phosphorus pentasulfide and unsaturated hydrocarbons as well as alcohols and posttreatment with zinc oxide have proved to be effective additives for engine oils.

c. Detergents. Detergent additives are intended to suspend impurities, stop the formation of sludge-like products from solid and liquid impurities, and neutralize acid substances.

The detergent additives are oil-soluble or finely dispersed metal salts of organic acids, which are primarily used to prevent deposits on hot engine parts.

3. Physical and Chemical Improvers for Interacting Surfaces

Additives that improve the friction and wear properties of interacting surfaces are probably the most important of all the additives used in oil formulations [39]. Figure 21 shows a survey of the most important antiwear (AW) and extreme-pressure (EP) additives. These additives can be divided into the following groups: adsorption or boundary additives, antiwear additives, extreme-pressure additives. A common distinction can be made between physically adsorbent and chemically reactive compounds [60,61].

Physically active additives act by adsorption with the metal surfaces or metal oxides and form intermediate films. Because of their polarity, the molecules orient themselves in a brush-like fashion on the surface of the metal. The strength of the adsorption depends on the van der Waals forces of the hydrocarbon groups and the dipole moment of the polar group with respect to the metal surfaces. The load capacity is increased by layer formation and polarity. However, the shear strength of the adsorption layers is low. The layers sheared off

Chemical composition	Initial products	Name
R – (S)$_x$ – R	Unsaturated fatty acid esters, cyclic and aliphatic olefins, sulphur, hydrogen sulphide	Sulphurated fat oils and sperm oil, sulphurates terpenes and olefins
(RO)$_3$ PO, (RO)$_3$ P (RO)$_2$ P(O) OH	Alcohols, phosphorus oxychloride, phosphorus trichloride	Tertiary and secondary esters of phosphoric or phosphorous acid
R … (C H Cl)$_x$ … CH$_2$ Cl	Aliphatic hydrocarbons, chlorine	Chloroparaffins
[(RO)$_2$ P(S) S]$_2$ Me	Alcohol, phosphorus pentasulphide, metal oxide (metal usually zinc)	Metal dialkyl (diaryl) dithiophosphates
(RO)$_3$ P(S)$_2$ – S$_x$, x = 1 or 2	Alcohol, phosphorus pentasulphide	Dialkyl (diaryl) dithiophosphates sulphides
(R) P (S) … S … R*)	Olefins, phosphorus pentasulphide	Alkyl dithiophosphonates

R – alkyl (aryl) residue
* Chemical composition largely unknown. P/S ratio depends on manufacturing process

Figure 21 Survey of the most important AW and EP additives. (From Ref. 39.)

during mixed friction are continuously replaced. The most important representatives of these additives are metal soaps, esters, fatty oils, and organic acids.

Chemisorption is a mechanism that takes place when electron transfer occurs between an adsorbed molecule and contact surfaces. This produces a covalent or an electrovalent bond that is far stronger than the physical adsorption. Physical adsorption is always reversible, chemisorption is generally irreversible, so that chemisorbed molecules are only removed when the element of the surface to which they are attached is removed by wear. Chemically active additives include sulfur, phosphorus, chlorine, and lead compounds. The most effective are also combined compounds, an example of this type of additive is metal dialkyl dithiophosphate, based on zinc.

a. Adsorption Additives—Friction Modifiers.   Friction is defined as the resistance a body meets when moving over another body in respect of transmitting motion. The friction coefficient is defined as friction force/normal force. In the lubricated surface situation, the coefficient of friction will be determined by the lubrication regime. In simple terms, there are two lubricant regimes, boundary and hydrodynamic regimes, as shown in Fig. 22 [57].

The mechanical losses can be reduced by friction reduction. But it must be remembered that in the range of liquid friction, only reducing the viscosity achieves the desired effect, while the action of friction-reducing additives is restricted to the mixed friction range. Reducing the viscosity enlarges the mixed friction area. This disadvantage is equalized with friction modifiers that act most successfully in the boundary regime, reducing friction and wear [39]. Figure 23 gives a schematic representation of the combined effects of reducing viscosity and using friction modifiers [62].

Friction-modifier additives are generally long slender molecules. They normally have a straight hydrocarbon chain consisting of at least 18 carbon atoms. The polar head group is

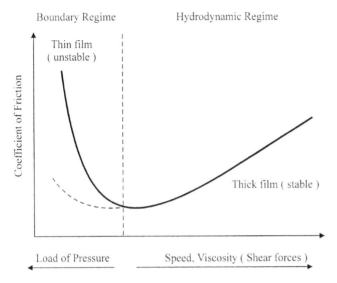

Figure 22 Effect of viscosity, speed, and load on friction and lubrication regimes. (From Ref. 57.)

the dominant factor in the effectiveness of the molecule as a friction modifier. Such polar groups (see Tables 1 and 2) consist of carboxylic acids or derivatives; phosphoric acids and their derivatives; amines, amides, and their derivatives [57]. Adsorption additives are very sensitive to the effects of temperature. They lose their effectiveness at temperatures between 80°C and 150°C, depending on the type of additive used. With increased temperature, there is sufficient energy input to the surface for the additive to desorb.

Fatty acids, as well as esters and amines of the same fatty acids, are normally used as friction modifiers. Specialized additives, which combine adsorption or boundary properties with other functions such as corrosion protection, are also in use. The most frequently used compounds are sulfurized fatty acid derivatives and phosphonic acids.

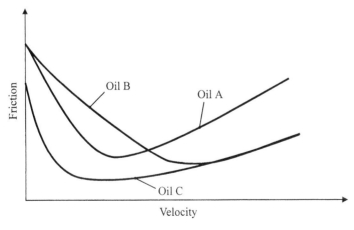

Figure 23 Combined effects of viscosity reduction and the use of friction modifiers. (A) Reference oil; (B) viscosity reduction; (C) viscosity reduction and friction modifiers. (From Ref. 39.)

b. Antiwear Additives. To protect interacting surfaces at temperatures that are above the effective range of adsorption additives, antiwear additives have been designed. There are several different types of antiwear additives that are currently used in oil formulations. The ideal molecule for reducing mild wear by forming a protective film on the interacting surfaces consists of a polar molecule for strong adsorption with a long nonpolar chain that will orient itself perpendicular to the surface and thus create a thick film. Such an additive is cetyl alcohol (hexadecanol), which is adsorbed on an oxidizing steel surface. Other substances showing similar behavior are long-chain fatty acids such as palmitic and stearic acids and their esters, such as ethyl stearate. This category of long-chain esters also includes vegetable oils and fats. Mineral base oils have inferior antiwear properties, and the technique of adding small amounts of natural vegetable oils or fats, or long-chain alcohols, acids, or esters has been practiced for many years.

The mildly adsorbed long-chain compounds described above are not very powerful antiwear additives. They tend to be removed from surfaces during strong rubbing and may be desorbed if the surface temperature rises much above 100°C [64].

More powerful antiwear additives have to chemically adsorb onto the interacting surfaces. Most of the common synthetic antiwear additives are compounds containing phosphorus, such as zinc dialkyl dithiophosphate (ZDDP), tricresyl phosphate (TCP), and trixylyl phosphate. Zinc dialkyl dithiophosphate additives have antiwear, antioxidant, and corrosion-preventing properties. Like stearic acid, ZDDP initially adsorbs on the contact surface and in this state has mild antiwear properties [64].

c. Extreme-Pressure Additives. Extreme-pressure additives are designed to react with interacting surfaces under extreme conditions of load and velocity, e.g., slowly moving, heavily loaded gears and bearings. Under these conditions, the contact temperatures increase and the simpler adsorbed molecules desorb so that a protective film is no longer present to prevent adhesion of the contacting surfaces. Furthermore, excessive mechanical stress will either tear away the iron oxide or wear away the metal together with any adsorbed films. The result is a surface containing areas of unoxidized metal, which has to be protected with powerful films that are created by reacting various chemicals with the contacting surfaces. Most of these chemicals, known as EP additives, contain one aggressive nonmetal such as sulfur, antimony, iodine, phosphorus, or chlorine [39]. They react with exposed metallic surfaces creating protective, low-shear-strength surface films that reduce friction and wear.

The reacted film is worn away by rubbing against another bearing surface, but then more sulfur will react to regenerate the surface film. The repeated cycle of reaction and wear represents a form of corrosive wear. All EP additives are in fact corrosive to the surfaces that they protect. The main aim when selecting EP additives is to maximize the protection against severe wear, while at the same time minimizing the rate of corrosion [39,64].

V. ADDITIVE REACTION MECHANISMS

To understand additive reaction mechanisms, it is useful to consider again the fundamental nature of wear and lubrication with fluids. In an ideal situation, lubricated surfaces are separated by a thick film of lubricant, and all the forces between the surfaces are transmitted by the lubricant. If, for any reason, the thickness of the lubricant film decreases, a point will be reached when the contact stresses are increasingly carried by a direct solid/solid contact between the real surfaces of the tribosystem. This process was originally investigated by Stribeck and has been already discussed in Sec. I.

Additive reaction mechanisms are a very complex phenomena. They depend on many parameters, but the following three factors have the main influences: the frictional heating-contact temperature, the velocity of the triboelements, and the bearing loads. Friction occurs whenever two solid bodies slide against each other. It takes place by a variety of mechanisms in and around the real contact area between the sliding or rolling/sliding bodies (see Fig. 4). It is through the frictional processes that velocity differences between the bodies are accommodated. In these processes, mechanical energy is transformed into internal energy or heat, which causes the temperature of the sliding bodies to increase [67–69]. The exact mechanism by which this energy transformation occurs may vary from one sliding situation to another. Most tribologists agree that nearly all of the energy dissipated in frictional contacts is transformed into heat [70]. This frictional heating is responsible for the increase of the temperature within the contact region of the triboelements. For the purposes of this discussion, it will be assumed that all frictional energy is dissipated as heat that is conducted into the contact bodies at the actual contact interface.

There are three levels of temperature in sliding or rolling contacts. The highest contact temperatures (T_C) occur at the small contact spots between surface roughness peaks or asperities on the sliding surfaces. These temperatures can be very high—over 1000°C in some cases—but last only as long as the two asperities are in contact (Fig. 24) [71]. This contact time could be less than 10 thousandths of a second [71–74]. The asperity contacts are often confined to a small region of the surface of the contact bodies; this region can be referred to as the nominal area. At any instant, there are usually several short-duration flash-temperature rises at particular asperity contact spots within the nominal contact area. The integrated average of the temperatures of all points within the contact area can be referred to as the nominal or mean contact temperature (T_{nom}). The nominal contact temperature can be over 500°C for severe sliding cases, e.g., brakes, but it is usually much lower. The temperature diminishes at short distances from the contact area, and decreases to a modest bulk volumetric temperature (T_b) several millimeter into the contact bodies. These temperatures are generally less than 100°C [71,73,74].

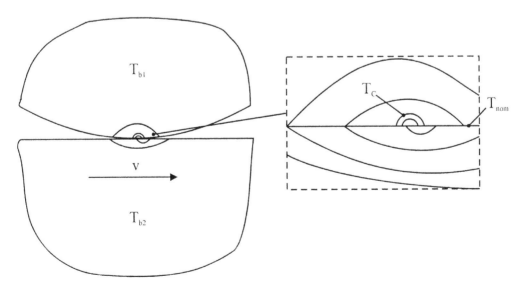

Figure 24 Schematic diagram of temperature distribution around sliding contact. (From Ref. 71.)

The effectiveness of additives is strongly dependent on contact temperature (T_C). Friction and wear with lubricants containing physically or chemically adsorbed additives rapidly deteriorate when the contact temperature reaches a critical value at which the additive molecules desorb (Fig. 25) [53].

The additive reaction mechanisms can be classified in terms of additive reaction mechanisms in the volume of the carrier oil and additive reaction mechanisms on the interaction surfaces, as shown in Table 7.

A. Additive Reaction Mechanisms in the Volume of the Carrier Oil

1. Mechanisms of Viscosity-Index Improvers

Viscosity-index improvers must work over a large temperature range. A further factor of interest is how the viscosity varies with temperature. There are many ideas as to how viscosity-index improvers function [75–79]. However, to improve the viscosity–temperature behavior, the additives that have better solubility at high temperatures, rather than at low temperatures, are used. This means that they increase the viscosity of the base oil less at low temperatures than at high temperatures. This function is performed by oil-soluble polymers with high molecular weights. The extent of their influence depends on the structure of the polymer molecules, their average molecular weight, and the molecular weight spectrum. The mechanism of the function is schematically shown in Fig. 26 [39,80].

At low temperatures, the long-chain molecules of the polymer are tightly coiled and increase the viscosity of the base oil, which to some extent acts as a solvent. As the temperature rises, the molecules of the polymer gradually uncoil, their spatial expansion increases at an increasingly faster rate than that of the solvent molecules, and this causes a progressive flow impedance. The thickening effect of such polymeric additives is several times greater at high temperatures than at low temperatures and thus causes a reduction in the temperature dependence of the viscosity [58].

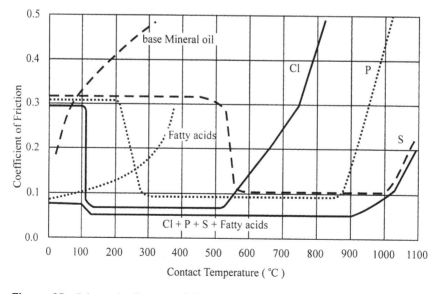

Figure 25 Schematic diagram of the temperature and additive influences on the coefficient of friction. (From Ref. 53.)

Table 7 Type of the Additive Reaction Mechanisms

Additive reaction mechanisms					
Additive reaction mechanism on the interaction surfaces					
State of the tribosystem	Temperature in the contact of interaction surface	Additive reaction mechanisms	Additive reaction mechanism in the volume of the carrier oil		
Steady or run-in condition	Ambient or very low contact temperature	Additives adsorption mechanisms— multimolecular layer	Mechanism of viscosity index improvers	Detergents and dispersants additive	Oxidation process and mechanism of antioxidation
Normal working condition	Operating temperature High operating	Adsorption additives reaction			
Extra heavy working conditions	temperature High operating and contact temperature	mechanisms— monomolecular layer Antiwear reaction mechanisms EP additive reaction mechanisms			

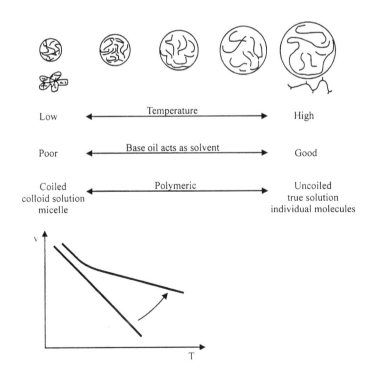

Figure 26 Mechanism of polymeric additives used as viscosity-index improvers. (From Ref. 39.)

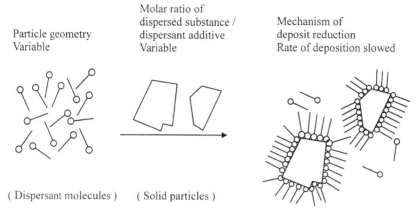

Figure 27 Peptization process. (From Ref. 63.)

2. Mechanisms of Detergent and Dispersant Additive

The function and the action of detergents and dispersants can generally be divided into three groups: peptization solubilization, and neutralization [39].

In the process of peptization, solid dirt particles are coated with ashless dispersants or metal detergents, as shown in Fig. 27 [63].

While both metal detergents and ashless dispersants can peptize solid particles, solubilization can only be carried out by ash-free dispersants (Fig. 28) [63]. Oil-soluble foreign substances, such as acids or fuel particles, are coated by the dispersant and their effects are hindered or diminished. This coating looks almost the same as for peptization, but the nucleus is a liquid and not a solid.

The neutralization process is a chemical neutralization of acid constituents in the oil (Fig. 29) [63].

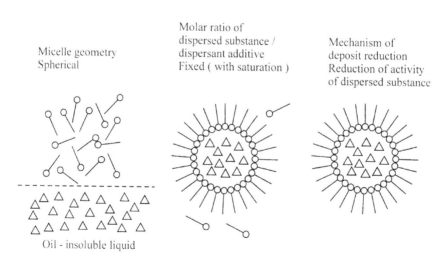

Figure 28 Solubilization process. (From Ref. 63.)

Molar ratio of
dispersed substance /
dispersant additive
One to one

Mechanism of
deposit reduction
Formation of
stable salts

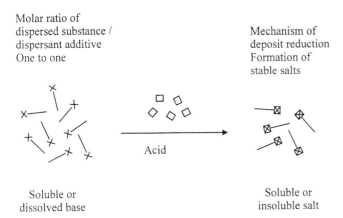

Acid

Soluble or
dissolved base

Soluble or
insoluble salt

Figure 29   Neutralization process. (From Ref. 63.)

3.   The Oxidation Process and the Mechanisms of Antioxidation

In this part, the reaction mechanism for the degradation processes of lubricants resulting from oxidation and the factors that influence them is presented. The following parameters play a role in oil oxidation:

- temperature;
- catalysts, e.g., metal parts, metal abrasion;
- combustion products, aging products, acids;
- O_2 content.

While oil oxidation relatively slowly proceeds at temperatures below 100°C, it accelerates at higher temperatures. It is well known that hydrocarbons oxidize via a free radical mechanism. The oxidation process is very complex, but it can be divided into four steps: initiation, propagation, branching, and termination [7,81–86].

- Initiation

$$RH \rightarrow R^* + H$$

- Chain propagation

$$R^* + O_2 \rightarrow ROO^*$$
$$ROO^* + RH \rightarrow ROOH + R^*$$

- Chain branching

$$ROOH \rightarrow RO^* + HO^*$$
$$RO^* + RH \rightarrow ROH + R^*$$
$$HO^* + RH \rightarrow H_2O + R^*$$

- Chain termination

$$2R^* \rightarrow R - R$$
$$R^* + ROO^* \rightarrow ROOR$$
$$2ROO^* \rightarrow ROOR + O_2$$

It is clear that this reaction, once initiated, is a chain reaction; the reactions are discussed in detail in Ref. 82. In successive reactions, the peroxides give rise to organic acids, esters, ketones, etc., which lead to souring of the oil. Polymerization results in oil-insoluble products and precipitates. The effectiveness of the oxidation inhibitors relies on termination of the oxidation chain, i.e., the exclusion of the peroxides and radicals. Various mechanisms could be responsible for this. Because of their chemical reduction potential, they counteract the formation of organic peroxides, and, as radical traps, they lead to the termination of particularly undesirable chain reactions by trapping those radicals, which, as intermediate products, are responsible for propagating the aging reaction. Finally, as deactivators and passivators, they prevent reactions that take place under the influence of homogeneous or heterogeneous catalysis by coating catalytic metallic surfaces or reacting with metallic salts dissolved in the oil.

B. Additive Reaction Mechanisms on the Interaction Surfaces

1. Adsorption Additives Reaction Mechanism—Multimolecular Layer

Mechanical systems such a gears, transmission systems, engines, wet brakes, etc. are normally stationary before the start of operation. In this situation, the bulk and contact temperature of the interacting surfaces of the tribosystem are equal to the ambient temperature. When the tribosystem starts to run-in, the contact temperature rises, but in any state of run-in conditions, this temperature has to be very low. Under certain conditions, the contact between the interacting surfaces of the bodies of a tribosystem can be prevented by a multimolecular layer, produced by the reaction of the adsorption additives. In this case, friction-modifier (FM) adsorption additives are used. These additives are dissolved in the base oil and adsorbed onto the metal surfaces. The resulting adsorption forces that act on the metal surface are very strong.

The polar head of a friction modifier is attracted to the iron oxide surface in a physical adsorption mechanism and the long hydrocarbon tail is left solubilized in the carrier oil. The anchoring of the molecule to the iron oxide surface results in the hydrocarbon tail being perpendicular to the surface. At normal concentrations, the FM's hydrocarbon tails will line up with each other, and as a result of the hydrogen bonding and the Daby orientation forces, their polar groups can be attracted with a force in dimer clusters. van der Waals forces will cause the molecules to align themselves such that they form multimolecular clusters that are parallel to each other. This orienting of the adsorbed layer can also induce further clusters to position themselves with their respective terminal methyl groups stacking onto the methyl groups of the adsorbed molecules being formed. The interactive forces between methylene carbon atoms and between methyl radicals are weak but positive. Under normal stress conditions, such forces lead to simple breaking and shear. Such a layer of molecules is hard to compress but very easy to shear; thus it is easy to appreciate the slippery nature of the metal surface due to the friction modifier [57].

A detailed structural representation of the physisorption of a friction modifier on an iron/steel–iron oxide substrate as a multimolecular layer is shown in Fig. 30 [57]. On a molecular scale, it is easy to imagine that for two contact surfaces moving against each other, both surfaces would have the FM adsorbed onto them. In a stationary situation, friction-modifier molecules are attracted to the contact surface [Fig. 31(a)] [57]. When a shear force is applied to the adsorbed surface layers, then they shear easily and appear in the oil [Fig. 31(b)].

The effectiveness of this mechanism of additive reaction is limited to low temperatures and low bearing loads. It was found that contact pressures have to be lower than 1 MPa and that even relatively low operating temperatures of about 50°C can result in a severe reduction in the film thickness.

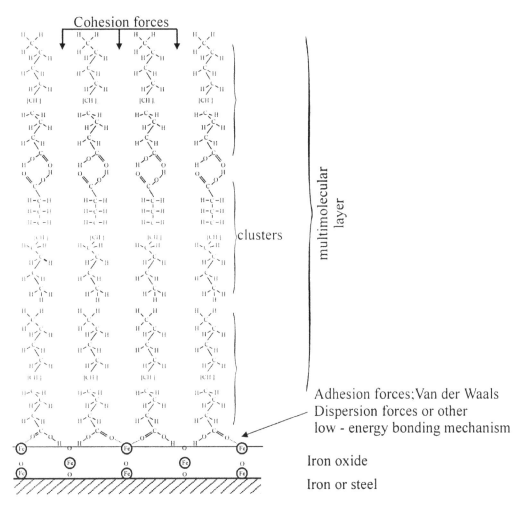

Figure 30 Detailed structural representation of the physisorption of a friction modifier on an iron/iron oxide substrate as multimolecular layer. (From Ref. 57.)

2. Adsorption Additives Reaction Mechanism—Monomolecular Layer

The additive reaction mechanisms at low contact temperatures (100–150°C) and bearing contact pressures up to 1 GPa is of considerable practical importance. Organic polar molecules, such as fatty acids, alcohols, and other friction modifiers, chemically adsorb onto metallic oxide surfaces and are not easily removed. A monomolecular layer separates the contacting surfaces. This layer is so thin that the mechanics of asperity contact are identical to that of a dry surface in contact. The mechanisms of the chemical adsorption of stearic acid onto oxide contact surfaces moving against each other are shown in Fig. 32 [65,87,88].

For example, one end of a molecule of stearic acid, which is the carboxyl group of the fatty acid -COOH, is chemically attracted to the oxide surface with adhesion forces, while the other end, which is an alkyl group -CH$_3$, is repellant to almost any other substance. Strong adsorption ensures that almost every available surface site is occupied by the fatty acid so as to produce a dense and robust film. The repulsion or weak bonding between the

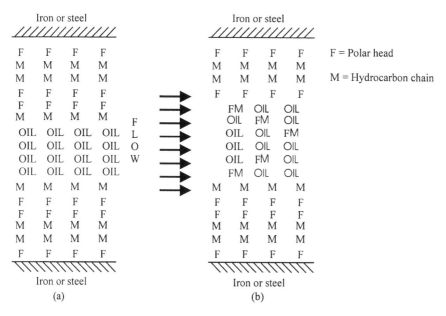

Figure 31 Schematic illustration of the adsorption of friction modifiers (FM) to metal surfaces. (a) Stationary situation; (b) application of force leading to shear and appearance of FM in oil. (From Ref. 57.)

contacting alkyl groups ensures that the shear strength of the interface is relatively low. This is the adsorption model of the additive reaction first postulated by Hardy and Doubleday [90,91], and later developed by Bowden and Tabor [92].

   a. Influence of Different Parameters on Adsorption Additive Reaction Mechanisms. Physisorption of the molecules of the adsorbant may attach or detach from a surface without any irreversible changes to the interacting surface or the adsorbate. Physisorption is effective in reducing friction, provided that the temperature does not rise much above ambient temperature. This effect is illustrated in Fig. 33 [92].

   Chemisorption is an irreversible or partially irreversible form of adsorption. The strength of the chemical bonding between the adsorbate and substrate, which affects the friction transition temperature, depends on the reactivity of the substrate material as shown in Table 8 [93]. It can be seen from Table 8 that zinc, cadmium, copper, and magnesium are comparatively well lubricated by lauric acid. On the other hand, the inert metals such as platinum and silver show high friction coefficients and low transition temperatures. Glass, which is virtually inert, is also poorly lubricated by lauric acid. Other materials such as nickel, aluminum, and chromium also conform to the pattern of a high friction coefficient and a poor retention of lauric acid [65].

   The molecular structure or shape of the additives has a very strong influence on the effectiveness of the boundary film. In addition to the basic requirement that the adsorbing molecules be polar, preferably with an acidic end group for attraction to an oxide surface, the shape of the molecule must also facilitate the formation of close-packed monolayers. This latter requirement virtually ensures that only linear molecules are suitable for this purpose. The effects of linear and branched molecules are shown in Fig. 34 [65].

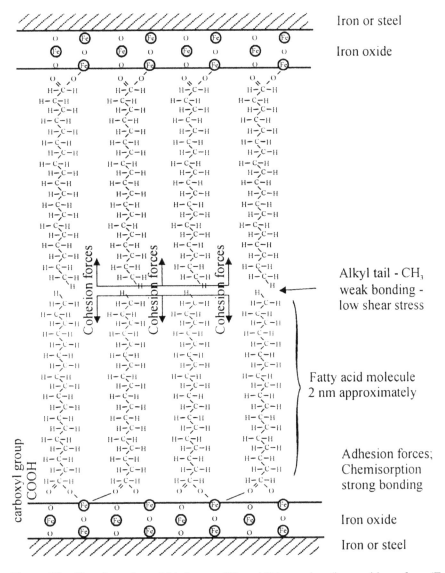

Figure 32 Chemisorption of friction-modifier additive on iron/iron oxide surface. (From Refs. 65 and 87.)

It was found, for example, that the friction transition temperature for fatty acids increased when their molecular weight was increased [92]. The minimum chain length for effective lubrication is $n = 9$. An increase in n from 9 to 18 raises the friction transition temperature by about 40°C. The explanation for the effect of chain length on the boundary film may lie in the relatively weak bonding between CH_2 groups [65].

Atmospheric oxygen and water are always present in lubricated systems unless they are actively excluded. These two substances are found to have a strong influence on adsorption additive mechanisms, as chemically active metals such as iron react with oxygen and water. A surface film of oxide is formed on the metallic surface immediately after contact is made with oxygen. This oxide film is later hydrated by water. Unless the

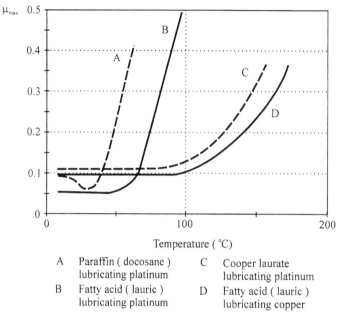

| A | Paraffin (docosane) lubricating platinum | C | Cooper laurate lubricating platinum |
| B | Fatty acid (lauric) lubricating platinum | D | Fatty acid (lauric) lubricating copper |

Figure 33 Effect of temperature on the friction of platinum and copper surfaces lubricated by docosane and lauric acid. (From Ref. 92.)

conditions of wear are severe, the oxide film usually survives sliding damage and forms a substrate for adsorbates. The removal of these oxide films by severe wear can result in the failure of adsorption additives.

The additive reaction on the interacting surfaces is ineffective in preventing severe wear in a steel-on-steel sliding contact without any ambient oxygen and water [94]. It has also been found that oxygen alone is more effective than water without oxygen. However, the

Table 8 Frictional Data for Lauric Acid Lubricating Metals of Varying Reactivity

Material	Coefficient of friction at 20°C	Transition temperature (°C)	Percentage of acid[a] reacting	Type of sliding at 20°C
Zinc	0.04	94	10.0	Smooth
Cadmium	0.05	103	9.3	Smooth
Copper	0.08	97	4.6	Smooth
Magnesium	0.08	80	Trace	Smooth
Platinum	0.25	20	0.0	Intermittent
Nickel	0.28	20	0.0	Intermittent
Aluminum	0.30	20	0.0	Intermittent
Chromium	0.34	20	Trace	Intermittent
Glass	0.3–0.4	20	0.0	Intermittent (irregular)
Silver	0.55	20	0.0	Intermittent (marked)

Source: Ref. 74.
[a] Estimated amount of acid involved in the reaction assuming formation of a normal salt.

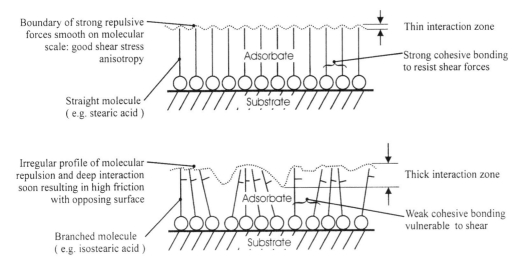

Boundary of strong repulsive forces smooth on molecular scale: good shear stress anisotropy

Thin interaction zone

Adsorbate

Strong cohesive bonding to resist shear forces

Straight molecule (e.g. stearic acid)

Substrate

Irregular profile of molecular repulsion and deep interaction soon resulting in high friction with opposing surface

Thick interaction zone

Adsorbate

Weak cohesive bonding vulnerable to shear

Branched molecule (e.g. isostearic acid)

Substrate

Figure 34 Disruption of adsorbate film structure by branched molecule. (From Ref. 65.)

combination of oxygen and water provides the lowest friction and wear. Studies of a range of lubricant additives revealed that ambient oxygen and water enhance the functioning of most lubricants except certain phosphorus additives for which water has a harmful effect [95–97].

In equipment such as high-speed gears, the repetition rate of frictional contacts may reach several hundred cycles per second. However, as more recently observed, it is extremely unlikely that adsorbate films can reach chemical and thermal equilibrium under such dynamic sliding conditions. The rate-limiting process in the formation of an adsorbate film under dynamic sliding conditions is believed to be a readsorption mechanism. A minimum concentration of friction-modifier additives is required for this process to occur within the time available between successive sliding contacts [65]. The model of adsorption kinetics under dynamic sliding conditions and the required minimum concentrations of some fatty acids are shown in Figs. 35 and 36 [65,98].

The effect of alloying and heat treatment to produce a specific microstructure also exerts a major influence on whether a low coefficient of friction can be obtained by additive reaction on the contact surfaces. The frictional characteristic for steel-on-steel contacts vs. temperature for two steels, a martensitic plain carbon steel and an austenitic stainless steel, lubricated by mineral oil, are shown in Fig. 37 [99]. Both steels are lubricated by mineral oil.

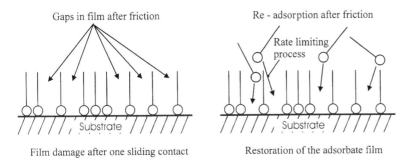

Gaps in film after friction

Re - adsorption after friction

Rate limiting process

Substrate

Substrate

Film damage after one sliding contact

Restoration of the adsorbate film

Figure 35 Model of adsorption kinetics under dynamic sliding conditions. (From Ref. 65.)

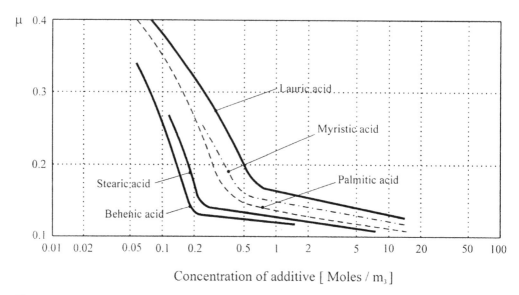

Figure 36 Effect of solute fatty acid concentration on friction coefficients. (From Ref. 98.)

It can be seen from Fig. 37 that the coefficient of friction of the austenitic stainless steel rises sharply at 160°C, reaching values greater than unity by 200°C. The differences between these two steels can be explained in terms of reactivity, as the austenitic steel is considered to be less reactive than the martensitic steel. The greater reactivity causes more rapid formation or repair of oxide films and readsorption of the additive reaction films under conditions of repeated sliding contact. In a more recent study [100], it was found that austenitic steels have lower friction transition temperatures than martensitic steels. Similar tests show that when conducted with additive-enriched oil, low-alloy steels exhibit a lower coefficient of friction than high-alloy steels [101].

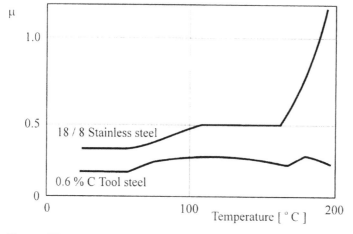

Figure 37 Frictional characteristics of plain carbon and stainless steel vs. temperature under mineral oil lubrication. (From Ref. 99.)

In practice, the base fluid can also influence the lubrication mechanism [65]. It was found that the heats of adsorption of stearic and palmitic acid on iron powder were up to 50% greater than heptane when hexadecane was the carrier fluid [102]. The heat of adsorption dictates the friction transition temperature. This aspect of adsorption additives has also been relatively neglected, partly because of the difficulty in manipulating mineral oil as a carrier fluid. With the adoption of synthetic oils, which offer a much wider freedom of chemical specification, systematic optimization of the heat of adsorption may eventually become practicable [65].

C. Antiwear Additive Reaction Mechanism

When solid/solid interactions occur between asperities, the asperities tend to adhere and the friction thus produced is called adhesive friction. The stresses on the asperities are sufficient to remove material, and this material loss is called adhesive wear. If the asperities of metal surfaces were purely metallic, the adhesion between them would be very strong, and would cause high friction and severe wear. In practice, the bearing surfaces of a tribosystem are normally covered with a film of metal oxide. Adhesion between the asperities that are oxide-coated tends to be much milder and leads to lower friction and milder wear. Mild wear is generally characterized by the removal of metal oxide and, if the process is taking place in the presence of oxygen, the surfaces will reoxidize. Oxygen is one of the most effective antiwear substances; only the antiwear additives that reduce mild wear and reduce the coefficient of friction are more effective. Two basic mechanisms involved in the case of the antiwear additive interaction on bearing surfaces are chain matching and the formation of thick films of soap layer [65,93].

1. Chain Matching

Chain matching refers to the improvement of lubricant properties that occurs when the chain lengths of the solute fatty acid and the solvent hydrocarbon are equal. An example of scuffing-load data vs. the chain length of various fatty acids is shown in Fig. 38 [103,109].

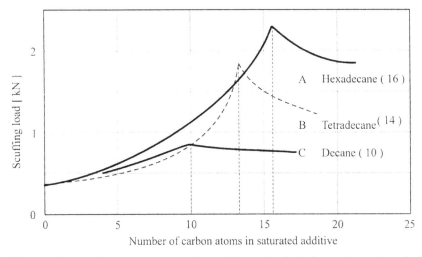

Figure 38 Scuffing loads as a function of fatty acid chain length for various hydrocarbon carrier oils. (From Ref. 103.)

Three carrier solvents were used in a number of "four-ball" tests: hexadecane, tetradecane, and decane with chain lengths of 16, 14, and 10 hydrocarbons, respectively. The maximum scuffing load occurred for a fatty acid chain length of 10 for decane, 14 for tetradecane, and 16 for hexadecane. This effect can be explained by the fact that a coherent viscous layer formed on the contact surface when the chain matching occurred. It was suggested that when chain matching occurs, a thin layer with an ordered structure forms on the contacting surfaces. If the chain lengths do not match, then a coherent surface structure cannot form and the properties of the surface structures remain similar to those of the disordered bulk fluid. This process is similar to the low temperature in the stationary state of a tribosystem discussed previously in Sec.V.B.1.

When the chain length of the fatty acid matched with the hexadecane, the highest friction transition temperature of about 240°C was recorded. For the other acids, the friction transition temperature was much lower, between 120°C and 160°C [65].

2. Soap Layer

Soap layers are formed by a reaction between a metal hydroxide and a fatty acid, which results in soap plus water. If the reaction conditions are favorable, there is also the possibility of soap formation between the iron oxide of a steel surface and the stearic acid, which is in the base oil as a additive. The iron oxide is less reactive than alkali hydroxides. Soap formation, promoted by the heat and mechanical agitation of the sliding contact, was proposed to model the frictional characteristics of stearic acid [105–107]. Figure 39 [87] shows the chemisorption of stearic acid to a steel contacting surface. It should be noted that stearic acid was mentioned earlier as an adsorbed additive. The probable mechanism involves a stearic acid molecule initially adsorbing onto the contact surface, but if the temperature sufficiently rises, electron transfer can take place producing an iron stearate salt. A monolayer of soap would form by chemisorption between the fatty acid and the underlying metal oxide, e.g., copper oxide and lauric acid form copper laurate. The soap formed by this reaction is believed to lubricate by providing a surface layer that is much more viscous than the carrier oil.

D. Extreme-Pressure Additive Reaction Mechanisms

1. Formation of the Extreme Pressure Film on the Freshly Worn Surfaces

If the sliding motion is too fast, or the contact stresses too high, there may not be enough time for adequate regeneration of the oxide film before the next asperity interaction takes place. The interactions will then be between unoxidized or, at best, underoxidized asperities. Adhesion will be strong, with high friction, and this will eventually lead to severe wear. If the severe wear continues, a point at which there is gross welding or seizure of the surfaces will usually be reached. If the surface friction is too great the contact temperature increases to 300°C or more, and the adsorbed molecules will desorb so the protective film is no longer present to prevent adhesion of the contact surfaces [93]. On the other hand, excessive mechanical stresses, up to 2 GPa, will either tear away the surface films, or wear away the metal together with its adsorbed films. In either case, the result is a surface containing areas of unoxidized or underoxidized metal, and more powerful films are required to protect it. At such high temperatures, as are likely to occur in asperity interactions at high load and velocity, the remaining organic structures will decompose, and only the inorganic compounds will be stable. These films result from the reaction of various additives with the metal surfaces, and most of these chemicals contain sulfur and phosphorus. These are the so-called extreme pressure (EP) additives.

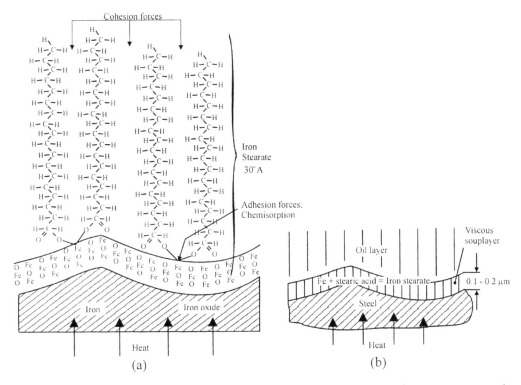

Figure 39 (a) Chemisorption of stearic acid, (b) formation of a viscous soap layer on contact steel surfaces. (From Refs. 87 and 105.)

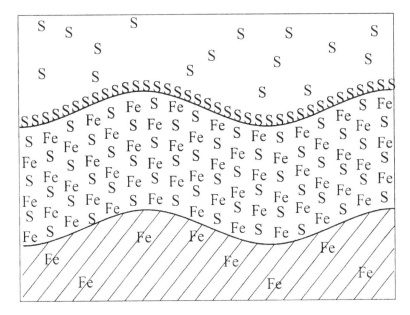

Figure 40 Reaction of sulfur with a steel surface. (From Ref. 93.)

The simplest EP additive is elemental sulfur, which can be dissolved in a mineral base oil to produce a so-called sulfurized oil. The free sulfur will react with a metal surface, especially when the metal has been freshly exposed by wear. Figure 40 shows an idealized picture of the surface film produced by free sulfur reacting with a steel surface [93]. The surface EP film, produced by sulfur or an other EP additive, will be worn away by rubbing against the another bearing surface, and more sulfur will then react to regenerate the surface EP film. It was found that the performance of a lubricant is proportional to its corrosiveness or film formation as shown in Fig. 41 [109]. As is evident from Fig. 41, sulfur is the most effective element providing the greatest lubricating effect for the least corrosion, this is followed by phosphorus and chlorine. The main aim in selecting an EP additive is to maximize the protection against severe adhesion, while at the same time minimizing the rate of corrosion [65].

A freshly worn metal contact surface has a higher surface energy than an unworn equilibrium contact surface. This surface is far more reactive than an oxidized contact surface because there is no oxide barrier between the metal and the reactants; the surface atoms release electrons known as "Kramar electrons" to initiate reactions [110]; the freshly worn surface formed by sliding contains numerous defects that provide catalytic sites for reactions [111].

These characteristics of a freshly worn surface are schematically illustrated in Fig. 42, and are discussed in detail in Ref. 112.

The release of electrons is critical to the initiation of a reaction between the EP additive and the freshly worn metal contact surfaces. Low-energy electrons emitted by the surface ionize molecules of the additive and then these ionic radicals (transformed additive molecules) adsorb onto positive points on the freshly worn contact surface [113]. The electron emission is associated with initial oxidation of the freshly worn surface by atmospheric oxygen [110,114] or by initial sulfurization. The positive points have only a

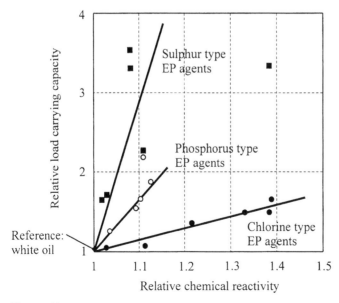

Figure 41 The relationship between corrosivity and lubricating effect on some EP additives. (From Ref. 109.)

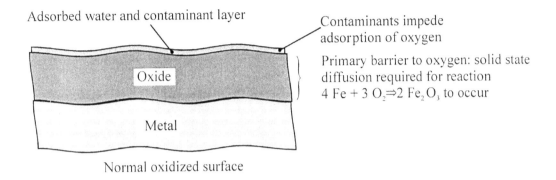

Adsorbed water and contaminant layer

Contaminants impede
adsorption of oxygen

Primary barrier to oxygen: solid state
diffusion required for reaction
$4 Fe + 3 O_2 \Rightarrow 2 Fe_2O_3$ to occur

Oxide

Metal

Normal oxidized surface

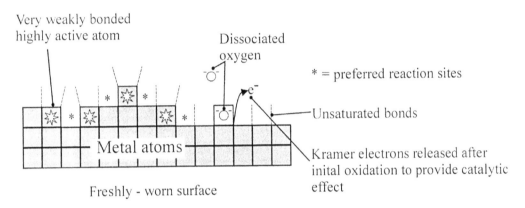

Very weakly bonded
highly active atom

Dissociated
oxygen

* = preferred reaction sites

Unsaturated bonds

Metal atoms

Kramer electrons released after
inital oxidation to provide catalytic
effect

Freshly - worn surface

Figure 42   Characteristics of freshly worn surfaces. (From Ref. 112.)

very short lifetime. On the fresh contact surface of a typically conductive metal, this is only about 10^{-13} sec, but this is sufficient to initiate chemical reactions. Therefore contact surface adsorption reactions are activated by the low-energy electrons and may very rapidly progress. The model of an ionic reaction mechanism between EP additives and a freshly worn metallic contact surface is shown in Fig. 43 [113]. Reaction rates between sulfur and a freshly worn steel surface are about 1000 times more rapid for a freshly worn surface than for an oxide surface. This velocity of the reaction ensures that a thin film of sulfidized material, about 5-nm thick, would form in a few milliseconds, rather than over several seconds. An EP film based on this thin film could be sustained even at high contact rates when the angular

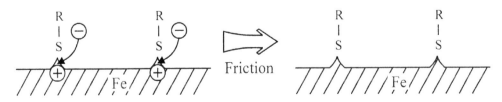

Friction

Figure 43   Ionic model of reaction between an additive and a freshly worn contact surface. (From Ref. 113.)

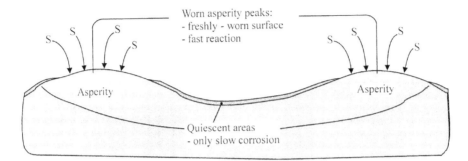

Figure 44 Preferential formation of EP film at localized areas of a freshly worn surface. (From Ref. 65.)

speed is several thousand revolutions per minute [115]. The freshly worn contact surfaces play a significant role in lubrication by the EP film through:

— raising the film formation rates to a level sufficient to sustain lubrication by the EP additive reaction mechanisms in the practical high velocity of a tribocontact;

— confining the corrosive attack by the EP additive, particularly at low temperatures, to asperity peak locations where a freshly worn surface is most often found. These characteristics are illustrated schematically in Fig. 44 [65].

2. Influence of Oxygen and Water on the Lubrication Mechanism by an Extreme-Pressure Film

Oxygen and water exert a strong influence on an EP additive reaction mechanism as well as on an adsorption mechanism. Oxygen competes with the EP additives for the freshly worn surface. The chemistry of oxidation is fundamentally similar to sulfidization and phosphidization. High concentrations of sulfur or phosphorus on freshly worn surfaces are found to coincide with scuffed regions, while smooth surfaces are covered with oxygen-rich layers [115]. An EP film rich in sulfur was only found if oxygen was deliberately excluded from the surrounding environment by imposing an atmosphere of pure nitrogen [116]. The requirement of thermodynamic equilibrium ensures that sulfides are eventually oxidized to sulfates and later to oxides in the presence of oxygen [65].

At the asperity peaks where a freshly worn surface is repeatedly formed, sulfidization occurs because this is by far the more rapid process. In between the asperity peaks, a slower form of corrosion attack occurs, which causes those compounds closest to thermodynamic equilibrium to form oxides or oxidation products of sulfides. This duplex structure and the formation of the EP film is schematically illustrated in Fig. 45 [65]. As shown in Fig. 45, a mixed oxide–sulfide film was found on the freshly worn asperities, which have a much higher load capacity than a pure sulfide film under certain conditions of load and sliding [117,118].

3. Milder Extreme-Pressure Additives Reaction Mechanisms

Most of the existing theories suggest that EP additives also function by a modified form of adsorption lubrication. This adsorption provides a useful lubricating effect or wear-reducing effect at moderate loads and is called the antiwear effect. An increase in loads, sliding velocity, or contact temperatures causes the adsorbed additive to decompose on the

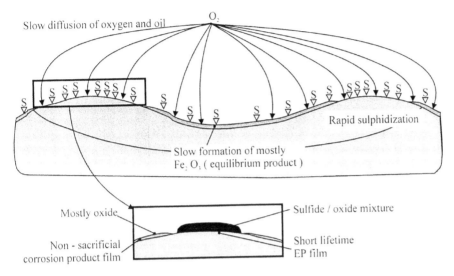

Figure 45 Formation and structure of the EP film in the presence of atmospheric oxygen. (From Ref. 65.)

worn surface leaving a sulfur atom, or any of the other active additives to react with the iron of the worn metal [119–121]. The mechanism of sulfur reaction is shown in Fig. 46 [119]. When the additive finally decomposes to produce a sulfide film, organic residue molecules such as alkanes and olefins are released.

4. Other Extreme-Pressure Additive Reaction Mechanism

Phosphorus is generally believed to provide a lubricating effect similar to sulfur. When a phosphorous compound containing a phosphate radical reacts with a metallic surface, a metallic phosphate film is formed on the worn surface [96,122]. Tricresylphosphate (TCP) and zinc dialkyldithiophosphate (ZnDDP) are usually used for this purpose. The final product from TCP will be iron phosphate or iron phosphide, depending on the availability of oxygen. Under similar conditions, ZDDP may generate iron phosphate, iron phosphide, iron sufide, iron sulfate, or zinc compounds.

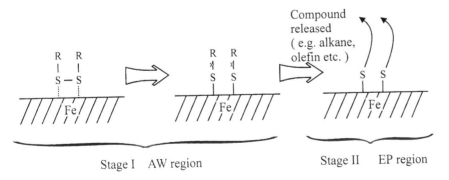

Figure 46 Reaction mechanism of milder EP additives. (From Ref. 119.)

It was mentioned earlier that if as a result of wear the oxide is completely removed, leaving a free metal surface, simple hydrocarbons or fatty acids will be severely broken down, leaving fragments that provide no protection against wear. If EP additives are present, similar reactions will release phosphorus and sulfur, which can rapidly react with a bare metal surface to form protective films. All of these compounds have high thermal stability, high enough to withstand any flash temperatures [108] generated in the sliding contacts. Mechanically, they will strongly adhere to the metal substrates but will themselves more readily shear, therefore providing lower sliding friction. At same time, they have only a very weak tendency to adhere to the opposite sliding surface.

5. Additive Reaction Mechanism with Two Active Additives

In practice, it has been found that the combination of two or more additives, e.g., sulfur and phosphorus, give a much stronger reactive film on the interacting surface than a single additive. The sulfur–phosphorus system is most widely used as a combination of two additives [123]. The effect of combining additives is demonstrated in Fig. 47, which shows data from Timken tests. The mineral base oil used in these tests was enriched with the additives dibenzyl disulfide and dilauryl hydrogen phosphate. By using these additives separately and together, the effect of phosphorus, sulfur, and a combination of both on the seizure load was determined [123]. It can be seen that although the phosphorous additive by itself is ineffective in comparison with the sulfur additive, the combination of phosphorus and sulfur is significantly better than either additive acting in isolation.

The chemistry of steel surfaces after lubrication with sulfur–phosphorus additives in combination with oil was also studied [123–126]. Films found on wear scars formed under several conditions, e.g., the Timken test, consisted mostly of sulfur. Under milder load and lower slide/roll ratios, which are characteristic for general machinery, it was found that

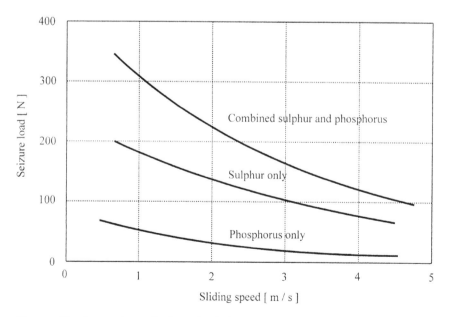

Figure 47 Comparison of seizure loads for sulfur, phosphorus, and combination-enriched lubricants. (From Ref. 123.)

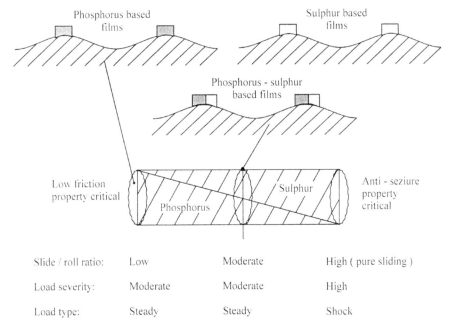

Figure 48 Dependence of sulfur–phosphorus wear scar film chemistry on severity of the sliding condition. (From Ref. 65.)

phosphorus predominates in the wear-scar films. The model of film chemistry vs. sliding severity is illustrated schematically in Fig. 48. It can be seen from Fig. 48 that a sulfur–phosphorus-based lubricant provides considerable versatility in lubricating performance. The sulfur is essential to prevent seizure under abnormally high loads and velocities, while phosphorus maintains low friction and wear rates under normal operating conditions.

REFERENCES

1. Meyer, K. Synergistische und antagonistische Wirkungen zwischen Schmierstoffkomponenten aus grenzflächenchemischer. Sicht. Tribol. Schmier.Tech. 1990, *37*, 1, 31.
2. Bartz, W.J.; Holinski, R. Synergistische Effekte bei der Anwendung von Festschmierstoffen in gebundener. Form. Tribol. Schmier.Tech. 1986, *140* (33), 4, 223.
3. Harmer, R.S.; Pantano, C.G. Synergetic Mechanisms of Solid Lubricants. Techn. Report AFMG TR-77-227, Wright-Patterson Air Force Base OH 45433-6563, USA, 1978
4. Meyer, K. Synergistische und antagonistische Wirkungen zwischen Schmierstoffkomponenten und ihr Einfluβ auf ausgewählte tribologische. Eigensch. Schmier.tech. 1989, *20* (68), 103.
5. Reynolds, O. On the theory of lubrication and its application to Mr Beauchamp Tower's experiments, including an experimental determination of the viscosity of olive oil. Philos. Trans. R. Soc. 1886, *117*, 157–234.
6. Dowson, D.; Higginson, G.R. A numerical solution to the elastohydrodynamic problem. J. Mech. Eng. Sci. 1959, *1*, 6–15.
7. Hsu, S.M.; Gates, R.S. boundary lubrication and boundary lubricating film. In *Modern Tribology Handbook* Bhushan, B Ed.; CRC Press: Boca Raton, 2001; 455–492.
8. Booser, R.E., Ed.; *CRC Handbook of Lubrication—Theory and Practice of Tribology*; CRC Press Inc.: Boca Raton, 1983.

9. Podgornik, B.; Vižintin, J.; Leskovšek, V. Tribological properties of plasma and pulse plasma nitrided AISI 4140 steel. Surf. Coat. Technol. 1998, *108–109*, 454–460.
10. Podgornik, B. Influence of substrate pre-treatment on the tribological properties of hard coatings during sliding. University of Ljubljana, Faculty of mechanical engineering, Ljubljana, 2000
11. Podgornik, B.; Vižintin, J. Wear Properties of Plasma Nitrided Steel in Dry Sliding Conditions. Trans. ASME 1999, *121*, 802–807.
12. Podgornik, B.; Vižintin, J.; Ronkainen, H.; Holmberg, K. Friction and wear properties of DLC-coated plasma nitrided steel in unidirectional and reciprocating sliding. Thin Solid Films 2000, *377–378*, 254–260.
13. Hutchings, I.M. *Tribology: Friction and Wear of Engineering Materials*; Edvard Arnold, 1992.
14. Bhushan, B. Surface roughness analysis and measurement techniques. In *Modern Tribology Handbook* Bhushan, B Ed.; CRC Press: Boca Raton: Florida, 2001; 49–114.
15. Greenwood, J.A.; Williamson, J.B.P. Contact of nominally flat rough surfaces. Proc. R. Soc. Lond., A 1966, *295*, 300–319.
16. Johnson, K.L. *Contact Mechanics*; Cambridge University Press: Cambridge, 1985.
17. McCool, J.I. Comparison of models for the contact of rough surfaces. Wear 1986, *107*, 37–60.
18. Whitehouse, D.J. *Handbook of Surface Metrology*; Institute of Physics Publishing: Bristol, 1994.
19. Bhushan, B. *Tribology and Mechanics of Magnetic Storage Devices,* 2nd Ed.; Springer: New York, 1996.
20. Thomas, T.R. *Rough Surfaces,* 2nd Ed; Imperial College Press: London, 1999.
21. Gatos, H.C. Structure of surfaces and their interactions. Interdisciplinary Approach to Friction and Wear; Ku, P.M Ed.; SP-181, NASA: Washington, DC, 1986; 7–84.
22. Buckley, D.H. *Surface Effects in Adhesion, Friction, Wear and Lubrication*; Elsevier: Amsterdam, 1981.
23. Buckley, D.H. Properties of surfaces. In *Handbook of Lubrication*; CRC Press: Boca Raton, Florida, 1989; Vol. II, 17–30.
24. Bhushan, B. Surface roughness analysis and measurement techniques. In *Modern Tribology Handbook*. Bhushan, B Ed.; CRC Press: Boca Raton, 2001; 49–114.
25. Bhushan, B.; Gapta, B.K. *Handbook of Tribology*; McGraw-Hill Inc.: New York, 1991.
26. Rabinowicz, E. *Friction and Wear of Materials*; John Wiley & Sons: New York, 1965.
27. Umehara, N.; Yamaguchi, Y.; Kato, K. New evaluation method for surface energy with μm size liquid droplets. Tokyo Metropolitane Institute, Japan, 2001.
28. Bhushan, B. Self-assembled manolayers for controlling hydrophobicity and for friction and wear. In *Modern Tribology Handbook*; Bhushan, B., Ed.; CRC Press LLC: Boca Raton, 2001; Vol. 2, 909–924.
29. Bhushan, B. *Handbook of Micro/Nanotribology,* 2nd Ed.; CRC: Boca Raton, 1999.
30. Haltner, A.J. The physics and chemistry of surfaces: surface energy, wetting and adsorption. In *Boundary Lubrication*. Ling, F.F., Ed.; ASME: New York, 1969; 39–60.
31. Bowden, F.P.; Moore, A.C. *Physical and Chemical Adsorption of Long Chain Polar Compounds* Research: London, 1949; 585–586.
32. Rehbinder, P.A.; Likhtman, V.I. Effects of surface active media on strains and rupture in solids. Proc. 2nd Int. Congress on Surface Activity, London. 1957; 563–580.
33. Brainard, W.A.; Buckley, D.H. preliminary Studies by Field Ion Microscopy of Adhesion of Platinum and Gold to Tungsten and Iridium. Washington D.C.: Report No. TND-6492, NASA, 1971.
34. Egert, B.; Panzner, G. Electron Spectroscopic Study of Phosphorus Segregated to α-Iron Surfaces. Surf. Sci. 1982, *118*, 345–368.
35. Mouttet, C.; Gaspard, J.P.; Lambin, P. Electronic structure of the Si–Au surface. Surf. Sci. 1981, *111*, L755–L758.
36. Jung, C.; Dannenberger, O.; Xu, Y.; Buck, M.; Grunze, M. Self-assembled monolayers from

organosulfur compounds: a comparison between sulfides, disulfides, and thiols. Langmuir 1998, *14*, 1103–1107.

37. Mori, S.; Suginoya, M.; Tamai, I. Chemisorption of organic compounds on a clean aluminum surface prepared by cutting under high vacuum. ASLE Trans 1982, *25*, 261–266.
38. Prince, R.J. Base oils from petroleum. In *Chemistry and Technology of Lubricants*; Mortiem, R.M. Orszulik, S.T. Eds.; Blackie Academic & Professional: Glasgow, 1993; 1–32.
39. Bartz, W.J. *Lubrication of Gearing*; Expert Verlag, Mechanical Engineering Limited: London, 1993.
40. Bartz, W.J. Reasons for using synthetic lubricants and operational fluids. The Use of Synthetic Lubricants and Operational Fluids in Industry and Vehicles. Course at Technische Akademie Esslingen, Ostfildern, 1981.
41. Schmid, W.A. Base fluids in synthetic lubricants—comparative consideration of their characteristics. The Use of Synthetic Lubricants and Operational Fluids in Industry and Vehicles. Course at Technische Akademie Esslingen, Ostfildern, 1981.
42. Szdywar, J. Synthetic hydrocarbons, e.g., polyaphaolefins. The use of synthetic lubricants and operational fluids in industry and vehicles. Course at Technische Akademie Esslingen, Ostfildern, 1981.
43. Kussi, S. Polyether in lubrication engineering. The Use of Synthetic Lubricants and Operational Fluids in Industry and Vehicles. Course at Technische Akademie Esslingen, Ostfildern, 1981.
44. Prey, G. Ester fluids for lubrication. The Use of Synthetic Lubricants and Operational Fluids in Industry and Vehicles. Course at Technische Akademie Esslingen, Ostfildern, 1981.
45. Szydywar, J. Aviation turbine lubrication. The Use of Synthetic Lubricants and Operational Fluids in Industry and Vehicles. Course at Technische Akademie Esslingen, Ostfildern, 1981.
46. Wunsch, F. The use of polyphenyl ethers and polyfuoroalkyl ethers in lubrication technology. The use of synthetic lubricants and operational fluids industry and vehicles. Course at Technische Akademie Esslingen, Ostfildern, 1981.
47. Bartz, W.J. Gear lubrication—current state of the art and future development (in German). Tribol. Schmier.Tech. 1984, *31*, 126–131, 198–203.
48. Bartz, W.J. *Gear Lubrication: Applications of Motive Power Engineering (in German)* Krausskopf-Verlag: Mainz, 1974; 391–455.
49. Bartz, W.J. Principles of fear lubrication (in German). Maschine 1970, *4*, 70–74; *5*, 35–39.
50. Bartz, W.J. Gear oil as a functional element (in German). Mineralltechnik 1971, *16*, 1–34.
51. Randles, S.J.; Stroud, P.M.; Mortier, R.M.; Orszulik, S.T.; Hoyes, T.J.; Brown, M. Synthetic base fluids. In *Chemistry and Technology of Lubricants*; Mortier, R.M., Orszulik, S.T. Eds.; Blackie Academic & Professional: Glasgow, 1994; 32–62.
52. Bartz, W.J. *Additive für Schmierstoffe. Band 433*; Expert Verlag: Renningen–Malmsheim, 1994.
53. Bartz, W.J. *Additive für Schmierstoffe. Reihe Tribotechnik, Band 2*; C.R. Hannover: Vincentz-Verlag.
54. Kajdas, C. Engine Oil Additives. Tribol. Schmier.Tech. 1990, *37*, 5, 250.
55. Bartz, W.J., Ed.; *Manual of Fuels and Lubricants for Vehicles, Part 2 (in German)* Expert-Verlag: Ehningen bei Böblingen, 1983; 45–127.
56. Mortier, R.M., Orszulik, S.T. Eds.; *Chemistry and Technology of Lubricants*; Blackie Academic & Professional: Glasgow, 1993.
57. Crawford, J.; Psaila, A. Miscellaneous additives. In *Chemistry and Technology of Lubricants*. Mortier, R.M. Orszulik, S.T. Eds.; Blackie Academic & Professional: Glasgow, 1984; 160–173.
58. Eckert, R.J.A. Pour point improvers, foam inhibitors and adhesion improvers. *Additives for lubricants (in German)* Vincentz-Verlag: Hanover, 1984; 219–231.
59. Kristen, U. Oxidation inhibitors, copper deactivators and rust inhibitors. In *Additives for Lubrications (in German)*; Vincentz-Verlag: Hanover, 1984; 19–46.
60. Lindstaedt, B. Ash-forming extreme-pressure and anti-wear additives. In *Additives for Lubricants (in German)*; Vincentz-Verlag: Hanover, 1984; 47–74.

61. Kristen, U. Ash-free extreme-pressure and anti-wear additives. In *Additives for Lubricants (in German)*; (pp. 75–92) Vincentz-Verlag: Hanover, 1984.

62. Bartz, W.J. Reducing friction with lubricants and additives. In *Additives for Lubricants (in German)*; Vincentz-Verlag: Hanover, 1984; 107–120.

63. Raddatz, J. Detergent/dispersant additives—manufacture, function and applications. In *Additives for Lubricants (in German)*; Vincentz-Verlag: Hanover, 1984; 139–160.

64. Lansdown, A.R. Extreme pressure and anti-wear additives. In *Chemistry and Technology of Lubricants*; Mortier, R.M., Orszulik, S.T., Eds.; Blackie Academic & Professional: Glasgow, 1993; 269–282.

65. Stachowiak, G.W.; Batchelor, A.W. *Engineering Tribology*; Tribology Series 24, Elsevier Science Publishers BV: Amsterdam, 1993.

66. Kennedy, F.E. Surface temperatures in sliding systems—a finite element analysis. ASME J. Lubr. Technol. 1981, *103*, 90–96.

67. Kennedy, F.E. Single-pass rub phenomena—analysis and experiment. ASME J. Lubr. Technol. 1982, *104*, 582–588.

68. Kennedy, F.E. Surface temperature measurement, friction, lubrication and wear technology. *Metals Handbook*; 10th Ed.; Blau, P.J. Ed.; ASM International: Metals Park, 1992.

69. Kennedy, F.E. Thermal and thermomechanical effects in dry sliding. Wear 1984, *100*, 453–476.

70. Uetz, H.; Föhl, J. Wear as an energy transformation process. Wear 1978, *49*, 253–264.

71. Kennedy, F.E. Frictional heating and contact temperatures. In *Modern Tribology Handbook*; Bhushan, B., Ed.; CRC Press: Boca Raton, 1999; 235–259.

72. Pezdirnik, J.; Vizintin, J.; Podgornik, B. Temperatures at the interface and inside an oscillatory sliding microcontact—theoretical part. Tribol. Int. 1999, *32*, 481–489.

73. Kalin, M.; Vizintin, J. High temperature phase transformations under fretting conditions. Wear 2001, *248*, 172–181.

74. Kalin, M.; Vizintin, J.; Vleugels, J.; Van Der Biest, O. Influence of mechanical pressure and temperature on the chemical interaction between steel and silicon nitride ceramics. J. Mater. Res. 2000, *15*, 1367–1376.

75. Selby, T.W. The non-Newtonian characteristics of lubricating oils. Trans. ASLE 1958, *1*, 68–81.

76. Tamai, T.; Toshikazu, Y.; Mogi, M. Flow activation quantities of VI-improver-blended mineral lubricating oils. Bull. Jpn. Pet. Inst. 1977, *19*, 131–134.

77. Jordan, E.F. Jr.; Smith, S. Jr.; Zabarsky, R.D.; Austin, R.; Wrigley, A.N. Viscosity index. II. Correlation with rheological theories of data for blends containing *n*-octadfecyl acrylate. J. Appl. Polym. Sci. 1978, *22*, 1529–1545.

78. Dare-Edwards, M.P.; Kempsell, S.P.; Barnes, J.R.; Craven, C.J.; Wayne, F.D. Nuclear magnetic resonance of lubricant-related systems. *6th International Colloquium, Industrial Lubricants-Properties, Application, Disposal*; Bartz, W.J. Ed.; Technische Akademie Esslingen: Ostfieldern, 1988; Vol II, 12.3-1–12.3-15.

79. Mueller, H.G. Mechanism of action of viscosity index improvers. Tribol. Int. 1978, *11*, 189–192.

80. Wincierz, C.; Müller, M.; Hedrich, K.; Neveu, C. Influence of viscosity index improvers on the tribological properties of modern transmission fluids. *in press*.

81. Hucknell, D.J. *Selective Oxidation of Hydrocarbons*; Academic Press: New York, 1974.

82. Rosberger, M. Oxidative degradation and stabilisation of mineral oil based lubricants. *Chemistry and Technology of Lubricants*. Mortier, R.M. Orszulik, S.T., Eds.; Blackie & Professional: Glasgow, 1993; 83–122.

83. Jensen, R.K.; Korcek, S.; Mahoney, L.R.; Zinbo, M. J. Am. Chem. Soc. 1979, *101*, 7574–7584.

84. Jensen, R.K.; Korcek, S.; Mahoney, L.R.; Zinbo, M. J. Am. Chem. Soc. 1981, *103*, 1742–1748.

85. Perez, J.M.; Kelley, F.A.; Klaus, E.E.; Bagrodia, V. SAE paper 872028, 1987

86. Dorinson, A.; Ludema, K.C. *Mechanics and Chemistry in Lubrication*; Tribology Series 9: Elsevier Science Publishers BV: Amsterdam, 1985.

87. Lansdown, A.R. Extreme-pressure and anti-wear additives. *Chemistry and Technology of Lubricants*. Mortier, R.M. Orszulik, S.T., Eds.; Blackie Academic & Professional: Glasgow, 1993; 269–281.

88. Lindsey, A.R. *Lubrication in Practice*; Robertson, W.S., Ed.; Marcel Dekker: New York and Basel, 1984.

89. Timmons, C.O.; Patterson, R.L.; Lockhart, L.B. Adsorption of carbon-14 labelled stearic acid on iron. J. Colloid Interface Sci. *26*, 120–127.

90. Hardy, W.B.; Doubleday, I. Boundary lubrication—The paraffin series. Proc. R. Soc. Ser. A 1922, *101*, 487–492.

91. Hardy, W.B.; Doubleday, I. Boundary lubrication—The temperature coefficient. Proc. R. Soc., Ser. A 1922, *101*, 487–492.

92. Bowden, F.P.; Tabor, D. *The Friction and Lubrication of Solids Part 1*; Clarendon Press: Oxford, 1950.

93. Bowden, F.P.; Gregory, J.N.; Tabor, D. Lubrication of metal surfaces by fatty acids. Nature 1945, *156*, 97–98.

94. Goldman, I.B.; Appleldoorn, J.K.; Tao, F.F. Scuffing as influenced by oxygen and moisture. ASLE Trans. 1970, *13*, 29–38.

95. Daniels, R.O.; West, A.C. The influence of moisture on the friction and surface damage of clean metals. Lubr. Eng. 1955, *11*, 261–266.

96. Godfrey, D. The lubrication mechanism of tricresyl phosphate on steel. ASLE Trans. 1965, *8*, 1–11.

97. Goldblatt, I.L.; Appeldoorn, J.K. The antiwear behaviour of tricresylphosphate (TCP) in different atmospheres and different base stocks. ASLE Trans. 1970, *13*, 203–214.

98. Okabe, H.; Masuko, M.; Sakurai, K. Dynamic behaviour of surface-adsorbed molecules under boundary lubrication. ASLE Trans. 1981, *24*, 467–473.

99. Grew, W.J.S.; Cameron, A. Role of austenite and mineral oil in lubricant failure. Nature 1968, *217*, 481–482.

100. Bailey, M.W.; Cameron, A. The effects of temperature and metal pairs on scuffing. ASLE Trans. 1973, *16*, 121–131.

101. Rounds, F.G. The influenced of steel composition on additive performance. ASLE Trans. 1972, *15*, 54–66.

102. Groszek, A.J. Heats of preferential adsorption of boundary additives at iron oxide/liquid hydrocarbon interfaces. ASLE Trans. 1970, *13*, 278–287.

103. Askwith, T.C.; Cameron, A.; Crouch, R.F. Chain length of additives in telation to lubricants in thin film and boundary lubrication. Proc. R. Soc., Ser. A 1966, *291*, 500–519.

104. Frewing, J.J. The heat of adsorption of long-chain compounds and their effect on boundary lubrication. Proc. R. Soc., Ser. A 1944, *182*, 270–285.

105. Cameron, A.; Mills, T.N. Basic studies on boundary, E.P. and piston-ring lubrication using a special apparatus. ASLE Trans. 1982, *25*, 117–124.

106. Cann, P.; Spikes, H.A.; Cameron, A. Thick film formation by zinc dialkyldithiophosphate. ASLE Trans. 1986, *26*, 48–52.

107. Blok, H. Les temperatures de surface dans des conditions de graissage sous pression extreme. Proc. Second World Petroleum Congress, Paris 1937, *4*, 151–182.

108. Blok, H. The flash temperature concept. Wear 1963, *6*, 483–494.

109. Sakurai, T.; Sato, K. Study of corrosivity and correlation between chemical reactivity and load carrying capacity of oils containing extreme pressure agent. ASLE Trans. 1966, *9*, 77–87.

110. Ferrante, J. Exoelectron emission from a clean, annealed magnesium single crystal during oxygen adsorption. ASLE Trans. 1977, *20*, 328–332.

111. Gulbransen, E.A. The role of minor elements in the oxidation of metals. Corrosion 1956, *12*, 61–67.

112. Meyer, K. *Physikalich-Chemische Kristallographie*; Buchdruckerei: Gutenberg, 1977.

113. Kajdas, C. On a negative-ion concept of EP action of organo-sulfur compounds. ASLE Trans. 1985, *28*, 21–30.

114. Gessel, T.F.; Arakawa, E.T.; Callcott, T.A. Exoelectron emission during oxygen and water chemisorption on fresh magnesium surface. Surf. Sci. 1970, 20, 174–175.
115. Bjerk, R.O. Oxygen, an extreme-pressure agent. ASLE Trans. 1973, 16, 97–106.
116. Masuko, M.; Ito, Y.; Akatsuka, K.; Tagami, K.; Okabe, H. Influence of Sulphur-Vase Extreme Pressure Additives on Wear Under Combined Sliding and Rolling Contact (in Japanese). Proc. Kyushu Conference of JSLE. 1983; 273–276.
117. Lauer, J.L.; Marxer, N.; Jones, W.R. Ellipsometric surface analysis of wear tracks produced by different lubricants. ASLE Trans. 1986, 29, 457–466.
118. Sakai, T.; Murakami, T.; Yamamoto, Y. Optimum Composition of Sulfur and Oxygen of Surface Film Formed in Sliding Contact. Proc. JSLE. Int. Tribology Conf., July 8–10, 1995; Tokyo, Elsevier, pp. 665–660.
119. Forbes, E.S. The load carrying action of organic sulfur compounds, a review. Wear 1970, 15, 87–96.
120. Forbes, E.S.; Reid, A.J.D. Liquid phase adsorption/reaction studies of organo-sulfur compounds and their load carrying mechanism. ASLE Trans. 1973, 16, 50–60.
121. Allum, K.G.; Forbes, E.S. The load carrying mechanism of some organic sulphur compounds— An application of electron microprobe analysis. ASLE Trans. 1968, 11, 162–175.
122. Godfrey, D. The lubrication mechanism of tricresylphosphate on steel. ASLE Trans. 1965, 8, 1–11.
123. Kubo, K.; Shimakawa, Y.; Kibukawa, M. Study on the Load Carrying Mechanism of Sulphur-Phosphorus Type Lubricants. Proc, JSLE. Int. Tribology Conf., 8–10 July, 1985 Elsevier: Tokyo, 1985; 661–666.
124. Masuko, A.; Hirata, M.; Watanabe, H. Electron probe microanalysis of wear scars of timken test blocks on sulfur-phosphorus type industrial gear oils. ASLE Trans. 1977, 20, 304–308.
125. Matveevsky, R.M. Temperature of the tribochemical reaction between extreme-pressure (E.P.) additives and metals. Tribol. Int. 1971, 4, 97–98.
126. Bollani, G. Failure criteria in thin film lubrication with E.P. additives. Wear 1976, 36, 19–23.

10

Surface Chemistry of Extreme-Pressure Lubricant Additives

Wilfred T. Tysoe
University of Wisconsin–Milwaukee, Wisconsin, U.S.A.

Peter V. Kotvis
Benz Oil, Inc., Milwaukee, Wisconsin, U.S.A.

I. INTRODUCTION

Commercial lubricants that are required to operate under severe conditions are composed of several components. The most abundant of these is the base fluid, which may be a mineral oil or, in some cases, water. The lubricants are "formulated" by adding various components. Some of these function, e.g., to stabilize the oil against oxidation or biological decay. Others improve the tribological performance. This chapter addresses the role of additives that thermally decompose at the lubricated surfaces to deposit reactively formed films so that the surface film arises from a chemical reaction between the additive and the surface. The key parameter in understanding the properties of these reactive additives is the temperature of the interface, because an increase in temperature of the rubbing surfaces correspondingly increases the rate at which the tribological film is formed. Additives that thermally decompose to form tribological films are therefore, in some sense, "smart" additives because they form films just in those regions of the interface that most require them [1]. This chapter therefore focuses on understanding the way in which these lubricant additives react at the tribological surface, the nature of the film that is formed, and the role of this film in improving tribological behavior. This improvement may be in the reduction of friction, in allowing an increase in the loads that can be sustained, or in generally providing a protective coating on the surface. Fully understanding these phenomena requires the synthesis of mechanical approaches with an understanding of the chemical and physical properties of the interface, and therefore requires a truly interdisciplinary strategy. The chapter will thus focus on an area where it has been well established that surface chemical reactivity plays a key role in forming tribological films, that of extreme-pressure (EP) lubricants that are used for processes such as machining, wire drawing, fineblanking, etc. A wide range of compounds have been claimed to be effective EP additives but the ones that are currently most commonly used generally contain chlorine, sulfur, or phosphorus [2–12]. These are most often added as organic compounds, which render them soluble in the base lubricating fluid. Because many of the compounds that are currently used for this purpose are either environmental pollutants or health hazards, or both, these will ultimately have to be replaced by more benign alternatives [13]. Because these additives are understood to thermally decompose at the lubricated interface, the nature of this interface is of crucial relevance to understanding the behavior of the additive. This means that the additive + surface combination must be considered as a whole because the lubricant film is formed by a chemical reaction between them. This clearly means that what may be a good lubricant additive for one metal may not be such a good additive for another. In addition, the presence of surface contaminants, e.g., oxide films or carbonaceous layers, can profoundly affect the reactivity at the surface. This problem is to some extent obviated in the EP regime because wear rates are usually sufficiently high that these surface contaminants are worn away.

II. SURFACE AND BULK ANALYTICAL METHODS

Analysis of the nature of the tribological film is central to understanding the chemical processes that give rise to these films. There are several key issues in deciding which technique is the most appropriate for film analysis. The first involves both the depth and spacial resolution of the techniques. Most tribological films range from a few ångströms to a few microns in thickness, although the composition of the film may vary considerably through this range. Spectroscopic techniques that are based on electrons (e.g., Auger and x-ray photoelectron (XPS) spectroscopies) are generally extremely surface sensitive because electrons interact strongly with matter and therefore generally penetrate only a small distance (ångströms) into the sample [14]. Auger spectroscopy, where electrons are used to excite the Auger transitions and are also detected, is extremely surface sensitive. In the case of x-ray photoelectron spectroscopy, which is excited by x-ray photons, but where electrons are detected, the surface sensitivity is somewhat lower and is primarily limited by the escape depth of the emitted electrons. The lateral spacial resolution of each of these spectroscopies is limited by the size of the electron or x-ray beam that is used to excite them. Electron beams can be relatively easily focused whereas focusing x-rays, although possible, is more complicated. In this case, good spacial resolution can be achieved by imaging the electrons emitted from the sample [15]. Both of these surface-sensitive techniques can be used to provide information on the depth variation in the structure of films when combined with ion bombardment (ion sputtering) where material is removed in a controlled fashion using a beam of energetic ions. Thus, a depth profile of a film can be obtained by sequentially removing controlled amounts by ion bombardment and by then analyzing the nature of the film after each of these steps using Auger or x-ray photoelectron spectroscopy. The main drawback of these methods is that they provide primarily elemental information, although XPS can also provide some information regarding the chemical environment of the elements through the "chemical shift" in which the electron binding energy depends on the oxidation state of the element. On this basis, XPS is also often called "electron spectroscopy for chemical analysis" (ESCA) [16].

Vibrational techniques such as infrared or Raman spectroscopy provide much more detailed information regarding the molecular nature of the film but are not inherently surface sensitive. Their spacial resolution is limited by the ease and extent to which the incident radiation can be focused. This is relatively straightforward for the incident laser beam used to excite Raman spectroscopy and somewhat more difficult for infrared radiation, although infrared microscopes are now commercially available. It should be emphasized that the list of experimental techniques for probing the nature of thin tribological films that are discussed below is by no means complete but includes some of the most powerful and widely used methods. In fact, surface science now includes in its armory almost an embarrassing array of analytical and imaging tools that have given us an unprecedented level of understanding of surface processes. For a more detailed discussion of these, the reader is referred to a number of monographs, books, and reviews in this area [17–20].

A. X-ray Photoelectron Spectroscopy

X-ray photoelectron spectroscopy is based on the photoelectric effect where a photon of frequency v, incident on a sample in which an electron is bound with an energy E_B emits an electron with kinetic energy E_K such that:

$$E_K = hv - E_B \tag{1}$$

where h is Planck's constant. Thus, for a fixed incident photon energy, the kinetic energy of the emitted electron is a measure of its binding energy in the material. The experiment consists of illuminating a sample with a monochromatic beam of x-ray photons and measuring the kinetic energy of the emitted electrons. Because electrons travel only short distances in air, these experiments are done in vacuo. A typical x-ray photoelectron spectrometer then consists of an x-ray source and an electron energy analyzer to measure the kinetic energy of the emitted electrons. Because electrons are emitted from the sample under the influence of the incident x-ray beam, it must be electrically grounded to avoid charging because any charge appearing on the sample will affect the kinetic energy of the emitted electron and distort the spectrum. Note that Eq. (1) emphasizes that the photon energy must be larger than the binding energy of the electron for it to be emitted; thus, to probe high-binding-energy core levels, high photon energies are required and x-ray photons are therefore used [14]. An analogous technique that probes the shallow valence levels using ultraviolet radiation is known as ultraviolet photoelectron spectroscopy (UPS).

X-rays are most often generated by having high-energy electron (~10 keV) impinge on an anode of known material. These high-energy electrons ionize core energy levels in the anode causing radiation to be emitted as these are filled by transitions from lower binding energy levels. The energy of the radiation emitted therefore depends on the anode material used. The most common materials are aluminum (emitted radiation at 1487 eV) and magnesium (emitting radiation at 1254 eV). These are selected because they provide relatively monochromatic radiation. Anode currents up to 40 mA are typically used to provide sufficient intensity to detect strong signals so that powers of ~400 W are typically dissipated. This requires that the anodes be water cooled. The anode is usually separated from the sample by a window (generally a thin aluminum foil) primarily to avoid contaminants that might emanate from the hot filament reaching the sample, although it also prevents the possibility of any secondary electrons from the source reaching the electron energy analyzer. It is also possible to use radiation furnished by a synchrotron to collect photoelectron spectra. This has the advantage that the photon energy can be tuned but suffers from the disadvantage that experiments must be performed at a remote location.

The emitted electron energies are analyzed and a plot of the rate of electron emission vs. kinetic energy constitutes the photoelectron spectrum. Because the incident photon energy is known, it is usual to directly convert kinetic energy into binding energy using Eq. (1). Electrostatic electron energy analyzers are most often used in x-ray photoelectron spectrometers where an electric field applied perpendicularly to the direction of motion of the electrons exerts a force on the electrons that is balanced by their centripetal force, so that the angle through which they are bent depends on their energy. The two main designs used for XPS are the hemispherical and cylindrical-mirror analyzers depicted in Fig. 1. In the

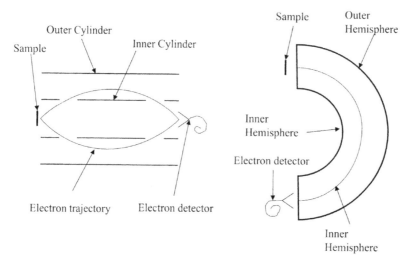

Cylindrical Mirror Analyzer   Hemispherical Analyzer

Figure 1 Schematic diagrams of hemispherical and cylindrical mirror electron energy analyzers.

hemispherical analyzer, a potential V is applied between the inner and outer hemisphere and the kinetic energy E_K of electrons entering the entrance slit and reaching the detector depend linearly on the voltage V. Similarly, for the cylindrical-mirror analyzer, a potential V applied between the inner and outer cylinder allows electrons of kinetic energy E_K to reach the detector. In principle, therefore, the spectrum can be collected simply by varying V and plotting the signal at the detector vs. this value. Unfortunately, however, the resolution of the spectrometer also varies with V (and therefore with E_K) so that this leads to a high-resolution spectrum at low kinetic energies, with the resolution decreasing at higher energies. To obviate this problem, the voltage V between either the inner and outer hemisphere (for the hemispherical analyzer) or the inner and outer cylinder (for the cylindrical-mirror analyzer) is kept constant. This allows only electrons of a single kinetic energy to pass through the analyzer, which now operates as an electron energy filter. This energy is known as the pass energy (E_P). The whole spectrometer is then raised to a variable negative retarding potential V_R so that the energy of the electrons that enter the analyzers is $E_K - V_R$. These will reach the detector when $E_K - V_R = E_P$. Now the spectrum is scanned by varying V_R. This mode of operation has the advantage that the resolution is constant throughout the whole spectral range, and this can be conveniently varied by changing the pass energy. As with all spectrometers, the signal intensity increases as the resolution is decreased.

Figure 2 shows a Cl 2p x-ray photoelectron spectrum of wear particles formed using a model lubricant composed of methylene chloride dissolved in a poly α-olefin (PAO) compared with $FeCl_2$, $FeCl_3$, and polyvinyl chloride (PVC), where peaks are evident corresponding to the electron binding energies in the sample. There are several points to note about this spectrum. The first is that the electron binding energies are typical for particular elements so that these peak positions allow the elements that are present in the sample to be identified. In this particular case, the peaks at the $\sim$198.5-eV binding energy are due to Cl^- and that at $\sim$199.5 eV due to Cl^0. In addition, the spectral background may be higher at the high binding energy side of each peak compared to the low binding energy side. This effect arises because the x-rays can penetrate much further into the sample than the

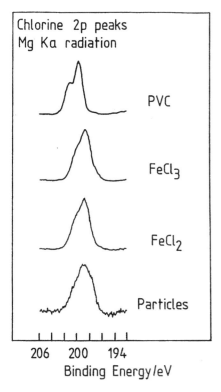

Chlorine 2p peaks
Mg Kα radiation

PVC

FeCl₃

FeCl₂

Particles

206   200   194
Binding Energy /eV

Figure 2 Chlorine 2p x-ray photoelectron spectra of wear particles formed when using methylene chloride dissolved in poly α-olefin as a model lubricant in a pin and v-block apparatus compared with the spectra for $FeCl_2$, $FeCl_3$, and PVC. (From Ref. 55.)

electron escape depth. This means that photoelectrons generated deep inside the material interact strongly with electrons in the sample as they emerge and thus lose energy. These appear as an increased intensity at lower kinetic energy and therefore at apparently higher binding energies. There are several methods for removing this background, the most common one being due to Shirley [21]. The finite electron escape depth, because it depends primarily on the electron energy and is relatively independent of the material, can also be used to measure film thicknesses. The XPS features can also be used to estimate relative concentrations using standard cross sections. As noted above, one of the main advantages of XPS lies in the fact that the binding energy is sensitive to the chemical environment; this shift in peak position is known as the "chemical shift." There are several origins for this shift. One of the main contributions arises because the energies of the core electrons are affected by the presence of all of the other electrons in the atom because of electron–electron repulsion. This means that if one of the electrons is removed from the atom to form an ion to change its oxidation state, the electron–electron repulsion correspondingly decreases and causes the binding energy to increase. This is even true if an outer valence electron is removed and the shift is generally of the order of a few electron volts. This shift can easily be measured using the types of electron energy analyzers described above so that precise measurement of the binding energy of a photoemission peak provides information on the chemical nature (most commonly the oxidation state) of the element. It is important to emphasize that the electron–electron repulsion effect described above is not the only contributor to this chemical shift.

However, because it is due to changes in the initial state of the atom (i.e., its chemical state), it is known as an initial state effect. Other effects, e.g., an interaction between the positively charged ion, formed by photoemitting an electron, and the photoemitted electron, can also cause binding energy changes (so-called final state effects) so that caution must be exercised in interpreting chemical shifts. Nevertheless, it remains an extremely useful and powerful analytical tool.

It also should be noted that Eq. (1) as applied to atomic and molecular systems is an approximation that essentially assumes that the electron is emitted so rapidly by the incident photon that the other electrons in the atom have no chance to relax. This is commonly assumed to be true in calculating ionization energies, and the energies measured using XPS are generally in very good agreement with this assumption. There are, nevertheless, some cases in which relaxation does occur. Such relaxations can cause additional features (satellites) to appear at higher binding energies to the main photoemission feature and are known as "shake-up" or "shake-off" peaks, depending on the nature of the relaxation that gives rise to them. The extent to which this happens depends on the oxidation state of the atom and can often be used to provide additional chemical information [16].

B. Auger Spectroscopy

The Auger process occurs when a core level has been ionized. These levels can be ionized using x-rays (as in Sec. II.A) or, more commonly, using an incident beam of electrons, because high fluxes of high-energy electrons are easier to obtain. This vacant core level is filled by electrons of lower binding energy. The excess energy from this transition can be dissipated in two ways. The first is by the emission of a photon, where the photon energy $h\nu$ is equal to the energy difference between the two levels. The energies of these photons can be measured, and this technique is used in scanning electron microscopy (SEM) as an analytical tool where the core levels are ionized by the incident electron beam used to collect the SEM image and the emitted x-rays analyzed (where the technique is known as EDAX or EDS). This process is also the basis of the x-ray sources described above. However, rather than the excess energy being removed by a high-energy photon, it can also be removed by emitting an electron. This process was first described by Pierre Auger in 1925 [22] and is schematically illustrated in Fig. 3 and involves three electrons. The first is the initial electron emitted to form a core hole. As noted above, this can be formed using either high-energy electrons (generally with energies ~3 keV) or x-rays, and indeed Auger features are often detected in x-ray photoelectron spectra. The second electron is one of lower binding energy, which fills the core hole, and the third is that which is emitted to remove the excess energy. The kinetic energy of this electron depends on the energies of the levels shown in Fig. 3 and is therefore characteristic of the element from which it arises; plotting these peak positions vs. kinetic energy of the electron yields the Auger spectrum. While shifts dependent on the chemical state of the atom are found with Auger spectroscopy, these are generally harder to measure and interpret than chemical shifts in XPS. The kinetic energy of the Auger electron is measured using electrostatic electron energy analyzers as shown in Sec. II.A. However, because spectra are excited using an incident high-energy electron beam, the small Auger features invariably appear on an intense background of inelastically scattered electrons. Because the background varies much more slowly with kinetic energy than do the Auger features, the small Auger peaks are more clearly distinguished from the background by differentiating the spectrum, so that Auger spectra are plotted as the first derivative of the number of emitted electrons, $dn(E)/dE$, as a function of kinetic energy E. An Auger spectrum of a typical tribological surface generated within an EP contact [23] is plotted in Fig. 4

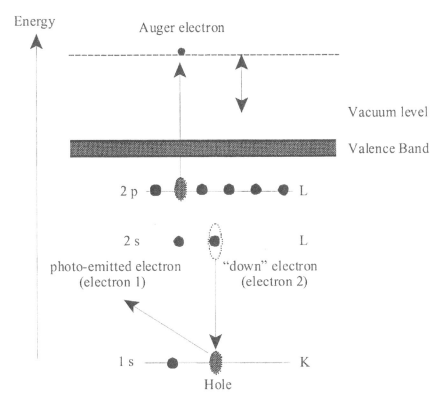

Figure 3   Schematic diagram illustrating the Auger process.

showing the first derivative behavior of the features. Note that, rather than operate the electron energy analyzer as an energy filter as was done for XPS (see Sec. II.A), the voltage V across the plates is generally varied. As noted above, this leads to variable resolution as the kinetic energy changes. Because high resolution is generally not required for Auger spectroscopy, this usually does not too badly degrade the spectral quality. The first derivative spectrum can be obtained in two ways. It can first be done by superimposing a small sinusoidal signal of amplitude δV on the voltage V applied to the plates of the analyzer. The a.c. component of the detected electron current is measured at the same frequency as the sinusoidal voltage applied to the electron energy analyzer plates using a lock-in amplifier. For sufficiently small values of δV, this a.c. signal is proportional to the slope of the electron current vs. voltage emanating from the sample. The a.c. signal measured using the lock-in amplifier is thus also proportional to the slope and therefore yields a first derivative spectrum directly when plotted vs. kinetic energy. More recently, it has become common to collect the $n(E)$ spectrum directly using a computer and to differentiate the data numerically.

Because the depth to which this technique probes depends both on the depth to which the incident exciting electron beam can penetrate as well as the distance the Auger electrons can escape, this technique is very surface sensitive. Using an electron source in which the electron beam is incident on the sample at a grazing angle can render it even more surface sensitive. Auger spectroscopy can also be used for quantitative measurements where Auger peak intensities are measured directly from the peak-to-peak amplitude of the first-

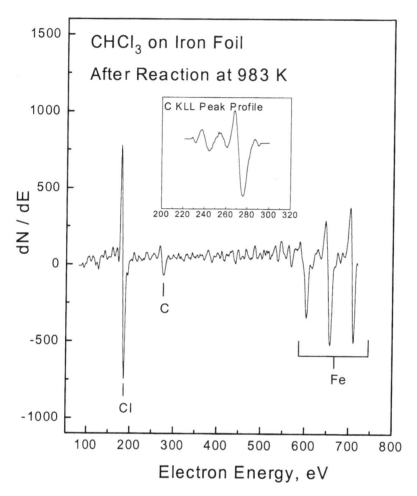

Figure 4 Typical Auger spectrum of an iron foil following reaction with chloroform at 983 K. Shown as an inset is the profile of the carbon KLL peak. (From Ref. 23.)

derivative spectrum, where it is generally accepted that this varies linearly with surface concentration.

Finally, a word of caution when using Auger spectroscopy for analysis is in order, particular when exciting with electron beams. The high-energy electron beam can cause electron damage either by dissociating or decomposing the surface species or by desorbing them from the surface. Halides are particularly susceptible to the latter effect through a process first described by Knotek and Feibelman [24] and arises directly from the Auger process. If a halide ion X^{-1} which is surrounded by positive ions is ionized by the incident electron beam, it forms a neutral species X^0. De-excitation and subsequent emission of the Auger electron (Fig. 3) leads to a positively charged ion X^{+1} that is surrounded by positive ions in the crystal lattice and is therefore unstable and can be ejected from the sample.

C. Infrared Spectroscopy

Both infrared [25,26] and Raman [27,28] spectroscopies (see Sec. II.D) probe the vibrational modes of the sample. Vibrations typically occur between 10^{13} and 10^{14} Hz and are absorbed by infrared radiation. Infrared radiation is particularly useful in identifying molecular species on surfaces because their vibrations typically occur in the region in which infrared radiation is easily available (between 400 and 3500 cm^{-1}). The lattice modes of inorganic materials generally appear at frequencies around or lower than 400 cm^{-1} because of the relatively large masses of the atoms in most inorganic lattices. This means that although these generally absorb strongly, it is often difficult to measure the infrared absorption spectrum of these species with commonly available spectrometers so that Raman spectroscopy, which is more easily capable of measuring these lower frequency vibrations, is often more useful in these cases.

Infrared spectroscopy is an absorption technique so that the infrared radiation must pass through the material for a spectrum to be collected. Because tribological films are often deposited on essentially planar substrates, spectra are best collected by reflecting infrared radiation from the surface. Conceptually, an infrared spectrometer is composed of four parts: the infrared source, a monochromator, the sample, and a detector. The infrared source is generally a filament heated to $T \sim 1000$ K. Assuming that this behaves like a black body, the maximum frequency λ_{max} in the spectral distribution is given by Wien's displacement law, $\lambda_{max}T = 2.9 \times 10^{-3}$ m K, so that at 1000 K, the maximum in the spectrum is at ~ 3500 cm^{-1}, which peaks at the high end of the desired spectral range. A better match can be obtained with the region of typical molecular vibrations (as noted above, between 400 and 3500 cm^{-1}) by lowering the source temperature. However, the total energy emitted is proportional to T^4 (from the Stefan–Boltzmann law), so that the total energy emitted diminishes substantially with temperature; thus 1000 K represents a good compromise between these two effects. In addition, heating a filament to higher temperatures can substantially affect its lifetime. Modern infrared sources generally are made from carbon and the surroundings may be water-cooled. Intense and polarized infrared radiation can also be obtained using a synchrotron but, because these are less accessible than laboratory sources, are generally reserved for more specialized experiments.

The monochromator separates the light into its component frequencies. Infrared spectra used to be collected using a monochromator with a dispersive element such as a diffraction grating to separate out a single frequency. Because the signals from thin films deposited on surfaces are generally rather small and because, by its nature, a dispersive spectrometer discards most of the radiation, it has become increasingly common to use Fourier transform spectrometers to collect infrared spectra because this analyzes all frequencies simultaneously. This uses a Michaelson–Morley interferometer (of essentially the same design as was used to try to measure the "ether drift" leading to the theory of relativity, Fig. 5). In this case, the infrared beam is split into two equal portions using a beam splitter. Each of these split beams is reflected back to the beam splitter and either constructively or destructively interfere depending on the path length difference of the two arms of the interferometer. If this difference is an integral number of light wavelengths, constructive interference occurs because the beams emerge in phase. On the other hand, if the beams emerge out of phase, they destructively interfere. This means that if one of the mirrors is moved, the output intensity oscillates and the frequency of this oscillation depends on how rapidly the mirrors are moved and on the wavelength of the radiation. If the mirrors are moved at an extremely constant velocity, the oscillation frequency due to the constructive and destructive interference is constant and inversely proportional to the light wavelength, and thus proportional to the frequency. If the light impinging on the interferometer now contains a range of frequencies (as in a spectrum), the oscillations emerging from the interferometer will be a superposition of *all* of the oscillations for each frequency in the spectrum. The original spectral distribution can be obtained by Fourier transforming this oscillation pattern, hence the name Fourier transform infrared spectrom-

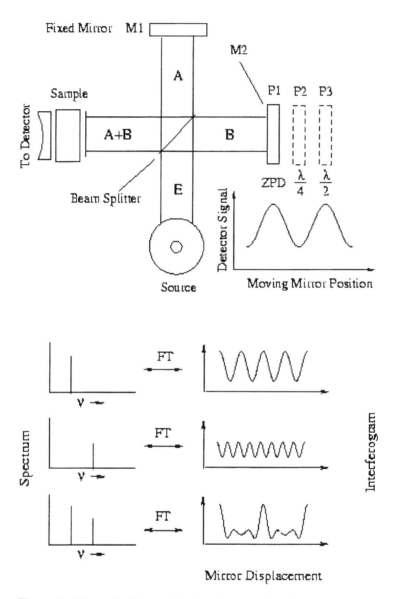

Figure 5   Schematic diagram showing the operation of a Fourier transform infrared spectrometer.

eter. The technology for such spectrometers has been significantly facilitated by the availability of inexpensive, high-speed computers capable of rapidly performing this mathematical operation. As noted above, this strategy has the advantage in that it allows all infrared wavelengths to be analyzed simultaneously, resulting in a significant improvement in spectral quality.The infrared beam is then reflected off the sample and infrared radiation is absorbed as it passes through the film. The optimal experimental geometry depends on the optical constants (the real and imaginary parts of the dielectric constant) for the substrate at infrared frequencies. This means that different substrates, e.g., insulators, semiconductors, or metals, will generally require different geometries. Metal substrates have been most commonly used for these types of experiments in a technique known as reflection-absorption infrared spectroscopy (RAIRS). The optimum experimental conditions for these experiments with metals was discussed some years ago by Greenler [29]. In this case, because metals are extremely good conductors at infrared frequencies, infrared radiation polarized parallel to the surface is effectively screened by the metal so that the only component of the infrared radiation that can be sustained at a conducting metal surface is that oriented perpendicularly to the surface. Because the polarization vector is oriented perpendicularly to the direction of propagation of the radiation, this means that the light should be incident on the surface at a grazing angle to it. However, the light flux also decreases as the incidence angle increases from normal (because the photons are distributed over a larger area) so that the optimum incidence angle is a compromise between these two effects. This situation can be analyzed using the Fresnel equations. The optimum geometry depends on the details of the optical constants for the substrate but, in general, an incidence angle of ~80° to the surface normal is likely to yield the best results [29]. Note that this is not likely to be true for insulators and semiconductors. Several detectors can be used.

Because the changes in infrared signal for thin films are generally rather small, it is most common to use a small band gap semiconductor as a detector. These are made either from mercury cadmium telluride (MCT) for the mid infrared region (from ~800 to 2000 cm^{-1}) or an indium phosphide detector for higher frequencies. These are cooled to liquid nitrogen temperatures to avoid thermal noise. Because the combination of spectral intensity variation and detector sensitivity yields a nonuniform background, spectra collected with the sample present is ratioed to a background to yield the infrared spectrum. As emphasized above, the signals obtained from thin films on surfaces is usually quite small; thus care must be taken to keep the delay between background and sample spectrum collection as small as possible to minimize the effects due to changes in the spectrometer, which will then change the background and possibly result in artefacts in the ratioed spectrum.

There are several strategies available for analyzing the resulting spectra. The most complete is to perform a full group theoretical analysis, or even a full vibrational analysis of the molecule, and to assign the most likely peaks, using group theoretical selection rules. This strategy is appropriate for small adsorbates but becomes less feasible as the molecule becomes larger.

Such a group theoretical analysis in gas phase indicates that homonuclear diatomic molecules do not absorb infrared radiation whereas heteronuclear ones do. This has substantial experimental implications because the most abundant components of air (O_2 and N_2) do not absorb infrared radiation so that, in many cases, infrared spectra can be collected in air. Unfortunately, air also contains small amounts of water and carbon dioxide, which do strongly absorb infrared radiation. Because the spectral features for adsorbed films are generally small, interference from these components in the atmosphere can be a problem so that it is common to enclose the light path and purge it with dry, CO_2-free air.

In the case of large molecules, it is often more useful to use the concept of "group frequencies." Although, in principle, a molecular vibration involves the simultaneous motion of all of the atoms in the molecule in a "normal mode," which vibrates at a characteristic frequency, in practice, vibrations are often more or less localized in one portion of the molecule. The approximation is often made that similar functional groups in various molecules have associated with them a characteristic vibrational frequency associated with that group. Thus, e.g., a $C=O$ group has a characteristic "group frequency" at ~1700 cm^{-1}. This is clearly an approximation because the $C=O$ group can differ both chemically and vibrationally depending on the nature of the rest of the molecule. Thus, group frequencies are generally quoted as a range of frequencies and different ranges are often given for different molecular environments. These are tabulated with varying degrees of detail in various sources [30], and this strategy provides a very powerful method for identifying molecular species adsorbed on surfaces.

D. Raman Spectroscopy

Raman spectroscopy [27,28] is a complementary technique to infrared spectroscopy because it similarly yields a vibrational spectrum but, because it operates on a different physical principle, it has different selection rules. This can mean that vibrational modes that are not detected in infrared spectroscopy can be monitored using Raman. It is also inherently capable of probing low-frequency vibrations (in favorable cases down to 20 to 50 cm^{-1}) and thus is particularly useful for probing lattice modes of inorganic (e.g., chloride or sulfide) films. It is a nonlinear optical scattering technique and therefore requires extremely intense light sources to excite the vibrational modes. In its early days, Raman spectra were collected using very powerful mercury arc sources. The more recent availability of intense, mono-chromatic lasers has meant that very good quality laser Raman spectrometers are commer-cially available.

    If an atom or molecule is exposed to an electric field E, a dipole moment μ is induced. For relatively moderate electric fields, the induced dipole moment is taken to be propor-tional to the applied field such that:

$$\mu = \alpha E \tag{2}$$

where α is known as the polarizability. For larger electric fields, higher-order terms can be included that give rise to nonlinear effects (e.g., second-harmonic generation). If the sample is illuminated by electromagnetic radiation oscillating with an angular frequency ω (where $\omega = 2\pi v$), the electric field is oscillatory and given by $E = E_0 \cos(\omega t)$. Substituting into Eq. (2) shows that this induces an oscillatory dipole moment in the sample:

$$\mu(t) = \alpha E_0 \cos(\omega t) \tag{3}$$

This oscillating dipole reradiates at a frequency ω so that light is scattered at the same frequency as the incident radiation. This is known as Rayleigh scattering. Because α increases with increasing frequency (as approximately v^3), high-frequency radiation is more effectively scattered than low. Such a frequency-dependent scattering cross section gives rise to a blue sky.

    Consider now a system vibrating with some characteristic angular frequency ω_{vib}. For a molecule or lattice, this is just the normal mode vibrational frequency. Because the molecule or lattice distorts at this frequency, its electron wave function also distorts so that the polarizability of the system varies in an oscillatory fashion as $\alpha = \alpha_0 + \delta\alpha \cos(\omega_{vib}t)$. Substituting into Eq. (3) and manipulating yields:

$$\mu(t) = \alpha_0 E_0 \cos(\omega t) + \frac{1}{2} E_0 \delta\alpha \cos(\omega + \omega_{vib})t + \frac{1}{2} E_0 \delta\alpha \cos(\omega - \omega_{vib})t \tag{4}$$

This reveals that the molecular dipole moment has components at ω, $(\omega + \omega_{vib})$, and $(\omega - \omega_{vib})$. This results in light being scattered at these angular frequencies. The component at a frequency ω is the Rayleigh scattering referred to above and, because the term α_0 is much larger than $\delta\alpha$, the scattered intensity at this frequency is the largest. Light is also scattered at $(\omega \pm \omega_{vib})$ and gives rise to the Raman peaks. Because peaks appear at both higher and lower frequencies compared to the Rayleigh line, it is usual to collect data only at lower frequencies to this peak because the same information is duplicated at higher frequencies. Note that more detailed quantum mechanical descriptions of the Raman scattering process have been developed and reveal that these Raman peaks are also the most intense.

A typical experimental arrangement for collecting Raman spectra is displayed in Fig. 6 and consists of a light source, the sample, a monochromator for analyzing the scattered light, and a detector. Because Raman scattering is a relatively low probability process, lasers are used as light sources, the most common being argon ion lasers (yielding light at 4880 and 5145 Å) or helium/neon lasers, emitting radiation at 6328 Å. This is incident on the sample, and the collected light is focused onto the entrance slit of the monochromator. Because the Rayleigh peak is much more intense than Raman features, care must be taken to avoid scattered light from the Rayleigh peak from overwhelming the Raman peaks. This is generally achieved using a tandem monochromator where two monochomators are placed in series. Because the Raman peaks are rather weak, the resulting light is detected by photon counting using a photomultiplier, and it is common to cool the photomultiplier used for this purpose to minimize the background (or dark) counts. A typical Raman spectrum is displayed in Fig. 7 and shows the Raman spectrum of an iron sulfide film formed by the thermal decomposition of dimethyl disulfide, a model extreme-pressure additive, on iron. The presence of FeS, rather than FeS_2, is confirmed by comparison with the Raman spectrum of a sample of ferrous sulfide [31].

A problem that is often encountered with Raman spectroscopy is the interference by fluorescence in the sample. This appears as a broad background to low frequencies of the exciting (Rayleigh) energy and can, in some cases, completely overwhelm the Raman peaks or, at the very least, make them difficult to detect on the large fluorescent background. It has recently been shown that this problem can be avoided by using ultraviolet radiation as the incident light source [32]. In this case, the incident energy is likely to be too large to cause electronic transitions that give rise to fluorescence and, in many cases, this is the case. Also, as noted above, the Raman scattering cross section increases rapidly with frequency of the incident radiation so that this provides an additional important advantage to using higher frequencies to excite the Raman spectrum. The ultraviolet Raman technique has recently been used to probe tribological interfaces in situ; a typical spectrum is displayed in Fig. 8

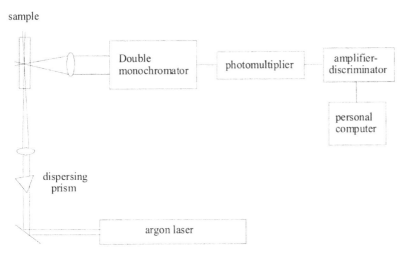

Figure 6 Typical experimental arrangement for collecting Raman spectra.

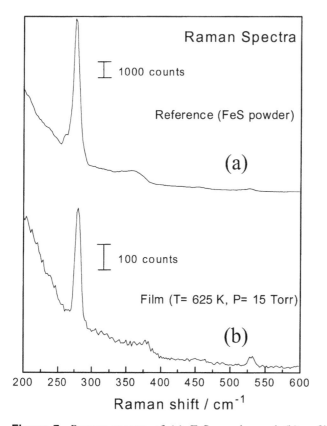

Figure 7 Raman spectra of (a) FeS powder and (b) a film grown on iron by the thermal decomposition of 15 Torr of dimethyl disulfide at 625 K. (From Ref. 65.)

showing the formation of graphitic species from 2001A [(C$_5$H$_9$(CH$_2$C(CH$_2$)$_7$H(CH$_2$)$_9$ CH$_3$)$_3$]$_3$ rubbed between steel and alumina [32].

E. Mössbauer Spectroscopy

Mössbauer spectroscopy is not generally regarded as a surface-sensitive technique [33]. It is, however, useful in tribological problems because it is very sensitive to iron and can distinguish simultaneously among several iron phases. Mössbauer spectroscopy exploits the nuclear transitions that are accessible in the iron nucleus and so is a nuclear spectroscopy. These transitions occur in the γ-ray region of the electromagnetic spectrum at ~14.4 keV. Such high-energy photons carry a large momentum so that it would be expected that the recoil of the nucleus caused by the emission of the γ-ray photon would sufficiently Doppler-broaden the line so that its uses as a spectroscopic tool would be limited. However, Mössbauer found that the transition was sufficiently fast that the recoil momentum was absorbed by the whole sample rather than by a single atom, thus essentially eliminating all Doppler broadening. This, therefore, allows the spectroscopic potential of this phenomenon to be exploited. The shifts caused in the frequency of the emitted γ-ray photon by changes in the environment around the emitting nucleus (^{57}Fe) are exceedingly small. This means that an extremely sensitive spectroscopic technique must be used to detect these spectral shifts.

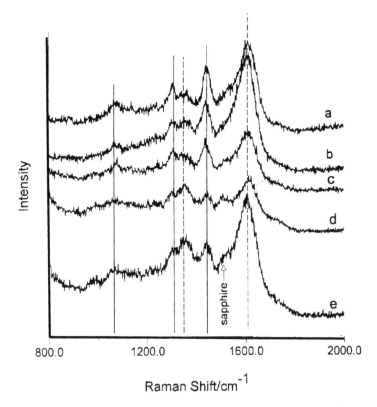

Figure 8 In situ ultraviolet Raman spectrum of 2001A $[(C_5H_9(CH_2C(CH_2)_7H(CH_2)_9CH_3)_3]_3$ rubbed between steel and alumina showing the effect of sliding at several loads. (a) No load, after sliding; (b) 43.3 N load, after sliding; (c) 66.9 N load, after sliding; (d) 90.5 N load, after sliding; (e) 114.1 N load, after sliding. (From Ref. 32.)

Gamma ray photons generated by the ^{57}Co-in-rhodium-foil source are extremely monochromatic because the recoil momentum is absorbed by the whole lattice. This radiation passes though the iron sample and is detected by a proportional counter. The nuclear absorption energies of the sample are affected by its environment primarily because of hyperfine interactions that shift and split the absorption slightly, and these effects are illustrated in Fig. 9. To bring the energy of the source photons into coincidence with the energy levels of the sample, the sample is oscillated back and forth, which then causes the emitted frequency to Doppler-shift slightly. The required shift is extremely small and requires the sample to be moved at velocities of only a few millimeters per second. As the sample is oscillated sinusoidally and reaches a velocity where the Doppler-shifted emitted frequency comes into coincidence with a nuclear transition in the sample, the transmitted energy changes. Rather than plotting this change in transmission vs. frequency, it is plotted vs. the velocity of the source. Figure 10 shows a typical Mössbauer spectrum of a film grown by the thermal decomposition of carbon tetrachloride on iron [34]. Because carbon tetrachloride is a model EP lubricant additive, this mimics the thermal reaction of this additive on iron (see below). These experiments were done on a thin iron foil so that any signals due to metallic iron did not completely overwhelm those of the film and could be subtracted out to yield features due to the film alone. This displays several features that are

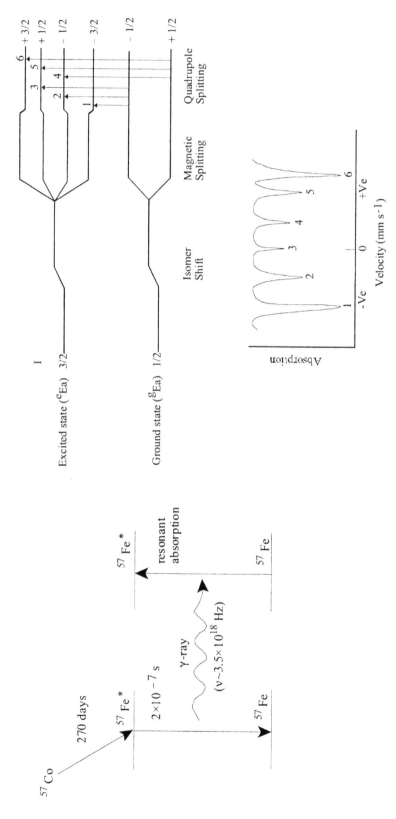

Figure 9 Energy level diagram and the corresponding Mössbauer spectra for the isomer shift, the quadrupole splitting, and the magnetic splitting in ^{57}Fe.

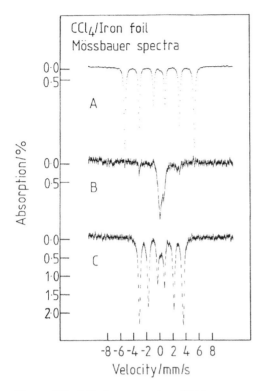

Figure 10 Mössbauer spectra of (A) an unreacted foil and iron foils that have been reacted in 55 Torr of CCl₄ at (B) 830 K and (C) 1045 K. (From Ref. 34.)

shifted from the zero point (0 mm/s), which corresponds to the energy of the incident gamma ray photon. These features are also split into multiplets. These effects are due to the hyperfine interactions. These can be modeled using two parameters, quadrupole splitting and magnetic splitting, which arise from extranuclear electric and magnetic fields. The fits are shown on the data and can be compared with standard spectra that reveal the presence of both iron chloride ($FeCl_2$) and iron carbide (Fe_3C) in the film.

F. X-Ray Diffraction

X-ray diffraction [35,36] has also proven useful in determining the structure of thin tribological films. Because the wavelengths of x-rays are typically of the same order of magnitude as interatomic spacings (~ångströms), when a coherent beam of x-rays is incident on a crystalline sample and scatters from adjacent parallel rows of atoms, these beams can either constructively or destructively interfere, depending on the difference in distance the beams travel. If the spacing between adjacent rows in the sample is d and the angle of incidence of the beam with respect to these rows is Θ, the path length difference between x-rays that are scattered from adjacent rows is simply given by $2d \sin \Theta$. If this path length difference is an integral multiple of the x-ray wavelength λ, then constructive interference will occur when

$$2d \sin \Theta = n\lambda \qquad (5)$$

This equation is known as the Bragg equation and is the fundamental equation behind analyzing x-ray diffraction data. In most cases, tribological films are polycrystalline. In this case, statistically all orientations of crystallites are equally likely (although, in some cases preferential orientations may be observed). In this case, the equivalent of a powder diffraction pattern is obtained in which the incidence and detection angles are scanned synchronously and peaks in the scattered intensity are given by the Bragg equation (Eq. 5). As noted above, the value d corresponds to the spacing between lattice layers and so the pattern of the diffraction peaks depends on the type of crystal lattice and the actual values on the value of the lattice constant. This suggests that the x-ray diffraction pattern is unique to each crystal structure and thus provides a very powerful method of analyzing the nature of a polycrystalline film. This is generally achieved by comparing the unknown pattern with a database of diffraction patterns of known materials [37]. These are now computerized so that the comparison between the unknown pattern and those in the database is made very rapidly and conveniently.

III. INTERFACIAL TEMPERATURE MEASUREMENTS DURING TRIBOLOGY

This chapter addresses the nature of the tribological films that are reactively formed on the lubricated surface by the thermal decomposition of the additive. Joule was the first to demonstrate the equivalence between mechanical energy and heat. Thus, the mechanical energy dissipated during friction ultimately appears as thermal energy that causes the surface temperature to rise. In molecular terms, the relative motion of the contacting bodies causes the atoms in the near surface region to vibrate with larger amplitudes. This first appears as a large displacement of the atoms from their equilibrium positions in a localized region where the two surfaces contact. This large local distortion is often described in terms of a high "flash" temperature [38], although caution must be used in ascribing a temperature to this effect because it is unlikely that the initial energy distribution arising from this localized interaction is anywhere close to a Boltzmann distribution. These localized high-temperature regions relatively quickly equilibrate to yield an average rise in the temperature near the surface. This section briefly discusses these interfacial temperature rises and ways in which they can be measured. It should be emphasized that it is a considerable experimental challenge to measure the temperature at a solid–solid interface because of the difficulty in accessing this region. Because it is assumed that heating at this interface can cause chemical reaction of lubricant additives, a discussion of the measurement and effect of temperature on chemical reaction rates is included.

A. Effect of Temperature and Concentration on Chemical Reaction Rates

Chemical reaction kinetics are measured from the rate at which reactants disappear or products are formed [39]. Consider a simple chemical reaction:

$$a\text{A} + b\text{B} \longrightarrow c\text{C} \tag{6}$$

where the rate of reaction is given by the rate of reactant (A or B) removal or product (C) formation, ensuring that the law of conservation of mass is maintained. The rate is therefore given by:

$$\text{Rate} = -\frac{1}{a}\frac{dA}{dt} = -\frac{1}{b}\frac{dB}{dt} = +\frac{1}{c}\frac{dC}{dt} \tag{7}$$

The measured rate of the reaction is often proportional to the concentrations of the reactants raised to some power so that the rate of the above reaction can be written as:

$$\text{Rate} = k[\text{A}]^n[\text{B}]^m \tag{8}$$

where [A] refers to the molar concentration of A. The constant k is known as the rate constant for the reaction because the larger the value of k, the faster the reaction proceeds. The exponents n and m are known as the orders of the reaction. These may, in rare cases, follow the stoichiometry of the reaction. In the gas phase, assuming that reactant A can be approximated as a perfect gas, $[\text{A}] = P_\text{A}/RT$ where P_A is the pressure of A, T the Kelvin temperature, and R the gas constant. In most cases, an overall chemical reaction arises from several sequential reactions (elementary steps) so that the reaction order does not follow the overall stoichiometry and must be determined experimentally. The rates of most reactions increase as the temperature is raised, and it is found empirically that many reactions have rate constants that follow the Arrhenius equation:

$$k = Ae^{-E_\text{a}/RT} \tag{9}$$

where A is called the pre-exponential factor for the reaction and E_a is the activation energy. The pre-exponential factor is formally independent of temperature although it may have a weak temperature dependence. This variation is, however, much slower than the exponential variation of the activation energy term so that, to a good approximation, it can be considered to be constant. This equation shows that the reaction rate can vary rapidly with temperature, this variation being given by the magnitude of the activation energy; large activation energies lead to more rapid variations than small ones. Thus, interfacial frictional heating leads to a more rapid thermal reaction there. It is important to note, however, that the validity of Eq. (9) fundamentally relies on the system being in thermal equilibrium so that the energy distribution at a temperature T is given by a Boltzmann function. As noted above, the temperature at individual asperity–asperity contacts at a rubbing interface are unlikely to result in Boltzmann energy distributions so that correlations between temperatures measured at rubbing interfaces and those in the bulk must be made with caution.

B. Measurement of Reaction Rates at Surfaces

Measuring the rate at which tribochemical additives decompose and react at surfaces is central to understanding the way in which they act. These experiments can conceptually be divided into those performed under high pressures or modest vacuums, and those performed in ultrahigh vacuum (UHV). For reactions at higher gas pressures, the thickness of the reactively formed film is measured as a function of reaction time. This can be done ex situ by removing the sample at intervals and measuring the resulting film thickness. A more convenient method is to monitor the film thickness in situ and a common way of doing this is to continually measure changes in the sample mass as a function of time using a microbalance [40]. The thickness of the film can be estimated using its density. However, because the structure of the film may be rather complex and the composition may vary as a function of depth in the film, this may result in systematic errors in film thickness measurements.

Kinetic experiments can be done in ultrahigh vacuum in two general ways. It should be emphasized that carrying out experiments under ultrahigh vacuum conditions offers several advantages, the major of which is that well-characterized (often single crystal) samples can be kept clean. Perhaps the most common method of measuring surface reaction kinetics is temperature-programmed desorption (TPD) [41–44]. In this technique, a controlled amount of reactant is adsorbed onto a sample where the sample is generally cooled to a temperature below which any surface reaction would occur, often liquid nitrogen temperatures ($\sim$80 K). The sample temperature is then ramped, generally linearly as a function of time. Species desorbing from the surface are simultaneously monitored as a function of temperature. This can be done in a fairly crude manner by measuring the pressure in the vacuum chamber or with greater precision using a mass spectrometer, allowing the nature of the desorbing species to also be identified in cases where the adsorbate undergoes chemical reactions. As noted above, reaction rates increase with increasing temperature [Eq. (9)], so that as the temperature is raised, the rates of reaction at and desorption from the surface increase, where this variation depends on the activation energy of these processes. Because the intensity of the signal of the desorbing species is proportional to the desorption or reaction rate, and this is proportional to $\exp(-E_{act}/RT)$, a plot of the ln(desorption rate) vs. $1/T$ for this initial desorption region is linear, with a slope of $-E_{act}/R$, and so provides a convenient method of measuring the surface reaction activation energy. As the surface reaction proceeds while the temperature is raised, and because a fixed amount of reactant was initially adsorbed onto the surface, it becomes depleted of reactants so that the desorption rate then decreases, ultimately to zero, as the surface becomes completely depleted. The temperature-programmed desorption data then generally consist of a series of peaks when the desorption rate is plotted as a function of sample temperature and is often also called a thermal desorption spectrum. The positions, shapes, and peak temperatures of these data can be analyzed to yield kinetic parameters such as reaction order, activation energies, and, in favorable cases, pre-exponential factors [41–44].

This method suffers from the drawback that it generally probes only reactions occurring in the first monolayer. In addition, high-temperature reactions are often not detected because all of the reactant is likely to have desorbed or decomposed before these reactions occurred. This problem can be avoided by performing a similar experiment while continually replenishing the surface with reactant using a molecular beam where the impingement flux is constant [45]. In this case, the reactant is beamed at the surface and the products that are formed are continually monitored as a function of sample temperature while this is raised. Unreacted species in the beam that reflect from the sample are also detected. However, as a surface reaction occurs, this reflected beam decreases in intensity. This variation in reflected intensity with sample temperature can be analyzed to yield the reaction activation energy and the formation of any reaction products can also be simultaneously monitored using the mass spectrometer.

C. Interfacial Temperatures During Rubbing

Consider two objects rubbing against each other with an applied normal load W and where the friction coefficient is μ. Assuming the sliding velocity is v, the rate at which heat is dissipated is proportional to $\mu W v$. The heat that is produced is removed at the hot interface either by thermal conduction through the solids or by cooling by the environment. Assuming that cooling by the environment obeys Newton's law, the rates of both conduction and convection are proportional to $(T - T_0)$ where T is the interfacial temperature and T_0 is the ambient temperature. This implies that:

$$(T - T_0) = A\mu W v \qquad (10)$$

where A is a proportionality constant that depends in detail on the assumptions used in the model, e.g., on the proportion of heat that flows through each of the rubbing objects, the contact geometry, the cooling coefficient, etc.

There are several methods that have been used to measure interfacial temperatures. The first is to measure the radiation emitted near the junction. This method, however, does not measure the temperature at the junction, but near to it. One of the most elegant methods for measuring the temperature *at the interface* was devised by Bowden and Ridler [46]. In this case the two rubbing materials were selected so that they formed a thermocouple. Thus, any heating at the interface causes a thermoelectric emf (electromotive force) to be generated which can directly measure the interfacial temperature. The results of these experiments (Fig. 11) reveal that the interfacial temperature does indeed vary linearly with sliding speed as implied by Eq. (10) where temperatures up to ~770 K were measured using this method. This confirms that cooling at the interface is dominated by linear processes (thermal conduction or cooling that follows Newton's law of cooling and other effects, such as radiative cooling, are not critical). It was also found that it was necessary to carry out the experiment on fresh portions of the sample because repeatedly rubbing on the same portion of the surface caused the thermoelectric emf to decrease substantially. This was ascribed to the effect of material being transferred from one surface to the other so that the thermocouple junction was no longer coincident with the contact region. This occurred for even

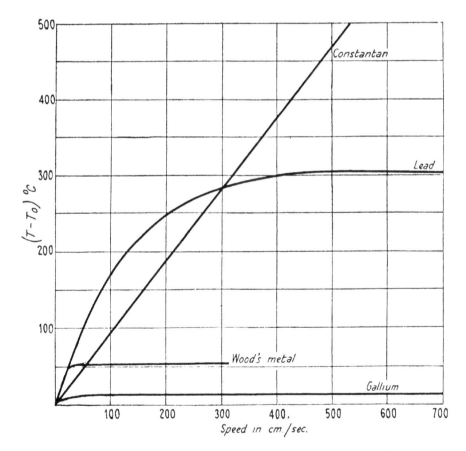

Figure 11 Maximum temperature reached when small cylinders of gallium, Wood's metal, lead, and constantan are slid on a steel surface (load = 100 g). The temperature did not exceed the melting point of the metal. (From Ref. 46.)

very thin films, indicating that the hot surface region is extremely localized. The temperature variation as a function of the distance x away from the interface for a cylindrical contact is given by:

$$T \sim e^{-\sqrt{\frac{2\sigma}{kr}}x} \tag{11}$$

where r is the contact radius, σ the cooling coefficient, and k the thermal conductivity [47]. Putting in typical values of k (0.05 cal/cm sec K) and σ (0.001 cal/cm^2 sec K) gives the scale length for the temperature variation x_0 as $x_0 \sim 5\sqrt{r}$, implying that the radii of the regions of contact where the temperature "flashes" occur are also small.

As the sliding speed is increased further, a point is reached at which the temperature no longer varies (at 304°C for lead, 57°C for Wood's metal, and 14°C for gallium). The thermocouple temperature at this maximum corresponded numerically to the melting point of the lowest-temperature component of the sliding interface [46]. It should be emphasized that this temperature corresponds to an *average* temperature. Measuring the "instantaneous" temperatures by rapidly monitoring the thermocouple voltage shows that this varies rapidly with a time scale of $\sim 10^{-4}$ sec where the maximum for this period significantly exceeds the average temperature. These are the "flash" temperatures referred to above. These have been discussed in detail by Blok [38]. Note however, that a physicochemical change, in this case melting of the surface, does not depend on the maximum value attained, i.e., the flash temperature, but on the average temperature measured by the thermocouple. This is an important observation because it implies that the most appropriate temperature to measure in order to probe the effect of surface temperature on reactivity is this average, not the flash, temperature.

Equation (10) above suggests that the temperature rise at the surface should be proportional to the applied load. The constant A, however, depends on the contact radius so that Eq. (10) is valid only for systems in which the contact area is constant. The contact area, of course, may vary with applied load, this variation depending on whether the contact is elastic or plastic. It is often assumed that the area of contact between two surfaces is determined by the yield pressure p_m of the softer of the two surfaces. In this case, it has been suggested that Eq. (10) should be modified as:

$$(T - T_0) = B\mu v\sqrt{W} \tag{12}$$

where the $\sqrt{W}$ dependence now occurs because the area of contact increases, thus simultaneously increasing cooling and energy dissipation. Some experimental verification of this model has been found [46].

The observation that melting occurs at the interface when the average surface temperature reaches the melting point at the interface can also be exploited to measure interfacial temperatures at high loads and in situations in which the thermoelectric emf cannot be measured [48]. This has been done using a conventional pin and v-block apparatus in which the pin was made of a material of known melting point. The wear rate was measured as a function of the applied load where it is anticipated that the wear rate will be come exceedingly large as the interfacial temperature approaches the melting point of the lowest melting point material as suggested above. The results of wear rate vs. applied load for a copper pin sliding against steel is shown in Fig. 12 [48]. This indeed shows a rapid increase in wear rate and the asymptote in this curve is taken to be due to the interfacial temperature reaching the melting point of copper. These data can be analyzed in greater detail. It is assumed that $(T - T_0) \propto W$ [Eq. (10)] rather than $\sqrt{W}$ because, as the surfaces are wearing,

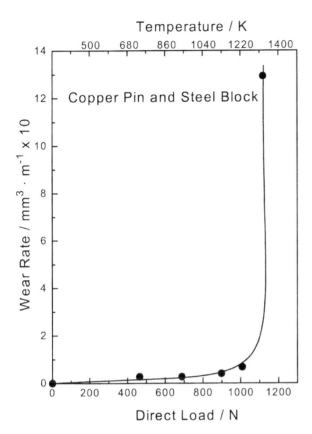

Figure 12   Plot of wear rate (measured from the amount of material removed in 600 sec of rubbing) vs. applied load using a copper pin immersed in a poly α-olefin. (From Ref. 48.)

any surface asperities will have been removed allowing the two surfaces to conform to each other. Archard has suggested that the wear rate R depends on the load W and interfacial shear strength S as [49,50]:

$$R \alpha \frac{W}{S} \qquad (13)$$

The shear strength has also been shown to be temperature dependent such that:

$$S \alpha \ln\left(\frac{T_m}{T}\right) \qquad (14)$$

so that as $T \rightarrow T_m$, $S \rightarrow 0$ [51]. From Eq. (13), this yields the large wear rate as the interfacial temperature approaches the melting point. The line shown plotted through the data of Fig. 12 is a theoretical plot from Eqs. (10), (13), and (14), where the agreement with the experimental data is very good. This method allows one to calibrate the interfacial temperature during the pin and v-block experiment. This apparatus is particularly well suited to studying EP additives because the loads are sufficiently high that it operates in the EP régime and also allows the interfacial temperature to be measured. As noted above, it does not measure the "flash" temperature but an average temperature at which a physicochemical change (in this case, surface melting) takes place. It is likely that this is also a reasonable measure of the temperature that controls the rate of chemical reaction at the surface.

IV. THE CHEMISTRY AND TRIBOLOGY OF EXTREME-PRESSURE ADDITIVES

We begin with a discussion of chlorinated hydrocarbon EP additives. As emphasized above, it has been suggested that these thermally decompose on the surface to form a lubricating film of an iron halide. Extreme-pressure lubrication provides a particularly appropriate system for study from a fundamental point of view because the high wear rates found at these high applied loads means that any surface contaminants that are initially present on the surface will be removed. In addition, as suggested by the data of Fig. 12, the extreme conditions will lead to high interfacial temperatures. In addition, although chlorinated hydrocarbon EP additives will ultimately have to be replaced by more environmentally benign and less toxic alternatives, these still represent the most commonly used EP additives. We will, in later sections, extend the understanding of the surface and tribological chemistry of chlorinated hydrocarbons to that of sulfur-containing molecules and finish with a discussion of molecules that contain phosphorus used as EP lubricant additives. These three additive types still clearly represent the most popular of all EP lubricant additives.

A. Chlorinated Hydrocarbons

1. Tribological Properties of Chlorine-Containing Additives

Because one of the main roles of EP additives is to prevent seizure between the sliding surfaces under high loads, the EP effectiveness of a range of small chlorinated hydrocarbons dissolved in a poly α-olefin (Gulf 4cSt) was gauged from the seizure load using a pin and v-block apparatus. The apparatus used for these experiments is displayed schematically in Fig. 13. The torque required to rotate the pin at a constant angular velocity (30 radians/sec) was measured as a function of load and varies extremely linearly with applied load [52]. Using the geometric factor of 117 m^{-1} for this apparatus along with the slope of this line yields the friction coefficient directly, giving a value of 0.110 ± 0.008 for a model EP additive consisting of 3.0 wt.% CH_2Cl_2 (methylene chloride) in PAO [53]. This linear variation continues up to the point at which the lubricant fails and is indicated by a rapid increase in torque. The load at which this occurs is designated the seizure load, and this is plotted in Fig. 14 for a range of small, chlorinated hydrocarbon model EP additives [54]. This experiment yields remarkably reproducible data and shows that the addition of even small amounts of chlorinated hydrocarbon results in a substantial increase in seizure load, the most dramatic effect being evident when CCl_4 is used as additive. This point will be discussed in greater detail below. Both methylene chloride and chloroform show similar trends when used as additives; there is an initial rapid increase in seizure load with the addition of up to ~2 wt.% chlorine and, as indicated in Fig. 14, the addition of further chlorinated hydrocarbon causes no further increase in seizure load.

An analysis of the interface and, in particular, the material that had been removed from the surface using XPS revealed the presence of a chloride on the surface. This is illustrated in the spectra shown in Fig. 2, which compares the Cl 2p photoemission feature with those for $FeCl_2$, $FeCl_3$, and PVC. The chemical shifts for the organic and inorganic chlorine are different and reveal that the surface film is present as a halide [55]. The corresponding iron 2p features are also in agreement with this conclusion.

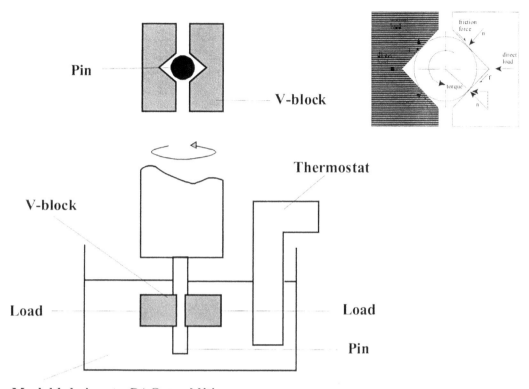

Model lubricant: PAO + additive

Figure 13   Schematic diagram of the pin and v-block apparatus. Inset shows a detail of the pin and v-block.

2.   Thermal Reactions of Extreme-Pressure Additives on Iron: The Kinetics of Films Grown from Methylene Chloride

The data above suggest that the chlorinated hydrocarbon thermally decomposes at the hot iron interface. The kinetics of this reaction were measured on clean iron foils using a microbalance [52,56,57]. A schematic of the equipment used for these experiments is displayed in Fig. 15. The iron foil sample is suspended from one arm of a microbalance that continually measures changes in mass of the sample. The sample is enclosed in an evacuated tube and can be heated by means of a furnace. The chlorinated hydrocarbon is introduced into the evacuated tube, the sample heated to the required temperature, and the reaction rate monitored from the change in mass as a function of time. The mass change is converted into film thickness by using the density assuming that the film consists entirely of $FeCl_2$. It will be found that this is not entirely true but this method nevertheless leads to reasonably reliable values. Figure 16 displays a series of results for film growth by the thermal decomposition of methylene chloride on iron measured in this way. It is evident that films initially grow rather rapidly and essentially reach a saturation thickness after some period of growth. It has been proposed that this cessation in growth is due to a poisoning of the surface by carbon [52]. The initial film growth rate can, nevertheless, be measured from the initial slope of these curves. This allows both the temperature and pressure dependencies of these reactions to be measured [52]. This shows that the initial rate of thermal

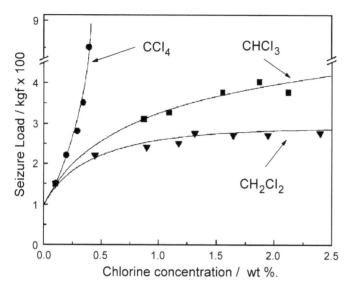

Figure 14 Plot of seizure load vs. additive concentration using a pin and v-block apparatus when using CCl_4, CH_2Cl_2, and $CHCl_3$ as additives dissolved in poly α-olefin. (From Ref. 54.)

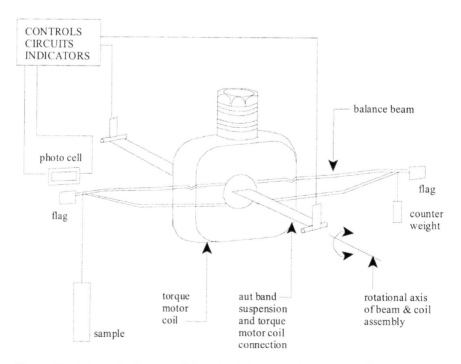

Figure 15 Schematic diagram of the microbalance used to measure film growth kinetics on iron.

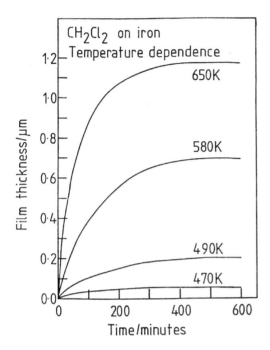

Figure 16 Plot of film thickness vs. time at various temperatures for the thermal decomposition of methylene chloride on iron measured using a microbalance. (From Ref. 52.)

decomposition of methylene chloride to yield films on clean iron in units of micrometers per minute is given by:

$$\text{Rate} = 1.5P(CH_2Cl_2)^1 \exp(-9700/RT) \qquad (15)$$

so that the reaction is first order in methylene chloride pressure P (in Torr) and proceeds with an activation energy of 9700 cal/mol.

The nature of the reactively deposited films can be measured using various surface analytical techniques. For example, Fig. 17 displays a Raman spectrum of the film, which reveals low-frequency modes that can be assigned to the presence of $FeCl_2$ [56]. In addition, features are also evident at ~1350 and 1600 cm^{-1}. These are due to the presence of carbonaceous particles. The ratio of the intensities of these two features has been related to the size of the small carbonaceous particles and indicates that particles of ~50-Å diameter are present in the film [58,59]. Note that the nature of this film is similar to that found during tribological experiments (Fig. 2).

3. Wear Rates During Extreme-Pressure Lubrication

Material is also being removed at the same time that it is reactively formed on the surface. The wear rate can be measured from the width of the wear scar in the pin and v-block experiment and the results are displayed in Fig. 18 [52]. The wear rate initially increases almost linearly with applied load but increases more rapidly at higher loads. The line shown through these data represents the calculated rate using Eqs. (10), (13), and (14). The asymptote in the wear rate then corresponds to the melting of the tribological interface and, from this fit, corresponds to a temperature of 940 ± 50 K, in good agreement with the

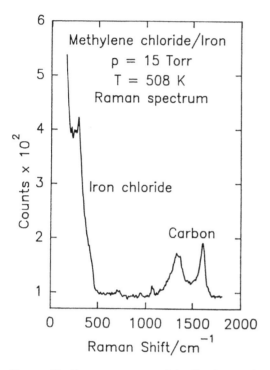

Figure 17 Raman spectrum of the film formed by the reaction of 15 Torr methylene chloride with an iron foil at 508 K. (From Ref. 56.)

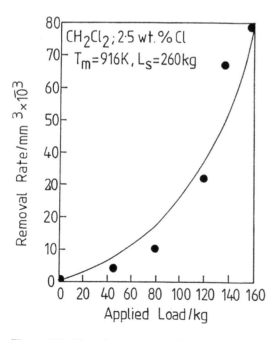

Figure 18 Plot of wear rate vs. load measured using the pin and v-block apparatus for 2.5 wt.% chlorine from methylene chloride dissolved in poly α-olefin. (From Ref. 52.)

melting point of $FeCl_2$. This is also in good agreement with the nature of the film found during tribological experiments (Fig. 2) or by the thermal reaction of methylene chloride (Fig. 17).

4. Model for Methylene Chloride as an Extreme-Pressure Lubricant Additive

The above discussions indicate that a film consisting of $FeCl_2$ that incorporates small carbonaceous particles is formed by a thermal reaction at the hot iron interface. This film is simultaneously worn from the surface, and the resulting film thickness arises from a balance between these two processes. In the seizure experiments shown in Fig. 14, the applied load is increased linearly with time after the completion of an initial "run-in" period. As noted above, the interfacial temperature increases as the load increases thereby increasing both the rate of reaction and the wear rate at the surface. The rates of both film growth and wear have been independently measured as described above. This allows the film thickness to be calculated at any point during the pin and v-block experiment. Because the role of the reactively formed ferrous chloride + carbon layer is to provide a solid lubricating film and to prevent seizure, it is reasonable to assume that if this film is removed at some load during the experiment, seizure will occur. This seizure load can then be simply determined as that at which the film thickness becomes zero. Because the film formation rate depends on reactant concentration [see Eq. (15)], the seizure load also depends on the methylene chloride concentration (if all the other parameters are held constant). The result of this calculation is displayed as the solid line in Fig. 19, and these calculated seizure loads are compared with experimentally measured values (●) as a function of additive concentration [52]. The agreement between the experimentally determined and calculated values is very good. This

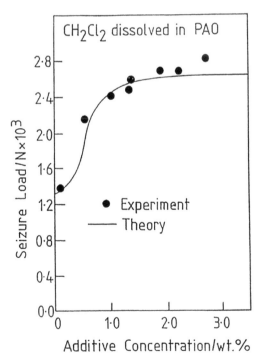

Figure 19 Comparison between theory (—) and calculation (●) for the variation of seizure load with additive concentration for methylene chloride dissolved in poly α-olefin. (From Ref. 52.)

indicates that the assumptions used in the model are correct, namely, that the methylene chloride model EP additive thermally decomposes at the hot interface to deposit a film that, in this case, consists of $FeCl_2$ and small carbonaceous particles. This strongly implies that the tribological substrate is sufficiently clean that, at least within the accuracy of these experiments, its reactivity closely resembles that of a clean iron sample. Because the assumption that the point at which this film is removed corresponds to seizure at the interface appears to be borne out experimentally, this confirms that the halide film indeed provides a lubricating layer. The picture that emerges is that the film is reactively formed on the hot, relatively clean, interface to form a lubricating film of ferrous chloride and carbon, and that the thickness of the film at any point during the reaction arises from a balance between the rate of its formation and removal.

The fundamental tribological assumptions made in this model are, first, that the wear rate is well represented by the Archard equation as long as this is appropriately corrected to take account of the temperature dependence of the interfacial shear strength. The second assumption is that the interfacial friction coefficient remains constant as a function of applied load up until the point at which seizure occurs. This assumption is implicit in the equations used to model the interfacial temperature (Fig. 12). A model proposed by Bowden and Tabor [60] and others [61] suggests that the friction coefficient of a film of shear strength S_f grown on a substrate of hardness H_s is given by $\mu = S_f/H_s$. This model simply and elegantly rationalizes many tribological properties. However, because S_f is assumed to be temperature dependent [see Eq. (14)], this correspondingly implies that the interfacial friction coefficient should also be temperature dependent for constant H_s. In fact, no such temperature dependence is detected. This apparent paradox clearly requires further investigation, perhaps by considering the pressure dependence of S_f.

5. Other Chlorinated Hydrocarbons as Extreme-Pressure Lubricant Additives

It is clear from the data of Fig. 14 that other chlorinated hydrocarbons are effective as EP additives. In particular, carbon tetrachloride appears to be outstandingly effective because the load that can be borne by the pin and v-block apparatus increases by a factor of 8 with the addition of only ~0.5 wt.% Cl as carbon tetrachloride. The asymptote in the wear rate vs. load curve has been shown above to correspond to an interfacial temperature that equals the melting temperature of the interface. This type of measurement can therefore be used to identify the nature of the surface antiseizure film [34]. This is illustrated when using carbon tetrachloride as an additive in the data of Fig. 20. This plots the wear rate vs. load and displays a relatively complicated behavior. Shown marked along the top axis of this plot is the interfacial temperature corresponding to the applied load plotted along the bottom axis. This reveals that the first asymptote corresponds to the melting point of $FeCl_2$, indicating that, at low loads, a similar tribological film is formed from carbon tetrachloride as from methylene chloride. This, however, does not lead to catastrophic failure. Instead, another asymptote is found at higher loads where the interfacial temperature corresponds to the melting/decomposition temperature of an iron carbide. Carbon may also be deposited at the interface at higher loads. This suggests that the remarkable EP behavior of carbon tetrachloride compared to other model additives (e.g., methylene chloride, CH_2Cl_2, or chloroform, $CHCl_3$ [62]) arises from its tendency to form more thermally stable carbide layers. The origin of this behavior will be discussed in greater detail below.

6. The Extreme-Pressure Chemistry of Model Additives in Ultrahigh Vacuum

The results described above show that a tribological interface is formed by the reaction of model chlorinated hydrocarbons with iron that consists of an iron chloride or an iron

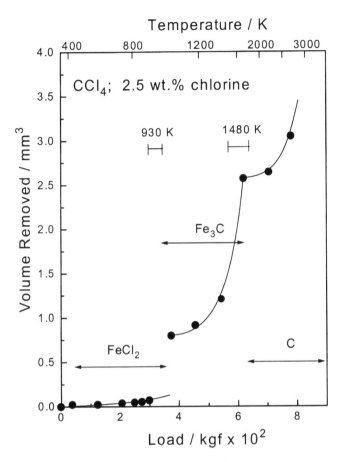

Figure 20 Plot of wear rate (measured from the amount of material removed in 600 sec) vs. applied load measured using the pin and v-block apparatus for a model lubricant consisting of CCl_4 dissolved in poly α-olefin at a concentration of 2.5 wt.% chlorine. (From Ref. 34.)

carbide, depending on the nature of the additive and tribological conditions. Similar surface reaction experiments can also be done in ultrahigh vacuum where the background pressure is $\sim 1 \times 10^{-10}$ Torr. The advantage of this type of experiment is that reactions can be done on clean surfaces because the pressure is sufficiently low to preclude adsorption of other substances. In addition, tribological experiments can be done in ultrahigh vacuum, again to allow frictional measurements to be made on contaminant-free surfaces. The reactivity of chlorinated hydrocarbons was monitored using a reactant beam with a constant flux incident onto an atomically clean iron surface. This allows the nature of the reaction products to be measured using a mass spectrometer, as well as the composition of the interface to be monitored using Auger or x-ray photoelectron spectroscopy (see Sec. II.A and II.B). The product fluxes are measured as a function of the sample temperature and typical results are displayed in Fig. 21 [23] for the reaction of methylene chloride with clean iron. These results show that the reflected flux of methylene chloride (O) from the surface decreases measurably as the sample temperature exceeds 700 K. This effect is due to the reaction of methylene chloride with the iron. The only gas-phase product detected from reaction with the surface is hydrogen; the signal of hydrogen is also plotted as a function of sample temperature (●) and its increase directly mirrors the decrease in methylene chloride

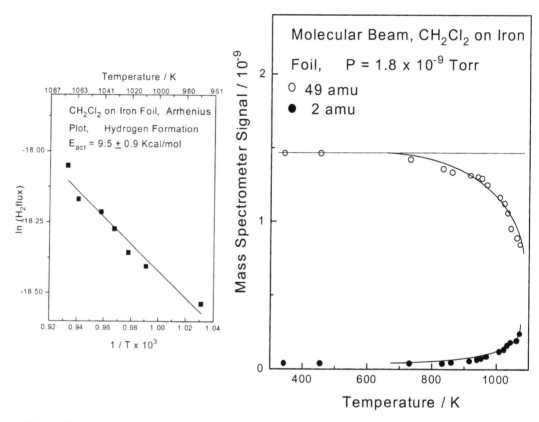

Figure 21 Plot of methylene chloride and hydrogen fluxes as a function of sample temperature for a beam of constant flux of methylene chloride on an initially clean iron foil. Shown as an inset is an Arrhenius plot taken from these data. (From Ref. 23.)

signal. This indicates that the reaction at the surface is relatively simple and can be represented by the equation:

$$Fe + CH_2Cl_2 \longrightarrow FeCl_2 + H_2 + C \tag{16}$$

where the composition of the carbon + $FeCl_2$ film that is formed is described above. The presence of these species on the iron surface after reaction has been confirmed spectroscopically. In addition, the activation energy of the reaction can be measured from the increase in hydrogen signal as a function of temperature, because this increase is proportional to the reaction rate at the iron surface. An Arrhenius plot of these data is shown as an inset in Fig. 21, and the slope yields a reaction activation energy of 9.5 ± 0.9 kcal/mol. This value is in good agreement with the measurements made using the microbalance [9.7 kcal/mol; Eq. (15)] and indicates that identical reactions are probed in each régime. Similar experiments have been performed for chloroform [23] and carbon tetrachloride [45,63]. These results reveal a similar film-growth chemistry where chloroform reacts as:

$$3Fe + 2CHCl_3 \longrightarrow 3FeCl_2 + H_2 + 2C \tag{17}$$

and carbon tetrachloride as:

$$2Fe + CCl_4 \longrightarrow 2FeCl_2 + C \qquad (18)$$

where, in each case, the ratio of the amount of surface halide to carbon depends on the stoichiometry of the reaction. This means that a larger proportion of chlorine is present in a film formed using carbon tetrachloride than when using either methylene chloride or chloroform. To probe the effect of chlorine on the behavior of carbon adsorbed on iron, surfaces were prepared by adsorbing either carbon tetrachloride or methylene chloride on clean iron at 300 K. This forms a combined overlayer of chlorine and carbon atoms where the ratio of each depends on the corresponding ratio in the reactant molecules. The surface is then heated to various temperatures and the amount of carbon remaining on the surface monitored using Auger spectroscopy. No gas-phase desorption products other than hydrogen are found as the sample is heated so that any loss of carbon from the surface is due to the diffusion into the bulk of the sample. The amplitude of the carbon Auger signal for surfaces formed by methylene chloride (●) and carbon tetrachloride (■) adsorption are displayed in Fig. 22 as a function of annealing temperature [63]. It is clear that carbon diffuses much more rapidly into the bulk of the iron when formed from carbon tetrachloride than when formed from methylene chloride. This result is clearly in accord with the tendency of carbon tetra-

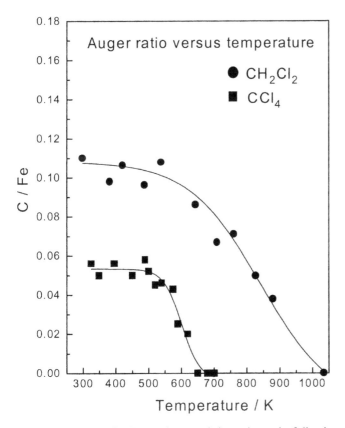

Figure 22 Plot of C/Fe peak-to-peak intensity ratio following saturation of a clean iron surface with methylene chloride (●) and carbon tetrachloride (■) as a function of annealing temperature. (From Ref. 63.)

chloride to form carbides discussed above and is the origin of its effectiveness as an EP additive. A possible molecular origin for this effect can be postulated. Both carbon and chlorine bond to the iron surface by accepting electrons into their vacant p-orbitals. Thus, the heat of adsorption of carbon will decrease when coadsorbed with an electronegative adsorbate such as chlorine. This lower heat of adsorption on the surface will correspondingly lower the activation energy for diffusion into the bulk of the sample and thus facilitate the formation of a carbide. This is apparently fast enough that, once the halide layer has been removed, sufficient carbon has diffused onto the bulk of the iron to form a hard carbide.

7. Summary of the Extreme-Pressure Chemistry of Chlorinated Hydrocarbons

Chlorinated hydrocarbons thermally decompose at the hot iron interface during lubrication to form a reactive film that consists of ferrous chloride that can incorporate small carbonaceous particles. Under suitable conditions (i.e., when using carbon tetrachloride or other additives at high enough concentrations [23,63]) carbon can diffuse into the bulk of the iron to form a carbide. This film is continually worn from the surface and the net film thickness is a result of a balance between its rate of formation and removal. This film ultimately fails, and seizure occurs, when the film is completely removed. This indicates that the highest interfacial temperature that can be sustained by any tribological film is its melting temperature because, at this point, its wear rate becomes essentially infinite. Some years ago [1], it was suggested that a parameter for gauging the effectiveness of an EP additive was T_m/μ, where T_m is the melting temperature and μ the shear strength of the interfacial material. The results described above are completely in accord with this view because low interfacial shear strengths lead to relatively lower interfacial temperatures [Eq. (10)] and high melting points to films that will withstand higher loads.

B. Sulfur-Containing Additives

Sulfur-containing molecules also provide good EP lubricating properties. For example, early work showed that the formation of a sulfide film on copper significantly reduced the interfacial coefficient of friction from ~1.3 for the surface without the film to ~0.5 for a film thickness between 1 and 2×10^{-5} cm [64]. The reduction in friction coefficient for the sulfide film is less marked than for a chloride film, but the sulfide films are not as sensitive to the presence of moisture and they retain their properties up to very high temperatures.

The seizure load vs. additive concentration response has been measured for several model compounds. The results for CS_2 and dimethyl disulfide [31] are displayed in Fig. 23, which reveal that the seizure loads increase with the addition of a small amount of additive, forming a plateau at an applied load of 2600 N for CS_2 and 3800 N for dimethyl disulfide. The interfacial temperatures in the plateaux regions of similar plots for chlorinated hydrocarbon additives correspond to the melting temperatures of the interfacial films. Similar analyses for the plateaux when using sulfur-containing additives lead to temperatures of ~1460 K, in good agreement with the melting point of FeS [31]. An analysis of thermally grown films on iron foils from carbon disulfide and dimethyl disulfide also reveal the presence of a slightly nonstoichiometric FeS film, in accord with the tribological behavior [65]. It is clear, therefore, that sulfur-containing EP additives operate in a similar way to chlorine-containing ones. Thus, they also thermally decompose at the hot iron interface to deposit a film which, in this case, consists of ferrous sulfide. This is simultaneously removed from the surface with seizure occurring when the film is removed.

Molecular beam experiments were also done in ultrahigh vacuum to further explore the nature of the surface chemistry of both dimethyl and diethyl disulfide [66]. It was found

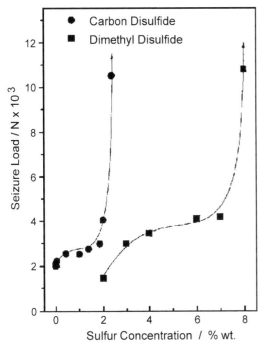

Figure 23 Comparison of the seizure loads measured using a pin and v-block apparatus plotted vs. sulfur concentration when using carbon disulfide (●) and dimethyl disulfide (■) as additives, both dissolved in poly α-olefin. (From Ref. 31, with permission.)

that reaction was initiated by S–S bond cleavage to form surface thiolates. Methyl thiolate species reacted to form methane and deposit the sulfide. In contrast, ethyl thiolate underwent a β-hydride elimination to evolve ethylene. The disulfides reacted at substantially higher temperatures than the halides, and a depth profile of the resulting surfaces revealed that the iron below the sulfide film contained substantial amounts of carbon [31]. It appears, therefore, that a similar diffusion of carbon into the bulk of the sample occurs for sulfur-containing additives as for chlorine-containing additives, particularly CCl₄. This process appears to be more important when using sulfur-containing additives because of the higher reaction temperatures. This effect may offer an explanation for the subsequent rise in seizure load at higher concentrations noted in the data of Fig. 23. This may be due to a carbide film remaining on the surface once the sulfide film has been removed along with subsequent, rapid carbon deposition from all available hydrocarbon moieties of the lubricant onto the smooth, hot ferrous metal surface. This antiseizure film would be directly analogous to that necessary for the highest load, and thus temperature, regime of Fig. 20 for carbon tetrachloride.

C. Phosphorus-Containing Additives

1. Introduction

Relatively low values of sliding friction coefficient (0.10 ± 0.05) for the EP films produced by phosphorus-containing additives [67–69], and their environmental benefits relative to the chlorine- and sulfur-based additives, make the phosphorus additive class, the third of the group of three, by far most commonly used, EP lubricant additive classes. These environmental benefits are well-heralded in discussions of vapor-phase lubrication and ceramic, low-heat-rejection engines (LHREs) [70]. More acidic and malodorous decomposition gases are produced by the chlorine- and sulfur-based additives than phosphorus-containing ones at the high-temperature surfaces in these ceramic engines—conditions that are similar to those of EP tribological interfaces. Treatment of iron surfaces using phosphate ester solutions at high temperatures is also viewed as environmentally preferable to standard phosphating procedures [71]. Clearly, a phosphate coating for a steel surface exposed to humid conditions is preferable over chloride or even sulfide films produced using chlorine- and sulfur-containing additives. However, the increased cost and seemingly more limited effectiveness of organophosphorus compounds as EP tribological additives relative to these others must be overcome by more efficient usage. This should be obtainable through a clear fundamental understanding of their tribological function and perhaps by some enhancement of their properties using structure modification and/or synergies with other additives.

2. Additive Structures

The hydrocarbon-based phosphorus compounds most analogous to the chlorinated and sulfurized hydrocarbon lubricants discussed above are called organophosphines. Like phosphine itself, these are toxic and unstable [72] and thus impractical to use in most ordinary lubricants, unlike sulfur- and chlorine-containing additives. However, just the addition of oxygen to mitigate this problem and to provide usable phosphorus compounds for lubricants adds considerable complexity to their molecular structures and the resulting surface chemistry. The correspondingly more complex structure of their tribologically important films than those from chlorine- or sulfur-based additives is then to be expected. However, a thorough understanding of the other two additive classes is likely to be essential to understanding phosphorus-based additive functions in tribology. As is demonstrated below, this appears to be the case. Although intramolecular sulfur is often used in combination with phosphorus (along with hydrocarbon functionalities [73,74]), oxygen is clearly the most common and is probably the most economically and environmentally desirable element to combine with phosphorus. Therefore, the general molecular structure for this additive class can be represented as:

where R, R', and R'' represent hydrogen (a maximum of two, each providing acid functionality) or a hydrocarbon moiety (possibly aromatic, substituted, and even cross-linked) [75]. The doubly bonded oxygen structure in parentheses represents that which makes it a phosphate rather than a phosphite ester additive.

3. Tribological Film Composition

Early work by Beeck et al. [76] showed that some phosphorus-containing additives function by reacting with the metal to form metal phosphides, and therefore appear to operate in a similar manner to the chlorine- and sulfur-containing additives discussed in the previous sections. Later studies contradicted this conclusion, indicating that a metal phosphate was formed instead [67,77]. However, recent tribological studies using an organic phosphite lubricant additive [78], as well as a phosphate additive under some conditions [71], where the resultant film was analyzed using XPS and depth profiling, clearly reveal the formation of phosphides, as well as other phosphorus compounds in higher oxidation states nearer the surface. Our own XPS analyses [79] of films on steel surfaces, produced by grinding the steel with a lubricant containing a phosphate ester additive to mineral oil, also reveal a small amount of metal phosphide.

Much of the disagreement relative to the composition of surfaces modified by the use of phosphate or phosphite esters as lubricant additives stems from differences in the severity of the tribocontact conditions and its environment, as well as the particular additive molecular structure and the composition of the tribopair. The significance of the latter is shown dramatically in differences in reactivity for ceramic surfaces compared to iron- or copper-containing ones [80]. It was shown that the addition of less than 1% by weight of aluminum to iron or copper "kills" the formation of metal phosphate film through formation of the very stable oxide of aluminum. A loose "carbonaceous" film is also formed in this case instead of the tenacious phosphate.

A thorough study of the composition of the initial monolayer film formed on iron at room temperature from a carefully chosen acid phosphate ester is given in a two-part analytical study: Part I using XPS angle-resolved depth analyses and ellipsometry to determine chemical states and Part II using reflection-absorption infrared spectroscopy (RAIRS) at grazing angle to determine specific bonding [81,82]. These studies clearly show the formation of the metal organophosphate corresponding to the starting phosphate ester, where bonding with the metal occurs *through the acid group*. No corresponding chemisorption was observed on a gold substrate. The presence of an ordered, 30-Å-thick monolayer film on the iron or steel substrates also indicated that the phosphate moiety combined ionically (i.e., formed the iron salt) with the approximately 20-Å-thick oxide layer on the polished iron-based substrate. This work supports an early contention that it was principally the acid form of the phosphate ester that reacts with steel surfaces to form tribologically significant films [77,83]. The hydrolysis of neutral phosphate esters to form acid phosphate esters occurs readily in some applications, such as engine oils, whether in the lubricant solution or on the reactive surfaces themselves. Reference 81 above does claim the presence of the hydrated iron(III) oxide on normal steel surfaces exposed to ambient conditions. However, some pin and v-block experiments suggest that dissolved water in the lubricant itself due to ambient humidity positively affects phosphate ester tribological performance [84]. Our own pin and v-block experiments [79] also show a direct correlation between published free acid content and increased seizure load for a series of commercial aromatic phosphate esters. Different film compositions are produced on tribological surfaces thermally activated by friction or heating in the presence of organophosphorus lubricant additives. Electrochemical acceleration of the otherwise spontaneous film-forming process on ferrous metal surfaces from dilauryl acid phosphate (R and R' a C_{12} saturated hydrocarbon group with the generic structure above) was shown to provide relatively clean iron(II) phosphates and release unsaturated hydrocarbons into solution—the former identified by x-ray diffraction and the latter by infrared spectroscopy [85]. A similarly acidic "mixed alkyl"-based phosphate ester also produced a film that Auger spectroscopic intensity ratios indicated as being iron(II) phosphate [71]. This appears to be consistent with the β-hydride elimination mechanism demonstrated in the diethyldisulfide molecular beam experiments on iron foils above. Lack of β-hydrogens in aromatic phosphate esters may then explain the up-to-90 (wt.)% carbon contents for films formed from the most popular of these, tricresyl phosphate (TCP) [86]. However, some carbon was also found in the film formed from the C_{12} saturated ester above [87]. It would be anticipated that under ambient conditions, some addition polymerization between the unsaturated hydrocarbon products would occur [88,89] to provide relatively low friction coefficients from the resulting polymeric hydrocarbon and somewhat oxygenated versions thereof (from air exposure) [89]. In any case, the relative stability of this iron(II) phosphate at high interface temperatures in this essentially reducing environment is seen in Ref. 90. This oxidation state for iron ($+2$) is also the same as that for the most tribologically important films grown from chlorinated and sulfurized additives: iron(II) chloride and iron(II) sulfide, as discussed above.

Higher oxidation states for iron also appear to be generated under these high temperature conditions. The possible formation of iron(III) chloride from chlorinated additives is often reported on tribological surfaces [91] and was detected in XPS studies mentioned above [55]. The nonstoichiometric iron sulfide noted above [65] can also be regarded as simply a mixture of iron(II) and (III) sulfide. In the discussion above for the growth process by diffusion under these conditions for chloride and sulfide films, one would expect a continuum of oxidation states for iron near the surface from zero (its bulk value) to three (its maximum). Thus, not considering the effect of carbon in these films, iron(III) chloride, sulfide, and phosphate would be expected to be found on the outer part of the corresponding iron(II) films. Because this uppermost layer is likely to be tribologically removed first, it would appear that the iron(II) compound would be the last line of defense against seizure in this case, and so more tribologically important. This was shown above to be the case with the chloride and sulfide additives. In any case, the many reports of the formation of iron(III) phosphate from various organophosphates as being *the* film formed by this additive class [67,92] are not inconsistent with those finding the iron(II) phosphate. The temperature, concentration of additive, and the other possible environmental factors should be expected to affect the ratio of the measurable iron(III) to (II) contents. The variety of experimental conditions represented by all those who have done research in this area almost certainly guarantees the formation of a variety of different products.

Exemplifying somewhat more severe conditions, thick iron phosphate films were produced on steel parts using heated forming dies and compared with plates of the same steel inductively heated to the same temperature while immersed in the same forming lubricant (of 10 (wt.)% phosphate ester in mineral oil [93]). Although the various esters were not identified, infrared spectroscopy was used to identify the films as iron(III) phosphate or pyrophosphate, both when grown thermally or by forming. The latter film composition indicates that temperatures were high enough to dehydrate the phosphate, the initial stage of the "phosphate glass" formation from TCP more thoroughly characterized for somewhat higher temperature conditions [94]. The molecular weight of such a phosphate film was also determined for similar conditions and found to be in the range of 6–60 kDa [95]. From infrared, Raman, and Auger analyses of the phosphate glass produced on ferrous metals at temperatures up to about 800 K from aromatic phosphate esters [96], the picture of the film composition that emerges is that of an iron-containing matrix of cross-linked phosphates having P–O–P (as in the pyrophosphate above) and P–O–C (graphite) linkages—the latter perhaps derived from the P–O–R bond of the original additive. This is illustrated schematically in Fig. 24. It is not clear, however, that such graphitic structures would be formed from *nonaromatic* R groups, except perhaps at such high interface temperatures that would provide facile C–H bond cleavage. Under mild sliding conditions, the low-friction films produced by such saturated hydrocarbon groups [85] suggest that β-hydride elimination and also, but probably limited, addition polymerization occur for them [68]; this polymerization would be analogous to the low-friction films at low loads produced in our study of higher molecular weight chlorinated hydrocarbons [97] and likely also produced by the intermediate methyl thiolate film derived from the dimethyl disulfide additive discussed above [65].

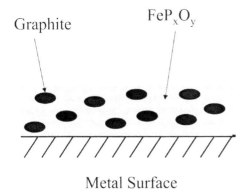

Metal Surface

Figure 24 Schematic depiction of the iron-containing phosphate glass with embedded graphite particles produced by an aromatic phosphate ester on the surface of one member of an iron tribopair under relatively mild conditions.

However, such superficial carbonaceous species on *top* of their corresponding inorganic metal salts are likely to be readily removed under more severe sliding conditions. The "embedded graphite" found in the above phosphate glass (Fig. 24), however, is believed to provide low-friction regions in the real contact area and thus be beneficial to their rolling contact [96]. Using atomic force microscopy, low-friction regions of MoS_2 produced by a molybdenum dithiocarbamate additive at the hot, real areas of a rubbing contact (where more complete thermal decomposition is likely to occur) were recently discovered [98]. (Other molybdenum-containing additives, together with sulfur in the lubricant, are also known to lower friction in common lubricants [99]). Graphite's extremely high sublimation point would also likely guarantee its existence at such hot spots of tribological interfaces [100] so that its low-friction property could also impart such a benefit to the above phosphate glass. At an EP tribological interface, however, we determined that the embedded graphitic particles in iron(II) chloride films produced by some chlorinated hydrocarbons were actually detrimental to preventing seizure under these conditions [53]. These graphitic particles appeared to create higher friction coefficients so that interface temperatures were greater and seizure loads lower. Their effect in a phosphate glass may then explain why phosphate esters are commonly regarded as having limited usefulness as EP additives, just as low concentrations of chlorinated hydrocarbons are [101]. At higher concentrations, at least the latter have the potential of creating the carbide underlayer necessary to prevent seizure at the highest EP loads encountered [62]. Finally, as discussed extensively by Forster [102], the diffusion of iron from the substrate through a phosphate glass binds it to the substrate. This important process would also appear to be analogous to the sulfide and chloride EP film-formation processes discussed above [34,63].

From its phase diagram, the decomposition products of the iron(II)/(III) phosphate glass at higher interface temperatures, in the absence of carbon or hydrogen, are the corresponding oxides of iron and P_2O_5 (g) [103]. Also, if there were no chemical interaction between any hydrocarbon moiety, or its decomposition products, with the deposit resulting from reaction of the organophosphate with an *inert* surface above 800 K, the phosphorus oxides or their acids appear to be driven off the surface to leave only a loose carbonaceous residue [104]. The sublimation point of P_2O_5 is ~600 K [100]. Thus, an active metal must be used to provide a tenacious, lasting film for high-temperature tribological surfaces. Our extreme-pressure sliding contact surfaces were determined to be above 600 K [48].

Because such high temperatures are known to cause a chemical reduction of stable iron oxides by carbonaceous materials [105], a thermodynamic study showing the sequential reduction of iron phosphates to a phosphide is not surprising [90]. One can then envision the partial loss at high temperatures of the phosphorus pentoxide from the derived iron(III) phosphate films of Godfrey and Bieber [67,83] consistent with an open system and with the Fe–O–P phase diagram [103], and then reduction of the rest of the film by *sufficient* carbon or hydrogen from the hydrocarbon moiety *and* time to allow the thermodynamically predicted phosphide to form [90]. This is then consistent with the above early study by Beeck reporting this phosphide [76]. Thus, we have gone full circle by varying conditions and materials to provide different phosphorus-containing tribological films. This also rationalizes the more recent discovery of iron phosphides in film-growth studies at up to ~1000 K [71,106], and laboratory [73] or commercial [78] EP tribological testing.

The formation of phosphides is also supported by recent well-controlled surface science studies of trimethylphosphite $[P(OCH_3)_3]$ (thus having no extra oxygen in the general structure above) on iron and nickel single crystals [107,108]. Here the carbon and oxygen leave the surface together as a volatile C_1 aldehyde, alcohol, or CO. Some evidence for P–O bonding on the surface is, nonetheless, detected in these experiments after initial production of atomic phosphorus from complete P–O bond cleavage. However, the oxygen preferentially recombines with carbon to leave as CO at higher temperatures. Although no analogous studies have been done on the corresponding organophosphate $[OP(OCH_3)_3]$, this recombination nonetheless implies that given temperatures high enough in the presence of sufficient carbon, a phosphide would most likely be produced. Thus, the significant effort made recently to understand and apply vapor phase lubrication science has substantially aided our understanding of organophosphorus chemistry as it relates to EP tribology. What no doubt has caused some of the confusion in this area is the facile oxidation of the phosphides all the way to phosphates by unintentional or unavoidable exposure of the hot tribological surfaces to air during or after test. The limited solubility of oxygen in oily films slows this process, but does not eliminate it [109]. This conversion to phosphate is also clearly thermodynamically allowed [100].

Under the reducing environment of the pin and v-block tribological interface at high loads, one then expects to eventually form a phosphide. If it is formed quickly enough, one expects to see a plateau in the seizure load vs. concentration curve such as that seen in Fig. 14 for chlorine-based additives and Fig. 23 for sulfur-based ones, corresponding to the melting points of the respective inorganic iron salts. Preliminary experiments in our laboratory with low phosphorus concentrations have indeed confirmed this (Fig. 25), and our calibration of this apparatus [110] indicates that the plateau is consistent with a melting point close to 1250 K for the film. Although this is close to the melting point for Fe_3P [111], it appears to be a little closer to the pseudobinary eutectic point for $Fe_3P + Fe_3C$ [103] and also about the same as that reported by Beeck: 1225 K [76]. Because the EP tribological importance of iron carbides has been shown above with both chlorine- and sulfur-based additives, these pin and v-block results would also show a consistent picture for the phosphorus-based additives together with the others. The film-growth experiments and subsequent surface analyses of Klaus et al. [106] also show the presence of both iron carbides and phosphides (as well as phosphates nearer the surface) from the reaction of vapor phase tricresyl phosphate (TCP; above R, R' and R''= aromatic C_7H_7) with polycrystalline iron foils at 973 K (Fig. 26). The carbides are present at larger concentrations than the phosphide and spontaneously oriented parallel to the surface, the tribologically most desirable orientation.

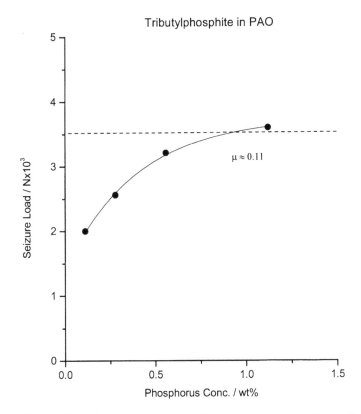

Figure 25 Seizure load vs. phosphorus concentration from tributylphosphite (mostly n-butyl) dissolved in a C_{30} poly α-olefin, using the pin and v-block apparatus (see Fig. 23). The plateau level appears to correspond to the melting of iron phosphide/carbide eutectic at ~1225 K [76] by using previous calibrations of the apparatus [48] and an average friction coefficient μ ~ 0.11.

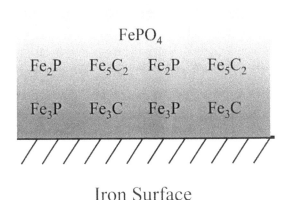

Figure 26 A schematic diagram of the model structure of TCP vapor deposition on an iron foil at 700°C for 20 min. (Adapted from Ref. 106.)

D. Kinetics of Film Growth

The time necessary to grow these tribologically desirable films from organophosphorus additives is especially important in EP tribology. In the dynamics of the growth and simultaneous removal of these films in the EP contact, there must be some film present at all times to avoid excessive, if not catastrophic, wear at the contact. The formalism for this has been developed and presented above for chlorinated hydrocarbons [52]. No analogous theory has yet been developed for tribological contacts in the presence of phosphorus-based lubricant additives although the need for such is clear [69,112]. This need is seen for both rolling contacts [102,104] and EP sliding contacts in metalworking [113]. For the latter type of contact especially, the importance of adhesion of the films to the substrate is often ignored but known to be essential [114,115].

Assuming delamination wear [116] due to poor film attachment is not the principal wear mode for these phosphorus-derived films on iron, the more traditional treatment using Archard's wear equation is invoked for neutral organophosphate additives [69] in much the same way as was applied to the chloride films above, except without explicit treatment of the temperature dependence of the shear strength of the film as it enters into this wear [34]. Although this temperature dependence can be used to explain higher wear rates at higher temperatures for any EP film, an interesting alternate explanation has recently been given for a vapor phase triarylphosphate ester (very similar to TCP) used on a steel tribopair [102]. Here the rate of wear for the steel surfaces was controlled by the rate of iron diffusion through the derived phosphate film—more specifically by the velocity of the Matano interface. The striking symmetry of iron and carbon in the Auger depth profiles also suggests the importance of carbon diffusion into the steel at the same time iron diffuses out of it into the film (Fig. 27). Note that this also suggests the likely presence of the carbide together with any possible phosphide that would form to provide the key eutectic mixture originally proposed by Beeck et al. [76] and the film composition by Klaus et al. [106]. The controlling cationic diffusion mechanism proposed is also consistent with that proposed for dimethyl disulfide above [65]. It is also consistent with the parabolic film growth discovered for the latter and found independently also for TCP by others [106] [see their results for iron in Fig. 28 ($\blacktriangle$)]. The small diffusion coefficient found for the films derived using Forster's neutral aromatic phosphate ester on ferrous metals (10^{-14} cm^2/sec [102]), as compared to that calculated at the same temperature for chlorinated or sulfurized additives where parabolic (i.e., diffusion-limited) growth kinetics are found (10^{-8} to 10^{-7} cm^2/sec), is believed to be because of the barrier represented by the P–O–P cross-linking of the phosphate glass produced by their additive. Alternatively, or in addition, their low value may be because of carbon poisoning of the reaction sites at the surface of the film to slow its growth; this has been shown to slow the growth of films on ferrous metals derived from chlorinated hydrocarbons not showing parabolic growth [52]. It is not clear that diffusion-controlled, parabolic growth of the derived phosphate film is actually occurring at the relatively low temperatures used [102]; such a small diffusion coefficient is also not typical of corrosion-type, parabolic growth on polycrystalline metals [117]. In any case, the lower diffusion coefficient for the phosphate glass is indicative of slower film growth and perhaps a reason for the less effective EP additive performance of neutral aromatic phosphate esters. As indicated above, this was seen in pin and v-block experiments [52].

The rate of this film growth must, of course, be greater than its removal under tribological conditions in order for the film to be of tribological importance. Delivery of the additive to the contact region and their mutual affinity are factors too often ignored in using additives for EP contacts. The latter factor is used to explain the relative performance of

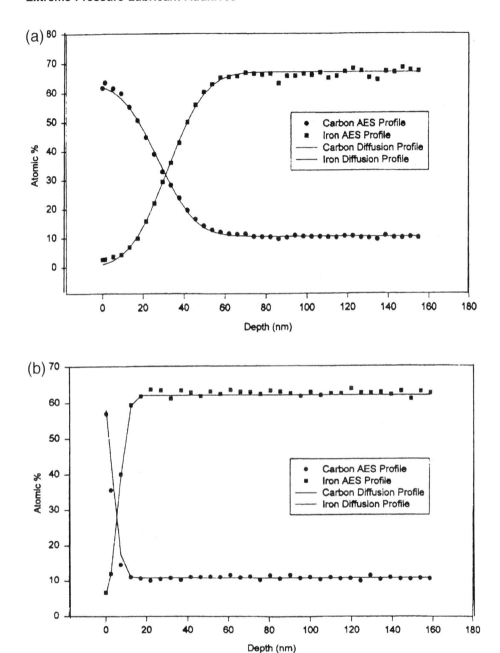

Figure 27 Iron and carbon surface concentration profiles found on ferrous tribological surfaces using Auger electron spectroscopy after long (a) and short (b) test durations at elevated temperatures, showing their relative symmetries in either case. (From Ref. 102.)

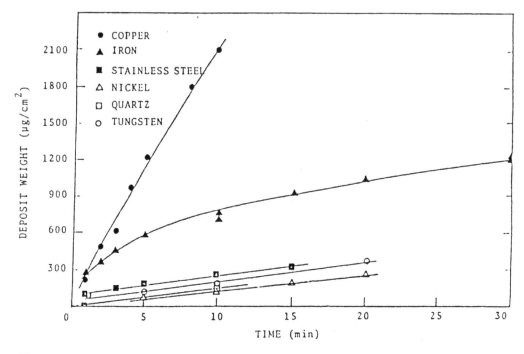

Figure 28 Deposition on various substrates with 1.6% TCP in nitrogen stream at 700°C. (From Ref. 106.)

neutral phosphate esters in mineral oil for a cross-pin-type lubricant tester [118]. A different, more polar fluid medium can compete with the additive for the surface or thermodynamically decrease the additive's availability to the surface. The latter can be expressed in terms of a "solubility parameter" presented recently [119]. Another method to increase additive surface concentration is the use of inactive particulates holding the EP additive and capable of being trapped within the EP contact [120,121].

Assuming adequate delivery of EP additives like these organophosphates into the tribocontact, their film-formation rate must then also be greater than their removal rate. Despite a dearth of film-growth experiments of the type presented above for chlorine- and sulfur-based additives, there are two of significance with which to compare such studies of the latter additive types. As mentioned above, of particular note is that of Klaus et al. [106], where vapor phase delivery of TCP at 973 K on iron clearly shows parabolic film growth on iron foils [Fig. 28 (▲)] indicative of the diffusion-controlled mechanism of Forster [102], and also seen at somewhat lower temperatures with some chlorine- [34] and sulfur-based [65] additive types. The better-developed film-growth models of the latter pair allow the comparison of their relative film growing propensities to TCP, clearly the most common and studied phosphorus-based additive. In particular, using temperature and concentration dependencies, the all-important initial (and greatest) growth rate for TCP can be estimated from its growth curve and is determined to be less than that for the chlorine- and sulfur-based additives under conditions where parabolic growth kinetics are found. However, at 973 K, the initial growth rate of TCP is significantly greater than chlorine or sulfur additives where linear growth kinetics are found [52,62]. Perhaps this explains why certain phosphate and phosphite esters can function as effective EP additives. It would also appear that they have greater potential than what is currently being utilized.

A somewhat later film growth study of TCP on "Cr–Ni alloy" (containing iron) [86], demonstrates in their Fig. 11 (Fig. 29) the linear-to-parabolic transition found and analyzed for the dimethyl disulfide above [65]. This figure also seems to demonstrate the substrate change (here claimed to be a phase change in the alloy) causing a significant jump in growth rate between 973 and 1073 K (700° to 800°C). This jump appears to be analogous to that seen with chlorinated hydrocarbon additives in microbalance studies [53,62] and verified by UHV experiments [23,122]. Interestingly, some of their growth curves at intermediate temperatures would also indicate reactive surface site poisoning, for which a formal model was developed using two chlorinated hydrocarbons above [52,62]. Although the Avrami kinetics applied here to determine the activation energy is suspect (the initial growth rate does not appear to be close to zero as expected for an Avrami "induction" period), their curve fitting happens to fit rather well the same first-order dependence on active surface sites and the site poisoning model of these chlorinated hydrocarbons. Their Fig. 4 (Fig. 30) showing film growth with TCP on various substrates at 973 K (700°C) demonstrates this more consistently, and their Fig. 8 (Fig. 31) shows the possible cleaning effect of oxygen on these surface sites. This would imply that carbon poisons these sites. The surface-reaction-controlled (initial) activation energy of 13.85 kcal/mol found for TCP on the above, less active alloy is perhaps predictably greater than those found for these chlorinated hydro-

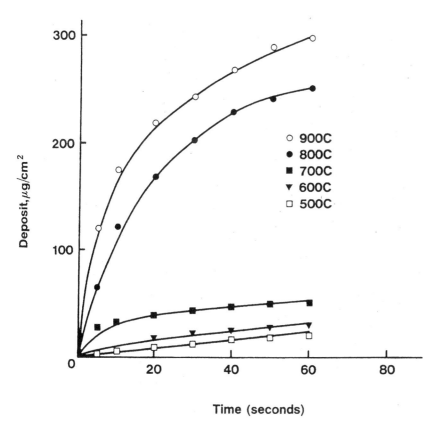

Time (seconds)

Figure 29 Deposit thickness vs. time of chromium–nickel substrates exposed to 1.5% TCP in 200 mL/min nitrogen carrier gas at different temperatures. (From Ref. 86.)

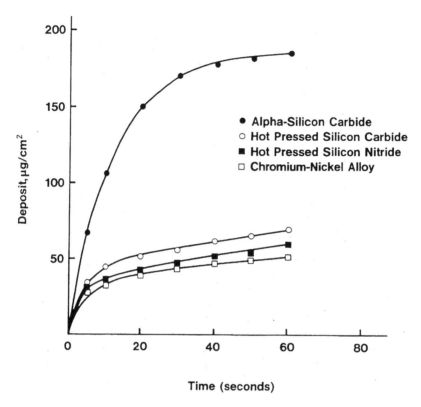

Figure 30 Deposit thickness vs. time of four different surfaces exposed to a vapor of 1.5% TCP in a 200 mL/min nitrogen carrier gas at 700°C. (From Ref. 86.)

carbons on iron (8–10 kcal/mol). The molecular details of this film growth for common organophosphorus additives is now being investigated in earnest, again mostly because of the interest in vapor phase lubrication.

A common thread for all three EP additive classes with regard to surface decomposition is the concept of β-hydride elimination. The loss of a hydrogen atom in the β position of the hydrocarbon chain of a chlorinated hydrocarbon is well known in even their uncatalyzed dehydrohalogenation reactions to give the corresponding binary acid (here HCl) [89]. The molecular beam experiments using $(CH_3CH_2S)_2$ clearly show such β-hydride elimination in its reaction with iron to give only ethylene (CH_2CH_2) and hydrogen [66]. There is no β hydrogen for $(CH_3S)_2$, which gives only methane (CH_4) and adsorbed carbon under the same conditions. This mechanism also appears to have been at play in forming the products of phosphate ester decomposition at varying temperatures in simple pyrolysis studies under helium [123]. The lack of a hydrogen atom in the β position to P for the aryl phosphates (such as TCP) perhaps explains their greater thermal stability [124]. Thus, at their higher required reaction temperatures, the corresponding aromatic hydrocarbon was lost instead because of C–O bond scission and recombination with hydrogen. This hydrocarbon was also claimed to have been formed in early, careful surface science studies of TCP on iron and its oxide [125] but has only very recently been confirmed by temperature programmed reaction spectroscopy (TPRS) of TCP on polycrystalline iron by its direct detection [126]. This latter study also confirms both the C–O and P–O bond

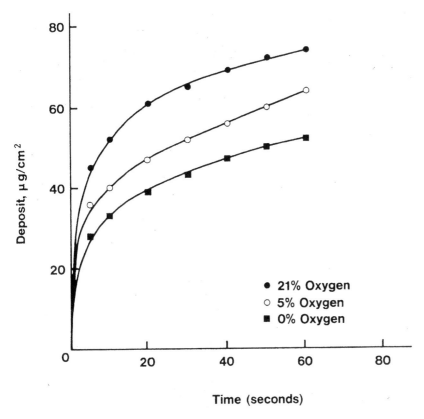

Figure 31 Chromium–nickel alloy substrates exposed to 1.5% TCP in a 200 mL/min nitrogen carrier gas with different oxygen concentrations at 700°C. (From Ref. 86.)

breakage in phosphate esters claimed by L'homme et al. [123] and that β-hydride elimination is indeed seen for the decomposition of those nonaromatic phosphate esters that have β hydrogens. In this case, C–O bond cleavage then produces almost entirely the corresponding unsaturated hydrocarbon. Studies by others on various metal single crystals with phosphite esters [107,108,127] and their corresponding alcohols [128,129] were also used to verify this β-hydride elimination mechanism and, furthermore, any carbon deposition. The acknowledgment that such β-hydride elimination produces double bonds has also recently been invoked to explain antiwear effectiveness for an acid nonaromatic phosphite ester in ferrous pin-on-disc tribological testing at relatively high temperatures under boundary, but not EP, conditions [130]. The corrosive wear caused by too rapid elimination and too little of the hydrocarbon moiety appeared to be controlled by poisoning of the active surface sites for film growth by any unsaturated hydrocarbon. Clearly, film growth was slowed by such poisoning in the experiments reported above on chlorinated hydrocarbons [52,62]. Nonetheless, the overall conclusion that the aromatic phosphate (and phosphite) esters leave more carbon on the surface than nonaromatic ones is consistent with the report that beneficial graphitic carbon inclusions are found in the phosphate glass films produced by TCP for vapor phase lubrication of rolling contacts [102]. As indicated above, these inclusions, however, may not be beneficial for EP sliding contacts.

V. SUMMARY

It is evident that the complete understanding of tribochemical properties requires an integration of a wide range of strategies from fundamental surface studies in ultrahigh vacuum on clean, well-characterized substrates to tribological experiments to measure wear rates, friction coefficients, and film properties under more realistic conditions. Even for relatively simple systems such as chlorinated hydrocarbons, the synthesis of all of these facets is a challenging problem. In the case of more complex additives such as phosphate esters, defining the physical parameters such as the nature and structures of the interface, the temperature at the tribological contact combined with the large number of reaction pathways available for the additives offers the tribochemist formidable challenges in understanding these systems. In spite of these challenges, the relatively recent availability of a wide range of analytical tools and experimental environments as well as more sophisticated theoretical tools suggests that these are now eminently tractable problems.

REFERENCES

1. Williams, G.C. Mechanism of action of extreme pressure lubricants. Proc. R. Soc. A. 1952, *212*, 512–515.
2. Gong, D.; Zhang, P.; Xue, Q. Studies on relationship between structure of chlorine-containing compounds and their wear and extreme-pressure behavior. Lubr. Eng. 1990, *46*, 566–572.
3. Davey, W. The extreme-pressure lubricating properties of some chlorinated compounds as assessed by the four-ball machine. J. Inst. Pet. 1948, *32*, 590–595.
4. Dorinson, A. The additive action of some organic chlorides and sulfides in the four-ball lubricant test. ASLE Trans. 1973, *16*, 22–31.
5. Mould, R.W.; Silver, H.B.; Syrett, R.J. Investigations of the activity of cutting oil additives III. Oils containing both organochlorine and organosulfur compounds Wear 1973, *26*, 27–37.
6. Kotvis, P.V. Overview of the chemistry of extreme-pressure additives. Lubr. Eng. 1986, *42*, 363–366.
7. Shaw, M.C. The metal cutting process as a means of studying the properties of extreme pressure lubricants. Ann. N. Y. Acad. Sci. 1950–1951, *53*, 962–978.
8. Lauer, J.L.; Benoy, P.A.; Vleck, B.L.; Calabrese, S.J. Formation of adsorbed and chemically reacted sulfur films on steel surfaces during sliding. Tribol. Trans. 1990, *33*, 586–594.
9. Sakurai, T.; Okabe, H.; Takanishi, Y. A kinetics study of the reaction of labeled sulfur compounds in binary additive system during boundary lubrication. ASLE Trans. 1967, *10*, 91–101.
10. Sakurai, T.; Sato, K. Study of the corrosivity and correlation between chemical reactivity and load carrying capacity of oils containing extreme pressure agents. ASLE Trans. 1966, *9*, 77–87.
11. Plaza, S. The studies of dibenzyl disulfide tribochemical reactions in the presence of other additives. Tribol. Trans. 1989, *32*, 70–76.
12. Komatsuzaki, S.; Nakano, F.; Uematsu, T.; Narahara, T. Antiweld properties of metal-forming lubricants using extreme-pressure agents. Lubr. Eng. 1983, *41*, 98–102.
13. Fifth Annual Report on Carcinogens, National Toxicology Program (51 FR 36607). US Dept. of Health and Human Services, Washington DC, 1986.
14. Keller, D.V. Jr. Adhesion, friction, wear, and lubrication by modern surface science techniques. J. Vac. Sci. Technol. 1971, *9*, 133–142.
15. Tonner, B.P. Energy-filtered imaging with electrostatic optics for photoelectron microscopy. Nucl. Instrum. Methods., 1990, *A291*, 60–66.
16. Wagner, C.D.; Riggs, W.M.; Davis, L.E.; Moulder, J.F. Handbook of X-Ray Photoelectron Spectroscopy; Physical Electronic: Eden Prairie, MN, 1978.
17. Rivière, J.C. *Surface Analytical Techniques*; Oxford University Press: New York, 1990.

18. Christmann, K. *Introduction to Surface Physical Chemistry*; Springer-Verlag: New York, 1991.
19. Ertl, G.; Küppers, J. *Low Energy Electrons and Surface Chemistry*; 2nd Ed; VCH: Deerfield Beach, FL, 1985.
20. Woodruff, D.P.; Delchar, T.A. *Modern Techniques of Surface Science*; Cambridge University Press: New York, 1986.
21. Shirley, D.A. ESCA. Adv. Chem. Phys. 1973, *23*, 85–159.
22. Auger, P. Sur l'effect photoélectrique composé. J. Phys. Radium 1925, *6*, 205–208.
23. Javier, Lara; Tysoe, W.T. The interaction of effusive beams of methylene chloride and chloroform with clean iron: tribochemical reactions explored in ultrahigh vacuum. Langmuir 1998, *14*, 307–312.
24. Knotek, M.L.; Feibelman, P.J. Stability of ionically bonded surfaces in ionizing environments. Surf. Sci. 1978, *90*, 78–90.
25. Griffiths, P.R. *Chemical Infrared Fourier Transform Spectroscopy*; John Wiley and Sons: New York, 1975.
26. Suëtaka, W.; Yates, J.T. *Surface Infrared and Raman Spectroscopy: Methods and Applications*; Plenum Press: New York, 1995.
27. Weber, W.H.; Merlin, R. *Raman Scattering in Materials Science*; Springer: New York, 2000.
28. Long, D.A. *Raman Spectroscopy*; McGraw-Hill: New York, 1976.
29. Greenler, R.G. Infrared study of adsorbed molecules on metal surfaces by reflection techniques. J. Chem. Phys. 1996, *44*, 310–316.
30. Colthup, N.B.; Daly, L.H.; Wiberley, S.E. *Introduction to Infrared and Raman Spectroscopy*; Academic Press: New York, 1964.
31. Lara, J.; Surerus, K.; Kotvis, P.V.; Contreras, M.E.; Rico, J.L.; Tysoe, W.T. The surface and tribological chemistry of carbon disulfide as an extreme-pressure additive. Wear 2000, *239*, 77–82.
32. Cheong Cheng, U.; Stair, P.C. In situ study of multialkylated cyclopentane and perfluoropolyalkyl ether chemistry in concentrated contacts using ultraviolet Raman spectroscopy. Tribol. Lett. 1998, *4*, 163–170.
33. Lang, G. Mössbauer spectroscopy of haem proteins. Q. Rev. Biophys. 1970, *3*, 1–60.
34. Kotvis, P.V.; Lara, J.; Surerus, K.; Tysoe, W.T. The nature of the lubricating film formed by carbon tetrachloride under conditions of extreme pressure. Wear 1996, *201*, 10–14.
35. Suryanarayana, C.; Norton, M.G. *X-Ray Diffraction: A Practical Approach*; Plenum Press: New York, 1998.
36. Warren, B.E. *X-Ray Diffraction*; Addison-Wesley Pub. Co.: Reading, MA, 1969.
37. Swanson, H.E., Ed.; (1960). *Standard X-Ray Diffraction Powder Patterns*; U.S. Dept of Commerce: Washington, DC, 1960.
38. Blok, H. General discussion on lubrication? Inst. Mech. Eng. 1937, *2*, 222–230.
39. Atkins, P.W. *Physical Chemistry*; 3rd Ed.; W. H. Freeman: New York, 1986.
40. Cabrera, N.; Mott, N.F. Theory of the oxidation of metals. Rep. Prog. Phys. 1949, *12*, 163–184.
41. Redhead, P.A. Thermal desorption of gases. Vacuum 1962, *12*, 203–211.
42. Carter, G. Thermal resolution of desorption energy spectra. Vacuum 1962, *12*, 245–254.
43. McCarrol, B. Analysis of thermal desorption spectra. J. Appl. Phys. 1969, *40*, 1–9.
44. Madey, T.E.; Yates, J.T. Jr. Desorption methods as probes of kinetics and bonding at surfaces. Surf. Sci. 1977, *63*, 203–231.
45. Kaltchev, M.; Celichowski, G.; Lara, J.; Tysoe, W.T. A molecular beam study of the tribological chemistry of carbon tetrachloride on oxygen-covered iron. Tribol. Lett. 2000, *9*, 161–166.
46. Bowden, F.P.; Riddler, K.E.W. The physical properties of surfaces: III. The surface temperature of sliding metals. The temperature of lubricated surfaces. Proc. R. Soc. A 1936, *154*, 640–656.
47. Bowden, F.P.; Tabor, D. *The Friction and Lubrication of Solids, Part 1*; Oxford University Press: London, 1971, Chapter 2.
48. Blunt, T.J.; Kotvis, P.V.; Tysoe, W.T. Determination of interfacial temperatures under extreme pressure conditions. Tribol. Lett. 1996, *2*, 221–230.

49. Archard, J.F.; Hirst, W. The wear of metals under unlubricated conditions. Proc. R. Soc. A 1956, *236*, 397–402.

50. Rabinowitz, E. *Friction and Wear of Materials*; Wiley: New York, 1965.

51. Ernst, H.; Merchant, M.E. Proceedings Special Summer Conference on Friction and Surface Finish. MIT Report 1940; Vol. 15, 76–91.

52. Kotvis, P.V.; Huezo, L.A.; Tysoe, W.T. The surface chemistry of methylene chloride on iron: a model for chlorinated hydrocarbon lubricant additives. Langmuir 1993, *9*, 467–474.

53. Huezo, L.; Kotvis, P.V.; Crumer, C.; Soto, C.; Tysoe, W.T. Surface chemistry and extreme pressure lubricant properties of chloroform on iron. Appl. Surf. Sci. 1994, *78*, 113–122.

54. Kotvis, P.V.; Millman, W.S.; Huezo, L.; Tysoe, W.T. The surface decomposition and extreme pressure tribological properties of highly chlorinated methanes and ethanes on ferrous surfaces. Wear 1991, *147*, 401–419.

55. Kotvis, P.V.; Tysoe, W.T. The surface chemistry of chlorinated hydrocarbon lubrication additives. Appl. Surf. Sci. 1989, *40*, 213–221.

56. Huezo, L.A.; Soto, C.; Crumer, C.; Tysoe, W.T. Growth kinetics and structure of film formation by the thermal decomposition of methylene chloride on iron. Langmuir 1994, *10*, 3571–3576.

57. Kotvis, P.V.; Huezo, L.; Millman, W.S.; Tysoe, W.T. Surface decomposition and extreme pressure tribological properties of highly chlorinated methanes and ethanes on ferrous surfaces. Applications of Surface Science and Advances in Tribology: Experimental Approaches, ACS Symposium Series 1992, No. 485. American Chemical Society, Washington, DC, 144–155.

58. Lespade, P.; Al-Jishi, R.; Dresselhaus, R. Model for light scattering from incompletely graphitized carbons. Carbon 1982, *20*, 427–431.

59. Knight, D.S.; White, W.B. Characterization of diamond films by Raman spectroscopy. J. Mater. Res. 1989, *4*, 385–393.

60. Bowden, F.P.; Tabor, D. *The Friction and Lubrication of Solids, Part 1*; Oxford University Press: London, 1971, Chapter 5.

61. Merchant, M.E. The mechanism of static friction. J. Appl. Phys. 1940, *11*, 230.

62. Tysoe, W.T.; Surerus, K.; Lara, J.; Blunt, T.J.; Kotvis, P.V. The surface chemistry of chloroform as an extreme pressure lubricant additive at high concentrations. Tribol. Lett. 1995, *1*, 39–46.

63. Lara, J.; Tysoe, W.T. The surface and tribological chemistry of carbon tetrachloride on iron. Tribol. Lett. 1999, *6*, 195–198.

64. Constable, F.H. Sulphide colours on metallic copper. Proc. R. Soc. A 1929, *125*, 630–637.

65. Lara, J.; Blunt, T.; Kotvis, P.; Riga, A.; Tysoe, W.T. The surface chemistry and extreme-pressure lubricant properties of dimethyl disulfide. J. Phys. Chem. 1998, *B102*, 1703–1709.

66. Kaltchev, M.; Kotvis, P.V.; Lara, J.; Blunt, T.J.; Tysoe, W.T. A molecular beam study of the tribological chemistry of dialkyl disulfides. Tribol. Lett. 2001, *10*, 45–50.

67. Godfrey, D. The lubrication mechanism of tricresyl phosphate on steel. ASLE Trans. 1965, *8*, 1–11.

68. Wang, S.S.; Tung, S.C.; S.C. Using electrochemical and spectroscopic techniques as probes for investigating metal–lubricant interactions. Tribol. Trans. 1990, *33*, 563–572.

69. Groeneweg, M.; Hakim, N.; Barber, G.C.; Klaus, E. Vapor delivered lubrication of diesel engines—cylinder kit rig simulation. Lubr. Eng. 1991, *47*, 1035–1039.

70. Rao, A.M.N. Vapor-phase lubrication: application-oriented development. Lubr. Eng. 1996, *52*, 856–862.

71. Placek, D.G.; Shankwalkar, W.G. Phosphate ester surface treatment for reduced wear and corrosion protection. Wear 1994, *173*, 207–217.

72. Quin, L.D. *A Guide to Organophosphorus Chemistry*; Wiley-Interscience: New York, 2000.

73. Kawamura, M.; Moritani, H.; Esaki, Y.; Fujita, K. The mechanism of synergism between sulfur- and phosphorus-type EP additives. ASLE Trans. 1986, *29*, 451–456.

74. Rizvi, S.Q.A. Lubricant additives and their function. In: *ASM Handbook. Friction, Lubrica-*

tion, and Wear Technology; Blau, P.J., Ed., ASM International: Metals Park, OH, 1992; Vol. 18, 98–112.

75. McDonald, R.A.; Burt, G.D. Phosphorus Amine Lubricant Additives. US Patent 5,348,670, Sept. 20, 1994

76. Beeck, O.; Givens, J.W.; Williams, E.C. On the mechanism of boundary lubrication: II. Wear prevention by addition agents. Proc. R. Soc. A 1941, *177*, 103–118.

77. Barcroft, F.T.; Daniel, S.G. The action of neutral organic phosphates as EP additives. J. Basic Eng. 1965; 761–770.

78. Riga, A.; Cahoon, J.; Pistillo, W.R. Organophosphorus chemistry structure and performance relationships in FZG gear tests. Tribol. Lett. 2000, *9*, 219–225.

79. Kotvis, P.V. Unpublished results

80. Hanyaloglu, B.F.; Graham, E.E.; Oreskovic, T.; Hajj, C.G. Vapor phase lubrication of high temperature alloys. Lubr. Eng. 1995, *51*, 503–508.

81. Schuetzle, D.; Carter, R.O.; Shyu, J.; Dickie, R.A.; Holubka, J.; McIntyre, N.S. The chemical interaction of organic materials with metal substrates: Part I. ESCA studies of organic phosphate films on steel. Appl. Spectrosc. 1986, *40*, 641–649.

82. Carter, R.O. III; Gierczak, C.A.; Dickie, R.A. The chemical interaction of organic materials with metal substrates: Part II. FT-IR studies of organic phosphate films on steel. Appl. Spectrosc. 1986, *40*, 649–655.

83. Bieber, H.E.; Klaus, E.E.; Tewksbury, E.J. A study of tricresyl phosphate as an additive for boundary lubrication. ASLE Trans. 1968, *11*, 155–161.

84. Nakayama, K. Tribochemical behaviour and lubrication characteristics of aromatic compounds. Lubr. Sci. 1993, *5*, 113–127.

85. Wang, S.S.; Tung, S.C. A reaction mechanism for producing low-friction iron phosphate coatings. Tribol. Trans. 1991, *34*, 45–50.

86. Makki, J.F.; Graham, E.E. Vapor phase deposition on high temperature surfaces. Tribol. Trans. 1990, *33*, 595–603.

87. Wang, S.S.; Tung, S.C. A new technique to enhance film-coating processes in oil-based media. Tribol. Trans. 1994, *37*, 175–181.

88. Gauthier, A.; Montes, H.; Georges, J.M. Boundary lubrication with tricresyl phosphate. ASLE Trans. 1982, *25*, 445–455.

89. Morrison, R.T.; Boyd, R.N. *Organic Chemistry*; Allyn and Bacon: Boston, 1959.

90. Kaell, J.C.; Jeannot, F.; Gleitzer, C. Étude de la reduction managee de $Fe_3(PO_4)_2$ et $Fe_9(PO_4)O_8$. Ann. Chim. Fr. 1984, *9*, 169–180.

91. Shaw, M.C. On the reaction of metal cutting fluids at low speeds. Wear 1958/1959, *2*, 217–227.

92. Morales, W.; Hanyaloglu, B.; Graham, E.E. Infrared analysis of vapor phase deposited tricresylphosphate (TCP). NASA Tech. Memo, Scientific and Technical Information; Hanover, MD, Jan. 1994; *106423*.

93. Komatsuzaki, S.; Narahara, T. An examination of antiweld film formed by reaction between metal and extreme-pressure agents in metal forming. Lubr. Eng. 1985, *41*, 543–549.

94. Forster, N.H. Rolling contact testing of vapor phase lubricants: Part III. Surface analysis. Tribol. Trans. 1999, *42*, 1–9.

95. Hanyaloglu, B.; Graham, E.E. Vapor phase lubrication of ceramics. Lubr. Eng. 1994, *50*, 814–820.

96. Forster, N.H.; Trivedi, H.K. Rolling contact testing of vapor phase lubricants: Part II. System performance evaluation. Tribol. Trans. 1997, *40*, 493–499.

97. Blunt, T.J.; Kotvis, P.V.; Tysoe, W.T. The surface chemistry of chlorinated hydrocarbon Lubricant additives: Part II. Modeling the tribological interface. Tribol. Trans. 1998, *41*, 129–139.

98. Miklozic, K.T.; Graham, J.; Spikes, H. Chemical and physical analysis of reaction films formed by molybdenum dialkyl-dithiocarbamate friction modifier additive using Raman and atomic force microscopy. Tribol. Lett. 2001, *11*, 71–81.

99. Rounds, F. Effects of organic molybdenum compounds on the friction and wear observed with ZDP-containing lubricant blends. Tribol. Trans. 1990, *33*, 345–354.

100. Weast, R.C.; Lide, D.R. *CRC Handbook of Chemistry and Physics*; CRC Press: Boca Raton, 1989–1990.

101. Antara Lubricants for Metalworking Fluids. GAF Corp.: New York, 1974; Werner, J. "Lubrhophos" Phosphate Ester Surfactants Replace Chlorine in Metal Working. *Techn. Info. CR 510 SRF 12/90*; Rhone-Poulenc: Cranbury, NJ, 1990; 1–4.

102. Forster, N.H. Rolling contact testing of vapor phase lubricants: Part IV. Diffusion mechanisms. Tribol. Trans. 1999, *42*, 10–20.

103. Raghavan, V. *Phase Diagrams of Ternary Iron Alloys, Part 3: Ternary Systems Containing Iron and Phosphorus*; Indian Institute of Metals: Calcutta, 1988.

104. Makki, J.; Graham, E. Formation of solid films from the vapor phase on high temperature surfaces. Lubr. Eng. 1991, *47*, 199–206.

105. Nebergall, W.H.; Schmidt, F.C.; Holtzclaw, H.F. *General Chemistry*; Heath: Lexington, MA, 1968.

106. Klaus, E.E.; Phillips, J.; Lin, S.C.; Wu, N.L.; Duda, J.L. Structure of films formed during the deposition of lubrication molecules on iron and silicon carbide. Tribol. Trans. 1990, *33*, 25–32.

107. Holbert, A.W.; Batteas, J.D.; Wong-Foy, A.; Rufael, T.S.; Friend, C.M. Passivation of Fe(110) via phosphorus deposition: the reactions of trimethylphosphite. Surf. Sci. 1998, *401*, L437–L443.

108. Ren, D.; Zhou, G.; Gellman, A.J. The decomposition mechanism of trimethylphosphite on Ni(111). Surf. Sci. 2001, *475*, 61–72.

109. Cutiongco, E.C.; Chung, Y. Prediction of scuffing failure based on competitive kinetics of oxide formation and removal: application to lubricated sliding of AISI 52100 steel on steel. Tribol. Trans. 1994, *37*, 622–628.

110. Lara, J.; Kotvis, V.; Tysoe, T. The surface chemistry of chlorinated hydrocarbon extreme pressure lubricant additives. Tribol. Lett. 1997, *3*, 303–310.

111. *Baker, H., Ed.; (1992). ASM Handbook, Alloy Phase Diagrams*; ASM International: Metals Park, OH, 1992.

112. Placek, D.G.; Freiheit, T. Progress in vapor phase lubrication technology. J. Eng. Gas Turbine Power 1993, *115*, 700–705.

113. Komatsuzaki, S.; Nakano, F.; Uematsu, T.; Narahara, T. Antiweld properties of metal-forming lubricants using extreme-pressure agents. Lubr. Eng. 1985, *41*, 98–102.

114. Saiki, H.; Ngaile, G.; Ruan, L. Characterization of adhesive strength of phosphate coatings in cold metal forming. J. Tribol. 1997, *119*, 667–671.

115. Ohmori, T.; Kitamura, K.; Danno, A.; Kawamura, M. A cold forging oil containing phosphorus type EP additives. Tribol. Trans. 1991, *34*, 458–464.

116. Suh, N.P. *Tribophysics*; Prentice-Hall: Englewood Cliffs, NJ, 1986.

117. Reed-Hill, R.E. *Physical Metallurgy Principles*; PWS Publishers: Boston, MA, 1973; 421 pp.

118. Kawamura, M.; Fujita, K. Organic sulphur and phosphorus compounds as extreme pressure additives. Wear 1981, *72*, 45–53.

119. Han, D.H.; Masuko, M. Elucidation of the antiwear performance of several organic phosphates used with different polyol ester base oils from the aspect of interaction between additive and base oil. Tribol. Trans. 1998, *41*, 600–604.

120. Komatsuzaki, S.; Narahara, T. Cold forming of steel with lubricating oil. Lubr. Eng. 1996, *52*, 259–266.

121. Komatsuzaki, S.; Uematsu, T. Lubricating oils for cold backward extrusion of aluminum. Lubr. Eng. 1997, *53*, 29–34.

122. Smentkowski, V.S.; Cheng, C.C.; Yates, J.T., Jr. The interaction of carbon tetrachloride with Fe(110): a system of tribological importance. Langmuir 1990, *6*, 147–158.

123. L'homme, V.; Bruneau, C.; Soyer, N.; Brault, A. Thermal behaviour of some organic phosphates. Ind. Eng. Chem. Prod. Res. Dev. 1984, *23*, 99–102.

124. Marino, M.P.; Placek, D.G. Phosphate esters. In *CRC Handbook of Lubrication and Tribology, Monitoring, Materials, Synthetic Lubricants, and Applications*; Booser, E.R. , Ed.; CRC Press: Boca Raton, 1994; Vol. III, 269–286.

125. Wheeler, D.R.; Faut, O.D. The adsorption and thermal decomposition of tricresylphosphate (TCP) on iron and gold. Appl. Surf. Sci. 1984, *18*, 106–122.
126. Sung, D.; Gellman, A.J. Thermal decomposition of tricresylphosphate isomers on iron. Tribol. Lett. 2002, *13* (1), 9–14.
127. Ren, D.; Gellman, A.J. Initial steps in the surface chemistry of vapor phase lubrication by organophosphorus compounds. Tribol. Lett. 1999, *6*, 191–194.
128. Ren, D.; Gellman, A.J. Reaction mechanisms in organophosphate vapor phase lubrication of metal surfaces. Tribol. Int. 2001, *34*, 353–365.
129. Ren, D.; Gellman, A.J. The carbon deposition mechanism in vapor phase lubrication. Tribol. Trans. 2000, *43*, 480–488.
130. Minami, I.; Hong, H.S.; Mathur, N.S. Antiwear Mechanism of Dialkyl Hydrogen Phosphites. STLE Presentation 99-NP. Society of Tribologists and Lubrication Engineers, Park Ridge, IL.

11
Tribochemistry of Boundary Lubrication Processes

Ilia A. Buyanovsky and Zinaida V. Ignatieva
Mechanical Engineering Research Institute, Russian Academy of Sciences, Moscow, Russia

Ruvim N. Zaslavsky
Venchur-N Ltd., Moscow, Russia

I. BOUNDARY LUBRICATION AND BOUNDARY LAYERS IN TRIBOLOGICAL CONTACT

Boundary lubrication practically occurs in any lubricated friction units of machines and mechanisms. This mode of friction can be realized either in particular operating periods (such as start–stop, dead points in reciprocal motion, etc.) or permanently (at low speeds and high pressures and temperatures, and when contact geometry rules out of hydrodynamic lubrication). According to Gee and Rowe, "boundary lubrication is a condition of lubrication in which the friction and wear between two surfaces in relative motion are determined by the properties of the surfaces and by the properties of the lubricant rather than bulk viscosity" [1].

Boundary lubrication takes place when rubbing surfaces are not completely separated by a thick film of an original lubricant, but the immediate contact of rubbing elements is prevented or at least restricted by boundary lubricating layers of a different origin. These lubricating layers are formed as a result of the interaction (physical–chemical, colloidal–chemical, or chemical) between the rubbing surfaces and the lubricant, and promote decrease in wear of rubbing parts and prevent seizure of friction units.

W. B. Hardy found out that boundary films consist of oriented layers of absorbed molecules of lubricant active components, which are formed and strongly influenced by the solid phase field of force. Friction in this case is controlled both by the ability of adsorbed films to screen off the solid phase field of force and by the strength of boundary film molecules bonding to contacting surfaces [2].

He revealed that these boundary layers are not squeezed from the contact zone in a rather wide range of loads; their properties substantially differ from the bulk properties of the lubricating material and are determined in the same degree by the properties of rubbing materials and that of lubricants. The viscosity of boundary layers essentially differs from the bulk viscosity of lubricating media [3]; these layers possess some real elasticity and can be considered as quasi solid and quasi crystalline bodies [4].

On the other hand, some metals (such as copper, cadmium, zinc), while rubbing in fatty acids or in the solutions of these acids, form boundary layers consisting of soaps of corresponding metals [5]. These layers possess high longitudinal cohesion and high adhesion to metal surfaces, which result in their effective lubricating properties under moderate loads and temperatures. In the last 30–40 years, a trend appeared to introduce in oils for hypoid gears some chemically active additives that improve their antiseizure properties under high loads and temperatures (extreme pressure additives, EP). The mode of lubrication that resulted from the specific action of EP additives was thus named EP lubrication. This term, opposite to boundary lubrication by adsorbed layers, describes the lubrication of friction units by a chemically modified layer, which is formed as a result of chemical interaction between active components of EP additives (sulfur, chlorine, etc.) and metal surfaces. The modified layer also restricts or prevents immediate metallic contact of rubbing elements. It should be pointed out that the modified layers ensure the transition from severe adhesive wear to a milder chemical one [6].

Lately, the phenomenon of selective transfer (ST) for steels rubbing against some bronzes and brasses in glycerin and a number of other reducing lubricating media was found. Under conditions of ST, the value of friction coefficient decreases by a factor of 10 and that of wear rate by a factor of 10^2 and more as compared to typical heavy loading conditions of boundary lubrication [7]. Although many authors consider the ST process to be totally unlike boundary lubrication, there is a good reason to consider it as one of the modes of boundary lubrication. Really, the selectively transferred films that formed as a result of the interaction between contacting materials and a lubricant fit well the above-mentioned definition of boundary lubrication.

Friction under boundary lubrication is always accompanied by wear of rubbing parts. The wear rate is usually in the range of 10^{-9}–10^{-10}, but for some modes of boundary lubrication, the rate decreases up to 10^{-11}–10^{-12}. The wear debris may include both the products of lubricant–contacting metals interaction and particles of the latter. The friction coefficients under boundary lubrication, depending on contacting materials, lubricants, and operating conditions, vary in rather wide ranges. The values of friction coefficient are usually in the 0.05–0.15 range, whereas in particular cases these values can approach the values typical for fluid lubrication (0.01–0.005).

The areas of boundary lubrication that exist for distributed and concentrated contacts are schematically given in the Hersey–Striebeck diagram and in the Begelinger transition diagram, respectively (Fig. 1) [8].

The capability of boundary layers to diminish friction losses, to decrease wear, and to prevent seizure can be explained by the following:

- As a rule, boundary films separate the rubbing surfaces to a distance higher than the adhesion forces radius of action. This results in a significant decrease of adhesion interaction between contacting elements, as the adhesion forces tend to sharply decrease with the increase in distance between bodies (to the power 3 or 4). Even a boundary layer of one molecule thickness essentially diminishes adhesion. As the real thickness of these layers is one-tenth of a micron, their effect on adhesion interaction is obvious.
- Boundary layers are characterized by an essential anisotropy of mechanical properties. Very thin ones are capable of bearing high pressures without failure. At the same time, even relatively small tangential efforts cause shearing in boundary layers at the planes of easy sliding. It provides both minimal energy losses and high load-bearing capacity of boundary films.

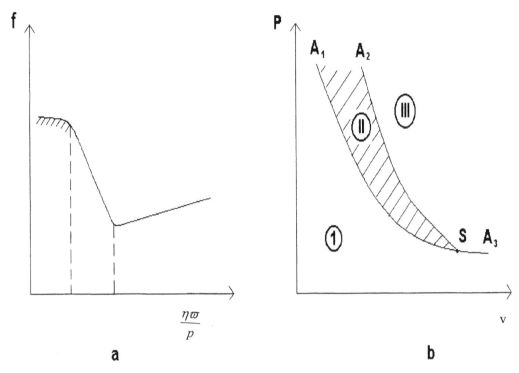

Figure 1 Regions of boundary lubrication existence: (a) on the Hersey–Streebeck diagram; (b) on the Begelinger and de Gee transition diagram (f = coefficient of friction; ω = angular velocity; η = dynamic viscosity; p = nominal contact pressure; P = load; v = speed). (Part b is reprinted from Ref. 8.)

- Active components of lubricating materials interacting with rubbing surfaces cause the adsorption plasticization of these surfaces (the so-called Rhebinder's effect), the selective dissolution of some alloying components in thin surface layers, as well as the reverse transfer of the material of softened surface layers from one contacting element to another. All these processes result in the decrease of shear strength of rubbing materials.

The listed factors give rise to tangential stresses (shearing deformation) localization in the boundary layers or in the finest surface layers of rubbing elements, which prevents failure in the body of these elements.

Boundary layer formation occurs because of a number of interconnected tribochemical processes. In other words, these layers arise as a result of activation by frictional interactions between components of a lubricant and contacting materials. It should be outlined that chemical reactions and adsorption–desorption processes under friction significantly differ from analogous thermoinduced processes [9].

Elastic–plastic deformations of surface layers under friction are the key activation factors. The work of friction forces is mainly spent for heat generation, which itself is a very powerful activation factor, and for physical–chemical and structural changes in surface layers of contacting elements. Among these changes, we can mention dispersion of the surface structure up to its amorphism, increase of different defects because of distortion and partial breakdown of crystalline lattices, oxidation, appearance of chemical compounds of nonstoichiometric formula, etc. Microcracks form in surface layers and fresh juvenile areas of metal surfaces arise, which result in emission of exoelectrons and high-energy electrons.

Nonuniform deformation and heating during friction superimposed on initial non-uniformity of rubbing surfaces result in their heterogeneity and correspondingly to activation of electrochemical processes. Friction leads to intensification of diffusion processes. Metal oxides and especially freshly naked metal because of wear have a catalytic action on chemical transformations in the zones of contact.

Under friction, mechanical activation of chemical processes in lubricants occurs, which leads to serious changes in them. So, in the contact zones, which are characterized both by high concentration of energy and anomalous state of material, the chemical reactions can take place, which usually require higher activation energies and higher temperatures for their realization under static conditions. Moreover, in many cases under static conditions, such reactions are thermodynamically improbable or even impossible. Such reactions under friction are called tribochemical reactions [9–11]. The mechanical activation usually taking place under friction could activate chemical reactions of various natures, including metathesis, decomposition, and fusion reactions [12].

II. THE SUCCESSION OF TRIBOCHEMICAL PROCESSES AND KINETIC MODELS DESCRIBING THEIR REALIZATION

Spikes [13] wrote "despite its great practical importance, boundary lubrication remains by far the least well understood lubricating regime." Nevertheless, long-term experimental works in the field of boundary lubrication allowed studying of the phenomenology of the process, formulating some concept overviews about its mechanism, and developing a number of its models. It should be taken into account that the changes in operating regimes of a friction unit under boundary lubrication condition in some cases can result in an abrupt change in friction and wear behavior. Let us consider the succession of the processes occurring under boundary lubrication, including both the transitions from normal wear to surface damages and the transitions from one mechanism ensuring the normal wear to another. The structural scheme of the process of boundary lubrication in the wide ranges of loading conditions of a friction unit (Fig. 2) is rather clearly shown [14].

The given scheme demonstrates the progressive stages of the process, which can be characterized by subsystems from the first to the seventh order. It is possible to subdivide these subsystems to the following main groups: O—interaction of friction surfaces before the beginning of the tribological process; A—low duty conditions of friction, i.e., low loads and moderate temperatures on friction contact; B—heavy duty conditions, i.e., high contact loads and temperatures; C—catastrophic conditions, i.e., severe loading regimes resulting in seizure, scuffing, and failure of a friction unit. When operation conditions correspond to the groups O and A (subsystems 2 and 3), the boundary films are forming because of adsorption (tribosorption) of active components of a lubricant by contacting surfaces.

The boundary films separate the surfaces and eliminate, or at least localize, metal contact in single contact points. In the last case, the friction coefficient would be determined both by the portion of lubricated contact and that of immediate metallic contact [6]:

$$f = \alpha f_m + (1 - \alpha)f_1, \tag{1}$$

Here f is the friction coefficient under boundary lubrication; α is a portion of metallic contact (fractional film defect), which tends to increase with temperature growth; and f_m and f_1 are

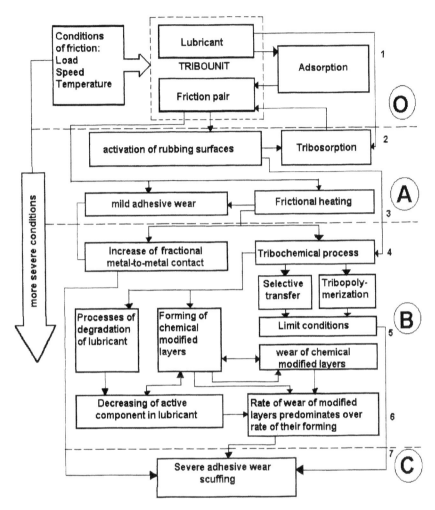

Figure 2 Structural scheme of the interaction of friction-determinative parameters under boundary lubrication. (From Ref. 14.)

the friction coefficients during metallic contact and contact through an undamaged boundary lubricating film, respectively.

In the regions of metallic contact, a moderate adhesive wear is observed in single contacting microasperities tips. The smaller the portion of metallic contact, the smaller the friction losses and wear.

When the parameters of operating regime exceed some critical level, the absorbed boundary layers become invalid to provide antiwear and antifriction effects because of a sharp increase of the share of metallic contact. According to Fig. 2, there is a transition to group B (subsystems 3 and 4), wherefrom under further regime, intensifying the transition to catastrophic wear occurs (to group C).

In groups A and B, a remarkable lubricating effect can be achieved from the realization of tribochemical processes that promote such phenomena as selective transfer [7], formation of friction polymers (tribopolymers) [15], and organometallic compounds, which separate rubbing surfaces [16], etc. The effect of the said compounds on friction and wear properties depends on the composition of the lubricant and rubbing materials, as well as on the operating regime. The above-mentioned phenomena provide normal operation of a friction unit in a certain range of loads and temperatures. Beyond the stated ranges, boundary layers of the mentioned origin break down and the transition to group C takes place.

On the other hand, there is a trend nowadays to introduce in the composition of lubricants some chemically active components, which used to be decomposed under operating loads and temperatures. The products of decomposed compounds react with the surface metallic layers, which leads to the formation of modified layers with decreased shear strength. So the modified layers having the thickness high enough to separate contacting surfaces ensure the reduction in friction and the transition from the severe adhesive wear to the mild corrosive mechanical one. Here the friction coefficient can be written as

$$f = \beta f_{cm} + (1 - \beta) f_m, \tag{2}$$

where β is the share of the contact through a modified layer (at the first approach, it is a probability of the contact through this layer); f_{cm} is the friction coefficient when both rubbing surfaces are covered by a modified layer providing EP and AW lubrication; and f_m is the share of metallic contact.

Normal operation of tribounits will depend on the correlation between the rate of formation of modified layers and that of their breaking down, i.e., the wear rate.

When operating conditions become difficult or when the depletion of chemically active additives in a lubricant occurs, the wear rate can exceed the rate of formation of modified layers. It will result in the transition to catastrophic wear resulting to seizure, scoring, and scuffing, which can be attributed to group C.

However, the transition to group C can be immediately done from subsystem 4, when some critical share of metallic contact is achieved because of the desorption of surface-active components of a lubricant and in the absence of any other components, which are able to ensure the separation of contacting surfaces and minimize the share of metallic contact.

So, in all considered cases of boundary lubrication (by an adsorbed layer, by a chemically modified layer, or by a layer of another origin), under certain operating conditions, the transition to severe wear can take place. It is accompanied by an essential decrease in the energy of activation of wear process.

Let us consider the transition conditions from normal friction process to scuffing resulting in catastrophic wear and failure of a friction unit. It is possible to suggest the following succession in the stages of development of seizure (Fig. 3):

- Breaking down of lubricating boundary layers followed by baring of the active sites, which were till then the adsorption centers [Fig. 3(a)].
- Formation of adhesive bonds between naked metallic contact spots, which were activated because of the plastic deformation of rubbing bodies, and their breaking down resulting from the relative displacement of friction surfaces [Fig. 3(b)].
- The seizure of contacting elements when the number of adhesion bonds exceeds some critical value (this stage can be combined with the previous one).

Seizure may be considered as a kinetic process. To analyze the process of adhesive bond formation, it is possible to use a static approach (i.e., without taking into account the forced current of interacting materials in various directions) and describe it by kinetic equations of the first order. If we assume that the number of adhesive bonds x in the contact

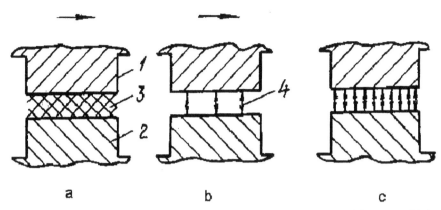

Figure 3 The stages of seizure under boundary lubrication: (a) initial stage; (b) boundary layer failure, adhesive bonds formation, and breaking down of moving surfaces; (c) formation of a critical number of adhesive bonds between surfaces and seizure arising (1,2 = contacting bodies; 3 = boundary layer; 4 = adhesive bonds). (From Ref. 17.)

zone with damaged boundary layer is proportional to the real area of contact, it is possible to calculate it at a temperature T_x using the following equation [14,18]:

$$x - \alpha^* a^* (p_a/H) t^m k_0^i \exp\left(-\frac{E_x}{RT_x}\right) \tag{3}$$

where α is a share of metallic contact; a is a number of active sites on a unit of the real area of contact; k_0^i is a preexponential factor; p_a is the specific load; t is the duration of the process ($t = d/v$, where v is the velocity of relative displacement and d is a characteristic size); m is a constant; E_x is the activation energy of the adhesive bond formation; R is the universal gas constant; T_x is the temperature of the process; and H is the hardness of the milder element of the contacting pair at the test temperature.

When some critical number of adhesive bonds $x = x_{cr}$ is achieved [Fig. 3(c)], the temperature runs up its critical values too ($T_x = T_{cr}$), which is accompanied by a sharp rise in the friction coefficient. Apparently, for any friction unit, there is a constant value x_{cr}, which does not depend on the loading regime. The value T_{cr} is suggested to be a constant for the given combination of rubbing materials and a lubricant.

The critical temperature T_{cr} can be determined from Eq. (3):

$$T_{cr} = \frac{E_x}{R\ln\left[\alpha B v^{-m}(p_a/H)\right]} \tag{4}$$

where $B = (ak_0^i(d)^m)/x_{cr}$. This value is assumed to be constant for the given combination of contacting materials and a lubricant.

A normal friction process usually occurs at temperatures below T_{cr}. It may be illustrated by the temperature dependence of friction coefficient shown in Fig. 4. One can see that there are three transition temperatures characterizing the given dependence: T_{cr1}, T_{cr2}, and T_{cm}, i.e., the temperature of chemical modification of surface layers. As mentioned above, when the temperature reaches its critical values, T_{cr1} and T_{cr2}, an abrupt rise in friction is observed accompanied by the transition to severe adhesive wear and seizure. When T_{cm} is achieved, a modified boundary layer is formed because of the chemical

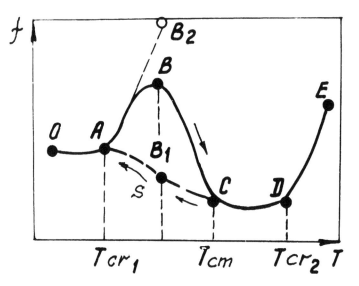

Figure 4   A generalized dependence of friction coefficient f upon the temperature T. (From Ref. 17.)

interaction between active compounds of the lubricant and the friction surfaces. Having the decreased shearing strength, this layer contributes to friction lowering and ensures the transition from severe adhesive wear to more mild corrosive mechanical wear.

The first critical temperature T_{cr1} results from the competition of adsorption and desorption processes of active compounds of a lubricant. The second one is the result of the competition between the processes of formation and failure of modified layers. So, in Eq. (4), the parameter E_x characterizes the interaction between rubbing bodies, whereas the parameter α characterizes the interaction of lubricating oil with the rubbing metals and properties of the modified layer. One of the first stages of that interaction is the adsorption of active lubricant compounds on the phase interface.

In the frames of the model under consideration, it is important to evaluate the ability of the lubricant molecules to minimize the parameter a (the so-called fractional film defect according to Kingsbury [19]). The computation of this parameter for the first critical temperature is based on the representation of a lubricant as the solution of its active components in some inactive medium. To compute the first critical temperature, the modified equation can be used:

$$\alpha = \left[b_0 C^{\nu} \exp\left(\frac{\Delta\mu}{RT_a}\right) \right]^{-1} \tag{5}$$

where b_0 is a preexponential factor in a formula to compute for the constant of exchange reaction, taking into account the change of the presentation of the active component concentration from mole portions to mass concentration; ν is an exponent; $\Delta\mu$ is the chemical potential difference of surface-active compounds in the boundary layer and in the body of the lubricant; T_a is the temperature of the surface that adsorbed the surface-active molecules. When desorption begins to prevail over adsorption, T_a becomes equal to T_{cr1}.

Submitting α from Eqs. (5) and (4) when $x = x_{cr}$, we obtain the equation for computing the first transition temperature, when the failure of the adsorbed boundary layer takes place (i.e., the first critical temperature T_{cr1}):

$$T_{cr1} = \frac{E_x + \Delta\mu}{R\ln\left[B_1 p_a (Hv^m C^v)^{-1}\right]} \qquad (6)$$

Here $B_1 = ad^m k_0^i/(x_{kr} b_0) = B/b_0$.

Analysis of Eq. (6) allows to conclude that the first critical temperature linearly depends on the value $E_x + \Delta\mu$. It appears that for the case $x = x_{cr}$ and for all factors except one variable, there are linear dependences between $\ln C$, $\ln v$, $\ln p_a$, $\ln H$ from one side, and $(E_x + \Delta\mu)/RT_{cr1}$ from another.

Under heavy-duty operating conditions, the use of EP additives can provide normal tribological characteristics of friction units. As mentioned above, EP additives form chemically modified layers on friction surfaces, which prevent immediate metallic contact of these surfaces. Having low shear strength, the chemically modified layers ensure reduction of friction and transition from severe adhesive wear, accomplished by the failure of boundary layers of adsorption or of any other origin, to milder corrosive mechanical wear.

The formation of the modified layer is a complicated multistage process. It includes as a minimum the adsorption of molecules of chemically active components of a lubricant, their decomposition, and the following interaction of the active reaction products with metallic surfaces. Adsorption and chemical interaction under boundary lubrication are characterized by different rates. The duration of these processes is in the microseconds and milliseconds [20].

Analysis of two competitive processes—modified layer formation and breaking down—shows that the α value depends on the relation of the time necessary for the formation of a modified layer with the thickness sufficient for aversion of metallic contact and the time necessary for the removal of this layer because of wear.

This relationship can be evaluated using Kingsbury's modified equation [14,19]. The wear process is assumed to yield to the Archard's law, whereas the wear coefficient is considered to be a temperature-dependent constant of the reaction rate.

Assuming the exponential law of the growth of a modified layer and computing the rate constants of competitive processes from the Arrhenius equation, it is possible to obtain the following expression to compute a share of the metallic contact [14], if both processes—formation and breaking down—are developed at the same temperature:

$$\alpha = \frac{k_a v^y}{C^n} \exp\left(\frac{E_w - E_m}{RT_p}\right) \qquad (7)$$

where k_a and y are the constants; n is the order of the process; E_m and E_w are the activation energies of the modified layer formation and breaking down, respectively.

Submitting a from Eq. (7) to Eq. (4) when $x = x_{cr}$, we obtain the equation to compute the second transition temperature, when the failure of the modified boundary layer takes place (i.e., T_{cr2}):

$$T_{cr2} = \frac{E_x - E_p}{R\ln\left[B_3 v^{y-m} C^{-n} (p_a/H)\right]}. \qquad (8)$$

Here $B_3 = ak_a/x_{cr}$; $E_p = y(E_m - E_w)$.

It is known that hardness tend to fall with increasing temperatures in friction contact. According to Eq. (8), this process causes decreasing T_{cr2}. That is why for materials rubbing under heavy loading conditions, which could result in seizure and scoring, the rate decrease in hardness in relation to temperature should be minimized. The seizure resistance of contacting materials can be improved by increase in the parameter E_x [Eqs. (6) and (8)].

However, the interaction of juvenile surfaces even of such materials under high loads and temperatures can give rise to seizure and failure of the friction unit. To prevent, or at least to postpone for some time, this interaction is made possible by the right choice of materials and chemically active lubricants, which should ensure high rate of surface modified layer formation (the value E_m is small) and high wear resistance of these layers (the value of E_u is large). The concentration C of an active component in a lubricant should be high enough up to the temperatures close to T_{cr2}.

Let us evaluate the temperature range in which modified layers prevent the immediate interaction of juvenile surfaces under boundary lubrication. The upper limit of this range is T_{cr2}. At this temperature, the modified layers are not capable anymore to prevent the formation of strong adhesive bonds. On the contrary, at the temperature T_{cm}, the modified layers are capable of separating the contacting surfaces and correspondingly to resist the intense adhesive interaction of surfaces (Fig. 4).

At the temperature in point B, the modified layer of the contacting surfaces is forming (if this layer is not formed, point B shifts to position B_2). As the temperature rises, the rate of interaction between metal and the active component of a lubricant increases. A share of the contact through the modified layer β increases, and the thickness of this layer becomes high enough to separate rubbing surfaces and to reduce friction. When the temperature T_{cm} is achieved, the value of β share becomes critical, so the metallic contact may not be taken into account.

As the modified layer is first formed on the tips of surface microasperities while these asperities are out of the contact, we assume that the probability of the contact through the modified boundary layer will be determined by the relation of the time necessary for its formation and growth up to a certain thickness at the temperature T_β to the time between successive contacts of microasperities, i.e., the time during which the layer can grow in thickness.

$$\beta = k_\beta C^n v^{-y} \left(\frac{H}{p_a} \right) \exp \left(-\frac{E_p^i}{RT_\beta} \right) \tag{9}$$

where k_β is the constant of the process, $E_p^i = yE_m$.

When $T_\beta = T_{cm}$, β reaches some critical value β_{cr}, which does not depend on the operating conditions [19]. Equation (9) can be solved for the temperature T_{cm} [14]:

$$T_{cm} = \frac{E_p^i}{R\ln(B_2 C^n v^{-y} (H/p_a)^y)} \tag{10}$$

where $B_2 = k_\beta/\ln\beta_{cr}$. The methods of computing the parameters for Eqs. (6), (8), and (10) based on laboratory test results are considered in Ref. 12.

Now we consider a simplified method for evaluating parameters of Eq. (10) based on the analysis of the descending branch BC of the curve f–T (Fig. 4). According to Eq. (2), the probability of contact through the modified layer can be expressed as:

$$\beta = (f_m - f)/(f_m - f_{cm}) \tag{11}$$

Submitting β from Eq. (9) to Eq. (11), one can obtain expression for the case when all parameters except temperature are constant:

$$\ln \beta = \ln\left[(f_m - f)/(f_m - f_{cm})\right] = [-E_p^i/RT] + \text{Const} \tag{12}$$

On the other hand, at constant values of v and T, it is possible to obtain:

$$\ln \beta = \ln \left[(f_m - f)/(f_m - f_{cm}) \right] = -n\ln C + \text{Const} \qquad (13)$$

At the constant values of C and T:

$$\ln \beta = \ln \left[(f_m - f)/(f_m - f_{cm}) \right] = -m\ln v + \text{Const} \qquad (14)$$

E_p may be considered as a tangent of the slope of the curve $\ln \beta - 1/T_{cm}$ multiplied by R, whereas n and m are the parameters of the slope of the curves $\ln \beta - \ln C$ and $\ln \beta - \ln v$, respectively.

Thus, in the range of temperatures $T_{cm}-T_{cr2}$, corrosive mechanical wear occurs. Analyzing Eqs. (8) and (10), one can conclude that to widen the range of operating temperatures without altering the said mode of wear, it is necessary to choose such combination of rubbing materials and a chemically active lubricating oil that would ensure high enough rates of tribochemical interaction to provide boundary modified layers formation. The triboreaction rate should be higher than that of modified layer destruction (i.e., low values of E_p and high C), but not as high to result in severe surface damage due to corrosive action. The surface hardness of contacting elements should be high enough too.

It is also necessary to take into account some aspects connected with the representation of tribochemical processes in the model under consideration. At first, if the mean surface temperature is lower than T_{cr1}, whereas the total surface temperature is higher than T_{cm} because of the high temperature flash, there are conditions for the formation of modified layer.

As noted above, the modified boundary layer ensures decrease in friction. Accordingly, the temperature flash decreases, which leads to the transition in the direction of the arrow s on curve CB_1A (Fig. 4). As the wearing of the modified layer occurs, friction and correspondingly the temperature flash increase again. So we deal with an oscillatory process in friction contact, if only the loading conditions are not heavy enough to provide for the transition to catastrophic wear and friction unit failure (i.e., transition to group C in Fig. 2).

The friction surfaces are as a rule covered by heterogeneous films. This heterogeneity is revealed even in the area of the same wear spot, which explains different wear resistances of local regions of surface films resulting in their various thickness. The dispersion processes along the same wear spot can obey dissimilar mechanisms depending on the wear area localization [17].

Really, in getting in contact, the temperature is relatively low, which results in the prevailing of wear rate over the rate of modified film formation. On coming out of contact, the temperature increases and, according to the theory of contact elasto-hydrodynamic lubrication, there is a peak of pressure here. This results in prevailing of the rate of modified film formation over the wear rate.

In our lab tests on the four-ball machine, we used the additive "Chloef-40"* to the base oil. The products of the interaction of active components of this additive with surface metallic layers are breaking out mainly because of a polydeformation process caused by a flood of lubricating oil, as it can be seen from the scanning electron microscope (SEM) photograph of the worn region (Fig. 5) and on the profilograms of the wear scar at different sections (Fig. 6).

---

*Dibutyl ether of trichlormethylphosphonic acid. Oil M-11 is a distillate purified oil without additives with viscosity 11.5 mm^2/sec at 100°C.

Figure 5 Microphotograph of wear scar on the lower ball of standard four-ball machine after tests in 1% solution of the additive Chloef-40 in M-11 oil; load = 200 N; time = 10 hr. (From Ref. 17.)

These conclusions are also proved by the data on composition of wear scar surface layers (Table 1), which significantly differ through the wear scar. Data on structure and composition heterogeneity of wear scar under boundary lubrication by chemically active lube oils are also presented in Ref. 21.

Together with the model of surface interaction based on data given in Fig. 5 and Eqs. (4) and (7), the presented data allow better understanding of the mechanism of lubricating action of chemically active media in the range of conditions corresponding to the BC region of the curve f–T in Fig. 4.

For example, according to Cameron's boundary lubrication theory and his experimental data, the thickness of the films protecting rubbing bodies from immediate metallic contact at the range of temperatures 95–230°C (which seems to correspond to the region AC on the curve f–t; Fig. 4) is of one and more microns. These films are organometallic-polymeric compounds in origin.

Nevertheless, the appearance of "thick films" does not contradict the model of tribochemical processes under consideration, of course, if it is possible to identify transition temperatures through the dependence of friction coefficient on temperature.

It should also be taken into account that while active metals are being tested in fatty acids or in oil solutions of fatty acids, seizure occurs at a temperature of softening of metal soaps formed because of the tribochemical interaction in the process of friction. This temperature can be considered for the given case as the first critical one [9]. This is not in

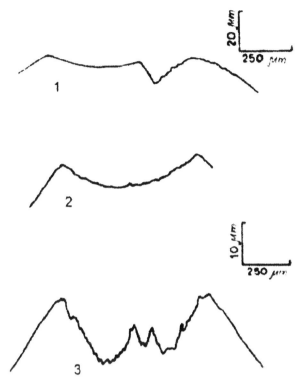

Figure 6 Profilograms of the wear scar presented in Fig. 5: (1) in the direction of movement of the ball that wears the lower ball; (2) across the worn-out area; (3) across the eroded area. (From Ref. 17.)

disagreement with the model under consideration and according to Fig. 2 leads to immediate transition of the friction unit to group C, i.e., to catastrophic wear.

At last, one has to bear in mind that under some definite circumstances, transition temperatures T_{cr1} and T_{cm} could not be realized. As seen from the data in Fig. 7, it can happen as the active component concentration grows, which leads to an increase in T_{cr1}. Under these conditions, according to Eq. (10), the value of T_{cm} decreases, so $T_{cr1} > T_{cm}$ and the transition from group A to group B occurs (Fig. 2) without the intermediate region of high friction and intense adhesive wear.

Table 1 The Content of Elements in the Surface Layer of Specimens Made of Steel ShKH-15

Surface	Lubricant	Fe (%)	Cr (%)	P (%)	Cl (%)
Initial surface	—	97.513	2.549	—	—
Wear scar	Oil	98.011	1.988	—	—
Wear scar	Oil + additive	97.559	1.889	0.545	0.035
Eroded area	Oil + additive	93.749	5.640	0.781	0.092

Tested in pure oil M-11 and in oil M-11 with 1% dibutyl ether of trichlormethylphosphonic acid (four-ball machine, load 200 N, test time—9 hours).
Source: Ref. 80.

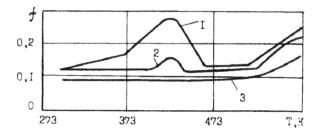

Figure 7 Coefficient of friction vs. temperature while testing in the solutions of the additive Lubrizol 5901 in ASV-10 oil: (1) 0.3%; (2) 0.7%; (3) 3.0 wt.% additive. (From Ref. 76.)

Below, we will consider in more details the mechanisms of tribochemical processes for various modes of boundary lubrication.

III. MECHANISMS OF LUBRICATION ACTION OF OILS, EXTREME PRESSURE AND ANTIWEAR ADDITIVES, AND TRIBOCHEMICAL PROCESSES THAT CONTRIBUTED TO THEIR ACTION

A. Tribochemical and Thermochemical Processes

Hardy, in his classical studies, did not take into account the possibility of changes in the characteristics of lubricant molecules adsorption during friction under boundary lubrication [2], although he considered the need of a system to ensure the optimum orientation of molecules on surfaces during latent period of boundary lubrication [22]. However, Deryaguin [23] has noted that if the orientation of molecules of active components of a lubricant on friction surfaces renders greater influence on tribological process, one can expect that the latter should strongly affect the molecule orientation. This standpoint is confirmed by the data on boundary lubrication atom force studies and computer simulation of boundary layer structures in dependence on temperature and relative displacement rate [24]. As noted above, friction activates clean rubbing surfaces to a very high level, which gives rise to hydrocarbon chemisorption on these surfaces and even to chemical interaction with them [9,10], which is impossible at the same temperatures under static conditions. Hsu and Klaus [25] studied wear products obtained when the steel balls were subjected to wear in ultrapure liquid paraffin. The authors had shown that under boundary lubrication, this oil interacted with metal in the presence of oxygen, which resulted in the formation of three types of reaction products: soluble in oil, soluble in pyridine, and insoluble in pyridine. It is of importance to outline that the wear products under study were formed at a significantly lower temperature (calculated according to the Block–Archard's equation) as compared to the temperature required for such products obtained under static tests, i.e., when we deal with the thermoactivated reaction only.

At low testing loads, the difference between the static temperatures required to obtain the same reaction products and the friction-induced temperature can reach 250°C. It is interesting to note that the authors [25] believe the catalytic effect of freshly naked rubbing metallic surfaces in this case is of no importance. Interesting data were obtained when 1% solution of dibenzyldisulfide in inactive liquid paraffin had interacted with low-chromium, ball-bearing steel [26]. The tests were performed on the four-ball machine at very low rotating speed (0.24 mm/sec), temperature of 300°C, and axial load of 108 N. The wear scars on the lower balls after 60-sec tests were subjected to examination by Auger spectroscopy.

The distribution of chemical elements through the depth of the worn area and beyond was obtained by successive sputtering of surface layers (5 nm/sec) in argon ion steam and corresponding Auger spectra examination. The analysis of the data presented in Fig. 8(a,b) shows that under friction at speeds undoubtedly insufficient to cause observable local rise of temperature, the depth of penetration of sulfur in surface layer is nearly an order more in the wear scar as compared to the surface areas beyond it, which are merely heated up to the same temperature in the lubricating medium under study. The wear scar is also characterized by an increased content of carbon and oxygen, whereas the unworn surface zones are impoverished in these elements (the content of these elements are practically very small—within the accuracy of the method).

It is an evidence of formation of thicker layers of ferrous sulfides and ferric oxides at tribochemical reactions as compared to thermochemical ones at the same temperature [Fig. 8(a,b)]. So the intensity of tribochemical reactions is higher than that of thermochemical ones. If it is possible to get rid of friction (or external) heating, friction itself does not ensure effective activation of some tribochemical reactions. When similar tests were performed without external heating, i.e., at room temperatures, the sulfur content in worn areas was remarkably less than that at 300°C [Fig. 8(c)], in spite of analogous friction conditions. At the same time, friction without heating has more activating effect on carbon- and oxygen-containing compounds formation than heating without friction [compare Fig. 8(b) and (c)]. So tribochemical reactions are usually realized as the result of a combination of a number of factors and it is very difficult to mark out a leading one among these factors.

The difference between tribochemical and thermochemical reactions was studied in Refs. 9 and 27–29. Thus, according to the data given by Buckley [28], x-ray photometric spectrums of steel surfaces exposed in chemically active media significantly differ from the ones obtained from surfaces subjected to machining or rubbing in the same media. Kajdas [27] pointed out that friction activates and accelerates chemical reactions in contact areas; he also paid attention to the fact that some tribochemical reactions, when contacting bodies are highly stressed, e.g., in vibration mills, may have no (or slight) dependence on temperature.

It is also known that the activation energies of many tribochemical reactions are less than that of thermochemical ones, whereas rates of tribochemical reactions exceed that of ordinary thermochemical reactions. Degradation of active components of a lubricant is intensified under friction as compared to merely heating (without friction) to the same temperatures [10]. So Koujarov and Rjabuchin [29] determined the output of tribochemical and chemical reaction products—complex compounds of bivalent copper for reactions of complex formation between copper and a number of salycilidenarilamines and N-benzo-ylarylamides. The rates of tribochemical reactions of complex formation exceed that of traditional chemical reactions between a compact metal and organic ligands in 2–4 orders of magnitude. At the same time, the main products of tribochemical interaction are identical to complexes produced in the process of chemical interaction. Heinicke [10] paid attention to the discrepancy of successive rows of thermal and tribochemical stability of some hydro-carbons. There is also a discrepancy of rows of radiation and tribochemical degradation of these compounds.

For this reason, in a general case, it is impossible to directly estimate the tribological behavior of lube media on the basis of the static tests results. It compels to refuse the simple models of lube action, according to which the temperature generated in the contact zone is the main and even the only factor responsible for development of tribochemical reactions (except the life of a single contact) [30,31], and to go on working out more adequate models of lubrication action. These models should take into account more acting factors and first of

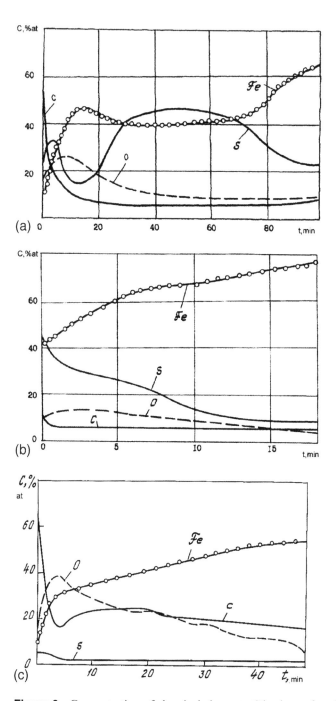

Figure 8 Concentration of chemical elements C in the surface layer of the specimen made of low chromium ball bearing steel tested in 1% solution of dibenzyldisulfide in a neutral oil at 300°C under friction (a), without friction (b), and under friction at 20°C (c) vs. the time of spraying t: Auger spectrometry, accelerating voltage of electron beam 10—kV, the diameter of the electron beam—5 µm. Figure 8 a,c is reprinted from Ref. 26 with permission from Elsevier Science.

all transition of electrons, friction heating, and catalytic effect of freshly formed surfaces on reactions under friction.

B. Mechanism of Tribochemical Processes Under Boundary Lubrication and the Negative Ion Radical Action Mechanism (NIRAM) Concept

As mentioned above, elastic–plastic deformation of contacting surface layers during friction should be attributed to the main activating factor in the tribological process. Indeed, all multiform phenomena, which are developed in the process of contact and affect both tribochemical reactions and surface layers fracture, result from the deformation processes [31–34].

It is necessary to point out the peculiarity of the stress–strain state under friction. Here the maximum stresses arise in deformed microvolumes of a very thin surface layer. According to the concept of dissipative heterogeneity proposed in Ref. 33, the friction system strives for concentration of all types of interaction in thin surface layers.

As a result, about 90–98% energy released during friction is accumulated and dissipated in secondary surface structures, in the formation of which tribochemical reactions play a significant role.* The study of the mechanism of tribochemical reactions is of fundamental importance as it allows clarifying the way by which mechanical energy generated under friction is supplied to the tribological contact that goes to work as a reactor, where these reactions come about. The present-day models connect the activation of contact zones under boundary lubrication leading to tribochemical reactions initiation mainly with transfer of electrons and formation of anions of organic compounds [9,27,37–39].[†]

In our opinion, the most accurate description of tribochemical processes can be obtained by considering the anion radical mechanism of lubricating action of oils and their components (the NIRAM concept) proposed by Kajdas. In accordance with this concept, interaction of lubricants with friction surfaces is determined mainly by two ways of friction energy transfer into the reaction zone: temperature rise in contact zones because of friction and emitting low-energy electrons (exoelectrons).[‡] It should be pointed out that a spontaneous exoelectron emission from solid surfaces occurs at their machining, work hardening, and friction, and even at desorption of previously adsorbed molecules of lubricants [28,39].[§] The emitted exoelectrons ionize the molecules of a lubricant in the immediate vicinity of exoelectron sources. This results in the formation of anions and anion radicals interacting with local zones of metal surfaces, which gain the positive charge because of exoelectron emission.

---

*According to Kostetsky et al. [33,34], secondary structures are thin surface films with thickness of 2–8 nm. These structures are formed due to kinetic phase transition resulting from the cooperative action of deformation and diffusion processes, friction heating, and chemical interaction under specific conditions of friction contact. The secondary structures result in passivating of rubbing surfaces. The boundary lubricating layer may be considered as a particular case of secondary structures.

†It is necessary to distinguish tribochemical reactions from that stimulated by catalysis. Kajdas [9] believed in the "link between tribochemistry and catalysis."

‡According to Refs. 12 and 36, the energy of exoelectrons is of 1–4 eV, whereas other authors believe that its value is of 10 eV. It should be noted for comparison that mechanoemission electrons emitted under disruption of dielectrics have the energy of tens of keV [12].

§Kajdas [9] pays lesser attention to emission of high-energy electrons, ions, components of destroyed crystalline lattice, etc. from rubbing surfaces.

Kajdas attaches particular significance to the second of the processes noted above.

The succession of boundary layer formation due to tribochemical reactions produced by friction may be presented as follows:

- Freshly formed, during friction process, local zones on rubbing metals surfaces emit exoelectrons: $Me \xrightarrow{friction} Me^+ + e^-$. The zones, which have emitted exoelectrons, obtain the positive charge.
- Exoelectrons interact with adsorbed (or with just desorbed) molecules, which leads to negative ions, ion radical, and radical creation.
- While the positively charged metal surface zones interact with anions and anion radicals, the latter are chemisorbed by these zones and the adsorption complexes are formed. The radicals produced by tribochemical reactions in turn react with the surface covered by a chemisorbed layer, which can result in the appearance of various organic products such as resins, peroxides, friction polymers, etc.
- If, in the composition of molecules of a lubricant, such chemically active elements as sulfur, phosphorous, etc. are introduced under heavy loading conditions when a chemisorbed complex is breaking down, these active elements begin to interact with the metal surface, forming an inorganic modified layer that protects metal surfaces from immediate metallic contact, severe wear, and seizure.
- The process of the modified layer renewal after breakdown develops in the same succession.

It is worthwhile mentioning that Kajdas had also analyzed the concept of acid–base interaction developed by Lewis and the principle of hard and soft acids and bases (HSAB) [40] with relation to the possibility of their application to tribochemical processes based on the NIRAM concept. He showed that in terms of the NIRAM concept, adsorption processes could be related to the typical interactions of soft acids with weak bases, whereas tribochemical processes could be attributed to the interactions of hard acids with hard bases [27]. Thereafter, the combined concept NIRAM–HSAB describing mechanism of tribochemical reactions was proposed but will not be considered in detail here.

Let us consider the application of the NIRAM concept to the process of boundary lubrication by various lubricants.

C. Tribooxidation and Tribochemical Processes in Hydrocarbon Lubricants

It is well known that tribooxidation is one of the widely spread tribochemical reactions strongly influencing the tribological behavior of contacting materials. Yet in 1957, Vinogradov observed that tribooxidation of a low-viscosity lube oil [naphtheneparaffin (NPF) fraction of diesel oil] during tests on the four-ball machine resulted in the appearance of a resin-like sediment on contact points of steel balls and in the accumulation of products of tribooxidation in NPF. The rate of formation of tribooxidation products grows with the increase of testing speed and loads. Both the products of tribooxidation and resin-like substances promote high running-in capacity and lead to degeneracy of seizure [41]. It should be outlined that NPF of residual oils reveals less turn for oxidation, possibly because of the presence of some intrinsic inhibitors of oxidation.

The resin-like sediments produced while testing steel specimens lubricated by a commercial fuel consist of oxygen-containing products such as carboxylic acids and their derivatives, complex ethers carboxylates, sulfoxides, sulfone acids, sulfates, and sulfones. Systematic studies of friction of lubricated bodies in various gaseous surroundings show that the effect of gases on antiwear and antiscoring characteristics of lubricants is mainly related with the presence of molecular oxygen and with oxidation processes activation or deceleration. Oxidation on surface layers depends on the presence of molecular oxygen in friction zones and lubricants and the capacity of rubbing materials to oxidize. As it is well known, oxidation leads to the formation of protective oxide layers resulting in friction and wear reduction and prevention of seizure and scoring.

So the content of oxygen in environments controls the process of modified layer formation and consequently the tribological characteristics of the system, which are determined by this process [42,43]. For every combination of rubbing materials, lubricant, and operation conditions, there is an optimal oxygen concentration providing minimal wear.

Resistance to scoring directly depends on oxygen transportation to contact zones. Highly active oxygen-containing compounds arising in oils during oxidation (e.g., organic peroxides) replace molecular oxygen and ensure lubrication even in vacuum. Under light loading conditions, intensive transport of oxygen to contact zones can lead to increasing wear [41,42]. At present, the main mechanism of bulk lube oils oxidation is considered to be a free radical one.

In tribocontact, the mechanism of oxidation of surfaces and oil boundary layers may differ. Grondkovsky et al. [44] studied the products of n-hexadecane destruction under chemical and tribochemical oxidation. They showed that these products, produced at static heating up to 180°C, significantly differ from the products of the same fluid oxidation during steel specimens friction under boundary lubrication (including the case when dispersed copper oxide was introduced in the oil). The said products of tribooxidation were precipitated on rubbing surfaces.

In the first case, the oxidation products were aliphatic hydrocarbons and oxygen-containing compounds, including carboxyl acids. The products of tribooxidation (without copper oxide additives) were mainly complex oxygen-containing compounds as well as organic acids, ketones, and ethers. In the presence of copper oxides, carboxyl acids and alcohols were deposited on rubbing surfaces and then were converted to ethers as a result of tribochemical reactions, providing high antiwear properties to the boundary layer. The products of static oxidation of hydrocarbons essentially differ from that of tribooxidation. It also concerns the metal oxides obtained by static heating and tribooxidation, which result in different tribological behavior of these oxides. Cavdar and Ludema [45], using the ellipsometry method, had established that during tests of a steel cylinder on a steel disk in plain mineral oil, a dual film was formed on rubbing surfaces, the bottom layer of which was a blend of Fe_3O_4, some small amount of iron carbides, and metallic iron, whereas the top layer consisted of ester-based and acid-based iron compounds. So in the process of tribochemical reactions, some organometallic compounds are produced and the boundary layer behaves itself like a solid rather than a viscous fluid.

At first, the film thickness tends to grow with increasing load, but further increase in load leads to a sharp decrease in the thickness of the dual layer and to surface damage due to the wear rate exceeding over that of film formation. The role that plastic deformation plays in such film formation should be mentioned, which can possibly be connected with activation of exoelectron emission. The maximum rate of the film formation was observed at the temperatures of 250–300°C.

Various hydrocarbon oils possess dissimilar lubrication action at different conditions [46–48]. In air, the hydrocarbons can be arranged by their lubrication action in the following series (in a descending line): aromatic, isoparaffin, naphtene, and paraffin hydrocarbons of usual structure.

In inert dry surroundings, saturated hydrocarbons of paraffin series demonstrate a sufficient bearing capacity, whereas polycyclic aromatic hydrocarbons of the same molecular mass under such conditions have rather poor lubrication action.

Accordingly [49], polyaromatic hydrocarbons in inert gases form negative ions: $M + e^- \rightarrow M^-$. The interaction of M^- ion with positively charged areas of the surface could lead to direct iron bonding with carbon. As a result of decomposition of this compound during friction, the particles of FeC appear to have an abrasive effect and provoke increased wear.

In air (or in inert gases with a touch of water vapor), negative ion formation can be described as the follows: $M + O^- \rightarrow (M-1)O^- + H^.$, so the reaction goes with the atom of hydrogen splitting off.

This ion is chemisorbed on the friction surfaces. After the chemisorbed layer breaks down during friction, ferrous oxide film is formed on the metallic surface, which is responsible for the high antiwear characteristics of polycyclic aromatic hydrocarbons in oxygen-containing gases.

Saturated hydrocarbons (paraffin and naphtene series) in similar surroundings form negative ions in accordance with the scheme $M + OH^- \rightarrow MOH^-$ with the subsequent splitting off of the fragments (first of all hydrogen and olefins). In absolutely inert surroundings, ions are not formed at all.

In oxygen-containing gases, MOH^- ions form a chemisorbed layer, the decomposition of which can result in oxidative wear [49].

Unsaturated hydrocarbons used as lubricating materials are exposed to serious tribochemical conversion during friction, resulting in the formation of boundary layers, which differ by the composition from the initial state. The unsaturated hydrocarbons are capable of creating chemical bonds with metals using oxygen bridges and possess better antiwear characteristics [49].

It is also worthwhile to note the ability of unsaturated hydrocarbons to acquire high enough lubricating properties in the process of friction due to polymer films formation on rubbing surfaces, which allows to replace the wear of contacting materials with that of a polymer film. The films tend to regenerate as they wear [46,49–53].

Friction polymers can be formed during friction in various hydrocarbons (both liquid and gaseous), in mineral oils, etc. Of course, it is of great importance to work on tailoring these polymer films for concrete friction units and operating conditions. However, it is rather difficult to control the process of friction polymer formation in commercial lubricants on rubbing surfaces. As petroleum oils are complex mixtures of hydrocarbons, there is no guarantee that any combination of these components even in the borders of the same oil grade would ensure realization of the process under required operating conditions.

That is why Zaslavsky et al. [15] and Furey [54] introduced in lubricants some additives (unsaturated monomers and oligomers) able to polymerize during friction and correspondingly to form films of friction polymers, which promote friction and wear reduction and prevent tribounits from seizure and scoring.

These films should have high adhesion to rubbing surfaces, an optimal thickness, and an optimal physical–chemical properties for given conditions, resulting in obtaining high-performance friction units reliably operating in required ranges of loads, temperatures, and speeds. If a tribopolymerizing additive is correctly chosen, a film of thickness 1–3 μm is formed, which has high adhesion to metal surfaces and has a trend to regenerate during wear.

The rate of the film regeneration is from 1 to 3.1 mg/hr. The rate of friction polymeric film formation from the base oil, which was used for tests, does not exceed 0.4 mg/hr.

Table 2 Effect of Some Functional Groups on Antiwear Properties of Olefin and Acetylene Compounds Introduced in Oil SU (Concentration 1%)

Acetylene compounds	d (mm)	Olefin compound	d (mm)
$HC \equiv C\text{-}C_6H_5$	0.53	$H_2C = CH\text{-}C_6H_5$	0.84
$HC \equiv C\text{-}COOH$	0.32	$H_2C = CH\text{-}COOH$	0.37
$HC \equiv C\text{-}CH_2\text{-}OH$	0.33	$H_2C = CH\text{-}CH_2OH$	0.48
$H_3COOC \equiv C\text{-}COOCH_3$	0.31	$H_3COOC\text{-}CH = CH\text{-}COOCH_3$	0.33
$HC \equiv C\text{-}CHO$	0.36		0.75

Note: The diameter of wear scar was measured after tests on four-ball machine during 4 hours under load 140 N and room temperature.
Source: Ref. 56.

The capacity of lubricants to form high-performance tribopolymeric films on rubbing metallic surfaces depends on structure and composition of monomers and oligomers using as friction polymer forming additives [15,54]. In the case of acetylene and olefin combinations, which are used for increasing antiwear properties of the base mineral oil SU ($v_{50} = 50$ mm^2/sec), joining the functional groups -OH, -COOH, -CHO, -COOR, etc. to the molecules of the said compounds has a marked effect on improvement of antiwear properties.

Some comparative data for olefin and acetylene compounds of similar structure are given in Table 2. As can be seen from these data, polymers of the acetylene series (polyenes) have more effect on the improvement of antiwear properties. It appears to account for the higher adhesion of the friction polymer films, forming in this case to the metallic substrate due to donor–acceptor complexes arising between polyenes and metals [55]. As was shown by tests on the standard four-ball machine (load 140 N, testing time 4 hr), the compounds forming metal–polymer complexes in friction zones really provide more antiwear effect as compared to ones that do not form such complexes (Table 3) [55]. The appearance of metal–polymer complexes in friction zones was proved by mass spectroscopy, IR and UV spectroscopy, x-ray analysis, and some other methods of surface analysis.

The presence in a lubricating composition of active polymerizing compounds and functional groups is a necessary but insufficient condition for obtaining friction polymer

Table 3 Antiwear Properties of Tribopolymerizing Additives That Form or Don't Form Metal-Polymeric Complexes

Lubricant	Friction polymer	Diameter of wear scar, mm
Oil I-20A	Is not formed.	0.78
Oil I-20A +1% derivative of 1,3-propylene glycol	Friction polymer is not bonded chemically with metal.	0.57
Oil I-20A +1% derivative +1% derivative of ethylene glycol	Friction polymer is not bonded chemically with metal.	0.59
Oil I-20A +1% derivative of methacrylic acid	Metal-polymeric complex is formed.	0.36
Oil I-20A +1% EF-357 (commercial additive)	Metal-polymeric complex is formed.	0.29

Note: Oil-20A is an industrial oil without additives, $v = 32$ mm^2/s.
Four-ball machine tests during 4 hours, load 140 N, base oil 1-20A.
Source: Ref. 56, pp. 158–165.

films with high antiwear and antiscoring characteristics. Monomers or oligomers themselves should have a certain configuration or conformation that would facilitate chemisorption of these compounds on metallic surfaces. The films should have an ordered plain structure and high adhesion to metal.

The introduction of bulk or branched radical substitutes of hydrogen results in a drastic change in antiwear action of friction polymer films for the worse. This effect does not depend on the character of polymer-forming bonds. Thus the hydrocarbon chain increase from C_1 to C_3 supplies ethers of acetylenedicarboxylic acids with high antiwear properties, whereas the further chain growing above C_5 leads to deterioration in antiwear properties. The evidence of this fact is a considerable increase in wear-scar diameter during tests of these compounds. It is of interest to note that some of the monomer additives under study reveal high antiwear activity in a wide range of concentrations in the base oil (0.01–1.0 mass%). High antifriction properties of polymer films deposited on rubbing surfaces are demonstrated in Fig. 9. Introducing a friction polymer forming additive of acetylene series in petroleum oil SU results in polymer film appearance, which reduces the friction coefficient by four times. Practically, reduction of friction coefficient immediately begins after the additive was introduced, which makes evident the high rates of friction polymerization. Numerous experimental data verify the formation of a protective antiwear film on metallic surfaces during the process of polymerization under friction with the effect of after-effect [54,56]. The formation of protective friction polymeric films can be demonstrated by using as examples highly effective commercial friction polymer forming additives EF-7, EF-262, and EF-357 [55], which were developed on the basis of some model compounds. The tests were performed on a friction machine, where a steel ball was sliding on the thrust-bearing ring in *n*-tetradecane. Immediately after the test, scoring of ball surface was started. Introduction in *n*-tetradecane of friction polymer forming additive EF-7 (mass 1%) leads to protective film appearance on scoring surface, to abrupt fall in friction, and healing of damaged zones as well (Fig. 10). At the same time, at the moment of the additive introduction, extreme activation of contact interaction occurs, which results in exposure of large zones of juvenile metal and provokes polymeric films formation. This provides reliable protection of a newly generated surface from seizure and scoring. In other words, the introduction of friction polymer forming additives contributes to fast running in of surfaces without large metal losses. It is of interest to mention an insular character of film distribution through the surface. The process of friction polymer formation was also considered in the works of Furey, Kajdas, and others [57] and is given in more detail in Chapter 25 by Furey and Kajdas.

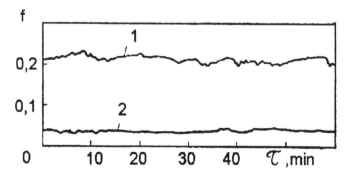

Figure 9 Changes of friction coefficient during tests on four-ball machine in oil SU (1) and in oil SU with 1% friction polymer-forming additive of acetylene series (2). (From Ref. 56.)

Figure 10 Scanning electron microscope photograph of the rubbing surface of the steel ball (testing scheme—a ball sliding on race track of thrust ball bearing ring; $v = 1.4$ m/sec; $P = 0.5$ N) after tests in n-tetradecane with 1 wt.% polymer-forming additive EF-7. (From Ref. 55.)

D. Tribochemical Processes Under Lubrication by Solutions of Surfactants

Lubricating materials practically always include surface-active compounds. Some of them get there as a result of insufficient refinement of row materials; others are introduced in lubricants as additives, e.g., carboxylic acids and their soaps and ethers, amines, imides, etc., which can be used as antifriction additives (friction modifiers). It should be mentioned that even chemically active additives possess some surface activity that can be rather high in some cases. Oxidation of hydrocarbons during lube oils storage and/or work also can lead to generation of surface-active compounds.

At last, molecules of some synthetic oils are surface-active compounds, e.g., molecules of polyglycols.

We dwell on some tribochemical aspects of action of surfactants incorporated into lubricating oils.

Among them, the most interesting for us are oxygen-containing surfactants and first of all fatty acids. The solutions of fatty acids in paraffin and some hydrocarbons may be used as very convenient models of lubricants for tribological studies. When friction of steel parts in solutions of surfactants occurs, the first transition temperature realization under boundary lubrication can be described well by the kinetic equation [Eq. (6)] given above. This is verified both by numerous experimental data obtained under tests of various lubricating oils at low-speed friction and heating of friction units by an external heat source and by literature data analysis as well [6,60–62].

As can be seen from data given in Fig. 11(a), curves 1 and 2, there is a linear dependence of the first critical temperature for fatty acids with the chain length C_{12}–C_{18} during friction of specimens made of mild steel [6] and during friction of steel against cast iron [60] on the value of differential heat of adsorption ΔQ of fatty acids from n-heptanes on α-Fe$_2$O$_3$ [61]. When passing from one fatty acid to another, the only parameter changed is $-\Delta Q$ (by which the value $\Delta \mu$ can be evaluated), whereas the rubbing materials are the same, i.e., E_x is a constant. Therefore we believe that the linearity of the first critical temperature via ΔQ is a sufficient evidence of the correctness of using Eq. (6) as applied to solutions of surfactants.

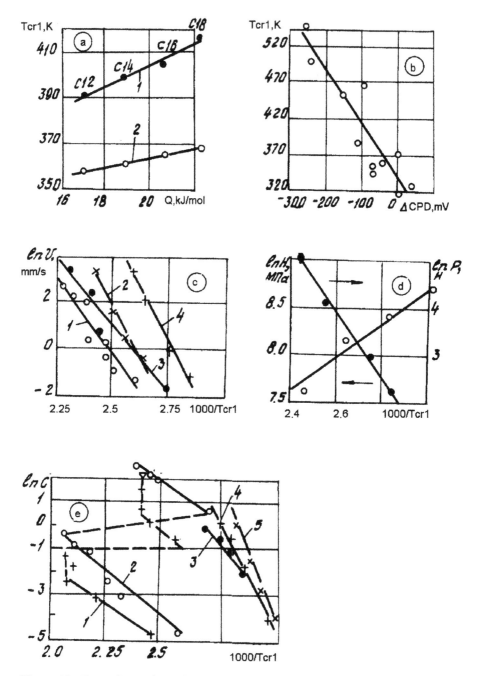

Figure 11 Dependence of experimental values of the first transition temperature T_{cr1} on (a) heat of adsorption of fatty acids, ΔQ, from n-heptane on α-Fe$_2$O$_3$ (1 = on steel; 2 = on cast iron); (b) contact potential difference, CPD, when a thin film of neutral Vaseline oil with 5% organoelemental surfactants [99] was placed on a measuring electrode; (c) speed v (for solutions in Vaseline oil: 1 = 0.1 wt.% stearic acid; 2 = 0.1 wt.% myristic acid; 3 = 1% dibenzyldisulfide; 4 = 1% commercial additive LZ-23k); (d) hardness H of surface layer and load P; (e) surfactant concentration C (1 = LiSt in oil c-220; 2 = LiOSt in spindle oil AU; 3 = diphenyldisulfide in Vaseline oil; 4 = commercial EP additive Sulfol (bistrichlormethyl sulfide) in Vaseline oil. (From Ref. 14.)

Equation (6) is also proved by data given in Fig. 11(b), where linear dependence of T_{cr1} via another characteristic of surfactant adsorption on metals—change in work function of electron $\Delta\varphi$ from steel surfaces covered by a thin layer of a lubricant under study—is shown. As lubricants, 5% solutions of a number of heteroorganic surfactants in medicinal Vaseline oil were used.

As seen in Fig. 11(c), tests also prove the linear dependence of inverse T_{cr1} on sliding speed (for the following compounds solution in Vaseline oil: 1 = 0.1% stearic acid; 2 = 0.1% myristic acid; 3 = 1% dibenzyldisulfide; 4 = 1% commercial additive LZ-23k). The similar dependences of the inverse T_{cr1} on hardness of steel specimens H and contact load P for a 0.1% solution of stearic acid in Vaseline is shown in Fig. 11(d). In Fig. 11(e), data are given as the linearity of the dependence of the inverse first critical temperature on the concentration of surface-active compound in oils: 1 = LiSt in oil C-220; 2 = LioSt in spindle oil AU; 3 = diphenyldisulfide DPHDS in Vaseline oil; 5 = commercial additive Sulfol* in Vaseline oil.

It should be noticed that Frewing [62] had established the linearity of dependence $\ln C - 1/T_{cr1}$ for low-concentration solutions of fatty acids in neutral oils. For both soap–oil systems (LiSt in oil S-220 and LioSt in spindle oil AU), linear dependences between $\ln C$ and $1/T_{cr1}$ have two branches with the break at the point of a critical concentration of structuralization peculiar to every system [14]. All dependences shown in Fig. 11 (besides the break in $C - 1/T_{cr1}$ mentioned above, which is not trivial and contains some new information concerning mechanism of lubrication action of greases) confirm Eq. (6) and dependences resulting from it under conditions when one factor is changing while others stay constant.

However, during friction of some metals (such as copper, cadmium, zinc) in fatty acids and their solutions, it is rather difficult to suppose the possibility of reverse adsorption of surfactants on rubbing surfaces. At present, it is established that the lubricating ability of fatty acid solutions is connected with active metal soaps production due to tribochemical reactions. It was first postulated in Ref. 5 and was proved with reference to copper on the basis of solubilization-effect studies by Kreuz et al. [63]. Next, by studying the orientation and binding energy of stearic acid monolayers, it was found that these monolayers after friction are similar to bulk cupric stearate, whereas the same lubricant adsorbed by copper surface without friction possesses all features of bulk stearic acid [64].

The chemical composition of rubbing bodies seriously affects friction and wear behavior under boundary lubrication by surface-active compounds. It can be explained first of all by the difference in the adsorption of molecules of surfactants on different materials.

The presence of oxides on rubbing surfaces and the possibility of their regeneration during wear are important for effective lubrication. It is established that the formation of soaps responsible for the lubricating action of fatty acids during friction of active metals is possible only on oxidized surfaces [5]. Nevertheless, not all oxides grant effective lubrication by fatty acid solutions. Therefore Vinogradov et al. [65] showed that deformation of more plastic metals as compared to their oxides can result in the breaking of the latter. In turn, an immediate metallic contact occurs, which cannot be prevented by boundary layers as they are removed from the contact zone.

On the other hand, it is right to expect that oxides generated on metallic surfaces exposed on air should differ from oxides arising during friction [43]. This is illustrated by Fig. 12 [59], where the results of temperature stability studies of neutral Vaseline oil with

*Sulfol is an antiscoring additive by strichlormethylsulfide.

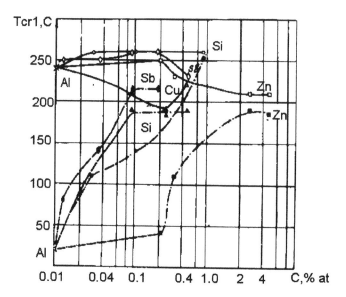

Figure 12 Dependence of the critical temperature T_{cr1} for Vaseline oil VM-1 with 0.1% stearic acid during friction of steel against binary aluminum alloys vs. the content of alloying elements in these alloys: (1) preliminary worn oxide film; (2) oxide film obtained by exposure on air. (From Ref. 59.)

stearic acid are presented as applied to the tests of a ball made of low-chromium steel on rings made of binary aluminum alloys in the said lubricant.

When oxides were preliminarily worn under an oil layer, T_{cr1} was higher as compared to the value in the case of friction on oxide layers formed under static conditions (exposed on air). The dependences of the first critical temperature, T_{cr1}, on the concentration of alloying element in aluminum for these cases significantly differ. All studied alloying elements increase the value of T_{cr1} to a greater or lesser extent in the presence of oxide films, but when the oxide film is preliminarily removed (and on its place under an oil layer before and during tests some other oxide films with properties distinct from that of removed ones were formed), such alloying elements as Si and Sb practically have no effect on temperature stability of oil under study, whereas Zn and Cu decrease it in some definite range of concentrations.

We have mentioned above that fatty acids are usually chemisorbed by metallic surfaces during friction. It appears that this process is activated due to ionization of molecules of a surfactant by exoelectron emission. In any case, while studying temperature stability of model lube oils during friction of some pure metals (Fig. 13) and alloyed steels (Fig. 14), it was found that the lower the work function of electron, the higher the first critical temperature T_{cr1} of the fatty acid and its solution in neutral oil.

In other words, the lesser the energy necessary for emission of electrons responsible for negative ion appearance, the stronger the bonds between the fatty acid molecules and metallic surfaces. For instance, Kramer [66] has assumed that exoelectrons emitted by surfaces of deformed metals interact with molecules of stearic acid, giving rise to metal stearates formation.

In accordance with the NIRAM concept, when the emitted electron is attached to the fatty acid molecule, most often the break of O–H bonds occurs. Further decomposition of

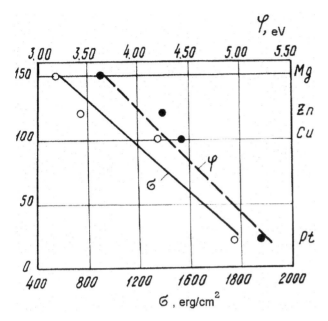

Figure 13 Values of the first critical temperature T_{cr1} for laurine acid during friction of some pure metals vs. their surface energy σ and the work function of electron φ. (From Ref. 59.)

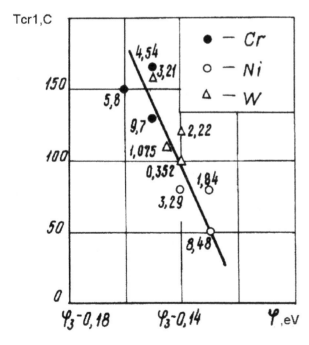

Figure 14 Dependence of the first critical temperature T_{cr1} of Vaseline oil with 0.1% stearic acid during friction of alloyed steels on the work function of electron φ determined by contact potential difference using Au as reference material. The content of alloying elements is shown on the points of the diagram (φ_3 is the work function of electron for Au).

the formed negative ion $R-CH_2-COO-$ goes by splitting two hydrogen atoms and carboxylate group formation. $-COO-$ group ensures strong bonding of molecules with positively charged surface of metal. The NIRAM concept allows to explain why lubrication properties of aliphatic acids and alcohols are similar; that is, it allows to consider the mechanism of chemisorption of aliphatic alcohols on rubbing surfaces.

According to Refs. 67 and 68, this process includes interaction of alcohol molecules with exoelectrons resulting in generation of anions, anions radicals, and hydrogen radicals, etc. Chemisorption of anions and anion radicals occurs on the positively charged sites of metal surface. Recombination of hydrogen radicals with molecules of hydrogen formation takes place as well as that of chemisorbed anion radicals with formation of cross-linked or unsaturated bonds. All these processes increase the strength of boundary layers. At high temperatures, the breaking of bonds is observed in molecules of chemisorbed alcohols resulting in formation of diene, acetylene, and diacetylene hydrocarbons, which can be subjected to further tribopolymerization.

Tribochemical transformations of complex ethers of palmetine acid and aliphatic spirits are analyzed in terms of the NIRAM concept by the example of steel–steel friction pair. The film of type $(RCOO)_n Me$ [69] is formed on rubbing surfaces.

Some other surface-active compounds, which were used in this capacity in various tribological studies during friction of steels (for instance, amines), when another materials are rubbing (in this case Al–Si alloys), can be subjected to degradation or oxidation due to tribochemical actions and then to form stable complexes between Al and amines (e.g., H_2N-CH_2COOAl and $Al-NH-CH_2-CH_2NH_2$) as well as tribopolymers that result in high lubricating characteristics and correspondingly in high performance of friction units [70].

E. Tribochemistry of Oils with Chemically Active Additives

In considering the tribochemical action of lube oils with chemically active additives (CAA), it should be noticed that there are a great number of studies and reviews devoted to this problem [21,71–75]. At first, we would like to point out that the tribological behavior of solutions of any chemically active additive in oils under variable operating conditions can be described rather well by the theory given in Sec. II. In accordance with this theory, the linear dependences (in logarithmic scale) of the concentration of chemically active additives [Fig. 15(a)] and sliding speeds [Fig. 15(b)] on the inverse temperature of chemical modification are observed, as well as the linearity of additive concentration dependence via the inverse second critical temperature T_{cr2} (Fig. 16) for various additives. The relationships $\ln \beta - 1/T$, $\ln \beta - \ln C$, and $\ln \beta - \ln v$ in accordance with this theory are also linear, which allows calculating the values of the parameters mentioned above (Figs. 17–19).

Chemically active additives may be classified according to their destination as anti-scoring extreme pressure (EP), antiwear (AW), and high-temperature antifriction–friction modifiers (FM). Extreme pressure additives, in terms we have used above (Fig. 2), mainly prevent a friction unit under high loads and temperatures from transition to group C, i.e., to catastrophic wear, scoring, and failure due to modified film formation on rubbing metallic surfaces. The thickness of these films should be high enough for metallic contact prevention.

Antiwear additives provide high antiwear properties for a friction unit under conditions of enduring operation at moderate and heavy loads and temperatures. These additives ensure formation of surface films with increased strength from the products of chemical interaction between active compounds of the additive and rubbing metals. Anti-wear additives possess some antiscoring effect under extreme loading conditions, whereas EP additives can act as antiwear materials under relatively low loads and temperatures.

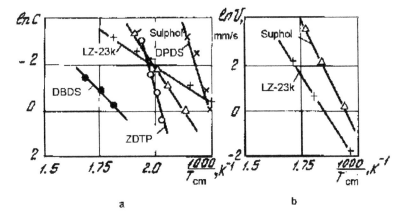

Figure 15 Concentration C of a chemically active additive (a) and sliding speed v (b) vs. the reverse temperature of chemical modification T_{cm} for additives insoluble in Vaseline oil: (1) LZ-23k; (2) Sulfol; (3) dibenzyldisulfide; (4) DF-11; (5) diphenyldisulfide. (From Ref. 14.)

Chemically active, oil-soluble FM ensure reduction of energy losses due to decrease in friction.

The tribochemical action of all listed types of additives consists in their decomposition due to the heat generated during friction, catalytic action of freshly formed metallic surface, exoelectron emission, etc. The active products of additive decomposition interacting with metallic rubbing surfaces form the modified layers, which impart to contacting elements anti-scoring, antiwear, or antifriction effects, or summary effect. Thus the efficiency of these additives depends on their capability to be adsorbed on the rubbing surfaces, which is followed by decomposition under particular conditions, resulting in the production of some chemically active reagents that interact with friction surfaces to form protective modified layers.

While testing on a standard four-ball machine, additives of three different types insoluble in oil M-11 (typical AW—zinc dialkylditiophosphate ZDTP; typical EP—dibenzyldisulfide, DBDS; and typical high-temperature antifriction additive—molybdenum

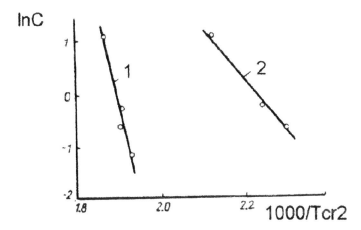

Figure 16 Concentration of additives insoluble in oil ASV-10 vs. the inverse second critical temperature T_{cr2}: (1) Lubrizol 5901; (2) Hightec E059. (From Ref. 76.)

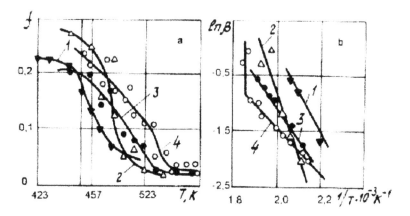

Figure 17 Coefficient of friction f vs. temperature T on the falling part of the friction coefficient temperature dependence (a) and $\ln \beta$ vs. reverse temperature $1/T$ (b) during friction of similar specimens made of steel with different chromium content in Vaseline oil with 1.5% chlorinated paraffin: (1) 1.5% Cr; (2) 4.5% Cr; (3) 5.6% Cr; (4) 9.5% Cr. (From Ref. 77.)

dialkylditiophosphate, MDTP) showed some antiwear effect. The more strongly pronounced effect was demonstrated by dialkylditiophosphates of zinc and molybdenum as compared to dibenzyldisulfide (Table 4). At the same time, for all comparable compositions, T_{cm} was about 200°C, which provides evidence of the efficiency of these additives at high temperatures. Dibenzyldisulfide demonstrated its high functional action only at extremely high temperatures and pressures. For DBDS as well as for other EP additives, the values of E_p are lower than for AW additives. It should be mentioned that DBDS does not cause the

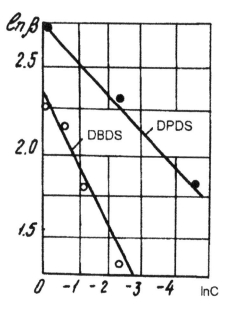

Figure 18 Relationship between $\ln \beta$ and $\ln C$ for solutions of diphenyldisulfide (1) and dibenzyldisulfide (2) in Vaseline oil under friction of similar specimens made of low chromium steel. (From Ref. 14.)

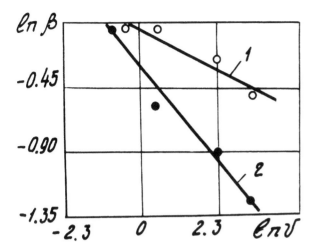

Figure 19 Relationship between ln β and ln v for 1% solutions in Vaseline oil: (1) Sulfol; (2) additive LZ-23k. (From Ref. 14.)

transition from regime 1 to regime 2 on the transition diagram [Fig. 1(b)] but enlarges zone 2; that is, it increases the bearing capacity of boundary lubricating layers in the field of heavy friction and enlarges zone 1 on the transition diagram (i.e., it shifts the transition from a partially elastohydrodynamic lubrication to a boundary one in the field of more heavy friction regime). It results from the fact that ZDTP additive forms protective boundary layers even under moderate operating regimes. When tests were performed in argon, the efficiency of ZDTP abruptly decreased, whereas that of DBDS remained constant [72].

Antifriction additives are usually identified by their capability to reduce friction during prolonged tests. In Fig. 20, some test results are given [79] where it may be seen that ZDTP has little effect on friction lowering as compared to the base oil, whereas MDTP ensures a significant decrease in friction coefficient even after 30 min of tests. The value of the friction coefficient is 0.065 in 60 min after the test began. The friction coefficient at the same testing time in zinc dithiophosphate solution is 0.108 [79]. Friction reduction due to the action of chemically active additives is not immediately observed after the start of test, because the processes that effects such reduction, such as optimal surface macro- and microroughness formation during tests and optimization of the structure and composition of modified layers formed as a result of tribochemical reactions, are time-dependent processes.

Table 4 The Values of the First Critical Temperature, T_{c1}, the Temperature of Chemical Modification, T_{cm}, and the Diameter of Wear Scar, d, under Tests of Oil M-11 with Two Chemically Active Additives, ZDTP (commercial additive DF-11) and MoDTP (commercial additive Molyvan-L), and Model EP additive, DBDS

Lubricant	T_{cr1} (°C)	T_{cm} (°C)	d (mm)
Oil M-11	120	—	0.69
Oil M-11 +1% ZDTP	125	195	0.35
Oil M-11 +1% MoDTP	110	215	0.32
Oil M-11 +1% DBDS	80	260	0.47

Note: Tests were conducted on four-ball machine with slow rotation of upper ball and bulk external heating of the friction unit (Matveevsky's method).

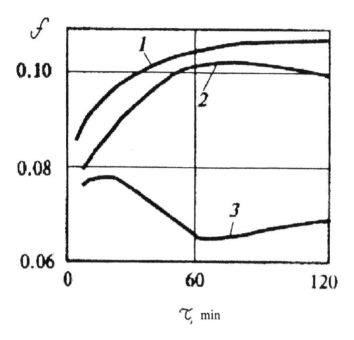

Figure 20 Friction coefficient f vs. testing time t (tests were performed on four-ball machine at load of 198 N): (1) base oil M-11; (2) base oil M-11 + 1% ZDTP (additive DF-11); (3) oil M-11 + 1% MDTP (Molyvan-L).

On the other hand, it is rather difficult to rank additives of various kinds by their capacity to smooth rubbing surfaces. For instance, the EP additive dibenzyldisulfide (DBDS) application results in rough surface of wear scar during tests on a four-ball machine. At the same time, the AW additive diphenyldisulfide, which is similar to DBDS by structure and composition, allows to obtain essentially smoother friction surfaces (Fig. 21). Comparing the AW additives ZDTP and MDTP, one can see that the first gives rather rough friction surface, whereas the latter ensures surface smoothing. Apparently, the key role in the ability of reagents to provide certain service performances belongs to modified layers formed on rubbing surfaces. The difference in the modified layers formed during friction in the presence of AW additives as compared with the surface subjected to wear in pure base oil may be seen in Figs. 22–24, where the distinction in the elemental composition of surface layers in solutions of DBDS and ZDTP in Vaseline is presented.

Chemical elements constituting the considered additives were found only on wear scars. Modified layers formed during tests in different lubricants essentially differ by composition and structure. For example, a solution of DBDS in Vaseline creates on rubbing surfaces a porous layer in the form of parallel strips located close to each other. These layers mainly consists of ferrous sulfide FeS, and ferrous and chromium oxides FeO, Cr_2O_3.

The friction surface, after tests in the solution of ZDTP, differs from that mentioned above. Here the distance between strips is greater; one can see the traces of delaminated films on protuberant zones. The thickness of delaminated films is of some microns. The modified layer composition in this case includes such compounds as FeS, ZnO, and/or ZnS, complicated compounds of phosphorus with oxygen. Oxygen occurs in two states with different energies of electron bonding (before ion etching of surface), which is an evidence of the presence of two ferrous oxides.

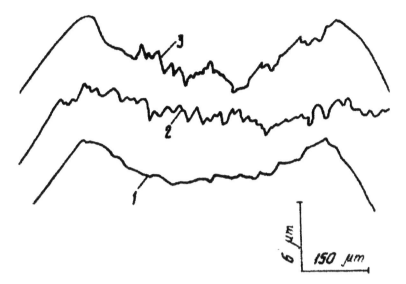

Figure 21 Profilograms of wear scars on balls made of ball-bearing steel ShKh-15, tested on a four-ball machine in 1% solutions in Vaseline oil: (1) DPDS; (2) DBDS (test duration 1 hr); (3) pure Vaseline oil (test duration 30 min). Load 198 N.

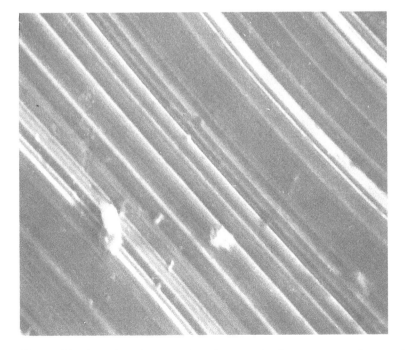

Figure 22 Scanning electron microphotograph of wear scar surface on the ball made of ball-bearing chromium steel ShKh-15 after 4-hr tests in Vaseline oil on a standard four-ball machine at the load of 196 N; magnification ×1100. (From Ref. 80.)

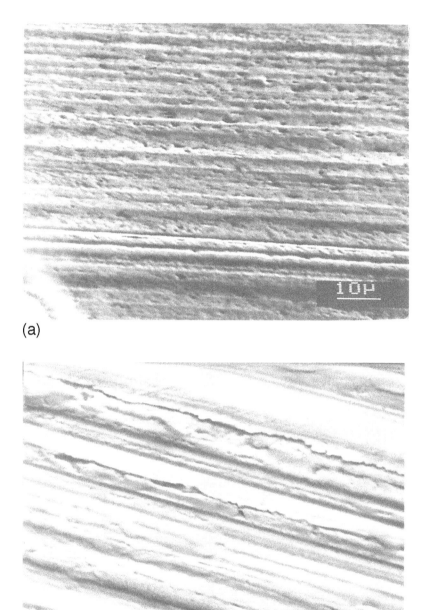

(a)

(b)

Figure 23 Scanning electron microphotograph of wear scar surface on the ball made of ball-bearing chromium steel ShKh-15 after 4-hr tests in Vaseline oil with additives on a four-ball machine at the load of 198 N: (a) 1% DBDS; (b) 1% ZDTP. Magnification ×1100 (From Ref. 80.)

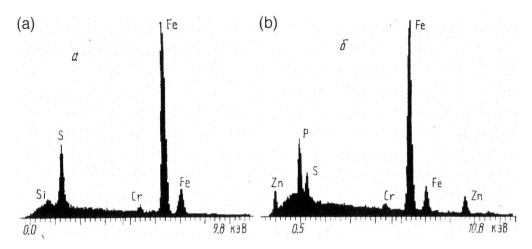

Figure 24 Spectrums of X-ray microanalysis of the surfaces of wear scar on the steel ball after tests in Vaseline oil with (a) 1% DBDS; (b) 1% ZDTP. (From Ref. 80.)

It appears that under moderate loading conditions, ZDTP layers adsorbed by rubbing surfaces ensure antiwear effect, whereas under heavier loading conditions, an initial decomposition of the additive occurs with the formation of primary chemically modified layers. During the transition to the EP regime, some changes in the primary modified layer composition takes place due to more complete decomposition of additives and secondary tribochemical reactions. These changes are directed to ensure surface strength of rubbing bodies and/or minimum friction. It could be realized by different ways, e.g., by secondary reactions on surfaces leading to MoS_2 formation in the solution of dithiocarbamate [75]. Tests on the Cameron–Plint machine under high loads and temperatures and low concentration of ZDTP lead to its decomposition. Higher concentration of this additive in combination with higher loads, moderate temperatures, and smooth surface result in long-chained polyphosphates formation. At temperatures higher than 150°C, the additive no longer stays in the initial state, and sulfur in the modified layer is in the form of sulfate.

The modified layer may be multilayered: Immediately on the friction surface of the steel, a layer is formed consisting of a blend of ferrous oxides and sulfides; then, a layer of polyphosphate glass with a mixture of zinc oxides and sulfides; lastly, the top layer consists of organic radicals. The total thickness of the protective layer is in the range of 100–1000 nm.

The AW and EP action of sulfur-containing additives may be explained by the NIRAM concept [81]. In accordance with this concept, the interaction of molecules adsorbed by metallic surfaces for monosulfides RSR with exoelectrons emitted during friction can be described as:

$$R - S - R + e^- \rightarrow RS^- + R^{\cdot}$$

For disulfides RSSR, this interaction can be written as:

$$R - S - S - R + e^- \rightarrow RS^- + RS^{\cdot}.$$

In the process of further interaction, another exoelectron may join the separated fragment:

$$RS^{\cdot} = e^- \rightarrow RS^-.$$

Therefore the initial decomposition of monosulfides is realized by the breakage of the R–S bond, whereas that of disulfides is realized by breaking the weaker bond S–S$^-$, although the decomposition of disulfides by hydrocarbon radical splitting with negative ion R–S–S$^-$ formation is possible too. It means that disulfides should possess better EP properties as compared to monosulfides. Further, the negative ions react with positively charged zones of metallic surface Me, forming chemisorbed layers on rubbing surfaces, which ensure AW action.

Under heavy loading conditions, the chemical bonds between sulfur and radicals break down (emitted exoelectrons possibly take part in this process). The sulfur bonds, which remained free, are saturated as a result of chemical interaction with metallic surfaces. Thus, on the surfaces, an inorganic modified layer is formed consisting of metal sulfides and disulfides (the metal is usually iron), which fulfills EP action. Released radicals in the presence of oxygen are subjected to oxidation; the products of oxidation are deposited on sulfide surface forming a durable adsorbed layer. Friction polymers are possibly formed on these surfaces to ensure effective prevention of metallic contact of surfaces. Surface-active compounds adsorption on ferrous sulfide is much better than on ferrous oxide, which determines better AW properties when tests were conducted in air. However, the reverse effect was observed on tests using argon.

The films are regenerated as they are worn out because there are conditions necessary for their reproduction. It is experimentally proven that in layers of rubbing steel surfaces tested in the solution of DBDS in the AW regime, there are less sulfur-containing compounds as compared to the EP regime. This may be explained by the fact that the modified layer at the EP regime consists of tens or even hundreds of molecular layers of iron sulfides and disulfides, whereas for the AW regime, only a thin layer of Fe(SR)$_2$ is needed with an organometallic film or tribopolymeric film deposited on this layer, etc.

As mentioned above, dibenzyldisulfide imparts higher EP properties to lubricating materials as compared to diphenyldisulfide [82].

From the NIRAM positions, it may be explained by some distinctive features of structure of radical DBDS and its transformations during tribochemical reactions:

$$R^i - S - S - R^i \rightarrow (C_7H_7)^- + R^i - S - S^-,$$

$$(C_7H_7)^- \rightarrow (C_5H_5)^- + HC{\equiv}CH.$$

Acetylene formed because of this reaction polymerizes in the friction zone, ensuring increase in EP performance of friction units. Synchronized formation of ferrous sulfide covered by tribopolymer and EP effect of their cooperative action was also observed in some other sulfur-containing additives having cyclic benzyl double bonds [83]. But the low value of binding energy in C–S bonds in DBDS as compared to that in DFDS may also be a cause of the better EP behavior of this additive.

Considering the tribochemical behavior of the additives having both sulfur and oxygen, we should note that such combination usually increases the effectiveness of lubrication action of additives to the extent defined by their structure and composition [83,84]. It is well known that the introduction of an atom of oxygen in a monosulfide results in C–S bonds weakening. Accordingly [83], among oxygen-containing additives based on monosulfides, better lubricating properties were demonstrated by sulfoxides, then by sulfones, and at last by sulfides. The difference in lubricating properties of organosulfur and organosulfuroxygen compounds lies in the different affinity to electron, i.e., different capacity to join exoelectron and correspondingly to be ionized [83].

The first stage of interaction of chemically active additives with metallic surfaces is adsorption. The capacity to be adsorbed by metals is directly related to the effectiveness of the additives. A set of tests conducted by Studt [86] on pin-disk and four-ball machines has shown that there is a direct correlation between the level of adsorption of various chlorine-containing organic compounds and their concentration in the base oil with load-carrying capacity of a friction unit, but effect of the level of adsorption is higher as compared to concentration. However, in 1940, Beeck et al. [87] showed that when the concentration of tricrezyl phosphate in white oil was increased up to 1.5%, a decrease in wear took place. On the contrary, the further increase in concentration of this compound resulted in increased wear. So there is an optimal concentration of chemically active reagents that ensures the minimal wear of contacting bodies. In Ref. 88, it was shown that operating conditions would determine the optimal concentration of active reagents.

Abzalov [89] explained the effects of additive concentration and optimal concentration of oxygen dissolved in lube oils as well as the influence of optimal film thickness (in air and argon) on wear behavior by competition between ion radicals of hydrocarbons and that of oxygen. They are both striving for tribochemical interaction with active sites on metallic surfaces and for modified layer formation with AW and/or EP action.

According to Nakayama and Sakurai [88], there is a minimum on the wear rate dependence upon temperature. This minimum exists for every studied concentration of the additive (elemental sulfur) in neutral oil. The authors had shown that two types of wear could be realized: When the sulfur concentration was lower than an optimal one, the adhesive wear occurred; at concentrations exceeding optimal, wear by delamination was observed.

It is possible to explain these data simply by the transition from conditions where the lubricating film only prevents metallic contact and adhesive wear, to conditions when higher temperatures and/or higher concentration of the reagent promote intensifying tribochemical and/or thermochemical interactions resulting in intense corrosive wear.

Besides, factors such as surface microroughness, type of base oil, and its viscosity have an effect on tribochemical reactions under boundary friction. According to Ref. 90, increase in oil viscosity leads to an increase in the constant of the rate of modified layer formation, but has no influence on the constant its wear rate.

The composition of base oils and additives of different action also influence the efficiency of EP and AW compounds under consideration. Sometimes, there is no direct connection with tribochemistry of lubricated contact. For instance, for effective running in of rubbing bodies, it is of importance to chose the right combination of the base oil and the structure of hydrocarbon radical of dithiophosphate additive. It is necessary to ensure such an interaction between them so that at the temperature of additive decomposition, its solvation by the oil was minimal and the rate of tribochemical interaction with rubbing surfaces was high enough. At lower temperatures, solvation should lead to durable colloidal particles arising from the molecules of the additive and the oil, which are capable to form a continuous boundary lubricating layer strongly adhered to friction surfaces [91].

Substantial improvement of EP and AW action of a number of sulfur-containing, chemically active additives can be obtained by using surfactants introduced in lube oils together with these additives. Surface-active compounds may be used, e.g., oxygen-containing surfactants such as fatty acids, complex ethers, and others [92]. At the same time, it should be noted that in a number of studies of combined tribological behavior of such AW additives as ZDTP and other components of lube oils (e.g., detergents) and the products of oil degradation (e.g., peroxide radicals) [21,92,93], it was pointed out that the antagonism of their action takes place. Interaction of these additives has a rather complicated character. So according to Kapsa et al. [92], the interaction of ZDTP and calcium sulfonate when used together depends on the level of detergent alkalinity and allows to consider the boundary layer as a complex colloidal system.

Finally, the tribological action of chemically active lubricants is strongly influenced by the composition of metals of rubbing parts. The capacity of steels to react with sulfur-containing compounds of lube oils during friction decreases as the content of chromium or tungsten in steels increases [an abrupt rise in T_{cm} in Fig. 17(a)], whereas the nickel content has no effect on the capacity of steels to interact with sulfur-containing compounds. On the contrary, chlorine-containing additives equally well reacts with steels alloyed by chromium, tungsten, or nickel [59].

Let us return to Fig. 17 given above. In Fig. 17(a), friction coefficient vs. temperature was shown for tests of steels with different content of chromium in the solution of chlorinated paraffin in Vaseline oil; in Fig. 17(b), a comparison was made of a share β of contact through the modified layer vs. reverse absolute temperatures. The data were obtained in the result of analysis of descending branches of the curves given in Fig. 17(a). According to Eq. (12), it is possible to evaluate the activation energy E_p of the process of modified layer formation by a slop of the straight line approximating the dependence of β upon $1/T$. In Fig. 25, data are presented on how changes in the value of E_p for CP depend on the chromium and tungsten contents in steel and how an increase in the value E_p corresponds to decrease of the diameter of wear scar d.

The observed results can be explained by the fact that even steels alloyed by chromium in the solution of chlorinated paraffin undergo corrosive wear, and the more actively metals interact with lube oils (i.e., the less the value E_p), the more rapidly the products of this interaction will be worn out.

Hence some tribochemical processes can be sufficiently described by the considered theory and the NIRAM concept allows to explain the mechanism of many tribochemical reactions.

F. Triboreactions Under Lubrication by Oils Containing Carbon as Additive

Iocebidze et al. [94,95] found that at friction in petroleum and some synthetic oils, with ultradisperse amorphous carbon used as an antipitting agent, EP, and antifriction additive,

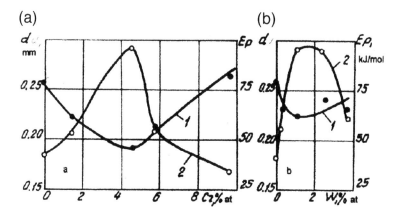

Figure 25 The diameter of wear scar d (1) and parameter E_p (2) vs. the content of chromium (a) and tungsten (b) in steels during tests in 1.5% solution of chlorinated paraffin in Vaseline oil.

formation of graphite and cubic and hexagonal diamond takes place, which together with amorphous carbon enter into the composition of a thin heterogeneous film formed on steel rubbing surfaces. A similar picture was observed during friction in low-molecular hydrocarbons.

The diamond–graphite structures are ordered in the form of quasi two-dimensional cells of lamellar-oriented graphite, which are bordered with a netting of fine, dispersed diamond particles (0.01–2.0 μm). The typical size of such a cell is of some microns and it is not conformed to structural parameters of friction surfaces, where these cells are formed. Time necessary for such graphite–diamond cell synthesis is of 10^{-4} sec and less, which corresponds to the duration of nonequilibrium fluctuation of "elemental" triboexcitement. The typical period of regeneration of such a structure would be determined by its wear rate.

The tribosynthesis of graphite–diamond structures on rubbing surfaces was observed both during sliding and rolling with sliding, and in all cases ensured friction reduction and increase in wear resistance. Because tribosynthesis of considered structures belongs to the so-called strongly nonequilibrium (synergetic) phenomena, its realization does not require high temperatures and pressures as well as the use of catalysts. The activation of this process may be achieved by increasing the level of nonequilibrium state of the system. Particularly, it may be realized by vibromodulation of normal load by high-frequency oscillations during friction. Such operating regime in conventional lube oils (with AW and EP additives) usually results in increase of surface failure. The use of low-viscosity oils with ultradisperse carbon under such conditions on the contrary provides smoothing of rubbing surfaces, improves adsorption ability of surface layers, and leads to metal catalytic activity decrease, contact strength increase, etc. All listed improvements in the case under consideration can be explained by intensification of tribosynthesis of graphite–diamond structures in the surface layers.

G. Tribochemistry of Lubricating Properties Fading Under Boundary Lubrication

It is well known that lubricating materials lose their effectiveness during their service life. This phenomenon was first investigated by Buche [96], who showed that films of various oils placed on surfaces of Babbitt bearing inserts contacting to steel shaft ensured satisfactory lubrication during some certain time for all studied oils. The friction coefficient during all this time was low and practically invariable, whereas a further time increase lead to an abrupt rise in friction and in wear rate, i.e., to transition to seizure and scoring. It is an evidence of breakage of the boundary lubricating layer. Some researchers later discussed the possibility of lubricating layer wearing out and estimated the values of specific wear of lubricating films. The obtained values were three orders lower than specific wear of metals at conditions of elastic deformation [97]. The rate of lubricating properties fading depends on the load, speed, and contact temperatures [98–100]. The higher the values of listed parameters, the higher the rate of lubricating properties fading. Deterioration of lubricating properties can also be caused by oxygen introduced in oil, whereas under friction in nitrogen, lubricating layers have higher resistance to wear as compared to friction on air [101]. Lubricating film life is strongly affected by the nature of lubricants and rubbing parts of materials [98–108]. For instance, when rubbing a steel ball against the monomolecular layer of a number of fatty acids deposited on a glass plate, the longer the length of the hydrocarbon chain, the longer the life of a boundary layer [104]. Surface-active compounds introduced in lube oils also ensure increase in life of lubricating layers [98,99]. Lastly, the thickness of the boundary layer essentially affects its lifetime. The thinner is the boundary layer separating the rubbing surfaces, the smaller time is required for this layer to lose its ability to separate surfaces (Fig. 26), without the possibility to get some additional feedback from the body of lubricant [99].

A boundary layer loses its lubricating properties when surface microasperities are worn out. The smoothing of surfaces leads to oil elimination of microreservoirs, which have ensured some additional feeding of boundary layers by oil [107].

Specialists in boundary lubrication usually consider two causes of lubricating properties fading. The first one is concerned with the adsorption of active compounds of lube oils by wear products. Really, wear products possessing large surface per unit of volume are able to actively affect the total process of adsorption and assist in removing active regents from contact zones. Therefore depletion of active compounds of a lubricant (at a real boundary layer thickness, wear products are able to passivate the volume of a lubricant exceeding that of wear products up to 20 times) resulted in degradation of lubricating characteristics [106]. Some researchers believe that one of the main causes of degradation of lubricating ability during friction are irreversible multistage transformations in lubricating materials under high contact pressures and temperatures combined with catalytic action of freshly naked metal. As a result of these transformations, a number of gaseous products as well as that of hydrocarbon condensation are formed, e.g., carbon, graphite, metal–carbon compounds, etc. These products, possessing low lubricating ability, lead to seizure and scoring in contacting machine parts, when a significant amount of them would be accumulated in contact zones [100].

According to some other data, the content of these products includes the products of oil primary oxidation—high molecular compounds, which seriously increase viscosity of an oil under study [99]. The results of oil microprobe studies by gel permeation chromatography (GPC) after tests on a diminutive four-ball machine allow to conclude that during process of wearing, some small amounts of insoluble products of oil destruction are formed,

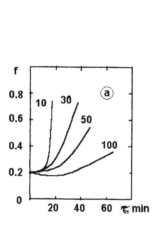

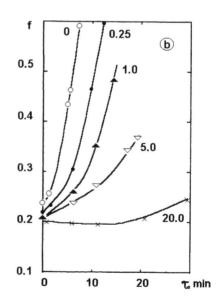

Figure 26 Life of boundary layers vs. test duration under a speed of 450 mm/sec and load 7N: (a) bright stock of different thickness; (b) Vaseline oil with different content of oleic acid. (From Ref. 98.)

but these products interacting with oil form grease-like mixtures that prevent fresh oil penetration to friction surfaces and thus stimulate seizure of friction units.

These products of oil degradation are similar to the products of oil oxidation at $400\,^\circ C$ and higher temperatures, whereas the oil dose that does not penetrate into contact zones is heated only up to $150\,^\circ C$.* The volume of oil involved in the formation of grease-like products ranges from 4% (for polyalphaolefins) up to 35% (for polyglycol ethers) from initial oil volume in the friction unit. Lifetime of an oil dose changes at rotation frequency 600^{-1} from 8 min/μl for polyphenylether up to 55 min/μl for mineral oil.

IV. SELECTIVE TRANSFER

One of the most effective tribochemical processes under boundary lubrication is the so-called selective transfer (SF), first described by Kraghelsky and Garkinov in 1957. Selective transfer in its classical variant can be realized under friction of steels against some bronzes and brasses in glycerin, alcohol–glycerin, and water–glycerin blends, seawater, some oils, and lithium lubricants. For its realization, certain operation conditions of friction units (including kinematics of conjunction, loads, temperatures, speeds, etc.) are needed [32,110–115]. Realization of this effect allows to obtain the values of friction coefficients 0.01–0.001, the values of wear rate 10^{-9}–10^{-11}, which provided reasons to consider this phenomenon as "effect of wear absence [7]."

The main idea of selective transfer mechanism is a thin metallic film formation on rubbing surfaces, which differs from the base metal both by structural state and elemental composition resulting in shear strength reduction (as compared with that of bulk base metal), deformation facilitating, and high activity concerning lubricants. On this metallic film, lubricating boundary layers with high load-carrying capacity are formed. The lubricants are also subjected to serious changes, which essentially increase their lubricating properties and particularly the ability to ensure adsorption plasticization of surface metallic layers.

The distribution of parameter β_{hkl} and the concentration C of alloying elements through the depth of the deformed zone of a copper alloy under ST conditions is schematically presented in Fig. 27.

As a rule, at the first stage of the process of selective transfer, friction coefficient is high enough to ensure the necessary triboactivation of ST.

A smearing of copper on the mate steel surface is possible [114]. At the same time, as this smearing appears, intensive selective anode dissolving of alloying elements of bronzes or brasses (Sn, Al, etc.) begins. Atoms of alloying elements pass away to the lubricant, so a very thin surface layer of a bronze or brass rubbing part would consist of pure copper.

Triboactivation is one of the necessary conditions for this process. In particular, because of triboactivation of glycerin, some specific products of its decomposition appear, which ensure lubrication of the pair steel–brass. Simakov [112] identified the following products of tribochemical conversion in glycerin at the initial stage of ST: formaldehyde-30, methyl alcohol-32, ethyl alcohol-46, formic acid-46, acrolein-56, propionic aldehyde-73, monoglycerin acid-105, diglycerin acid-118, glycerin dimer-166, etc.

---

*The products of oil degradation during rolling tests on a five-ball machine are also high molecular compounds (in air–resin and coke-like products; in vacuum–mainly coke-like ones). The degree of deterioration of lube oils is about 45-50% until the friction unit fails to operate [109].

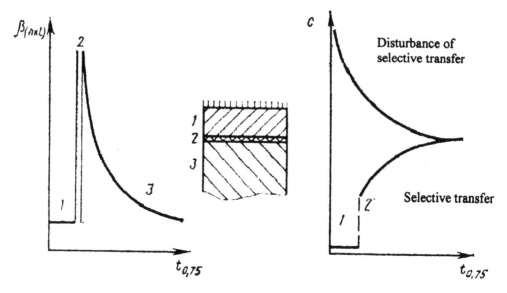

Figure 27 The data characterizing selective transfer structural criteria changes through the depth of deformation zone: (1) copper film; (2) oxide layer; (3) base metal. (From Ref. 32.)

Thus there is an interaction of the active components of surface layers of copper-based alloys not with glycerin itself but with the products of its conversion. At some stages of ST in the tribochemical interaction, such products as carboxylic acids and their salts (metallic soaps, the ions of which were present in a lube fluid) are involved. A remarkable decrease in glycerin surface tension may serve as an evidence of the presence of these compounds [Fig. 28(a)]. The changes in the lube fluid composition may be also proven by decrease in its viscosity [Fig. 28(b)] and by reduction of fluid acidity (initial value of pH was 5.45, whereas for the steady state ST regime, the pH is 6.40) [112]. Oxides and hydroxides of active metals (such as Zn, Sn, Fe) under conditions of trioactivation (dispersion of metals, catalytic effects of freshly formed metallic surfaces, exoelectron emission and that of high-energy electron, etc.) reacting with carboxyl acids produce salts, whereas copper compounds are reduced by iron and aldehydes up to the state of pure metal, which ensures the appearance of a thin film of pure copper on the alloy surface. During friction, this film being unoxidized and correspondingly rather active is transferred easily from one rubbing surface to another and cover this surfaces by a thin layer (the thickness is of 1–2 μm). So, under the steady state ST process, we have friction of copper on copper [7].

Surface-active compounds arising due to tribochemical interactions squeeze glycerin out the contact zones and wear particles and form a durable adsorbed layer, which protects metal from immediate contact with oxygen and hydrogen. These compounds lead to further dispersion of wear particles. That is why the size of copper wear particles tends to diminish [Fig. 28(c)].

In the steady state ST regime, the size of wear particles is similar to that of colloidal particles so micelles arise, which take part in boundary layer formation.

On the other hand, in the lube fluid, there are a lot of copper ions that form complex compounds with ligands of the lubricant. In the electrochemical system copper alloy–lube fluid–steel, electrical field influences on micelles and complex copper compounds in such a way that they have to migrate to the steel and copper surfaces. These phenomena are also beneficial for forming a film of pure copper on rubbing surfaces, which has abnormal

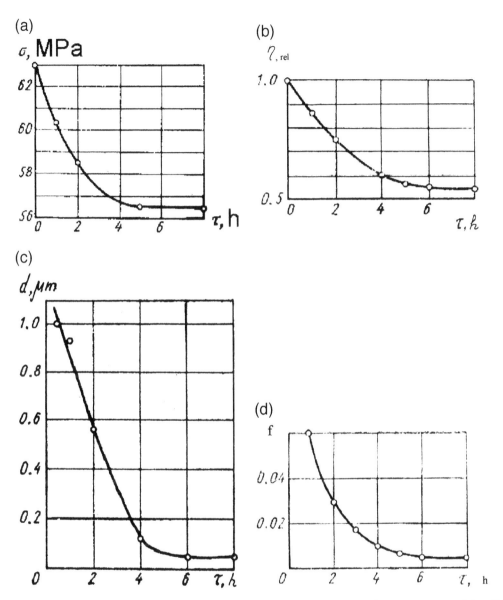

Figure 28 Glycerin surface tension σ (a); glycerin viscosity η (b); size of copper wear particles d (c); and friction coefficient f (d) vs. duration of friction τ during testing of steel on copper alloy in the regime of selective transfer. (From Ref. 112.)

properties [113–117]. A large amount of lattice vacancies arises in the copper film as the process of selective dissolving of alloying elements develops. The lattice vacancies partially coagulate in pores, so the crystalline lattice of the thin copper film under study contains more than 10% lattice points free of atoms. According to Bernall's theory, this leads to the situation when a solid body may behave like a fluid. On the other hand, dislocations in the surface layer of copper can be easily discharged because surface-active compounds formed during the process reduce the energetic barrier. Besides, annihilation of dislocations and lattice vacancies may occur accompanied by energy E liberation [115].

The copper layers formed as a result of selective transfer during friction significantly differ from that obtained by galvanic or by some other methods, which do not use friction as a technology procedure. It is not hardening, has abnormal plasticity, and being in loaded contact ensures pressure reducing in the points of real contact up to the values suitable for polymolecular lubrication realization and correspondingly low values of friction coefficients typical for this lubricating mode. At the same time, this film has significantly lesser shear strength as compared to the base metal, which is also of importance to ensure low friction and absence of wear effects. Thus the main advantages of the process of selective transfer— low friction and wear—are realized because of specific copper films formation, which are in great extent responsible for these characteristics. These films were named "servovite" (i.e., life supporting). The films have high catalytic activity resulting to polymerization of monomers and oligomers that arise due to glycerin tribodestruction. A friction polymer layer cover the top of the film (Fig. 29), which also contributes to low friction and wear. Returning to data presented in Fig. 28, it is possible to see that all processes under study are coincident and interrelated. Really, comparing the characteristics of friction coefficient decrease [Fig. 28(d)] during steel–bronze pair tests in glycerin with that of wear particles size change [Fig. 28(c)] and the characteristics of glycerin surface tension and viscosity [(Fig. 28(a,b)] changes, one can see that there is direct correlation between the reduction of these parameters. So the process of selective transfer clearly demonstrates its tribochemical character. One of the specific features of the process is self-organization at the steady stage of operation, when worn-out dispersed particles of copper getting into the lube fluid again form complex compounds and micelles, which migrate to the rubbing surface ensuring copper film regeneration. Because wear particles separating and worn-out material recurring to surface have a periodic character, superlow oscillations of contacting parts and wear rates are observed and correspondingly friction force oscillations under steady ST process [113,114].

In spite of the very substantial tribological effect of the ST process at present, it finds limited use because of the rather narrow ranges of operating conditions under which it can be developed, and a limited number of lubricating fluids and contacting materials for which this effect can be realized (Fig. 30). Besides, ST is very sensitive to the structural state of alloys. For example, it was shown that structural instability of solid solutions of copper alloys and their heterogeneity are factors restricting the ST process [111,113].

Studies made by Kuksenova et al. [111] showed that in some cases, there is a possibility of tailoring operating properties and structural state of surface layers of copper alloys. It may be done both by changes in copper alloy composition, which will prevent the de-

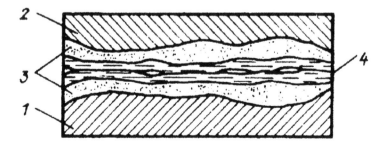

Figure 29 Scheme of friction contact at selective transfer regime realization (according to Garkunov): (1) steel; (2) bronze; (3) servovite films; (4) tribopolymer films.

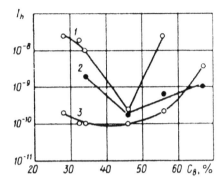

Figure 30 Wear rate I_h of some bronzes rubbing on steel vs. water concentration C_w in poly-ethylenglycol solution in water: (1) BrSP5-2,5; (2) BrSP10-1; (3) BrSP4-0,25. (From Ref. 32.)

composition of its structural constituents in the process of diffusion under friction, and by the control of lube fluid capacity to realize the ST process.

It should also be noted that by introducing in oils for steel–steel pair lubrication some complex copper compounds analogous to those formed during tribochemical interactions at ST, it is possible to maintain for a long time a process similar to ST, accompanied by servovite film formation on rubbing surfaces and low friction and wear rate values [114,116].

V. WEAR PROCESS CONDITIONED BY HYDROGEN

As distinct from the above-considered tribochemical processes, wear conditioned by hydrogen or "hydrogen wear," being itself without doubt a tribochemical process, is nevertheless unwanted as this results in a serious decrease in reliability and service life of friction units operating in hydrogen-containing fluids, gases, etc., or in lubricants where hydrogen can appear because of some tribochemical processes development. Hydrogen wear manifests itself as the failure of rubbing metallic elements resulting from hydrogen adsorption by metals.

Outwardly, this mode of wear is revealed in some as rather unusual damages of surface layers. For example, a transfer of steel particles on the mate bronze surface takes place while rubbing in kerosene or a steel transfer on mild Babbitt (pair steel crankshaft–Babbitt bearing); increased wear of titanium and its alloys in mineral oils is observed (as compared to wear in the absence of oils); increased wear is also observed in metallic friction elements lubricated by water, etc. Hydrogen wear usually extends to a depth of the surface layer deformed during friction—some tenth of a millimeter—but it is enough for irreversible damage of rubbing surfaces. The process of hydrogen wearing was first described by Poljakov et al. [117]. The process differs from the well-known hydrogen embrittlement by the origin. Really, hydrogen liberation and the corresponding reactions passing on rubbing surfaces are the results of friction. Moreover, friction initiates hydrogen adsorption by surface layers because of an abrupt diffusion rate increase resulting from surface layer deformation and temperature increase. All these processes finally lead to surface layer embrittlement, dispersion, or damage [118].

The process of hydrogen wear includes several stages, the final of which is the failure of contacting surfaces. The first stage is hydrogen transport from the source of its generation to the rubbing surfaces. In the process of boundary lubrication, hydrogen is generated because of lube oil destruction in the friction zone and first of all that of molecules with polar groups. Hydrogen liberation arises or is considerably accelerated because of mechanical–chemical and thermal factors, ionization, and catalytic action of metallic surfaces freshly prepared during friction, especially under heavy loading conditions.

One more source of hydrogen in the friction zone is the decomposition of water presenting in oil [118–120].

Hydrogen can be also liberated because of electrochemical corrosive processes in liquids with increased electrical conductivity, e.g., in oils in the presence of water and electrolytes. The second stage of the process of hydrogen wear is its adsorption on rubbing surfaces. For instance, while the selective-transfer hydrogen wear was not observed, the transition to it can occur at a temperature rise above 80–90°C.

The third stage of the process under study is hydrogen diffusion to the deformed layer of steel to some "hydrophile" zone, which soaks it up according to Garkunov [118].

The volume of hydrogen diffused in surface layers depends on the operating conditions and the metal state and composition. For example, such alloying elements as chrome, vanadium, and titanium decrease penetration of hydrogen in steel. Cold hardening in its turn increases hydrogen adsorption. Steel after cold working adsorbs in 1000 times more hydrogen as compared with the tempered one. According to Bronstein et al. [120], hydrogen-charged layers of steel are not formed during friction in base oils having insufficient lubricating properties because at these conditions, maximum of temperatures and stresses are localized in the contact zone, so hydrogen does not diffuse in metal.

The final phase of the process is the failure of surface layer of steel. Practically simultaneously in a subsurface layer, numerous confluent cracks arise, which result in surface layer destruction and wear products appearance. There are two types of hydrogen wearing: wearing by dispersion when hydrogen intensifies steel and cast iron dispersion and wearing by failure, which manifests itself in instant failure of surface layer to the depth of 1–2 μm.

VI. CONCLUSION

Tribochemical aspects of friction under boundary lubrication are considered as applied to lubrication of contact by adsorbed, chemically modified, and tribopolymeric layers as well as that under conditions of selective transfer. Analysis of these processes allows making the following conclusions:

- Tribochemical processes are one of the main mechanisms that ensure lubricating action under boundary lubrication.
- The succession of tribochemical processes under boundary lubrication was considered for lubrication by adsorbed and chemically modified boundary layers (AW and EP lubrication). The equations for computing the conditions of boundary lubricating layers formation and breaking down were given, including the computation of transition temperatures from lubricating by boundary layers of adsorption and chemical origin (i.e., chemically modified layer) to boundary layer failure and seizure arising. The computation of temperatures of chemically modified layer formation was also given.
- The comparison of thermochemical (static) and tribochemical processes under boundary lubricating layer formation was made, and the difference in these processes and the possibilities of their realization were discussed. The difference

in the rate of boundary layer formation at thermochemical and tribochemical processes was revealed. It was shown that one of the most appropriate explanations of the mechanism of tribochemical processes may be given by the ion radical concept developed by Kajdas (the NIRAM concept).

- The mechanism of tribochemical processes under boundary lubrication in hydrocarbon liquids including low-molecular ones was considered. The role played by oxidation and tribopolymerization in their lubricating properties was shown. The lubricating action of hydrocarbons and surface-active compounds contained in hydrocarbons is explained from the position of the NIRAM concept. Test results are presented, which proved the given concepts and equations.

- The tribochemical processes were considered to ensure the lubricating action of EP and AW additives using solutions of some typical EP and AW additives and friction modifiers in lube oils as an example. The influence of operating conditions on the efficacy of lubrication action of these additives was demonstrated. The equations describing processes of modified layer formation and breaking down were experimentally proven. The lubricating ability of sulfides and disulfides was explained using the NIRAM concept.

- The mechanism of the decline of lubricating ability during friction was discussed. It is shown that this is caused by carrying away from the contact zone of a share of active reagents of lube oils and complex tribochemical processes, resulting in formation of compounds without the lubricating ability.

- The phenomenon of selective transfer, which is a unique variety of boundary lubrication as it allows obtaining the values of friction coefficient similar to that at hydrodynamic lubrication and extremely low wear rates, was described. The mechanism of selective transfer and tribochemical processes leading to its realization were considered. The conditions of realization of selective transfer were considered including lubricating fluids (glycerin, seawater, some synthetic and petroleum oils), contacting materials, and operation regimes.

- It was shown that during the process of selective transfer, some deep tribochemical conversion occurred in the third body consisting of the boundary lubricating layer and very thin surface layers of rubbing materials. The tribochemical processes here include selective dissolution of alloying elements of copper alloys, formation of the so-called servovite copper film, with an abnormal plasticity, which ensures real contact pressures reducing and lubrication by a polymolecular film. Surface-active compounds formed because of tribochemical processes provide this film with effective adsorption on activated friction surfaces.

- The phenomenon of wear of steels conditioned by hydrogen was considered to result from adsorption of hydrogen by surface layers of rubbing bodies. Hydrogen is liberated as a result of a set of tribochemical reactions in lube oils. Saturation of surface layers of steel parts by hydrogen results in their failure. The role of tribochemical processes in hydrogen wear was considered.

REFERENCES

1. Buyanovsky, I.A.; Fouks, I.G.; Shabalina, T.N. *Boundary Lubrication: Stages of Tribology Development*; Publishing House "Oil and Gas", Russian State University of Oil and Gas: Moscow, 2002, 230 pp.
2. Hardy, W.B.; Doubleday, I. Boundary lubrication—The paraffin series. Proc. R. Soc. Lond. A 1922, *100*, 550–574.

3. Deryaguin, B.V.; Karassev, V.V.; Zakhavaeva, N.N.; Lazarev, V.P. The mechanism of boundary lubrication and the properties of the lubricating film. Wear 1957/58, *1* (4), 277–290.
4. Akchmatov, A.C. *Molecular Physics of Boundary Friction*; Physmftgiz: Moscow, 1963; 472 pp.
5. Bowden, F.P.; Gregory, J.N.; Tabor, D. Lubrication of metal surfaces by fatty acids. Nature 1945, *156* (3952), 97–99.
6. Bowden, F.P.; Tabor, D. *Friction and Lubrication of Solids*; University Press: Oxford, 1950; 176–246.
7. Garkunov, D.N. Selective transfer (effect of wearlessness) and its applications in triboengineering. In: *Tribology Grounds (Friction, Wear and Lubrication)*, Chichinadze, A.V., Ed.; The Center "Science and Engineering:" Moscow, 1995; 225–248.
8. Salomon, G. Failure criteria in thin film lubrication—the IRG program. Wear 1976, *36*, 1–6.
9. Kajdas, C. Tribochemistry. In: *Tribology*, Franek, F., Bartz, W.J., Pauschitz, A., Eds.; Scientific Achievements, Industrial Applications, Future Challengers: Plenary and Key Papers from 2nd World Tribology Congress, Vienna, Austria, 3–7 Sept. 2001: 39–46.
10. Heinicke, H. Tribochemistry. *Translation from English*; Mir: Moscow, 1987; 584 pp.
11. Poljakov, A.A.; Garkunov, D.N.; Simakov, Yu S.; Ya Matuchenko, V.; Vasiliev, I.I.; Gorelov, L.R. *Protection from Hydrogen Wearing*; Machinostroenie: Moscow, 1980; 43–52.
12. Avvakumov, E.G. *Mechanical Methods of Chemical Processes Activation*; Nauka (Siberian Department): Novosibirsk, 1979; 74–165.
13. Spikes, H. Advances in the study of thin lubricant films. In *New Directions in Tribology*; Hutchings, I.M. Ed.; MEP: Bury St. Endmunds and London, 1997; 355–369.
14. Buyanovsky, I.A. Temperature-kinetic method of evaluation of temperature ranges of efficiency of lubricating materials under heavy conditions of boundary lubrication. Friction Wear 1993, *14*, 129–142.
15. Zaslavsky, Yu S.; Berlin, A.A.; Zaslavsky, R.N.; Cherkashin, M.I.; Belozerova, K.E.; Rusakova, V.A. Antiwear, extreme pressure and antifriction action of friction polymer formation additives. Wear 1972, *20*, 287–297.
16. Raiko, M.V.; Dmytrychenko, N.F. Some aspects of boundary lubrication in the local contact of friction surfaces. Wear 1988, *126*, 69–78.
17. Buyanovsky, I.A. The role of surface interactions in tribological process. Chem. Technol. Fuels Oils 1992, *11*, 7–13.
18. Vasiliev, Yu N.; Buyanovsky, I.A.; Matveevsky, R.M.; Shepilov, Yu P. A model of scoring under boundary lubrication. In *Calculating-Experimental Methods of Friction and Wear Estimation*, Kraghelsky, I.V., Ed.; Nauka: Moscow, 1980; 65–69.
19. Kingsbury, E.P. Some aspects of thermal desorption of a boundary lubricant. J. Appl. Phys. 1958, *29* (6), 888–891.
20. Cameron, A. On a unified theory of boundary lubrication. In *Mixed Lubrication and Lubricated Wear*, Dowson, D., Godet, M., Eds.; Butterworth: London, 1985; 135–146, 170–171.
21. Sakurai, T. Role of chemistry in the lubrication of concentrated contacts. ASME Trans. J. Lubr. Technol. 1981, *103*, 473–485.
22. Hardy, W.B.; Doubleday, I. Boundary lubrication—The latent period and mixtures of two lubricants. Proc. R. Soc. Lond. A 1923; Vol. 104, 25–39.
23. Deryaguin, B.V. The problems of boundary lubrication. In; *Lubricating Materials—Efficiency Increasing and Application*; Gostoptechizdat: Moscow, 1957; 5–17.
24. Bhushan, B.; Israelachvili, J.K.; Landman, U. Nanotribology: friction, wear and lubrication at the atomic scale. Nature 1995, *344*, 607–616.
25. Hsu, S.M.; Klaus, E.E. Estimation of the molecular junction temperatures in four-ball contacts by chemical reaction rate studies. ASLE Trans. 1978, *21*, 201–210.
26. Matveevsky, R.M.; Buyanovsky, I.A.; Karaulov, A.K.; Mischuk, O.A.; Nosovsky, O.I. Transition temperatures and tribochemistry of the surfaces under boundary lubrication. Wear 1990, *136*, 135–149.
27. Kajdas, C. Tribochemical and thermochemical reactions of tribological additives. Proceedings of Symposium INTERTRIBO'90. Zbornic prednasok. Vysoke Tatry, Strbske Pleso. Ceskoslovensko 1990, *17-20.04*, 40–48.

28. Buckley, D.H. *Surface Effects in Adhesion, Friction, Wear and Lubrication*; Elsevier Science: Amsterdam-Oxford-New York, 1981.

29. Koujarov, A.S.; Rjabuchin, YuI. Triboactivation of complexes forming of compact copper. Friction Wear 1991, *12*, 99–107, *in Russian*.

30. Sanin, P.I. Chemical aspects of boundary lubrication. Friction Wear 1980, *1*, 45–57.

31. Postnikov, S.N. *Electrical Phenomena Under Friction and Machining*; Volgo-Vyatskoe izdatelstvo: Gorky, 1975; 38–106.

32. Rybakova, L.M.; Kukcenova, L.I. *Structure and Wear Resistance of Metals*; Machinostroenie: Moscow, 1975; 3–57.

33. Kostezky, B.I.; Nosovsky, I.G.; Bershadsky, L.I.; Karaulov, A.K. *Surface Strength of Materials Under Friction*; Technika: Kiev, 1976; 128 pp.

34. Kostezky, B.I. Evolution of structural and phase state and mechanisms of material self-organization under external friction. Friction Wear 1993, *4*, 773–783.

35. Thiessen, P.A.; Meyer, K.; Heinicke; G. Grundlagen der Tribochemie. K1. *Chemie, Geologie, Biologie, 1966, Nr. 1*; Akademie Verlag: Berlin, 1967; 160s.

36. Butiaguin, PYu. Kinetics and nature of mechano-chemical reactions. Prog. Chem. 1971, *40*, 1935–1959.

37. Ravikovitch, A.M.; Borstcheevsky, S.B.; Belinskaia, R.M. Influence of organo-sulfur compounds on their antiseizure performance. *VNIINP Trans., XXI, Advances in Additives for Lubricating Oils*; Chimia: Moscow, 1977; 74–80.

38. Goldblatt, I.L. Model for lubrication behavior of polynuclear aromatics. Ind. Eng. Prod. Res. Dev. 1971, *10*, 270–278.

39. Kajdas, C. Ion-radical mechanism of lubrication action of oil components. Friction Wear 1986, *7*, 626–633. *in Russian*.

40. Pearson, R.G. *Hard and Soft Acids and Bases*; Dowden, Hutchinson and Ross: Stroudsburg, PA, 1973.

41. Vinogradov, G.V.; Podolsky, Y. Y. Mechanism of antiwear and antifriction action of lubricants under heavy regimes of boundary lubrication. In *To the Nature of Friction of Solids*, Ishlinsky, A.Y., Ed.; Nauka i technika: Minsk, 1971; 281–304.

42. Vinogradov, G.V. Some experience of EP characteristics of hydrocarbon lubricating media studies. In *Methods of Antiscoring and Antiwear Characteristics of Lubricating Materials Evaluation*, Kchrutshev, M.M., Matveevsky, R.M., Eds.; Nauka: Moscow, 1969; 3–24.

43. Tomaru, M.; Hironaka, S.; Sakurai, T. Effect of some chemical factors on film failure under EP conditions. Wear 1977, *41*, 117–140.

44. Grondkovsky, M.; Makovska, M.; Kajdas, C. Studies of tribochemical reactions in n-hexadecane. *Proc. International Scientific and Practical Symposium "SLAVIANTRIBO-5" Land and Aerospace Tribology-2000: Problems and Achievements*; Pogodaev, L.I., Ed.; Rybinsk State Academy of Aviation Technology: Rybinsk, 2000; 100–103.

45. Cavdar, B.; Ludema, K.C. Dynamics of dual film formation in boundary lubrication of steels. Part 1. Functional nature and mechanical properties. Wear 1991, *148*, 305–327.

46. Aksenov, A.F.; Beliansky, V.P.; Shepel, A.Y. To some ways of metal wear resistance improving under friction in hydrocarbon oils. Friction Wear 1980, *1*, 70–78. *in Russian*.

47. Chaikin, S.W. On frictional polymer. Wear 1967, *10*, 49–60.

48. Goldblatt, I.L. Surface fatigue initiated by fatty acids. ASLE Trans. 1973, *16*, 150–159.

49. Kajdas, C. About a negative-ion concept of the antiwear and antiseizure action of hydrocarbons during friction. Wear 1985, *101*, 1–12.

50. Molenda, Y.; Grondkovsky, M.; Kajdas, C. Tribochemical behavior of vinyl compounds studies. *Proc. International Scientific and Practical Symposium "SLAVIANTRIBO-5" Land and Aerospace Tribology-2000: Problems and Achievements*; Pogadaev, L.I., Ed.; Rybinsk State Academy of Aviation Technology: Rybinsk, 2000; 153–156.

51. Hermance, H.W.; Egan, T.F. Organic deposits on precious metal contacts. Bell Syst. Tech. J. 1958, *37*, 739–776.

52. Fein, R.S.; Kreuz, K. Chemistry of boundary lubrications of steel by hydrocarbons. ASLE Trans. 1965, *8*, 29–38.

53. Campbell, W.E.; Lee, R.E. Polymer formation on sliding metals in air saturated with organic vapors. ASLE Trans. 1962, 5, 91–103.

54. Furey, M.J. The formation of polymeric films directly on rubbing surfaces to reduce wear. Wear 1973, 26, 369–392.

55. Zaslavsky, R.N.; Zaslavsky, Y.S.; Vasilieva, E.M.; Asrieva, V.D.; Aguievsky, D.A. To the mechanism of running-in action of friction-polymer-forming additives to oils. Doklady AN SSSR 1980, 251, 863–865.

56. Zaslavsky, Y.S.; Zaslavsky, R.N. *Mechanism of Action of Antiwear Additives to Oils*; Chimia: Moscow, 1978; 123–173.

57. Zaslavsky, R.N.; Zaslavsky, Y.S. Application of friction-polymer-forming additives for the running in of rubbing surfaces. Wear 1987, 118, 1–26.

58. Furey, M.J.; Kajdas, C.; Ward, T.C.; Hellgeth, J.W. Thermal and catalytic effect on tribo-polymerization as a new boundary lubrication mechanism. Wear 1990, 136, 85–97.

59. Matveevsky, R.M.; Buyanovsky, I.A.; Lozovskaya, O.V. *Antiscoring Resistance of Lube Oils Under Friction at Boundary Lubricating Regime*; Nauka: Moscow, 1978; 62–167.

60. Huges, T.P.; Wittingam, G. The influence of surface films on the dry and lubricated sliding of metals. Trans. Faraday Soc. 1942, 38, 9–17.

61. Allen, T.; Patel, R.M. The adsorption of long-chain fatty acids on finely divided solids using a flow microcalorimeter. J. Colloid Interface Sci. 1971, 35, 647–655.

62. Frewing, J.J. The heat of adsorption of long-chain compounds and their effect on boundary lubrication. Proc. R. Soc. Lond. A 1944, 181, 270–285.

63. Kreuz, K.L.; Fein, R.S.; Rand, S.J. Solubilization effects in boundary lubrication. Wear 1973, 23, 393–407.

64. Fisher, D.A.; Hu, Z.S.; Hsu, S.M. Molecular orientation and bonding of monolayer stearic acid on a copper surface prepared in air. Tribol. Lett. 1997, 3, 41–45.

65. Vinogradov, G.V.; Kusakov, M.M.; Sanin, P.I.; Morozova, O.V.; Bezborodko, M.D.; Ulianova, A.V.; Razumovskaya, E.A.; Zaslavsky, Y.S.; Riabova, D.V. Organo-phosphorous and thioorganic compounds as antiwear additives to lube oils. In *Efficacy Improvement and Lubricating Materials Application,* Locikov, B.V., Krein, S.E., Foucks, G.I., Eds.; Gostoptrch-izdat: Moscow, 1957; 51–72.

66. Kramer, I.R. The effect of surface-active agents on the mechanical behavior of aluminum single crystals. AIME Trans. 1961, 22, 989–993.

67. Kajdas, C. About an anionic-radical concept of the lubrication mechanism of alcohol. Wear 1987, 116, 167–180.

68. Kajdas, C.; Obadi, M. The effect of aliphatic alcohols C_4–C_{18} on wear of the steel-on-steel mating element. Tribologia 1998, 159(3), 367–380. *in Polish.*

69. Kajdas, C.; Shuga'a, A-K. The investigations of antiwear properties and tribochemical reactions of esters of palmitic and aliphatic alcohols in the steel-on-steel system. Tribologia 1998, 159(3), 389–402. *in Polish.*

70. Hu, L.; Chen, J.; Liu, W.; Xue, Q.; Kajdas, C. Investigation of tribochemical behavior of Al–Si alloy against itself lubricated by amines. Wear 2000, 243, 60–67.

71. Forbes, E.S. Antiwear and extreme pressure additives for lubricants. Tribology 1970, 3, 145–152.

72. Zaslavsky, Y.S. *Tribology of Lubricating Materials*; Chimia: Moscow, 1991; 6–210.

73. Liu, W.; Xue, Q. Tribochemistry and the development of AW and EP oil additives: a review. Lubr. Sci. 1994, 7, 81–92.

74. Papey, A.G. Friction Reducers for Engine and Gear Oils—Review of the State of the Art. In *Additive for Lubricants and Operational Fluids*; Bartz, W.J., Ed.; Technishe Akademie Esslingen Druck: Ostfildern, 1986; Vol. 1, 5.1-1–5.1-10.

75. Mitchell, P. Oil-soluble Mo–S compounds as lubricant additives. Wear 1984, 100, 405–432.

76. Buyanovsky, I.A.; Karaulov, A.K.; Lukinuk, L.M.; Pereverzeva, E.A. Evaluation of an upper limit of chemically active additives efficacy by lab tests data. Chem. Technol. Fuels Oils 1989, 12, 34–35. *in Russian.*

77. Matveevsky, R.M.; Buyanovsky, I.A.; Lashkchi, V.L.; Vipper, A.B. Evaluation of activation

energy of friction surfaces chemical modification process under boundary lubrication. Chem. Technol. Fuels Oils 1976, *2*, 50–52. *in Russian.*

78. Buyanovsky, I.A.; Kloss, H. Estimation of the order of tribochemical processes by temperature dependences of coefficients of friction. Friction Wear 1985, *6*, 527–531. *in Russian.*

79. Buyanovsky, I.A.; Kukcenova, L.I.; Rybakova, L.M.; Foucks, I.G. Methods of lubricating action improving by two-layered film creation. Vestn. Machinostr. 2000, *4*, 6–17.

80. Buyanovsky, I.A.; Zaslavsky, R.N. Concerning to usage of the method of antiwear properties estimation of lube oils with additives on four-ball machine. Friction Wear 1991, *12*, 113–117. *in Russian.*

81. Kajdas, C. On a negative-ion concept of EP action of organo-sulfur compounds. ASLE Trans. 1985, *28*, 21–30.

82. Hironaka, S.; Yahagi, Y.; Sakurai, T. Effect of adsorption of some surfactant on antiwear properties. ASLE Trans. 1978, *21*, 231–235.

83. Azouz, A.; Rowson, D.M. A comparison of technique for surface analysis of extreme pressure films formed during wear. In: *Microscopic Aspects of Adhesion and Lubrication.* Jeorges, J.M., Ed.; Tribology Series, Elsevier Scientific Pub.: Amsterdam, 1982; Vol. 7, 763–778.

84. Plaza, S. Some chemical reactions of organic disulfides in boundary lubrication. ASLE Trans. 1987, *30*, 493–500.

85. Anand, O.N.; Malik, V.P.; Neemla, K.D. Studies on extreme pressure films-reactions of sulphurized vegetable oils and alkyl phenols with iron. Tribol. Int. 1986, *19*, 128–132.

86. Studt, P. Die Adsortion von Schmierolzusatzen an Stahloberflachen und ihre tribolgishe Bedeuutung. In: *Additives for Lubricants and Operation Fluids,* Bartz, W.J., Ed.; Technishe Akademie Esslingen Druck: Ostfildern, 1986; Vol. 1, 3.6-1–3.6-16.

87. Beeck, O.J.; Givens, J.W.; Williams, E.C. On the mechanism of boundary lubrication. II. Wear prevention by addition agents. Proc. R. Soc. Lond 1940, *A177*, 103–118.

88. Nakayama, K.; Sakurai, T. The effects of surface temperature on chemical wear. Wear 1974, *29*, 373–385.

89. Abzalov, P.N. Kinetics of wear of metals under boundary lubrication taking into account lubricating action of ion-radicals of hydrocarbons. Prediction and Control of Behavior of Lube Materials for Mobile Machines, Abzalov, P.N., Ed.; Izdatelstvo "Fan" AN RU: Tashkent, 1993; 70–78.

90. Okabe, H.; Masuko, M.; Oshino, H. Effects of viscosity and contact geometry on tribochemical surface reactions. ASLE Trans. 1982, *25*, 39–43.

91. Shore, G.I. Mechanism of action and express-estimation of performance of lube oils with additives. *Review*; ZNIITEneftechim: Moscow, 1996; 1–66.

92. Kapsa, Ph.; Martin, J.M.; Blane, C.; Georges, J.M. Antiwear mechanism of ZDDP in the presence of calcium sulfonate detergent. ASME Trans. J. Lubr. Technol. 1981, *103*, 486–496.

93. Yin, Z.; Kasrai, M.; Bancroft, G.M.; Fife, K.; Colianni, M.L.; Tan, K.H. Application of soft X-ray absorption spectroscopy in chemical characterization of antiwear films generated by ZDDP. Part II: Effect of detergents and dispersants. Wear 1997, *202*, 192–201.

94. Iocebidze, Kutelia, E.R.; Bershadsky, L.I.; Zamansky, L.S. To the phenomenon tribosynthesis in dissipative-ordered heterophase systems on the base of carbon. Dokl. Akad. Nauk GSSR 1987, *128*(1), 29–32.

95. Bershadsky, L.I.; Iocebidze, D.S.; Kutelia, E.R.; Zamansky, L.S. Effect of tribostimulation on graphite-diamond dissipative structures tribosynthesis. In *Proc. of All Union Symposium on Mechanochemistry and Mechanoemission of Solids, Theses of Reports, Tchernigov*, September 11–14, 1990; Vol. 1, 12–14.

96. Buche, W. Untersuchungen uber molekularphysikalishe Eigenschaften des Shcmiermittel und ihre Bedeutung bei halbflussiger Reibung. –Petroleum 1931, *27*,s. 587.

97. Kraghelsky, I.V. *Friction and Wear in Machines*; Machgiz: Moscow, 1962; 253–255.

98. Deryaguin, B.V.; Zachovaeva, N.N.; Kusakov, M.M.; Lazarev, V.P.; Samyguin, M.O. To the nature of oilness of lube materials and methods of its numerical estimation. In *All Union Conference on Friction and Wear in Machines. Reports*; Izdatelstvo AN SSSR: Moscow-Leningrad, 1939, 519–534.

99. Chandrasekaran, M.; Batchelor, A.W.; Loh, N.L. Lubricated seizure of mild steel observed by X-ray imaging. Proc. Inst. Mech. Eng. 2000, *214*(Part J), 359–374.

100. Klimov, K.I. Antiscoring oil characteristics as a function of the rate of their tribochemical transformations in the friction zone. In: Methods of Estimation of Antiscoring and Antiwear Characteristics of Oils, Kchrutshev, M.M., Matveevsky, R.M., Eds.; Nauka: Moscow, 1969; 26–34.

101. Deryaguin, B.V.; Zachovaeva, N.N. To the effect of air oxygen on the lube film wear under kinetic friction. In *Proc. of the Second All Union Conference on Friction and Wear in Machines*; Isdstelstvo AN SSSR: Moscow-Leningrad, 1947; Vol. 1,96–102.

102. Koinkar, V.N.; Bhushan, B. Micro/nanoscale studies of boundary layers of liquid lubricants for magnetic discs. J. Appl. Phys. 1996, *79*, 8071–8075.

103. Hsu, S.M.; Zhang, X.H. Lubrication: traditional to nano-lubricating films. In *Micro/Nanotribology and its Applications*. Bhushan, B., Ed.; NATO ASI Series, Series E: Appl. Sci. 1996, *330*, 399–414.

104. Zisman, W.A. Friction. Durability and wettability properties of monomolecular films on solids. In Friction and Wear: Proc. Symp. on Friction and Wear, Detroit, 1957, Davies, R., Ed.; Elsevier: Amsterdam, London, NY, Princeton, 1959; 110–148.

105. Holzhauer, W.; Ling, F.F. In-situ SEM study of boundary lubricated contacts. Tribol. Trans. 1988, *31*, 360–369.

106. Karacik, I.I.; Cherny, A.S. Greases lubricating ability deterioration due to wear passivating action. Friction Wear 1984, *5*, 1045–1050. *in Russian.*

107. Li, H.; Chao, K.K.; Duda, J.L.; Klaus, E.E. A study of wear, chemistry and contact temperature using microsample four-ball wear tests. Tribol. Trans. 1999, *42*, 529–534.

108. Bernikova, N.B. Development and research of technological process of lubricating and washing of diminutive assembled friction units. *Thesis of Ph.D. dissertation*; NII Chasovoy promyshlennosti: Moscow, 1988; 28 pp.

109. Sosnulina, L.N.; Scriabina, T.G. Studies of products of greases tribochemical transformation under rolling. Friction Wear 1984, *5*, 923–929. *in Russian.*

110. Garkunov, D.N.; Kraghelsky, I.V. To the corpuscular seizure of materials under friction. Dokl. Akad. Nauk SSSR 1957, *113*, 326–327.

111. Kuksenova, L.I.; Poljakov, A.A.; Rybakova, L.M. Lubricating materials and the selective transfer phenomenon under friction. Vestn. Machinostr. 1990, *1*, 35–40.

112. Simakov, Y.S. Physical–chemical phenomena under selective transfer. In *Selective Transfer in Heavy Loaded Friction Units*, Garkunov, D.N., Ed.; Machinostroenie: Moscow, 1982; 88–111.

113. Koujarov, A.S.; Marchak, R.; Guzik, Y.; Kravchik, K.; Zadoshenko, E.G. Tribological evidence of self-organization in the system brass–glycerin–steel. Friction Wear 1996, *17*, 113–122. *in Russian.*

114. Koujarov, A.S. Complexes formation under friction and selective transfer regime. Vestn. Machinostr. 1990, *9*, 27–30.

115. Poljakov, A.A. Dislocation–vacation mechanism of selective transfer. Effect Wearlessness Tribotechnol. 1992, *3–4*, 3–10. *in Russian.*

116. Chigarenko, G.G.; Ponomarenko, G.G.; Burlov, A.S.; Garnovsky, A.D. Studies of tribochemical reactions resulting in lubricating layer formation under selective transfer. Effect Wearlessness Tribotechnol. 1994, *3–4*, 64–75. *in Russian.*

117. Poljakov, A.A.; Kraghelsky, I.V.; Garkunov, D.N. To hydrogen wear. Dokl. Akad. Nauk SSSR 1970, *195*, 658–666.

118. Garkunov, D.N. Self-organizing processes in tribological system under friction interaction. In Handbook on tribology, Hebda, M., Chichnadze, A.V., Eds.; Machinostroenie: Moscow, 1989; Vol, 1. 288–323.

119. Kchrustalev, Y.A. Physical–chemical concept of metals hydrohenization. Effect wearlessness Tribotechnol. 1997, *2*, 19–35. *in Russian.*

120. Bronstein, L.A.; Shechter, Y.N.; Pashkov, E.V.; Furman, A.Y. Hydrogen factor of metals wear in hydrocarbons. Friction Wear 1991, *12*, 838–847. *in Russian.*

12

Electrochemistry, Corrosion, and Corrosion-Wear

Matgorzata E. Ziomek-Moroz and Jeffrey A. Hawk
U.S. Department of Energy, Albany, Oregon, U.S.A.

I. INTRODUCTION

Corrosion affects many sectors of the nation's economy, either directly by material degradation or through the design choices made by engineers [1]. The annual cost of degradation of metallic materials due to corrosion and wear–corrosion interactions is approximately 3–4% of the gross national product (GNP) of the United States, with approximately 20% of this cost being avoidable through better utilization of existing materials and protection technologies (e.g., proper design, selection of materials, coatings and linings, cathodic protection, inhibitors, etc.). Table 1 summarizes the cost of metallic corrosion in the United States [1,2]. However, more important than the cost alone, metallic corrosion and the costs to control corrosion result in the excess utilization of materials, energy, labor, and technical expertise that would otherwise be available for alternative uses [1]. Table 2 summarizes the elements of these costs of corrosion [1,3].

The never-ending search for better and more efficient technologies presents a unique problem to the material scientist. Technologists are constantly looking to improve existing processes and initiate innovative ones, and as such, the demand for new and better materials for our economy is insatiable. To utilize a new technology, existing materials must be used in its construction. In many instances, the existing materials cannot withstand the severe conditions of the service environment. This is particularly so when a material must exhibit good chemical stability in a corrosive environment. An awareness of the limits of the chemical stability of these materials of construction, and how to affect such limits of chemical stability are needed. In the case of corrosion, the immense variety of conditions under which it occurs, and the number of forms it may take, make it hard to properly design structures and components to withstand the environmental conditions beforehand. This is usually compounded by a general lack of knowledge of the environment that the structure or component will operate in. As such, the design engineer makes the best choice of materials for the application given the general conditions of operation. Once the structure is built or the component goes into operation, it can be monitored, and if there are corrosion problems, the situation can be analyzed in light of the new information, and a more appropriate material selected for repair or remediation of the part.

Table 1 Metallic Corrosion in the United States

	1975 (Billions of current dollars)	1995
All Industries		
Total	82.0	296.0
Avoidable	33.0	104.0
Motor Vehicles		
Total	31.4	94.0
Avoidable	23.1	65.0
Aircraft		
Total	3.0	13.0
Avoidable	0.6	3.0
Other Industries		
Total	47.6	189.0
Avoidable	9.3	36.0

As with most engineering problems, corrosion control is a systems problem. By that, each corrosion event is unique in terms of materials of construction, the way the materials are placed in contact, the overall macroenvironment the materials experience, and the localized microenvironment in which many of the complex ionic processes occur. In trying to understand corrosion and to control its deleterious effects, the fundamentals of electro-chemistry and corrosion of metals must be understood. When this basic knowledge is used in conjunction with information found in handbooks and textbooks, it is possible to make better design decisions where corrosion is identified as a potential problem. As such, it is essential to have a basic understanding of the fundamentals of electrochemistry so that in-service corrosion problems can be analyzed as they occur and proper solutions devised when needed.

Table 2 Some Elements of the Cost of Corrosion

Capital Costs
Replacement of equipment and buildings
Excess capacity
Redundant equipment
Control Costs
Maintenance and repair
Corrosion control
Design Costs
Materials of construction
Corrosion allowance
Special processing
Associated Costs
Loss of product
Technical support
Insurance
Parts and equipment inventory

The basics of electrochemistry, the types of metallic corrosion occurring in corrosive service environments, an understanding of how these problems can be simulated in the laboratory, and some simple approaches to corrosion control are discussed. Because the nature of corrosion is based on electrochemistry, fundamental electrochemical concepts of metallic corrosion are discussed through the use of thermodynamics and kinetics. This discussion will be followed by brief descriptions of the forms of corrosion and the basic experimental methods used in electrochemical testing. The chapter concludes with a brief overview of surface modification for corrosion control.

II. BASICS OF ELECTROCHEMICAL CORROSION

A. Electrochemical Principles

To understand the basis for various approaches to tribological surface modification in aqueous solutions where corrosion may be a concern, it is necessary to develop a basic understanding of the thermodynamics and kinetics of electrode reactions. Thermodynamics can be applied to determine which processes can occur and how strong the relative tendency is for these changes to take place (i.e., for the metals to corrode). Its principle value is in yielding information on intermediate products of the complementary anodic and cathodic partial reactions that take place at the electrodes and constitute a complete process. The structures and characteristics of the intermediate products often control the resistance of the metal surface to corrosion attack. The Nernst equation, or the fundamental thermodynamic expression, is the primary tool used in this approach and it relates reversible, or single electrode, potentials to the effective concentration of species in the solution. However, thermodynamics tells us nothing about the rate at which corrosion occurs. Kinetics, on the other hand, provides the necessary information to determine the rate at which a corrosion reaction occurs.

The stability potential–pH diagrams, more commonly known as Pourbaix diagrams, are discussed in this section. Pourbaix diagrams are the electrochemical analog to metallurgical-phase diagrams and show phase stability as a function of electrode potential and pH. However, Pourbaix diagrams do not give any information about the rates at which reactions occur; they only provide information as to whether or not a reaction can take place. Thus, the use of Pourbaix diagrams with reaction rate theory allows the corrosion of a material in a particular aqueous environment to be better understood.

In general, any approach to understanding electrochemistry and corrosion makes use of both thermodynamics and kinetics. Thermodynamics provides an indication of whether an electrode reaction will occur, while kinetics provides information as to the relative rate of occurrence of these reactions. They are useful in the discussion of electrode/solution interface reactions and thus provide the basis for the development of a rate expression for the exchange current density at the reversible potential.

Although real-life processes are irreversible in nature, it is always easier to talk about reversible reactions. With that in mind, in aqueous solutions deviation from reversible conditions leads to anodic or cathodic net currents that exceed the exchange current density. The difference between potentials with and without current is called polarization. There are three kinds of electrochemical polarization: activation, concentration, and resistance.

The corrosion potential and corrosion rate in terms of active–active and active–noble metal couples are presented using polarization diagrams, also called Evans diagrams.

Some metals, such as Fe, Al, Cr, Ti, and their alloys, form a protective thin, adherent surface film. Polarization diagrams involving the presence of the passive films will also be discussed.

B. Thermodynamics Basics [4–6]

The degradation of metallic materials due to corrosion in aqueous environments usually exhibits an electrochemical nature. Therefore, it is necessary to develop an understanding of both the thermodynamics and kinetics of electrode reactions.

The basic thermodynamic functions are summarized below, where the terms listed have the following meanings [4–6]:

U—internal energy
H—enthalpy, where $H = U + pV$ (p—pressure, V—volume)
S—entropy
G—free energy, also called free enthalpy, Gibbs potential, or thermodynamic potential, and where: $G = U + pV - TS$ (T is the absolute temperature)

The case where $\Delta G = 0$ represents the equilibrium condition for isothermal (T is constant) and isobaric (p is constant) conditions.

1. Electrochemical Potential

For a substance, the chemical potential, μ, is defined as its free energy per mole, i.e.,

$$\mu = \frac{\partial G}{\partial n}$$

For ions, chemical and electrical potentials are considered separately by introducing the electrochemical potential, $\bar{\mu}$. This potential is represented by the following equation:

$$\bar{\mu} = \mu + zFE$$

where μ is the chemical potential, z is the change of ionic charge in electrode process, F is the Faraday's constant $= 96,484.56$ C/equivalent, and E is the electrical potential.

For the change of the electrochemical potential, $\Delta\bar{\mu}$, of an ion participating in an electrode reaction, the following equation applies:

$$\Delta\bar{\mu} = \Delta\mu + zF\Delta E$$

In the system, when a metal is immersed in a solution containing ions of that metal, a dynamic equilibrium is reached. Therefore, the electrochemical potential of the ions is equal in both phases, i.e., $\Delta\bar{\mu} = 0$, or

$$\Delta\mu + zF\Delta E = 0$$

This indicates that the chemical and the electrical potential differences at the phase boundary counterbalance each other. As a result, no net transfer of ions occurs. The electric potential difference, ΔE, can then be thermodynamically expressed by the following equation:

$$\Delta E = -\frac{\Delta\mu}{zF} \quad \text{or} \quad e = -\frac{\Delta G}{zF}$$

Here, ΔG is the free energy change in electrode reaction, and e is the electrode potential (also called half-cell potential, or redox potential).

For standard conditions, the electrode potential is called the standard potential, e°, and is represented by the following equation:

$$e^\circ = -\frac{\Delta G^\circ}{zF}$$

where ΔG° is the standard free energy.

For spontaneous electrochemical processes, e is positive, resulting in a negative free energy change, ΔG.

2. Physical Concept of Electrode Potential [7–10]

A conducting metal containing mobile electrons immersed in an aqueous solution has a tendency to form a complex interface. Unsymmetrical, polar water molecules, with the hydrogen atoms positively charged and the oxygen atoms negatively charged, are attracted to the conductive surface forming an oriented solvent layer that prevents close approach of charged species, e.g., ions in solution from the bulk electrolyte. Charged ions also attract their own sheath of polar water molecules resulting in further insulation of the conducting metal surface. In the case of a metal with a great tendency to dissolve, and therefore, give off its positively charged cations, the metal is negatively charged relative to the electrolyte. The plane of closest approach of the cations to this surface is called the outer Helmholtz plane, as shown in Fig. 1. An interfacial structure of separated charge is formed, called the electrical double layer. The thickness of the double electric layer is approximately 10^{-7} cm. Also, this layer behaves similarly to that of a charged capacitor (Fig. 2).

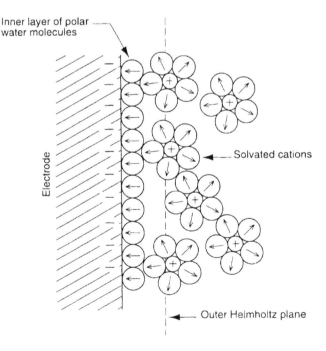

Figure 1 A simple model of the electric double layer at the electrode interface.

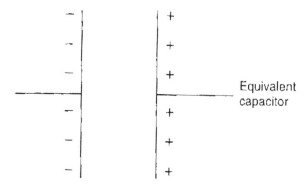

Figure 2 A schematic equivalent electric capacitor of the electrode interface.

Some metals have a lower tendency to ionize in aqueous electrolytes, and therefore acquire a positive charge relative to the electrolyte. Also, in these cases, an electric double layer is formed at the interface of the metal and electrolyte.

The electric field of the electric double layer prevents easy charge transfer. This results in limiting electrochemical reactions at the metal surface.

3. Types of Electrochemical Cells [4]

An electrochemical cell is a combination of electronic (metallic) and ionic (electrolytic) conductors, in which electrochemical processes occur by the passage of electric current. If the electrochemical cell produces electric energy by consuming chemical energy, it is called a *galvanic cell*. If the electrochemical cell consumes current from a power supply and stores chemical energy, it is called an *electrolytic cell*. A corrosion cell is a galvanic cell in which the electrode reactions lead to corrosion.

A corrosion cell with dimension ≤ 0.1 mm is called a local cell. These cells frequently occur on multiphase alloys, metals with electrically conducting coatings, and on materials with inclusions, such as oxides, sulphides, or carbides. The presence of local cells and their corrosion action results in localized attack, such as pitting.

4. Definition of Anode and Cathode

A galvanic cell with an electrolyte bridge, also called a liquid junction and denoted by the symbol //, is schematically represented by the following example [11]:

$$Zn/Zn^{2+}//Cu^{2+}/Cu$$

The zinc electrode is negative, whereas the copper electrode is positive. Usually, the positive pole is placed on the right hand side of the expression.

For the notation of the two electrodes in an electrochemical cell, either galvanic or electrolytic consisting of two electrodes, the following general definition is valid [12]: *the anode is the electrode through which positive current passes into the electrolyte.*

In accordance with the above definition, the following rules apply to electrochemical cells:

- In an electrolytic cell, the positive electrode is the anode and the negative electrode is the cathode.

- In a galvanic cell, the negative electrode is the anode and the positive electrode is the cathode.

Also, in an electrochemical cell:

- The anode reaction is an oxidation and the cathode reaction is a reduction;
- Anions migrate toward the anode and cations migrate toward the cathode.

5. Standard Electrode Potentials

The electrode potential of a metal in an electrolyte containing metal ions of the same species is reversible. If the species participating in the electrode process are present in certain standard states as pure metals, pure gases of 0.1 MPa pressure, electrolytes of ionic activity 1, and at a temperature of 298 K, the potentials obtained are called standard potentials. These potentials are measured vs. the standard hydrogen electrode (SHE), the potential that is arbitrarily assumed to be zero at all temperatures [10,13].

The standard electrode potential e^o is assumed to be the thermodynamic quantity $[\Delta G^o/zF]$; and the sign of the potential depends on the procedure used to write the electrode reactions. For example, e^o for the zinc electrode is -0.76 V if the reaction is written as the reduction reaction shown below:

$$Zn^{2+} + 2e = Zn \qquad e^o_{Zn2+/Zn} = -0.76 \text{ V}$$

On the other hand, if the above reaction is written as the oxidation reaction, e^o equals $+0.76$ V:

$$Zn = Zn^{2+} + 2e \qquad e^o_{Zn/Zn2+} = +0.76 \text{ V}$$

Note that reduction potentials have the major advantage in that they match the polarity of experimentally measured potentials. Standard potentials for selected electrode reactions are shown in Table 3 [14].

6. Nernst Equation

Another way of expressing an electrode reaction is shown below [15]:

$$ox + ze = red$$

According to the thermodynamics of chemical equilibrium, a relationship between the free energy change ΔG and the standard state free energy change ΔG^o is expressed by the following equation [4]:

$$\Delta G = \Delta G^o + RT \ln \frac{a_{red}}{a_{ox}}$$

where R is the gas constant, a_{red} is the activity of reductant, a_{ox} is the activity of oxidant, and T is the absolute temperature.

Because

$$\Delta G = -ezF \qquad \text{and} \qquad \Delta G^o = -e^o zF$$

$$e = e^o - \frac{RT}{zF} \ln \frac{a_{red}}{a_{ox}}$$

Table 3 Standard Potentials for Selected Electrode Reactions

	Reaction	Standard potential, e° (V vs. SHE)
Noble	$Au^{3+} + 3e^- = Au$	+1.498
	$Cl_2 + 2e^- = 2Cl^-$	+1.358
	$O_2 + 4H^+ + 4e^- = 2H_2O$ (pH 0)	+1.229
	$Pt^{3+} + 3e^- = Pt$	+1.2
	$O_2 + 2H_2O + 4e^- = 4OH^-$ (pH 7)	+0.82
	$Ag^+ + e^- = Ag$	+0.799
	$Hg_2^{2+} + 2e^- = 2Hg$	+0.788
	$Fe^{3+} + e^- = Fe^{2+}$	+0.771
	$O_2 + 2H_2O + 4e^- = 4OH^-$ (pH 14)	+0.401
	$Cu^{2+} + 2e^- = Cu$	+0.337
	$Sn^{4+} + 2e^- = Sn^{2+}$	+0.15
	$2H^+ + 2e^- = H_2$	0.000
	$Pb^{2+} + 2e^- = Pb$	−0.126
	$Sn^{2+} + 2e^- = Sn$	−0.136
	$Ni^{2+} + 2e^- = Ni$	−0.250
	$Co^{2+} + 2e^- = Co$	−0.277
	$Cd^{2+} + 2e^- = Cd$	−0.403
	$Fe^{2+} + 2e^- = Fe$	−0.440
	$Cr^{3+} + 3e^- = Cr$	−0.744
	$Zn^{2+} + 2e^- = Zn$	−0.763
	$2H_2O + 2e^- = H_2 + 2OH^-$	−0.828
	$Al^{3+} + 3e^- = Al$	−1.662
	$Mg^{2+} + 2e^- = Mg$	−2.363
	$Na^+ + e^- = Na$	−2.714
Active	$K^+ + e^- = K$	−2.925

This is the Nernst equation, which shows the way potential varies with the activities of the participating species. The activity of a dissolved species is equal to its molal concentration multiplied by the activity coefficient γ. Because

$$2.303 \frac{RT}{F} = 0.059 \text{ (at } 25^\circ C)$$

the Nernst equation can be rewritten in the following form:

$$e = e^\circ - \frac{0.059}{z} \log \frac{a_{red}}{a_{ox}}$$

The dependence of electrode potentials on hydrogen ion activity, or pH, is of utmost importance. For the hydrogen electrode [8]

$$2H^+ + 2e = H_2$$

The electrode potential is recalculated from pH = 0 to pH = 14 according to the following Nernst equation

$$e_{H^+/H_2} = e^\circ_{H^+/H_2} - \frac{0.059}{2} \log \frac{p_{H_2}}{a_{H^+}^2}$$

where p_{H2} is the fugacity of hydrogen in atmospheres, a_{H+} is the activity of hydrogen ions.

For the case where the fugacity of hydrogen is equal to 1 atm and where $pH = 14$

$$e_{H^+/H_2} = e^o_{H^+/H_2} - \frac{0.059}{2} \, pH = 0.000 - 0.059 \times 14 = -0.83 \text{ V}$$

The value -0.83 V is the standard electrode potential for the following reaction:

$$2H_2O + 2e = H_2 + 2OH^-$$

For the oxygen electrode

$$O_2 + 4H^+ + 4e = 2H_2O$$

the electrode potential is recalculated from $pH = 0$ to $pH = 14$ according to

$$e_{O_2/H_2O} = e^o_{O_2/H_2O} - \frac{0.059}{4} \, \log \frac{a^2_{H_2O}}{p_{O_2} a^4_{H^+}}$$

and where p_{O_2} is the fugacity of oxygen in atmospheres. For the case where the fugacity of oxygen is equal to 1 atm and where $pH = 14$, the Nernst equation is as follows:

$$e_{O_2/H_2O} = e^o_{O_2/H_2O} - 0.059 \, pH = 1.23 - 0.83 = 0.40 \text{ V}$$

which is the value of the standard electrode potential for the reaction:

$$O_2 + 2H_2O + 4e = 4OH^-$$

The electrode potentials for the hydrogen and oxygen electrodes as a function of pH are shown in Fig. 3. The slope for both lines is -0.059 V.

Figure 3 is divided into three distinct regions. These regions highlight the stability of different species, as follows:

1. The upper region shows the stability of oxygen;
2. The intermediate region shows the stability of water;
3. The lower region shows the stability of hydrogen.

7. Potential–pH Diagrams [16–18]

Most metals have a tendency to corrode as a result of reactions with surrounding environments. Thermodynamics can be applied to predict under which circumstances a metal can and cannot corrode. Marcel Pourbaix [19] pioneered the construction and use of these diagrams, which are graphical representations of thermodynamic information appropriate to electrochemical reactions [20]. They are available for almost all metals. The objective of diagram construction is to represent the relative stabilities of the solid phases and the soluble ions that are produced in the reaction between the metal and water. The parameters are the electrode potential, or the redox potential, which is plotted along the y axis and the pH that is plotted along the x axis [20]. Using thermodynamic data and the Nernst equation for electrode potentials, it is possible to determine the region boundaries between the phases that form in the metal–water system. Within these regions, the metal or its compounds are considered to be thermodynamically stable. A simplified diagram of the potential–pH diagram for the copper–water system is shown in Fig. 4 [21]. For a particular potential and

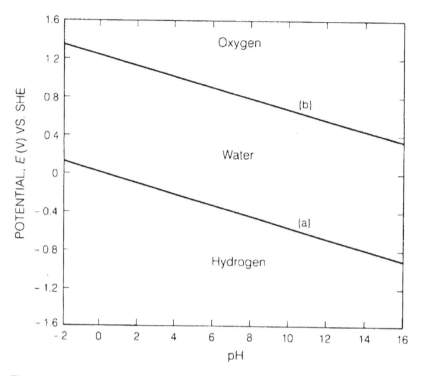

Figure 3 Potential–pH diagram indicating stability of water and its decomposition products: oxygen and hydrogen.

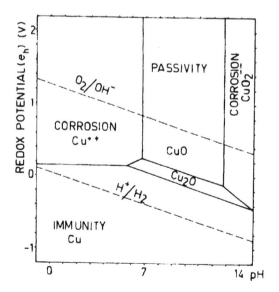

Figure 4 A scheme of the potential–pH diagram for copper–water system for copper species activity 10^{-6}.

pH, a point is defined on the diagram that suggests the likely response of the metal by the region, or domain, in which it lies. There are three kinds of domain response [20].

1. Domain of immunity in which the metal is the stable species and is immune to corrosion (in the terms for which the diagram is drawn).
2. Domain of corrosion in which the stable species is a soluble ion and the metal is expected to corrode *if the kinetics is favorable.*
3. Domain of passivity in which the stable species is insoluble.

In general, all sloping lines in the diagram represent pH-dependent redox equilibrium boundaries. Horizontal straight lines represent redox equilibrium boundaries that do not depend on pH, such as Cu^{2+}/Cu. Vertical lines represent equilibrium boundaries that do not involve an oxidation state change, and therefore, do not depend on potential. An example of this is the solubility equilibrium between CuO and Cu^{2+}.

If conditions are such that the metal is a stable phase, a state of immunity exists. In regions where oxide films, hydroxides, or other sparingly soluble compounds form, regardless of their protective properties, the metal may be passive. Where ions are stable, corrosion is possible.

Figure 4 also indicates that the potential for the hydrogen electrode (line *a* in the figure) falls within the region of copper immunity at all pH values. This means that deaerated solutions of nonoxidizing and noncomplex-forming acids, bases, and salts do not attack copper, whereas in aerated solutions copper is attacked.

An activity of 10^{-6} is often selected as a reasonably low value for calculations used in Pourbaix diagram constructions.

Because the Pourbaix diagrams are based on thermodynamics, they do not provide any kinetic data, e.g., corrosion rates, which is one of the Pourbaix diagram's limitations [22].

Pourbaix diagrams are very useful in identifying corrosion problems, but they only apply for the conditions of their construction, and they are not infallibly predictive because of the following limitations [20].

1. The diagrams are derived from thermodynamic considerations and yield no kinetic information.
2. Domains in which the solid substances are considered to be the stable species relative to arbitrary soluble ion activities $<10^{-6}$ give good indications of conditions in which a metal may be passive. However, whether or not passivation actually occurs depends on the nature, adherence, and coherence of the solid substance film.
3. The diagrams yield information only on the reactions considered in their construction. They do not take into account any known or unsuspected impurities in the solution or in the alloy components of the metal that may modify the reactions.
4. The form and interpretation of a Pourbaix diagram are both temperature-dependent. Pourbaix diagrams by the nature of their construction represent isothermal conditions.

Impurities in the aqueous solution and multicomponent alloys often result in the improper use of Pourbaix diagrams. This often leads to false conclusions, or the belief that thermodynamic considerations may be of little value generally in terms of aqueous corrosion. When dealing with these more complicated systems, it is important to know at the outset whether the additional elements are associated with the corrosive media or with the alloy. The former places constraints on the chemical potentials or concentrations; the latter places constraints on mass balance [17].

III. ELECTROCHEMICAL PRINCIPLES

A. Polarization

The difference between the potential of an electrode with current (e_i) and one without current (e_0) is called electrochemical polarization (η) [23]. For cathodic polarization, electrons are supplied to the surface, and a buildup in the metal due to a slow reaction rate, causes the potential e_i to become negative to e_0. Therefore, cathodic polarization is always negative. For anodic polarization, electrons are removed from the metal. Their deficiency results in a positive potential change due to the slow liberation of electrons by the surface reaction. Therefore, anodic polarization is positive. There are three different types of polarization [4,23]:

1. Concentration (η_{conc})
2. Activation (η_{act})
3. Resistance (η_r)

1. Concentration Polarization

Concentration polarization is caused by a deviation of the concentration on the electrode surface from that of the bulk solution [24]. This results in a concentration gradient at the metal/electrolyte interface. For example, during electrolytic copper deposition, the cathodic reduction reaction depletes the adjacent solution of Cu^{2+}. The concentration profile of Cu^{2+} as a function of electrode distance is shown in Fig. 5. C_B represents the Cu^{2+} concentration of the bulk solution, and δ is the thickness of the Nernst diffusion layer, within which the concentration gradient occurs. In an unstirred solution, the diffusion layer has a thickness of approximately 0.1 mm, whereas, the electric double layer has a thickness of 10^{-6} mm.

The potential of the electrode without polarization is given by the Nernst equation.

$$e_0 = e^o_{Cu^{2+}/Cu} + \frac{0.059}{2} \log C_B$$

The potential of the polarized electrode is also determined by the Nernst equation:

$$e_i = e^o_{Cu^{2+}/Cu} + \frac{0.059}{2} \log C_e$$

Therefore, the concentration polarization is equal to

$$e_i - e_0 = \frac{0.059}{2} \log \frac{C_e}{C_B}$$

Concentration polarization is also quantified by the limiting current density I_L, which is the highest current density possible for a given electrode reaction.

The limiting current density is inversely proportional to the thickness of the diffusion layer and can be evaluated from the following equation:

$$I_L = \frac{DzFC_B}{\delta}$$

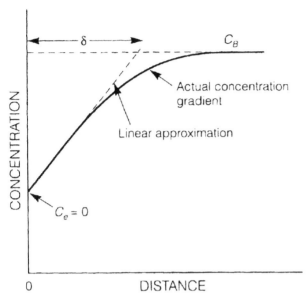

Figure 5 Concentration of H^+ in electrolyte near a surface controlled by concentration polarization.

In this equation, D is the diffusion coefficient for the ions participating in an electrode reaction. If I_L is the limiting current density for a cathodic reaction and I is the applied current density, the concentration polarization is given by

$$e_i - e_0 = \frac{2.303RT}{zF} \log \frac{C_e}{C_B} = \eta_{conc} = \frac{2.303RT}{zF} \log\left(\frac{1-I}{I_L}\right)$$

The concentration polarization decreases with increasing temperature because of an increase in diffusion velocity. Also, stirring reduces the concentration polarization, because the diffusion layer becomes thinner and the limiting current becomes higher. The concentration polarization disappears after the applied current is disconnected.

2. Activation Polarization

Activation polarization is caused by a slow step that is controlled by the rate of charge (electron) flow in the electrode reaction [4]. Because some value of activation energy is needed for the reaction to proceed, the reaction is said to be under activation or charge-transfer control. If a metal is immersed in an electrolyte containing its own ions, a dynamic equilibrium is reached. Therefore, the current density for the forward reaction is equal to the current density for the reverse reaction, as shown in Fig. 6. This forward/reverse current density at electrode equilibrium is called the exchange current density, i_0. If the electrode is not at equilibrium due to its anodic polarization, the cathodic partial current density (i_c) decreases while the anodic partial current density (i_a) increases. This situation is shown in Fig. 7(a). The net impressed current density is thus equal to

$$I_a = i_a - i_c$$

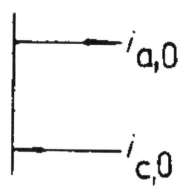

Figure 6 Exchange current density i_0 for dynamic equilibrium of the anodic and cathodic reactions.

When the electrode is polarized cathodically, the net impressed current density becomes

$$I_c = i_c - i_a$$

and this situation is shown in Fig. 7(b).

The relationship among the reversible equilibrium potential (e_0), exchange current density (i_0), cathodic current density (I_c), and anodic current density (I_a), for polarization

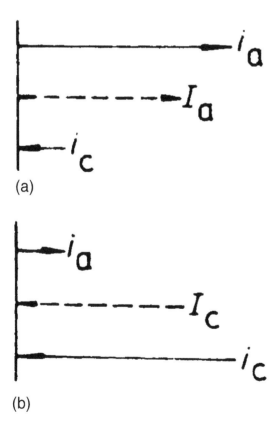

Figure 7 (a) Net current I_a, partial anodic current i_a, and partial cathodic current i_c during anodic electrode dissolution process. (b) Net current I_c, partial anodic current i_a, and partial cathodic current i_c during cathodic electrode process.

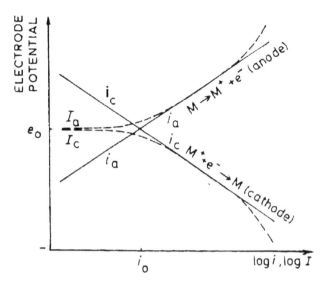

Figure 8 Polarization conditions at noncorroding electrode; solid line—i_a, i_c: real anodic and cathodic current, respectively; dashed line—I_a, I_c: experimental anodic and cathodic current, respectively; i_0—exchange current; e_0—reversible equilibrium potential.

conditions for electrode M is shown graphically in Fig. 8. If the electrode is polarized by an anodic current, I_a, for example, the activation polarization η_{act} can be expressed as follows:

1. For values of $I_a < 10i_0$, a linear relationship exists between η_{act} and I_a:

$$\eta_{act,a} = \frac{RTI_a}{zFi_0}$$

    This relationship corresponds to a curved line in Fig. 8 (dashed lines).

2. For values of $I_a > 10i_0$, a logarithmic relationship is obtained according to the Tafel equation:

$$\eta_{act,a} = \beta_a \log \frac{I_a}{i_0}$$

    where β_a is the is the anodic Tafel constant equal to $2.3(RT/(1-\alpha)zF)$ and α is the symmetry coefficient, which describes the shape of the rate-controlling energy barrier. Note, that for cathodic polarization, the Tafel equation is written as:

$$\eta_{act,c} = \beta_c \log \frac{I_c}{i_0}$$

    where β_c is the cathodic Tafel constant equal to $2.3(RT/\alpha zF)$. In Fig. 8, the straight lines, called Tafel lines, represent Tafel behavior. Also, for anodic polarization, $\eta_{act,a}$ and β_a are positive, whereas, for cathodic polarization, $\eta_{act,c}$ and β_c are negative.

3. For $I_a i_0$, other types of polarization, such as concentration and/or resistance polarization (described below), may cause deviations from the Tafel lines.

The activation polarization depends upon the chemical composition of an electrolyte, in particular, its content of anions and corrosion inhibitors.

Stirring of an electrolyte does not affect the activation polarization. However, it decreases with increasing temperature. The activation polarization disappears within a few milliseconds after the current is interrupted.

B. Resistance Polarization

Resistance polarization is a result of an ohmic resistance in a film, present on the electrode surface and/or a portion of the electrolyte surrounding the electrode, causing an ohmic potential drop [4]. The ohmic polarization can be expressed by the following equation:

$$\eta_r = iR = \frac{il}{\rho}$$

where i is the current density, R the ohmic resistance, l is the length of the resistance path, and ρ is the specific conductivity.

Resistance polarization is not affected by stirring of an electrolyte. It disappears simultaneously when the current is interrupted.

C. Electrochemical Aspects of Corrosion

Any metal surface is a combination of positive and negative electrodes, electrically short-circuited through the metal body, as shown in Fig. 9 [25]. As long as the metal is dry, corrosion is not observed because there is no current flow through the electrodes. However, if the metal is immersed in an aqueous solution or exposed to a moist environment, the electrodes form local galvanic cells that are able to function as an electric circuit. This condition may lead to the conversion of the metal to corrosion products.

Impurities in a metal, constituting the electrodes in a local cell, adversely affect its corrosion resistance because of the formation of dissimilar electrode cells. Therefore, processes to remove impurities from metals may improve their corrosion resistance. However, they can sill corrode because local cells can also form because of variations in the environment and/or temperature.

One of the environmental variations is caused by a difference in the solution concentration. The formation of a local cell on the metal surface due to solution concentration difference is called a salt concentration cell. The electrode exposed to a less concentrated solution is an anode, whereas the electrode exposed to a more concentrated solution is a cathode.

Another concentration cell can be formed as a result of the difference in the nonuniform distribution of air/oxygen at the metal/solution interface. This cell is called a differential aeration cell. The anodic process, i.e., oxidation, takes place on the electrode (anode)

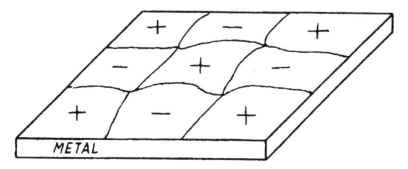

Figure 9 A schematic of local cells on a metal surface.

exposed to the solution with a lower fugacity of oxygen f_1; while a cathodic process, i.e., reduction, takes place on the electrode (cathode) exposed to the solution with a higher fugacity of oxygen f_2.

Some oxidation reactions occurring on the anode sites of the electrode can be written:

$$4OH^- \rightarrow O_2 + 2H_2O + 4e$$

$$Fe \rightarrow Fe^{2+} + 2e$$

Reduction reactions taking place on the cathode sites could be:

$$O_2 + 2H_2O + 4e \rightarrow OH^-$$

$$Fe^{2+} + 2e \rightarrow Fe$$

Local cells can also form as a result of the nonuniform distribution of temperature through the metal surface that is exposed to the same solution. These types of cells are called differential temperature cells. Depending on the corrosion system, the electrode at the higher temperature could be the cathode, while the electrode at the lower temperature could be the anode. Such polarity was found for the $Cu/CuSO_4$ system. However, for Fe exposed to dilute aerated NaCl, the electrode at higher temperature was the anode, while the electrode at the lower temperature was the cathode. After a certain period of time, the electrode polarity could reverse itself, as a result of, for example, aeration and stirring rate.

Local cells responsible for corrosion can form as a result of combinations of dissimilar electrode cells, concentration cells, and differential temperature cells. Taking into account that corrosion in these systems is electrochemical in nature, determining electrode potentials, in addition to corrosion currents, are of utmost importance.

D. Types of Cell Potentials

1. Reversible Cell Potentials

The principle of a reversible cell potential applied to an electrochemical corrosion system can be explained by considering the displacement reaction between copper and zinc [26]:

$$Zn + Cu^{2+} = Zn^{2+} + Cu$$

The equal sign indicates that the above reaction takes place at equilibrium. As a redox reaction, the reaction can be expressed as the sum of two half-reactions, which are conceptual reactions indicating the loss and gain of electrons. Therefore, the half-cell reactions can be written as follows:

Reduction of Cu^{2+} : $\quad Cu^{2+} + 2e = Cu$

$$\underline{\text{Oxidation of Zn}: \quad Zn = Zn^{2+} + 2e}$$
Overall reaction(sum) : $Cu^{2+} + Zn = Cu + Zn^{2+}$

If the above half-cell reactions are written as reductions, the overall reaction is the difference of the two, namely:

Reduction of Cu^{2+} : $\quad Cu^{2+} + 2e = Cu$

$$\underline{\text{Reduction of } Zn^{2+}: \quad Zn^{2+} + 2e = Zn}$$
Overall reaction (difference) : $Cu^{2+} + Zn = Cu + Zn^{2+}$

As described in a previous section, each half-cell reaction has a free-energy change of ΔG and a corresponding potential equal to e. The overall reaction also has the free energy change analogous to ΔG and a corresponding electrochemical potential equal to E. At equilibrium, the relationship between ΔG and E is:

$$\Delta G = -nFE$$

E is also called the electromotive force (EMF) and is associated with half-cell potentials in the following way:

- if $e_{r,c}$ is the half-cell reduction potential for a reduction (cathodic) reaction and $e_{o,a}$ is the half-cell oxidation potential for the oxidation (anodic) reaction, E is a sum of these potentials, namely:

  $$E = e_{r,c} + e_{o,a}$$

- if $e_{r,c}$ is the half-cell reduction potential for a reduction (cathodic) reaction and $e_{r,o}$ is the half-cell reduction potential for the oxidation (anodic) reaction, E is a difference of these potentials, namely:

  $$E = e_{r,c} - e_{r,o}$$

Currently, reduction potentials are used for any half-cell reactions, which is in agreement with a convention established by the International Union of Pure and Applied Chemistry (IUPAC) in 1953.

A reversible cell representing the overall reaction $Cu^{2+} + Zn = Cu + Zn^{2+}$ is shown in Fig. 10. The cell consists of copper and zinc electrodes in equilibrium with their ions

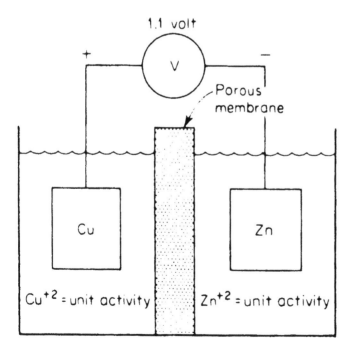

Figure 10 A schematic of reversible cell for copper and zinc electrodes.

separated by a porous membrane. For the overall reaction $Zn + 2H^+ = Zn^{2+} + H_2$, a reversible cell is shown in Fig. 11.

According to the IUPAC convention, the half-cell reaction and their potentials are [4]:

$$2H^+ + 2e = H_2, e_{r,c}$$

$$Zn^{2+} + 2e = Zn, c_{r,a}$$

Overall reaction (difference) $2H^+ + Zn = H_2 + Zn^{2+}, E = e_{r,c} - c_{r,a}$

Pt is used as an inert substrate for the hydrogen electrode.

In these reversible cells, where equilibrium is reached, electrical conductors do not carry current, and they have uniform and constant potentials. Hence, thermodynamics can be applied. Corroding systems are not at equilibrium, and therefore, equilibrium thermodynamics cannot be applied. Thus, reversible cell potentials cannot be used for characterizing the system. The potentials, which are considered for studying corrosion mechanisms, are called corrosion potentials.

2. Corrosion Potential

When the anode electrode and the cathode electrode are short-circuited in a cell, and net oxidation and reduction processes occur at the electrode interfaces, the electrodes will not be in equilibrium [26]. An example of the short-circuited cell with zinc and hydrogen electrodes is shown in Fig. 12. In this cell, the zinc electrode corrodes (dissolves), and simultaneously

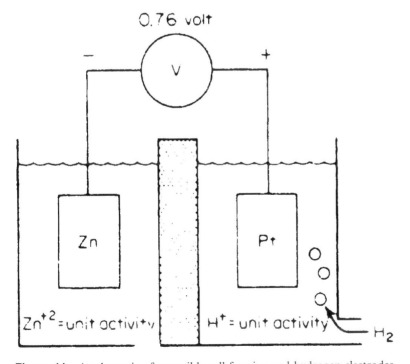

Figure 11 A schematic of reversible cell for zinc and hydrogen electrodes.

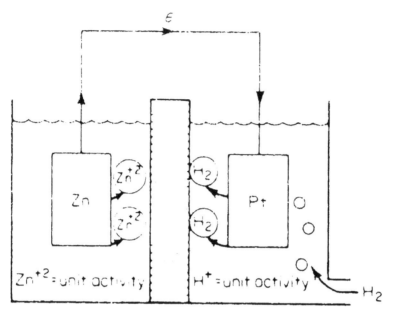

Figure 12 A schematic of short-circuited cell for zinc and hydrogen electrodes.

hydrogen evolves at the platinum surface. Electrons released from the zinc-oxidation reaction are transferred through the external conductor to the platinum where they are consumed during the hydrogen-reduction reaction. In this process, both electrodes undergo polarization, i.e., the displacement of electrode potential resulted from a net current. The process that occurs in Fig. 12 is exactly the same as what occurs while zinc is immersed in a hydrogen-saturated acidic solution containing Zn^{2+}. In both cases, the overall reaction is the dissolution of zinc and the evolution of hydrogen:

$$Zn \rightarrow Zn^{2+} + 2e \text{ anodic (oxidation)}$$

$$2H^+ + 2e \rightarrow H_2 \text{ cathodic (reduction)}$$

The potential of the Zn metal measured by a voltmeter (V) with respect to a reference electrode (REF) is called corrosion potential, E_{corr}. A schematic simple cell for measuring corrosion potentials is shown in Fig. 13. The corrosion potential falls between the half-cell electrode potentials. Therefore, E_{corr} is a combination of at least two half-cell reactions, which cause electrode polarization as shown in Fig. 14. E_{corr} is also called a mixed potential, which can be explained by the mixed-potential theory based on the following hypothesis:

Any electrochemical reaction can be divided into two or more partial oxidation and reduction reactions.

There can be no net accumulation of electric charge during the electrochemical reaction.

3. Kinetics of Electrochemical Corrosion

Thermodynamic considerations of corrosion indicate whether or not a tendency toward corrosion exists. If there is a thermodynamic tendency toward corrosion, then the kinetics of

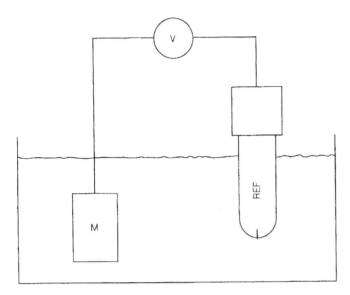

Figure 13   A simple cell for measuring corrosion potential.

the processes involved must be examined to determine how fast the corrosion reaction takes place [27].

Taking into account the electrochemical nature of corrosion, the rate of corrosion is measured by the current of the metal ions leaving the metal surface in the anodic region [28]. The flux of these ions gives rise to a corrosion current I_{corr} that can be identified with anodic current I_a. Because any current emerging from the anodic region must find its way to the

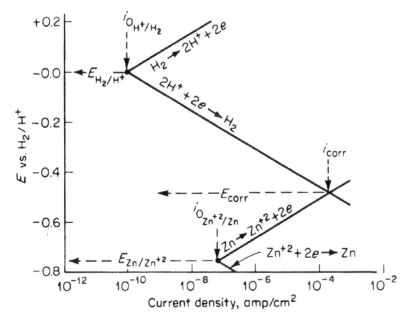

Figure 14   Polarization diagram for zinc immersed in acidic solution.

cathodic region, the cathodic current I_c must also be equal to the corrosion current. In terms of current densities at the anodic (oxidation) i_a and cathodic (reduction) i_c sites

$$I_{\text{corr}} = i_a A = i_c A'$$

where A is the area of the oxidation (anodic) process and A' is the area of the reduction (cathodic) region.

$$I_{\text{corr}} = (i_a \times A \times i_c \times A')^{1/2}$$

When $i_a i_c$ is i and AA' is A

$$I_{\text{corr}} = \underline{iA}$$

Using the Tafel equations described in the previous section, the half-cell electrode potential for the anodic process changes according to

$$\eta_a = \frac{RT}{(1 - \alpha)F} \ln \frac{i_a}{i_{o,a}} = \beta_a \log \frac{i_a}{i_{o,a}}$$

and for the cathodic process

$$\eta_c = -\frac{RT}{\alpha F} \ln \frac{i_c}{i_{o,c}} = \beta_c \log \frac{i_c}{i_{o,c}}$$

until they become equal to E_{corr}. The overpotentials in the two regions are:

$$\eta_a = E_{\text{corr}} - e_{\text{eq},a}$$

and

$$\eta_c = E_{\text{corr}} - e_{\text{eq},c}$$

At E_{corr}, the rates of the anodic and cathodic reactions are equal. Therefore, for $\alpha = 1/2$

$$I_{\text{corr}} = A i_o e^{EF/4RT}$$

where

$$i_o = (i_{o,a} i_{o,c})^{1/2}$$

$$E = e_{\text{eq},c} - e_{\text{eq},a}$$

A number of conclusions can be drawn from the last equation: The corrosion rate depends on the surfaces exposed, namely if either A or A' is zero, the corrosion current is zero. Therefore, to reduce corrosion, covering the surface with a protective coating is one method for achieving corrosion protection.

For corrosion reactions with similar exchange current densities, the rate of corrosion is high when E is large. This indicates that rapid corrosion can be expected when the oxidizing and reducing couples have widely differing electrode potentials. The exchange current density affects corrosion rates. For example, the exchange current density of the reaction $4H^+ + O_2 + 4e \rightleftarrows 2H_2O$ on iron is 10^{-14} A/cm^2, and for the reaction $2H^+ + 2e \rightleftarrows H_2$ is 10^{-6} A/cm^2. This indicates that the latter dominates kinetically, and iron corrodes with hydrogen evolution in acidic solutions [29].

As the electrode (corrosion) reaction is heterogenous it is natural to express its corrosion rate as the amount of material reacted per unit area of the electrode surface per unit time [30]. For electrochemical reactions, current (I) is the parameter used for measuring reaction rate. The mathematical dependence that exists between current and mass reacted (m) is given by Faraday's Law:

$$m = \frac{ItM}{zF}$$

where t is the time, M is the atomic weight, z is the change of ionic charge in an electrode process, and F is the Faraday's constant.

Dividing the above equation through by t and surface area (A) gives corrosion rate:

$$r = \frac{m}{tA} = \frac{iM}{zF}$$

where i is the current density.

This equation shows proportionality between mass loss per unit area per unit time and current density.

Current density can be precisely measured to values as low as 10^{-9} A/cm^2 and up to several A/cm^2. Therefore, electrochemical measurements are very sensitive and they make convenient tools for investigating corrosion in the laboratory and in the field.

Corrosion rate in terms of penetration results from dividing $r = m/tA = iM/zF$ by density ρ of the material. For a corrosion rate in mils (0.001 in.) per year, or mpy:

$$r = 0.129 \, \frac{iM}{z\rho}$$

for the following units: i in [μA/cm^2], ρ in [g/cm^3].

The mpy unit is the most commonly used corrosion rate unit in the United States. A substitute expression is required to facilitate the conversion of mpy to the metric system. Some equivalent metric corrosion penetration rates are:

1 mpy = 0.0254 mm/yr = 25.4 μm/yr = 2.90 nm/hr = 0.805 pm/sec

The corrosion rates of usefully resistant materials generally range between 1 and 200 mpy.

4. Passivity [31–33]

All metals and alloys possess a thin protective corrosion product film on their surface resulting from the reaction with the environment [32]. Without such films, materials exposed to the environment would revert back to their thermodynamically stable condition, the ores used to produce them. Some of these films—*the passive films*—on some, but not all, metals and alloys have special characteristics that enhance their resistance to corrosive attack [32]. These protective "passive" films lead to the phenomenon of passivity [31–33].

Two types of passivity have been defined by Uhlig and Revie [32,34]:

1. Type 1—"A metal is passive if it substantially resists corrosion in a given environment resulting from marked anodic polarization" (low corrosion rate, noble potential).
2. Type 2—"A metal is passive if it substantially resists corrosion in a given environment despite a marked thermodynamic tendency to react" (low corrosion rate, active potential).

Passivity results in a strongly reduced tendency for metals to corrode as a consequence of the formation of thin, protective surface films composed of corrosion products

[31]. Some metals spontaneously passivate in water if the pH is within the range corresponding to potential-independent domains of stability for oxides and hydroxides [33].

For many metals, such iron, chromium, and nickel, the corrosion rate decreases above the so-called primary passive potential E_{pp} despite the fact that high anodic polarization is applied. This situation is shown in Fig. 15. This region of corrosion immunity is called "passive." Below E_{pp}, the metal corrodes at a relatively high corrosion rate. This region is called "active." Corrosion rates in the passive regions are much lower than in the active regions. Depending on the potential, or oxidizing power of the corrosive environment, an alloy may exist in the passive state above E_{pp}, or may undergo dissolution in the active region below it. For example, Fig. 16 shows the effect of deaeration on corrosion of stainless steel in neutral saltwater. As can be seen, the material is passive in aerated salt water and active in deaerated salt water.

Passive films are usually thin and often fragile. Therefore, they can break down very easily. This could result in the occurrence of different forms of corrosion discussed in the next section. Breakdown of the passive film can occur in a variety of ways, the most usual being general breakdown, local breakdown, and mechanical breakdown. Metal and alloys that rely on passive films for corrosion protection are vulnerable to corrosion failure should the passive film break down. The nature of the any subsequent corrosion damage yields valuable information as to the nature of the corrosive attack.

General breakdown of passivity is usually not a problem, except for those instances where depassivation of the metal or alloy is associated with the inadvertent failure to replenish the oxygen as it becomes depleted [32]. Local breakdown of the passive film is more of a problem because it leads to localized corrosion such as crevice and pitting corrosion. In these cases, the solution in the crevice or pit can become locally depleted of oxygen, and as such, the passive film breaks down in this area, and corrosion in the pit or crevice accelerates as a result because of the small local anode. Penetration of the metal or alloy can be swift in these regions, leading quickly to failure. Mechanical breakdown of the film can occur

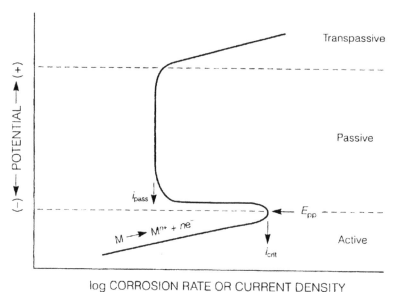

Figure 15   Anodic polarization curve for active–passive metal.

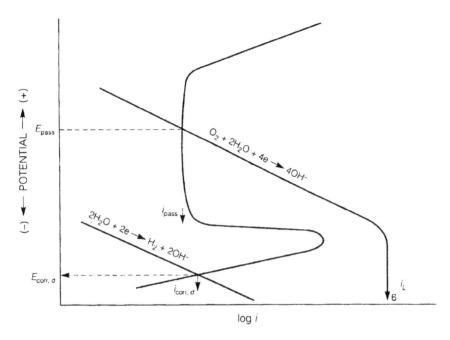

Figure 16   Effect of dearation and aeration on corrosion of active–passive stainless steel neutral salt water.

through a variety of means, such as a constant applied stress (stress corrosion cracking), cyclic stresses such as rotating–bending (corrosion fatigue), exploding bubbles from rapid part rotation (cavitation erosion), or by the rubbing away of the film by particles in a fluid (erosion–corrosion) [32]. These forms of corrosion are discussed in greater detail in the next section, but suffice it to say that each mechanical situation can damage the passive film, and if the film cannot repair itself as quickly as it is damaged, then failure will eventually occur.

IV. FORMS OF CORROSION

Over the course of time, various forms of corrosion have been studied in great detail [35]. Certain similarities between these various forms have been noted, and as such, it is possible to group them accordingly. One grouping of corrosion forms is shown in Table 4. In this group, the most familiar forms are uniform or general corrosion, crevice corrosion, pitting corrosion, intergranular corrosion, selective leaching, erosion–corrosion, stress corrosion, and hydrogen damage. It is not possible to talk about all the various forms of corrosion that have been identified to date, or to talk about them in detail. Only the most generally recognized forms of corrosion are discussed in this section.

Taking into account that the corrosion of metals and alloys in aqueous environments is electrochemical in nature, corrosion processes occur as electrode reactions [36]. In the case of localized corrosion attack, anodic and cathodic surfaces of existing corrosion cells can be usually found. These cells can be macroscopic or microscopic in size. The occurrence of these cells is caused for a variety of reasons: e.g., varying composition in the metal or of the corroding electrolyte can cause localized corrosion cells to form. The presence

Table 4 Forms of Corrosion

General Corrosion
Atmospheric corrosion
Galvanic corrosion
Stray-current corrosion
General biological corrosion
Molten salt corrosion
Corrosion in liquid metals
High temperature corrosion
   Oxidation
   Sulfidation
   Carburization
   Other forms
Localized Corrosion
Filiform corrosion
Crevice corrosion
Pitting corrosion
Localized biological corrosion
Metallurgically Influenced Corrosion
Intergranular corrosion
Dealloying corrosion
Mechanically Assisted Corrosion
Erosion corrosion
Fretting corrosion
Cavitation and water drop impingement
Corrosion fatigue
Environmentally Induced Cracking
Stress-corrosion cracking
Hydrogen damage
Liquid metal embrittlement
Solid metal induced embrittlement

of active–passive cells on the metal surface are another way a cell can form, and one way this may occur is through differences in lattice strain on a metal surface formed during cold-working.

For localized corrosion, separate corrosion cells exist on the metal surface and can be distinguished by variation in the electrode potential or by the appearance of separate anodic and cathodic corrosion products. In the case of general corrosion, the corroding metal is considered as a single electrode upon which both anodic and cathodic reactions simultaneously occur.

During general and localized corrosion, polarization occurs. An example of this is the Evans polarization diagrams for general corrosion, as shown in Fig. 17. The equilibrium potential at the anode ($e_{a,o}$) is that potential the metal of the anode would reach if it did not corrode, i.e., if it were in equilibrium with its own ions in the electrolyte. The equilibrium potential at the cathode ($e_{c,o}$) is that potential the metal of the cathode would reach if it were inert, i.e., its potential was determined by the cathodic reaction, such as hydrogen evolution, under equilibrium conditions. When a metal corrodes, a displacement of the individual electrodes from their equilibrium values takes place. The anode potential is displaced in a positive direction, while the cathode potential is displaced in a negative direction. If the

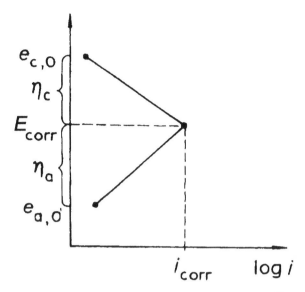

Figure 17   A schematic polarization diagram for uniform electrochemical corrosion; $\eta_c = E_{corr} - e_{c,o}$, $\eta_a = E_{corr} - e_{ac,o}$.

corrosion process is controlled by the activation polarization, the point of intersection between the Tafel lines, i.e., i_{corr} and E_{corr}, characterizes the state of the corroding metal. Therefore, the corrosion current is assumed to be the same for the anodic or the cathodic processes. Also, because of the submicron dimensions of the corrosion cells, potential drops in the electrolyte can be disregarded; thus, the corroding metal is able to reach one potential value, E_{corr}.

An example of the Evans polarization diagram for localized corrosion is shown in Fig. 18. In this case, the anodes are separated from the cathodes and have different potentials. The difference arises because an internal potential drop in the electrolyte for this corrosion cell exists. Therefore, it is impossible to determine one corrosion potential value for the entire metal surface. For example, if the potential of the pitted metal surface covered with an oxide film was measured, a value (e_c) that corresponds to the cathodic reaction taking place on the intact part of the oxide would be obtained. The potential value (e_a), corresponding to the anodic reaction that takes place at the bottom of a pit, could be lower than the e_c potential.

Taking into account the electrochemical nature of corrosion with emphasis on the size of the corrosion cells, the most prevalent forms of corrosion could be classified as follows:

Uniform corrosion
Galvanic corrosion
Pitting corrosion
Crevice corrosion
Intergranular corrosion
Dealloying
Stress corrosion cracking
Corrosion fatigue
Erosion corrosion
Cavitation corrosion

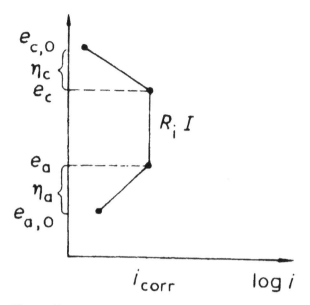

Figure 18 A schematic polarization diagram for localized electrochemical corrosion; $\eta_c = e_c - e_{c,o}$, $\eta_a = e_c - e_{ac,o}$, $E = R_i I + \eta_c + \eta_a$.

A. Uniform Corrosion [37–40]

Many corrosion problems involve uniform or general corrosion of metal surfaces. Uniform corrosion is usually defined as corrosion attack dominated by uniform thinning without appreciable localized attack. For uniform corrosion to occur, the corrosive environment must have the equivalent access to all parts of the metal surface, and the metal must be metallurgically and compositionally uniform [39,40]. Weathering steels and copper alloys are good examples of materials that undergo uniform corrosion. The corrosion of an 18-8 stainless steel pipe in hot water, contaminated with 0.5% H_2SO_4 and 0.5% HNO_3, is another example of uniform corrosion. Typically, materials such as stainless steels and nickel–chromium alloys usually undergo localized corrosion of one form or another.

B. Galvanic Corrosion [41–44]

Galvanic corrosion, also called contact corrosion, occurs at the contact point between two different metals or alloys possessing different electrode potentials. In most real-life situations, a potential difference exists between two dissimilar metals when they are immersed in a corrosive conductive environment. If these metals are electrically connected, galvanic corrosion takes place such that the cathodic areas are protected at the expense of the anodic ones. The metal that possesses the more negative potential undergoes corrosion. For galvanic corrosion to occur, three conditions must be met: (1) the contacting materials possess different surface potentials; (2) a common electrolyte exists; and (3) there is a common electrical path. For determining galvanic activity of alloys and pure metals, it is convenient to use values of their corrosion potentials measured in seawater [44]. A listing of these corrosion potentials, constituting the galvanic series, is shown in Table 5. The most active metal or alloy in a material couple, i.e., the one with the largest negative potential, is always attacked preferentially through galvanic corrosion.

Table 5 Galvanic Series for Some Common Metals and Alloys in Seawater at 25°C

Corroded End (Anodic, or Least Noble)
Magnesium
Magnesium alloys
Zinc
Galvanized steel or galvanized wrought iron
Aluminum alloys
Low carbon steel
Wrought iron
Cast iron
Type 410 stainless steel (active)
Type 304 stainless steel (active)
Type 316 stainless steel (active)
Lead
Tin
Copper alloys (e.g., naval brass)
Nickel 200 (active)
Inconel 600 (active)
Hastelloy alloy B
Copper alloy (e.g., yellow brass)
Copper alloy (e.g., red brass)
Copper
Copper alloy (e.g., silicon bronze)
Copper alloy (e.g., leaded tin bronze M)
Type 410 stainless steel (passive)
Type 304 stainless steel (passive)
Type 316 stainless steel (passive)
Incoloy alloy 825
Inconel alloy 625
Hastelloy alloy C
Silver
Titanium
Gold
Platinum
Protected End (Cathodic, or Most Noble)

C. Pitting Corrosion [45–50]

Pitting corrosion is a serious form of corrosion damage because of the rapidity in which the metal is penetrated. Pitting is the localized type of corrosion that produces pits that are relatively small in area compared to the overall exposed surface area, and generally occurs on an oxide-covered metal surface. The reason pitting is more prevalent on oxide covered surfaces is due to a stimulation of the anodic reaction by activating anions, and of the cathodic reaction by the presence of oxidizing agents and by effective cathode surfaces with low polarization. For a pitting to occur, a pitting potential, also called a breakdown potential, must be reached, as shown in Fig. 19. This potential is lower than the trans-passive potential and located in the passive region. A pit is initiated on defective sites in the oxide film, such as slag inclusions or precipitates of secondary phases [49]. In commercial steels, pits are usually initiated around sulphide inclusions [50]. An example

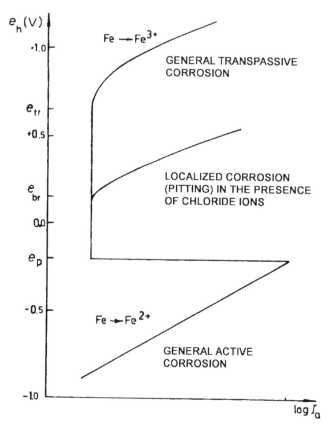

Figure 19 Effect of pitting on the anodic polarization curve of active–passive metals; e_p = passivation potential, e_{br} = breakdown or pitting potential, e_{tr} = transpassive potential.

of pitting for 18-8 stainless steel in sulfuric acid containing ferric chloride is shown in Fig. 20.

D. Crevice Corrosion [45,46,51–55]

Crevice corrosion occurs in the presence of narrow openings or spaces between metal-to-metal or nonmetal-to-metal components, and gives rise to localized corrosion at those sites. Severe localized corrosion frequently occurs within these spaces and at other shielded areas on metal surfaces exposed to corrosive environments [53,54]. This type of attack is usually associated with small volumes of stagnant solution caused by holes, gasket surfaces, lap joints, surface deposits, and crevices under bolt and rivet heads [53,54]. The mechanism of crevice corrosion, sometimes called deposit or gasket corrosion, is very similar to that of pitting [55]. An example of crevice corrosion found on a stainless steel pipe flange is shown in Fig. 21.

E. Intergranular Corrosion [25,56–60]

Intergranular corrosion, also known as intergranular attack, occurs at the grain boundaries, or areas adjacent to the grain boundaries, which are anodic relative to the grain interior

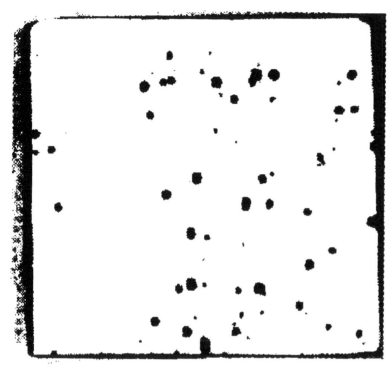

Figure 20 Pitting corrosion of 18-8 stainless steel in sulfuric acid containing ferric chloride.

[25,58,59]. The attack can be fast, penetrating deeply into the metal, sometimes causing catastrophic failures. The best-known form of intergranular corrosion occurs in austenitic stainless steel when heat treatments deplete the grain boundaries of chromium through conversion with carbon to form chromium carbides (primarily $Cr_{23}C_6$). The grain boundaries become "sensitized." The sensitized areas do not have sufficient chromium in solution to develop a stable passive film necessary to protect the underlying metal from corrosive attack. These areas become anodic with respect to the surrounding passive material [60]. An example of intergranular corrosion of sensitized austenitic stainless steel is shown in Fig. 22.

F. Dealloying [56,60,61]

Dealloying, which is also called selective leaching or parting, is a corrosion process in which one constituent of an alloy corrodes preferentially, leaving an altered residual structure in its wake. This preferentially corroded alloy constituent is active, e.g., negative electrochemically, compared to the major solvent constituent [60,61]. The dealloying of brass, known as dezincification, is an example of this form of corrosion degradation. An example is shown in Fig. 23. Zinc is active relative to copper, and leaches out of brass, leaving behind relatively pure porous copper with poor mechanical properties.

G. Stress Corrosion Cracking [59,62–68]

Stress corrosion cracking refers to cracking caused by the simultaneous presence of a tensile stress and a specific corrosive environment [66]. The stress in the metal may be residual in

Figure 21 Crevice corrosion on a large stainless steel pipe flange.

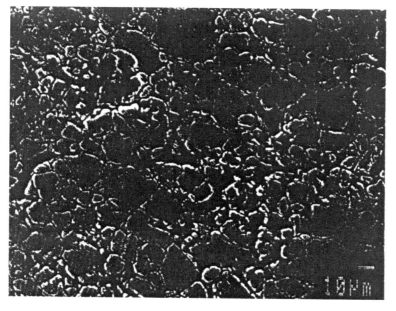

Figure 22 Scanning electron micrograph (SEM) of intergranular attack in sensitized stainless steel.

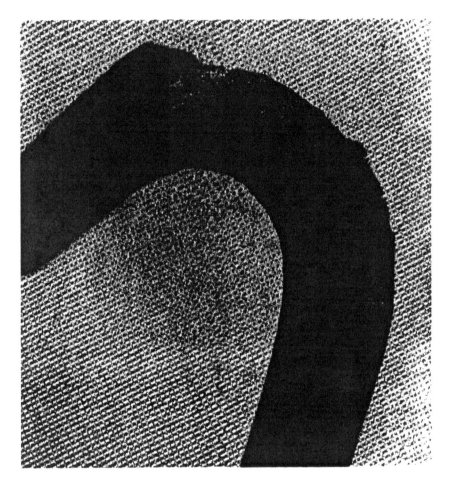

Figure 23   Dealloying of brass bolt (30% Zn).

nature, an artifact from cold working or heat treatment. Alternatively, the tensile stress may be the result of an externally applied load. The observed cracks are intergranular or transgranular, depending on the metal and the corrosive environment [59]. Intergranular stress corrosion cracking of carbon steel in hot 8.5 M NaOH is shown in Fig. 24 [67], and transgranular stress corrosion cracking for austenitic stainless steel in hot chloride detailed in Fig. 25 [68].

H. Corrosion Fatigue [59,63,64,69–71]

If a metal cracks when subjected to repeated or alternate tensile stresses in a corrosive environment, it is said to fail by corrosion fatigue [59]. In the absence of a corrosive environment, a metal stressed similarly but with a value below a critical applied stress, called the fatigue limit or endurance limit (shown in Fig. 26), will not fail by fatigue even after a large, or infinite, number of cycles. A true endurance limit does not commonly exist in a corrosive environment, and the metal fails after a prescribed number of stress cycles regardless

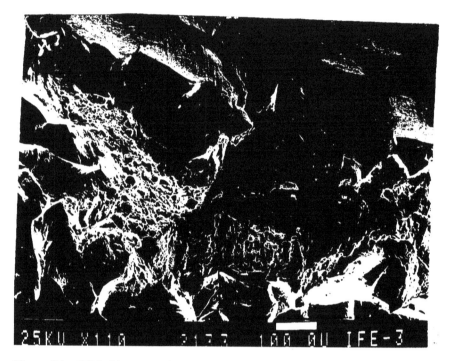

Figure 24 SEM of intergranular stress corrosion crack in carbon steel exposed to hot concentrated sodium hydroxide.

of a stress level [71]. Corrosion fatigue cracking in carbon steel boiler tube is shown in Fig. 27.

I. Erosion Corrosion [63,72–76]

Erosion corrosion occurs when the velocity of a fluid is sufficient to remove protective films from the metal surface [74]. The same stagnant or slow-flowing fluid will cause a low or modest corrosion rate, but rapid movement of the corrosive fluid physically erodes and removes the protective film, exposes the reactive alloy underneath, and accelerates the corrosion attack [75]. Figure 28 shows how the corrosion rate may change with the fluid flow velocity. The corrosion rate increases slowly while in the laminar range flow region. It then increases rapidly in the turbulent range [76]. The physical manifestations of corrosive attack generally follow the direction of localized flow and turbulence around surface irregularities, as shown in Fig. 29 [75].

J. Cavitation Corrosion [63,73,75,77]

Cavitation corrosion, or cavitation, is a result of the combined action of high liquid flow rates and corrosion action [77]. Cavitation corrosion occurs where fluid velocity is so high that pressure reductions in the flow are sufficient to nucleate water vapor bubbles, which then collapse on the metal surface [75]. The result is a strong hammering action, which destroys the passivating oxide film and also damages the underlying metal [77]. The surface becomes pitted and appears to be spongy. An example of cavitation corrosion is shown in Fig. 30 [75].

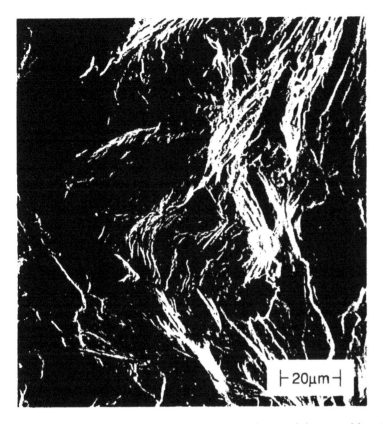

Figure 25 SEM of transgranular stress corrosion crack in austenitic stainless steel exposed to hot chloride solution.

V. LABORATORY TEST METHODS

A. Test Methodologies for Electrochemical Measurements

Most forms of metallic corrosion can be studied by using accelerated electrochemical laboratory test methods. The forms of corrosion that are usually studied in the laboratory include uniform corrosion, galvanic corrosion, pitting corrosion, crevice corrosion, dealloying, stress corrosion cracking, and hydrogen-induced failure. In addition to these forms of corrosion, the mechanisms of passivation, anodization, cathodic and anodic protection, as well as the performance sacrificial coatings on metals, can all be investigated by using accelerated electrochemical laboratory procedures. It is not the intention of this discussion to describe in any detail the wide variety of electrochemical test procedures currently in use. For more information on the actual electrochemical test procedures, the reader is referred to Refs. 78–80. These references provide a good starting point for understanding the test procedures available. Most of these electrochemical test methods can be classified into one of the following groups [81].

1. Recording of Anodic and Cathodic Polarization Curves

a. Polarization Resistance Method. For small polarizations, i.e., up to 20 mV from the corrosion potential, there is a linear relationship between the applied current (i_{app}) and

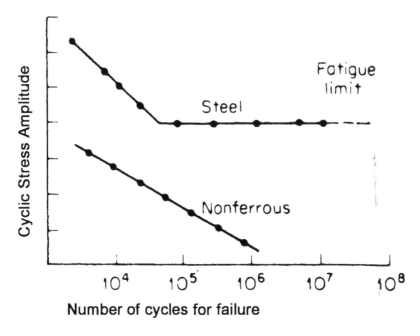

Figure 26 Schematic fatigue curves for ferrous and nonferrous alloys.

Figure 27 Optical macrograph of corrosion fatigue crack in carbon steel boiler tube.

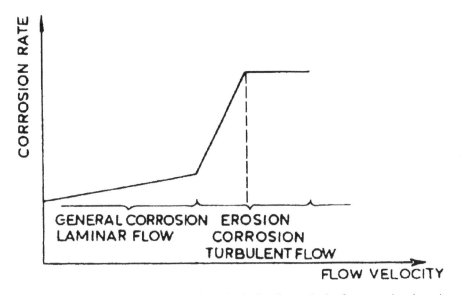

Figure 28 Corrosion rate as a function of solution flow velocity for general and erosion corrosion.

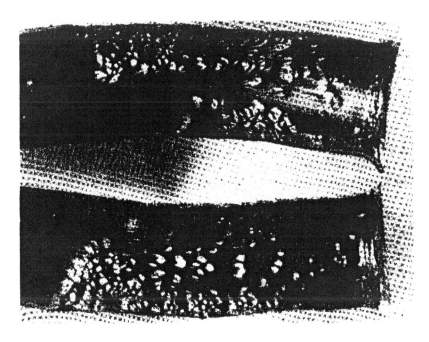

Figure 29 Erosion corrosion in brass condenser tube exposed to salt water.

Figure 30 Cavitation of the cast iron suction bell from a low pressure water pump.

polarization (η). The slope of this straight line is

$$\frac{d\eta}{di_{app}} = \frac{\beta_a \beta_c}{2.3(\beta_a + \beta_c)i_{corr}}$$

where β_a and β_c are the slopes of the anodic and cathodic Tafel lines. An experimental determination of $d\eta/dI$, shown in Fig. 31, allows the calculation of i_{corr}. This method may be applied to the determination of the corrosion rate even for passive metals ($\beta_a = \infty$), or if the cathodic reaction is controlled by diffusion (for the case where $\beta_c = \infty$).

   b. Extrapolation of Tafel Lines.   For polarizations of greater magnitude, i.e., $\eta > 50$ mV, the corrosion rate may be determined by extrapolation of the Tafel line for the anodic or cathodic reaction to the corrosion potential. The corrosion current may also be directly obtained from the point of intersection of the Tafel lines. Figure 32 illustrates this method [80].

   The methods described above are applicable for general corrosion corresponding to an even distribution of charge on the test electrode.

2. Recording of Anodic Polarization Curves for Passivating Metals

Electrochemical methods for investigating passive metals and alloys depend on the use of the potentiostatic techniques. These methods are based on recording anodic polarization curves of the type shown in Fig. 14. This approach is particularly useful when developing new alloys and for determining the influence of various alloying constituents and impurities on the corrosion properties of such alloys. The current peak at the passivation potential defines the tendency of the alloy to become passive in a given corrosion domain, whereas the passivity current gives the corrosion rate in the passive range. The passivation potential and the transpassive potential define the limits of the passive range. By recording the anodic

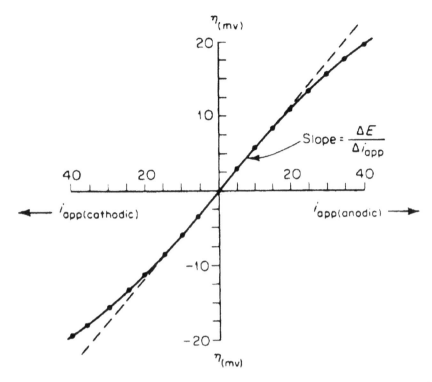

Figure 31   Schematic linear polarization curve showing experimental determination of $\Delta E/\Delta I$.

polarization curves of passive metals in environments containing chloride ions, or other activating ions, it is possible to determine pitting potential as shown in Fig. 33 [82].

   For recording polarization curves, it is possible to use a potentiodynamic method with a continuous increase of the potential, or a potentiostatic method with a stepwise increase of the potential. In each case, the potential increase should be slow in order to obtain reproducible results.

3.   Maintaining a Constant Anodic Potential

This method is suitable for accelerated laboratory tests, e.g., investigations of intergranular corrosion attack. For intergranular corrosion, one or more structural constituents at the grain boundaries dissolve at a faster rate than the interior of the grains. This phenomenon can be investigated by using potentiostatic testing where the applied potential is maintained so that the less noble areas corrode at a high rate, while those areas susceptible to general corrosion (i.e., the grain interior) corrode at a negligible rate. For testing the susceptibility of austenitic stainless steel to intergranular corrosion, the applied potential is maintained at a potential value where the austenite matrix passivates, while the chromium-depleted grain boundaries are active.

B.   Test Methodologies for Wear

New materials are continually being developed for a variety of applications, including those in the pulp and paper industry, the mining and mineral-processing industry, the metal-

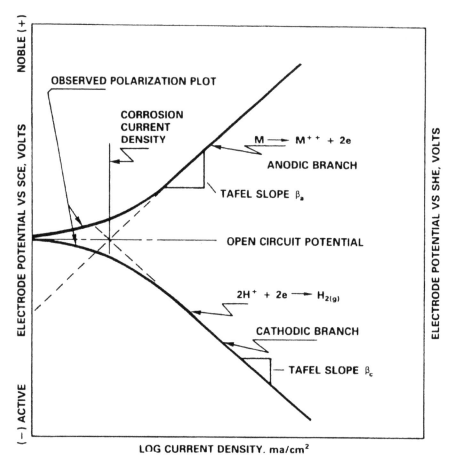

Figure 32 Illustration of experimental determination of anodic Tafel slope β_a from the anodic branch and cathodic Tafel slope β_c from the cathodic branch of a polarization curve.

lurgical industry, and the electric industry. Because of their applications, some components made from these materials will work under combined action of corrosion and wear [83–85].

The literature indicates that the wear process can be easily enhanced by corrosion when materials are exposed to aqueous solutions where the potential for accelerated corrosion exists [86–88]. This is likely to happen for metals and alloys that show relatively good corrosion resistance as a result of passive film formation.

If the wear is continuous, repassivation may be difficult, and active/passive cells will further enhance degradation of the material. Therefore, materials that possess good corrosion resistance in the presence of passive film formation may deteriorate rapidly as a result of wear. In the case of the proposed materials, chromium is added to increase their corrosion resistance because of the formation of passive films [58–60]. To fully understand the deterioration process and provide the best means for its prevention, it is necessary to determine the role of corrosion and wear in the deterioration process. This is best accomplished by determining the wear–corrosion synergism of the material in a corrosive media that simulates some industrial environment, and in electrolytes that can provide useful information with respect to pH changes of the industrial environment [89–92].

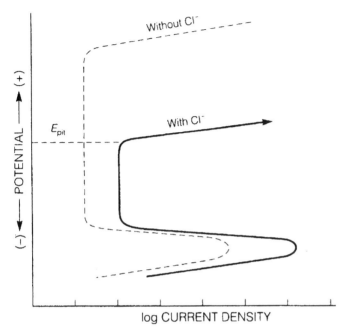

Figure 33 Schematic determination of critical pitting potential, E_{pit}, from an anodic polarization curve.

C. Determination of the Synergistic Effect of Corrosion and Wear [93]

The synergistic effect between corrosion and wear can be calculated by using the following equation:

$$S = T - W_0 - C_0$$

where S is the synergistic component when corrosion and wear act simultaneously, T is the corrosive wear rate, i.e., when corrosion and wear act in concert, W_0 is the wear rate in the absence of corrosion (wear component), and C_0 is the corrosion rate in the absence of wear (corrosion component).

Furthermore, the augmentation of wear by corrosion (S') and the augmentation of corrosion by wear (S) can be determined by using the following equations:

$$S = S' + S''$$

$$S' = T - C_w - W_0$$

$$S'' = C_w - C_0$$

where the corrosion rate (C_w) is augmented by wear.

Determination of the synergistic effect is important, because the sum of pure wear and pure corrosion may be less than the total corrosive wear. The difference, which is defined as the synergistic effect, results from the coupling effect of corrosion and wear. The magnitude of the synergistic effect may be large, even when the corrosion component relative to the wear component is small.

To determine the combined effect of corrosion and wear, corrosion, wear, and corrosion–wear experiments need to be performed. Because of the electrochemical nature of the corrosion processes occurring during wear under wet and/or corrosive conditions, electrochemical techniques can be used to quantify this corrosion–wear interaction.

To determine the wear component (W_0) in terms of mass loss, the specimen is weighed before and after the wear test. During the wear experiment, a potential from the cathodic region is applied to provide cathodic protection, i.e., to eliminate the corrosion effects.

The corrosion rate (C_0) and the corrosion rate augmented by wear (C_w) can be determined by using two electrochemical methods: (1) when the specimen is under static, and (2) wear conditions. In the first method, a constant potential is applied on the specimen. This potential is applied to accelerate the corrosion process. During the experiment, the change in current density with time is monitored. This dependence permits the calculation of the total charge that passes during the specimen. The charge is used to calculate the mass loss, and using Faraday's law, the corrosion rate. In the second method, the linear polarization technique is employed. As described in the previous section, the current density is measured as a function of potential in the region about ± 20 mV about the open-circuit corrosion potential. From the straight portion of the curve, the slope ($\Delta E / \Delta i_{aap}$) as shown in Fig. 31 is used to calculate the corrosion current. The Stern–Geary equation is then used to determine the corrosion rate.

D. Wear–Corrosion Testing

Corrosion, erosion, and abrasion are frequent problems in chemical and mineral processing plants, leading to failures of equipment operating under severe hydrodynamic condition such as pumps, turbines, centrifuges, agitators, and ball, mills. The combined and simultaneous action of these deterioration processes results in various types of surface wear of the equipment, which are defined according to the Standard Definitions of Terms Relating to Corrosion and Corrosion Testing, American Society for Testing and Materials (ASTM) G 15-83a [94], and the Standard Terminology Relating to Erosion and Wear, ASTM G 40-83 [95].

To understand the causes and mechanisms of corrosive wear phenomena and to measure its magnitude, many industries apply different laboratory and plant devices, which simulate service conditions and sometimes increase the severity of the chemical and mechanical factors involved in corrosive wear. Many apparatuses for studying corrosion–wear interaction through its synergistic effect have been used and are reported in the literature [90]. The Standard Guide For Determining Synergism Between Wear and Corrosion, ASTM G 119-83 [93], from slurry abrasive wear and sliding wear experiments is widely used for quantifying the corrosion and wear processes as well as its synergistic effect. For example, Kotlyar et al. [89] studied the abrasive–corrosive wear of high-carbon, low-alloy steel under grinding conditions in 15% quartz slurries using the rotating cylinder-anvil apparatus shown in Fig. 34. The apparatus consists of a rotating specimen in the shape of a cylinder (i.e., the working electrode) and two opposing abrasive anvils made either of the same material as the test specimen or of some chemically inert material, e.g., quartz. The apparatus also contains two platinum electrodes (i.e., the counter electrodes) and a Luggin capillary that serves as the reference electrode when carrying out electrochemical experiments. Pure corrosion, pure wear, and total wear–corrosion can be determined in terms of mass loss. The synergistic effect between corrosion and wear was specified as the mass loss as

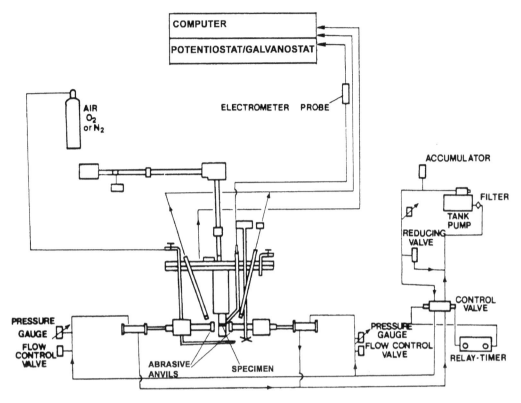

Figure 34 A schematic of cylinder-anvil apparatus.

determined by the difference between the total wear and the sum of the mass loss caused by pure abrasion plus corrosion. The results indicate that wear intensity is magnified by the corrosive environment. Also, the magnitude of the synergistic effect is considerable, even if the corrosion component of the overall material loss is small.

Friedersdorf and Holcomb [92] used an electrochemical pin-on-disk test technique to study the corrosion–wear synergism between 314 stainless steel in both 1 N H_2SO_4 and 1 N Na_2SO_4 solutions. The electrochemical pin-on-disk apparatus is shown in Fig. 35 [96]. The pin-on-disk apparatus is a modified ball-on-disk wear apparatus allowing wear experiments to be performed in corrosive environments with electrochemical control and continuous monitoring. Pure wear intensity was determined by sliding a test specimen in the shape of a pin across the surface of an abrasive disk. The electrochemical cell has three electrodes with the pin inserted in the circuit as the working electrode. The corrosion and corrosion–wear behavior were determined from potentiodynamic experiments. The material loss due to pure corrosion was determined from the potentiodynamic experiments, whereas the mass loss due to corrosion during wear was determined from the potentiostatic experiments. The material loss due to wear alone was gravimetrically determined under conditions of cathodic protection. The total material loss resulting from the combined action of corrosion and wear was also gravimetrically determined. The calculations of the synergistic effect of corrosion and wear indicate that the mechanical removal of material via abrasive wear was responsible for 95% of the total wear–corrosion losses in 304 stainless steel.

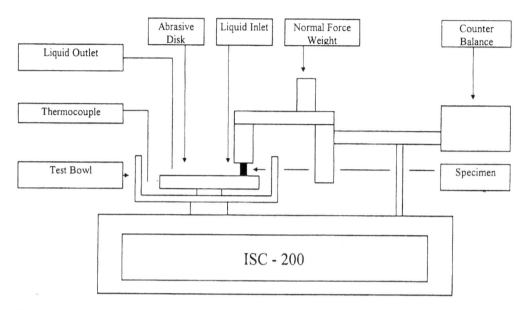

Figure 35 A schematic of pin-on-disk apparatus.

VI. SURFACE MODIFICATION FOR CORROSION CONTROL

This chapter has dealt with the basics of electrochemistry, corrosion, and corrosion-related testing techniques that are in general use. These techniques can be used to determine the electrochemical parameters for corrosion in any material. The rest of this section will briefly describe the two historical methods of modifying the surface of a material and will present some corrosion results. This section is based on the material found in Ref. 97.

Surface modification, in a historical sense, is the alteration of surface composition or structure by the use of energy or particle beams. A surface modified by one of these means is distinguished from conventional coatings by the greater similarity to the underlying bulk material. The utility of modifying the surface of a material with an energy or particle beam allows the surface composition to be changed by alloying on a microscale, through the development of metastable phases, or by development of an amorphous surface layer as a result of the high cooling rate (lasers) or disordered structure (particle implantation) [97,98].

In general, the advantages for corrosion resistance for modified surfaces include:

1. Alteration of the surface without sacrifice of bulk properties.
2. Conservation of scarce, critical, or expensive alloying elements.
3. Production of novel surface alloys (unattainable by conventional metallurgical techniques) with superior properties.
4. Avoidance of coating adhesion problems.

Attendant with corrosion and property improvements, there are disadvantages to using energy and particle beams to modify surfaces. These include:

1. Processing materials having deep or hidden contours.
2. Cost of processing equipment.
3. Substrate sensitivity to high-energy input (e.g., thermal instability).
4. The thinness (susceptibility to damage) of the coatings.

A. Ion Implantation

Ion implantation is the bombarding of a surface with an ionized species. The species impacts the surface and can knock atoms from equilibrium lattice positions into interstitial sites, and/or the ionized species can occupy the interstitial positions in the lattice. Ion implantation for wear resistance is quite widespread. Ion implantation for corrosion resistance is mainly confined to enhancing the passivation characteristics of the bulk composition and/or creating a novel material at the surface. The primary ion implantation techniques are:

1. Ion beam sputtering
2. Ion beam mixing
3. Plasma ion deposition
4. Ion beam assisted deposition

The advantages of using ion implantation to modify the surface of a material include [97]:

1. No sacrifice of bulk properties.
2. Solid-solubility can be achieved.
3. Alloy preparation is independent of diffusion constants.
4. No coating adhesion problem, because there is no interface.
5. No significant change in sample dimensions.
6. Depth concentration distribution is controllable.
7. Compositions can be changed without affecting grain size.
8. Precise location of implanted area(s).
9. Clean vacuum process.
10. Allows screening of the effects of changes in alloy compositions.

1. Corrosion Control

In using ion implantation for corrosion resistance, enhancing the passivation characteristics of the bulk composition and/or creating a novel material at the surface is the major motivation. In the former instance, implantation creates a surface composition based on conventional metallurgical experience that reacts to form a passive layer. In the latter case, ion implantation creates a surface composition/structure that may be difficult or impossible to create using conventional metallurgical approaches.

There are two problems associated with using ion implantation for corrosion resistance. The first is obtaining a modified layer of sufficient thickness to last the design lifetime of the part. As such, it is important to have a corrosion rate for the modified surface that is low and, if possible, self-repairing. Elements such as chromium and molybdenum have been used to modify the surface structure of conventional alloys. These two elements, when implanted into iron or steel, provide additional material at the surface to create a passive layer. Alternative approaches have been to use elements when alloyed with the base metal have good corrosion resistance, e.g., tantalum in iron forms a corrosion resistant intermetallic phase [97].

A second problem, i.e., creating a surface that is self-healing, is approached by introducing elements that are not soluble in, or removed by, the corrosion environment. For example, palladium in titanium is a very corrosion-resistant alloy system. Palladium strongly promotes titanium passivation, and when alloyed, provides lasting self-passivation properties in acidic media. However, palladium is only needed at the interface with the corrosive media. Because palladium is a very expensive element, it is more economical and effective to modify the titanium surface with palladium, i.e., where it is needed most, rather than alloy the entire part [97].

B. Laser Surface Processing

Lasers with a continuous output of 0.5–10 kW can be used to modify the metallurgical structure of a surface and to tailor the surface properties without unduly affecting the underlying material structure or composition. Laser surface modification can take the following forms [97,98]:

1. Transformation hardening, in which the surface is heated so that thermal diffusion and solid-state transformations can take place.
2. Surface melting, which results in the refinement of the surface structure because of rapid cooling from the melt temperature.
3. Surface alloying, in which alloying elements are added to the melt pool to change the composition of the surface layer.

Laser surface modification results in several important microstructural changes. These include the following: (1) a redistribution of major alloying components, (2) a redistribution of any second-phase particles or precipitates, (3) a change in the crystalline character of the surface, and/or (4) a change in the composition. In laser surface modified materials, two changes occur that can be used to modify corrosion behavior [97]. The first one is a change in the composition of the surface layer augmented by a refinement of the grains and phases found therein. The localized melting of the surface results in a leveling of any large-scale compositional variations, and a possible enrichment of the surface with a desired element or phase. For example, the rapid solidification of the melt pool (i.e., 10^7 K/sec) can extend the solid-solubility limits of the surface layer with attendant benefits for corrosion resistance (i.e., easier passivation). Concomitant with extended solid-solubility comes the refinement or elimination of undesirable second phases and/or grain size. At the very least, the phase and grain size is much reduced by the rapid cooling and the steep temperature gradient in the melt pool (i.e., 10^5 K/cm). The phases are also dispersed more uniformly throughout the surface layer. The microstructural modification would tend to promote general corrosion of the surface as opposed to possible localized corrosion of the second phases or severe intergranular attack at the grain boundaries [97–99].

1. Corrosion Control [97–100]

Laser modification has been used on a number of materials. For example, 304 stainless steel and ferritic Fe–13Cr–xMo (x = 0, 2, 3.5, or 5) stainless steel have been so processed. The critical current density needed for passivating 304 stainless steel in a 1 N sulfuric acid solution decreased by 2 orders of magnitude compared to that of the unprocessed steel. In the case of the ferritic stainless steel, the width of the active peaks decreased compared to the unprocessed material. Pitting potential increased by approximately 150 mV for the 304 and 5% Mo ferritic laser modified stainless steels. The other unprocessed ferritic stainless steels showed decreases in their pitting potentials [97,100].

Laser surface alloying was used to incorporate chromium into the surfaces of SAE 1018 steel and molybdenum into type 304 stainless steel. The 1018 steel passivated similarly to corresponding bulk alloys with 5–80% chromium. The 304-3Mo material was similar in pitting resistance to that of 316 stainless steel. The 304-9Mo material was superior to 316

Table 6 Effect of Laser Surface Alloying with Molybdenum on Pitting Potentials of Austenitic Stainless Steels in 0.1 M NaCl

Sample	Composition (%)			Pitting potential E_{pit} (V vs. SCE)
	Cr	Ni	Mo	
Type 304	18–20	8–10	0	0.300
Type 316	16–18	10–14	2–3	0.550
304-3Mo	18.9	9.1	3.7	0.500
304-9Mo	19.2	11.7	9.6	did not pit

stainless steel and showed no pitting up to oxygen evolution potentials. Table 6 shows some results of this research [97].

VII. SUMMARY AND CONCLUDING REMARKS

The intent of this chapter was to provide some background information on electrochemistry, corrosion, and corrosion testing techniques to understand how surface modification has been and can be used to improve the corrosion performance of materials. The basic principles for good corrosion resistance for bulk materials can be applied to designing modified surfaces and the corrosion resistance enhanced with proper processing. However, care must be taken to make sure that any modified surface has enough thickness to perform throughout the lifetime of the part, or that the corrosion behavior of the modified surface is enhanced to the point where corrosion is virtually nonexistent. Much work has been carried out, but advances in surface modification technology present unique opportunities to develop new materials specially designed for specific applications where corrosion resistance is needed.

REFERENCES

1. Kruger, J. Cost of metallic corrosion. In *Uhlig's Corrosion Handbook*; 2nd Ed.; Revie, R.W., Ed.; John Wiley & Sons, Inc.: New York, NY, 2000; 3–10 pp.
2. Bennett, L.H.; Kruger, J.; Parker, R.I.; Passaglia, E.; Reimann, C.; Ruff, A.W.; Yakowitz, H.; Berman, E.B. Economic effects of metallic corrosion in the United States—a three part study for Congress, Part I. Natl Bur Stand Spec Publ, U.S. Government Printing Office: Washington, DC, 1973; Vol. 511–1.
3. Economic effects of metallic corrosion in the United States—a 1995 update. Report to the Specialty Steel Industry of North America. Columbus, OH: Batelle Memorial Laboratories, April 1995.
4. Wranglen, G. *An Introduction to Corrosion and Protection of Metals*; Chapman and Hall: London, 1985; 1–17.
5. Guminski, K. *Physical Chemistry*; National Scientific Publisher: Warsaw, 1980; 21–74. *in Polish*.
6. Marek, M.I. Thermodynamics of aqueous corrosion. In *Metals Handbook,* 9th Ed., *Corrosion*; Korb, L.J., Olsen, D.L. Eds.; ASM International: Materials Park, OH, 1987; 18 pp.
7. Pomianowski, A. *Physical Chemistry*; State Scientific Publisher: Warsaw, 1980; 1017-1041 *in Polish*.

8. Jones, D.A. *Principles and Prevention of Corrosion*; Macmillan Publishing Company: New York, NY, 1992; 35–39.

9. Natalie, C.A. Electrode processes. In *Metals Handbook*, 9th Ed.; Korb, L.J., Olsen, D.L., Eds.; ASM International: Materials Park, OH, 1987; Vol. 13, 18–19.

10. Natalie, C.A. Electrode potentials. In *Metals Handbook*, 9th Ed.; *Corrosion*; Korb, L.J., Olsen, D.L., Eds.; ASM International: Materials Park, OH, 1987; Vol. 13, 18–21.

11. Atkins, L.P.W. *Physical Chemistry*, 4th Ed.; W.H. Freeman and Company: New York, NY, 1990; 257–258.

12. Uhlig, H.H.; Revie, R.W. *Corrosion and Corrosion Control*, 3rd Ed.; John Wiley & Sons, Inc.: New York, NY, 1985; 8–10.

13. Shreir, L.L. *Corrosion*; John Wiley & Sons, Inc.: New York, NY, 1965; 1.23–1.25.

14. Jones, D.A. *Principles and Prevention of Corrosion*; Macmillan Publishing Company: New York, 1992; 42–43.

15. Atkins, L.P.W. *Physical Chemistry*, 4th Ed.; W.H. Freeman and Company: New York, NY, 1990; 251–255.

16. Verink, E.D., Jr. Simplified procedure for constructing Pourbaix diagrams. In *Uhlig's Corrosion Handbook*, 2nd Ed.; Revie, R.W., Ed.; John Wiley & Sons, Inc.: New York, NY, 2000; 111–124.

17. Thompson, W.T.; Kaye, M.H.; Bale, C.W.; Pelton, A.D. Pourbaix diagrams for multi-component systems. In *Uhlig's Corrosion Handbook*, 2nd Ed.; Revie, R.W., Ed.; John Wiley & Sons, Inc.: New York, NY, 2000; 125–136.

18. Piron, D.L. Potential versus pH (Pourbaix) diagrams. In *Metals Handbook*, 9th Ed.; *Corrosion;* Korb, L.J. Olsen, D.L., Eds.; ASM International: Materials Park, OH, 1987; Vol. 13, 24–28.

19. Pourbaix, M. *Atlas of Electrochemical Equilibria*; Pergamon Press: London, 1964.

20. Talbot, D.; Talbot, J. *Corrosion Science and Technology*; CRC Press: Boca Raton, FL, 1998; 80–95.

21. Wranglen, G. *An Introduction to Corrosion and Protection of Metals*; Chapman and Hall: London, 1985; 52–95.

22. Flis, J. *Selected Actual Problems of Chemical and Mechanical–Chemical Degradation of Metals: Potential–pH Diagrams*; Ossolineum National Publishing Company: Warsaw, 1975; 49–70.

23. Palczewska, W.; Siedlecka, Z. *Physical Chemistry*; State Scientific Publisher: Warsaw, 1980; 1076–1097. *in Polish.*

24. Jones, D.A. *Principles and Prevention of Corrosion*; Macmillan Publishing Company: New York, NY, 1992; 83–85.

25. Szklarska-Smialowska, Z. *Novel Approaches in Corrosion Science*; Ossolineum National Publishing Company: Warsaw, 1967; 5–64. *in Polish.*

26. Fontana, M.G. *Corrosion Engineering*, 3rd Ed.; McGraw-Hill Publishing Company: New York, NY, 1986; 447–452.

27. Shoesmith, D.W. Kinetics of aqueous corrosion. In *Metals Handbook*, 9th Ed.; *Corrosion*; Korb, L.J., Olsen, D.L., Eds.; ASM International: Materials Park, OH, 1987; Vol. 13, 29–36.

28. Atkins, L.P.W. *Physical Chemistry*, 4th Ed.; W.H. Freeman and Company: New York, NY, 1990; 926–928.

29. Fontana, M.G. *Corrosion Engineering*, 3rd Ed.; McGraw-Hill Publishing Company: New York, NY, 1986; 464–468.

30. Jones, D.A. *Principles and Prevention of Corrosion*; Macmillan Publishing Company: New York, 1992; 75–76.

31. Jones, D.A. *Principles and Prevention of Corrosion*; Macmillan Publishing Company: New York, 1992; 115–120.

32. Kruger, J. Passivity. In *Uhlig's Corrosion Handbook*, 2nd Ed.; Revie, R.W., Ed.; John Wiley & Sons, Inc.: New York, NY, 2000; 165–171.

33. Talbot, D.; Talbot, J. *Corrosion Science and Technology*; CRC Press: Boca Raton, FL, 1998; 110–115.

34. Uhlig, H.H.; Revie, R.W. *Corrosion and Corrosion Control*, 3rd Ed.; John Wiley & Sons, Inc.: New York, NY, 1985; 61 pp.

35. Craig, B.; Pohlman, S.L. Forms of corrosion: introduction. In *Metals Handbook*, 9th Ed.; Korb, L.J., Olsen, D.L., Eds.; ASM International: Materials Park, OH, 1987; Vol. 13, 79 pp.

36. Wranglen, G. *An Introduction to Corrosion and Protection of Metals*; Chapman and Hall: London, 1985; 62–65.

37. Pohlman, S.L. General corrosion. In *Metals Handbook*, 9th Ed.; Korb, L.J., Olsen, D.L., Eds.; ASM International: Materials Park, OH, 1987; Vol. 13, 80–87.

38. Natalie, C.A. Evaluation of uniform corrosion. In *Metals Handbook*, 9th Ed.; Korb, L.J.; Olsen, D.L., Eds.; ASM International: Materials Park, OH, 1987; Vol. 13, 229–230.

39. Jones, D.A. *Principles and Prevention of Corrosion*; Macmillan Publishing Company: New York, NY, 1992; 11 pp.

40. Wranglen, G. *An Introduction to Corrosion and Protection of Metals*; Chapman and Hall: London, 1985; 84–85.

41. Zhang, X.G. Galvanic corrosion. In *Uhlig's Corrosion Handbook*, 2nd Ed.; Revie, R.W., Ed.; John Wiley & Sons, Inc.: New York, NY, 2000; 137–164.

42. Hack, H.P. Evaluation of galvanic corrosion. In *Metals Handbook*, 9th Ed.; Corrosio; Korb, L.J., Olsen, D.L., Eds.; ASM International: Materials Park, OH, 1987; vol. 13, 234–238.

43. Wranglen, G. *An Introduction to Corrosion and Protection of Metals*; Chapman and Hall: London, 1985; 85–87.

44. Jones, D.A. *Principles and Prevention of Corrosion*; Macmillan Publishing Company: New York, NY, 1992; 168–169.

45. Böhni, H. Localized corrosion of passive metals. In *Uhlig's Corrosion Handbook*, 2nd Ed.; Revie, R.W., Ed.; John Wiley & Sons, Inc.: New York, NY, 2000; 173–190.

46. Dexter, S.C. Localized corrosion. In *Metals Handbook*, 9th Ed.; Korb, L.J., Olsen, D.L., Eds.; ASM International: Materials Park, OH, 1987; Vol. 13, 104–108.

47. Asphahani, A.I.; Silence, W.L. Pitting corrosion. In *Metals Handbook*, 9th Ed.; Korb, L.J., Olsen, D.L., Eds.; ASM International: Materials Park, OH, 1987; Vol. 13, 113–114.

48. Sprowels, D.O. Evaluation of pitting corrosion. In *Metals Handbook*, 9th Ed.; Korb, L.J., Olsen, D.L., Eds.; ASM International: Materials Park, OH, 1987; 94–95.

49. Wranglen, G. *An Introduction to Corrosion and Protection of Metals*; Chapman and Hall: London, 1985; 94–98.

50. Fontana, M.G. *Corrosion Engineering*, 3rd Ed.; McGraw-Hill Publishing Company: New York, NY, 1986; 64–66.

51. Kain, R.M. Crevice corrosion. In *Metals Handbook*, 9th Ed.; Korb, L.J., Olsen, D.L., Eds.; ASM International: Materials Park, OH, 1987; Vol. 13, 108–113.

52. Kain, R.M. Evaluation of crevice corrosion. In *Metals Handbook*, 9th Ed.; Korb, L.J., Olsen, D.L., Eds.; ASM International: Materials Park, OH, 1987; Vol. 13, 303–310.

53. LaQue, F.L. The problem of localized corrosion. Proceedings of International Conference on Localized Corrosion, Williamsburg, Virginia; 1971; i47–i53.

54. Fontana, M.G. *Corrosion Engineering*, 3rd Ed.; McGraw-Hill Publishing Company: New York, NY, 1986; 51–52.

55. Wranglen, G. *An Introduction to Corrosion and Protection of Metals*; Chapman and Hall: London, 1985; 98–99.

56. Steigerwald, R. Metallurgically influenced corrosion. In *Metals Handbook*, 9th Ed.; Korb, L.J., Olsen, D.L., Eds.; ASM International: Materials Park, OH, 1987; Vol. 13, 123–135.

57. Corbett, R.A.; Saldanha, B.J. Evaluation of intergranular corrosion. In *Metals Handbook*, 9th Ed.; Korb, L.J., Olsen, D.L., Eds.; ASM International: Materials Park, OH, 1987; Vol. 13, 239–241.

58. Kaesche, H. *Metallic Corrosion*, 2nd Ed.; National Association of Corrosion Engineers: Houston, TX, 1985; 360–362.

59. Uhlig, H.H.; Revie, R.W. *Corrosion and Corrosion Control*, 3rd Ed.; John Wiley & Sons: New York, NY, 1985; 15 pp.

60. Jones, D.A. *Principles and Prevention of Corrosion*; Macmillan Publishing Company: New York, NY, 1992; 19 pp.
61. *Dealloying in Basic Corrosion Study Manual*; NACE International: Houston, TX, 1997; 5:76–5:77.
62. Parkins, R.N. Stress corrosion cracking. In *Uhlig's Corrosion Handbook*, 2nd Ed.; Revie, R.W., Ed.; John Wiley & Sons, Inc.: New York, NY, 2000; 191–204.
63. Glaeser, W.; Wright, I.G. Mechanically assisted corrosion. In *Metals Handbook*, 9th Ed.; Korb, L.J., Olsen, D.L., Eds.; ASM International: Materials Park, OH, 1987; Vol. 13, 136–144.
64. Craig, B. Environmentally induced cracking. In *Metals Handbook*, 9th Ed.; Korb, L.J., Olsen, D.L., Eds.; ASM International: Materials Park, OH, 1987; Vol. 13, 145–163.
65. Sprowels, D.O. Evaluation of stress–corrosion cracking. In *Metals Handbook*, 9th Ed.; Korb, L.J., Olsen, D.L., Eds.; ASM International: Materials Park, OH, 1987.
66. Fontana, M.G. *Corrosion Engineering*, 3rd Ed.; McGraw-Hill Publishing Company: New York, NY, 1986; 109–112.
67. Ziomek-Moroz, M. The influence of carbon on passivation and stress corrosion cracking in caustic soda solutions. Ph.D. Dissertation, Institute of Physical Chemistry of the Polish Academy of Sciences, Warsaw, 1986.
68. Jones, D.A. *Principles and Prevention of Corrosion*; Macmillan Publishing Company: New York, NY, 1992; 240 pp.
69. Sprowels, D.O. Evaluation of corrosion fatigue. In *Metals Handbook*, 9th Ed.; Korb, L.J., Olsen, D.L., Eds.; ASM International: Materials Park, OH, 1987.
70. Wang, Y.-Z. Corrosion fatigue. In *Uhlig's Corrosion Handbook*; 2nd Ed.; Revie, R.W., Ed.; John Wiley & Sons, Inc.: New York, NY, 2000; 221–232.
71. Fontana, M.G. *Corrosion Engineering*, 3rd Ed.; McGraw-Hill Publishing Company: New York, NY, 1986; 1139–1141.
72. Postlethwaite, J.; Nesic, S. Erosion–corrosion in single and multiphase flow. In *Uhlig's Corrosion Handbook*; 2nd Ed.; Revie, R.W., Ed.; John Wiley & Sons, Inc.: New York, NY, 2000; 249–272.
73. Evaluation of erosion and cavitation. In *Metals Handbook*, 9th Ed.; Korb, L.J., Olsen, D.L., Eds.; ASM International: Materials Park, OH, 1987; Vol. 13, 311–313.
74. *Erosion Corrosion in Basic Corrosion Study Manual*; NACE International: Houston, TX, 1997; 5:58 pp.
75. Jones, D.A. *Principles and Prevention of Corrosion*; Macmillan Publishing Company: New York, NY, 1992; 21–24.
76. Wranglen, G. *An Introduction to Corrosion and Protection of Metals*; Chapman and Hall: London, 1985; 82–98.
77. Wranglen, G. *An Introduction to Corrosion and Protection of Metals*; Chapman and Hall: London, 1985; 118 pp.
78. Silverman, D.C. Practical corrosion prediction using electrochemical techniques. In *Uhlig's Corrosion Handbook*, 2nd Ed.; Revie, R.W., Ed.; John Wiley & Sons, Inc.: New York, NY, 2000; 1179–1225.
79. Scully, J.R. Electrochemical methods of corrosion testing. In *Metals Handbook*, 9th Ed.; Korb, L.J., Olsen, D.L., Eds.; ASM International: Materials Park, OH, 1987; Vol. 13, 212–220.
80. Scully, J.R. Corrosion methods for laboratory corrosion testing. In *Corrosion Testing and Evaluation: Silver Anniversary Volume*; Baboin, R., Dean, S.W., Eds.; American Society for Testing and Materials: Philadelphia, PA, 1990.
81. Wranglen, G. *An Introduction to Corrosion and Protection of Metals*; Chapman and Hall: London, 1985; 247–249.
82. Jones, D.A. *Principles and Prevention of Corrosion*; Macmillan Publishing Company: New York, NY, 1992; 208–209.
83. Doan, O.N.; Hawk, J.A. Abrasion resistance of in situ Fe–TiC composites. Scr Metall Mater 1995, *33*, 953–958.

84. Doan, O.N.; Hawk, J.A. Microstructure and abrasion resistance of Fe–TiC–M_7C_3 composite. Microstruct Sci 1996, *23*, 251–257.

85. Doan, O.N.; Alman, D.E.; Hawk, J.A. Wear resistant, powder processed in-situ iron–matrix TiC composite. In *Advances in Powder and Particulate Materials*; Philips, M., Porter, J., Eds.; MPIF: Princeton, NJ, 1996; 83–97.

86. Buchanan, R.A.; Turner, G.D.; Gray, P.D.; Melendez, J.G.; Talbot, T.F.; McDonald, J.L. A new apparatus for synergistic studies of corrosive wear. Corrosion 1983, *39*, 377–379.

87. Abuzriba, M.B.; Dodd, R.A.; Worzela, F.J.; Conrad, J.R. Wear–corrosion: separation of the components of corrosion and wear. Corrosion 1992, *48*, 2–4.

88. Ziomek-Moroz, M. Wood solution for studies of corrosive wear of woodcutting tools. Corrosion 1994, *50*, 276–279.

89. Kotlyar, D.; Pitt, C.H.; Wadsworth, M.E. Simultaneous corrosion and abrasion measurements under grinding conditions. Corrosion 1988, *44*, 221–228.

90. Watson, S.W.; Friedersdorf, F.J.; Madsen, B.W.; Cramer, S.D. Methods of measuring wear–corrosion synergism. Wear 1995, *181–183*, 476–484.

91. Zhou, S.; Stack, M.M.; Newman, R.C. Characterization of synergistic effects between erosion and corrosion in an aqueous environment using electrochemical techniques. Corrosion 1996, *52*, 934–946.

92. Friedersdorf, F.J.; Holcomb, G.R. Pin-on-disc corrosion–wear test. J Test Eval 1998, *26*, 352–357.

93. Madsen, B.W. Standard guide for determining amount of synergism between wear and corrosion. 1994 Annual Book of ASTM Standards: ASTM International, West Conshohocken, PA, 03.02 ASTM G 119-93; 494–499.

94. Schorr, M.; Weintraub, E.; Andrasi, D. Erosion-corrosion measuring devices. In *Corrosion Testing and Evaluation: Silver Anniversary Volume*; Baboin, R., Dean, S.W., Eds.; American Society for Testing and Materials: Philadelphia, PA, 1990; 151–159.

95. Metal Corrosion, Erosion, and Wear in 1984 Annual Book of ASTM Standards: ASTM International, West Conshohocken, PA, 03.02 ASTM G 40-83; 426 pp.

96. Holmes, D.R. Corrosive wear behavior of zirconium in hot sulfide containing electrolytes. M.S. Thesis, Oregon Graduate Institute of Science and Engineering, 2001.

97. Granata, R.D.; Moore, P.G. Surface modification. In *Metals Handbook*, 9th Ed.; Korb, L.J., Olsen, D.L., Eds.; ASM International: Materials Park, OH, 1987; Vol. 13, 498–505.

98. Moore, P.G. Laser surface processing: an overview. In *Fundamental Aspects of Corrosion Protection By Surface Modification*; McCafferty, E., Clayton, C.R., Oudar, J., Eds.; The Electrochemical Society, Inc.: Pennington, NJ, 1984; 102–111.

99. McCafferty, E.; Moore, P.G. Corrosion behavior of laser alloyed stainless steels. In *Fundamental Aspects of Corrosion Protection By Surface Modification;* McCafferty, E., Clayton, C.R., Oudar, J., Eds.; The Electrochemical Society, Inc.: Pennington, NJ, 1984; 112–121.

100. Lumsden, J.B.; Gnanamuthu, D.S.; Moores, R.J. Electrochemical properties of laser processed surfaces. In *Fundamental Aspects of Corrosion Protection By Surface Modification*; McCafferty, E., Clayton, C.R., Oudar, J., Eds.; The Electrochemical Society, Inc.: Pennington, NJ, 1984; 122–129.

13

Corrosion and Its Impact on Wear Processes

Einar Bardal
Norwegian University of Science and Technology, Trondheim, Norway

Asgeir Bardal
Hydro Aluminium Technology Center Årdal, Årdal, Norway

I. INTRODUCTION

Corrosion has been defined in different ways, but the usual understanding of the term is "chemical attack on a metallic material by reaction with the environment." This definition is also used in the present chapter. Corrosion reactions sometimes strongly interact with mechanically induced local deformation of material, and combinations of various forms of wear and corrosion are among the most important and widespread such chemical–mechanical interactions.

The chapter consists of three main parts: General corrosion theories and examples; interaction between wear and corrosion; and characterization of surface topography and structure.

The first part is divided in aqueous (wet) corrosion and corrosion (oxidation) in dry gases. Both forms of deterioration are electrochemical in nature, but in somewhat different ways. Aqueous corrosion is characterized by anodic (oxidation) and cathodic (reduction) reaction(s), both taking place at the interface between a metallic material and an electrolyte. On the other hand, when corrosion occur in a dry gas, oxidation of metal is localized at the interface between the metal and a film of oxidation (corrosion) product, while reduction of an oxidant (e.g., oxygen) occurs at the interface between the oxidation product film and the gas. Deviation from this pattern may occur under extremely efficient removal of corrosion product formed in dry gas.

Aqueous corrosion is limited to a temperature range where water is in liquid form, while corrosion (oxidation) in dry gases primarily occurs at high temperature. Under certain special circumstances, significant oxidation can also take place at ambient temperature of the gas. This is of vital interest in the present work because wear occurring in an environment of dry oxidizing gases is a typical situation where considerable oxidation may occur at ambient temperature. This is possible because thin protecting surface films can be worn off or prevented from being formed so that a fresh and active metal surface is continuously exposed.

There is also a third group of corrosion processes, namely those occurring in other liquid media than aqueous solutions, e.g., salt melts and liquid metals. This group is not dealt with in the present chapter.

The chapter gives a brief and concentrated presentation of corrosion theory, with the intention to focus on fundamentals of considerable significance for the interaction between corrosion and wear. Of particular importance are the properties and conditions for formation of solid corrosion products on metal surfaces, the mechanical (e.g., flow) effects on reactions, and the electrochemical methods to describe and measure corrosion rate. These methods can also be used in many cases where corrosion is combined with wear.

The chapter distinguish between two groups of corrosion-affected wear, mainly from a practical point of view: (1) erosion and abrasion due to flow of hard mineral particles in a corrosive fluid, and (2) combined wear and corrosion under sliding, rolling, or oscillating contact between machine components.

The third main part of the chapter concentrates on the physical surface characterization methods that are useful for the study of both topography and structure of the surface material before and after exposure to corrosion and/or wear.

II. AQUEOUS CORROSION

A. Description of a Typical Wet Corrosion Process

Figure 1 is a schematic picture of corrosion of a divalent metal M in an aqueous solution containing oxygen (aerated). The corrosion process consists of an anodic and a cathodic reaction. The anodic reaction is oxidation of metal M to metal ions M^{2+}, i.e., dissolution of metal. The cathodic reaction is reduction of oxygen and water, which is by far the most common cathodic reaction in natural environments.

It is seen that the process makes an electric circuit without accumulation of charges anywhere: The electrons released by the anodic reaction are conducted through the metal to the cathode where they are consumed in the cathodic reaction. A prerequisite for such a corrosion process is that there is a conductive liquid (an electrolyte) in contact with the metal and covering both an anodic and a cathodic area. The electric circuit is closed by ion conduction through the electrolyte. In the shown example, metal ions are conducted to meet OH^- ions, and a metal hydroxide is formed, either dissolved or possibly more or less as a deposit on the metal surface.

Several metals corrode in principle such as that shown in the figure. For iron, the process is modified because divalent iron (ferrous) hydroxide is not stable and oxidizes under access of oxygen and water to a trivalent (ferric) hydrated oxide, $FeOOH \cdot H_2O$, which is well

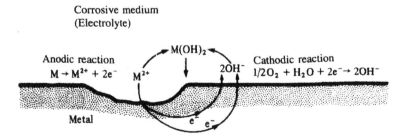

Figure 1   Aqueous corrosion of a divalent metal M in an electrolyte containing oxygen.

known as red-brown rust. With limited access of oxygen, magnetite (Fe_3O_4) is formed, which is black (without water) or green-black (with water) [1].

Because wet corrosion is characterized by an electric current across the metal–electrolyte interface at the electrodes, we can express the corrosion rate in terms of current or current density. The local corrosion current density on the anode is the anodic current divided by the anodic area. More commonly used is the average corrosion current density, which is the sum of anodic current per square centimeter of the total surface. The (average) corrosion current density can be transferred to thickness reduction (from one side of the material) per unit of time by Faraday's law:

$$ds/dt = \frac{i_{corr}M}{z\,F\rho}\,cm/s \tag{1a}$$

or

$$\frac{\Delta s}{\Delta t} = 3268\frac{i_{corr}M}{z\rho}\,mm/yr \tag{1b}$$

where i_{corr} is given in A/cm^2; z is the number of electrons in the anodic reaction equation (per atom of the dissolving metal); M is the mol mass of the metal (g/mol atoms) (atomic weight); F is Faraday's number $= 96,485$ C/mol electrons $\approx 96,500$ C/mol $e^- = 96,500$ As/mol e^-; and ρ is the density of the metal (g/cm^3).

Table 1 shows the factors for transfer of corrosion rate from one unit to another.

Figure 1 illustrates an electrochemical cell. The driving force of the process is the cell voltage, i.e., the potential difference between the cathode and the anode (the electrodes).

To understand under which conditions corrosion is possible and how various factors affect the corrosion rate, a study of the concept of electrode potential and its relation to reaction rates is necessary. A brief presentation of these subjects is given in Secs. II.B–II.E.

For a more thorough study, readers are referred to corrosion textbooks, e.g., Ref. 3 for introduction to corrosion theory and Refs. 4–6 for more comprehensive descriptions of corrosion conditions, corrosion forms and descriptions, materials, prevention, testing, etc.

B. Reversible (Equilibrium) Electrode Reactions: Pourbaix Diagram

If there is no externally impressed voltage, the maximum cell voltage, or the maximum potential difference between two electrodes with given electrode reactions, is the *reversible*

Table 1 Factors for Conversion Between Corrosion Rate Units

Material/reaction	Corrosion current density ($\mu A/cm^2$)	Weight loss per unit of area and time (mdd)	Average attack depth increment per unit of time	
			mm/yr	mpy
$Fe \rightarrow Fe^{2+} + 2e^-$	1	2.51	1.16×10^{-2}	0.46
$Cu \rightarrow Cu^{2+} + 2e^-$	1	2.84	1.17×10^{-2}	0.46
$Zn \rightarrow Zn^{2+} + 2e^-$	1	2.93	1.5×10^{-2}	0.59
$Ni \rightarrow Ni^{2+} + 2e^-$	1	2.63	1.08×10^{-2}	0.43
$Al \rightarrow Al^{3+} + 3e^-$	1	0.81	1.09×10^{-2}	0.43
$Mg \rightarrow Mg^{2+} + 2e^-$	1	1.09	2.2×10^{-2}	0.89

mdd $=$ mg/dm^2/day; mpy $=$ mils/yr.
Source: Ref. 2.

cell voltage. A reversible cell with the electrodes 1 and 2 is shown in Fig. 2. The electrode reactions may for instance be the same as in Fig. 1, but they are now in dynamic equilibrium, e.g., the metal is in equilibrium with its ions in the solution:

$$M^{2+} \Leftrightarrow M$$

To realize the reversible cell, i.e., to maintain the reversible cell voltage, the voltmeter must have a voltage so high that the current flowing from one electrode to the other can be neglected.

Generally, a reversible electrode reaction can be written as

$$aA + bB + \cdots + ze^- = lL + mM + \cdots \tag{2}$$

meaning that there are a moles of species A, etc. It is shown by the electrochemical theory [1–7] that the reversible (equilibrium) potential of the reaction is expressed by Nernst's equation

$$E_0 = E_0^0 - \frac{RT}{zF} \ln \frac{a_L^l a_M^m \cdots}{a_A^a a_B^b \cdots} \tag{3}$$

where a is the activity, usually replaced by concentration and E_0^0 is the standard reversible electrode potential, i.e., the reversible potential when all activities = 1.

Activity a of dissolved species is inserted as a concentration given in mol/L. For solvents and solid metals, a is expressed in mole fraction (usually ≈ 1), and for gases, $a = p$ (pressure) in atm.

The electrode potential, which can be physically considered as the potential difference between a metal and the adjacent electrolyte, can only be measured relative to the potential

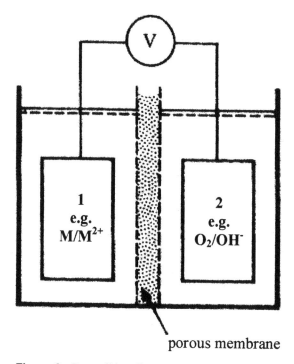

Figure 2   Reversible cell.

(difference) of another electrode, the reference electrode. Standard reversible potentials (relative to *standard hydrogen electrode*, SHE, with reaction $2H^+ + 2e^- \Leftrightarrow H_2$) for different reactions are given in Table 2. The equilibrium (reversible) potentials of the various reactions make the basis for determination of possible directions of the electrode reactions of given cells, i.e., electrode combinations. In other words, they can show if a reaction in a certain direction, e.g., dissolution of a metal, is possible or not.

From the list in Table 2, it can be seen that several reactions involve H^+. This means that the equilibrium potentials depend on pH ($= -\log a_{H+}$), which is a very important variable for corrosive media. On this background, Marcel Pourbaix derived equilibrium potential–pH diagrams for various metals, also called Pourbaix diagrams, which have become an important tool for illustration of the corrosion possibilities [1,7]. The Pourbaix diagram for iron in water at 25°C is shown in Fig. 3. The lines of the different reactions involving Fe^{2+} are based on activity $a_{Fe^{2+}} = 10^{-6}$, which is considered to be an actual lower value under corrosion of practical significance. For reaction (a), lines for activities 10^{-4} and 10^{-2} are also indicated. The dominating product of a reaction depends on whether the potential is above or below the equilibrium potential line. For instance, for the reaction (a)

Table 2 Standard Electrode Potentials E_0^0 at 25°C vs. Standard Hydrogen Electrode (SHE)

Electrode reaction	E_0^0 (V)
$Au^{3+} + 3e^- = Au$	1.50
$Cl_2 + 2e^- = 2Cl^-$	1.36
$O_2 + 4H^+ + 4e^- = 2H_2O$	1.23
$Pt^{2+} + 2e^- = Pt$	1.20
$Fe_3O_4 + 8H^+ + 2e^- = 3Fe^{2+} + 4H_2O$	0.98
$HNO_3 + 3H^+ + 3e^- = NO + 2H_2O$	0.96
$Ag^+ + e^- = Ag$	0.80
$Hg^{2+} + 2e^- = Hg$	0.79
$Fe^{3+} + e^- = Fe^{2+}$	0.77
$Fe_2O_3 + 6H^+ + 2e^- = 2Fe^{2+} + 3H_2O$	0.73
$O_2 + 2H_2O + 4e^- = 4OH^-$	0.40
$Cu^{2+} + 2e^- = Cu$	0.34
$AgCl + e^- = Ag + Cl^-$	0.22
$S + 2H^+ + 2e^- = H_2S$	0.14
$2H^+ + 2e^- = H_2$	0
$Fe_3O_4 + 8H^+ + 8e^- = 3Fe + 4H_2O$	−0.085
$Pb^{2+} + 2e^- = Pb$	−0.13
$Sn^{2+} + 2e^- = Sn$	−0.14
$Ni^{2+} + 2e^- = Ni$	−0.25
$Cd^{2+} + 2e^- = Cd$	−0.40
$Fe^{2+} + 2e^- = Fe$	−0.44
$Cr^{3+} + 3e^- = Cr$	−0.74
$Zn^{2+} + 2e^- = Zn$	−0.76
$2H_2O + 2e^- = 2OH^- + H_2$	−0.83
$Ti^{2+} + 2e^- = Ti$	−1.63
$Al^{3+} + 3e^- = Al$	−1.66
$Mg^{2+} + 2e^- = Mg$	−2.37
$Na^+ + e^- = Na$	−2.71

Source: Ref. 7

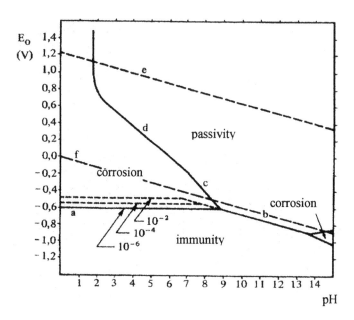

Figure 3 Pourbaix diagram of iron in water at 25°C. Reactions: (a) $Fe^{2+} + 2e^- = Fe$; (b) $Fe_3O_4 + 8H^+ + 8e^- = 3Fe + 4H_2O$; (c) $Fe_3O_4 + 8H^+ + 2e^- = 3Fe^{2+} + 4H_2O$; (d) $Fe_2O_3 + 6H^+ + 2e^- = 2Fe^{2+} + 3H_2O$; (e) $O_2 + 4H^+ + 4e^- = 2H_2O/O_2 + 2H_2O + 4e^- = 4OH^-$; (f) $2H^+ + 2e^- = H_2/ 2H_2O + 2e^- = H_2 + 2OH^-$.

$Fe^{2+} + 2e^- = Fe$, the dominating product is Fe^{2+} when the potential is above the reversible potential, i.e., when the metal is made more positive than that corresponding to equilibrium. In other words, above the line for reaction (a), iron corrodes. In this way, certain corrosion, passivity, and immunity regions are determined, as shown in the diagram. The passivity region is characterized by a thin dense and strong oxide film with good adhesion to the substrate, while the immunity region is characterized by a stable pure metal surface. The validity of the diagrams is, strictly speaking, limited to pure metals in water. Small amounts of impurities and alloying elements in the metals and dissolved species in the water do not necessarily affect the diagrams considerably. However, for instance, larger amounts of Cr as alloying element and Cl^- in water strongly change the diagram.

From the above description, we can see that the equilibrium (reversible) potentials and Pourbaix diagrams can show if corrosion is possible under the given conditions, i.e., at different pH–potential values. However, these tools do not tell us anything about the corrosion rates. This subject is treated in the next section. Extensive description and thorough explanation of reversible electrochemical reactions are given in Pourbaix's publications [1,7].

C. The Kinetics of the Reactions on an Active Metal

As shown in Sec. II.A, the rate of an electrochemical reaction can be expressed as reaction current or current density. The rate of each electrode reaction depends on the electrode potential. Usually, this relationship is expressed in a potential–log current diagram (E vs. log I diagram) or a potential–log current density diagram (E vs. log i diagram). An E vs. log I diagram for the electrode reactions in Fig. 1 is shown in Fig. 4. An active metal is assumed in this case; that is, the anodic reaction is not hindered by a passivating film on the surface.

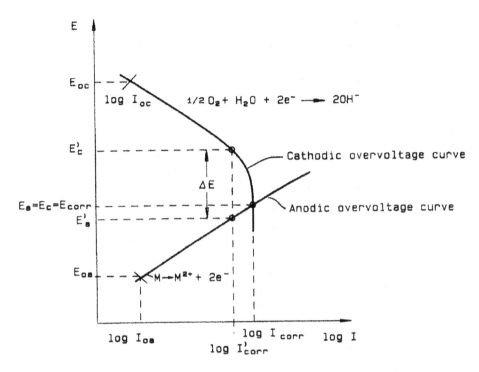

Figure 4 Potential–log current diagram for an active metal corroding in an aerated aqueous solution.

At the potential E_{oa}, the reaction $M \Leftrightarrow M^{2+} + 2e$ is in equilibrium; that is, the potential is identical with the equilibrium potential or reversible potential of this reaction. The oxidation rate of the reaction equals the reduction rate at this potential value. This rate can be expressed in terms of a current density named the exchange current density, generally denoted i_o, and in our case i_{oa}. For a certain electrode area, there is an exchange current I_o. In the case shown in Fig. 4, we have an exchange current I_{oa} as a basis for our anodic reaction. Similarly, we have the equilibrium potential E_{oc} and the exchange current I_{oc} for the cathodic reaction (oxygen reduction).

When the electrode potential of the M/M^{2+} reaction is more positive than the equilibrium potential E_{oa}, there is a net reaction rate in anodic direction, and this rate increases with increasing potential distance from the equilibrium. The total anodic current depends on the potential as shown by the lower curve, which is named the anodic overvoltage curve. The difference between an arbitrary potential E and the equilibrium potential E_{oa} is the so-called anodic overvoltage η_a ($\eta_a = E - E_{oa}$).

When the potential deviates from the equilibrium potential, the electrode is said to be polarized. The degree of polarization is expressed by the overvoltage. In Fig. 4, the anodic reaction is subject to *activation polarization*. This means that the rate of the reaction is determined by the resistance against charge transfer across the interface between the metal and the aqueous solution. The relationship between potential and anodic current density is given by the Tafel equation

$$\eta = E - E_{oa} = b \cdot \log \frac{i}{i_{oa}} = b \cdot \log \frac{I}{I_{oa}} \qquad (4)$$

i.e., a straight line in the diagram in Fig. 4.

If we look at the cathodic overvoltage curve, we see that the cathodic overvoltage is negative: $\eta_c = E - E_{oc}$. The actual cathodic reaction, reduction of oxygen, is under activation polarization at lower current values; hence the overvoltage curve is a straight line in this range. However, when the potential is lowered and the current increases, the reaction rate reaches a level where it is limited by limited access of oxygen. The reason is that oxygen, which is consumed at the electrode, has to be supplied from the bulk of the liquid. In a thin layer close to the metal, oxygen transport occurs by diffusion, and the diffusion rate is limited. Consequently, the current reaches the so-called diffusion-limiting current I_{lim}, and the current density correspondingly reaches the diffusion-limiting current density i_{lim}:

$$I_{lim} = ADzF\frac{c_B}{\delta} \tag{5a}$$

$$i_{lim} = DzF\frac{c_B}{\delta} \tag{5b}$$

In these equations, A is the electrode area, D is the diffusion coefficient of oxygen in water, z is the number of electrons per oxygen molecule in the reaction equation, F is Faraday's number, c_B is the bulk concentration of oxygen, and δ is the thickness of the layer where the transport occurs by diffusion, i.e., the *diffusion boundary layer*.

I_{lim} is graphically expressed by the vertical part of the cathodic curve in Fig. 4. In other words, the cathodic current is independent of potential in this potential range. The actual polarization mechanism is named *concentration polarization*. It is also said that the reaction is mass transfer controlled or diffusion controlled.

As already mentioned in Sec. II.A, the anodic current must equal the cathodic current for a metal under free corrosion, i.e., $I_a = I_c$. If we assume that there is practically no ohmic potential drop between the anode and the cathode, neither in the liquid nor in the metal, the electrode potentials at the anode and the cathode are equal too, i.e., $E_a = E_c$. These two conditions, $I_a = I_c$ and $E_a = E_c$, are satisfied only at one point in Fig. 4, namely at the intersection point between the anodic and the cathodic overvoltage curve. This point defines both the corrosion potential (open circuit potential, rest potential) E_{corr} and the corrosion current I_{corr}. This is the so-called mixed potential theory, which is extremely important for the understanding of corrosion behavior. In the case of diffusion control, $I_{corr} = I_{lim}$.

If there is a potential drop ΔE in the electrolyte between the cathode and the anode, the mixed potential theory has to be modified, as illustrated by the dotted line in Fig. 4. Now the anode potential E_a' is more negative than the cathode potential E_c' ($E_c' - E_a' = \Delta E$). The anodic and cathodic currents are still equal ($= I_{corr}'$), but lower than in the case of no potential drop in the solution.

The diffusion layer thickness δ decreases by increasing the flow velocity v of the liquid, i.e., I_{lim} increases with increasing flow rate. At low velocities, with a laminar boundary layer, the relationship is typically $I_{lim} \propto v^{0.5}$. At higher velocities giving turbulent boundary layer, we normally have $I_{lim} \propto v^{0.8-0.9}$.

On a corroding metal, a layer of corrosion products is usually formed on the surface. This layer is often rather porous. Nevertheless, it hinders diffusion of oxygen, sometimes to a relatively high extent. The total diffusion resistance is the sum of the resistance within the diffusion boundary layer (of thickness δ) in the liquid and the resistance within the corrosion product layer.

While the diffusion-limiting current density on a clean surface can be expressed by Eq. (5b),

$$i_{lim} = DzF\frac{c_B}{\delta} = \frac{zFc_B}{\delta/D}$$

this equation has to be modified when a deposit is developed on the surface. A modified version is

$$i_{lim} = \frac{zFc_B}{\delta/D + \delta_s/P_s} \tag{6}$$

where δ_s is the thickness of the surface deposit and P_s is the permeability of oxygen in the surface deposit. The ratio δ_s/P_s expresses the extra diffusion resistance due to the surface layer [8,9].

Both addends in the denominator of Eq. (6) are sensitive to the flow velocity of the corrosion medium, as well as to other erosive or abrasive factors. Therefore they play very important roles in all flow-dependent corrosion, including erosion corrosion. The effect of flow velocity on the cathodic reaction and subsequently on the corrosion rate of an active metal is schematically shown in Fig. 5.

D. Behavior of Active–Passive Materials

As mentioned above, the situations in Figs. 4 and 5 represent corroding metals that are active in the whole potential range in question. However, many of the corrosion-resistant materials, typically stainless steels, are passive due to an oxide film on the surface. This oxide film prevents the anodic reaction. The potential range where the metal or alloy is passive depends on several factors, of which pH, chloride concentration, and temperature are of particular importance. Schematic anodic overvoltage curves for an active–passive material at different pH values and chloride concentrations are shown in Fig. 6. Pitting corrosion occurs when the potentials is above a certain critical potential, the pitting potential E_p, which is characteristic for the actual material–environment combination. Pits are initiated as a result of very local breakdown of the oxide. Aggressive species such as chloride and other halogenide ions play an important role in this process. They are adsorbed on the surface and assumed to penetrate the oxide, preferably at sites with oxide defects. Small anodes and pits are formed at these points. As a result of the reactions, the electrolyte

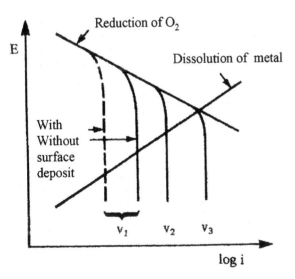

Figure 5 Schematic overvoltage curves for an active metal in an aerated aqueous solution.

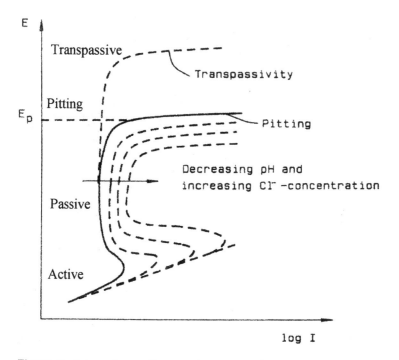

Figure 6 Schematic anodic overvoltage curves for an active–passive metal or alloy. Effects of pH and chloride concentration. E_p = pitting potential.

in the pits becomes acidic and more aggressive, which in turn accelerates the pitting process (autocatalytic process). The mechanism and influence by various factors have been described in several textbooks [3–6,10].

In Fig. 7, it is shown by anodic and cathodic overvoltage curves how a stable passive state (1) may be established in certain neutral solutions, while pitting may occur in a solution with higher Cl^- concentration and/or lower pH, or possibly at a higher temperature (state 2).

The experimentally determined pitting potential depends more or less on the test procedure. Therefore these data are not exact corrosion characteristics, but they indicate the potential levels at which pitting can be expected.

A typical passive oxide film such as that on stainless steel is highly resistant against erosion corrosion in pure liquid flow, but it is attacked by particle erosion. The repassivation rate of the oxide is an important quantity in this case.

E. Experimental Determination of Polarization Curves and Corrosion Rate

The methods dealt with in this section are described in more detail in most modern corrosion textbooks [3–6].

As we have seen, overvoltage curves can be used for the determination of corrosion rate. However, to draw the overvoltage curves, one must know E_o, i_o, b for each reaction, and possibly i_{lim}. E_o can be calculated by thermodynamics (or possibly measured on a reversible cell). The other quantities must be experimentally determined. That has already been performed for many reactions, but both b and particularly i_o and i_{lim} strongly depend on the conditions, and there is always a need for new data. The most important method for determining these is to record polarization curves with a *potentiostat*. As the name indicates,

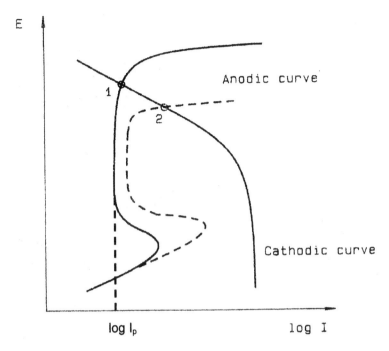

Figure 7 Anodic and cathodic overvoltage curves resulting in a stable passive state (1) with passive current I_p in a neutral solution, and pitting (2) at lower pH and higher Cl^- concentration.

the potentiostat is an instrument that keeps a set electrode potential (on the material we are investigating) constant, and it delivers the current that is instantaneously needed to maintain the potential constant. A potentiostatic setup is shown in Fig. 8. There are three electrodes:

1. The working electrode W (the specimen to be investigated).
2. The counter electrode C.
3. The reference electrode R*.

The reference electrode is usually electrolytically connected to the working electrode through a salt bridge with a luggin capillary close to the working electrode as shown in Fig. 8.

   As mentioned above, the potentiostat keeps the working electrode potential constant at a set value; that is, it keeps a constant voltage on the cell consisting of the working electrode and the reference electrode by delivering the necessary current between the counter electrode and the working electrode. For the working electrode, this is experienced as an *external current*.

   For the sake of simplification, we assume now that there is only one cathodic and one anodic reaction on the working electrode and that both these reactions are activation polarized.

   If we successively set different working electrode potentials by the potentiostat and measure the external current I_{ou} for each potential, and then plot the corresponding pairs of potential and current, we obtain the two dotted curves (shown in Fig. 9), i.e., one cathodic

---

* In the laboratories, a saturated calomel electrode (SCE) consisting of Hg, Hg_2Cl_2/KCl (saturated) is commonly used. At 20°C, this electrode has a potential 245 mV more positive than the standard hydrogen electrode potential.

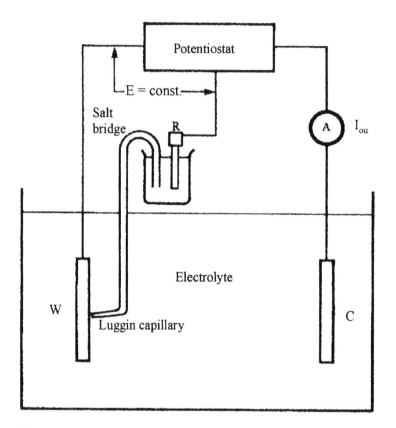

Figure 8 Setup for potentiostatic experiments.

and one anodic polarization curve. In a potential range far from the free corrosion potential E_{corr} (e.g., >50–100 mV from E_{corr} depending on the Tafel constants), there is a linear part on each of the curves. On the anodic side, the anodic reaction is completely dominating in this potential range (the cathodic current is negligible), which means that the external anodic current I_{ou} equals the real anodic current I_a on the electrode; in other words, the external polarization curve equals the overvoltage curve in the actual potential range. On the cathodic side, we have the same relationship between polarization and overvoltage curve. Therefore we can determine the complete overvoltage curves by extrapolating the linear parts of the external polarization curves. The corrosion current is determined by the intersection of anodic and cathodic overvoltage curve, as explained in Sec. II.C.

Another method of electrochemical determination of corrosion rate is the so-called LPR (linear polarization resistance) method. In this method, the working electrode is polarized within a small potential range, e.g., ± 10 mV from the corrosion potential E_{corr}. In this limited potential range, an increment of the electrode potential gives a proportional increment of the external current I_{ou}. The linear polarization resistance dE/dI_{ou} is expressed by the *Stern–Geary equation*

$$\frac{dE}{dI_{ou}} = \frac{b_a b_c}{2.3(b_a + b_c)I_{corr}} \tag{7}$$

where b_a and b_c are the anodic and cathodic Tafel constants, respectively. After measuring the polarization resistance dE/dI_{ou}, I_{corr} can be calculated by Eq. (7) if we know the Tafel constants. This method has the benefit that the natural corrosion conditions are not considerably disturbed, because the potential is only slightly moved from the free corrosion potential. Because knowledge of the Tafel constants is a prerequisite, the method can favorably be combined with the first method described above, the extrapolation of the linear parts of the polarization curves. It is often reasonable to determine linear polarization resistance at some times during the exposure period and record full polarization curves just before the exposure is completed.

One problem is that many reactions are not only activation polarized; therefore there may not be any clear linear part of the polarization curve. At least one of the polarization curves (anodic or cathodic) must have a linear part for the extrapolation method to be used. The curves maybe difficult to interpret and much experience is often needed to obtain reliable results.

The described methods for determination of corrosion rates can also be used in cases of combined wear and corrosion in aqueous media (Sec. IV.B.).

F. Erosion Corrosion

1. Characteristic Features and Occurrence

When there is a relative movement between the corrosive medium and a metallic material, the material surface is exposed to mechanical effects leading to an increased extent of corrosion that we call erosion corrosion. The usual mechanism is that corrosion products are worn off, or dissolved, or possibly prevented from being formed, so that fresh and

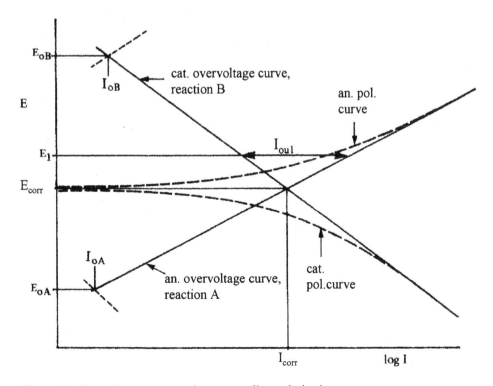

Figure 9 Overvoltage curves and corresponding polarization curves.

therefore more active metallic surface is exposed. It should be emphasized that—in spite of the erosion effect—all the material loss can be considered as electrochemical deterioration, because the corrosion products eroded away mechanically are replaced by new products from the electrochemical reactions. The results of erosion corrosion are grooves or pits with a pattern determined by the flow direction and local flow conditions (Figs. 10 and 11).

Reasonably, the corrosion form is typical at high relative velocities between the material surface and the fluid, and it is particularly intensive in cases of two-phase or multiphase flow, i.e., liquid–gas and liquid–solid particle flow. Components often liable to erosion corrosion are propellers, pumps, turbine parts, valves, heat exchanger tubes, nozzles, bends, and equipment exposed to liquid sputter and jets. Most sensitive materials are those normally protected by corrosion products with inferior strength and adhesion to the substrate, e.g., lead, copper and copper alloys, steel, and, under some conditions, aluminum/aluminum alloys. Stainless steel, titanium, and nickel alloys are much more resistant due to passive films with high strength and adhesion.

2. Various Mechanisms

Two types of erosion corrosion is often distinguished, but they may overlap each other:

1. Impingement corrosion, often found in systems with two-phase or multiphase flow, particularly where the flow is forced to change direction. Numerous impacts from, e.g., liquid drops in a gas stream, or particles or gas bubbles in a liquid flow lead to pits with a direction pattern as shown in Fig. 10(a). In cases with solid particles, the situation can be illustrated as in Fig. 12, where corrosion products are removed and the surface locally activated.
2. Turbulence corrosion, which occurs in areas with particularly strong turbulence, e.g., in the inlet end of heat exchanger tubes [Fig. 10(b)].

As mentioned, the removal of corrosion products may occur by impacts, i.e., where we have a force component normal to the material surface. In some cases, deposits may also be removed by high shear stresses (force components parallel to the surface). The shear stresses may considerably vary as a consequence of flow fluctuations or repeated impacts. Therefore it is possible that deposits of corrosion products are destroyed either by extreme values of shear stress or by fatigue.

The effects of high flow velocity and erosion on the reaction kinetics are in principle described in Fig. 13. Both the effect on the cathodic curve (increased transport of oxygen to

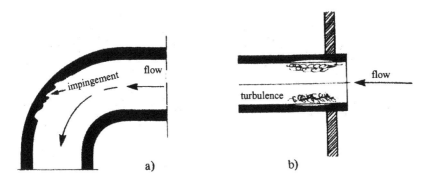

Figure 10   (a) Impingement and (b) turbulence corrosion.

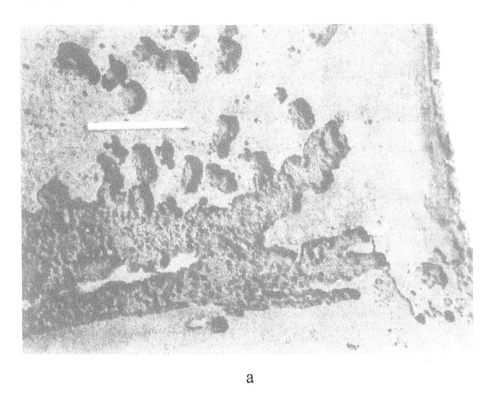

a

b

Figure 11 Flow-affected corrosion in a pipeline for oil and gas. (a) A picture from the bottom of the pipe. (b) Typical pit shapes. (Photo: J.M. Drugli, SINTEF Corrosion Center. From Ref. 2.)

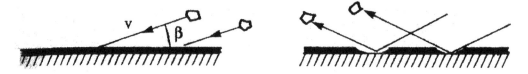

Figure 12 Impacts by solid particles in a flowing liquid causing removal of corrosion products from the metal surface. Erosion corrosion.

the surface) and on the anodic curve (increased activation of the metal in the corrosion potential range) are shown. The indicated passivation (case d) is possible only under certain conditions, like for instance on steel in fresh water without particles at very high velocity, e.g., as frequently occurring on parts of water turbines.

3. Influencing Factors and Protective Measures

The effect of flow velocity on the corrosion rate of various metallic materials in seawater in the most usual velocity range is shown in Table 3. The materials can roughly be divided into three groups based on their behavior:

1. Carbon steel and iron are active in the whole actual range of flow rates. The corrosion rate steadily increases, mainly because of more efficient supply of oxygen. There is some corrosion product layer intact on the steel surface at all the three velocities.

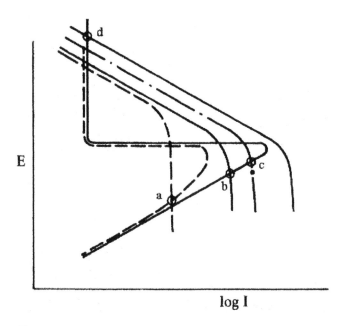

Figure 13 Overvoltage curves at (a) low flow velocity and formation of corrosion product deposits on the surface (- - -), (b) high flow velocity and removal of corrosion products from the surface (—) with (c) increased cathodic reaction due to galvanic contact with a more noble metal, and (d) flow velocity sufficiently high to cause passivation.

Table 3 Corrosion Rates in Seawater at Different Flow Velocities

Typical corrosion rates (mdd)[a]

	0.3 m/sec[b]	1.2 m/sec[c]	8.2 m/sec[d]
Carbon steel	34	72	254
Cast iron	45	—	270
Silicon bronze	1	2	343
Admiralty brass	2	20	170
Hydraulic bronze	4	1	339
G bronze	7	2	280
Al bronze (10% Al)	5	—	236
Aluminum brass	2	—	105
9010 CuNi (0.8% Fe)	5	—	99
7030 CuNi (0.05% Fe)	2	—	199
7030 CuNi (0.5% Fe)	<1	<1	39
Monel	<1	<1	4
Stainless steel AISI 316	1	0	<1
Hastelloy C	<1	—	3
Titanium	0	—	0

[a] For the listed materials except titanium, 1 mdd equals 4–5×10^{-3} mm/yr; that is, values about
  200–300 mdd in the right column corresponds to ca. 1–1.5 mm/yr.
[b] Immersed in tidal current.
[c] Immersed in seawater flume.
[d] Attached to immersed rotating disk.
Source: Ref. 11, from International Nickel Co.

2. All the copper alloys are reasonably well protected by surface oxides and hydroxides at the two lower velocities. When the velocity exceeds a critical value, the protective surface layers are dissolved. The critical velocity for erosion corrosion of most copper alloys is in the range 1–5 m/sec [12]. Special grades of copper–nickel and nickel–aluminum bronzes can tolerate higher flow rates.
3. Stainless steel, nickel alloys (Monel and Hastelloy C), and titanium are satisfactorily passive at all the three velocities. (However, it should be emphasized that most stainless steels are liable to crevice corrosion in seawater.)

A ranking of the materials listed in Table 3 would be similar for seawater and fresh water. Other environments may give a quite different picture. The pH value of the fluid is one of the most important parameters; another one is its oxidation or reduction character. Sometimes, the temperature is of considerable importance. Galvanic coupling to another material can reduce or increase the erosion corrosion to high extents, depending on the relative polarity of the materials.

An example of corrosion in a gas/condensate well is shown in Fig. 14 [13]. The aggressive oxidizer is in this case CO_2. Depending on the temperature, the steel is protected by a layer of primarily $FeCO_3$ at low and moderate velocities [14].

Erosion corrosion can in principle be prevented by various means, but they are not always easy to effectively apply in practice. It follows from the description above that appropriate selection of material is important. Protective coatings can be applied in many cases; inhibitors can be applied in others (e.g., as in Fig. 14). Separation of phases is important in cases of two-phase or multiphase flow (filters, separators) (Sec. IV.B). Last but

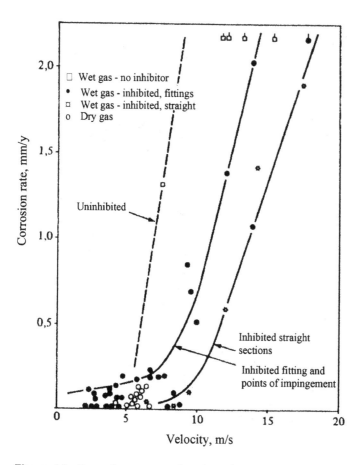

Figure 14 Corrosion rate as a function of gas velocity in gas/condensate well. (From Ref. 13.)

not the least, many problems of erosion corrosion can be solved by appropriate design. Examples are the following: Pipe dimensions large enough to avoid velocities above the critical values everywhere. Suitable cross section and flow direction changes, and smooth joints and surface, which minimize turbulence and its effect [Fig. 15(a,b)]. Critical areas localized where corrosion makes less harm [Fig. 15(c)]. Increased material thickness. Corrosion-resistant or easily replaceable linings [Fig. 15(d)].

III. CORROSION IN DRY GASES (OXIDATION)

A chemical reaction between oxygen in dry atmosphere and an engineering metallic material usually requires a high temperature to occur to a practically significant extent. However, under special conditions (e.g., wear conditions), oxidation of practical significance may also happen at ambient temperature. Reaction with oxygen causes an oxide film on the surface. Other oxidants that can react with metal in a similar way are, e.g., sulfur and halogens (chlorine, fluorine, bromine, iodine).

The oxidation product may form a dense and relatively strong film, which is usually the case for oxide films on the most common engineering metals and alloys (iron and steel,

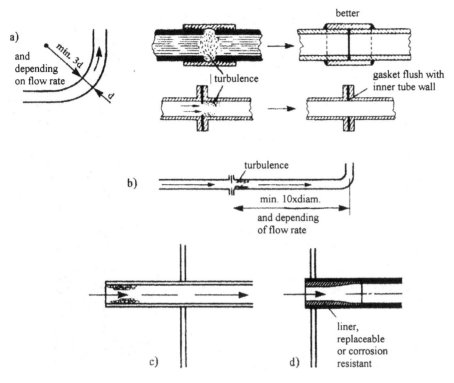

Figure 15 Design for prevention of erosion corrosion. (From Ref. 15.)

aluminum, chromium, copper, nickel, and their alloys). A divalent metal M may react with oxygen O_2 to the final product MO, but the reaction goes in steps.

At the interface between metal and oxide film:

$$M \rightarrow M^{2+}(\text{film}) + 2e^-(\text{metal})$$

and at the gas/oxide film interface

$$1/2 O_2(g) + 2e^-(\text{film}) \rightarrow O^{2-}(\text{film})$$

which in principle make a process similar to the electrochemical process in wet corrosion.

For most metals, the process is spontaneous; that is, it proceeds by itself in the atmosphere. This is expressed by a negative change of Gibbs free energy, which is the driving force of the reaction. Examples of standard values of free energy change during the oxidation process (standard free energies of formation of oxides) are shown in Table 4. The higher the value of $-G$, the stronger is the tendency of the metal to react with oxygen. As an example, the oxidation tendencies of aluminum and chromium are very strong, much stronger than that of iron. However, experience shows that Al and Cr can withstand high-temperature oxidation much better than Fe. The reason for this is that the kinetics of the oxidation reaction is not determined by $-\Delta G$, but by the barrier properties of the oxide film.

The first phase of oxidation of a fresh metal surface is a very rapid formation of a monomolecular layer of oxidation product. The process so far can occur at high as well as ambient temperatures. However, the further growth of the film depends to a high extent on the temperature. Some typical thicknesses after oxidation in air at various temperatures and

Table 4 Standard Free Energy ($\Delta G°$) of Formation of Some Oxides per Mole of Oxide at 300K

	$-kJ\ mol^{-1}$ of oxide
Ag_2O	10.9
Cu_2O	145
PbO	188
NiO	215
FeO	255 (at 227°C)
ZnO	319
MgO	570
SiO_2 (quartz)	824
Cr_2O_3	986
Al_2O_3	1578

Source: Ref. 16, after Smithells, *Met. Ref. Book*, Butterworths: London, 1955; 590.

times are shown in Table 5. The strong temperature dependence can be explained from the growth mechanism.

To make an oxide film, metal cations and/or oxidant anions have to be transported through the film. For thinner films, the transport mechanism is considered to be a combination of diffusion due to the concentration differences across the film and migration due to the electric field between the outer layer of anions and the inner layer of cations. For thicker films, diffusion is usually the major rate-determining mechanism of oxide film growth. The diffusion rate (mass transported through unit area per unit time) can be expressed by Fick's first law:

$$\mathrm{d}m/\mathrm{d}t = -D\frac{\mathrm{d}c}{\mathrm{d}x} \tag{8}$$

From this, the oxide film thickness Δx after a certain time t can be derived:

$$(\Delta x)^2 = kt + C \tag{9a}$$

Table 5 Some Typical Oxide Film Thicknesses After Oxidation in Dry Air

Metal	Temperature (°C)	Time	Thickness
Iron	20	100 h	1–2 nm
		8000 h	5 nm
	400	3 min	70 nm
	900	24 h	0.6 mm
Aluminum	20	1500 h	3 nm
	600	60 h	200 nm
Copper	20	100 h	2–3 nm
	500	30 min	500 nm
	800	6 h	20 μm
Nickel	20	100 h	ca. 2 nm
	800	1 h	450 nm

Source: Ref. 17.

and the weight increase per unit surface area $\Delta m/A_o$ can be expressed by [18]:

$$(\Delta m/A_o)^2 = k_g t + C^1 \tag{9b}$$

where k, k_g, C, and C^1 are constants. k and k_g includes both the diffusivity D and the concentration difference Δc of the diffusing element across the film.

Because, in principle, $\Delta x = 0$ for $t = 0$, one may set C and $C^1 = 0$. However, positive values of these constants may be used to fit experimental data that are affected by additional mechanisms that contribute to the film growth (e.g., migration due to electric field) [19].

The diffusivity D is expressed by

$$D = D_0 e^{-Q/RT} \tag{10}$$

where D_0 is a constant, Q is activation energy, R is the universal gas constant, and T is temperature in Kelvin. Equation (10) explains the strong dependence of oxide growth on temperature.

The rate constants strongly vary from one metal to another, as shown for k_g in Fig. 16. The figure shows that alloys forming chromium oxide, silicon oxide, and aluminum oxide films are much more suited for high-temperature service than those forming oxides of iron, cobalt, or nickel.

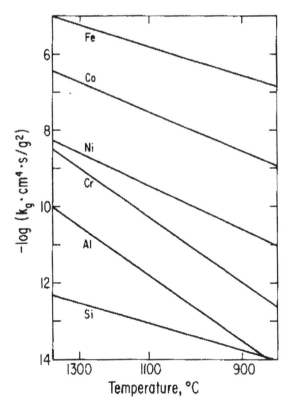

Figure 16 Representative values of the parabolic rate constants for the growth of several oxides. Data collected, oxidation conditions and forms of oxide specified by Yurek. (From Ref. 18.)

Equations (9a) and (9b) express the *parabolic law* of oxide film growth. Experiments have shown that this is valid at high-temperature corrosion. The explanation of film growth as given above is based on the assumption that diffusion occurs via point defects (vacancies) in the oxide lattice. The concentration of point defects may be large because of non-stoichiometric composition of the film, e.g., $Fe_{2.21}O_3$ instead of Fe_2O_3 on steel [19]. However, other mechanisms are also possible, e.g., grain boundary diffusion, and it should be noticed that other types of oxide growth than the parabolic one are possible. Figure 17 shows different "laws" of oxidation. A *linear film growth* can be the result when a parabolically grown film is periodically fracturing. The *logarithmic law* corresponds better to low-temperature oxidation.

The curve denoted "no inhibition of oxidation by oxide layer" may, e.g., indicate oxidation rate under corrosive wear conditions.

The thickness of a dense surface film is reflected by its appearance. Films of thickness 10 nm are transparent, while in the range 100–1000 nm, they attain various colors because of the interference (tempering colors). In the latter thickness range, the films are often important for friction and wear.

A combination of two or more oxidants often makes the surface film less strong and durable. This is the case when sulfur dioxide and oxygen are present. Sulfide film composition is often more nonstoichometric than oxides; therefore the diffusivity in sulfides may be higher and the film growth more serious, which is the case for iron and steel. Chlorine and Cl-containing compounds combined with oxygen is also destructive for iron and steel surfaces. Exposure of steel to hydrogen can cause decarburization and lead to inferior surface properties.

There are several interactions between surface film formation and mechanical stress and strain in the surface material. On one hand, a stressed and strained surface is more liable to oxidation (as it is to wet corrosion): Oxidation is accelerated because of the dynamic deformation and the temperature dependence of the oxidation rate may be even stronger

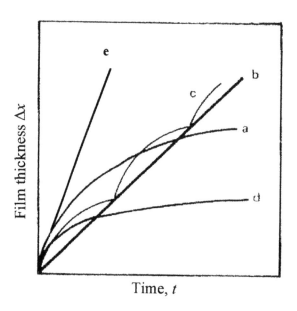

Figure 17   Laws of oxide film growth. (a) Parabolic, (b) linear, (c) quasi linear, (d) logarithmic, and (e) with no inhibition of oxidation by oxide layer.

than for a nondeformed surface. On the other hand, the film growth often generates considerable stresses. One reason is that there is frequently a misfit between the metal lattice and the lattice of the film. If this is not compensated for in some way during film growth, the interfacial stresses between metal and surface film increases by increasing film thickness, and at a critical thickness fracture occurs and the surface film may be partly detached. Cracks in the film can also be generated because outward diffusion of metal ions leads to tensile stresses in the film. Traditionally, mechanical stability of oxide films has been characterized by means of the *Pilling–Bedworth ratio*:

$$\phi = \frac{\text{the molecular volume of oxide}}{\text{the atomic volume of metal}}$$

Because a difference in molecular volume between oxide and metal causes stresses, Pilling and Bedworth stated that a ϕ value close to unity indicates a mechanically stable film, and a value far from unity indicates an unstable film. Experimentally, it has been found that most oxides roughly follow this rule, but there exist also some exceptions. Particularly high Pilling–Bedworth ratios make the oxide liable to cracking and detachment. This may for instance be the result if transport of oxygen ions through the oxide film is the predominating diffusion process, which causes relatively large compressive stresses at the metal/oxide interphase. If $\phi < 1$, the oxide is found not to give effective protection, and the growth rate tends to be linear. Oxides of most engineering metals have ϕ values in the range 1–2.

Metallic materials that are normally protected by an oxide film or another product of corrosion in dry gases are very sensitive to wear processes. However, both the mechanical strength of the film, its frictional properties and the intensity of wear, are important factors for the performance of materials and components. This subject is dealt with in Sections IV and V.

For further reading of corrosion in dry gases, see, e.g., Ref. 20.

IV. EFFECTS OF CORROSION ON EROSION AND ABRASION

A. Introduction

Sec. II.F comprised wet corrosion accelerated by removal of corrosion products from the metal surface (erosion corrosion), in which the removed corrosion products are replaced by new ones by electrochemical processes. Thus the total loss rate of metallic material can be measured by electrochemical methods. However, if the erosive action of the flowing medium is sufficiently increased, not only the corrosion products, but also small particles of the material itself, will be removed. This contribution to the material loss is a pure mechanical process, i.e., pure erosion. In many cases, we have a combination of pure erosion and erosion corrosion. It may occur by liquid drops in gas flow at very high velocities (e.g., $>$ ~100 m/sec for steel). It may also occur in pure liquid flow on material of low strength. However, the more typical and frequent cases of combined erosion and corrosion take place when solid particles are transported by gas or liquid flow.

Solid particle erosion at impact angles approaching zero is closely related to abrasion by solid particles, such as occurring in slurries of minerals. The latter deterioration form is also called *corrosive–abrasive wear*. Correspondingly, the combined erosion and corrosion may reasonably be denoted corrosive–erosive wear. The corrosion mechanism and the effect of corrosion on the deterioration rate is also quite similar for these two forms. Another well-known and special form of combined mechanical effects and electrochemical corrosion is cavitation erosion and corrosion.

B. Corrosive–Erosive Wear in Liquid Flow with Solid Particles

In strongly erosive environments (high particle concentration and/or high flow velocity), erosion is usually the dominating deterioration mechanism, while the corrosion contribution makes a smaller proportion, although the absolute corrosion rate may be high. The relative importance of corrosion increases with decreasing erosivity of the environment, and of course with decreasing corrosion resistance and increasing erosion resistance of the material. In principle, combined erosion and corrosion covers the whole spectrum between pure erosion and pure erosion corrosion.

The pure erosion rate can be expressed by the following formula

$$W \text{ (mm/yr)} = K_{mat} \cdot K_{env} \cdot c \cdot v^n \cdot f(\beta) \tag{11}$$

where K_{mat} is a material factor depending in a complex manner on (among other properties) hardness and ductility of the base material, K_{env} is an environmental factor that includes effects of size, shape (sharpness), density, and hardness of the particles, c is concentration of particles, n is the so-called velocity exponent, v is particle velocity, and β is the impact angle shown in Fig. 12.

For a certain geometrical element, it is convenient to replace $f(\beta)$ by a geometrical function, which, e.g., for pipe bends include pipe diameter and bend radius. Such functions as well as empirical values of the constants in Eq. (11) have been determined for fluids containing silica sand [21].

For pure particle erosion, a value of n in the order of 3 is frequently found. Sometimes, it is a little less and sometimes higher. $n = 3$ can be explained by a consideration of the kinetic energy of the particles hitting the material surface. The effect of impact angle is very strong, but quite different for ductile and brittle base materials, as shown, e.g., in Ref. 22.

Equation (11) is also of interest when erosion is combined with corrosion. It can be used in modified form for limited ranges of the involved parameters. The limitation is particularly caused by the fact that the velocity exponent of erosion corrosion is much less than that of pure erosion ($n_{corr} < 1.5$, usually $n_{corr} \leq 1.0$) [23]. In addition, erosion corrosion increases less than proportionally with particle concentration. Together, these relationships mean that for corrosive–erosive wear, $W \propto c^x \cdot v^y$, where $x < 1$ and $y < 3$.

A significant complication is that both x and y depend on the relative importance of erosion and corrosion, which in turn directly depends on both v and c. Finally, the factor K_{env} depends of course on the corrosivity of the environment (composition, temperature, pressure). K_{mat} is changed because corrosion results in different surface materials properties. The angle function $f(\beta)$ may also be affected by corrosion.

Because of these relationships, one needs to be very careful when using Eq. (11) on combined erosion and corrosion. To make such calculations reasonably accurate, they must be combined with and based on several experiments under actual conditions.

More thorough analyses and mechanism studies show that there is often a considerable synergy effect of erosion and corrosion. Generally, the total material loss rate W_T for such material deterioration can be expressed by

$$W_T = W_E + W_C + W_{EC} + W_{CE} \tag{12}$$

where W_E is pure erosion rate, i.e., mass loss rate when corrosion is eliminated, W_C is corrosion rate in the absence of sand erosion, and W_{EC} and W_{CE} are both synergy effects of erosion and corrosion: W_{EC} is the increase in erosion rate due to corrosion and W_{CE} is the increased corrosion rate due to erosion.

It is possible to determine the four contributions to the total material loss rate by the following experimental principles. The total mass loss rate W_T is determined by weighing the specimens before and after experiments under combined erosive and corrosive conditions. The sum of W_C and W_{CE} (the corrosion components) can be measured by electrochemical methods during the same experiments. (The methods described in Sec. II.E can also be used under erosive conditions.) W_E is determined by weighing the specimen before and after exposure in separate tests where corrosion is eliminated by cathodic protection (or possibly by other means), but otherwise under the same conditions as in the former experiments. W_C can be electrochemically measured in tests like the original ones but with all solid particles excluded. And finally, the synergy components W_{CE} and W_{EC} can be derived by Eq. (12) and the mentioned measurements.

In many cases reported in the literature, surface materials for combined erosive and corrosive conditions have been evaluated on the basis of separate erosion and corrosion studies, with the consequence that the synergistic effects are left out from the evaluation. Because one or both of these effects are frequently large, the conclusions may be quite wrong. For materials that usually are passive due to a dense oxide film, such as stainless steels, W_C is by definition very low. But because sand erosion more or less destroys the passive film, the corrosion rate strongly increases and may reach very high levels, i.e., the W_{CE} contribution in Eq. (12) may be particularly high for these materials. The other synergy effect, W_{EC}, is most pronounced for ceramic–metallic materials in which the metal phase has inferior corrosion resistance. This may be the case even if the measured corrosion rate under erosive conditions makes a small percent of the total mass loss rate. These relationships and the corresponding mechanisms are illustrated by a few examples below [24,25].

The experiments were carried out with specimens fixed to a rotating disk and arranged so that corrosion, erosion, and synergistic deterioration rates could be separately determined as mentioned above. In Fig. 18 [24], total material loss rate W_T of various materials are shown for two different peripheral velocities of the rotating disk. The corrosion contribution ($W_C + W_{CE}$) in percent of W_T is given in Fig. 19 for three of the materials. At the highest velocity the corrosion contribution is relatively low, and the erosion rate W_E is dominating for all materials (their synergy effects W_{EC} were also shown to be small or insignificant). Therefore the total material loss rate at the highest velocity reflects mainly the erosion resistance of the materials. However, at low velocity, the corrosion resistance of stainless steel makes these materials much better than the non-corrosion-resistant steels (Fig. 18). The corrosion percentage is only 7% on the stainless steel, while for the other steels it is >40% at the lowest velocity tested (Fig. 19).

The results are in agreement with the general rule that increasing velocity (increasing erosivity) increases the relative contribution of erosion and reduces the corrosion percentage. However, the absolute corrosion rate of unalloyed or low-alloy steel is very high at both velocities shown in Fig. 18 (3–8 mm/yr), which shows that the cathodic reaction (reduction of oxygen) must be strongly accelerated and corrosion products effectively carried away because of the relative movement between specimens and the sand/liquid. The major part of the corrosion rates on the unalloyed/low alloy steels is the synergy effect W_{CE}, but W_C is also significant (see the figures for steel in Table 3).

The absolute corrosion rates of stainless steel were 0.2 and 1.2 mm/yr at 8.9 and 22.8 m/sec, respectively. Because this steel is passive at such velocities in pure water flow, the corrosion rate measured is the synergy contribution W_{CE}. For stainless steel, this is an unusually high corrosion rate (except for cases of marked localized corrosion). But at the same time, the measured corrosion rate of stainless steel was much lower than that of the other steels. Altogether, the measurements show that stainless steel is activated by

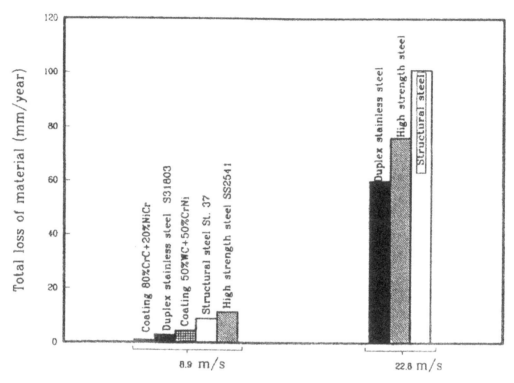

Figure 18 Total material loss in mm/yr for various materials at two peripheral velocities. (From Ref. 24.)

mechanical removal of the oxide film, but it is repassivated to some extent by formation of new oxide. There is a competition between activation and repassivation. Most other homogeneous engineering metals and alloys will in principle behave either like unalloyed steel or stainless steels depending on their normal corrosion resistance and passivation tendency. Deviating behavior can be found in metallic materials with inclusions and grain boundaries sensitive to local corrosion.

The interaction between erosion and corrosion for a ceramic–metallic material may be quite different from that described above. This is particularly so when the corrosion resistance of the metallic phase is relatively low, which is the case, e.g., for cemented carbides with a metallic binder phase of Co. An example of results with the same experimental technique as dealt with above illustrates the case.

In Table 6 [25], coating 1, consisting of 83 wt.% of WC grains in a binder phase of Co, shows particularly high corrosion rate ($W_C + W_{CE}$) as well as synergy effect (W_{EC}) and total material loss, while the pure erosion contribution W_E is roughly at the same level as for the other coatings. It is noticed that the synergistic deterioration rate (W_{EC}) of coating 1 makes nearly half of the total material loss rate. Coating 2 corrodes much more slowly and has a smaller although significant synergy effect W_{EC}. Coating 3, which shows no significant synergy effect, has the most corrosion-resistant metal binder phase of the three coatings.

With support in several erosion and corrosion experiments and surface studies , it is assumed that the deterioration process in cases like coating 1 and to some extent coating 2 (Table 6) can be briefly described as follows: The metal around the carbide particles corrodes

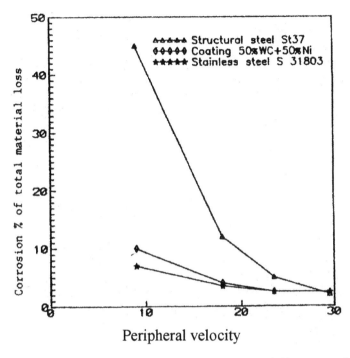

Peripheral velocity

Figure 19   Corrosion rate in percent of total material loss rate as function of peripheral velocity for three different materials. (From Ref. 24.)

Table 6   Total Material Loss Rate and the Relative Contribution of Erosion, Corrosion, and Synergy Effects on Specimens Exposed in Air-Saturated Synthetic Seawater at 20°C and with a Relative Velocity Between Specimens and Water of 14.3 m/sec

Material/ coating	Sand concentration	Total material loss	Erosion W_E mm/yr	%	Corrosion $W_C + W_{CE}$ mm/yr	%	Synergy effect W_{EC} mm/yr	%
Coating 1	0.25	5.292	0.836	16	1.905	36	2.551	48
	2.5	10.928	3.604	33	2.513	23	4.811	44
Coating 2	0.25	1.225	0.831	66	0.201	16	0.223	18
	2.5	4.157	2.983	72	0.166	4	1.008	24
Coating 3	0.25	0.855	0.888	104	0.051	6	−0.084	−10
	2.5	3.453	3.387	98	0.138	4	−0.072	−2
UNS 31803	0.25	0.658	0.625	95	0.059	9	−0.026	−4
(BalFe	2.5	8.611	8.455	98	0.172	2	−0.016	0
22 Cr 5Ni 3Mo)								

Source: Ref. 25.

away so that the WC particles are more easily removed by erosion, hence the large synergy effect W_{EC} (increased erosion due to corrosion). The transfer from unexposed state to a corroded and eroded state is schematically illustrated by Fig. 20.

Under extremely erosive conditions or if the corrosion rate is very low, erosion is highly dominating and the effect of the shown mechanism may be insignificant.

Erosion and erosion corrosion of coatings, steels, and other materials at elevated temperatures are described in Ref. 26.

1. Protective Measures

It follows from the description above that the selection of materials is very important. The cemented carbides are of great interest under conditions with sand erosion. As shown above, the composition of the metallic binder phase is particularly important when we have a corrosion component. Under the most extreme erosion and corrosion conditions, like, e.g., in oil/gas production choke valves, considerable improvement of monolithic cemented carbides has been obtained by selection of superfine-grain carbides with a relatively high corrosion resistance of the metal matrix [27]. Similar improvements have been reached with coatings within the same material group, which are actual for other types of valves, turbine parts, and various flow system components [28]. Several pure ceramic materials have very high erosion and corrosion resistance, but they are more brittle than the cemented carbides, which make them more sensitive to geometrical irregularities as well as impacts and concentrated load. In general, hardness, ductility, structure, and corrosion resistance are vital properties. This is the case both for monolithic materials and coatings. Further development of materials and coatings may give large improvements, particularly for more extreme conditions. A comprehensive description of coatings and surface treatment for corrosion and wear resistance is given in Ref. 29.

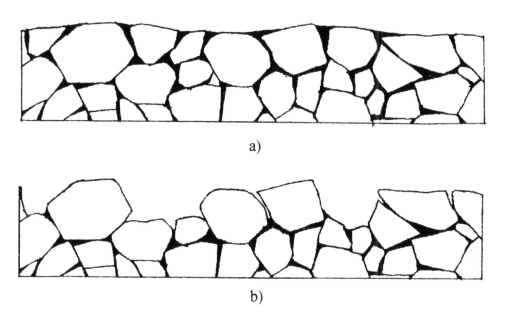

a)

b)

Figure 20 Schematic view of cemented carbide (a) before testing and (b) after exposure under corrosive and low-erosive conditions. (From Ref. 25.)

In addition, combined erosion and corrosion is minimized by the same methods as listed for erosion corrosion. Reduction of average and local velocity is an even more effective measure when the erosion component is large because the velocity exponent is so high. A specific protective measure against solid particle erosion and corrosion is of course reduction of particle content by filtration/separation or other methods when this is possible. Separation should be carried out as early as possible in the process, i.e., at an upstream position in the flow system (Fig. 21).

C. Corrosive–Abrasive Wear

Here, and in some other literature, the term corrosive–abrasive wear is primarily used for cases where abrasion is caused by particles and there is a significant corrosion component in addition, such as in slurries. (Combined abrasion and oxidation/corrosion between moving machine parts are dealt with in Sec. V.)

Particle abrasion with low pressure between the particles and the material surface is closely related to particle erosion at low impact angles. Therefore corrosive–abrasive wear has several features in common with combined erosion and corrosion. The corrosion mechanism and activation is based on abrasive removal of corrosion products, which often have much lower abrasion resistance than the base material, and this may cause large synergy effects. This is particularly the case at slow and moderate wear combined with rapid corrosion. On the other hand, rapid abrasion is not significantly accelerated by moderate or slow corrosion (the corrosion is slower than the wear of base material; thus no corrosion product film is established).

In agreement with the above-mentioned relationships, we have the following rough guidelines for materials selection. For cases where slow abrasion is combined with corrosion, high molecular polymers or elastomers may give lower wear rates than hard and corrodable materials. On the other hand, strongly abrasive conditions should be met with hard materials with a corrosion resistance matching the corrosivity of the environment. It should be noticed that abrasion resistance is more closely related to hardness and less dependent on ductility than is erosion resistance. These guidelines also apply for selection of coatings, for which good adhesion to the base material and sufficient thickness are other important properties.

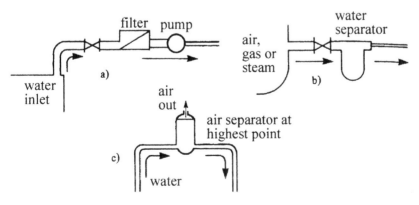

Figure 21 Separation of different phases for prevention of erosion and erosion corrosion. (From Ref. 15.)

D. Cavitation Erosion and Corrosion

Cavitation erosion is a special form of wear that is different from the above-described forms as regards both the mechanism and the appearance of the attack. The phenomenon occurs at large flow velocities and fluid dynamic conditions causing extensive pressure fluctuations at certain positions or pressure variation from one position to another in a flow system. Gas bubbles that are formed at low pressure collapse very rapidly when they suddenly come into a high-pressure zone or high-pressure moment of time. When this happens close to a metal surface, the bubble collapse causes a concentrated and intense impact against the metal, with the induction of high local stress and possibly plastic deformation. Parts of protecting films are removed. Repeated impacts may lead to microscopic fatigue and crack formation and subsequent removal of particles from the material itself. The material becomes more active and the corrosion rate increases. Also, the microfatigue processes are accelerated by corrosion.

    Cavitation erosion, probably in most cases affected by corrosion, is a frequent problem in water turbines, on propellers and pump rotors, and it is well known, e.g., on external wet surfaces of cylinders in diesel engines.

    The appearance of the attacks differs from ordinary erosion or erosion corrosion attacks. While the latter have a direction pattern reflecting the flow direction, cavitation

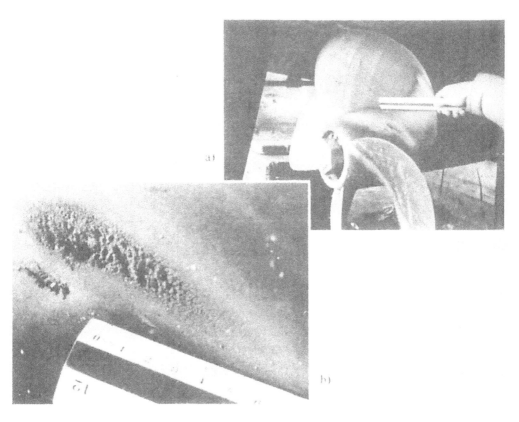

Figure 22 (a) Cavitation attack on propeller of a high speed boat. (b) Close-up of attacked area (the material is nickel–aluminum bronze). (Photo: Tor Erik Hammervold and Oddvar Sætre; from Ref. 2.)

attacks are deep pits perpendicular to the surface. The pits are often localized close to each other or are grown together, making a rough, spongy surface (Fig. 22).

Both high hardness and good corrosion resistance increases the cavitation erosion/corrosion resistance of materials and coatings. Relative resistances of various steels, stainless steels, irons, and copper and nickel alloys are reported in Ref. 4. In addition to appropriate base material selection, application of coatings (plastic, rubber, or hard inorganic) with good adhesion to the substrate, inhibitors, and design that prevents bubble formation are other useful protection methods [4].

V. OTHER WEAR PROCESSES AFFECTED BY CORROSION

A. Surface Reactions and Corrosive Wear Under Sliding or Rolling Contact

The term *corrosive wear* is in the present chapter used for cases where two machine parts or other structural components are in sliding or rolling contact with each other and exposed to a chemically reactive environment, which is responsible for a significant contribution to the wear rate. The most common basic process is that corrosion products are worn off so that fresh and active metal surface is exposed and rapid corrosion takes place.

However, a chemical surface reaction may have a positive or a negative effect on wear, depending on several factors. Sometimes, there is a positive effect because a dense, strong, and durable film is formed on the surface, and this film prevents adhesive wear by isolating the metal surfaces from each other. This mechanism is utilized when reactive antiwear and extreme pressure additives are used in lubricants. The additive reacts with the surface of the component and forms a layer of corrosion products. The extent of reaction depends on the reactivity and concentration of additive, and of the severity of wear. Increased effect of the additive reduces adhesive wear but at the same time increases corrosive wear, and for a certain wear intensity there is an optimum reactivity or concentration. The relationships are schematically shown in Fig. 23 [30].

Another example of a positive effect is the formation of a strong oxide on iron and steel surfaces exposed in concentrated sulfuric acid (acid concentration >90%). The oxide reduces adhesive wear as well as friction compared with wear in neutral water [19].

In other cases, the corrosion products between two components with sliding contact may be relatively soft and with a consistency that markedly reduces friction, but this will usually be at the expense of increased wear rate, because these corrosion products are easily removed.

In most other situations, comprising different environments, movements, and wear mechanisms, chemical reactions on the surface of components is only detrimental.

A frequent wear mechanism under sliding contact is that adhesive and abrasive wear imply high local pressure and friction temperature at the contact asperities, which induces rapid oxidation of metal particles and forms hard oxide particles that contribute to further abrasive wear. This may also happen for instance if the press fit between a collar and the shaft is insufficient to prevent relative motion between the parts. The increased abrasion rate due to oxide particles may cause more serious relative motion.

However, the mechanism of a typical corrosive wear in an environment containing oxygen is that corrosion products on the surface of the component are more easily worn off than is the metal itself. Similarly, in environments containing sulfur or chlorine, relatively weak and brittle sulfides or chlorides are easily removed from the surface mechanically. A model of the kinetics of corrosive wear under *cyclic* removal and formation of corrosion product film is illustrated in Fig. 24 [19].

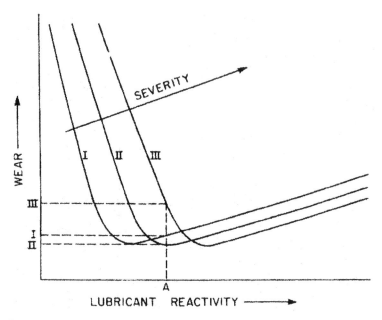

Figure 23 Schematic plot illustrating effect of severity of sliding on adhesive–corrosive wear balance. (From Ref. 30.)

Slip sometimes occurs between a driven and a braking roll. This leads to plastic deformation of the surface material, higher reactivity of the material, and penetration of oxygen down the grain boundaries. Increased surface corrosion as well as integranular corrosion accelerates the corrosive wear [31].

At rolling contacts exposed to atmosphere, e.g., between railway wheel and rail, corrosion products are periodically formed and detached. This may also happen in rolling bearings with insufficient lubrication.

Corrosive wear occurring in atmospheres at higher or ambient environment temperature is often called oxidative wear to distinguish between this form and corrosive wear in aqueous environment. The latter may cause the most extreme cases of corrosive wear, because it normally involves the highest corrosion rates. Also, in cases where we have a nominally corrosion-resistant material, such as stainless steel, the corrosive wear may be intensive, similar to that illustrated for erosion and corrosion. It has been shown that wear of stainless steel in an aqueous solution is reduced by cathodic protection and increased at high potential where the corrosion rate is high when the surface oxide is scraped off. The corrosion rate in such a case increases with increasing load and moving speed of the sliding contacts (as does abrasive wear rate) because the fresh area increases with these parameters [31].

In accordance with the description above, corrosive wear is generally prevented by adequate material selection, hard and corrosion-resistant coatings, reducing temperature, load or speed if possible, and changing environment. A brief presentation of actual coatings and surface treatment is given in Ref. 32; comprehensive descriptions within this topic are published in Ref. 29.

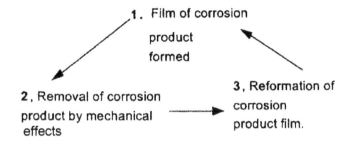

CYCLIC PROCESS OF CORROSIVE WEAR

1. Film of corrosion product formed

2, Removal of corrosion product by mechanical effects

3, Reformation of corrosion product film.

MODEL OF KINETICS OF CORROSIVE WEAR

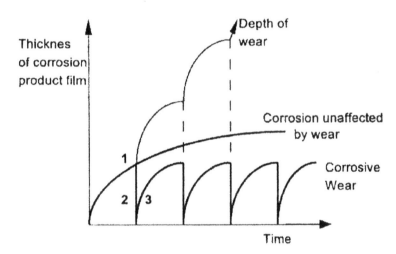

Figure 24 Corrosive wear by cyclic removal and formation of passivation films; mechanism and effect on corrosion kinetics. (From Ref. 19.)

B. Effect of Oxidation on Fretting Wear

As dealt with elsewhere, fretting wear occurs between components that are nominally static in relation to each other, but with microscopic and repeated relative motion. The amplitudes may be of the order of 1 μm, in other cases up to tens of micrometers. Typical sites are between shafts and bushes, on collars, inner ring of bearings, or other components with some press fit on the shaft. A critical position is at the end of a collar, as shown in Fig. 25. The attack is sometimes serious, in particular because it may lead to macroscopic motion between the parts, or a fatigue crack can develop in the shaft.

The mechanism of fretting wear may include elements of adhesive wear, microscopic fatigue crack development, and delamination that result in removal of small particles from the metal lattice. The particles make a debris that may partly adhere to the relatively moving surfaces and are trapped between these, and may partly escape from the fretting area. When exposed to air, the fretting process is accelerated compared with when it proceeds in vacuum

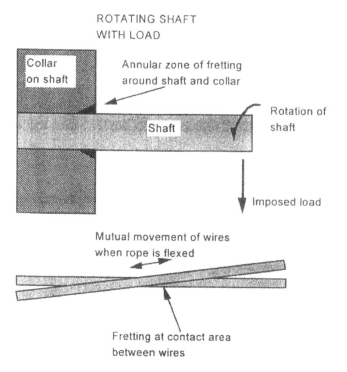

Figure 25   Location of fretting wear in some common engineering components. (From Ref. 19.)

or an inert environment. The acceleration effect depends on access of oxygen but not necessarily on humidity. So this is an example of significant oxidation at ambient temperature. The debris formed on steel interfaces is a mixture of α-Fe_2O_3 and iron particles.

The exact mechanism of combined fretting and oxidation is not fully understood. The following process has been proposed. The relative motion between the parts may promote oxidation of the surface, parts of the oxide is worn off, the fresh metal surface is highly active and oxidizes again, and this circular process is repeated. A probably more vital process is that metal particles are first released from the crystal structure by adhesive wear, microfatigue, and delamination, then the particles oxidize, making a debris consisting of brittle and friable oxide particles. It has been suggested that this debris is more easily moved out from the interfacial area between the components, and thus allow the deterioration process to continue at a higher rate. Possible effects in this process is that basic microfatigue contribution is accelerated by the environment and that sharp oxide particles, which are much harder than the metal particles, cause some abrasive wear in spite of the small size of particles. The friction coefficient is affected by the kind of debris. For further study, see Refs. 5, 6, 31, and 33. The visual result of fretting oxidation of steel interfaces is, in addition to a red-brown debris, pits or intrusions, which may lead to fatigue cracks. The attacked surfaces may also be discolored.

According to Ref. 5, the intensity of combined fretting and oxidation is increased by reduced (!) temperature and air humidity, and by increased pressure and amplitude of movement. However, there are different opinions about the effect of amplitude. Uhlig has expressed the material loss rate by the expression

$$W = (k_0 L^{1/2} - k_1 L)\frac{C}{f} + k_2 lLC \tag{13}$$

where L is the load, C is the number of cycles, f is the frequency, l is the sliding length, and k_0, k_1, k_2 are constants. The first two addends on the right side express the chemical contribution and the third one the mechanical effect.

Fretting oxidation is reduced or prevented by use of lubricants, which hinder the access of oxygen and at the same time suppresses adhesive wear. Other prevention methods are to prevent oxygen access by use of gaskets or sealing materials, and to change the mechanical, geometrical, or material parameters affecting the factors in Eq. (13). In general, see Refs. 4–6.

VI. SURFACE CHARACTERIZATION

A. Motivation

As pointed out already, corrosion behavior sensitively depends on factors such as the presence and state of passivating oxide layers or protective coatings, as well as on the composition and metallurgical state of the metal alloy itself. When sample and surface layers are a mixture of various phases with different chemical, electrochemical, and mechanical properties, the relative importance of corrosion, erosion or other wear processes, and synergetic effects may not at all be obvious. To understand or verify wear mechanisms in complex multiphase materials, a detailed knowledge is required about the structural and chemical state of the materials and their surfaces.

To be more specific, the important characteristics may include:

–Macroscopic defects, e.g., cracks, pits, or delamination of oxide layers or coatings.
–The microstructure, i.e., phase distribution, interfaces, structural defects (e.g., dislocations or stacking faults).
–The crystallographic structure of the different phases.
–The chemical composition of the phases
–The chemical (bonding, oxidation) state of the phases.
–The roughness of the surface.

Methods that probe larger volumes of the material and yield averaged information, be it structural (e.g., x-ray diffraction) or chemical (e.g., atomic spectrometry), may indeed be useful at the starting point of an analysis. Still, it is clear that the full picture needed to understand complex corrosion and wear mechanisms can only be obtained by means of methods that probe the material locally, typically on lengthscales below a few tens of micrometers. Therefore we concentrate here on microscopy methods and microanalysis with good lateral resolution (atomic scale to a few micrometers), surface analytical techniques with good depth resolution (monolayer to a few nanometers), and methods by which roughness can be measured on nanometer to millimeter lengthscales.

For some materials, characterizing the bulk microstructure is highly relevant also in the context of surface properties and wear behavior. For other materials, near-surface (a few micrometers) microstructures may differ from those of the bulk, as a result of surface modification or coating, or as a result of mechanical, chemical, or electrochemical processing/exposure. Although the microscopy or microanalytical methods for studying the near-surface region may be identical to those used for bulk characterization, the sample preparation may represent practical challenges. This particularly applies to cross-section geometries, in which the surface region is observed edge-on. The main advantage of cross-section microscopy samples is that structural and chemical variations with distance from the surface can be directly imaged. Plan-view samples, on the other hand, very well reveal how the microstructure varies across the surface. With plan-view samples, variations with depth can be studied by depth profiling, commonly used for surface analytical methods such as x-ray photoelectron spectroscopy (XPS), Auger spectroscopy, and secondary ion mass spectrometry (SIMS).

B. Characterization Methods

1. Microscopy and Microanalysis

Optical microscopy is the natural starting tool when setting out to characterize the microstructure of a material. Overview images will reveal inhomogeneities in microstructure over micrometer to centimeter lengthscales, and areas of special interest can be located. However, information on phase composition cannot be extracted, and the lateral spatial resolution is limited to about 1 μm. Hence microstructural information on submicrometer lengthscales and spatial variations of sample chemistry must be explored using electron microscopy and associated spectroscopic methods such as energy-dispersive x-ray spectroscopy (EDS), wavelength-dispersive x-ray spectroscopy (WDS), and electron energy loss spectroscopy (EELS). Comprehensive coverage of electron microscopy can be found, e.g., in Refs. 34 and 35.

In scanning electron microscopy (SEM), two imaging modes are commonly utilized, in which secondary electrons (SE) or backscattered electrons (BE) are detected. Secondary electron images carry information from the uppermost few tens of nanometers of the surface, and are primarily sensitive to surface topography. Fine surface detail can be imaged at a lateral resolution of a few tens of nanometers (conventional SEM) to a few nanometers (FEG-SEM with field-emission electron source). The field of view (and magnification) can be rapidly and easily changed from ~1 cm (20×) to ~1 μm (200,000×), making identification and closer study of interesting surface features straightforward. The specimen is observed in plan view, and specimen preparation is normally not needed in case of conducting samples.

Backscattered electron (BE) images are sensitive to sample composition as well as topography. When recorded from polished samples, image contrast is proportional to atomic number and the images display the phase distribution of the sample. The probing depth and lateral spatial resolution depend on the accelerating voltage for the electron beam and the type of material, but is typically a few hundred nanometers. Modern FEG-SEMs can operate at low accelerating voltages, which improves resolution in the BE mode. Because polishing is required for easy image interpretation in terms of atomic number contrast, BE imaging is mainly useful for learning about bulk phase distribution. By polishing cross-section samples, microstructural variations with distance from the surface can furthermore be studied. However, information from the ~1 μm closest to the surface may not be easily retrieved. Preparation-induced topography such as cracks or rounded edges between the surface and the embedding medium will obscure the contrast. This is difficult to avoid, even with careful preparation procedures.

Compositional information from EDS in the SEM can be recorded from submicrometer volumes, with a practical resolution limit of a few hundred nanometers using modern FEG-SEMs operating at low (3–5 keV) accelerating voltages. However, quantification of composition is more difficult at these low voltages. Electron probe microanalyzers (EPMA) have much in common with SEMs, but are optimized for microanalysis and are equipped with a set of wavelength dispersive spectrometers. Wavelength-dispersive x-ray spectroscopy normally utilizes higher-beam currents than EDS, and hence permits more accurate measurements of composition and better sensitivity limits for trace constituents (<0.1%). Spatial resolution with EPMA/WDS is typically several micrometers. Crystallographic information at submicrometer resolution can be obtained from the SEM specimen by the electron backscattering diffraction (EBSD) technique [36]. The main application is mapping of the crystallographic microtexture (spatial distribution of gain orientations) of known phases, and measurement/imaging of grain-boundary geometry.

Conventionally, samples for SEM and TEM must be compatible with the high-vacuum conditions of the analysis chambers. With modern low-vacuum SEMs, pressures up to ~1 Torr can be tolerated in the chamber, relaxing the demands on sample cleanliness. In environmental SEMs (ESEM), the practical pressure limit may be ~20 Torr, opening possibilities for in situ experiments in the microscope that involves liquids, corrosive media, and elevated temperatures [37].

To obtain detailed information from the uppermost ~1 μm of the near-surface region, with the surface viewed edge-on, transmission electron microscopy (TEM) is a very powerful method. Preparation of cross-sectional TEM specimens containing the surface has been considered challenging. The surface structure must be preserved at the same time as the TEM specimen is thinned to electron transparency (less than a few hundred nanometers). However, with the right choice of preparation technique, and by carefully establishing preparation procedures, it is clear that good-quality TEM specimens can be produced from a variety of materials surfaces. In particular, preparation routes involving ion milling or ultramicrotomy in the final stages have proven successful [38].

With the excellent spatial resolution of TEM, the finest microstructural detail can be studied. In the context of wear mechanisms, important fine-scale detail may be, e.g., the presence of wetting metal layers between the ceramic grains of a cermet, reaction products at the ceramic metal interface, or thin surface oxide layers. Features with sizes down to the lower nanometer range can be imaged, and even the atomic lattices of the phases if required. Moreover, phase identification by electron microdiffraction and measurement of chemical composition by microanalysis (EDS and EELS) can be performed on very small volumes. The typical lateral resolution for these analytical techniques is a few tens of nanometers for conventional TEMs and an order of magnitude better for FEG-TEMs. The fine-structure details of EELS spectra contain information on bonding and oxidation state of the analyzed elements, in addition to the purely compositional information. Element-specific imaging or mapping techniques based on the spectroscopic signal are available both for EDS and EELS, with a spatial resolution approaching the figures quoted for point analyses.

2. Surface Analytical Methods

The methods traditionally termed as surface analytical methods include x-ray photoelectron spectroscopy (XPS, also known as ESCA), Auger electron spectroscopy (AES), and secondary ion mass spectroscopy (SIMS) [39,40]. All of these methods are highly surface sensitive. The information depth of an XPS or AES recording is a few nanometers, whereas the SIMS information depth even may be at the monolayer level. The high surface sensitivity yields unique information that cannot be obtained by electron microscopy methods. On the other hand, the methods are very sensitive to surface contamination, which necessitates special specimen handling precautions, or even disqualifies samples that have been subject to a dirty environment during exposure. Surface contamination can, to some extent, be removed by light sputter cleaning, but bombardment with sputter ions tend to break atomic bonds and obscure bonding information in the spectra. Samples for the surface analytical methods must be compatible with the ultrahigh vacuum conditions in the analysis chamber.

An electron probe is used in AES, which yields the best lateral resolution of the surface analytical techniques. A few tens of nanometers can be achieved in favorable cases with modern instruments with a field-emission electron source. High-resolution compositional maps can be recorded by scanning the probe across the sample. Chemical composition can be quantitatively determined by means of sensitivity factors and standards, reaching accuracies of 5% with good standards. Sensitivity limits are typically around 1%. The energy shifts and shapes of the characteristic peaks in the AES spectra can in some cases provide information on the chemical state of the elements in the sample. Variations in sample composition with depth can be investigated by combining AES with ion beam sputtering and recording depth profiles. Samples for AES must be conducting to avoid charging.

X-ray photoelectron spectroscopy uses an x-ray source, and the lateral spatial resolution is more limited, but can reach 5 μm on modern instruments. Again, compositional mapping and quantification is available, with accuracy and sensitivity figures comparable to AES. The information on bonding and oxidation state found in the shape and position of characteristic elemental peaks is more complete than with AES. With XPS, insulating surfaces can also be studied. Depth information can be obtained by sputtering or by angle-resolved analysis. Like for AES, data acquisition may be time-consuming, and recording times may reach tens of hours for deep profiles or for compositional maps.

In secondary ion mass spectrometry (SIMS), secondary ions are ejected from the surface after bombardment by a focused primary ion beam, and subsequently analyzed by a mass spectrometer. Modern instruments with time-of-flight spectrometers (TOF-SIMS) combine extreme surface sensitivity (monolayer) and unique sensitivity for trace constituents (few parts per million) with capabilities for rapid depth profiling and very good lateral resolution [41]. Although not quite reaching the lateral resolution figures of AES, a few hundred nanometers can be achieved using a liquid-metal ion gun. Information on bonding can be obtained by detailed analysis of mass spectra. Quantification can be very challenging because of matrix effects; that is, secondary yields for a species may vary with several orders of magnitude, depending on its bonding state in the material.

Another very efficient method for depth profiling with high sensitivity that should be mentioned is glow-discharge optical emission spectrometry (GD-OES) [42]. Profiles tens of micrometers deep can be recorded in a few minutes, sensitivity can again be in the lower parts per million range, and quantification is generally much easier than with SIMS. A drawback is that the probe is several millimeters large, and hence profiles specific to microstructural features cannot be obtained.

3. Measuring Surface Roughness

With SEM images recorded in the SE mode, surface roughness is easily visualized, but quantitative measurements of surface height variations are not straightforward. The method of choice for obtaining quantitative roughness data depends on the material and the scale of the roughness. Contact measurements with stylus profilometers are commonly used for recording line profiles of roughness on lateral scales in the submillimeter to 100-mm range. The method is time-consuming and one must be aware of the possible damage to the surface by the sharp stylus, which is in full and moving contact with the surface.

A range of optical methods are available that give full $z(x,y)$ topography information in short recording times. In particular, modern white-light interferometers (WLI) have proven powerful in yielding accurate fine-scale roughness data at recording times of typically 10 sec per image (50×50 μm^2 to 5×5 mm^2). Height differences as small as 3 nm and as large as 500 μm can be resolved. White-light interferometry and other optical methods require some optical reflectance (typically > 2–10%) from the surface. Black surfaces can still be measured after coating with a thin layer of gold. The lateral resolution of the optical methods are limited by the wavelength of light to 0.5–1 μm.

With atomic force microscopy (AFM) [39,43], quantitative 3-D images of surface topography with very high resolution can be obtained. By means of a piezoelectric scanner, the sample is scanned in the xy-plane relative to a very thin etched tip mounted to a flexible cantilever. Weak interatomic contact forces between the sample surface and the tip cause vertical deflection of the cantilever. Measurements of the deflection are used for feedback to the control system for the scanner and measurement of $z(x,y)$. The lateral scan range spans from well into the submicrometer (with lateral resolution at the 1-nm level) to a maximum of typically 150×150 μm. Thus it overlaps with optical methods at the larger scan ranges, but also covers the finer details that cannot be resolved by optical methods. Height variations larger than 5–10 μm cannot be measured by AFM, which is sometimes a severe limitation. On the other hand, subnanometer height resolution can be achieved. Measurement accuracy can reach 1% laterally and a few percentage vertically, limited by the quality of standards.

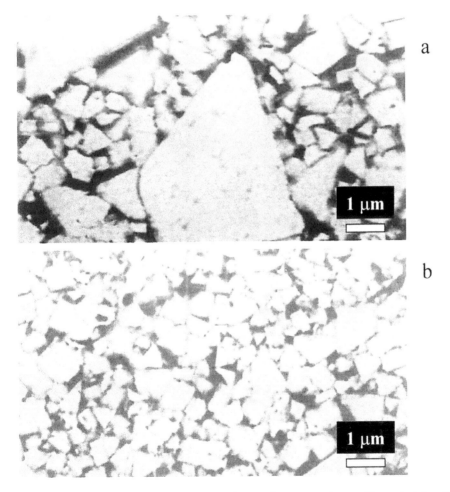

Figure 26 Scanning electron microscopy images from ceramic–metallic materials with WC grains and 6 wt.% cobalt as metallic binder. (a) Material with uneven grain size distribution, large grains, and inferior wear properties. (b) Material with more even grain size distribution.

C. Examples: Characterization of Ceramic–Metallic Materials

The examples are taken from the study of the combined erosion and corrosion properties of ceramic–metallic materials referred to in Sec. IV.B, illustrating how characteristics of the bulk microstructure in this case are found to be decisive for surface properties [44,45]. The most important materials parameters were expected to be the weight fraction of the metallic binder, the corrosion resistance of the metallic binder, and the size distribution of carbide grains.

Figure 26 shows SEM images from two different materials that exhibited very different wear properties. The metal binder content was in both cases 6 wt.%, and the binder composition was unalloyed Co. Scanning electron microscopy images were taken from polished sections of the bulk microstructure of these materials but contrast differences between carbide grains and binder were not all too clear. In particular, it was not easy to distinguish between larger carbide grains and clusters of grains separated by thin metal films. However, using an etchant that removes the metallic binder, distinct differences in carbide size distributions were strikingly revealed. The strongly increased wear rate of the sample shown in Fig. 26(a) was ascribed to the presence of the large WC grains along with larger volumes of metal binder between grains. The major wear mechanism is believed to be combined erosion and corrosion, where corrosion of the metal phase around carbide grains facilitates removal of the grains by erosion. Figure 27 shows an AFM image and roughness profiles of a surface after combined erosion and corrosion testing. Grooves around carbide grains are observed (profile 1), indicating local corrosion attack of the binder phase.

For finer and more complex microstructures, it may not be easy to extract enough information from SEM images, and TEM may yield valuable new information. Figure 28

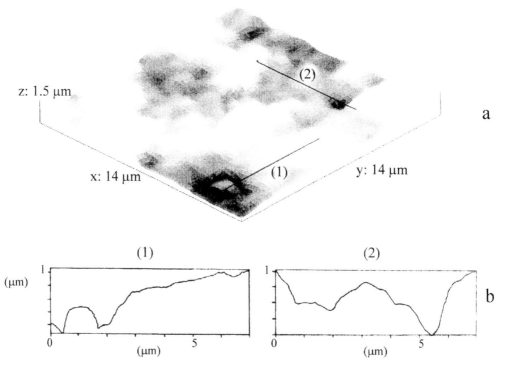

Figure 27 Atomic force microscopy 3-D image (a) and corresponding roughness profiles (b) from worn ceramic–metallic surface.

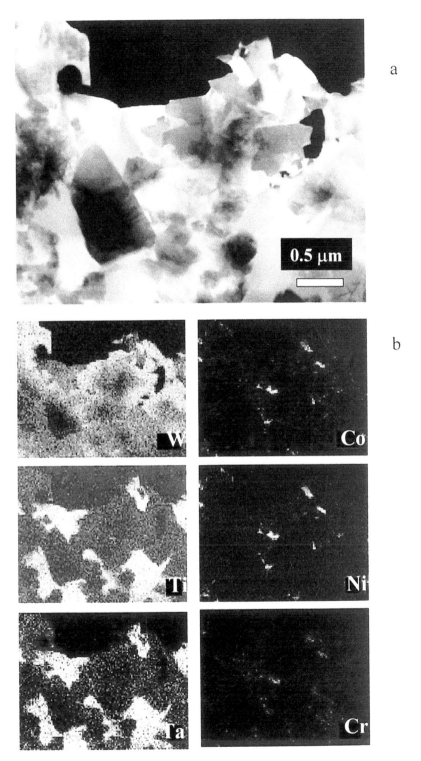

Figure 28   Dark-field STEM image (a) and elemental maps (b) from complex ceramic–metallic material.

shows a dark-field scanning TEM (DF-STEM) image and elemental maps from a material in which the metallic binder contained Ni, Co, and Cr, and the carbide ceramic contained Ti, Ta, and W. The elemental maps, based on the x-ray EDS signal, very efficiently reveal the phase distribution, and are helpful for interpreting the STEM image contrast. A single intermetallic phase containing Ni, Co, and Cr is present, along with two carbide phases: Ti–Ta–W–carbide and WC. The WC has the darker appearance in the STEM image, the Ti–Ta–W–carbide is medium grey, whereas the metallic binder yields the brighter contrast.

Details on the finer lengthscales may be more easily studied in conventional bright-field TEM images. In Fig. 29(a), the intermetallic binder is bright, whereas the carbide phases cannot easily be distinguished on the basis of image contrast alone. However, it is clearly

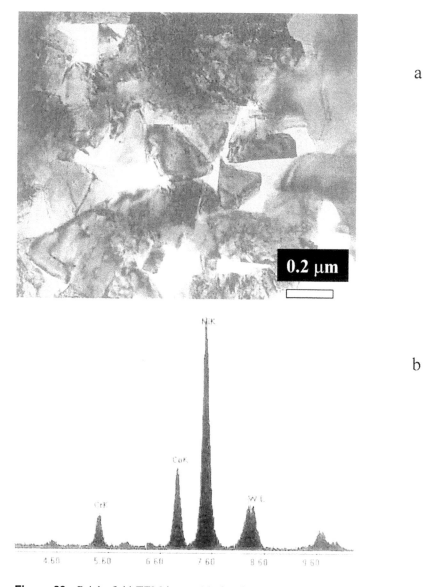

Figure 29 Bright-field TEM image (a) showing carbide grains (grey) and intermetallic Ni–Co–Cr binder (bright), and EDS spectrum (b) from intermetallic phase.

revealed that the material contains a significant fraction of very fine (~100–200 nm) carbide grains. It can also be seen that the intermetallic is found as thin (tens of nanometers) layers along grain boundaries, and as ~100–200 nm large junctions between grains. An EDS spectrum from the intermetallic phase is shown in Fig. 29(b). The probe size used was approximately 50 nm, and hence the information volume is well inside the junction between carbide grains. Quantification based on the EDS spectra yielded an approximate composition of the intermetallic phase of 67 Ni–24 Co–9 Cr (wt.%). The EDS spectra imply that some tungsten has dissolved in the intermetallic phase. The TEM studies of these materials did not reveal any reaction products at the interfaces between carbide and metal. The material shown in Figs. 28 and 29 exhibited very good wear resistance when subjected to combined erosion and corrosion testing. The mixture of fine (~1 μm) and very fine (~100–200 nm) carbide grains, and the presence of two different carbide phases made the material distinctly different from other materials that were tested.

REFERENCES

1. Pourbaix, M. *Lectures on Electrochemical Corrosion*; Plenum Press: New York–London, 1973.
2. Bardal, E. *Korrosjon og korrosjonsvern*; Tapir: Trondheim, 1985, 1994; 10 pp, *in Norwegian*.
3. Wranglen, G. *An Introduction to Corrosion and Protection of Metals*; Chapman and Hall, 1985.
4. Fontana, M.G.; Greene, N.D. *Corrosion Engineering*; McGraw Hill: New York, 1967, 1978, 1986.
5. Uhlig, H.H. *Corrosion and Corrosion Control*. 2d ed.; John Wiley and Sons: New York-London, 1971.
6. Shreir, L.L.; Farman, R.A; Burstein, G.T. *Corrosion*. Vol. 1 and 2; 3d ed.; Newnes–Butterworths: Oxford, 1994.
7. Pourbaix, M. *Atlas of Electrochemical Equilibria*, 2d ed.; NACE: Houston, Texas, 1974.
8. Gartland, P.O. Effects of flow on cathodic protection of a steel cylinder in seawater. SINTEF report STF16 F80007, Trondheim, 1980.
9. Bardal, E. Effects of flow conditions on corrosion. *Lecture Series on Two-phase flow*; NTH: Trondheim, 1984.
10. Kaesche, H. *Die Korrosion der Metalle*; Springer Verlag: Berlin-Heidelberg-New York, 1966; 251–284.
11. Fontana, M.G.; Greene, N.D. *Corrosion Engineering*, 2d ed.; McGraw Hill: New York, 1978; 76 pp.
12. Efird, K.D. Effects of Fluid Dynamics on the Corrosion of Copper-Base Alloys in Sea Water. Corrosion 1977, *33*, 3–8.
13. Duncan, R.N. Materials performance in Khuff gas service. Mater. Perform July 1980, 45–53.
14. de Waard, C.; Lotz, V. Prediction of CO_2 corrosion of carbon steel. *NACE CORROSION/93*; NACE Houston: Texas, 1993; Paper No 69.
15. Pludek, V.R. *Design and Corrosion Control*; The Mac Millan Press: London, 1977; 100–110, 144–158.
16. Scully, J.C. *The Fundamentals of Corrosion*. 3d ed.; Pergamon Press: Oxford, 1990; 1–53, 7 pp.
17. West, J.M. *Basic Corrosion and Oxidation*. 2d ed.; Ellis Horwood LTD: Chichester, 1986; 174–192, 178 pp.
18. Yurek, G.J. Mechanisms of diffusion controlled high-temperature oxidation of metals. In *Corrosion Mechanisms*; Mansfeld, F., Ed.; Marcel Dekker: New York, 1987; 397–446.
19. Batchelor, A.W.; Lam, L.N.; Chandrasekaran, M. *Materials Degradation and its Control by Surface Engineering*. Imperial College Press/World Scientific Publishing: Singapore, 1999; 124–143, 189–196, 43–46.
20. Kofstad, P. *High Temperature Corrosion*; Elsevier: Amsterdam, 1988.
21. DNV. Recommended Practice RP 0501. *Erosive Wear in Piping Systems*; Det Norske Veritas: Oslo, 1996.
22. Ives, L.K.; Ruff, A.A. Transmission and scanning electron microscopy studies of deformation at erosion impacts sites. Wear 1978, *46*, 149–162.

23. Ellison, B.T.; Wen, C.J. Hydrodynamic effects on corrosion. AIChE Symp. Ser.: The American Institute of Chemical Engineers, 1981; 161–169.
24. Bardal, E.; Eggen, T.G.; Stølan Langseth, Aa. Combined erosion and corrosion of steels and hard-metal coatings in slurries of water and silica sand. Proceedings of 12. Scandinavian Corrosion Congress/Eurocorr 92; Helsinki University of Technology: Helsinki, 1992; 393–400.
25. Bardal, E.; Rogne, T.; Bjordal, M.; Berget, J. Rates and mechanisms of combined erosion and corrosion of ceramic–metallic coatings/surfaces. Proceedings of Conference Organic and Inorganic Coatings for Corrosion Prevention, Nice, 1996, publ. Institute of Materials: London, 1997; 278–290.
26. Levy, A.V. *Solid Particle Erosion and Erosion Corrosion of Materials*; ASM International: Materials Park, OH, 1995.
27. Ahlen, C.E.; Bardal, E.; Marken, L.; Solem, T. New ceramic–metallic materials for choke valves in oil production. Proceedings of Eurocorr '97, Trondheim; NKF-NTNU-SINTEF: Trondheim, 1997, Vol. 1, 103–108.
28. Berget, J.; Rogne, T. A summary of recent developments of HVOF sprayed ceramic–metallic coatings for corrosion and wear resistance. Proceedings of the International Thermal Spray Conference, Singapore 2001; 1157–1163.
29. Strafford, K.N., Datta, P.K., Googan, C.G., Eds.; *Coatings and Surface Treatment for Corrosion and Wear Resistance*; Ellis Horwood LTD: Chichester, 1984.
30. Rowe, C.N. Lubricated wear. In *Handbook of Lubrication, Theory and Practice of Tribology, Vol. II Theory and Design*; Booser, E.R., Ed.; CRC Press: Boca Raton, 1983; 209–225.
31. Waterhouse, R.B. The effect of environment in wear processes and the mechanisms of fretting wear. In *Fundamentals of Tribology*; Suh, N.P., Saka, N., Eds.; The MIT Press: Massachusetts, 1978; 567–584.
32. Frank Murray, S. Wear resistant coatings and surface treatments. In *Handbook of Lubrication, Theory and Practice of Tribology, Vol. II Theory and Design*; Booser, E.R., Ed.; CRC Press: Boca Raton, 1983; 623–644.
33. Sproles, E.S. Jr; Gaul, D.J.; Duquette, D.J. A new interpretation of the mechanism of fretting and fretting corrosion damage. In *Fundamentals of Tribology*; Suh, N.P., Saka, N., Eds.; The MIT Press: Massachusetts, 1978; 585–596.
34. Goldstein, J.I., et al. *Scanning Electron Microscopy and X-ray microanalysis*. 2d Ed.; Plenum Press: New York, 1992.
35. Williams, D.B.; Carter, C.B. *Transmission Electron Microscopy*; Plenum Press: New York, 1996.
36. Dingley, D.J.; Field, D.P. Electron backscatter diffraction and orientation imaging microscopy. Mater. Sci. Technol. 1997, *13*, 69–78.
37. Danilatos, G.D. Environmental Scanning Electron Microscopy. In *In-Situ Microscopy in Materials Research*; Gai, P.L., Ed.; Kluwer Academic Publishers: Dordrecht, 1997; 14–44.
38. Lindseth, I.; Pettersen, G.; Nordlien, J.H.; Andersen, S.J.; Walmsley, J.C.; Bardal, A. Preparation of TEM specimens from aluminium surfaces: Methods and applications. 2nd International Symposium on Aluminium Surface Science and Technology (ASST 2000), Manchester, 2000.
39. Vickerman, J.C., Ed.; *Surface Analysis: The Principal Techniques*; Wiley: Chichester, 1997.
40. Watts, J.F. *An Introduction to Surface Analysis by Electron Spectroscopy*; Oxford Science Publications: New York, 1990.
41. Hagenhoff, B. High resolution surface analysis by TOF-SIMS. Mikrochim. Acta 2000, *132*, 259–271.
42. Payling, R., Jones, D.G., Bengtson, A., Eds.; *Glow Discharge Optical Emission Spectrometry*; Wiley: Chichester, 1997.
43. DiNardo, N.J. *Nanoscale Characterization of Surfaces and Interfaces*; VCH: Weinheim, 1994.
44. Bardal, E.; Ahlen, C.H.; Nøkleberg, L.; Solem, T.; Bardal, A. Erosion and corrosion of ceramic–metallic materials. *Proceedings of 13th International Corrosion Congress, Melbourne*; Australasian Corrosion Association, Melbourne, 1996, paper 194, pp. 1–8.
45. Rogne, T.; Bardal, A.; Solem, T. *Undersøkelse av cermet materialet VC808 (in Norwegian)*; SINTEF Materials Technology: Trondheim, 1998.

14

Heat Treatment: Tribological Applications

Lauralice Campos Franceschini Canale and Ovídio Richard Crnkovic
University of São Paulo, San Carlos, Brazil

I. INTRODUCTION

Wear performance is a complex matter that depends on hundreds of factors that define the tribological conditions of the system. The performance of the component is dependent on the wear conditions, which can be severe.

In tribological conditions, the microstructure of the material used has an important role in these tribosystems and sometimes can have strong influence on the wear resistance under determined conditions. A microstructure may be appropriate for one type of wear; the same cannot be true for others.

Different types of wear have been recognized. Adhesive wear is caused by the bonding of contact points on the mating surfaces, such that plastic deformation and fracture of these regions occur with material adhering to the surfaces. Abrasive wear involves the removal of material from a surface, with debris generated entering into the wear process.

Among these different modes of wear, abrasive wear is responsible for more than 50% of all wear problems and has been recognized as the most severe and the most common in the industry [1].

Abrasive wear is normally classified as two-body and three-body, depending on the interactions between the surfaces in contact. For these abrasive tribosystems, parameters related to abrasive characteristics such as hardness, size, and morphology influence the wear performance. Khruschov defined the K factor relating hardness of the abrasive and the material hardness showing three different phases:

$$K = H \text{ (abrasive)}/H \text{ (material)}.$$

When K is ≤ 0.7–1.0, wear resistance is infinite because there is no wear. In the range between 1.3 and 1.7, wear resistance and wear are independent of abrasive hardness. Between these ranges, the governing factor of abrasive wear remains the hardness of the abrasive, but the relationship is very dependent on the physical properties of the materials under test [2].

In measuring actual wear, the problem is how to characterize the process. Measurements of the volume of the wear debris, the depth of the wear track, and the power to move one surface past the other can be made; the results obtained are dependent on the type of wear test and whether a lubricant is used. Thus, it is necessary to compare the behavior of materials just as if they underwent the same kinds of tests and conditions.

Results from abrasive tests are normally expressed in terms of weight losses, volume losses, or abrasion rate (W) [3]. W can be calculated using the following expression:

$$W = \Delta m / \rho.A.S,$$

where Δm = weight loss, ρ = specimen density, A = cross-sectional area of worn specimen, and S = wear length.

To resist abrasive wear, iron-based alloys, whose microstructural characteristics can be changed through heat treatments, are usually used.

Sometimes the abrasive wear performance of alloys is related to their bulk hardness, but the rate of increase is quite low compared to pure metals. However, several investigations have shown iron alloys having the same bulk hardness but with different microstructures showing varied abrasion performance [2,4–9].

Some models have been developed that tried to analyze the abrasion behavior of iron-based alloys, such as the "equivalent hardness" model [9].

Equivalent hardness is equal to the sum of the products of the volume fraction of each phase [matrix (f_m) and carbide (f_{hp})] and its hardness (H_{hp}). The hardness of the work-hardened matrix after wear test is used (H_m). Thus:

$$H_{eq} = f_m H_m f_{hp} H_{hp}$$

This model showed good results, as seen in Fig. 1, where abrasion resistance is plotted against equivalent hardness. However, it is not possible to use it for predicting wear behavior.

Even for abrasion wear, the process is complex and, depending on the type of tribological contact, a work hardening zone can be formed during wear processes. Consequently, significant modification can occur in the microstructure (in the contact zone and below that), which affects the wear performance, making it difficult to predict the abrasion resistance of a microstructure [10–12].

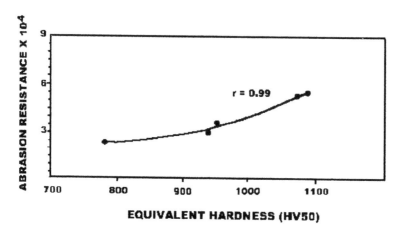

Figure 1   Abrasion resistance of Ni-hard (2.86% C/1.78% Si, 0.52% Mn/8.78% Cr/5.12% Ni/0.19% Mo) as a function of equivalent hardness. Results obtained after two-body abrasion test. (From Ref. 9.)

II. MICROSTRUCTURAL MODIFICATIONS DURING WEAR PROCESSES

Work hardening in the surfaces in contact can sometimes occur during the wear process, provoking alterations in the microstructure, because modifications in the surface hardness arise even until phase transformation.

Investigations made with iron-based alloys show significant hardness modifications after two-body abrasion tests. Quenched and tempered steels were tested using a pin abrasion test with flint abrasive (80 mesh, 900 HV). Hardness values were obtained before and after the tests demonstrating the strong effect of the work hardening [9]. Table 1 shows some partial values from this investigation. It is interesting to observe that the best abrasion resistance occurs in the steel with the highest work hardening capacity.

The superior wear resistance of perlitic structures compared with tempered martensitic structures of equal hardness has been measured on rail steel containing 0.72% C, 0.79% Cr, and 0.21% Mo. The superior wear resistance of the perlitic structure is attributed to its high work hardening during the wear processes [13,14].

Hadfield steel and medium manganese steels are used in applications where wear resistance is required. The main characteristic of this kind of steel is its high work hardening capacity. Wear results under impact abrasive tests were obtained with Hadfield steel (Mn13) and 6Mn–2Cr steel. Table 2 summarizes values of weight loss (W) and wear resistance (W^{-1}) under two impact conditions. These results are the average values of three specimens.

It can be seen that wear resistance of 6Mn–2Cr is considerably higher than that of Mn13 steel because the work-hardening exponent is about 20% higher than that of Mn13 steel.

Usually these kind of steels are submitted to heat treatment consisting of solution treatment following water quenching. In this condition, their microstructure is formed by austenite. Then during the wear processes, strain-induced martensite is produced with the wear resistance increased [15].

During the wear processes, most of the frictional work is turned into heat. When two surfaces slide against each other, a significant rise in temperature can cause changes in the mechanical properties, depending on the working conditions [3]. Thus, two aspects related to the microstructure are important to observe: thermal stability and thermal conductivity.

Investigations made with microstructures of high-carbon steel concluded that the perlitic microstructure has higher thermal conductivity than martensite, which has higher thermal conductivity than cementite. It means that the perlitic microstructure can transmit

Table 1 Microhardness of the Surface Before and After the Test in the Worn Surface and Abrasion Resistance

Steel	Before the test (HV_{30})	Worn surface (HV_{30})	Abrasion resistance ($\times 10^4$)
40 Ni Cr Mo 8 4	332	395	1.9
32 Mn Cr B 5 II	346	385	1.9
G S 42 Cr Mo 4	294	370	2.0
G S 50 Si Ni Mo 8 7 5	354	525	2.4

Source: Ref. 9.

Table 2 Results from Impact Abrasive Wear Tests

Steels	Condition (a)		Condition (b)	
	$W(g)$	W^{-1} (g^{-1})	$W(g)$	W^{-1} (g^{-1})
6 Mn 2 Cr	0.12294	8.13405	0.15673	6.33040
Mn 13	0.21143	4.72970	0.18337	5.45345

Impact conditions: (a) 5 kJ/m^2 and (b) 10 kJ/m^2.
Source: Ref. 14.

frictional heat into the substrate faster than the others. Because of this, the surface temperature is not very high [16].

If the frictional heat produced is high, the thermal stability of the microstructure is very important because it reflects in hardness changes with tempering temperature and/or time. The hardness of the perlitic structure does not change even when the tempering temperature increased to 700°C. The same is not true for the martensitic plus carbide or martensitic structure, whose hardness starts to decrease at 200°C, and bainite at 300°C. At 650°C, the hardness of the martensite becomes less than that of the perlite. It can be seen that the thermal stability of the structure affects the wear performance of the materials. The heat treatment chosen to obtain the best performance of the alloy is strongly related to the wear conditions.

Investigations made with 52100 and 1080 steels with different microstructures showed wear results related to the thermal stability of the microstructures. Spheroidized microstructures, lamellar perlite, bainite, and tempered martensite were analyzed by dry sliding for severe wear; results show that the wear resistance is increased in the following order: martensite + carbide + retained austenite; spheroidized structure; martensite; bainite, lamellar perlite. On the other hand, if the wear conditions are mild, there are no significant differences among the microstructures studied [17].

Microstructural effects in wear behavior of materials are very complex because during wear processes, different structures show different dynamic changes and energy consumption. If a structure consumes more energy during the process of changing, the structure will show better wear resistance [16].

To achieve the best abrasive wear performance, changes in the materials' properties take place using heat treatments, and most varied hardening surface-hardening treatments and other hardening techniques. Although hardness is not the only factor that determines wear performance, the first approach to improve wear resistance is to increase hardness. Because of this, heat treatment that causes surface hardening is often used.

III. HEAT TREATMENT

Heat treatment can be defined as an operation or set of operations in a metal involving heating and controlled cooling. Transformation in the solid state can be obtained using heat treatment procedures, which cause changes in microstructure resulting in materials with a wide range of hardness and mechanical properties.

Although an exact relationship between hardness and wear performance does not exist, hardness has traditionally been the property used for quality control, selection of steel, and heat treatment and performance evaluation.

A. Quenching

Hardened microstructures in steels require the generation of the parent phase austenite, the formation of crystals by diffusionless, shear-type martensitic transformation, and adjustment of final strength and toughness by tempering.

Martensite can be obtained using fast cooling rates that avoid the formation of diffusion microstructures, which would cause loss of hardness in the component. There is a limiting rate that allows the transformation from austenite to martensite and this rate changes from steel to steel determined by their hardenability, cooling conditions, and geometry and dimensions of the component.

Hardness values of the hardened components in the as-quenched state depend on the amount of martensite present after the heat-treating process. Figure 2 shows the variation in hardness with martensite percentage and carbon content [18].

As can be seen, the maximum hardness of any steel occurs when a fully martensitic microstructure is present. This hardness is dependent on carbon content and thus higher hardness is obtained for higher carbon contents. Martensitic morphologies can differ mainly according to the carbon content. Martensitic crystals that form at low temperature (high-carbon steels) have three-dimensional plate geometry and are called plate martensite. The martensite that forms in austenite of medium or low carbon content assumes a completely different morphology. These crystals appear to have a lath or board-shaped geometry [19]. Figure 3 shows the variation of morphology with varying carbon content and Ms temperatures.

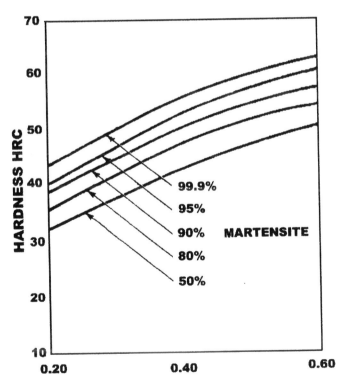

Figure 2   Hardness variation with martensite percentage and carbon content after quenching. (From Ref. 18.)

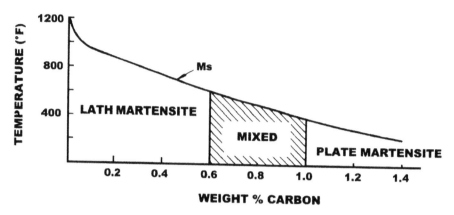

Figure 3 Martensitic morphologies varying with carbon contents. (From Ref. 19.)

For low-carbon steels, martensite crystals are relatively thin and flat with one long dimension and adjacent crystals form parallel to each other in stacks or packets.

For high-carbon steels and iron alloys with Ms below the ambient, the orientation of parallel groups can change. Its substructure consists of high densities of dislocation arranged in cells. Their crystal structure may be either (bct) or (bcc). However, medium-carbon steels, because they can contain a mixture of lath and plate martensite, have a more complex structure [20].

The effectiveness of the austenite to martensite transformation is dependent on the ability of the steel to resist the diffusion-controlled transformation of austenite during cooling to microstructures of lower hardness such as perlite, bainite, and ferrite. The ability of a steel of given chemistry, and of parts of a given size and geometry to form martensite under a given set of cooling conditions is related to hardenability. Generally, higher carbon content, alloying elements such as manganese, nickel, molybdenum, and chromium, coarse austenitic grain sizes, higher cooling rates, and small section sizes favor more complete transformation to martensite [21].

Although the alloying elements increase hardenability in steels, they do not exercise influence on the hardness of the martensite, as shown in Fig. 4 [22].

The rate increased up to about 0.60% C because of retained austenite increasing with higher carbon content as shown previously.

There are other microstructural factors affecting hardness besides retained austenite.

Fine grain size can improve the hardness, although toughness is the mechanical property more benefited with fine structure. The fine structure may have an appreciable effect on the wear resistance as well. The literature reports an increase of 2–5 times in wear resistance and a 15% reduction in the average value of the coefficient of friction when the grain of quenched and tempered steel was reduced. Thus, much finer microstructures are desirable to obtain better performance with respect to wear resistance [23].

High-carbon steels can present carbides in the martensitic matrix, which were not dissolved during austenitization before quenching. These can occur because the time in the austenitization temperature was not enough to dissolve the particles, or because the austenitization temperature used was intercritical in order to keep these undissolved carbides spread in the martensitic matrix. Thus, the results of hardness are related to the amount of carbon dissolved to harden the martensite vs. the amount of carbide present in the microstructure. Because of this, microstructure hardness values are not so easy to predict [24].

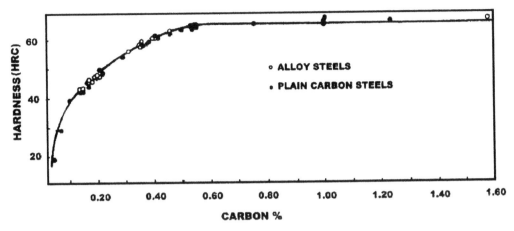

Figure 4   Hardness of martensite as a function of carbon content of the martensite. (From Ref. 22.)

Although high hardness values can be achieved with the quenching heat treatment, an as-hardened steel, whether this is a product of conventional quenching or martempering, is not a suitable structure for working. The structure in this condition may be a heterogeneous mixture of untempered martensite, retained austenite, and carbides. Retained austenite is present in small amounts in quenched plain carbon and low-alloy steels, but in the highest alloyed steels it is present in significant amounts. The tendency to retain austenite in the structure at room temperature depends on the austenitizing temperature. Also, as additions of alloying elements are made, which tends to stabilize the austenite, Mf temperature drops and it becomes more difficult to transform the austenite. Posthardening subzero treatment before the temper allows the steel to achieve the Mf temperature, thus transforming the retained austenite. It also reinforces the secondary hardening effect, and maximum hardness and wear resistance are achieved. Figure 5(a) and (b) demonstrates the effect of subzero treatment decreasing the retained austenite amounts.

Quenching temperatures affect the mean grain size of austenite, volume fraction of carbides, and mean size of primary carbides, and hence the mechanical properties, mainly wear resistance and toughness. High temperature makes it possible to obtain a partial dissolution of carbides and to saturate the austenite with carbon and alloying elements. Too high austenitizing temperature causes a significant increase in austenite grain size and retained austenite content. Similar effects are observed when too long austenitizing times are used [25].

High-speed steel (AISI M2) investigations show variations in those parameters at different quenching temperatures as demonstrated in Table 3.

Carbides in the microstructure considerably affect the properties of tool steels in two ways [26]:

- Partial dissolution of carbides in the matrix during austenitizing affects properties that depend on the matrix composition (e.g., hot hardness).
- The carbides themselves affect the mechanical properties especially the wear resistance.

The dissolution carbides during the austenitization could be measured during induction heating where the heating time is rapid and short. Experiments were performed with the tool steel AISI M2. The amount of dissolved carbide was obtained from the difference of the

(a)

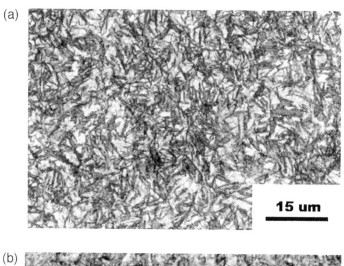

(b)

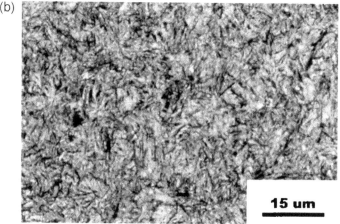

Figure 5 Microstructure of SAE 8620 case carburized (etched with nital 2%) showing martensite (dark areas) and retained austenite (light areas): (a) after quenching; (b) after quenching and subzero treatment.

Table 3 Variations in the Austenite Grain Size and Volume Percent of Carbides (M_6C and MC) at Different Quenching Temperatures

Quenching temperature (°C)	Grain size (μm)	M_6C (%)	MC (%)	Total %
1160	6.0 ± 0.1	7.8	0.8	8.6
1185	7.0 ± 0.1	7.7	Not detected	7.7
1210	8.0 ± 0.2	4.2	0.3	4.5
1235	10.0 ± 0.7	4.8	0.1	4.9

Source: Ref. 26.

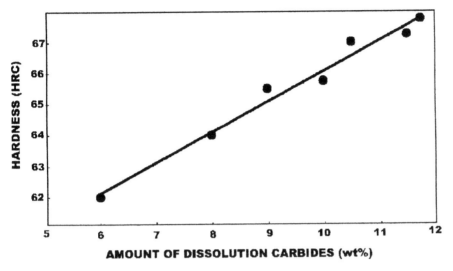

Figure 6   Amount of dissolved carbide (weight) × hardness. Holding time of 200 sec. (From Ref. 27.)

amount of extracted carbide in the spheroidized and the quenched state. Figure 6 shows the results of hardness with the variation of the amount of dissolved carbide.

To clarify the growth of matrix composition, transmission electron microscopy/energy dispersive X-ray (TEM-EDX) analysis was performed; the results are presented in Table 4. The increasing alloying elements in the matrix achieve the high hardness.

Matrix composition is increased during the heat treatment, but for long holding times the composition does not change because a balanced state is achieved.

For plain steels, but mainly for alloy steels, selection of the correct time and temperature of austenitizing determines the success of the hardening process.

B. Tempering

Microstructures in the as-quenched condition are brittle, and thus tempering is necessary. With this heat treatment, toughness is increased whereas hardness is decreased. The intensity of this effect is related to the temperature and holding time but the effect of the former is more pronounced than that of the latter. For carbon and low-alloy steels, decrease in hardness as tempering proceeds is greater the higher the tempering temperature. The reduction in hardness shown by the curves is associated with the martensite decomposition (ferrite and carbides) and with the coarsening of the carbides. These modifications are

Table 4   Results of TEM-EDX Analysis (at.%) for Different Holding Times

Holding time (sec)	V	Cr	W	Mo
1	0.36	4.01	0.20	0.66
1,000	0.94	4.50	0.82	1.66
1,000,000	0.92	4.45	0.84	1.63

Source: Ref. 27.

associated with the steel composition. Figure 7 shows these effects for AISI 4340 (0.39C/0.24Si/0.61Mn/1.46Ni/0.67Cr/0.17Mo) alloy steel [28].

Tempering temperature and holding time affect the mechanical properties and microstructural features of AISI 4340 steel. During the tempering at different temperatures, various types of carbide precipitate in different shapes and sizes in the matrix. Carbide distribution is directly affected by the tempering conditions. At low tempering temperature, t-Fe_3C (cementite) forms a platelike structure, which is replaced in two stages by Cr_7C_3, which has a spheroid-like structure, when the tempering temperature reaches 650°C. Cr_7C_3 carbide is probably formed due to the reaction of the Fe_3C with the matrix [28].

Although alloy elements do not affect the hardness of the martensitic microstructure as quenched, they can hardly affect during the tempering. In general, the main effect is on the rate of softening during tempering, so that alloy steels require a higher tempering temperature to obtain a given hardness. These effects are dependent on the type and amount of the alloy. Addition of Ni, Al, and Mn, which are not strong carbide formers, has little effect on the hardness of the tempered steel. Si retards the softening of low-carbon steel. Thus, it is possible to use higher tempering temperatures without significant decrease in the tensile strength [29]. Si influences microstructural modifications, inhibiting dissolution or conversion of ε-carbide to cementite. It means that Si retards the formation and growth of cementite [30]. Figure 8 demonstrates Si effects in the yield strength as a function of tempering temperature for three steel grades: Steel A: 0.60C/1.02Si/0.47Mn/0.51Cr/1.77Ni/0.18V/Fe bal.; Steel B: same composition but with 1.78Si; Steel C: same composition but with 2.44Si.

Microstructural observations obtained from this work have demonstrated that the maximum yield strength occurred when the microstructure presented a dense distribution of ε carbides within martensite laths. In the steel with high Si content, the temperature at which the phenomenon occurs is high as well [31,32].

Elements such as Cr, Mo, and V form part of the carbide phase and bring about a retardation of softening at the higher tempering temperatures. In high-alloy steel, as in some tool steels, increase in hardness occurs during the tempering, when these elements are themselves associated with the carbide phase provoking the secondary hardening. The use of Si as an alloying element in high-speed steels (Mo steels) has attracted interest because Si

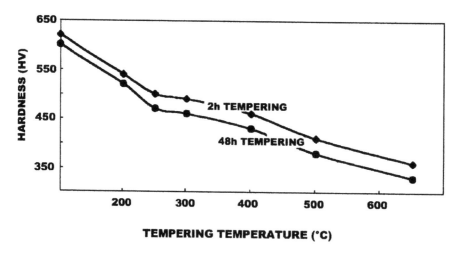

Figure 7 Variations of hardness with tempering temperature. (From Ref. 28.)

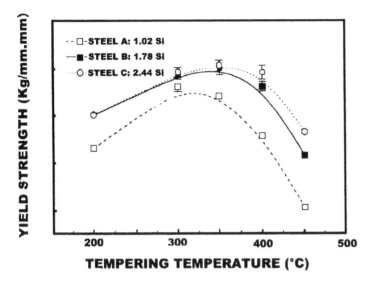

Figure 8   Yield strength as a function of tempering temperature. (From Refs. 31 and 32.)

additions may accelerate the transformation of M_2C carbide, a needle-shaped carbide harmful to properties, to M_6C. So the hardness temper is expected [33]. Figure 9 shows the tempering curve for 6W3Mo2Cr4V steel hardened at 1160°C, demonstrating the secondary hardening and the Si effect.

The presence of carbides in the martensitic matrix increases the wear resistance. If they are much finer than some abrasive particles, the material behaves essentially as a continuum.

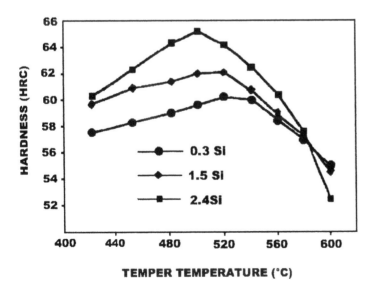

Figure 9   Secondary hardening and Si effect on the temper hardness of the high-speed steel 6W3Mo2Cr4V (approximate chemical composition: 0.60% C/3.15% W/2.10% Mo/1.62% Cr/4.1% N/–Si). (From Ref. 33.)

Additions of Mo and/or W elements result in a fine distribution of tempered carbide particles because of the decrease in the rate of coarsening of the particles (above 400 °C) [34]. The increase in wear resistance is related to the total volume and the hardness of the carbide.

Table 5 shows the general classification and properties of carbides found in tool steels [35].

The chemical composition of the carbides depends on the alloy additions. High-speed steels such as 9-2-2 (9W2Cr2V0.7Si), 11-2-2 (11W2Mo2V0.7Si) and 11-0-2 (11W2V0.7Si) had their compositions modified with different contents of Nb or Ti and a reduction in the V content in order to study if the former elements can be used as partial substitutions for V. The composition of the MC-type primary carbides changed, although the M_6C remained the same in all the steel compositions [36]. It was verified that MC primary carbides begin to originate as very rich in Nb, and only after local saturation with this element they gradually get enriched with V. For Ti additions, the same thing occurs but with Ti instead of Nb. Because these MC carbides practically do not dissolve in the solid solution during the heat treatment, it is possible to ensure sufficient concentration of V in the solid solution after quenching in the steels. This makes possible a sufficiently high precipitation hardening of the martensitic matrix during tempering and results in an important secondary hardness effect. These additions increase the hardness, about 0.5–2 HRC depending on the steel and the alloying element related to the steel grades without these additions and heat treated in the same conditions. Figure 10 shows the conclusions from that investigation.

Another important question about the addition of elements in steels, mainly tool-steels, is related to the volume of primary carbides not dissolved in the solid solution. Table 6 demonstrates that in high-speed steels previously mentioned the primary carbide amounts varied with the chemical composition and with the type of steel.

Similar steels were modified with 5% Co. The effect of this element in high-speed steels is interesting because it is not in high concentration in the MC and M_6C. Otherwise, it appears in high concentration mainly in solid solution. Even then, it can provoke changes

Table 5 General Classification and Properties of Carbides

Type of carbide	Lattice type	Remarks	Hardness (Knoop)
M_3C	Orthorhombic	M may be Fe, Mn, Cr with a little W, Mo, V	950
M_7C_3	Hexagonal	M is normally Cr. Hard and abrasion resistant	1600
$M_{23}C_6$	Face-centered cubic	M is normally Cr. It can be replaced with Fe to yield carbides with W and Mo	—
M_6C	Face-centered cubic	M is W or Mo. May contain small amounts of Cr, V, Co. Extremely abrasion resistant	1300
M_2C	Hexagonal	M is W or Mo. Appears after temper	—
MC	Face-centered cubic	M is V or Nb. Resists dissolution. Small amount that dissolves reprecipitates on secondary hardening	2300

Source: Ref. 35.

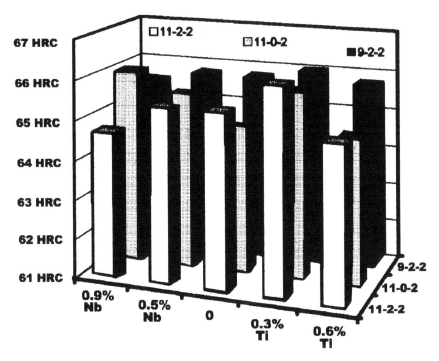

Figure 10 Effect of the Nb and Ti contents on hardness for three types of high-speed steels. (Austenitizing temperature: 1120–1300°C. Tempering temperature: 480–630°C). (From Ref. 36.)

during the heat treatment, and the final hardness after quenching and tempering may increase 2 HRC depending on the steel type. Austenitizing temperatures as high as 1270°C can promote carbide dissolution without increase in grain size. As a result, during the tempering the secondary hardening is raised and consequently the abrasive wear resistance is improved [37].

Although the hardenability of the matrix, and consequently its hardness, is related to the stability of primary carbides, their presence offer wear resistance to steel, mainly in tribological applications where abrasion wear is a most important requirement. However, excessive primary carbides are detrimental to toughness. When very severe impact load occurs, these carbides can cause steel parts to crack and break. In this case, excessive primary carbide should be strictly restrained [38].

In some tribological systems, the carbide originally acting as an antiwear phase can effectively resist microcutting of the abrasive particle and plastic deformation. However, if

Table 6 Carbide Percentage in Tool Steels (11-0-2 and 11-2-2) with Different Nb and Ti Additions

	0.9% Nb	0.5% Nb	0	0.3% Ti	0.6% Ti
11-0-2 steel	11	10	10	13	15
11-2-2 steel	15	10	—	15	18

Source: Ref. 36.

the matrix material has been partly worn away, carbide particles protruding out of the matrix surface come loose. When this occurs the carbide particle changes, from enhancing wear resistance to inducing further damage. These factors are related to the binding strength between the carbide and matrix material and the preferred shape geometry [39].

Additions of microalloying in the steel have an effect on the mechanical properties after tempering. Elements such as V and Nb can modify the hardness and yield strength. Spring steels SAE 5160 and 9260 had their composition modified by the addition of V, Nb, and V + Nb [40].

Results have shown that for tempering temperatures lower than 550°C, there was no effect. However, there was a definitive secondary hardening effect at 600°C, as shown in Fig. 11, since hardness values were higher for microalloyed steels.

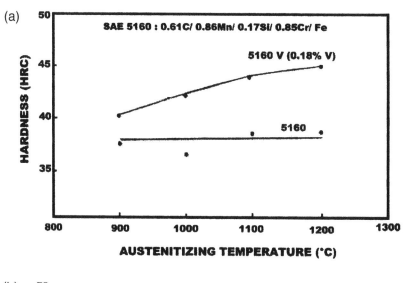

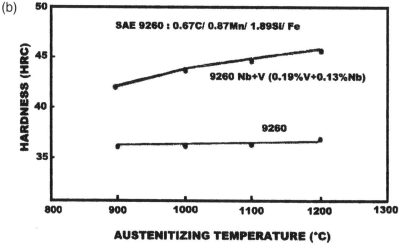

Figure 11   Hardness variation with chemical composition of the steels and austenitizing temperature. (a) Base steel: SAE 5160. (b) Base steel: SAE 9260. (From Ref. 40.)

As already observed for high-alloy steels, the general heat treatment response in this case was determined by the austenitizing temperature. Higher temperatures can dissolve precipitates, promoting their new precipitation as small particles during the tempering.

When the retained austenite amounts are large before the tempering, increase in hardness can be obtained from the transformation of this austenite during the tempering. Because of this multiple tempering is usual.

If the steel is tempered at a temperature to get the maximum hardness, good abrasion wear resistance can be expected [41].

1. Embrittlement in Tempered Steels

In general, for plain carbon and low-alloy steels hardness decreases when the holding time and temperature of tempering increase. Consequently, the ductility increases with increasing tempering temperature, but it does not mean that the toughness is continuously increased.

Sometimes, during the tempering, embrittlement can occur when the impact energy measured at 25°C decreases. This temper embrittlement is due to the retained austenite decomposition, the morphology of carbides, which form during tempering, and the segregation of elements in the austenite grain boundaries. The effects of these last two factors are not necessarily independent.

The temperature at which this occurs can change from steel to steel, but it would be in the 200–500°C range.

The segregation, mainly of P to austenite grain boundaries, has an important effect on the embrittlement, provoking intergranular fracture. Figure 12 shows that for the lower-P steel the fractures are less intergranular [42].

Studies of the mechanical properties of the AISI 4340 steels (low-alloy steel) were done as functions of tempering temperature and holding time. There was a loss of toughness after tempering at 300°C for 2 hr due to retained interlath austenite and the formation of interlath carbide films. These aspects are evident in works from the literature [28].

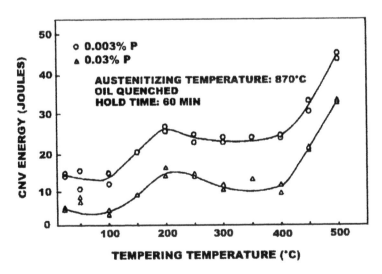

Figure 12   Charpy impact energy × tempering temperature for 4340 steel. Tests performed at room temperature. (From Ref. 42.)

Addition of Mo is an effective method to suppressing temper embrittlement caused by P segregation in the steels. Mo forms with the P, Mo-rich phosphates, or Mo–P–Fe clusters. Another effect is that the Mo segregation to grain boundaries enhances the grain boundary cohesion by itself. Studies were realized with the 2.25Cr1Mo steel using field emission gun scanning transmission electron microscopy (FEGSTEM). This high-resolution, grain boundary microanalysis technique allowed one to verify mean values for grain boundary Mo concentrations. Results showed this segregation exists as quenched condition and increases with the tempered holding time. Then proper addition of Mo in low-alloy steel is an effective approach to eliminate P-induced temper embrittlement [43,44]. Silicon has an important effect on the embrittlement associated with the carbide formation. It is possible to raise the temperature range at which the tempered martensite embrittlement occurs using Si additions.

The effect of Si content on the mechanical properties of 0.6C(1.0–2.5)Si2Ni0.2V (wt.%) was studied and impact energy (25°C)–tempering temperature curves were obtained (Fig. 13). It was verified that they showed distinctive behavior. All of them presented tempered martensite embrittlement, when the minimum impact energy value occurs, but for the higher Si content steel this minimum occurs at higher temperature ranges. That embrittlement is related to the formation and growth of cementite at boundaries [32]. Silicon acts as an inhibitor retarding the conversion of ε carbide to cementite within martensite laths and coarsening of cementite. Silicon is a useful steel alloying element, increasing the strength of low-alloy steels and raising the temperature range in which tempered martensite embrittlement occurs [31–34].

The mechanism of tempering and the complex changes in martensite have been the subject of much research. Tempering progressively transforms the austenite athermally on cooling from the tempering temperature. The product of the austenite transformation is not tempered martensite, which must be tempered by a second tempering cycle. Thus, the

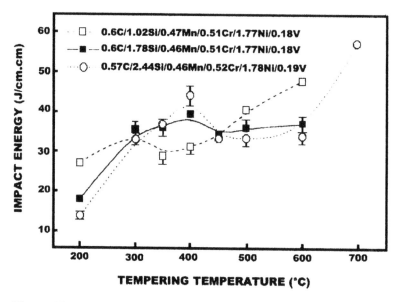

Figure 13 Charpy impact energy measured at room temperature as a function of tempering temperature. (From Refs. 31 and 32.)

tempering is not just a stress-relieving treatment. It is an important heat treatment to give the component the required properties of hardness, wear resistance, and toughness, thus increasing the durability of the component.

IV. INTENSIVE QUENCHING

Although increasing the quenchant severity produces higher core hardness and depth of the hardened layer, it provokes an increase of the crack and distortion risks and because of this the scrap is increased. As seen in Fig. 14, which shows the relationship between percentage of cracks and quenchant severity, the probability of quench cracking increases with cooling rate up to a maximum value, then decreases, achieving again low values for extremely high cooling rates [45].

Intensive quenching processes have explored this phenomenon. In contrast to the conventional quenching techniques, intensive quenching uses very fast cooling usually involving high-pressure sprays or very strong agitation. The part cooling rate is several times greater than that of agitation oil or even water. In contrast to the conventional quenching technique, fast cooling continues through the steel martensite stage. Table 7 shows different types of quenchant media and agitation compared with intensive cooling.

High quench severity factors are obtained because this process provides "direct convection cooling" without film formation and nucleate boiling that normally occur in conventional cooling processes. Intensive quenching produces a hard martensite case or shell that forms simultaneously over the whole part surface area producing surface compressive stresses due to fast martensite formation. The combination of strong martensite shell and high compressive stresses prevent the steel part from cracking and distorting [46,47].

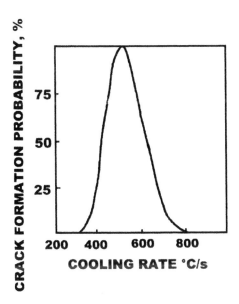

Figure 14 Effect of cooling rate on the probability of cracking. (From Ref. 45.)

Table 7 Classification of the Severity of Different Quenchant Media

Cooling rate classification	Quenchant type and agitation	Quench severity (*H*)
Slow cooling	Oil—none	0.2
	Oil—violent	0.7
Fast Cooling	Brine—none	2.0
	Brine—violent	5.0
Intensive Cooling	Brine or aqueous polymer—high-pressure spray	≥5

Source: Ref. 45.

Additionally, this process produces improves the mechanical properties of different steels. Table 8 shows comparative mechanical properties between oil quench and intensive quenching for different steels. These results are part of the intensive quench method validation tests.

Results from Table 8 are obtained after all parts (quenching in oil and intensive quenching) had been tempered to the same hardness. The improvements observed are caused by the increase of the fully hardened depth, the surface compressive stresses, and the finer martensite structure formation.

Wear is improved and the durability of the steel part is increased. The surface compressive residual stresses produced during the intensive quenching have a positive effect on the fatigue strength and superficial hardness of the specimen [49]. Compressive stress can improve the fatigue limit resistance up to 50% [50]. Intensive quenching can compete with other surface hardening such as induction hardening. Table 9 shows the hardness results in the relative depth between intensive quenched pins and induction hardened pins.

Intensive quenching process can be used in several steel compositions. Experiments performed with different steels showed that by applying the intensive quenching process the desired properties of the part could be obtained using less expensive steels, which contain 2 or 3 times less alloying elements than conventional alloy grades.

Among the benefits of intensive quenching are the optimum combination of high surface compressive stresses and a high-strength, wear-resistant, and fully quenched layer of optimum depth, all resulting in longer part service life at lower cost.

These combinations are ideal for applications requiring high strength and resistance to static, dynamic, or cycling loads [46].

Table 8 Mechanical Properties at the Core of Conventional (Oil) and Intensively Quenched Steels

Part	Property	Oil quench	Intensive quench
1340 Steel M 22 × 123 mm bolt	Tensile strength (MPa)	1093	1176
S5 Steel diameter 38 × 56 mm punch	Impact strength (N m)	6.8	12.2
4140 Steel hand tool socket	Torque to failure (N m)	168.6	223.3
4140 Steel diameter 45 mm king pin	Ultimate strength (kN)	313.6	414.8

Source: Ref. 48.

Table 9 Hardness in the Relative Depth Between Induction Hardened and Intensive Quenched Pins

Treatment	R (HRC)	0.75R (HRC)	0.5R (HRC)	0.25R (HRC)
Induction hardening	60	35	32	30
Intensive quenching	59	52	52	48

Source: Ref. 48.

V. SURFACE HARDENING

For many applications, it is necessary to get only a good performance of the surface of the part. Surface hardening includes processes involving only thermal processing that produces, by quench hardening, only a surface layer that is harder or more wear resistant than the core. There is no significant alteration on the chemical composition of the surface layer. The core properties can be adjusted by heat treatment before the surface heating and no change occurs during the surface thermal process.

In contrast to carburizing, nitriding, and other thermochemical processes, surface hardening does not require heating a whole piece. It is localized only in the areas where metallurgical changes are desired [51,52].

The surface-heating techniques frequently used are flame (flame hardening), high-frequency current (induction hardening), lasers, and electron beams, among others.

During the surface-hardening process, the surface zone with martensitic transformation cannot expand because of the central zone. As a result, high compressive stress is generated in the superficial hardened case. This stress state improves the fatigue resistance of the component and increases the superficial hardness [50–52].

These characteristics obtained allow substituting highly alloyed steels with inexpensive carbon steel or low-alloyed steel [51].

The surface-hardening treatment provokes the fatigue life, increased depending on the compressive stress level of the surface. In some situations, it is possible to verify that the fatigue resistance is better for induction hardening than carburizing at the same depth case.

Compressive stress can improve the fatigue limit resistance up to 50%. These results are obtained by Bertini and Fontanari [50] using low-alloy steel.

Although the surface-heating process is different in each case, the microstructural effects are similar. The microstructure developed depends on the hardenability at this location and the cooling rate. Usually the hardenability will be lower than it would be for the same steel conventionally austenitized. It happens because in the surface hardening, the heating time is necessarily short and the austenite may not be homogeneous. The nucleation and growth of austenite is of more concern than in conventional austenitizing where the time is long. As the distance from the surface increases, the cooling rates decrease and the hardenability decreases as well because of the lower austenitizing temperature. Then the hardness profile depends on the heating and the cooling cycles [52].

Austenitizing temperature and delay time resulting of the surface heating have an effect on the composition of the austenite that will be transformed and consequently on the hardenability, which determines the hardness profile.

Steels like AISI 52100 that are Cr alloyed are usually classified as high-hardenability steel. But in induction hardening the Cr carbides do not dissolve in the austenite modifying the hardenability. This can be seen in Fig. 15 where the 52100 steel hardness behaviors are worse than the expected [54].

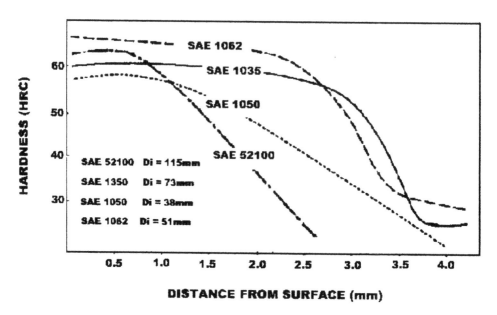

Figure 15 Relative induction hardenability of some steels. Parameters used: power, 50 kW (9600 Hz); scan rate, 11 mm/sec; quenchant, oil. (From Ref. 54.)

The prior microstructure before surface hardening has a significant effect on the final structure after the surface hardening. Microstructures with very small carbides dispersed can be quickly austenitized. When fast cooling rates are used it is possible to obtain thin hardened layers showing high hardness; consequently, high wear resistance is expected.

Tempered and quenched microstructures show very good response to induction hardening with the smallest shape/size distortion and minimum grain size growth. The hardness obtained is high with almost 100% of martensite in the layer. Normalized prior microstructure attains its final hardness value after surface hardening dependent on free ferrite phase present in the former microstructure. The more ferrite phase present, the less hardness and, worse, less wear resistance is achieved. Ferritic and perlitic microstructures require heating to higher temperatures in order to achieve the austenitic transformation.

Coarse carbides in the microstructure need high austenitizing temperatures to dissolve causing large grain size, coarse martensite, and greater distortion after induction heating. Retained austenite contents are high as well, provoking detrimental effects on toughness, fatigue, wear resistance, and hardness data scatter [51].

In surface hardening the second particle dissolution and the phase transformation can occur under different conditions when compared with tempered and quenched conventional process. Consequently, the expected microstructures are different in both cases although the hardness value sometimes can be the same.

Microalloying elements modify the characteristics of the hardened layer. Steels microalloyed with vanadium have carbides, and their effectiveness is related to the stability, size, and distribution of these carbides that have an influence on the final hardness of the layer [55,56].

The final hardness value through the layer is affected by the prior microstructure. Figure 16 shows the hardness behavior through the layer for different prior microstructures.

In flame and laser hardening the surface is the only location of energy input to the steel. In induction heating, heating of the steel occurs not only on the surface but also on the

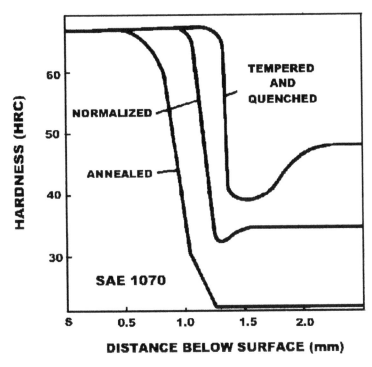

Figure 16 Hardness behavior after induction hardening obtained from different prior micro-structures. Material used: SAE 1070. (From Ref. 18.)

surfaces layers because induced currents produced cause the heating in the steel. The depth of penetration is related to the frequency and the power density used. As the frequency increases, the current is concentrated more in the surface layers.

The rate of energy transfer to the surface is lower in flame hardening, resulting in a larger heat-affected zone and slower cooling if only self-quenching is used. It is customary to water-quench the surface following heating to the desired temperature.

Laser surface transformation hardening is obtained when light radiation instanta-neously heats the surface and the heat is restricted to the optically defined area. A thin layer is rapidly achieved at the austenitizing temperature, which is cooled at a very fast rate due to self-quenching by the conduction of heat into the bulk body to produce a martensitic structure. This martensite phase and the volume contractions associated with it increase the hardness and the compressive stresses at the surface, resulting in wear resistance and increased fatigue life [57,58].

Since only a thin layer is formed, the total energy input is relatively small. The heat-affected zone is small and less distortion is observed. The laser beam can be controlled well, so good control of the hardening is achievable, improving the tribological properties and increasing the service life of the steel part [52,59].

Laser surface hardening was performed in medium-carbon steel samples (AISI 5135) and the wear behavior was evaluated. Using the pin-on-disc wear machine (sliding wear test) tests were performed without lubrication [59]. Figure 17 presents some results comparing treated and nontreated specimens for different loads: 10N, 20N, and 30N.

In this case, the laser surface hardening produced a wear-resistant and highly hardened surface layer consisting of undissolved martensite and carbides that were formed by a fast

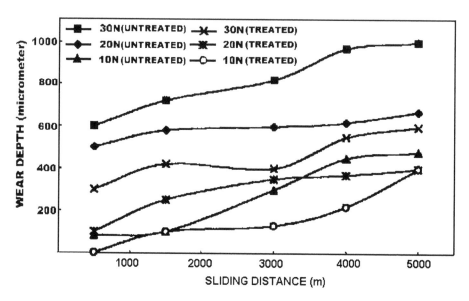

Figure 17 Wear results from treated and nontreated specimens. (From Ref. 59.)

self-quenching process, where cooling rate was estimated to be 5000°C/sec. After treatment the hardness could increase from 250 to 900 $HV_{0.2}$.

The hardness of the laser-quenched zone is substantially higher than that resulting from conventional high-frequency quench and tempering treatment [60]. Figure 18 shows the hardness profile of 40CrNiMoA steel samples under laser surface hardening (laser quenching) and high-frequency quenching (induction hardening) conditions. This type of

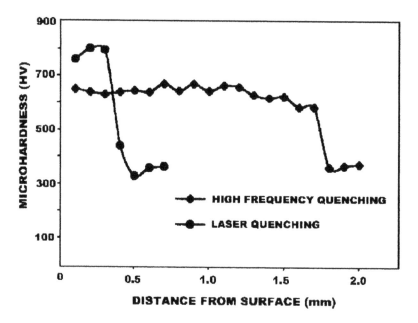

Figure 18 Hardness profile of the cross section of specimens resulting from laser quenching and high-frequency quenching processes. (From Ref. 60.)

steel (0.38C/0.72Cr/1.26Ni/0.19Mo/0.23Si/0.60Mn) is used for piston heads of large diesel engines where wear resistance is one of the most important requests.

The best performance of laser treatment samples is due to formation of unusually fine martensite structures being composed of high-density dislocation and little retained austenite. As such it can be expected to provide a very high wear resistance of the laser quenched specimens.

VI. THERMOCHEMICAL TREATMENT

A. Carburizing

Although high hardness does not always means better wear resistance, surface hardening by thermochemical process is used mainly to improve this property.

Carburizing is one of these treatments promoting high surface hardness, strength, fatigue resistance, and wear resistance. Therefore, it is used in highly stressed components such as shafts, bearings, and gears.

Carburizing process includes pack-carburizing, gas-carburizing, and salt bath. It is applied to low-carbon steels, about 0.2% C. The steel part is heated achieving the austenite region in an environment that allows carbon to be added to the steel at the surface. Depending on atmosphere control and carburizing temperature and time, surface carbon contents of 0.8% or more are obtained.

The carbon diffuses in from the surface producing a carbon gradient. This gradient has a strong influence on martensitic transformation, morphology, and properties [19].

The steel part is then quenched and tempered forming a complex microstructure with high-carbon tempered martensite, retained austenite, and massive carbides in the surface layer; the core may consist of tempered martensite, bainite, or ferrite and perlite, depending on the hardenability [61]. A typical microstructure of the carburized case is shown in Fig. 19.

Typically, the retained austenite content of the near-surface microstructure of direct-quenched carburized steel is between 20% and 30%.

Excessive retained austenite (40% or more) may be detrimental to the performance of carburized steels. Austenite has lower strength than tempered martensite; hardness, wear resistance, and fatigue resistance may be reduced [62]. The last property is related to the stresses of the surface, and large retained austenite volume may reduce superficial compressive residual stresses. In addition, it can cause unacceptable dimensional changes for precision parts in service.

Excessive retained austenite is related to too high carbon content caused by high carburizing temperatures that allow the austenite to became saturated with carbon; this is more intensive in steels directly quenched after carburizing [63].

The abrasive wear performance is related to the retained austenite amounts. Recent investigations about the abrasive wear resistance related to retained austenite amounts in carburized SAE 8620 steel were made by Silva and coworkers [15]. Microstructures with 37%, <6%, and 23% of retained austenite were tested using a pin-on-disk machine. Results of mass loss/area × time showed that the best wear performance occurs with the microstructure presenting 37% of retained austenite, as demonstrated in Fig. 20. The improvement in wear resistance may be related to both the hardening effect of the retained austenite and/or the strain-induced ability of retained austenite to transform into martensite during abrasion.

A survey of the literature indicates that the influence of retained austenite on wear resistance, in the case of carburized steel or hardened steel, has long been a controversial subject. This situation occurs because of the complexity of the mechanisms involved in the

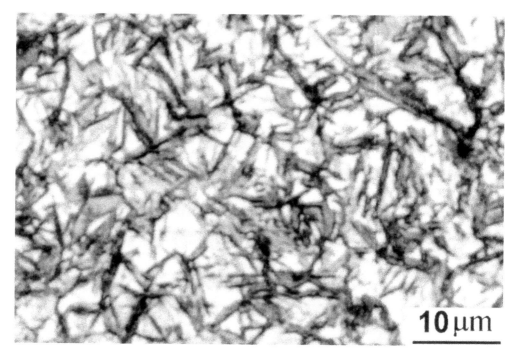

Figure 19 Typical microstructure of the carburized case. Dark areas formed by martensite and retained austenite in the light areas.

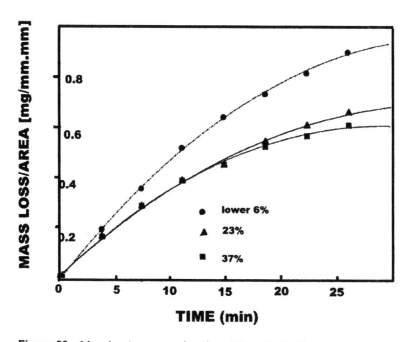

Figure 20 Mass loss/area × testing time. (From Ref. 15.)

wear of materials. In addition, there is the interaction of factors such as hardness, residual stress, surface finish, microstructure grain size, globular and network carbides, and intergranular oxidation that influence the wear and fatigue resistance.

Investigations on sliding wear behavior of carburized steels with retained austenite were performed using AISI 8620 and 4140 steels. Results concluded that for low load (20 Kg) the influence of retained austenite is negligible. Wear resistance is decreased as the retained austenite is increase from 6% to 30% when 40 Kg is used. For amounts higher than 30%, the wear resistance again becomes large [64].

Retained austenite reduction can be obtained by tempering, subzero cooling, inter-critical-temperature reheating, and shot peening.

Typical tempering temperatures for carburized steels are between 150° and 200°C, a range in which the retained austenite is stable. In order to transform it, higher temperatures can be used. However, care should be taken because changes in the microstructure can reduce toughness, hardness, and strength.

On the other hand, subzero treatments can be performed, but tempering is advisable both before and after cold cooling to reduce the detrimental effects on fatigue.

Heating the steel part at intercritical temperatures reduces austenite carbon content, increases the density of carbides, and reduces retained austenite content.

Shot peening can be effective to decrease austenite contents because the surface deformation causes the mechanical transformation from austenite to martensite, producing compressive residual stresses in the surface as well as increase of fatigue life [63]. Figure 21 demonstrates the effect of shot peening in the retained austenite amounts.

Different retained austenite amounts were obtained in the SAE 8620 carburized steel using heat-treatment-controlled parameters. After carburizing at 930°C for 3 hr, the specimens were treated as follows:

1. Austenitization to 840°C for 20 min, followed by martempering at 160°C for 20 min and tempering at 160°C for 2 hr.

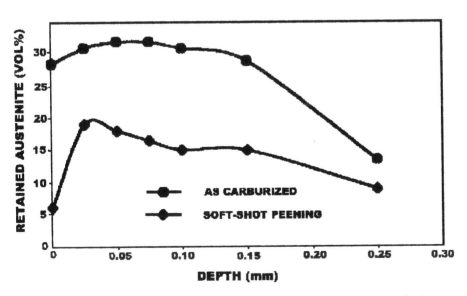

Figure 21 Retained austenite amounts through carburized case before and after shot peening. (From Ref. 63.)

2. The same [1] but cooled in liquid nitrogen at $-196\,^{\circ}$C followed by new tempering at $160\,^{\circ}$C for 1 hr.
3. Austenitization to $800\,^{\circ}$C for 20 min, followed by martempering at $160\,^{\circ}$C for 20 min and tempering at $160\,^{\circ}$C for 2 hr.
4. Austenitization to $840\,^{\circ}$C for 20 min, followed by martempering to $160\,^{\circ}$C for 20 min and tempering at $240\,^{\circ}$C for 2 hr.
5. Martempering at $160\,^{\circ}$C for 20 min directly from $930\,^{\circ}$C, tempering at $160\,^{\circ}$C for 2 hr, new austenitization at $840\,^{\circ}$C for 20 min, and second tempering at $160\,^{\circ}$C for 2 hr.

Retained austenite contents obtained from these heat treatments are summarized in Table 10.

The chemical composition of the steel has influence on the retained austenite amounts. Alloying elements (such as Mn, Ni, Cr and Mo) lower Ms and Mf temperatures and contribute to retention of case-retained austenite [65]. On the other hand, it was found that niobium refines martensite and reduces the formation of retained austenite in the case of carburized and hardened low-carbon steels. Additions of about 0.04% can improve the surface hardness, case hardness, and core hardness. Associations with nitrogen (0.04% Nb + 0.001% N) are more effective than niobium alone in decreasing the retained austenite content, although the presence of NbC or Nb(C, N) particles reduces the impact energy absorbed [66].

The effects of rare earths (REs) on the carburizing of steel were examined using additions in the 0.024% to 0.130% range. It was found that the presence REs may accelerate the carburizing process. REs in steel and in the carburizer both facilitate the carbon supply on the surface so as to have more carbon diffusing into the samples [67].

Another microstructural characteristic that may affect the case performance is internal oxidation. During gas carburizing, because of the atmosphere used, oxygen is introduced onto the surface of steel parts, promoting oxide formation [68,69]. Figure 22 shows internal oxidation of SAE 8620 steel after gas carburizing. The thickness of this affected zone has influence on the mechanical properties, mainly on the short crack growth. Larger zone thickness cases hasten the growth.

The chemical composition of steel affects the properties of the carburized case, mainly fatigue properties. Elevated sulfur levels produce MnS that is detrimental because these inclusions act as stress concentrators promoting fatigue crack initiation. Phosphorus stabilizes cementite. Thus, it promotes intergranular fatigue crack initiation [69]. In this manner, wear mechanisms evolving microfatigue can be affected by high levels of sulfur and phosphorus.

Table 10 Retained Austenite Amounts
Obtained from Different Heat Treatments

Heat treatment	Retained austenite (%)
I	37
II	<6
III	23
IV	15
V	32

Source: Ref. 15.

Figure 22 Internal oxidation of the SAE 8620 steel after gas carburizing.

Recently, plasma carburizing using glow discharge technology has been industrially applied. This technique has permitted stainless steel carburizing because the surface oxide layer can be removed by ion sputtering. High surface hardness of AISI 316L stainless steel was obtained using plasma carburizing over 90 min at 800°C. The specimens have surface hardness above 735 HV_{50g} and an effective hardening depth above 40 μm. In this process, the cooling rate had little effect on the hardness profile of the carburized specimen. This is probably because carburized stainless steel is hardened by precipitation hardening, whereas low-alloy steels are hardened by martensite transformation during quenching [70].

B. Nitriding

Nitriding is an important method to improve the tribological properties of several components. In this method the surface of the steel is chemically altered. Nitrogen is a relatively small atom and dissolves interstitially in iron. Because ferrite cannot dissolve large amounts of nitrogen, it forms nitrides with the iron. The hardening is not associated with the formation of hard martensite but with the formation of nitrites that provoke strong increase of the hardness, promoting wear resistance and improving the fatigue properties of the component. Heat treatments such as quenching and tempering are used to adjust the bulk properties as required; they are usually done before the nitriding.

The nitriding process is considered a low-temperature method, because in steel the addition of nitrogen is made below the eutectic temperature. This is one advantage because processes using high temperature lead to quick softening of the matrix.

Nitriding can be done using a mixture of gaseous nitrogen, hydrogen, and ammonia; it can also be done in liquid salts. At present, the most popular method is ion nitriding, also called plasma nitriding, mainly because in this case the pollution is absent and the nitriding time is reduced.

Plasma source ion nitriding is being used as a low-temperature, low-pressure nitriding promoting ion implantation and its diffusion into iron and steel.

These processes can increase the microhardness, decrease the coefficient of friction, and improve wear resistance. During the process, the diffusion of nitrogen into the substrate is a function of the time and temperature. When the nitrogen solubility of the materials is exceeded, nitrides are formed; first the iron nitride and then other alloying element nitrides or carbonitrides are formed in the matrix of nitride layer. Thus, the composition of the steel has an influence on the results obtained and many types of iron-based alloys undergo plasma ion nitriding in the matrix of nitride layer.

Steels recommended for nitriding have elements present that are strong nitride formers, permitting the development of high hardness, as shown by studies performed by K. E. Thelning [52]. Figure 23 demonstrates the influence of alloying elements on the hardness after nitriding of steel.

Tool steels can have their wear performance improved by nitriding. AISI H13 steel is often nitrided before use as hot-metalworking dies. When traditional gas nitriding is performed, although the hardness obtained is high, the compound layer produced on the surface can lead to spalling during working. This layer is usually removed before nitriding, but there is a detrimental effect on the adhesive resistance. However, with plasma nitriding the thickness of the compound layer can be controlled, and for some special conditions it can be suppressed. When the compound layer thickness is increased, the wear rate also increases by spalling of this layer. Thus, the wear behavior depends on the thickness of this layer. Long-time plasma nitriding at 550°C decrease the core hardness but increases the total case depth. A wide range of time and temperatures available with plasma nitriding can be used to obtain the required material condition [71].

Low alloy tool steel (0.90C/1.5Si/0.50Mn/1.10Cr and Fel bal) was submitted to the plasma source ion nitriding process. Before the superficial treatment the samples were

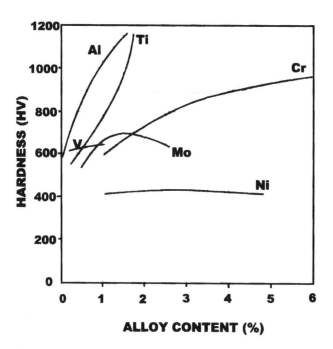

Figure 23 Alloy element additions effects on the hardness of a nitrided steel. (From Ref. 52.)

quenched and tempered to produce bulk hardness of 58–59 HRC. After nitriding at 250°C for 4 hr the superficial hardness increased by a factor of 2, showing positive effect on the wear resistance during the wear test, as demonstrated in Fig. 24. ε-Fe_2N phase on the outer surface with 1200 $HV_{0.25N}$ was responsible for the wear performance of the nitrided samples [72].

Similar tests were performed with other low-alloy tool steels (L7') with the composition 0.46C/0.50Mn/0.78 Si/2.5–2.7 Cr/ <0.01V and Fe bal. In the quenched and tempered condition, the samples showed values of microhardness of 560 $HV_{0.1}$ for the tempered martensite and 640 $HV_{0.1}$ for the network phase. After plasma nitriding at 520°C for 16 hr, the microstructure of the bulk changed because of the overtempering during nitriding treatment. Although the hardness of the case had become high, about 900 $HV_{0.1}$, hardness of the bulk decreased to about 380–400 $HV_{0.1}$, decreasing the lifetime of this tool. Otherwise under the same plasma nitriding conditions H13 and D2 tool steels improved their lifetime by 2 or 3 times, depending on the type of tool steel [73].

Because of the considerable increase in the use of stainless steels for engineering components, plasma nitriding in this kind of steel has become important. The conventional gas nitriding process is not efficient in stainless steel because the superficial oxide layers (Cr_2O_3) act as a barrier disturbing the nitrogen diffusion. However, in plasma nitriding the sputtering action removes all oxide layers before the actual plasma nitriding process starts [74].

Nitrogen is a well-known austenite-stabilizer and affects the matrix strength by solution strengthening mechanism. Thus, depending on the parameters used for the nitriding process, significant modifications can occur in stainless steels, mainly in austenitic stainless steel.

Plasma nitriding has been particularly important for austenitic stainless steels because despite showing excellent corrosion resistance, these steels have low hardness and therefore

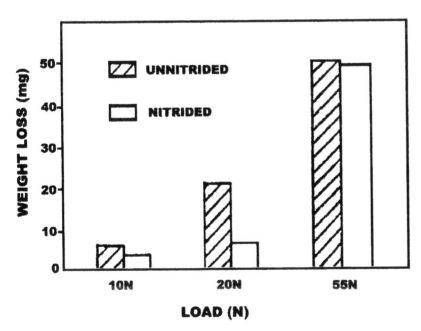

Figure 24 Wear resistance (weight loss) of an unnitrided and nitrided 9SiCr tool steel for different loads during wear test. (From Ref. 72.)

poor wear resistance leading to a short lifetime in industrial applications with intensive wear. These kinds of materials can be hardened by cold forming or using surface treatments. In nitriding above 500°C, the nitrided layer formed is accompanied by CrN precipitation causing decrease in corrosion resistance [75].

Investigations made with 304 stainless steel show different characteristics of nitrided layers at different nitriding parameters. Wear tests results (depth scar) were obtained from treated samples and samples without nitriding for the best parameters. Results demonstrate the beneficial effect of the plasma nitriding, increasing the microhardness and consequently the wear resistance [76,77]. Figure 25 shows these test results.

Similar investigations were performed in the same steel using pulsed-d.c. plasma with different N_2–H_2 gas mixtures and treatment times at a temperature range of 375–475°C [78]. The microhardness is increased because of the formation of a new phase and the wear is strongly reduced even at higher loads. Other mechanical properties were improved, as well as the fatigue strength.

The new nitrogen-rich phase obtained by the interaction between nitrogen and austenite at temperatures below 500°C has been called S phase, ε' phase, m phase, or expanded austenite. It is formed by the incorporation of nitrogen into interstitial positions in the fcc structure of stainless steel. Expanded austenite characterization can be found in the literature [75,79,80].

C. Boriding

Boriding is a thermochemical process in which boron atoms diffuse into a metal surface to form hard boride layers, which can exceed 2000 HV. These very hard borides may be used to protect against wear, mainly abrasive and adhesive wear.

The borided layer has good resistance to heat up to 1000°C; its resistance to heat is excellent after 650°C because oxidation on the borided surface produces protective effect on metal–metal contact, decreasing the friction coefficient as well [81–83].

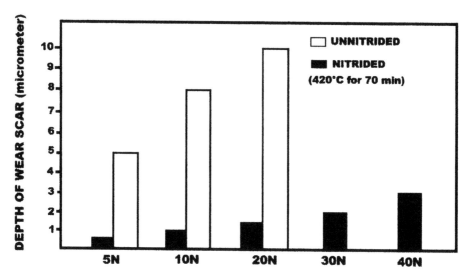

Figure 25 Depth of wear scar on AISI 304 specimen nitrided compared with unnitrided samples. (From Ref. 77.)

Solids, liquids, and gases are media for diffusion boron, but power-pack borizing is the most used because it can produce practical and economic results [84].

The structure of the borided layer depends on the boronizing type, temperature, and time, which are related to the activity of the boron-carrying agent. Two phases can be identified in this layer, one rich in boron, FeB, and the other with low boron content, Fe_2B [81].

The FeB phase is brittle and forms a surface that is under high tensile stress. It can be removed by diffusion annealing. The Fe_2B phase is preferred because it is less brittle and it forms a surface with high compressive stress [85–87].

Borides grow as columnar morphology during treating of low-alloyed steels. This particular morphology improves the adherence of the layer to the matrix. The diffusion layer is composed of the phases (Fe, M)B and $(Fe, M)_2B$ (M = Cr, Ni, Mn, etc.), which are solid solutions, derived from FeB and Fe_2B by partial substitution of iron with metallic atoms [88].

Boriding process can be successfully applied to steels; cast iron and nonferrous alloys may be borided as well [85] and the resulting properties, e.g., thickness layer, may vary with the chemical composition and with the process (temperature and time). Figure 26 illustrates the variation of depth boriding with the type of steel and with the time process. The temperature and time required depend on the material to be boronized and the thickness of the boride layer. Process cycle times range from one-quarter hour to over 10 hr.

Borizing is usually performed at 850° to 950°C and because of this hardening and tempering before the borizing are not done. Heat treatment before borizing can include stress relief annealing, which is useful for work pieces in which considerable surface stress may cause problems such as distortion or the boride layer to flake off after borizing. Usually

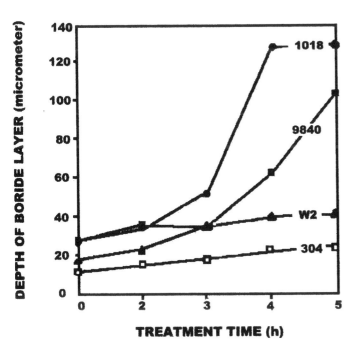

Figure 26   Influence of treatment time on the depth of boride layer at 920°C. (From Ref. 87.)

it is done at about 600°C for 1 hr. After borizing, diffusion annealing can be done to remove FeB, which forms only in alloy steel. Hardening and tempering are usually performed directly after borizing at the borizing temperature or after cooling. During hardening there is small boron diffusion to below the borided layer, but there is a carbon concentration in the same region as well. Therefore, it is possible to obtain a martensitic structure below the borided case even when using oil quenchant or martempering [81,84,85].

Wear performance of the borided steels depends on several factors such as the diffusion coefficient of the boron whose value decreases inversely with the alloying element content, which in turn restricts the layer growth. Process parameters may have influence on the wear behavior as well [87]. Figure 27 presents weight loss values for different borizing times.

Borided steels are extremely resistant to abrasion because of their great hardness. Borided layer thickness varies with applications from 20 to 300 μm and increases the service life by severalfold, showing better performance when it is compared with different coatings [81]. Results from abrasive wear test using 42CrMo4 steel are illustrated in Fig. 28.

Although plasma borizing has been studied for 20 years, only pack borizing has been used widely on a commercial basis. Investigations on plasma borizing were performed at 750°C using BCl_3 as precursor. Duplex layers are formed during plasma borizing, and the compound layer that is almost pore-free. This is one advantage of plasma borizing compared with other processes, promoting in the near future the growing application of plasma borizing [89,90].

D. Combined Processes

Surface modifications are highly desirable to solve the problems involving friction and wear.

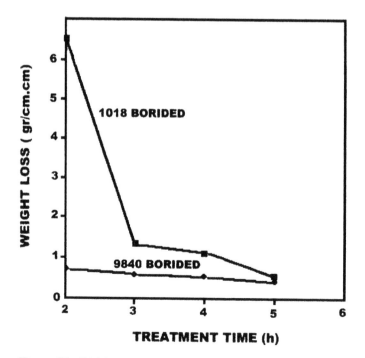

Figure 27   Weight loss obtained in wear test with different boriding times. (From Ref. 87.)

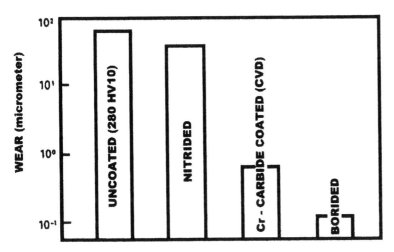

Figure 28 Results from abrasive wear of 42CrMo2 steel with different coatings (steel against alumina abrasive paper, grain size: P220, t = 3 min, P = 0.019 N/mm^2). (From Ref. 81.)

Nitrocarburizing is a thermochemical process that consists in diffusing nitrogen and carbon into the surface of ferrous materials, usually at 550–580°C, and can be done through different media: solids, liquids, gases, or gas glow discharge. It is applied widely to various materials such as carbon steels, alloy steels, tool steels, stainless steels, cast iron, and sintered materials [91,92]. In this process the case produced is formed by one compound layer (ε and/or γ' (Fe$_4$N) phases) and one diffusion zone where N and C are dissolved interstitially in the ferritic matrix. The former case is responsible for the wear resistance and the diffusion zone leads to improvement of fatigue properties [91].

Other treatments were developed to simultaneously improve the mechanical properties and corrosion resistance of nitrocarburized layers.

Among them, there is the nitrocarburizing-quenched duplex treatment. The surface of the steel part in this process is very different from that observed on the nitrocarburized specimen, because after the usual nitrocarburizing at about 570°C and air cooling the steel part is reheated to the austenitic region. The compound layer begins to decompose and the nitrogen and carbon penetrate into the austenite. After cooling, the martensite formed contained nitrogen and carbon along with retained austenite. Great changes have taken place in the microstructure of the surface and core and grain refinement is observed. The diffusion zone changes into fine acicular martensite and retained austenite and the core is mixed martensite. The process promotes characteristics of both nitrocarburizing in the surface and quenching in the core, which ensures that specimens will have higher surface hardness and a superior hardness profile. Investigations with AISI 1045 steel show best wear performance of the nitrocarburizing–quenching duplex when compared with the samples only nitrocarburized [93].

Nitrocarburizing followed by oxidation is another treatment combination that has been widely used. Investigations using AISI 1045 steel show comparative performance of these kinds of treatments.

Nitrocarburizing treatment before induction hardening is a combined process to find good abrasion and antiwear properties (obtained by induction hardening) and better corrosion qualities and antiscuffing properties (obtained by nitrocarburizing) [94]. Hardness results from this combined process and each one separated process was obtained using SAE 1010/3115/1040/4140 steels. In all steels there was increase in the hardness values in the samples that underwent the combined process. Thus it can be supposed that they have better mechanical properties.

VII. HEAT TREATMENT OF CAST IRON

Although the heat treatment of steels has been emphasized, cast iron is a special class of ferrous alloys with innumerous wear-resistant applications. Thus heat treatments are very useful for cast irons.

Tempering and quenching are used to increase wear resistance and other mechanical properties. Complex microstructures of cast iron, mainly gray cast iron with lamellar graphite, may act as stress concentrator during the heat treatment. Metallurgical reactions during the austenitizing are more complex for cast iron than for steel. So, the temperature control during the heat treatment is important as well. Martempering and oil as quenchant are frequently used whereas water quenchant is rare.

As in the case of steel, cast iron must be tempered after hardening. The effects of the tempering on the cast iron matrix are similar to that on steel [95].

Perlitic matrix is preferred to ferrite and graphite as the prior microstructure before hardening because the wear resistance and hardness after tempering and quenching are the best.

White cast iron is specially used in the mining industry, for grinding balls, rolling mills, earthmoving equipment, shredder hammer tips of sugar cane, etc. These applications require hardness and a very high abrasion wear resistance. Chemical compositions can contain high Cr and C and other hard carbide forming elements.

Microstructures as cast of this type of cast iron may present retained austenite and diffusion phases, usually perlite. The volume of those microstructures is related to Cr/C proportion, other elements added, and the presence of ledeburite and M_7C_3, M_3C (duplex carbides) [96].

Martensitic microstructures in white cast iron can be obtained by new austenitizing (considering the perlitic regions) and by austenite unstabilization (considering the retained austenite regions). Retained austenite after quenching can be transformed by heating at 450–550°C and by subzero treatment [97].

Large amounts of retained austenite are obtained when high temperatures are used dissolving large amounts of Cr and C in the matrix. On the other hand, if low temperatures are used, low-carbon martensite or ferrite can be transformed. Between these extremes, the 850–1100°C range is more usual for high Cr and high C white cast iron [97].

Elements such as Mn, Ni, Cu, and Mo are added to increase the hardenability. However, they reduce the Ms temperature, increasing retained austenite contents. In some cases they are intentionally added to promote stable austenitic matrix [98].

When the white cast iron microstructure is composed of continuous matrix (austenitic or martensitic) with M_7C_3 dispersed carbides, the best abrasive wear resistance can be achieved. Complex microstructures with spheroidized carbides and perlite do not have wear performance as good as the former microstructure [99].

Wear tests performed in different tribosystems demonstrated large mass losses due to microcracking of the carbides when the matrix was perlitic or austenitic. In hard matrix containing martensite the cracking is not so intense and the wear resistance becomes high.

Increase of hardness after heat treatment obtained in a high-Cr white cast iron (Nb alloyed) is discussed in Ref. 100. Austenitizing at 1100°C followed by subzero treatment and tempering at 450°C increased the hardness from 50 to 64 HRC. Microstructures formed by

M_7C_3 carbides and MC carbides dispersed in the martensitic matrix (Fig. 29) showed higher abrasion wear resistance than M_7C_3 continuous carbides in austenitic matrix [100].

Carbide precipitation in martensitic matrix can have a positive effect on the wear resistance. Additions of 0.38% Ti in a 15Cr–3Mo white cast iron increased the wear resistance by 30% through precipitation of alloy carbides [101].

Volume, distribution, and shape carbides are very important in wear performance. Small Bo additions can modify carbide morphology, providing white cast iron more toughness and wear resistance [102].

Carbide eutectic as network can undergo microcracking under impact-abrasion conditions, whereas carbides dispersed in a matrix can work better when wear resistance and toughness are required [103].

Usually, high M_7C_3 carbide percentage in white cast iron is not beneficial to wear characteristics. Although this can be observed, when carbide orientation with regard to the worn surface is achieved, abrasion wear may be improved. Carbide orientation can be achieved during solidification process by cooling control. Best wear resistance is shown when carbides are parallel with regard to worn surface. Hardness of carbide is affected by carbide orientation as well. Longitudinal hardness was 18% greater than transversal plan [104]. Similar results were obtained with Cr–Mo white cast iron. M_7C_3 carbides showed 1444 $HV_{0.05}$ in the parallel direction of growth and 950 $HV_{0.05}$ in the perpendicular direction [105].

Martensitic ductile iron is preferred when toughness is important in the wear process. Several different microstructures in ductile cast iron that have strong influence on the wear behavior may be obtained through heat treatment [106].

Austempering of ductile iron has been one of the most recent developments. Control of chemical composition, amount of perlite, and graphite shape allows austempering heat

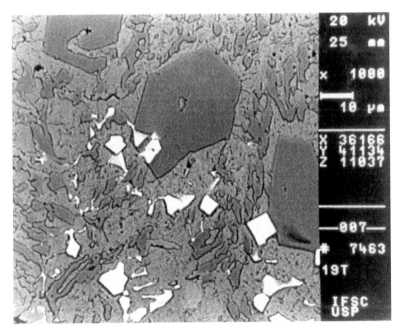

Figure 29 High-Cr white cast iron microstructure. Martensitic matrix with M_7C_3 and MC carbides dispersed. (Ref. 100.)

treatment to be made in a temperature range that is above perlitic transformation and below martensitic transformation. In this case the microstructure formed is called ausferrite. Austempered ductile iron (ADI) with hardness between 30 and 40 HRC has wear resistance similar to tempered and quenched steel at 60 HRC. In some wear processes without lubrication (dry abrasion) ADI components show better performance than tempered and quenched ductile iron, mainly for high loads. These high loads provoke deformation in the ADI component surface, increasing surface hardness [107].

Field tests with bucket tips made from ADI have demonstrated about 35% more abrasion resistance than the same component made from tempered and quenched low-carbon steel [108].

Ausferrite, which is a duplex ferritic–austenitic microstructure, has excellent mechanical properties including wear resistance, which is due to the strain-induced transformation of unreacted high-carbon austenite. This change occurs when the ADI surface is under deformation, provoking significant increase of the surface hardness and keeping the bulk ductile. Such phenomenon allows continuous maintenance of high surface hardness insofar as the worn layer is removed [109]. Figure 30 shows the hardness profile in an ADI sample under abrasion test.

The austempering temperature through microstructural changes mainly determines mechanical property variations. Usually a high austempering temperature produces a strong casting with excellent ductility and dynamic properties, whereas a low temperature results in a part with very high strength and wear resistance. The difference between low and high austempering temperature can be about 50°C. Time is another parameter that should be controlled. The presence of martensite is related to short times [109].

Austenitizing temperatures also affect the structure and properties of the austempered casting. A higher carbon content of the austenite, obtained when high austenitizing temperatures are used, increases the hardenability and stability affecting the austempering reaction [109].

Austempered ductile iron wear resistance can be improved by chilling during solidification. In this case eutectic carbides are together with the austempered structure. Ductile

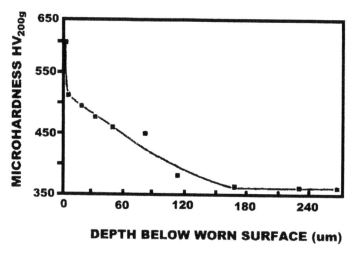

Figure 30   Microhardness profile of an abraded ADI sample. (From Ref. 109.)

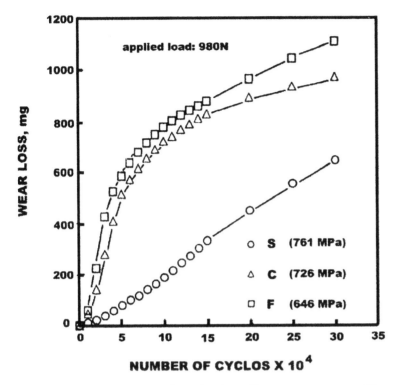

Figure 31   Wear loss results from dry slip–rolling contact wear test. (From Ref. 112.)

cast iron with Mo and Ni was processed under these conditions and the wear resistance increased without large toughness reduction [110].

Although ADI microstructure looks good for some wear types, in the case of rolling contact fatigue resistance the same is not observed. Nucleation crack resistance is low (graphite are nucleus) whereas the behavior of crack propagation is fine [111].

As the graphite can be a crack nucleus, its shape is very important. Three kinds of ADI samples were studied varying the graphite shapes [from spherical (S) to compact vermicular(C) and to flake (F)] by changing the amount of graphite spheroidizing agent. Dry slip rolling wear test results are shown in Fig. 31. Severe test conditions provoke nucleation of cracks that propagate and connect to each other to produce pitting or spalling traces. Spherical shapes (S) have better behavior because the matrix continuity is larger in this condition [112].

Other heat treatments may also be used in cast iron to increase hardness such as surface hardening by induction or flame, nitrating, and others.

REFERENCES

1. Eyre, T.S. Wear characteristics of metals. Tribol. Int. 1976, 6, 203–212.
2. Bradley, A.; Bingley, M.S.; Pitmann, A.N. Abrasive wear of steels in handling of bulk particulates: an appraisal of wall friction measurement as an indicator of wear rate. Wear 2000, 243, 25–30.

3. Zum Gahr, K.H. *Microstructure and Wear of Materials;* Elsevier: Amsterdam, 1987; Tribology Series 10, 560.

4. Larsen-Basse, J. The abrasion resistance of some hardened and tempered carbon steels. Trans. Metall. Soc. AIME 1966, *336*, 1461–1466.

5. Moore, M.A. The relationship between the abrasive wear resistance, hardness and microstructure of ferritic materials. Wear 1974, *28*, 59–68.

6. Mutton, P.J.; Watson, J.D. Some effects of microstructure on the abrasion resistance of metals. Wear 1978, *48*, 385–398.

7. Bhansali, K.; Miller, A.E. Resistance of pure metals to low stress abrasive wear. Wear 1981, *71*, 375–379.

8. Gore, G.J.; Gates, J.D. Effect of hardness on three very different forms of wear. Wear 1997, *203–204*, 544–563.

9. Al-Rubaine, K.S. Equivalent hardness concept of two-body abrasion of iron-base alloys. Wear 2000, *243*, 92–100.

10. Axen, N.; Jacobson, S. A model for the abrasive wear resistance of multhiphase materials. Wear 1994, *174*, 187–199.

11. Moore, M.A.; Douthwaite, R.M. Plastic deformation below worn surfaces. Metall. Trans. A 1976, *7*, 1833–1839.

12. Alpas, A.T.; Hu, H.; Zhang, J. Plastic deformation and damage accumulation below worn surfaces. Wear 1993, *162–164*, 188–195.

13. Kalousek, J.; Fegredo, K.M.; Laufer, E.E. The wear resistance on worn metallography of perlite, bainite and tempered martensite rail steel structures of high hardness. Wear 1985, *105*, 199–222.

14. Tianfu, J.; Fucheng, Z. The work-hardening behavior of medium manganese steel under impact abrasive wear condition. Mater. Lett. 1997, *31*, 275–279.

15. Silva, V.F.; Canale, L.F.; Spinelli, D.; Bose, W.W.; Crnkovic, O.R. Influence of retained austenite on short fatigue crack growth and wear resistance of case carburized steel. J. Mater. Eng. Perform. 1999, *8*, 543–548.

16. Wang, Y.; Lei, T.; Liu, J. Tribo-metallographic behavior of high carbon steels in dry sliding. III. Dynamic microstructural changes and wear. Wear 1999, *231*, 20–37.

17. Wang, Y.; Lei, T.; Liu, J. Tribo-metallographic behavior of high carbon steels in dry sliding. II. Microstructure and wear. Wear 1999, *231*, 12–19.

18. American Society for Metals. *Metals Handbook*; 9th Ed.; Heat Treating; ASM International: Metals Park, OH, 1981; Vol. 4.

19. Krauss, G. *Steels: Heat Treatment and Processing Principles*. ASM International; Materials Park, OH.

20. Shimizu, K.; Nishiama, H. Electron microscopic studies of martensitic transformation in iron alloys and steels. Metall. Trans., A 1972, *3*, 1055–1066.

21. Krauss, G. Understanding the heat treatment, microstructure and performance of high-strength hardened steels. Proceedings of 1st ASM International Automotive Heat Treating Conference, Puerto Vallarta, Mexico; ASM International: Metals Park, OH, 1998; 125–133.

22. Brooks, C.R. *Heat Treatment of Ferrous Alloys*; Hemisphere Publishing Corporation/McGraw-Hill Book Company: New York, 1979.

23. Farhat, Z.N.; Ding, Y.; Alpas, A.T.; Northwood, D.O. The processing and testing of new and advanced materials for wear resistance surface coatings. J. Mater. Process. Technol. 1997, *63*, 859–864.

24. Krauss, G. Martensite in steel: strength and structure. Mater. Sci. Eng., A 1999, *273–275*, 40–57.

25. Dobrzanski, L.A. Structure and properties of high-speed steels with wear resistant cases or coating. J. Mater. Process. Technol. 2001, *109*, 44–51.

26. Ambrosio, F.; Nogueiro, R.A.; Nevei, M.D.M.; Limo, L.F.C.P. Influence of heat treatment on microstructures of conventional and sintered AISI M2 high speed steel. Proceedings of the 20th ASM Heat Treating Society Conference, St. Louis, MO, ASM International: Metals Park, OH, 2000; 1006–1009.

27. Ozaki, K.; Urita, T.; Matsuda, Y.; Koga, H. Austenitization and carbide dissolution of high speed tool steel AISI-M2 in short time heating. Proceedings of the 20th ASM Heat Treating Society Conference, St. Louis, MO, ASM International: Metals Park, OH, 2000; 1046–1050.

28. Lee, W.S.; Su, T.T. Mechanical properties and microstructural features of AISI 4340 high-strength alloy steel under quenched and tempered conditions. J. Mater. Process. Technol. 1999, 87, 198–206.

29. Altstetter, C.J.; Cohen, M.; Averbach, B.L. Effect of Silicon on tempering of AISI 43xx steels, A.S.M. Trans 1962, 55, 287–300.

30. Horn, R.M.; Ritchie, R.O. Metall. Trans., A 1978, 9A, 439–443.

31. Nam, W.J.; Choi, H.C. Effect of Si on mechanical properties of low allow steels. Mater. Sci. Technol. 1999, 15, 527–530.

32. Nam, W.J.; Choi, H.C. Effect of silicon, nickel and vanadium on impact toughness in spring steels. Mater. Sci. Technol. 1997, 13, 568–574.

33. Pan, F.; Ding, P.; Zhou, S.; Kang, M.; Edmonds, D.V. Effects of silicon additions on the mechanical properties and microstructure of high speed steels. Acta. Mater. 1997, 45, 4703–4712.

34. Nam, W.J.; Lee, C.S.; Ban, D.Y. Effects of the alloy additions and the tempering temperature on the sag resistance of Si–Cr spring steels. Mater. Sci. Eng., A 2000, 289, 8–17.

35. Wilson, R. *Metallurgy and Heat Treatment of Tool Steels*; 3rd Ed.; McGraw-Hill Book Company: London, 1975.

36. Dobrzanski, L.A.; Zarychta, A.; Ligarski, M. High-speed steels with addition of niobium or titanium. J. Mater. Process. Technol. 1997, 63, 531–541.

37. Dobrzanski, L.A.; Kasprzak, W. The influence of 5% cobalt on structure and working properties of 9-2-2-5, 11-2-2-5 and 11-0-2-5 high-speed steels. J. Mater. Process. Technol. 2001, 109, 52–64.

38. Luo, X.; Cheng, X.; Jiang, F. Heat treating and mechanical properties of 5Cr8Mo2WSiV and its application to wood-chip cutting knives. Proceedings of the 20th ASM Heat Treating Society Conference, St. Louis, MO, ASM International: Metals Park, OH, 2000; 272–277.

39. Lau, K.H.; Mei, D.; Yeung, C.F.; Man, H.C. Wear characteristics and mechanisms of a thin edge cutting blade. J. Mater. Process. Technol. 2000, 102, 203–207.

40. Boyd, J.D.; Malis, T.F.; Nadkami, G. Tempering of micro alloyed spring steels, Proceedings of the SPEICH Symposium; ISS–AIME: Warrendale, PA, 1992; 276 pp.

41. Dobrzanski, L.A. Structure and properties of high-speed steels with wear resistant cases or coating. J. Mater. Process. Technol. 2001, 109, 44–51.

42. Krauss, G. Tempered Martensite Embrittlement in AISI 4340 steel. Metall. Trans., A 1979, 10 a(11), 1643–1649.

43. Song, S.H.; Faulkner, R.G.; Flewitt, P.E.J. Quenched and tempering-induced molybdenum segregation to grain boundary in a 2.25Cr1Mo steel. Mater. Sci. Eng., A 2000, 281, 23–27.

44. Raoul, S.; Marini, B.; Pineau, A. Effect of microstructure on the susceptibility of a 533 steel to temper embrittlement. J. Nucl. Mater. 1998, 257, 199–205.

45. Totten, G.E.; Sun, Y.H.; Kobasco, N.I.; Arunov, M.A. Intensive quenching practices for automotive parts production—an overview, Proceedings of the ASM 1st International Automotive Heat Treating Conference, Puerto Vallarta, Mexico, ASM International: Metals Park, OH, 1998; 99–105.

46. Aronov, M.A.; Kobasko, N.I.; Powell, J. Practical application of intensive quenching process for steel parts, Proceedings of the 12th International Federation for Heat Treatment and Surface Engineering Congress, Melbourne, Australia, The Institution of Engineers: Australia, 1999; 51–57.

47. Kobasco, N.I. Three types of intensive water quenching and their future applications, Proceedings of the 20th ASM Heat Treating Society Conference, St. Louis, MO, ASM International: Metals Park, OH, 448–454.

48. Aronov, M.A.; Kobasco, N.I.; Powell, J.A. Practical application of intensive quenching process for steel parts. Proceedings of the 20th ASM Heat Treating Society Conference, St. Louis, MO, ASM International: Metals Park, OH, 2000; 778–784.

49. Liss, R.B.; Massieon, C.G.; McKloskey, A.S. The Development of Heat Treat Stresses and Their Effect on Fatigue Strength of Hardened Steels; SAE Midyear Meeting: Chicago, 1965.

50. Bertini, L.; Fontanari, V. Fatigue behaviour of induction hardened notched components. Int. J. Fatigue, ASM International: Metals Park, OH, 1999, *21*, 611–617.

51. Rudvei, V.I.; Loveless, D.; Marshall, B. Gear heat treatment by induction, Proceedings of the 20th ASM Heat Treating Society Conference; St. Louis, MO, ASM International: Metals Park, OH, 2000; 862–871.

52. Brooks, C.R. Chapter 2: Surface treatments involving only thermal processing. *Principles of the Surface Treatment of Steels*; Technomic Publishing Company, Inc.: Lancaster, 1992; 3–197.

53. Xu, D.; Kuang, Z.B. A study on a distribution of residual stress due to surface induction hardening. J. Eng. Mater. Technol. 1986, *118*, 571–575.

54. Siebert, C.A.; Doane, D.V.; Breen, D.H. The Hardenability of Steels—Concepts, Metallurgical Influences and Industrial Application. Metals Park, OH: ASM, 176–211.

55. Stevens, N. Induction heating and tempering. *ASM Metals Handbook*; 1984, Vol. 8, 451–483.

56. Rivas, A.L.; Michal, G.M. Microstructural evolution during induction hardening heat treatment of a vanadium micro alloyed steel. Mat. Sci. Forum 1998, *284–286*, 403–410.

57. Devgun, M.S.; Molian, P.A. Experimental study of laser heat treated bearing steel. J. Mater. Process. Technol. 1990, *23*, 41–54.

58. Yang, L.J.; Jana, S.; Tam, S.C. Laser transformation hardening of tool-steel specimens. J. Mater. Process. Technol. 1990, *21*, 119–130.

59. Selvan, J.S.; Subramanian, K.; Nath, A.K. Effect of laser surface hardening on En18 (AISI 5135) steel. J. Mater. Process. Technol. 1999, *91*, 29–36.

60. Qingbin, L.; Hong, L. Experimental study of the laser quenching of 40CrNiMoA steel. J. Mater. Process. Technol. 1999, *88*, 77–82.

61. Krauss, G. Microstructure and performance of carburized steel. Part I: Martensite. Adv. Mater. Process. 1995, *147*, 40BB–40Y.

62. Krauss, G. Microstructure and performance of carburized steel. Part III: Austenite and fatigue. Adv. Mater. Process. 1995, *148*, 42EE–42II.

63. Krauss, G. Microstructure and performance of carburized steel. Part II: Austenite. Adv. Mater. Process. 1995, *148*, 48U–48Y.

64. Kim, H.J.; Kweon, Y.G. The effects of retained austenite on dry sliding wear behavior of carburized steels. Wear 1996, *193*, 8–15.

65. Krauss, G. Microstructure of carburized steels. Heat Treat. 1993, *25*, 12–15.

66. Islam, A.; Bepari, M.A. Structure and properties of carburized and hardened niobium micro-alloyed steels. J. Mater. Process. Technol. 1998, *74*, 183–189.

67. Yuan, Z.X.; Yu, Z.S.; Tan, P.; Song, S.H. Effect of rare earths on the carburizing of steel. Mater. Sci. Eng., A 1999, *267*, 162–166.

68. Krauss, G. Microstructure and performance of carburized steel. Part IV: Oxidation and inclusions. Adv. Mater. Process. 1995, *148* (6), 36Z–36DD.

69. Wise, J.P.; Krauss, G.; Matlock, D.K. Microstructure and fatigue resistance of carburized steels, Proceedings of the 20th ASM Heat Treating Society Conference; St. Louis, MO, ASM International: Materials Park, OH, 2000; 1152–1161 pp.

70. Suh, B.S.; Lee, W.J. Surface hardening of AISI 316L stainless steel using plasma carburizing. Thin Solid Films 1997, *295*, 185–192.

71. Karamis, M.B. An investigation of the properties and wear behavior of plasma-nitrided hot-working steel(H13). Wear 1991, *150*, 331–342.

72. Lei, M.K.; Wang, P.; Huang, Y.; Yu, Z.W.; Yu, N.L.J.; Zhang, Z.L. Tribological studies of plasma source ion nitrided low alloy tool steel. Wear 1997, *209*, 301–307.

73. Devi, M.U.; Chakraborty, T.K.; Mohanty, O.N. Wear behavior of plasma nitrided tool steels. Surf. Coat. Technol. 1999, *116–119*, 212–221.

74. Sun, Y.; Bell, T.; Wood, G. Wear behavior of plasma-nitrided martensitic stainless steel. Wear 1994, *178*, 131–138.

75. Xu, X.L.; Wang, L.; Yu, Z.W.; Hei, Z.K. Microstructural characterization of plasma nitrided austenitic stainless steel. Surf. Coat. Technol. 2000, *132*, 270–274.

76. Liang, W.; Bin, X.; Zhiwei, Y.; Yaqin, S. The wear and corrosion properties of stainless steel, nitrided by low-pressure plasma-arc source ion nitriding at low temperature. Surf. Coat. Technol. 2000, *130*, 304–308.

77. Zhu, X.; Huang, H.; Xu, K.; He, J. Structure and properties of plasma nitrided austenitic stainless steel, Proceedings of the 20th ASM Heat Treating Society Conference; St. Louis, MO, 2000; 217–221.

78. Menthe, E.; Bulak, A.; Olfe, J.; Zimmermann, A.; Rie, K.T. Improvement of the mechanical properties of austenitic stainless steel after plasma nitriding. Surf. Coat. Technol. 2000, *133–134*, 259–263.

79. Picard, S.; Memet, J.B.; Sabot, R.; Grosseau-Poussard, J.L.; Rivièri, J.P.; Meilland, R. Corrosion behavior, microhardness and surface characterization of low energy, high current ion implanted austenitic stainless steel. Surf. Coat. Technol. 2001, *303*, 163–172.

80. Fewel, M.P.; Mitchell, D.R.G.; Priest, J.M.; Short, K.T.; Collins, G.A. The nature of expanded austenite. Surf. Coat. Technol. 2000, *131*, 300–306.

81. Hunger, H.J.; Trute, G. Boronizing to produce wear-resistance surface layers. Heat Treat. Met. 1994, *2*, 31–39.

82. Quinn, T.F.J. Oxidation wear modelling. Wear 1992, *153*, 179–180.

83. Selçuk, B.; Ipek, R.; Karamis, M.B.; Kuzucu, V. An investigation on surface properties of treated low carbon and alloyed steels boriding and carburizing. J. Mater. Process. Technol. 2000, *103*, 310–317.

84. Pohlmann, K. Boretação em metais ferrosos. XXXI Congresso Anual da ABM—Ouro Preto e Belo Horizonte, ABM: Saõ Paulo-Brazil, Brazil, 1976; 232 pp.

85. Stewart, K. Boronizing protects metals against wear. Adv. Mater. Process. 1997, *3*, 23–25.

86. Capan, L.; Alnipak, B. Erosion-corrosion resistance of borided steels. Mat. Sci. Forum 1994, *163–165*, 329–334.

87. Melendez, E.; Campos, I.; Rocha, E.; Barron, M.A. Structural and strength characterization of steel subject to boriding thermochemical process. Mater. Sci. Eng., A 1997, *234–236*, 900–903.

88. Özmen, Y.; Can, A.C. An effective method to obtain reasonably hard surface on the steel X210 Cr12 (D3): Boriding. Mat. Sci. Forum 1994, *163–165*, 335–340.

89. Rie, K.T. Recent advances in plasma diffusion processes. Surf. Coat. Technol. 1999, *112*, 52–62.

90. Küper, A.; Stock, H.R.; Mayer, P. Plasma-assisted boronizing using trimethyl borate, Proceedings of the 12th International Federation for Heat Treatment and Surface Engineering Congress; Melbourne, Australia, The Institution of Engineers: Australia, 2000; 177–183.

91. Qiang, Y.H.; Ge, S.R.; Xue, Q.J. Sliding wear behavior of nitrocarburized bearing steel. Mater. Sci. Eng. A278, 261–266.

92. Nicoletto, G.; Tucci, A.; Esposito, L. Sliding wear behavior of nitrided and nitrocarburized steel. Wear 1996, *196*, 38–44.

93. Qiang, Y.H.; Ge, S.R.; Xue, Q.J. Microstructure and tribological behavior of nitrocarburizing-quenched duplex treated steel. Tribol. Int. 1999, *32*, 131–136.

94. Halabi, J. Improvement of mechanical properties combining ferritic nitrocarburizing and induction hardening, Proceedings of the ASM 1st International Automotive Heat Treating Conference; Puerto Vallarta, Mexico, ASM International: Materials Park, OH, 1998; 201–208.

95. Senai. Depto Regional de Minas Gerais. *Ferros Fundidos de Grafita Esferoidal*; 2a Ed. Publicações Técnicas—Fundição 5: SENAI, Senai MG: Belo Horizonte, 1987.

96. Sinatora, A.; Albertin, E. Considerações técnicas e econômicas sobre a fabricação e utilização de bolas de moinho fundidas. *Seminário sobre Materiais Resistentes ao Desgaste, 2, Uberlândia*, ABM: São Paulo, Brazil, 1991; 85–122.

97. Pattyn, R.L. Tratamento térmico de ferros fundidos brancos de alto cromo. Fundição e Serviç os, 1996, *Ano VII* (4b), 47–57.

98. Maratray, F.; Usseglio-Nanot, R. *Atlas: Courbes de Transformation de Fontes Blanches au Chrome et au Chrome–Molybdéne*; Climax Molybdenum S/A: Paris, 1970; 198 pp.
99. Guesser, W.L.; Costa, P.H.C.; Pieske, A. Nióbio em ferros fundidos brancos ligados ao cromo para aplicações em desgaste abrasivo. Metalurgia-ABM 1989, *45*, 768–776.
100. Farah, A.F. Desenvolvimento de Uma Liga de Ferro Fundido Branco Alto Cromo com Nióbio Tratada Termicamente para Resistência ao Desgaste Abrasivo. Tese (mestrado), São Carlos, EESC—Universidade de São Paulo, SP, Brasil, 1997.
101. Arikan, M.M.; Cimenoglu, H.; Kayahi, E.S. The effect of titanium on the abrasion resistance of 15 Cr–3 Mo white cast iron. Wear 2001, *247*, 231–235.
102. Izciler, M.; Celik, H. Two- and three-body abrasive wear behavior of different heat treated boron alloyed high chromium cast iron grinding balls. J. Mater. Process. Technol. 2000, *105*, 237–245.
103. Qian, M.; Chaochang, W. Impact-abrasion behavior of low alloy white cast irons. Wear 1997, *209*, 308–315.
104. Dõgan, O.N.; Hawk, J.A. Effect of carbide orientation on abrasion of high Cr white cast iron. Wear 1995, *189*, 136–142.
105. Maratray, F.; Usseglio-Namot, R. *Factors Affecting the Structure of Chromium–molybdenum White Irons*; Climax Molybdenum Co, 1970; 32 pp.
106. Luo, Q.; Xie, J.; Song, Y. Effects of microstructures on the abrasive wear behavior of spheroidal cast iron. Wear 1995, *1184*, 1–10.
107. Hasseb, A.S.M.A.; Islam, Md.A.; AliBepari, Md.M. Tribological behavior of quenched and tempered and austempered ductile iron at the same hardness. Wear 2000, *244*, 15–19.
108. Dommarco, R.; Galarreta, I.; Ortiz, H.; Daid, P.; Maghieri, G. The use of ductile iron for wheel bucket tips. Wear 2001, *249*, 101–108.
109. Trudel, A.; Gagné, M. Effects of composition and heat treatment parameters on the characteristics of austempered ductile irons. Metall. Q. 1997, *36*, 289–298.
110. Hemanth, J. Wear characteristics of austempered chilled ductile iron. Mater. Des. 2000, *21*, 139–148.
111. Dommarco, R.C.; Bastias, P.C.; Dael'O, H.A.; Hahn, G.T.; Rubin, C.A. Rolling contact fatigue (RCF) resistance austempered ductile iron (ADI). Wear 1998, *221*, 69–74.
112. Hatate, M.; Shiota, T.; Takahashi, N.; Shimizu, K. Influences of graphite shapes on wear characteristics of austempered cast iron. Wear 2001, *251*, 885–889.

15
Plasma-Based Surface Modification

Paul K. Chu and Xiubo Tian
City University of Hong Kong, Kowloon, Hong Kong

I. INTRODUCTION

Plasma surface modification has attracted much attention as both a research and production technology. Plasma nitriding is a relatively established technique in the metal industry, but new applications such as plasma nitriding of aluminum are being refined. Recently, the use of metal plasmas generated by vacuum arc plasma sources and gaseous plasmas in concert has greatly enhanced the versatility of the plasma implantation–deposition technique. It is fair to say that a single process most likely cannot satisfy all the requirements, and multiple processes may be necessary. For example, a titanium nitride coating fabricated on a plasma-nitrided surface by physical vapor deposition (PVD) exhibits more superior wear resistance. In this chapter, plasma-based surface modification techniques including plasma nitriding, plasma implantation, plasma-assisted vapor deposition (PAVD), and hybrid processes are discussed with the focus on the enhancement of tribological properties of materials and industrial components. The advantages, limitations, and typical applications of each technique are described.

II. PLASMA NITRIDING

A. Introduction

Plasma nitriding is an economical method to improve the hardness as well as fatigue and wear resistance of ferrous and nonferrous materials [1,2]. The technique was first introduced in the late 1930s [3,4] and industrial acceptance began in the 1970s. Plasma nitriding can be performed either at high or low pressure. High-pressure plasma nitriding is usually carried out at a pressure of 100 to 1000 Pa and a moderate d.c. substrate voltage of 0.3 to 1 kV. The substrate temperature is dominated by the discharge parameters and typically cannot be varied independently. The minimum current density necessary to maintain an abnormal glow discharge can be high and this phenomenon frequently leads to sample overtempering. The substrate voltage can also give rise to surface finish degradation and undesirable materials modification. Low-pressure plasma nitriding is conducted at a pressure of less than

10 Pa. However, the plasma must be sustained by an external source and the treatment voltage can be as high as several kilovolts. Generally speaking, during plasma nitriding, active particles from the plasma adsorb onto or bombard the sample surface and deep diffusion because of thermal and ion-irradiation effects entails.

B. Active Plasma Species

The first step in plasma nitriding is to generate active particles in a nitrogen environment. A considerable amount of the research activities have concentrated on N_2 and N_2–H_2 plasma characterization and the mechanism [5] although a common consensus has not been reached. Excited states of N_2^+, N_2, NH, H, N, and N^+ have been detected [6]. Assuming that N^+ is the dominant ion species [7], an increased ratio of nitrogen atoms to nitrogen molecules and N^+ to N_2^+ when the distance from the cathode decreases has been observed [8]. It has also been revealed that the excited vibrational states of N_2 (X, V = 10) are the most populated and they are 10^3–10^5 higher in density than N_2^+ ions and electrons [9]. Successful nitriding performed in nitrogen microwave postglow discharge shows the essential roles played by the long-living N_2 (X, V) and N neutral species in the plasma nitriding processes [10]. Note that the working gases are mostly mixed gases (with hydrogen or argon) in postglow discharge nitriding experiments [11,12] and a unified theory on the exact mechanism of plasma nitriding has not been established.

In practice, hydrogen is usually added to the nitrogen atmosphere to enhance the process. Similar to pure nitrogen plasma nitriding, the species active in N_2–H_2 plasma nitriding have not yet been identified, but there is some agreement. The concentration of almost all of the active species decreases with increasing hydrogen concentration above 10% regardless of the type of discharge [13]. This includes both electronically and vibrationally excited N_2, N atoms and atomic ions, and metastable $N(^2D)$ atoms. Nitrogen molecular ions are effectively quenched even in the presence of low levels of hydrogen [14,15]. The evidence of NH is varied, and decreased concentrations are observed at high percentages of hydrogen, although increased concentrations have been noted in the afterglow of a microwave discharge [15]. It has also been reported that there is no spectral evidence of N–H species at low hydrogen concentration whereas the concentration of species consisting of NH_{1-4} is highest at 50% hydrogen [13]. This discrepancy may be because of the lower pressure used in the experiments. The widely observed quenching effect at high hydrogen concentrations correlates well with known reaction dynamics. For example, the very strong quenching of N_2^+ ions is due to the large reaction coefficient of hydrogenation [15], and detailed calculation of the reaction kinetics has confirmed the strong quenching effect of hydrogen [16]. Generally, the addition of hydrogen alters the plasma characteristics and consequently affects the nitriding dynamics.

C. Energy Spectrum of Incident Particles

During plasma nitriding, the active particles arrive at the surface under the influence of an electric field or via gas phase diffusion. The dynamics of plasma nitriding is thus related to the energy distributions of the ions and neutrals because they affect the particle flux, surface state, and sputtering. The energy distribution is governed by the cathodic fall length, also known as the cathodic sheath thickness or the dark space distance, and the mean free path for charge exchange collision λ.

1. High-Pressure Nitriding

The ion and neutral energy distributions in the dark space have been described by Davis and Vanderslice [17]. It is assumed that (1) all the ions originate from the negative glow; (2) in a charge exchange collision, an energetic ion interacts with its neutral counterpart producing an equally energetic neutral and an ion with zero energy; (3) the collisional cross section for the charge transfer is independent of the incident ion energy; and (4) the electric field decreases linearly from the cathode to the edge of the negative glow. A more general form encompassing distributions other than the linear one has been formulated [18], and the normalized ion-energy distribution is given by [19]:

$$f = \frac{1}{m} \frac{L}{\lambda} (1 - E)^{(1/m)-1} \exp\left[-\frac{L}{\lambda}(1 - E)^{1/m}\right] \tag{1}$$

where E defines the ion energy relative to the maximum and lies between zero and unity. When $m = 2$, the equation is equivalent to the model proposed by Davis and Vanderslice [17]. When $m = 4/3$, the field corresponds to the more realistic space charge limited (free fall) case, although it has been shown that the chosen value of m has a minimal effect on the shape of the distribution. For a collisional sheath at an intermediate pressure, the dark-space length L is determined by Child law:

$$J = \left(\frac{2}{3}\right)\left(\frac{5}{3}\right)^{3/2} \varepsilon_0 \left(\frac{2e\lambda}{\pi M}\right)^{1/2} \frac{V_c^{3/2}}{L^{5/2}} \tag{2}$$

where J denotes the current density, ε_0 is the vacuum permittivity, e is the unit charge, and M is the ion mass.

The discharge voltage and working pressure have a different influence on the energy distribution. As L/λ diminishes with increasing voltage and current, the number of ion-neutral collisions per ion in the sheath is reduced, thereby shifting the ion-energy distribution toward higher values. Hence, the effects of higher discharge voltage are twofold, scaling with the ion energies and increasing the fraction of the high-energy ions [19]. With respect to the working pressure, no major change can be observed. If the discharge voltage is held constant, the pressure–dark space thickness product is nearly constant for a d.c. discharge. The change in the pressure (or mean free path) must thus be counterbalanced by a change in the sheath thickness L. Hence, there are no substantial variations in the ion-energy distribution with the working pressure.

2. Low-Pressure Nitriding

In low-pressure glow discharge such as intensified glow discharge [8,20], the dark space consists of energetic ions and neutrals due to collisions as well as low-energy (thermal) neutrals. Some ions do not encounter collisions and arrive at the cathode with full energy, but others suffer collisions resulting in an energy distribution exhibited in Fig. 1 [21]. Another model [22] has been proposed to describe the ion or neutral velocity distribution by assuming that (1) the electric field in the sheath decreases linearly from the electrode to sheath edge, (2) the ion motion is highly collisional and charge exchange is the dominant ion-neutral collision mechanism, and (3) fast neutrals generated in the sheath will not undergo collisions before reaching the electrode. The neutral energy distribution at the cathode can be expressed by:

$$f(\varepsilon, s) = \frac{s}{2\lambda_i \varepsilon_m} E_1\left(\frac{s\varepsilon}{2\lambda_i \varepsilon_m}\right) \tag{3}$$

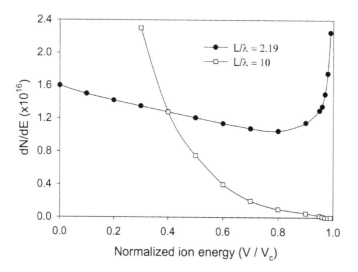

Figure 1   Ion energy distribution in high-energy plasma nitriding. (From Ref. 21.)

where $\varepsilon = Mu^2/2$ is the neutral kinetic energy, $\varepsilon_m = Mu_m^2/2$ is the maximum kinetic energy, M is the neutral mass, λ_i is the ion-neutral mean free path, and s is the sheath thickness.

D. Nitrogen Incorporation Mechanism

Plasma nitriding of ferrous materials is common in the industry [23]. The process involves bombardment of energetic particles (ions and excited or unexcited neutrals), radiation damage, surface adsorption, heating, surface sputtering, and diffusion of the incorporated species. According to Hudis [24], high-energy ion bombardment is mainly responsible for the high nitrogen-diffusion rates that cannot be explained by conventional diffusion theories. Tibbetts' results [25] indicate that neutral nitrogen atoms alone are responsible for ion nitriding, as nitriding is observed for samples protected against nitrogen ion bombardment with a biased grid. It has subsequently been suggested that the main nitriding mechanism is a thermal chemical reaction in which a dissociative adsorption reaction of nitrogen assisted by the addition of hydrogen, $N_2 \rightarrow N_2$ (ad) $\rightarrow 2N$ (ad), takes place resulting in the subsequent diffusion of nitrogen atoms [26].

The incorporation of nitrogen into the materials is a necessary process for successful plasma nitriding, and it has been proposed that the incorporation of nitrogen into the sample surface is due to ion sputtering. Sputtered iron atoms react with the excited nitrogen to form iron nitride (FeN) that redeposits onto the substrate surface. The (FeN) phase may decompose into Fe_2N, $Fe_{2-3}N$, and Fe_4N phases due to energetic ion bombardment. Deep nitrogen diffusion arises from thermal effects and the concentration gradient [27] and the mechanism is illustrated in Fig. 2 [28].

Surface adsorption in conjunction with low-energy ion bombardment plays an important role in plasma nitriding. This phenomenon has been used by Baba and coworkers [29] to explain their data on Ar ion bombardment of transition metals exposed to nitrogen at a partial pressure of 0.001 Pa, and also by Lancaster and Rabalais [30]. Lincoln [31] and Harper [32] have also observed that concurrent bombardment by inert or reactive gas ions

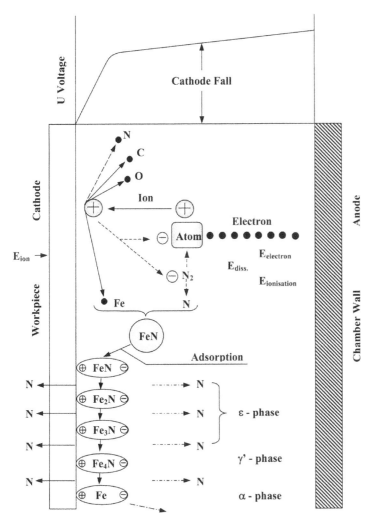

Figure 2   Ion plasma-nitriding mechanism. (From Ref. 28.)

increases the chemical reaction rate of reactive gas incorporation into the surface. Hubler et al. [33] show that the sticking coefficient of the nitrogen is increased by an order of magnitude [34] with ion bombardment. Other experiments have shown that sputtering and/or ion-induced desorption during plasma nitriding can enhance the nitriding rate if the materials have a strong tendency to form oxides with residual oxygen in the reactor, such as aluminum nitriding [35].

Ion implantation is an important nitrogen-incorporation mechanism. It is particularly important in low-pressure plasma nitriding at a high voltage. The large ion energy as a result of either the high voltage or small number of collisions makes it possible for the nitrogen to go beyond the top surface. In fact, the top surface of most materials is usually passivated by a dense oxide that tends to hinder nitrogen incorporation and subsequent diffusion. This explains why low-pressure plasma nitriding or elevated-temperature plasma immersion ion implantation (PIII) is more preferred for stainless steel or aluminum alloy nitriding.

Interestingly, by using hydrogen and a hydrogen/nitrogen mixture and placing the heated samples at the cathode, anode, and discharge zone, it has been observed that nitriding takes place on all samples [36]. Contrary to other reports, this suggests that ionized species and sputtering effects may not be necessary for the formation of the nitrided layers, and so the crucial reactions appear to be the surface adsorption of reactive species and subsequent diffusion. Successful nitriding thus requires a fresh surface rendered by either sputtering or reactive etching [10]. Regardless of whether the proposed mechanisms are neutral controlled or ion controlled [24,25,37], all the models use an electric field to transport the particles, including ion and energetic neutrals to bombard the sample surface. In fact, the λ/L ratio is so large that the neutrals are responsible for most of the energy transfer in spite of the use of a biased grid to shield the ions [25]. It seems that the neutrals play a role in transporting the energy, but surface cleanliness is an important factor.

E. Diffusion Mechanism

1. Iron

Plasma nitriding has been widely applied to iron-based alloys. It is usually carried out in the ferrite region of the iron–cementite diagram. At elevated temperature, the lattice vibration increases, enabling the nitrogen atoms to pass through the interstices to react with alloying elements to form nitrides to a relatively large depth. The nitrided layer comprises a compound layer consisting predominantly of ε and/or γ' (Fe_4N) phases and a diffusion zone where nitrogen is dissolved interstitially in the ferritic matrix. The diffusivity in the different layers varies and the concentration decreases gradually with depth. A typical nitrogen depth profile is illustrated in Fig. 3. The nitrogen diffusion coefficient in α-Fe is the largest compared to those in the other phases including the plasma/solid surface [38]. The thicknesses of the compound layer and diffusion zone generally follow a parabolic law with some deviations when the nitriding time is long [38–41] and the discrepancy can be ascribed to sputtering effects [42,43].

2. Stainless Steel

The nitrogen depth profile in austenitic stainless steel depicted in Fig. 4 deviates from the standard analytical solution of concentration-independent diffusion in a semi-infinite solid with a constant surface concentration. It has been observed that the retained dose increases from 16Cr to 19Cr steels [44]. This is inconsistent with internal nitriding theory in which the rate of nitriding and evolution of the nitrided layer are inversely proportional to the concentration of nitride-producing element in the steel. It has been proposed that the presence of Cr in the fcc structure is crucial to the process [45]. Between 350° to 450°C and for a time shorter than that taken by Cr to diffuse a distance of a few lattice constants, Cr effectively creates a metastably high solubility of N producing the γ_N phase due to the strong N–Cr bond. The N diffusivity in stainless steel is reduced relative to that in pure fcc Fe because of the retardation and periodic trapping by a high concentration of Cr (e.g., about 19 at.% in 304 SS). However, if a high concentration of N is introduced rapidly into the materials, all the Cr sites in the near surface quickly become occupied by one or more nitrogen atoms. The nitrogen diffusivity in this region then becomes much higher and the excess nitrogen can subsequently diffuse rapidly thereby tying up deeper Cr to increase the nitrogen diffusivity further down the depth. This model assumes that the nitrogen-diffusion mechanism remains interstitial in nature.

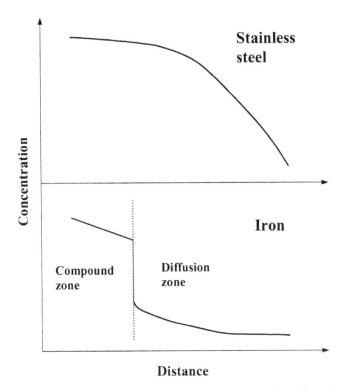

Figure 3 Schematic illustration of nitrogen depth profile on iron and stainless steel.

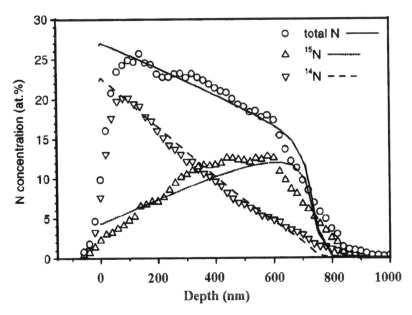

Figure 4 Experimental N depth profiles and N depth profiles calculated using the proposed transport model. (From Ref. 51.)

According to the model proposed by Williamson et al., when all the trap sites become occupied, additional N diffuses quickly through the highly saturated layer to reach the front of the layer where unoccupied trap sites are available. Experiments conducted under low-energy, high-flux conditions first with the ^{15}N isotope and subsequently with the ^{14}N isotope show that the ^{14}N isotope is not preferentially located at the diffusion front [46]. On the contrary, the nitrogen introduced in the second step is near the top surface, as shown in Fig. 4. Therefore, a refined model has been suggested including untrapping of nitrogen [47]. This model draws an analogy between the ^{14}N/^{15}N depth profiles in austenitic stainless steel and the hydrogen/deuterium profiles in amorphous silicon. Nitrogen depth profiles simulated using this model with diffusion and untrapping activation energies of 1.1 and 1.45 eV, respectively, are in good agreement with the experimental data.

3. Aluminum

Aluminum and its alloys are attractive to the industry because of the high strength-to-weight ratio, good corrosion resistance, and good formability. On the other hand, hardness and wear resistance are not satisfactory. Nitriding of Al is of interest because AlN has excellent physical and chemical properties, e.g., high hardness and increased wear resistance. However, effective Al nitriding is difficult. First, aluminum possesses a dense native surface oxide layer that acts as a barrier to nitrogen incorporation and diffusion, and nitriding must be preceded by successful removal of this oxide layer. Second, owing to the different sputtering rates of Al_2O_3 and Al, the sputtering conditions and gases must be optimized. Third, a relatively high substrate temperature between 400° and >500°C is required to produce thick enough AlN coatings within a treatment time of several hours. Fourth, there is high residual stress between AlN and Al, and good layer adhesion is difficult to achieve. Lastly, a thicker AlN layer possesses low electrical conductance.

Investigation on the mechanism of aluminum nitriding is more challenging because of the high affinity of nitrogen to aluminum. A model that is different from that of conventional plasma nitriding has been proposed based on experimental evidence [48]. Gold is implanted into Al to a depth of 200 nm as a depth marker. After nitriding, the Au marker is observed to move to a deeper region with an unchanged distribution. In contrast, both the location and depth profile remain the same after annealing at the nitriding temperature alone, as shown in Fig. 5. The results thus suggest that thermal effects can be eliminated and the dominant process is Al out-diffusion rather than N in-diffusion. That is, the nitride layer is formed on the surface with aluminum atoms supplied from the bulk. However, metallurgical dynamics such as the affinity of N to Au have not been considered in this work.

F. Low-Pressure Nitriding

Low-pressure nitriding has some advantages compared to conventional plasma nitriding conducted at high pressure because of the higher effective ion energy. Ion beam nitriding experiments performed at low pressure show that the nitriding efficiency increases linearly with the ion flux and implantation voltage (less than 1 kV), particularly in austenitic stainless steel materials as shown in Fig. 6 [23,49]. It implies that the amount of incorporated nitrogen available for diffusion increases linearly with ion energy. In experiments using a relatively constant power input to maintain the same sample temperature, an optimal implantation voltage (about 1 kV) is observed [50]. At low pressure, a higher voltage can be applied and there are fewer collisions because of the smaller L/λ. The incident particles thus possess a

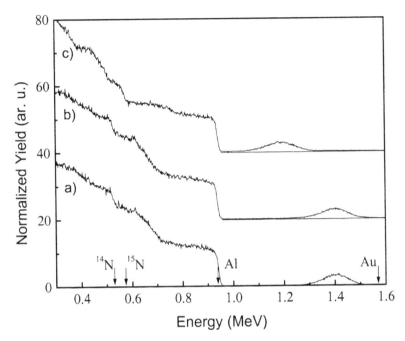

Figure 5 Rutherford backscattering spectra obtained after: (a) ^{14}N ion nitriding and Au ion implantation, (b) additional heat treatment at 400°C, and (c) additional ion nitriding with ^{15}N. The arrows denote the surface backscattering energies of Al, N, and Au. (From Ref. 48.)

higher energy and can penetrate the surface barrier to facilitate diffusion. Consequently, the heating effects by useless particles can be mitigated. Low-pressure, plasma-based nitriding has attracted much attention [35], as the process can treat samples at relatively low temperature and the advantage is appealing to some stainless steel [51–54] and aluminum alloys [55–57] applications. It also makes nitriding possible without hydrogen, even if the surface barrier (e.g., oxide) has not been removed.

Owing to the low pressure, external power must be supplied to sustain the plasma. The common external sources are thermionic electron sources [58,59], electron cyclotron resonance (ECR) plasma sources [60,61], radio frequency (rf) sources [62], and thermionic arc discharges [63,64]. During the process, surface phenomena such as surface adsorption, sputtering, ion implantation, and chemical reactions take place simultaneously.

G. Applications

Plasma nitriding is commonly used in the metallurgical industry to improve the surface hardness of structural alloy steels, hot and cold work steels (especially those that do not experience a reduction in the core hardness at the nitriding temperature), high-speed steels such as M2 and 9% W grades, as well as steels with special properties such as acid resistance, heat resistance, and creep resistance. It can also be used to treat titanium, molybdenum, and their alloys. As a result, plasma nitriding is used to extend the lifetime of machine components and tools such as injection mold screws and cylinders, dies, gears, punches, injection nozzles, and ultrasonic waveguides [65].

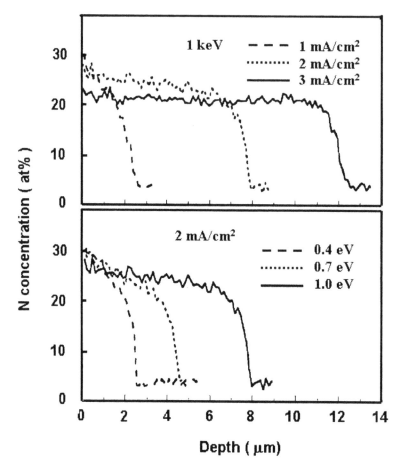

Figure 6 Influence of ion flux or implantation energy on the processing efficiency. The processing temperature and time are fixed at 400°C and 60 min, respectively. (From Ref. 49.)

1. Tool Steels

The lifetime of industrial components made of H13 and D2 steels can be increased by a factor of 2 to 4 using plasma nitriding, depending upon the type of steel and the geometry of the components. Reengineering of the surface/subsurface by plasma nitriding coupled with a better substrate can enhance the lifetime by a factor of 7. Adhesion-related failure is one of the important causes of wear in tool steels. Delamination occurs in H13 steel and both delamination and abrasive wear are observed in D2 steel. Plasma nitriding can reduce the adhesive-related failure substantially [66,67]. The surface hardness of AISI H-12 tool steel increases with the nitriding time, but a really long treatment time may reduce the surface hardness [68]. 9SiCr low-alloy tool steel treated by plasma nitriding at 250°C exhibits improved abrasive wear resistance. The formation of nitrides and nitrogen-containing solid solution is postulated to be the main cause of the improved tribological properties [69].

2. Titanium Alloys

The surface properties of pure titanium and Ti6Al4V alloy can be improved using intensified plasma-assisted processing (IPAP) at low pressure [70]. The surface structure consists of a surface titanium nitride (TiN) layer, a sandwiched Ti_2N layer, and a zone containing diffused nitrogen. The treated materials exhibit a threefold improvement in hardness as well as better wear and corrosion resistance. The surface roughness is observed to be a function of the degree of plasma intensification. The thicknesses of the nitrided and diffusion zones depend on the ion flux and energy, with a higher energy favoring TiN formation. Nano-indentation experiments performed on cross sections to an indentation depth of 400 nm demonstrate the effectiveness of IPAP at higher cathode current densities. The hardness is improved, whereas there is no functional dependence of the modulus on the treatment parameters. Tribological experiments show that the nitrided surface possesses significantly better wear resistance and lower friction coefficients [71].

3. Aluminum Alloys

Nitriding has been shown to improve the poor tribological properties of aluminum and aluminum alloys at a temperature of less than 500°C [72]. However, nitriding of aluminum and its alloys is difficult because of the high affinity of oxygen to aluminum resulting in a surface aluminum oxide layer that acts as a diffusion barrier and precludes nitriding even at high pressure (100–1000 Pa). Low-pressure (<10 Pa), high-density plasma nitriding processes such as plasma immersion ion implantation [73], arc-assisted nitriding [56], and intensified glow discharge nitriding [59] have been used to treat Al. It is observed that sufficient surface cleaning and the use of high enough incident ion energy are necessary to nitride aluminum alloys [74]. A high treatment temperature is also required if a thick AlN layer is desired. For example, a 5-μm AlN layer is produced at 450°C in 20 hr by conventional plasma nitriding [72], whereas 3 μm can be produced at 480°C in 3 hr using a pulsed d.c. dicharge, and 7 μm can be fabricated using arc-assisted nitriding at 450°C in 15 hr [56]. By conducting PIII at approximately 500°C, an AlN layer thickness of 12 μm (compared to 1.5 μm for conventional plasma nitriding at the same temperature) can be achieved, and the resulting wear resistance is also significantly better [75]. Thick AlN layers are also observed using d.c. presputtering in an rf plasma [74] or ECR plasma [76]. With a d.c. pretreatment before rf nitriding, a nitriding time of 2 hr yields a layer thickness of 18 μm. This surface layer is composed of three different regions: interfacial, compact AlN, and porous zones. A 15-μm AlN layer has been reported using ECR plasma treatment and the Vickers hardness is about 1400 without pretreatment. The treated region exhibits a bilayer structure that is related to surface melting from ion bombardment.

4. Stainless Steel

Both high- and low-pressure plasma nitriding can improve the surface properties of austenite stainless steels. However, the corrosion resistance may be compromised by temperature exceeding 440°C and a long treatment time because of precipitation of chromium nitride and ferrite in the near-surface region of the diffusion layer and the corresponding decrease in the chromium content in the γ phase [2]. In comparison, after low-temperature plasma nitriding/implantation, the wear resistance of austenite stainless steels can be ameliorated without degrading (sometimes even improving) the corrosion resistance [77–80]. Published results show that a layer of metastable fcc solid solution supersaturated in nitrogen sometimes known as "expanded austenite" (γ_N) formed between 310° and 420°C is observed by x-ray diffraction and transmission electron microscopy (TEM). This phase is very hard and has good wear properties while retaining the corrosion resistance of the austenite stainless steel. In fact, two expanded austenite phases have been observed. The formation mechanism is postulated to be a slow cooling process leading to the secondary expanded austenite [2]. Another school of thought is stress-assisted diffusion [81]. Of all the processing methods, PIII is preferred based on temperature consideration, and successful treatment has been demonstrated at 150°C [82].

III. PLASMA IMPLANTATION

A. Introduction

Conventional beam-line ion implantation is an effective surface modification and materials synthesis tool and has evolved into a standard technique in the semiconductor industry [83,84]. Even though it is a line-of-sight process, implantation into planar specimens such as silicon wafers is quite straightforward if beam scanning or sample translation/rotation is adopted. On the other hand, uniform implantation into large industrial components possessing an irregular shape can pose practical difficulties. Plasma immersion ion implantation is a radical departure from the beam-line concept [85–88]. In PIII, the sample is immersed in a plasma produced by one or multiple external sources and pulse-biased to a high negative potential. A plasma sheath forms around the specimen and ions bombard the entire sample in a non-line-of-sight fashion. To achieve uniform implantation into a sample with an irregular geometry, the processing conditions must be optimized to produce good plasma sheath conformality. The plasma sheath and experimental set up is schematically shown in Fig. 7.

At first glance, a nitrogen PIII system is similar to a pulsed plasma nitriding instrument, and the difference indeed becomes smaller when pulsed high-voltage glow discharge implantation is performed. The major difference lies in the voltage and average power density. Plasma nitriding utilizes a relatively low voltage but very high current (power density) to heat the samples to higher than 500°C in situ. In comparison, PIII is normally conducted at a much higher voltage and low duty cycle. Although the peak current in PIII may be quite large, the pulse modulator limits the average current and power density. However, to make nitrogen PIII commercially attractive, the bias duty cycle must not be too small, and the average power may become comparable to that used in plasma nitriding thereby blurring the border between the two techniques [89]. PIII at elevated temperature is a very good example combining the advantages of ion implantation and conventional nitriding.

In PIII, ions are implanted through the plasma sheath. The formation and evolution of the sheath is critical to PIII processes and have been investigated extensively using one- or multidimensional models. The evolution of the plasma sheath is composed of several continuous steps. When a negative voltage $-V_0$ is applied to the sample immersed in a plasma sustained by an external source, electrons near the surface are driven away on the timescale of the inverse electron plasma frequency ω_{pe}^{-1} leaving ions behind to form an electron-depleted ion sheath. Subsequently, on the time scale of the inverse ion plasma frequency ω_{pi}^{-1}, ions within the sheath are accelerated into the sample. The consequent drop in the ion density in the sheath drives the sheath-plasma edge further away, exposing new ions on the way and causing them to be accelerated toward the sample and implanted. The time evolution of the transient sheath determines the implantation current and ion energy distribution. On a longer timescale, the system evolves into a steady-state Child law sheath.

B. One-Dimensional Modeling

1. Plasma Sheath Evolution

One-dimensional (1-D) plasma sheath models have been extensively investigated due to the relative simplicity and flexibility. A 1-D plasma sheath derived from the Child–Langmuir

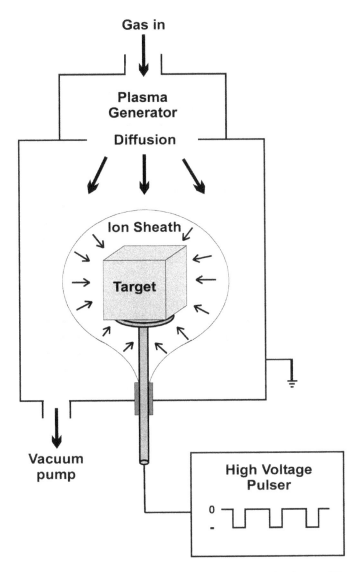

Figure 7 Schematic of a plasma immersion ion implanter. The target is surrounded by the ion sheath so that all the surfaces of the target are implanted simultaneously.

law is usually based on the following assumptions: (1) The ion flow is collisionless. (2) Electrons have zero mass and respond instantaneously to the applied potential. (3) The full voltage V_0 is applied at $t = 0$ and is much greater than the electron temperature T_e. (4) A quasi-static ion matrix sheath forms instantaneously and the current demanded by this sheath can be described by Child law and is supplied by the uncovering of ions at the propagating sheath edge. (5) The ion transit time across the sheath is zero. That is, the implant current is equal to the charge uncovered by the expanding sheath, and ions are singly charged. The current density j_c for a voltage V_0 across a sheath of thickness S is given by:

$$j_c = \frac{4}{9} \varepsilon_0 \left(\frac{2e}{M} \right)^{0.5} \frac{V_0^{1.5}}{S^2} \tag{4}$$

where ε_0 is the free space permittivity, e is the electronic charge, and M is the ion mass. Equating j_c to the charge per unit time crossing the sheath boundary $e n_0 ds/dt$ and integrating, the time-dependent sheath position becomes:

$$S = S_0 \left(1 + \frac{2}{3} \omega_{pi} \right)^{1/3} \tag{5}$$

where $S_0 = (2\varepsilon_0 V_0/e n_0)^{0.5}$ is the ion matrix sheath thickness, $\omega_{pi} = (e^2 n_0/\varepsilon_0 M)^{0.5}$ is the ion plasma frequency, ε_0 is the permittivity of the free space, and n_0 is the plasma density. Therefore, the plasma dynamics is dependent on the applied voltage, pulse duration t_p, plasma density, plasma species, and so on.

In typical PIII experiments, $2/3(n_0 q^2/\varepsilon_0 m)^{1/2} t_p$ is much greater than unity [90], and so the total ion dose per pulse (assuming a zero pulse voltage rise time) can be estimated by the following equation:

$$D_{total} = \left(\frac{2}{3} \right)^{\frac{1}{3}} 2^{\frac{1}{2}} \varepsilon_0^{\frac{1}{3}} q^{-\frac{1}{6}} n_0^{\frac{2}{3}} m^{-\frac{1}{6}} V_0^{\frac{1}{2}} t_p^{\frac{1}{3}} \tag{6}$$

In practice, no power modulators can deliver pulses with a zero rise or fall time because of hardware capacitance [91,92], and a model has been derived to account for the finite rise and fall times for a planar sample geometry based on the assumptions that the quasi-static Child law sheath forms instantaneously, the voltage rises and falls linearly, and the ions have no directed velocity toward the expanding sheath [93].

2. Sheath Extinction

The sheath automatically collapses at the end of each voltage pulse and the rate is related to the voltage fall time, plasma density, and plasma (including electron and ion) temperature. As the sheath collapses, the electron density reverts to the local ion density leaving a step gradient in the quasi-neutral plasma. For a collisionless sheath, the flow of nondrifting ions into the depleted region will be at the ion thermal velocity with an ambipolar enhancement because of the presence of electrons at a temperature much higher than that of ions. By modeling the plasma density using a spatially constant density n_s in the depleted region and a plasma density of n_0, the following relationship has been derived [94]:

$$\frac{n_s(t)}{n_0} = 1 - \left(1 - \frac{n_{s0}}{n_0} \right) e^{-\mu t} \tag{7}$$

where the fill-in rate (per second) is $\mu = (T_i T_e)(1/S)(KT_i/2\pi M)^{0.5}$ and n_{s0} is the initial density in the depleted region.

Experiments have been conducted to investigate the transient evolution/collapse of the sheath by measuring the electron saturation current using a Langmuir probe [95]. A spherical target biased to a 10-μsec, -10-kV voltage pulse is immersed in a uniform nitrogen plasma with a plasma density between 10^9 and 10^{10} cm^{-3} and an electron temperature of 1.4 eV. The experimental results are exhibited in Fig. 8. The sheath expansion speed measured by the Langmuir probe is higher than the ion acoustic speed until the sheath approaches the steady-state value governed by the Child–Langmuir law. As soon as the pulse voltage is switched off, the electron current begins to recover. When the pulse fall time is longer than the plasma transit time that is approximately the propagating time of an ion acoustic wave, the ion sheath shrinks in accordance with the reduction in the negative voltage. During the pulse fall time, which is shorter than the plasma transit time, the electron saturation current

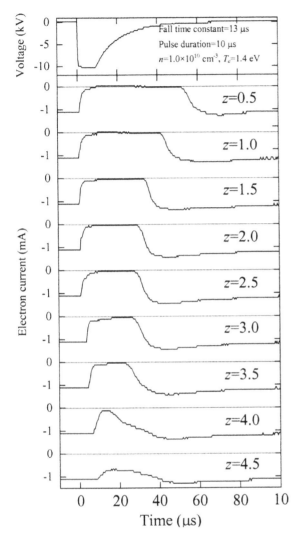

Figure 8 Time-dependent electron density during PIII processes with different pulse fall times. (From Ref. 95.)

overshoots to up to approximately twice the steady-state saturation current. This tidal wave propagates into the plasma at the ion acoustic speed and its amplitude decreases exponentially with the distance from the target [96–99].

3. Ion Energy Distribution

In general, the average energy of the incident ions is less than eV_0 (where V_0 is the voltage applied to the sample). This is due to the finite rise and fall times of the implantation voltage and ions present in the expanding sheath not accelerated to the full potential. In general, under collisionless conditions (small sheath width or low pressure), about 90% of the incident ions have the full energy [100]; on the other hand, when the sheath thickness is greater than the mean free path, the average incident energy will diminish.

The pulse voltage waveform affects the ion energy, particularly for a long voltage-rise time, as a large number of ions are implanted during this rise time period [101–104]. A model with simple analytical expressions indicates that the ion energy distribution is determined by the ratio of the total pulse duration to the sample voltage rise time but independent of the plasma composition, ion species, and implantation voltage [101]. As shown in Fig. 9, the ion energy spectrum has two superimposed components, a high-energy one for the majority of the ions implanted during the plateau region of the voltage pulse as well as a low-energy one encompassing ions implanted during the finite rise time of the voltage pulses. The lowest energy component is attributed to a small initial expanding sheath obeying the Child–Langmuir law.

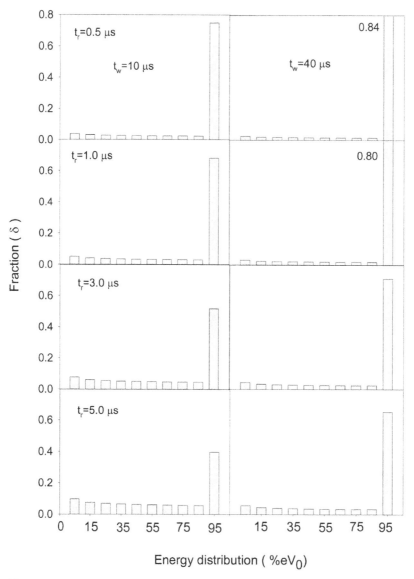

Figure 9 Ion energy spectra for different rise times for pulse durations of 10 or 40 µsec.

C. Two-Dimensional Sheath and Implant Uniformity

A big advantage of PIII over beam-line ion implantation is the capability to effectively implant objects with an irregular shape without beam rastering or target manipulation. Although a 1-D dynamic plasma sheath provides the basic knowledge, it does not adequately describe PIII into real samples, and 2-D models that are mathematically less complex than 3-D or quasi 3-D ones have been explored for different target geometries.

1. Theoretical Simulation Results

a. Square Targets. Corners and edges present special challenges to PIII researchers for the sheath edge is not smooth around corners [105]. For a 2-D corner (i.e., an edge), numerical simulation reveals that the influence of the corner extends roughly to a planar sheath width away from the vertex [106]. Ions impact at an oblique angle near a corner and the flux, which is enhanced above its planar value, peaks near, but not at, the corner. Far from the corner, the sheath is planar, the dose is uniform, and the impact angle is normal. It has been shown that the time-dependent sheath expansion away from a corner is essentially self-similar [107]. That is, the sheath moves quasi-statically through a sequence of Child–Langmuir-like configurations, and so the flux and impact angle are given approximately by those of the static sheath with the same width [105]. The processing parameters are crucial to the dose uniformity. When the sample size relative to the ion-matrix sheath is large, the incident dose peaks near the corner and decreases at the corner [108], but the maximum incident dose is found at the center of a square bar in the case of a relative small target or very high implantation voltage. The effects of the target dimension on the dose distribution are illustrated in Fig. 10.

b. Trenches. Trenches possess both convex and concave geometries. A 2-D fluid model has been used to compute the ion dynamics of a target consisting of an infinite array of rectangular trenches (or teeth) [109]. Ion pulses sweeping from the top to the bottom are observed along the sidewall of the trench and ions are focused and diverted by the potential structure. The sweep is quite abrupt (less than several inverse ion plasma frequency) and may happen during the rise time. In addition, once these pulses have elapsed, the ion flux to the sidewalls decreases precipitously, and ions follow a near-parallel trajectory not favoring implantation.

Particle-in-cell (PIC) simulation has been conducted on a trench structure for a longer timescale close to that in real applications [110–112]. The influence of the pulse length, rise time, ion mass, and trench aspect ratio on the ion dose distribution along the surfaces has been investigated. In Fig. 11, the dose on the sidewall is much less than those on the top and bottom surfaces, and the bottom surface receives a slightly smaller dose than the top surface. The sidewall ion dose remains almost constant for all pulse lengths probably because of the rapid propagation of the plasma sheath from the trench. Owing to glancing incidence, the projected range on the sidewall is smaller than those on the top or bottom surfaces. A more flexible model developed on a Windows platform has recently been developed to simulate PIII into trenches and other shapes using the PIC algorithm [113,114].

c. Inner Surfaces. Compared to outer-surface PIII, PIII into an inner surface such as the inside of a cylindrical bore poses interesting challenges [88]. If the dimension of the bore is smaller than several Debye lengths, the plasma will not be uniform and calculation of the characteristic sheath width will be difficult. Plasma replenishment between voltage pulses will take longer time because of the restricted geometry. In fact, it may be difficult to attain axial uniformity along the bore and the primary electrons and secondary electrons must exit the bore longitudinally, otherwise the local electron density is altered thereby affecting the

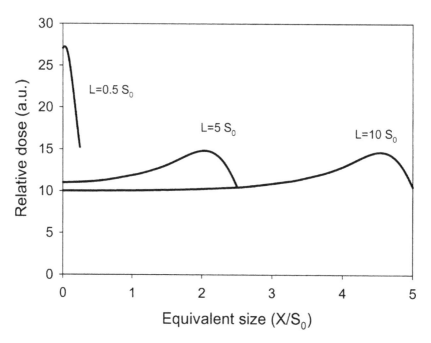

Figure 10 Effects of square-target size on the uniformity of incident ions.

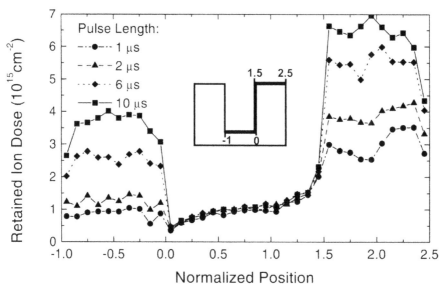

Figure 11 Distribution of incident ions along the exposed surfaces of the trench sample. (From Ref. 111.)

plasma and implantation. In an interior geometry, the sheaths grow toward each other from opposite sides and may overlap if the bore dimension is too small. Recent work shows that for inner-surface PIII, the implantation energy is less than the applied potential. Furthermore, unlike exterior-surface PIII in which the sheath propagation velocity decreases monotonically with time, an expanding interior sheath reaches a minimum velocity and then accelerates because of convergent effects near the axis [115], provided that the voltage pulse is long enough and assuming that the equilibrium Child–Langmuir sheath width is larger than the cylindrical radius [115]. The maximum normalized ion impact energy is [116]:

$$u_{max}^2 = 0.367 P^2 \qquad \text{for} \qquad P < 1$$

or   (8)

$$u_{max}^2 = 1 - \frac{0.8790}{P^2} + \frac{0.2468}{P^2} \qquad \text{for} \qquad P > 1$$

where $P = R/d$ is the dimensionless cylindrical radius, R is the cylindrical radius, and $d = (4\varepsilon_0 V/en_0)^{0.5}$. To improve the impact energy in cases where $P < 1$, an external electrode can be coaxially inside the bore [117,118]. As shown in Fig. 12, the auxiliary electrode improves the potential distribution and electric field inside the cylindrical bore significantly. Consequently, the impact energy can be improved. Simulation has recently been conducted on PIII into the inner and outer surfaces of ball bearings and specimens resembling a ring [119].

2. Experimental Results

a. Spherical Samples. Experimentally, ion dose and energy nonuniformity can be minimized by optimizing the implantation parameters, adjusting the specimen placement, and altering the location/dimension of the sample holder. The variation in the implantation

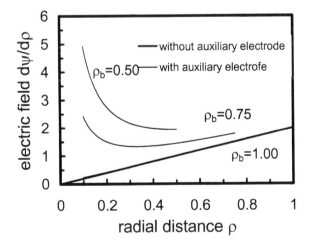

Figure 12 Electric field, $d\psi/dr$ in the sheath in a small cylindrical bore with and without an auxiliary electrode. The radius of the bore is marked on each curve. The radius of the auxiliary electrode is $r_a = 0.1$. Note: $\rho = r(4\varepsilon_0 V_0/en_0)^{-0.5}$.

dose on a spherical sample has been shown to be less than 20% in some earlier work [120]. The nonuniformity can be attributed to the relative position of the target to the chamber wall and also related to the nonuniform plasma density in the chamber [121]. More recent work on 10-mm-diameter 304 stainless steel balls discloses better uniformity [122].

   b. Triangular Samples. Samples with triangular shape and edges have been investigated [123–126]. Figure 13 shows the results on the surface of a wedge. The edges in the neighborhood of the 30° angle have a smaller retained dose as compared to the 90°-angle edge, and the wedge face with a smaller area has a higher average dose. The smaller dose retention near the acute angle may be attributed to more severe sputtering and shallower depth of the implanted ions because of glancing incidence. For multiple wedge-shaped samples, proper sample placement is important, as the relative shape of the sample and plasma sheath determines the experimental outcome. The sample surfaces near the edges are not parallel to the plasma sheath edge and, consequently, the ion trajectories near the sample edges are not parallel to the surface normal. The relative implantation dose on the edges is smaller, as shown in Fig. 14. The implantation voltage also affects the lateral implantation dose distribution. A lower voltage improves the conformality of the electric field lines with respect to the sample surfaces and may lead to better ion dose uniformity.

   c. Trenches. The same trend as observed on triangular (convex) samples is found on V-shaped (concave) trenches [125,126]. A first-order linear dependence on the angle is observed. The small implant dose on the sidewall of the wedge-shaped trenches is due to the fact that only ions from a small solid angle reach the interior sidewalls. The ion incident angle is more oblique, giving rise to more severe sputtering. The increase in the implant dose toward the upper edge is caused by the deviation in the ion trajectories due to the shape of the plasma sheath. The incident angle has a big effect on the ion dose on the inner surfaces. The dose decreases with the incident angle, and the average dose is much smaller than that on the top surface. When the angle is zero, the sample becomes a U-shaped trench [126–128]. In this case, the incident dose on the side surface decreases to about one-tenth of that on the

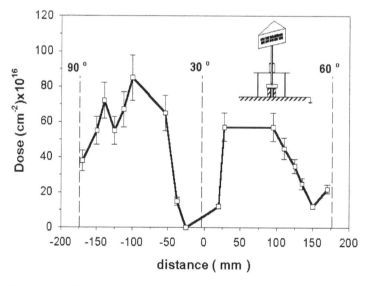

Figure 13   Measured dose on the wedge surface (implantation conditions: 50 kV/21 μsec at a dose of $8.0 \times 10^{17}/cm^2$). The surface is conceptually unfolded at the 30° angle. (From Ref. 123.)

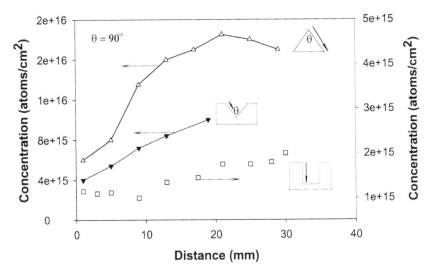

Figure 14 Influence of target shape on the retained dose. (From Ref. 126.)

top surface. The overall trend is, however, the same, and the dose increases toward the bottom of the trench along the side surface.

  d. Inner Surfaces. As aforementioned, the plasma density in a small hole or bore is not uniform [129] and ions cannot receive the full potential even if there is a grounded auxiliary electrode [130,131]. Experiments disclose that the implantation energy and retained dose are lower in inner-surface PIII when compared to outer-surface PIII. Without the auxiliary electrode, the implantation depth and retained dose are 33% and 19% of those achieved on the outer surface, respectively. Positioning a grounded electrode along the center of the hollow bore improves the implantation depth by 43% and retained dose by 71%. The nonuniformity in both the implantation energy and retained dose increases with a larger bore length. The overall variation is about 30% with or without the auxiliary electrode. To produce a uniform plasma in the inner hole, an external plasma source must be used. Radio frequency biased or grounded coaxial rods positioned inside the cylindrical bore have been proposed [132]. The central rod serves as both the rf antenna and sputtering cathode, and uniform TiN and diamond-like carbon (DLC) films have been successfully deposited on the inner surfaces. A moving solenoidal coil can also be used to produce a microwave plasma inside the cylindrical sample to improve the implant dose uniformity [133,134].

D. Ion Heating

In addition to the plasma sheath, ion heating affects the outcome of the PIII experiments. Heating due to ion bombardment can be mitigated using a sample cooling mechanism. One can also exploit the heating phenomenon to achieve better results. For example, elevated temperature nitrogen PIII is preferred for aluminum and stainless steel to enhance diffusion. The deposited energy can be calculated using the sheath width at the end of each pulse, and for a planar sample, the ion heating power is [135,136]

$$P_{\text{ave}} = \left(\frac{2}{3}\right)^{\frac{1}{3}} 2^{\frac{1}{2}} e^{\frac{5}{6}} \varepsilon_0^{\frac{1}{3}} t^{\frac{1}{3}} M^{-\frac{1}{6}} f V_0^{1.5} n^{\frac{2}{3}} \tag{9}$$

A simple time-dependent, lumped capacity thermal model has been proposed [136]:

$$mc_p \frac{dT}{dt} = A_h q_h - A_r q_r - Q_c \tag{10}$$

where m is the total mass of the sample platen and target, c_p is the mass-averaged heat capacity, A_h is the total heated area, q_h is the cycle-averaged heat flux applied to that area, A_r is the total radiation area, q_r is the radiation heat flux lost from that area, and Q_c is the total heat lost due to conduction. Theoretical simulation using the fluid model and classical thermal equations can disclose the 2-D sample temperature variation [137]. It is found that the lumped-capacity thermal model is adequate for a small sample or when the heating rate is low, whereas a 2-D model is more accurate for a rapid heating process, e.g., during the voltage rise time phase.

E. Applications

Many good examples can be found in the literature to demonstrate the effectiveness of PIII with regard to the enhancement of the tribological, chemical, and electrical properties of materials. Three general PIII techniques are used to enhance the tribological properties: (1) Implant N_2 or NH_3 at a dose between 10^{17} and 10^{18} ions/cm^2 and voltage exceeding 30 kV. (2) Deposit DLC films on the substrate using acetylene (C_2H_2)/CH_4 gaseous PIII plasma or carbon vacuum arc plasma. (3) Implant metallic ions using vacuum arc in PIII/ ion-beam-assisted deposition (IBAD).

1. Gas Ion Plasma Immersion Ion Implantation

Gas ion PIII was first used to improve the wear resistance and surface hardness of materials. The hardness enhancement can be attributed to the formation of hard, second-phase precipitates in the near-surface region. The excellent hardening effect of titanium plasma implanted by nitrogen is largely due to the formation of titanium nitride precipitates in the ion-implanted layer. Solid–solution strengthening may also be responsible for the surface hardening. For instance, elevated-temperature nitrogen plasma implantation of stainless steel results in expanded austenite with higher surface hardness without precipitation. Nitrogen PIII has also been applied to other materials such as aluminum alloy, titanium alloy, tool steel, carbon steel, and stainless steel with good results.

a. Aluminum Alloy. Aluminum and its alloys plasma implanted by nitrogen or oxygen form a surface nitride or oxide layer. Because of the strong reaction between oxygen and the aluminum surface and the non-ultrahigh vacuum (UHV) conditions in most PIII machines, some degree of oxidation is always observed in nitrogen PIII [138,139]. Compared to the untreated samples, the wear rate of the treated 7075 aluminum alloy decreases by approximately a factor of 10, although enhancement in the surface hardness is observed only for small test loads [140]. Improvement has been noted in nitrogen-plasma-implanted 2024 aluminum alloy and the degree of enhancement depends on the implant dose. For a nitrogen dose between 0.2 and 1.0×10^{18}/cm^2, the coefficient of friction decreases and the wear resistance increases. The mechanism is mainly adhesive, as adhesive wear becomes smaller at a higher implant dose [141]. Nitrogen ion implantation also enhances the hydrophilic properties [142].

Oxygen plasma has been used to improve the surface properties of aluminum alloys. Stronger Mg segregation (up to ~100% MgO) is found in 7075 Al alloy after oxygen plasma implantation. The nanohardness and Young's modulus are enhanced by factors of 2 and 4 to

1.5 GPa and about 300 GPa, respectively. Surface hardening is consistent with the formation of very fine oxide precipitates [143,144]. The surface hardness of pure aluminum after oxygen plasma implantation is improved by about a factor of 2 to 3 [145]. To increase the AlN layer thickness, elevated temperature ($>500\,^\circ$C) is used to enhance thermal diffusion [75]. An AlN layer thickness of 15 μm is obtained on an AlMg4.5Mn sample after 6 hr, 40 kV PIII at $500\,^\circ$C. The hardness of the nitrided AlMg4.5Mn alloy increases to 1200 HV that is similar to the value of compact AlN ceramics, as compared to a bulk hardness of 100 HV for the untreated alloy.

b. Titanium Alloy. Titanium alloys are attractive in the aerospace and biomaterial industries. Experimental results show that PIII improves the tribological properties of titanium surfaces. The wear rate of Ti6Al4V samples decreases substantially after 50 kV, 3×10^{17} ions/cm^2 PIII [146]. The wear rate follows a trend similar to the depth profile of the implanted species. A low friction is observed for the Ti6Al4V sample treated by nitrogen PIII when tested against an ultrahigh-molecular-weight polyethylene (UHMWPE) stylus [147,148]. A smaller buildup and change in the morphology of the wear debris of polyethylene are observed when it is rubbed against the implanted sample. The surface properties of the treated titanium depend on the sample temperature [149], and $550\,^\circ$C is necessary for an appreciable improvement. After the treatment (40 kV, 5 hr), the wear rate diminishes by four orders of magnitude. This is thought to be associated with the increased mobility of nitrogen in α-Ti at high temperature thereby producing a thicker hardened zone. Oxygen plasma implantation improves the surface properties of titanium alloys [150]. Microindentation tests show an increase of up to 100% in the surface hardness, and dry pin-on-disk tests with spherical-ended UHMWPE pins reveal a very significant increase in the wear resistance of oxygen-implanted Ti6Al4V samples.

c. Tool and Stainless Steels. Plasma immersion ion implantation is an effective surface-modification technique for tool steels [151] and stainless steels [152–155]. For example, a factor of 2 enhancement in the friction coefficient has been achieved in nitrogen-plasma-implanted tool steel samples. A duplex subsurface structure consisting of an outer nonequilibrium layer typical of nitrogen implantation (containing over 20 at.% nitrogen) on top of diffused zone of a lower nitrogen content can be produced on H13 tool steel by a single PIII process at elevated temperature [156].

Elevated-temperature PIII is usually preferred for stainless steels despite the high implantation voltage allowing deep ion penetration through the sample surface. Nitrogen plasma implantation at elevated temperature (less than $450\,^\circ$C) leads to the formation of the expanded austenite phase, where the fcc austenite lattice is expanded by a few percent due to the presence of nitrogen. This reduces the wear by one to two orders of magnitude without degrading the corrosion resistance [46,157,158]. Recent results show that the implantation voltage in elevated-temperature PIII may not be very crucial [159], and even a lower voltage can give rise to good surface properties [160–162]. The observed enhancement is probably caused by the high ion density promoting fast diffusion of the implanted species. Experimental results indicate that a higher ion density is the most important parameter to achieve rapid surface nitriding, followed by the implantation energy, and so low-voltage PIII is desirable in some applications.

d. Diamond-Like Carbon. Diamond-like carbon films have unique properties between those of diamond and graphite. PIII-based DLC formation has been reported on silicon [163,164], chromium [165,166], titanium [167], aluminum [168,169], and stainless steel [170]. External or pulsed glow discharge plasmas in addition to carbon vacuum arc have been used to deposit DLC films [163,164,171,172]. To improve adhesion between the DLC coating and substrate, PIII is used to produce a compositionally graded interface. The improved adhesion can also result from the formation of carbide phases in the matrix of the substrate surface acting as anchors for the DLC coating. However, it is necessary to remove the surface amorphous carbon layer to attain excellent adhesion [173] and fabrication of large-area DLC films ($\sim$3m^2) has been demonstrated [174].

By using a proper gas ratio in the PIII process, fluorinated DLC films have nonwetting properties similar to those of Teflon and are 10 times harder [164]. Based on nanoindentation measurements and simulation, the surface Young's modulus of the plasma-treated amorphous polyolefin (APO) plastic sample is increased to 25 GPa from the untreated value of 1.8 GPa [175]. A duplex oxide/DLC coating deposited by PIII has excellent tribological prosperities. The hardness of the duplex alumina/DLC coatings is over 2000 HK$_{10g}$ [176]. An intermediate alumina layer is instrumental to withstanding the sliding wear at high contact loads and minimizing the impact wear damage. A C$_2$H$_2$ to Ar flow ratio of 0.25–0.35 is shown to produce hard amorphous carbon films with low hydrogen content, good adhesion, and low internal stress. Diamond-like carbon layers with a surface hardness of 30 GPa and compressive stress of 9 GPa have been produced using rf plasma containing C$_2$H$_2$ and Ar [177]. Carbon thin films have also been fabricated on 9Crl8 AISI440 stainless bearing steel by C$_2$H$_2$ PIII. The friction coefficient of the untreated sample reaches a relatively high value of 0.8–1.0 quickly, whereas the friction coefficients of all the treated samples remain at a low value of 0.1–0.2 throughout the test. It was also observed that there is an optimal process window in which the synthesized films have superior properties. Pulsing rates and rf power that are too high increase the sp^2 contents and give rise to worse tribological characteristics [178].

2. Metal Ion Implantation–Deposition

Metal ion implantation in conjunction with gas ion PIII can yield more superior tribological properties in some materials that do not form hardened phases after gas ion implantation alone. Although solid strengthening using nitrogen or carbon may be responsible for the improvement in the surface tribology, precipitation strengthening is more effective. Therefore, PIII using vacuum arc metal plasma sources is an effective way to implant and deposit metals. Pure metal ion implantation [179–181] and plasma ion mixing (a hybrid process combining deposition and implantation) [182–185] can be readily conducted, and this technique will be described in detail in a latter section.

IV. PLASMA-BASED VAPOR DEPOSITION

A. Introduction

Unlike plasma nitriding or implantation, plasma-based vapor deposition (PBVD) can produce a surface layer or film that is different from the bulk materials. A PBVD process generally consists of three steps: (1) generation of the deposition species, (2) transport of the species to the substrate, and (3) film deposition. All processes can be characterized by two sets of parameters: plasma and process parameters. Examples of the plasma variables include the electron density, electron energy, and ion distribution, whereas process variables include the evaporation/sputtering rate, gas composition, pressure, gas flow rate, substrate bias, and temperature [186]. These parameters are coupled to each other to a certain extent. The two most common forms of PBVD are PVD and chemical vapor deposition (CVD). Typical plasma-based physical vapor deposition (PBPVD) processes are evaporation, ion plating, sputtering, and plasma/ion beam based and/or assisted deposition techniques.

Plasma-based chemical vapor deposition (PBCVD) involves chemical reactions on the substrate surface and most likely reactions in the gas phase as well. The treatment temperature in PBCVD is usually higher to facilitate the chemical reactions. Current PBVD processes allow for the deposition of metals, alloys, ceramic, and polymer thin films onto a myriad of substrate materials [187–189].

B. Plasma-Based Physical Vapor Deposition

1. Introduction

Plasma-based physical vapor deposition processes can be classified according to the means of ionization, source of deposition elements, coating method, and so on [65]. For example, magnetron sputtering and "nonenhanced evaporation" typically use a weakly ionized plasma with an ionization fraction of <<10%. On the other hand, processes such as "enhanced" evaporation, cathodic arc deposition, ionized magnetron sputtering, and ECR plasma deposition have ionized fractions of over 10% [184].

2. Evaporation

In the evaporation process, vapors are produced from a source heated by direct resistance, radiation, eddy currents, electron beam, laser beam, or an arc discharge. The process is usually carried out in vacuum (typically 10^{-5} to 10^{-6} Torr) so that the evaporated atoms travel in an essentially collisionless line-of-sight trajectory before condensation onto the substrate that is usually held at a ground potential. The film thickness variation usually follows the cosine rule, and more uniform film deposition can be accomplished by rastering the sample or by operating at a higher pressure to enhance collisions [190]. A reasonable uniformity of $\pm 10\%$ can be attained in gas-scattering evaporation or pressure plating.

Ion plating processes can be conducted using a glow discharge [191,192]. The material is vaporized in a manner similar to that in the evaporation process but passes through a gaseous glow discharge on its way to the substrate ionizing some of the vaporized atoms (Fig. 15). The glow discharge is produced by biasing the substrate to a high negative potential (several kilovolts) in the presence of a gas, usually argon or reactive gases, at a pressure of 5 to 200 mTorr. In this simple mode, which is known as diode ion plating, the substrate is bombarded by energetic ions leading to constant cleaning of the substrate (i.e., removal of surface impurities by sputtering) thereby yielding better adhesion and lower-impurity contents. Ion bombardment also causes modification of the microstructures as well as residual stresses. It can produce undesirable effects such as reduction in the deposition rate because some of the deposited materials are inevitably sputtered off, as well as unintentional sample heating. The latter problem can be overcome using the supported discharge ion-plating process in which the substrate is no longer subject to a high negative potential. The electrons required to sustain the discharge are typically emitted from an auxiliary heated tungsten filament. Binary alloy compound layers are often fabricated using evaporation ion plating and the film composition can be optimized relatively easily. For example, $(Ti_{1-x}Cr_x)N$ coatings can be deposited in a reactor with two evaporating sources, Ti and Cr heated by an electron beam and resistively [193]. The Ti and Cr concentrations in the coating are controlled by the Ti/Cr evaporation ratio. The coating hardness increases with higher Cr concentration (x) and exhibits a maximum value of 6000 HK at $x = 0.8$. The $(Ti_{1-x}Cr_x)N$ coatings possess compressive residual stress that increases with a larger Cr content.

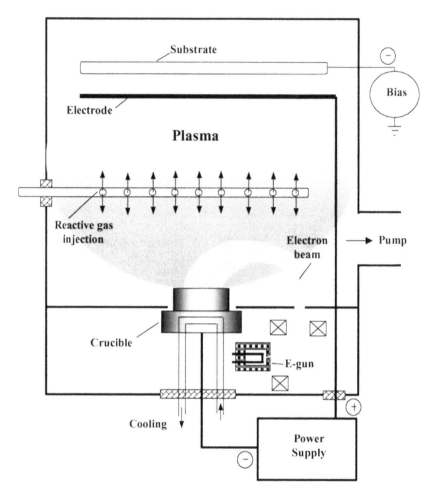

Figure 15 Reactive evaporation using an EB evaporation source. (From Ref. 65.)

3. Sputtering

a. Basics. Sputtering is an effective tool although the deposition rate is typically smaller than that of the evaporation and ion plating. The simplest sputtering technique is d.c. glow discharge sputtering (planar diode sputtering). An important parameter is the energy distribution of the sputtered species that usually has a maximum at about 4–7 eV and a long tail to over 50 eV. Theses energies are more than an order of magnitude higher than those of the evaporated species and allow the depositing species more penetration and surface mobility to form dense, well-bonded films [194]. Under typical metal or semiconductor thin film deposition conditions, most sputtered materials are ejected in the neutral atomic state. The fraction of the charged particles sputtered from clean metal and semiconductor surfaces is quite small and about 10^{-4}, and the ion yield increases if the surface is contaminated by strongly electropositive or electronegative species [195–198].

The sputtering yield depends on the target species and nature, energy, and angle of incidence of the bombarding species [199], but is relatively insensitive to the target temperature [200]. The yield is also independent of whether or not the bombarding species is ionized. In fact, incident ions have a high probability of being neutralized by field-emitted electrons before impact [201]. Molecular species behave as if individual atoms arrive separately with energies partitioned according to their mass ratios [200]. The sputtering yield tends to be largest when the mass of the bombarding particle is of the same order of magnitude or larger than that of the target atoms. The use of inert gases avoids chemical reactions in the target and substrate and argon is often used because of its mass and relatively low cost [199].

The dependence of the sputtering yield on the ion energy exhibits a threshold of 20–40 eV [297] followed by a nearly linear region that can extend to several hundred electron volts. In the regime between 0.5 and 2.0 keV [202],

$$Y(E) = Y(1\text{keV}) \left(\frac{E}{1\text{keV}} \right)^{0.5} \tag{11}$$

At higher energies, the yield vs. ion energy relationship becomes sublinear. However, when the ion energy increases such that the ion range exceeds several atomic layers, ion penetration is too deep and the sputtering yield diminishes. The sputtering process is most effective in the linear range. The sputtering yield also depends on the ion angle of incidence and the maximum yield is reached at angles at 60–80° from normal [202,203]. An empirical expression can be used to describe the overall dependence of the sputtering yield on the incident angle [204]:

$$Y = \frac{Y(0)}{\cos^f \theta} \exp \left[f \cos \theta_{\max} \left(1 - \frac{1}{\cos \theta} \right) \right] \tag{12}$$

where $Y(0)$ is the sputtering yield at normal angle, and $\theta_{\max}$ is the incident angle corresponding to the maximum ion yield, $f = \sqrt{U_{\text{sb}}} (0.94 - 1.33 \times 10^{-3} \times m_r/m_p)$, and m_r and m_p are the mass of the target and sputtering ions, respectively. The maximum normalized yield, $Y(\theta)/Y(0^0)$, is higher for a lighter projectile, higher projectile energy as well as larger surface binding energy of the target materials. The sputtering yield also depends on the crystallographic orientation of a single crystal target. The energy dependence of the yield on selected incident directions and angular dependence for different ion energies can be explained in terms of channeling [203].

There is an optimal sputtering pressure in practical applications. At a low pressure, the number of ions striking the target per unit time is small, and so a high pressure is preferred for high rate deposition. However, an overly high gas pressure introduces excessive collisions in the gas phase reducing the ion bombarding energy and sputtering rate. The biggest limiting factor at high pressure is the higher probability that sputtered atoms return to the cathode by scattering. For instance, at a sputtering pressure of 100 mTorr, only 10% of the atoms ejected from the target are able to diffuse beyond the sheath [203]. Sputtering of a composite target at low ion energies under normal ion incidence causes the lighter atoms to be preferentially ejected in a direction normal to the target surface resulting in the enrichment of the lighter species in the deposited film [205]. This phenomenon is due to reflective collisions of lighter atoms by the heavier ones [206].

b. Enhancement of Ionization and Sputtering Rate. A large sputtering rate is usually preferred in industrial applications and the means to achieve it has been investigated extensively. The sputtering rate can be increased by raising the discharge voltage but the limitation is that the ionization cross section decreases with increasing electron energy for energy greater than 100 eV. Another way to increase the sputtering current (rate) at a given voltage is to increase the pressure. However, if the pressure is increased too much, a reduction in the sputtering rate occurs as the ions are slowed by collisions and scattering with the gas atoms. For a d.c. glow discharge, the typical pressure is between 20 and 100 mTorr [206]. Adjusting the distance between the electrodes to minimize collisions helps.

Enhancing the ionization efficiency is the most effective approach, particularly at low pressure. The longer mean free path under low pressure leads to both higher ion bombarding energy and larger number of sputtered particles impinging on the sample surface. Consequently, the sputtering yield (proportional to ion energy) and deposition rate can be higher at low pressure. However, to compensate for the reduced number of available ions at low pressure, more efficient ionization is required. Many techniques such as d.c. triode, rf, and magnetron sputtering have been used to promote the ion density.

c. Thermionic Electron Assisted Sputtering. In a diode glow discharge system, the use of a heated filament can very effectively increase ionization and lower the operating pressure [207]. Higher deposition rates (several hundred nanometers per minute) can be achieved with triodes in lieu of planar diodes. In this technique, electrons for the discharge are generated by thermionic emission from a hot filament and attracted to an anode at about 100 to 200 V d.c. relative to the hot filament. When a negatively biased substrate is introduced in the sputtering system, it is possible to control the substrate voltage or current density independent of the gas pressure. This is in contrast to diode ion plating in which the current density is influenced by the substrate bias and gas pressure. Perhaps the most important benefit of the thermionically assisted triode system is the capability to increase the substrate current density even at a low bias or gas pressure [188]. However, the triode system has some drawbacks. One of the problems is the nonuniform plasma density distribution that varies with the chamber size, characteristics of the thermionic filament, location of the electrodes with one another, potentials at the electrodes, and pressure [208]. For example, the ion density is higher near the filament leading to nonuniform sputtering, and the target must be inclined relative to the axis of the electron beam to compensate for it.

d. High-Frequency Sputtering. A glow discharge can be created and sustained at low pressure using high-frequency or rf electromagnetic field. The effect is similar to that created by the thermionic electrons in a triode system providing a longer effective electron lifetime to collide with other particles. Thus, sputtering can be performed at a pressure of about 2 mTorr [203] and even lower than 0.1 mTorr [188]. The electrical coils can be installed inside or outside the vacuum chamber, with the latter configuration being more practical when reactive sputtering gases are used. Pinhole free coatings of In_2O_3, TiN, and TaN have been fabricated on glass substrates [188]. Transparent MgO films have been deposited using rf plating, and both the MgO (111) and (220) x-ray diffraction patterns are observed, whereas for the triode-type ion plating process, MgO (200) has the preferred orientation [209].

e. Magnetron Sputtering. In spite of its success to deposit thick coatings, a wider use of the triode configuration has been hampered by the difficulties in scaling and vulnerability of the thermionic emitter to reactive gases. As a result, magnetron sources are more common in high deposition rate devices [199]. The development of high-performance magnetron sputtering sources enables relatively high deposition rates, large deposition area, and low substrate heating, thus revolutionizing the sputtering process [210]. The combination of electric and magnetic fields (50–500 Gauss) illustrated in Fig. 16 causes the secondary electrons to drift in a closed circuit, or magnetron tunnel, in front of the target surface [199,202,211]. The magnetic field concentrates the plasma in the space immediately above the target trapping electrons near the target surface. This electron confinement significantly increases the efficiency and consequently, a magnetron can be operated at low pressure (e.g., 1–3 mTorr) and low voltage (e.g., 350 V). Because of the magnetic field line shape, the erosion of the target is nonuniform, but this problem can be overcome by moving the target

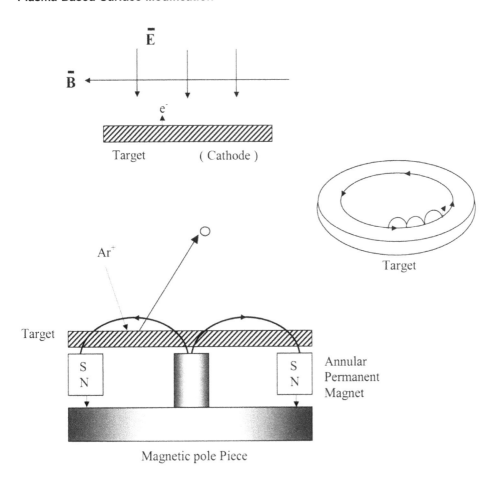

Figure 16 Schematic of magnetron sputtering arrangements. (From Ref. 202.)

with respect to the magnets. Magnetrons work well in either the d.c. or rf mode. Modern magnetrons have characteristic dimensions ranging from a few centimeters to more than several hundreds of centimeters. The power range is from 1 W to 120 kW [211], and a power density as high as 200 W/cm^2 can be accomplished [212].

Many variations of the magnetron sputtering technique have been developed [194] such as unbalanced magnetron [213] in which the magnetic fields are arranged so that some flux lines flow toward the substrate. This can be achieved by using magnets of different strengths [Fig. 17(a)] or applying an additional electromagnetic field using current-carrying coils [Fig. 17(b)]. Some electrons will spiral along these field lines toward the substrate. Ions can be formed along the electron path and because there is a net movement of negative charge from the magnetron region, positive ions will be attracted to the substrate. A pragmatic configuration is to arrange two or four or six magnetrons to face each other in pairs having opposite magnetic polarities [Fig. 17(c)] and place the substrate between them, forming a closed-field unbalanced magnetron sputtering system (CFUBMS) [214,215].

   f. Hybrid Processes. Different techniques can be combined to yield better results. For example, rf reactive sputtering and ion plating have been used to fabricate Ti–Al–O films [216], silicon carbide films [217], highly tetrahedral, dense amorphous carbon (ta-C)

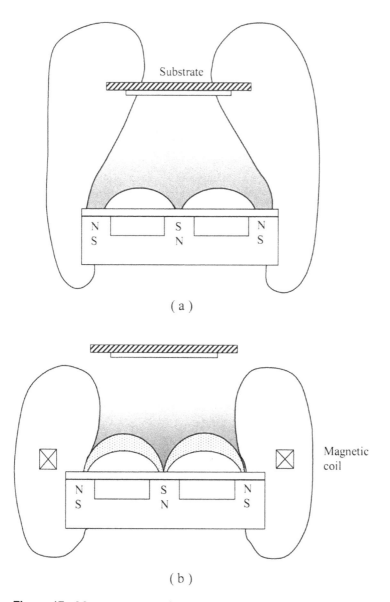

Figure 17 Magnetron sputtering systems: (a) unbalanced magnetron, (b) unbalanced magnetron with controlled field. (From Ref. 194.)

films [218], and cubic boron nitride films [219]. Superhard materials such as nanocrystalline cubic boron nitride (c-BN), beta-silicon carbide (β-SiC), amorphous boron carbide (B_4C), and highly tetrahedral amorphous carbon (ta-C) are produced by rf unbalanced magnetron sputtering in combination with intense ion plating in an argon discharge [220]. Implantation, surface mobility enhancement due to ion plating, and thermal crystallization are the three important factors affecting the formation of superhard phases with strong covalent bonding.

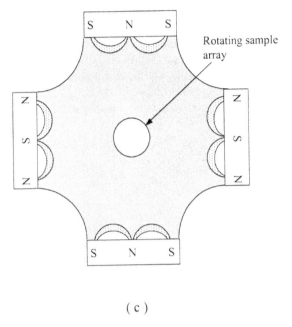

(c)

Figure 17 (c) closed-field unbalanced magnetron. (From Ref. 194.)

4. Cathodic Arcs

Energetic ion bombardment induces ion mixing that can improve film adhesion and remove loosely bonded surface atoms by sputtering to enhance the film quality. From this standpoint, cathodic arc deposition has some advantages because the average energy of the ions is about 40 eV per particle compared to 0.1 eV in evaporation and 5–10 eV in typical sputtering processes [221]. The ionization efficiency (approaching 1.0) in cathodic arc processes is high (approaching 1.0), whereas the ionization efficiency of ion plating (~0.2) and sputtering (<0.1) is much smaller. In cathodic arc deposition, one can easily optimize the processes by adjusting the substrate bias, pulse duration, and other parameters. The high ionization efficiency provides highly reactive metallic and gaseous species to form compounds such as nitrides, carbides with excellent properties [222].

The characteristics of vacuum arcs such as the triggering voltage, macroparticle removal, charge state, and ion velocity have been described in a previous section. The evaporation rate W that is crucial in cathodic arc deposition depends on the arc current and voltage [221]. In the case of Cu, the evaporation rate in grams per second at an arc voltage of 20 V and current (I_a) less than 150 A is given by:

$$\log W = -4.19 + 1.32 \log I_a - 0.147 (\log I_a)^2 \tag{13}$$

Cathodic arc deposition can be implemented in either the d.c. or pulsed mode. A d.c. discharge (10 to 300 A) is a well-established PVD technique to form hard coatings such as TiN, CrN, TiCN, and AlTiN on cutting and forming tools to reduce friction and wear. Industrial instruments automatically carry out the essential steps including heating, cleaning, and deposition. An a.c. arc discharge is sustained by the repetition of short current pulses up to several kiloamperes. The average current is usually higher and bodes well for applications in which some sort of filtering is required to reduce or eliminate droplet (macroparticle) deposition onto the substrates. Another useful characteristic of the pulse mode is the higher ionization efficiency that is especially beneficial to the deposition of hard amorphous carbon and metal interconnects in the semiconductor industry. An a.c. system can be installed on industrial d.c. equipment relatively easily to offer more manufacturing flexibility [223], even though filtered cathodic arc deposition has mostly been limited to experimental studies and precommercial demonstrations because macroparticle filters reduce the deposition rate [224].

C. Plasma-Based Chemical Vapor Deposition

1. Introduction

Chemical vapor deposition is a versatile and economical method to produce coatings, fibers, microelectronics, and monolithic components. It is possible to produce coatings of almost any metallic and nonmetallic elements, including carbon or silicon, as well as compounds such as carbides, nitrides, oxides, and intermetallics [225]. The basic processing steps consist of the following [226]: (1) transportation of the reactants from the reactor inlet to the deposition zone, (2) gas-phase reactions leading to the formation of the film precursors and by-products, (3) transportation of the film precursors to the growth surface, (4) adsorption of the film precursors on the growth surface, (5) surface diffusion of the film precursors to deposition sites, (6) surface reactions and incorporation of film constituents, (7) desorption of by-products from the surface, and (8) transportation of by-products to the bulk gas flow region. Each of these process steps must be understood and controlled in order to fabricate films with the desired materials properties. The chemical processes used in CVD of the thin films can typically be classified as (1) thermal decomposition (pyrolysis), (2) reduction, (3) oxidation, (4) hydrolysis, (5) nitridiation, (6) chemical transport, (7) disproportionation, (8) catalysis, (9) synthesis, (10) photolysis, and (11) combined reactions.

2. Basic Chemical Vapor Deposition Process

The optimal deposition temperature in some thermal CVD reactions may be impractically high. For example, titanium nitride deposition at an acceptable rate using a gas mixture of $TiCl_4$, N_2, and H_2 requires a temperature of about 1000°C that is higher than the softening temperature of tool steels. The processing temperature can sometimes be reduced by using different gases and reactions such as $TiCl_4$ and NH3 to deposit titanium nitride at ~600°C [227], but the choices can be limited. An alternative is to create an electrical discharge in the reactant gases to produce a significant number of free radicals (i.e., $SiH_4 \rightarrow SiH_2$, SiH and $NH_3 \rightarrow NH$, NH_2, etc.) enabling successful deposition at a lower surface temperature [228]. The major advantage of plasma CVD over thermal CVD is the lower temperature as shown in Table 1 [225] thereby allowing deposition on substrates such as aluminum, organic polymers, and austenitic steels, which undergo structural changes at high temperature. Other advantages of plasma CVD include minimal diffusion, reduced stress due to small thermal expansion and mismatch, as well as higher deposition rates.

There are three common reactor designs [229]: (1) electrodeless reactor, (2) external electrodes or tubular reactors, and (3) internal electrodes or parallel plate reactors. The plasma is typically generated by d.c. glow discharge [230,231], rf [232,233], or microwave/ECR [234,235]. Proper selection of the reactor and plasma depends on the required deposition rate, thin film properties, and applications.

Table 1 Typical Deposition Temperature in Thermal and Plasma CVD

Materials	Deposition temperature (°C)	
	Thermal CVD	Plasma CVD
Epitaxial silicon	1000–1250	750
Polysilicon	650	200–400
Silicon nitride	900	300
Silicon dioxide	800–1100	300
Titanium carbide	900–1100	500
Titanium nitride	900–1100	500
Tungsten carbide	1000	325–525

3. Pulsed Glow Discharge

Pulsed d.c. glow discharges are common in CVD systems for hard coatings [236–239]. Compared to continuous d.c. discharges, pulsed d.c. discharges produce fewer arcs, and compared to rf discharges, there is less electromagnetic interference. It is also possible to control the substrate temperature by adjusting the pulse duty cycle without changing the plasma parameters. However, to achieve good deposition uniformity, the reactor must be designed properly. It has been demonstrated that a gas distribution system is required for a uniform thickness distribution and better film quality [240,241]. In addition, increasing the size of the reactor for higher capacity and sample throughput requires larger discharge currents that can increase the possibility of unipolar arcs, and so shorter pulses may be necessary. However, shorter pulses may compromise the thickness uniformity and film quality, and the problem is related to the dynamics of the plasma.

The glow discharge dynamics has been investigated, particularly in large reactors, and nonuniform deposition has been observed [242–244]. For instance, TiN PBCVD has been studied using video cameras monitoring the optical emission from the plasma, Langmuir probe measuring the plasma potential, and current probe determining the discharge current [242,243]. In the presence of TiCl4, the formation and propagation of the discharge in a large reactor are slow thereby causing time-dependent plasma density nonuniformity as shown in Fig. 18 [242]. The development of the discharge depends on the geometry of the cathode, placement of the substrates, gas mixture, and voltage waveform. The optical emission spectra acquired in the vicinity of the sample in pulsed $TiCl_4$–H_2–N_2–Ar discharge disclose the presence of 12 species, Ti, Ti^+, Cl, Cl^+, H_2, H, N_2, N_2^+, N, N^+, Ar, and Ar^+, in the TiN deposition process. Analysis of the waveforms of both the discharge current and optical emission spectroscopy (OES) signals shows that the temporal evolution of the discharge can be ascribed to the slow movement of the ionization front from the cathode to the anode [244]. The problem can be overcome by increasing the conductivity of the plasma at the beginning of each pulse, for instance, by applying additional synchronized short high-voltage pulses.

4. Applications

Plasma-based chemical vapor deposition has been used to fabricated hard films such as TiN, $Ti_{1-x}Al_xN$ [238,245–248], DLC [249–253], boron nitride or carbide [254–256], and so on. Successful deposition requires special processes, and deposition of TiN and DLC is described here as an example.

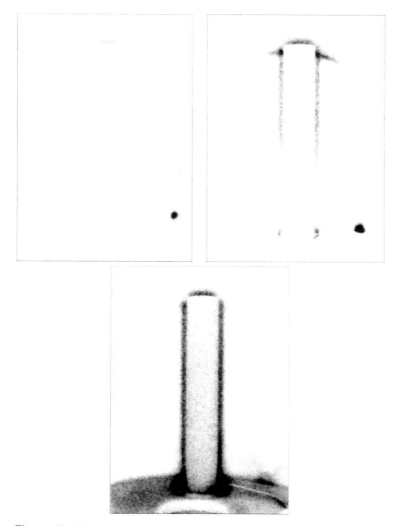

Figure 18 Pictures taken with the video system of the cylinder at 3, 55, and 100 μsec into the pulse for a plasma with a TiCl$_4$ flow of 1 slh. (From Ref. 243.)

a. Titanium Nitride. Titanium nitride is one of the first coatings successfully fabricated by PBCVD. The materials possess superior physical and mechanical properties including high hardness, wear resistance, and chemical inertness. A TiN coating can also lower the friction between the tool and the specimen, decrease heating, and increase the tool lifetime.

CHLORINE CONTENT. TiCl$_4$ is a common precursor in TiN deposition. The deposition temperature is generally between 600° and 800°C to reduce Cl incorporation when NH$_3$ is used [257]. TiN can also be deposited at 300°C by plasma CVD, but the lower the substrate temperature, the higher the chlorine content if TiCl$_4$ is used as the precursor. A high Cl content affects the corrosion resistance seriously. Figure 19 shows that a moderate temperature and high voltage are preferred [258]. Using hydrogen reduction, the Cl concentration can be reduced to 1 at.% at 200°C [259].

COATING UNIFORMITY. Nonuniform coating thickness is expected because sputtering and plasma phenomena depend on the object shape. The thicknesses of the coatings de-

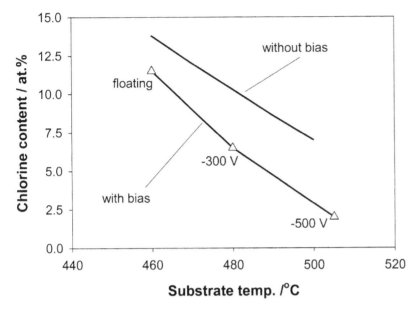

Figure 19 Influence of substrate temperature, bias voltage on chlorine content in PBCVD TiN. (From Ref. 258.)

posited onto 15 × 15 × 6 mm tool steel samples using various voltages are shown in Fig. 20 [258]. The black bars show the coating thicknesses using rf plasma and electrically floating substrates, whereas the white bars show the thicknesses using d.c. glow discharge plasma at a voltage of −800 V. The coatings fabricated using rf plasma and negative bias voltages (e.g., −300 V) exhibit intermediate values. The mechanism has been postulated as follows. In rf plasma, all the active species in front of the substrate react to form the film, and so the layer thickness depends on the solid angle. Inner edges have smaller solid angles, and the thickness

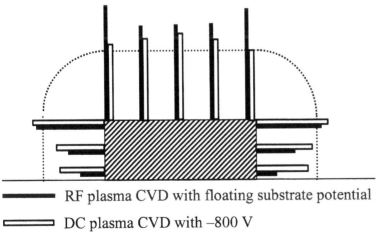

RF plasma CVD with floating substrate potential

DC plasma CVD with −800 V

Figure 20 Thickness distribution of PBCVD TiN fabricated under different processing conditions. (From Ref. 258.)

on the side surface decreases toward the bottom electrode. In d.c. plasma, sputtering occurs simultaneously with CVD. Sputtering occurs preferentially at the outer corners, while the inner surfaces gain additional thickness from backscattering effects more pronounced at high pressure. Nonuniformity in the coating thickness of a layer fabricated by rf plasma CVD can thus be mitigated to some degree by applying an appropriate negative bias voltage. Generally speaking, the thickness uniformity is relatively good for low-voltage CVD because the plasma sheath is thin and more conformal around the objects [260]. Large samples require more attention [242,243], and it is imperative that the glow discharge enshrouds the entire surface in order to deposit a uniform film. Unfortunately, glow discharge is always ignited between certain points and then propagates to other areas, and particular measures such as applying initial high-voltage pulses must be taken.

   b.   *Diamond-Like Carbon.*   Diamond-like-carbon is a collective name of amorphous carbon materials with hydrogen contents of less than 1% (ta-C) to about 50 at.% (a-C:H). DLC has attractive properties such as low friction coefficient, high wear resistance, chemical inertness, a relatively high optical gap, and high electrical resistivity. The structure and tribological characteristics of DLC films depend on the deposition process, hydrogen concentration, and chemical bonding.

   HYDROGEN CONTENT.   To fabricate DLC films by PBCVD, the substrates are usually biased to a negative voltage relative to the plasma to facilitate ion bombardment. Deposition is performed in an ambient containing 10–50% hydrogen that is required to obtain the diamond-like properties in the materials [261]. It determines the film structure, passivates the dangling bonds in the amorphous structures, gives rise to the optical and electrical properties, and modifies the internal stresses [262,263]. The stress in undoped DLC films increases with larger fractions of free (unbonded) hydrogen. A higher compressive stress is observed in films deposited with methane, whereas lower stress is found in films deposited using cyclohexane [263]. The stress can be lowered by introducing N, Si, O, or metals into the films [261]. The hydrogen contents also affect the tribological properties [264]. A high steady-state friction (friction coefficient of 0.6) is observed for the least hydrogenated and mostly sp^2 DLC films in UHV. On the other hand, very small steady-state friction coefficients are observed for the highly hydrogenated films in UHV and for the slightly hydrogenated film in hydrogen ambient. The high steady-state friction coefficient observed for the slightly hydrogenated film with dominant sp^2 hybridization is associated with a pi–pi* subband overlap. It is responsible for the increased across-the-plane chemical bonding with high shear strength similar to that observed in unintercalated graphite under the same UHV conditions. The extremely low friction of the other samples is due to hydrogen saturation across the shearing plane and weak van der Waals interactions between the polymer-like hydrocarbon top layers [264]. It has also been reported that under dry conditions, the ta-C has a high friction coefficient that is reduced by doping with hydrogen. The hydrogen content together with graphitization plays an important role in the friction properties of DLC films [265].

   BIAS AND PRESSURE.   The deposition and properties of DLC films are controlled by the substrate temperature and bias with the latter being more dominant [261]. The hardness, density, refractive index, and deposition rate increase with higher substrate bias. It has also been observed that even for films deposited in a helical resonator, the substrate must be biased negatively in order to obtain diamond-like properties [261]. The bias also affects the morphology of DLC films [266]. Under low pressure and low bias voltage conditions, the DLC film is rough due to surface particles. Increasing the pressure and bias voltage yields films that are smoother and more particle free. With increasing self-bias voltage (rf), the hardness of the film first increases, reaches a maximum at around 250 V, and then decreases.

The impact of pressure on the surface roughness has been studied [267]. At high pressure, the surface roughness increases with thickness, but the roughness of films deposited at medium pressure is almost constant regardless of the film thickness. The surface roughness decreases at pressure below 53 Pa and hard films are obtained. The experimental results can be explained by the different species in the plasma. At low pressure, ionic species accelerated toward sharp areas on the film surface lead to sputter removal of weakly bonded materials. At higher pressure, neutral species in the plasma dominate, and so the relative magnitude of sputtering is reduced. The hydrogen contents and sp^3/sp^2 ratios determined from infrared spectra suggest that the lower hardness of the rough surfaces can be explained by a higher content of sp^2 in the films deposited at high pressure. However, sputter removal of the sp^2 content leads to a significant increase in the ratio at low pressure [267].

DOPED DLC.   Some materials are similar structurally to DLC or ta-C but, in addition to carbon and/or hydrogen, contain nitrogen (NDLC or CN_x films), silicon (SiDLC), silicon and oxygen, fluorine (FDLC), and metal atoms (MeDLC) [261,268]. Compared to undoped DLC, these materials have smaller internal compressive stress (N, Si, metal incorporation), lower surface energy, smaller friction coefficients (F, Si–O incorporation), and better electrical properties. Nitrogen incorporation has been shown to improve the field emission properties of both DLC and ta-C, and Si incorporation reduces the etching rate of DLC in oxygen plasma making it useful as an etch stop for undoped DLC. These doped diamond-like carbon films are deposited by the same techniques as the regular films while adding

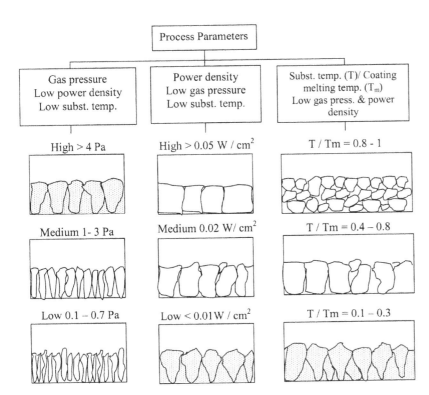

Figure 21   Effects of ion plating process parameters on coating structure. (From Ref. 271.)

species containing the modifying elements to the plasma, and details can be found in the references cited above.

D. Particle Bombardment During Coating Process

Ion bombardment during plasma-based vapor deposition affects the film density, grain size, crystallographic orientation, morphology, topography, and many other characteristics of the films [269]. In particular, the ion energy, ion/neutral flux, angle of incidence are of importance. The influence of the ion energy on the coating has been investigated [34,294,269,270]. This energy dependence varies with different coating materials and bombarding species. It should also be mentioned that many parameters affect the film formation, and the substrate temperature, gas pressure, and bombardment energy are most important. Fig. 21 illustrates the parameter-dependent film structures and a high discharge power and low gas pressure may lead to a layered structure [271].

V. HYBRID PROCESSES

A. Introduction

Plasma surface modification unambiguously enhances the tribological properties of materials and industrial components, but is a complex science because the mechanism depends on many factors. One single coating or surface modification technique may thus not yield the desirable mechanical, chemical, and tribological properties. For example, a multilayered structure can act as a crack inhibitor thereby increasing the fracture resistance of the coating. This effect can be the result of several different mechanisms such as crack deflection due to weak interfaces [272], crack tip shielding by plastic deformation in combination with strong interfaces [273,274], favorable residual stress distribution [275], crack deflection due to the large differences in the cohesive strength between adjacent lamellae [276], and differences in

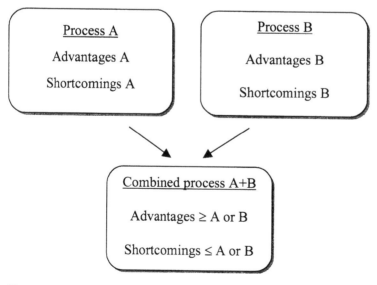

Figure 22 Optimizing the films using hybrid processes. (From Ref. 279.)

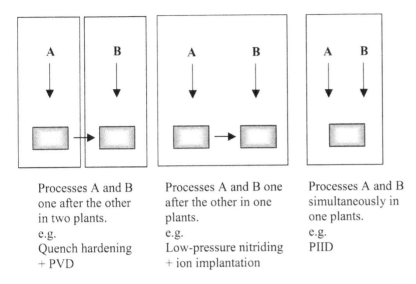

Processes A and B
one after the other
in two plants.
e.g.
Quench hardening
+ PVD

Processes A and B one
after the other in one
plants.
e.g.
Low-pressure nitriding
+ ion implantation

Processes A and B
simultaneously in
one plants.
e.g.
PIID

Figure 23   Schematic illustration of different hybrid processes. (From Ref. 279.)

the elastic properties and/or coating morphology between adjacent lamellae [277]. A well-designed hybrid process can combine the advantages of several independent techniques while minimizing their negative impacts. However, the optimal combination of different techniques tends to be quite complicated, and one must consider the cost as well as compatibility of equipment and processes. For instance, a pressure range of 10 to 1000 Pa is usually used in plasma nitriding but plasma implantation demands a much better vacuum in the regime of 0.1 to 0.01 Pa. Nonetheless, the successful implementation of hybrid processes can offer unmatched technical and economical benefits [278].

The principles on combining processes are illustrated in Fig. 22. Two single processes A and B are combined into treatment A with their advantages added but disadvantages eliminated [279]. For example, a hybrid PVD/ion implantation process not only achieves a thick layer (PVD) but also a highly sticking film (ion mixing from ion implantation). Combined processes can be realized in different sequences. Processes A and B can be implemented in two machines/plants or successively in the same machine/plant. The latter is usually preferred due to logistic reasons. The integration level increases from left to right as illustrated in Fig. 23. As an example shown in Fig. 24, hybrid ion implantation or hybrid plasma-based CVD is usually used to strengthen the surface properties.

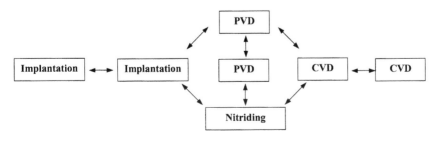

Figure 24   Examples of plasma-based hybrid processes.

B. Hybrid Plasma-Based Physical Vapor Deposition

An early application of the hybrid PVD concept is the combination of arc evaporation and magnetron sputtering to deposit homogeneous or graded (Ti,Al)N coatings [258,280,281]. The process uses an aluminum cathodic arc source and titanium sputtering electrode in a nitrogen/argon atmosphere to prevent oxidation. Coatings with graded Al concentrations are deposited by changing the magnetron power. $(Ti,M_2)N$ or $(Cr,M_2)N$ with M standing for Zr, V, Nb, Ta, Cr, or W can be deposited in an argon/nitrogen ambience [258]. Recently, combined steered cathodic arc–unbalanced magnetron deposition has been used to deposit CrN/NbN [282], TiAlN, and CrN films [283]. These coatings exhibit increased hardness and considerably reduced sliding and microabrasive wear compared to TiAlCrN and CrN. The TiAlN–VN films are exceptional in dry sliding wear tests and all TiAlN-based films perform well in microabrasive wear tests [283].

C. Plasma-Based Physical Vapor Deposition and Chemical Vapor Deposition

The low friction coefficient and high hardness compared to steel make DLC an attractive material for tribological applications. However, its excellent chemical inertness and compressive stress make good adhesion to a steel substrate difficult. It has been shown that a small amount of metals at a concentration below 10 at.% can lower the stress of DLC and increase film adhesion [258]. In fact, Me:C coatings composed of DLC and metal carbides exhibit similar tribological properties as DLC films [284]. The ideal Me:C coating should possess a layered structure with smooth transition from pure metal to Me:C at the interface. The process starts with magnetron sputtering in an argon discharge for Zr [285]. A PVD process is used to deposit the interfacial metallic layer giving the high film adhesion. In a second step, a hydrocarbon gas is mixed with argon. The hydrocarbon is then cracked in the sputtering discharge and carbon is incorporated into the deposited metal layer. Moreover, carbon is introduced into the sputtering target to poison the target. As a result, the metal sputtering rate decreases drastically and the deposited layer becomes carbon rich. At this point, the process is dominated by plasma CVD with the hydrocarbon/argon discharge because the sputtering rate of carbon from the poisoned target is low.

D. Plasma Nitriding and Plasma-Based Physical Vapor Deposition

Plasma nitriding and physical vapor deposition are sometimes used together to extend the lifetime of the industrial components. This hybrid process has been applied to tool steel [286], stainless steel [287], aluminum alloy [288], titanium alloy [289,290], and construction steel [291]. A hybrid process for Ti6Al4V is described here to exemplify the effectiveness [289]. Plasma nitriding is first carried out at 2 kW power, 1500 V, 750°C, and treatment time of 16 hr. The surface roughness after nitriding is 0.32 ± 0.04 µm. Subsequently, a 1.0-µm CN_x film is deposited on the plasma-nitrided Ti6Al4V sample using unbalanced rf magnetron sputtering at a base pressure of 5×10^{-5} Torr. A high-purity (99.99%) planar graphite target is used in a 99.999% nitrogen discharge. The gas flow rate is 40 sccm, the cathode discharge current is 1 A, the substrate temperature is about 200°C, and the substrate bias voltage is -300 V. The improvement in the tribological characteristics can be attributed to the combined benefits of plasma nitriding and CN_x films. Plasma nitriding produces a graded hardened layer serving as an underpinning and load-bearing layer for the hard CN_x film. The CN_x film deposited at low temperature yields a wear-resistant and low-friction surface without offsetting the benefits of the plasma-nitrided layer. The CN_x film effectively reduces both the interface and surface stress. Graphitization of the wear debris during dry sliding condition can help decrease the coefficient of friction and improve the wear resistance. As a second example, reduction of the friction coefficient by approximately one order of magnitude has been reported for AISI 4140 steel [292,293]. After the plasma treatment, the substrates are coated with TiN, TiAlN, and a hydrogen-free hard carbon film using commercial PVD processes. Test results reveal improved mechanical and wear properties of the plasma-nitrided and coated specimens compared to their untreated counterparts.

E. Nitriding and Plasma-Based Chemical Vapor Deposition

Like hybrid plasma nitriding/PVD, nitriding in conjunction with PBCVD has received attention. For example, duplex DLC [294,295] and TiN and TiC coatings [248,296,297] have been fabricated. A DLC coating has been deposited on steel substrate without an interlayer [294] using a laboratory-size plasma nitriding system (300 mm in diameter, and 400 mm tall) equipped with a 33-kHz unipolar power supply. After the nitriding treatment, Ar sputtering is conducted at 450°C for 30 min to clean and roughen the surface to enhance film adhesion. The DLC film is deposited by rf plasma-assisted CVD in a $CH_4:N_2:H_2 = 66:24:10$ (by volume) ambient at 80 W for 20 min at room temperature. The surface roughness after treatment is about the same as that of the surface that has undergone plasma nitriding alone, but film adhesion is noticeably improved because of the formation of the thin transition layer.

F. Nitriding and Ion Implantation

Hybrid processes combining nitrogen ion implantation and plasma treatment such as ion nitriding and ECR plasma nitriding have been combined for aluminum and Al–Mg alloys [298,299]. The samples exhibit better hardness, friction coefficient, and wear. In particular, ion implantation followed by ion nitriding is very effective in reducing the wear volume in aluminum.

Elevated-temperature nitrogen PIII is intrinsically a hybrid process in which the nitrogen implantation and nitrogen diffusion simultaneously take place. It has been demonstrated that nitrogen can be retained during implantation at elevated temperature even in unalloyed steels [156]. It thus enables controlled and deep diffusion of nitrogen thereby improving the load-bearing capacity of the implanted layer. Heating by the incident ions is sometimes adequate and no external heating is required. The temperature can be controlled by adjusting the frequency and width of the high-voltage pulses applied to the sample. Elevated-temperature PIII yields a surface layer typical of nitrogen implantation (containing in excess of 20 at.% nitrogen) on top of a region consisting of diffused nitrogen in stainless steel [154,158,300,301] and aluminum alloys [73,302]. Better corrosion and wear resistance are accomplished by performing plasma nitriding at elevated temperature and subsequent high-voltage PIII at lower temperature [303].

G. Plasma-Based Physical Vapor Deposition and Ion Implantation

Nitrogen PIII treatment of the TiN films deposited by magnetron sputtering increases the lifetime of the HSS (M series) cutting tools [304]. The sample having a 5-μm predeposited TiN film is implanted at 95 kV with a nitrogen dose of 2.5×10^{17} cm^{-2}. The wear rate of TiN-

(a)

(b)

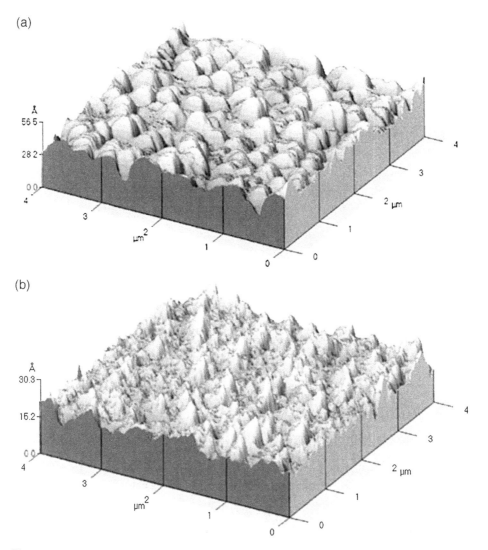

Figure 25   Surface morphology revealed by atomic force microscopy: (a) 8-kV sample, (b) 16-kV sample.

coated cutters is reduced by four to eight times compared to that of a pure TiN coating. An apparatus with combined PVD and PIII capabilities has been produced to synthesize TiN films [305]. The plasma is generated by an ECR plasma confined in the upper part of the vacuum chamber. In the lower part of the vacuum chamber, a magnetron sputtering cathode and resistively heated evaporator are mounted for film deposition. An instrument for in situ or sequential plasma immersion ion beam treatment in combination with rf sputtering deposition or triode d.c. sputter deposition has also been built [306]. In this apparatus, several coating processes in conjunction with PIII can be conducted. The first step is usually the precleaning of the substrate by low-voltage d.c. sputtering, followed by deposition using d.c. or rf/d.c. triode sputtering. PIII is carried out either during the coating process or afterward in one step or sequentially. In this way, a transition layer between the substrate and coating is first formed before the deposition and treatment of the final coating.

(c)

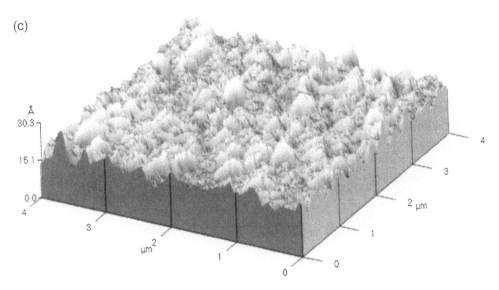

Figure 25   (c) 23-kV sample.

H. Plasma Immersion Ion Implantation–Deposition

Because of the capability to deposit films containing both metallic and gaseous elements, plasma immersion ion implantation–deposition (PIII-D) has received attention from materials scientists and engineers. In this mode, a vacuum arc plasma source in a pulsed or d.c. mode is operated in concert with a gas plasma typically comprising nitrogen [307, 308]. In the pulsed mode, the cathodic arc pulses are synchronized with the substrate bias pulses to achieve either deposition or implantation. Implantation occurs when the sample is biased whereas the deposition results when no bias is applied to the sample. The process efficacy can be enhanced by using a long vacuum arc pulse (or d.c.) in conjunction with continuous or gated sample bias pulses. The ratio of ion implantation to deposition is determined by the duty cycle of the bias voltage [309].

The coexistence of gas plasma enables the synthesis of compound films. For instance, TiN films can be fabricated using a Ti cathodic arc and nitrogen plasma [310]. Magnetron sputtering and PIII have also been combined to conduct PIII-D [311]. Stoichiometric TiN coatings are produced using Ti sputtering targets in an N_2/Ar atmosphere at a pressure of 1.12×10^{-6} bar. TiN layers with good quality and adherence can also be fabricated on 100Cr6 ball bearing steel by this method. Other PIII-D processes using dual plasmas have been suggested [312,313]. During the processes, the vacuum arc voltage is synchronized with the sample high voltage to conduct pure metal ion implantation, pure deposition, ion mixing (metallic ion implantation/deposition), and other hybrid processes under an inductively-coupled rf nitrogen plasma or hot filament glow discharge plasma. A deposition ratio of about 1 to 9 is shown to deliver good results [314].

In PIII-D, the implantation voltage affects the flux and energy of the incident ions and consequently the surface topography of the coatings. Figure 25 depicts the atomic force microscopy (AFM) images of three samples. The 8-kV sample shows a surface featuring small islands that are more or less uniformly distributed on the surface. The surface morphology of the other two samples is, however, quite different. The density of the islands decreases and the islands are higher and steeper at higher bombarding voltage. The hillocks on the 23-kV sample become less uniform and appear like clusters of arrows. The implantation voltage alters the growth dynamics of the coatings, and the difference is believed to be due to ion bombardment [315].

I. Interface Engineering

The characteristics of the interface between the coating and substrate affect the overall quality of the coating, as adhesion of hard coatings strongly depends on the surface structure and properties such as internal stress, surface roughness, and topography [316–319]. Different processing conditions tend to give different results. For instance, a compound layer formed by nitriding has a negative effect on the coating substrate adhesion and reduces the durability of the wear-resistant coating [320–324]. In contrast, a compound layer acting as an intermediate hard layer can give rise to superior sliding wear properties of the composite [292,293,325].The rule of thumb is that a transition layer or special interfacial treatment leads to better results.

Diamond-like carbon deposited on steel substrates is notorious for the high internal compressive stress and poor adhesion. Therefore, buffer layers have been added to enhance adhesion [326,327]. Addition of titanium to the interface between the nitride layer and Ti-based coating has attracted attention. It has been suggested that the Ti interlayer reacts with nitrogen in the nitrided substrate and chemically improves the adhesion between the substrate and Ti interlayer [292,321]. A thin Ti or Ti–TiN_x intermediate layer thus improves adhesion [328]. Some conflicting results have also been reported. Walkowicz et al. [329] have conducted an investigation on the impact of the interfacial structure on film adhesion. Substrates made of hot working steel EW X35CrMoV5 (0.4% C, 0.4% Mn, 1.0% Si, 5.0% Cr, 1.3% Mo, 0.3% V) undergo different thermal–chemical and finishing surface treatments to obtain nitrided layers with different structures: Fe-α(N), Fe-α(N) + "white layer"-ε, γ'. Four different coatings of nitrides, TiN, Ti(C,N), CrN, (Ti,Cr)N, are deposited using the vacuum arc method. The presence of titanium ions is found to degrade adhesion of the PVD coating to the nitrided substrate when the nitrided layer includes the compound zone. This effect is, however, not observed for a CrN coating deposited without Ti. In fact, it has also been observed that Ti ion bombardment increases the thickness of the black layer (see the next paragraph) and degrades the surface properties [330]. Therefore, the interface chemistry and structure are quite complex and have a crucial impact on the effectiveness of the hard coating deposited on nitrided substrates.

Hybrid processes are performed either in the same or different instruments. For the latter, exposure to air is usually inevitable. This can lead to surface contamination and oxidation that can degrade the coating and interface properties. Surface contaminants can be removed by sputtering before the subsequent coating step. However, iron nitride is not very stable and may decompose into ferritic α-iron during ion bombarding at high temperature producing the so-called "black" layer observed in the low alloy steel En40B after Ar ion bombardment at above 450°C [331]. Owing to decomposition, the hardness is lower (HV 400–500). Shearing of the black layer thus occurs, finally leading to delamination of the TiN films. It has been suggested that the coating temperature must be lowered to suppress the formation of this soft black layer, but if the intermediate treatment temperature is reduced, the adhesion strength must be sacrificed because of the inadequate chemical bonding between TiN and iron nitride [332]. Titanium ion bombardment of the nitrided AISI 4140 has been suggested to clean the surface, but the black layer is still observed. The adhesion strength can be somewhat improved by introducing nitrogen gas during the bombardment, but it does not match that of the duplex treatment that does not have α-Fe [290,330]. To overcome this problem, the TiN coating process and ion nitriding should be conducted in the same vacuum chamber, and there is no reduction in the adhesion strength [290,328].

VI. CONCLUDING REMARKS

Plasma-based surface treatment is a practical and relatively economical tool to enhance the surface properties of materials and industrial components. Even though plasma processes have been adopted for a long time, many of the processes are still not fully understood and sometimes quite complicated. It suffices to say that since the outcome depends very much on the plasma chemistry, a successful process requires the selection of the right technique(s) as well as optimization of the plasma parameters including excited species formation, reactant transportation, chemical and metallurgical interaction between the plasma and substrate, and behavior of the incorporated species on/beneath the surface. There are pros and cons with each technique, and to yield the best results, a process may encompass several techniques. It is important to exploit the pros while minimizing the cons. Plasma surface modification has diversified from conventional metallurgy into microelectronics and biomedical engineering, and new techniques and protocols are constantly being invented to satisfy the ever-growing demands of the industry and materials research.

REFERENCES

1. Hochman, R.F.; Hillary, S.L.; Legg, K.O. Ion implantation and plasma assisted processes. Proceedings of the Conference on Ion Implantation and Plasma Assisted Processes for Industrial Applications, Atlanta, Georgia, May 22–25, 1988, ASM International, Metals Park, OH, 1988.
2. Czerwiec, T.; Renevier, N.; Michel, H. Low-temperature plasma-assisted nitriding. Surf. Coat. Technol. 2000, *131*, 267–277.
3. Berghaus, B. Process for Surface Treatment of Metallic Elements. German Patent DRP 668 639, 1932.
4. Berghaus, B. Vacuum Furnace, Heated by Glow Discharge. German Patent DRP 851540, 1939.
5. Michel, H.; Czerwiec, T.; Gantois, M.; Ablitzer, D.; Ricard, A. Progress in the analysis of the mechanisms of ion nitriding. Surf. Coat. Technol. 1995, *72*, 103–111.
6. Petitjean, L.; Ricard, A. Emission-spectroscopy study of N_2–H_2 glow-discharge for metal-surface nitriding. J. Phys. D 1984, *17*, 919–929.
7. Leyland, A.; Fancey, K.S.; James, A.S.; Matthews, A. Enhanced plasma nitriding at low-pressures—a comparative study of DC and RF techniques. Surf. Coat. Technol. 1990, *41*, 295–304.
8. Rusnak, K.; Vlcek, J. Emission-spectroscopy of the plasma in the cathode region of N_2–H_2 abnormal glow-discharges for steel surface nitriding. J. Phys. 1993, *D 26*, 585–589.
9. Cernogora, G.; Hochard, L.; Touzeau, M.; Matos-Ferreira, C. Population of $N_2(A_3$-Sigma $U^+)$ metastable states in a pure nitrogen glow discharge. J. Phys. B. 1981, *14*, 2977–2987.
10. Ricard, A.; Oseguera Pena, J.E.; Falk, L.; Michel, H.; Gantois, M. Active species in microwave postdischarge for steel-surface nitriding. IEEE Trans Plasma Sci 1990, *18*, 940–944.
11. Ricard, A.; Deschamps, J.; Godard, J.L.; Falk, L.; Michel, H. Nitrogen atoms in Ar-N_2 flowing microwave discharges for steel surface nitriding. Mater. Sci. Eng. 1991, *A139*, 9–14.
12. Oseguera, J.; Salas, O.; Figueroa, U.; Palacios, M. Evolution of the surface concentration during discharge nitriding. Surf. Coat. Technol. 1997, *94/95*, 587–591.

13. Kumar, S.; Baldwin, M.J.; Fewell, M.P.; Haydon, S.C.; Short, K.T.; Collins, G.A.; Tendys, J. The effect of hydrogen on the growth of the nitrided layer in r.f.-plasma-nitrided austenitic stainless steel AISI 316. Surf. Coat. Technol. 2000, *123*, 29–35.

14. Renevier, N.; Czerwiec, T.; Collignon, P.; Michel, H. Diagnostic of arc discharges for plasma nitriding by optical emission spectroscopy. Surf. Coat. Technol. 1998, *98*, 1400–1405.

15. Bockel, S.; Amorim, J.; Baravian, G.; Ricard, A.; Stratil, P. Spectroscopic stud of active species in DC and HF flowing discharges in N_2–H_2 and Ar-N_2-H_2 mixtures. Plasma Sources Sci. Technol. 1996, *5*, 567–572.

16. Garscadden, A.; Nagpal, R. Non-equilibrium electronic and vibrational kinetics in H_2-N_2 and H_2 discharges. Plasma Sources Sci. Technol. 1995, *4*, 268–280.

17. Davis, W.D.; Vanderslice, T.A. Ion energies at the cathode of a glow discharge. Phys Rev 1963, *131*, 219–228.

18. Rickards, J. Energies of particles at the cathode of a glow-discharge. Vacuum 1984, *34*, 559–562.

19. Budtz-Jorgensen, C.V.; Bottiger, J.; Kringhoj, P. Energy distribution of particles bombarding the cathode in glow discharges. Vacuum 2000, *56*, 9–13.

20. Jiang, J.C.; Meletis, E.I. Microstructure of the nitride layer of AISI 316 stainless steel produced by intensified plasma assisted processing. J. Appl. Phys. 2000, *88*, 4026–4031.

21. Adjaottor, A.A.; Ma, E.; Meletis, E.I. On the mechanism of intensified plasma-assisted processing. Surf. Coat. Technol. 1997, *89*, 197–203.

22. Wang, D.Z.; Deng, X.L.; Ma, T.C. An analytic model for energy distribution of neutrals striking a planar cathode at low pressure glow discharge. Thin Solid Films 1999, *345*, 182–184.

23. Wei, R. Low energy, high current density ion implantation of materials at elevated temperatures for tribological applications. Surf. Coat. Technol. 1996, *83*, 218–227.

24. Hudis, M. Study of ion-nitriding. J. Appl. Phys. 1973, *44*, 1489–1496.

25. Tibbetts, G.G. Role of nitrogen atoms in "ion-nitriding". J. Appl. Phys. 1974, *45*, 5072–5073.

26. Szasz, A.; Fabian, D.J.; Hendry, A.; Szaszne-Csih, Z. Nitriding of stainless-steel in an RF plasma. J. Appl. Phys. 1989, *66*, 598–5601.

27. Edenhofer, B. Physical metallurgical aspects of ionitriding. Heat Treat Met 1974, *1*, 23–28.

28. Totten, G.E.; Howes, M.A.H. *Steel Heat Treatment Handbook*; Marcel Dekker, Inc.: New York, 1997.

29. Baba, Y.; Sasaki, T.A. Nitride formation at metal-surfaces by Ar^+ ion-bombardment in nitrogen atmosphere. Mater. Sci. Eng. A 1989, *115*, 203–207.

30. Lancaster, G.M.; Rabalais, J.W. Chemical-reactions of N_2^+ ion-beams with 1st-row transition metals. J. Chem. Phys. 1979, *83*, 209–212.

31. Lincoln, G.A.; Geis, N.W.; Pang, S.; Efremow, N. Large area ion beam assisted etching os GaAs with high etch rates and controlled anisotropy. J. Vac. Sci. Technol. 1983, *131*, 1043–1046.

32. Harper, J.M.E.; Cuomo, J.J.; Hentzell, H.T.G. Quantitative ion-beam process for the deposition of compound thin-films. Appl. Phys. Lett. 1983, *43*, 547–549.

33. Hubler, G.K.; Carosella, C.A.; Donovan, E.P.; Van Vechten, D.; Bassel, R.H.; Andredis, T.D.; Rosen, M.; Mueller, G.P. Physical aspects of ion beam assisted deposition. Nucl. Instrum. Methods. 1990, *B46*, 384–391.

34. Hubler, G.K.; Sprague, J.A. Energetic particles in PVD technology: Particle-surface interaction processes and energy-particle relationships in thin film deposition. Surf. Coat. Technol. 1996, *81*, 29–35.

35. Czerwiec, T.; Michel, H.; Bergmann, E. Low-pressure, high-density plasma nitriding: mechanisms, technology and results. Surf. Coat. Technol. 1998, *108/109*, 182–190.

36. Michalski, J. Ion nitriding of armaco iron in various glow discharge regions. Surf Coat Technol 1993, *59*, 321–324.

37. Strack, H. Ion bombardment of silicon in a glow discharge. J. Appl. Phys. 1963, *34*, 2405–2409.

38. Dimitrov, V.I.; D'Haen, J.; Knuyt, G.; Quaeyhaegens, C.; Stals, L.M. Modeling of nitride layer formation during plasma nitriding of iron. Comput. Mater. Sci. 1999, *15*, 22–34.

39. Yan, M.F.; Jiang, J.H.; Bell, T. Numerical simulation of nitrided layer growth and nitrogen distribution in ε-Fe$_{2-3}$ N, γ-Fe$_4$N and α-Fe during pulsed plasma nitriding of pure iron. Model. Simul. Mat. Sci. Eng. 2000, *8*, 491–496.

40. Dimitrov, V.I.; D'Haan, J.; Knuyt, G.; Quasyhaegens, C.; Stals, L.M. A diffusion model of metal surface modification during plasma nitriding. Appl. Phys. A 1996, *63*, 475–480.

41. Sun, Y.; Bell, T. A numerical model of plasma nitriding of low alloy steels. Mater. Sci. Eng. 1997, *A224*, 44–47.

42. Meletis, E.I.; Yan, S. Low-pressure ion nitriding of AISI 304 austenitic stainless steel with an intensified glow discharge. J. Vac. Sci. Technol. A 1993, *11*, 25–33.

43. Metin, E.; Inal, O.T. Formation and growth of iron nitrides during ion-nitriding. J. Mater. Sci. 1987, *22*, 2783–2788.

44. Gemma, K.; Taharam, T.; Kawakami, M.K. Effect of chromium content on remarkably rapid nitriding in austenite Fe–Ni–Cr alloys. J. Mater. Sci. 1996, *31*, 2885–2892.

45. Williamson, D.L.; Ozturk, O.; Wei, R.; Wilbur, P.J. Metastable phase-formation and en-hanced diffusion in fcc alloys under high-dose, high-flux nitrogen implantation at high and low ion energies. Surf. Coat. Technol. 1994, *65*, 15–23.

46. Parascandola, S.; Günzel, R.; Grötzschel, R.; Richter, E.; Möller, W. Analysis of deuterium induced nuclear reactions giving criteria for the formation process of expanded austenite. Nucl. Instrum. Methods. 1998, *B138*, 1281–1285.

47. Parascandola, S.; Möller, W.; Williamson, D.L. The nitrogen transport in austenitic stainless steel at moderate temperatures. Appl. Phys. Lett. 2000, *76*, 2194–2196.

48. Telbizova, T.; Parascandola, S.; Kreissig, U.; Gunzel, R.; Moller, W. Mechanism of diffu-sional transport during ion nitriding of aluminum. Appl. Phys. Lett. 2000, *76*, 1404–1406.

49. Williamson, D.L.; Davis, J.A.; Wilbur, P.J.; Vajo, J.J.; Wei, R.; Matossian, J.N. Relative roles of ion energy, ion flux, and sample temperature in low-energy nitrogen ion implantation of Fe–Cr–Ni stainless steel. Nucl. Instrum. Methods. B 1997, *127/128*, 930–934.

50. Wei, R. The effects of low-energy-nitrogen-ion implantation on the tribological and micro-structural characteristics of AISI 304 stainless steel. J. Tribol. 1994, *116*, 870–876.

51. Li, X.Y. Low temperature plasma nitriding of 316 stainless steel—nature of S phase and its thermal stability. Surf. Eng. 2001, *17*, 147–152.

52. Larisch, B.; Brusky, U.; Spies, H.J. Plasma nitriding of stainless steels at low temperatures. Surf. Coat. Technol. 1999, *119*, 205–211.

53. Gunzel, R.; Betzl, M.; Alphonsa, I.; Ganguly, B.; John, P.I.; Mukherjee, S. Plasma-source ion implantation compared with glow-discharge plasma nitriding of stainless steel. Surf. Coat. Technol. 1999, *112*, 307–309.

54. Collins, G.A.; Hutchings, R.; Short, K.T.; Tendy, S.J. PI³—A new nitriding process. Heat. Treat. Met. 1995, *22*, 91–94.

55. Shim, Y.K.; Kim, Y.K.; Lee, K.H.; Han, S. The properties of AlN prepared by plasma nitriding and plasma source ion implantation techniques. Surf. Coat. Technol. 2000, *131*, 345–349.

56. Renevier, N.; Czerwiec, T.; Billard, A.; Von Stebut, J.; Michel, H. A way to decrease the nitriding temperature of aluminium: the low-pressure arc-assisted nitriding process. Surf. Coat. Technol. 1999, *119*, 380–385.

57. Quast, M.; Mayr, P.; Stock, H.R.; Podlesak, H.; Wielage, B. In situ and ex situ examination of plasma-assisted nitriding of aluminium alloys. Surf. Coat. Technol. 2001, *135*, 238–249.

58. Fancey, K.S.; Matthews, A. Some fundamental-aspects of glow-discharges in plasma-assisted processes. Surf. Coat. Technol. 1987, *33*, 17–29.

59. Meletis, E.I.; Yan, S. Formation of aluminum nitride by intensified plasma ion nitriding. J. Vac. Sci. Technol. 1991, *A9*, 2279–2284.

60. Lei, M.K.; Zhang, Z.L. Plasma source ion nitriding—a new low temperature, low pressure nitriding approach. J. Vac. Sci. Technol. 1995, *A13*, 2986–2990.

61. LeCoeur, F.; Arnal, Y.; Burke, R.R.; Lesaint, O.; Pelletier, J. Ion implantation based on the uniform distributed plasma. Surf. Coat. Technol. 1997, *93*, 265–268.

62. Priest, J.M.; Baldwin, M.J.; Fewell, M.P.; Haydon, S.C.; Collins, G.A.; Short, K.T.; Tendys,

J. Low pressure r.f. nitriding of austenitic stainless steel in an industrial-style heat-treatment furnace. Thin. Solid. Films. 1999, *345*, 113–118.

63. Dhaen, J.; Quaeyhaegens, C.; Knuyt, G.; De Schepper, L.; Stals, L.M.; Van Stappen, M. An interface study of various PVD tin coating on plasma-nitrided austenitic stainless steel AISI 304. Surf. Coat. Technol. 1993, *60*, 468–473.

64. Renevier, N.; Collignon, P.; Michel, H.; Czerwiec, T. New trends on nitriding in low pressure are discharges studied by optical emission spectroscopy. Surf. Coat. Technol. 1996, *86/87*, 285–291.

65. Burakowski, T.; Wierzchon, T. *Surface Engineering of Metals. Principles, Equipment, Technologies*; CRC Press: Cleveland, OH, 1998; 506 pp.

66. Devi, M.U.; Mohanty, O.N. Plasma-nitriding of tool steels for combined percussive impact and rolling fatigue wear applications. Surf. Coat. Technol. 1998, *107*, 55–64.

67. Devi, M.U.; Chakraborty, T.K.; Mohanty, O.N. Wear behaviour of plasma nitrided tool steels. Surf. Coat. Technol. 1999, *116*, 212–221.

68. Albarran, J.L.; Juarezislas, J.A.; Martinez, L. Nitrided width and microhardness in H-12 ion nitrided steel. Mater. Lett. 1992, *15*, 68–72.

69. Lei, M.K.; Wang, P.; Huang, Y.; Yu, Z.W.; Yuan, L.J.; Zhang, Z.L. Tribological studies of plasma source ion nitrided low alloy tool steel. Wear. 1997, *209*, 301–307.

70. Muraleedharan, T.M.; Meletis, E.I. Surface modification of pure titanium and Ti6Al4V by intensified plasma ion nitriding. Thin Solid Films 1992, *221*, 104–113.

71. Meletis, E.I.; Cooper, C.V.; Marchev, K. The use of intensified plasma-assisted processing to enhance the surface properties of titanium. Surf Coat Technol 1999, *113*, 201–209.

72. Arai, T.; Fujita, H.; Tachikawa, H. In *Proceedings of the International Conference on Ion Nitriding,* Spalvins, T., Kovaxs, W.L., Eds.; ASM International: Cleveland, OH, 1986; 37 pp.

73. Manova, D.; Huber, P.; Mandl, S.; Rauschenbach, B. Surface Modification of aluminum by plasma immersion ion implantation. Surf. Coat. Technol. 2000, *128/129*, 249–255.

74. Stock, H.R.; Jarms, C.; Seidel, F.; Doring, J.E. Fundamental and applied aspects of the plasma-assisted nitriding process for aluminum and its alloys. Surf. Coat. Technol. 1997, *94/95*, 247–254.

75. Richter, E.; Günzel, R.; Parasacandola, S.; Telbizova, T.; Kruse, O.; Möller, W. Nitriding of stainless steel and aluminium alloys by plasma immersion ion implantation. Surf. Coat. Technol. 2000, *128/129*, 21–27.

76. Ebissawa, T.; Saikudo, R. Formation of aluminum nitride on aluminum surfaces by ECR nitrogen plasmas. Surf. Coat. Technol. 1996, *86/87*, 622–627.

77. Lei, M.K.; Zhang, Z.L.; Ma, T.C. Plasma-based low-energy ion implantation for low temperature surface engineering. Surf. Coat. Technol. 2000, *131*, 317–325.

78. Tian, X.B.; Zeng, Z.M.; Zhang, T.; Tang, B.Y.; Chu, P.K. Medium-temperature plasma immersion-ion implantation of austenitic stainless steel. Thin Solid Films 2000, *366*, 150–154.

79. Sun, Y.; Bell, T. Sliding wear characteristics of low temperature plasma nitrided 316 austenitic stainless steel. Wear 1998, *218*, 34–42.

80. Leyland, A.; Lewis, D.B.; Stevenson, P.R.; Matthews, A. Low-temperature plasma-diffusion treatment of stainless steels for improved wear resistance. Surf. Coat. Technol. 1993, *62*, 608–617.

81. Williamson, D.L.; Davis, J.A.; Wilbur, P.J. Effect of austenitic stainless steel composition on low-energy, high-flux, nitrogen ion beam processing. Surf. Coat. Technol. 1998, *103/104*, 178–184.

82. Li, X.; Samandi, M.; Dunne, D.; Collins, G.A.; Tendys, J.; Short, K.; Hutchings, R. Cross-sectional transmission electron microscopy characterization of plasma immersion ion implanted austenitic stainless steel. Surf. Coat. Technol. 1996, *85*, 28–36.

83. Ziegler, J.F. *Ion Implantation Technology*; North-Holland, Amsterdam, New York, 1992.

84. Ryssel, H.; Rige, I. *Ion Implantation*; John Wiley & Sons Ltd.: Chichester, 1986.

85. Conrad, J.R.; Radtke, J.L.; Dodd, R.A.; Worzala, F.J.; Tran, N.C. Plasma source ion-implantation technique for surface modification of materials. J. Appl. Phys. 1987, *62*, 4591–4596.

86. Tendys, J.; Donnelly, I.J.; Kenny, M.J.; Pollock, J.T.A. Plasma immersion ion implantation using plasmas generated by radio frequency techniques. Appl. Phys. Lett. 1988, *53*, 2143–2145.

87. Mizuno, B.; Nakayama, I.; Aoi, N.; Kubota, M.; Komeda, T. New doping method for subhalf micro trench sidewalls by using an electron resonance plasma. Appl. Phys. Lett. 1988, *53*, 2059–2061.

88. Anders, A. *Handbook of Plasma Immersion Ion Implantation and Deposition*; John Wiley & Sons: New York, 2000.

89. Conrad, J.R. *Introduction, in Handbook of Plasma Immersion Ion Implantation and Deposition*; John Wiley & Sons: New York, 2000; 1–26.

90. Tian, X.B.; Chu, P.K. Modeling of the relationship between implantation parameters and implantation dose during plasma immersion ion implantation. Phys. Lett. A 2000, *277*, 42–46.

91. Tian, X.B.; Wang, L.P.; Kwok, D.T.K.; Tang, B.Y.; Chu, P.K. Capacitance of high-voltage coaxial cable in plasma immersion ion implantation. J. Mater. Sci. Technol. 2001, *17*, 41–42.

92. Tian, X.B.; Tang, B.Y.; Chu, P.K. Accurate determination of pulsed current waveform in plasma immersion ion implantation processes. J. Appl. Phys. 1999, *86*, 3567–3570.

93. Stewart, R.A.; Lieberman, M.A. Model of plasma immersion ion implantation for voltages with finite rise and fall time. J. Appl. Phys. 1991, *70*, 3481–3487.

94. Wood, B.P. Displacement current and multiple pulse effects in plasma source ion implantation. J. Appl. Phys. 1993, *73*, 4770–4778.

95. Yatsuzuka, M.; Miki, S.; Morita, R.; Azuma, K.; Fujiwara, E. Spatial and temporal growth and collapse in a PBII plasma. Surf. Coat. Technol. 2001, *136*, 93–96.

96. Kim, Y.W.; Kim, G.H.; Han, S.H.; Lee, Y.H.; Cho, J.H.; Rhee, S.Y. Measurement of sheath expansion in plasma source ion implantation. Surf. Coat. Technol. 2001, *136*, 97–101.

97. Shamim, M.; Scheuer, J.T.; Conrad, J.R. Measurements of spatial and temporal sheath evolution for spherical and cylindrical geometries in plasma source ion-implantation. J. Appl. Phys. 1991, *69*, 2904–2908.

98. Goeckner, M.J.; Malik, S.M.; Conrad, J.R.; Breun, R.A. Laser-induced fluorescence measurement of the dynamics of a pulsed planar sheath. Phys. Plasmas 1994, *1*, 1064–1074.

99. Collins, G.A.; Tendys, J. Measurements of potentials and sheath formation in plasma immersion ion-implantation. J. Vac. Sci. Technol. 1994, *B12*, 875–879.

100. Mandl, S.; Brutscher, J.; Grunzel, R.; Moller, W. Ion energy distribution in plasma immersion ion implantation. Surf. Coat. Technol. 1997, *93*, 234–237.

101. Tian, X.B.; Kwok, D.T.K.; Chu, P.K. Modeling of incident particle energy distribution in plasma immersion ion implantation. J. Appl. Phys. 2000, *88*, 4961–4966.

102. Kellerman, P.L.; Bernstein, J.D.; Bradley, M.P. Ion energy distributions in plasma immersion ion implantation-theory and experiment. Conference on Ion Implantation Technology; IEEE Press: Alpbach, Austria, 2000; 484–487.

103. Fan, Z.N.; Zeng, X.C.; Kwok, D.T.K.; Chu, P.K. Surface hydrogen incorporation and profile broadening caused by sheath expansion in hydrogen plasma immersion ion implantation. IEEE Trans Plasma Sci 2000, *28*, 371–375.

104. Kwok, D.T.K.; Chu, P.K.; Takase, M.; Mizuno, B. Energy distribution and depth profile in BF_3 plasma doping. Surf. Coat. Technol. 2001, *136*, 146–150.

105. Sheridan, T.E. Effect of target size on dose uniformity in plasma-based ion implantation. J. Appl. Phys. 1997, *81*, 7153–7157.

106. Watterson, P.A. Child–Langmuir sheath structure around wedge-shaped cathodes. J. Phys. D: Appl Phys 1989, *22*, 1300–1307.

107. Sheridan, T.E. Self-similar sheath expansion from a segmented planar electrode. Phys. Plasmas 1996, *3*, 2461–2466.

108. Sheridan, T.E.; Alport, M.J. 2-Dimensional model of ion dynamics during plasma source ion-implantation. Appl. Phys. Lett. 1994, *64*, 1783–1785.

109. Sheridan, T.E. Pulsed-sheath ion dynamics in trench. J. Phys. 1995, *D 28*, 1094–1098.

110. Keller, G.; Paulus, M.; Mandl, S.; Strizker, B.; Rauschenbach, B. Modeling on plasma immersion implantation of trenches. Nucl. Instrum. Methods B 1999, *148*, 64–68.

111. Keller, G.; Rude, U.; Stals, L.; Mandl, S.; Rauschenbach, B. Simulation of trench homogeneity in plasma immersion ion implantation. J. Appl. Phys. 2000, *88*, 1111–1117.
112. Keller, G.; Mandl, S.; Rude, U.; Rauschenbach, B. Ion mass and scaling effects in PIII simulation. Surf. Coat. Technol. 2001, *136*, 117–121.
113. Kwok, D.T.K.; Chu, P.K.; Chan, C. Ion dose uniformity for planar sample plasma immersion ion implantation. IEEE Trans. Plasma. Sci. 1998, *26*, 1669–1679.
114. Sheridan, T.E.; Kwok, T.K.; Chu, P.K. Kinetic model for plasma-based ion implantation of a short, cylindrical tube with auxiliary electrode. Appl. Phys. Lett. 1998, *72 15*, 1826–1828.
115. Sheridan, T.E. Sheath expansion into a large bore. J. Appl. Phys. 1996, *80*, 66–69.
116. Sheridan, T.E. Analytical theory of sheath expansion into a cylindrical bore. Phys. Plasma. 1996, *3*, 3507–3512.
117. Zeng, X.C.; Tang, B.Y.; Chu, P.K. Improving the plasma immersion ion implantation impact energy inside a cylindrical bore by using an auxiliary electrode. Appl. Phys. Lett. 1996, *69*, 3815–3817.
118. Zeng, X.C.; Kwok, T.K.; Liu, A.G.; Chu, P.K.; Tang, B.Y.; Sheridan, T.E. Effects of the auxiliary electrode radius during plasma immersion ion implantation of a small cylindrical bore. Appl. Phys. Lett. 1997, *71*, 1035–1037.
119. Kwok, T.K.; Zeng, Z.M.; Chu, P.K.; Sheridan, T.E. Hybrid simulation of sheath and ion dynamics of plasma implantation into ring-shaped targets. J. Phys. 2001, *D 34*, 1091–1099.
120. Conrad, J.R.; Baumann, S.; Fleming, F.; Meeker, G.P. Plasma source ion-implantation dose uniformity of a 2×2 array of spherical targets. J. Appl. Phys. 1989, *65*, 1707–1712.
121. Qin, S.; Chan, C.; McGruer, N.E.; Browning, J.; Warner, K. The response of a microwave multipolar bucket plasma to a high-voltage pulse. IEEE Trans. Plasma. Sci. 1991, *19*, 1272–1278.
122. Baba, K.; Hatada, R. Nitrogen ion implantation into three-dimensional substrates by plasma source ion implantation. Mater. Chem. Phys. 1998, *54*, 135–138.
123. Malik, S.M.; Muller, D.E.; Sridharan, K.; Fetherson, R.P.; Tran, N.; Conrad, J.R. Distribution of incident ions and retained dose analysis for a wedge-shaped target in plasma source ion implantation. J. Appl. Phys. 1995, *77*, 1015–1019.
124. Ensinger, W.; Hochbauer, T.; Rauschenbach, B. Lateral implantation homogeneity of wedge-shaped samples treated by plasma immersion ion implantation. Surf. Coat. Technol. 1997, *94/95*, 352–355.
125. Hochbauer, T.; Ensinger, W.; Schrag, G.; Hartmann, J.; Strizker, B.; Rauschenbach, B. Homogeneity measurements of plasma immersion ion-implanted complex-shaped samples. Nucl. Instrum. Methods B 1997, *127*, 869–872.
126. Ensinger, W.; Hochbauer, T.; Rauchenbach, B. Treatment uniformity of plasma immersion ion implantation studied with three-dimensional model systems. Surf. Coat. Technol. 1998, *104*, 218–221.
127. Mandl, S.; Thorwarth, G.; Huber, P.; Schoser, S.; Rauschenbach, B. Comparison of trenches treated in PIII-systems. Surf. Coat. Technol. 2001, *139*, 81–86.
128. Wnsinger, W.; Volz, K.; Hochbauer, T. Plasma immersion ion implantation of complex-shaped objects: an experimental study on the treatment homogeneity. Surf. Coat. Technol. 2000, *128*, 265–269.
129. Sun, M.; Yang, S.Z.; Li, B. New method of tubular material inner surface modification by plasma source ion implantation. J. Vac. Sci. Technol. 1996, *A14*, 367–369.
130. Liu, A.G.; Wang, X.F.; Chen, Q.C.; Tang, B.Y.; Chu, P.K. Inner surface ion implantation using deflecting electric field. Nucl. Instrum. Methods 1998, *B143*, 306–310.
131. Liu, A.G.; Wang, X.F.; Tang, B.Y.; Chu, P.K. Dose and energy uniformity over inner surface in plasma immersion ion implantation. J. Appl. Phys. 1998, *84*, 1859–1862.
132. Malik, S.M.; Fetherston, R.P.; Conrad, J.R. Development of an energetic ion assisted mixing and deposition process for TINx and diamondlike carbon films, using a co-axial geometry in plasma source ion implantation. J. Vac. Sci. Technol. 1997, *A15*, 2875–2879.
133. Baba, K.; Hatada, R. Ion implantation into the interior surface of a steel tube by plasma source ion implantation. Nucl. Instrum. Methods B 1999, *148*, 69–73.

134. Baba, K.; Hatada, R. Ion implantation into inner wall surface of a 1-m-long steel tube by plasma source ion implantation. Surf. Coat. Technol. 2000, *128*, 112–115.

135. Tian, X.B.; Chu, P.K. Direct temperature monitoring for semiconductors in plasma immersion ion implantation. Rev. Sci. Instrum. 2000, *71*, 2839–2842.

136. Blanchard, J.P. Target temperature prediction for plasma source ion-implantation. J. Vac. Sci. Technol. 1994, *B12*, 910–917.

137. Tian, X.B.; Chu, P.K. Target temperature simulation during fast-pulsing plasma immersion ion implantation. J. Phys. D: Appl. Phys. 2001, *34*, 1639–1645.

138. Gunzel, R.; Wieser, E.; Richter, E.; Steffen, J. Plasma source ion-implantation of oxygen and nitrogen in aluminum. J. Vac. Sci. Technol. 1994, *B12*, 927–930.

139. Schhoser, S.; Brauchle, G.; Forget, J.; Weber, T.; Voigt, J.; Rauschenbach, B. XPS investigation of AlN formation in aluminum alloys using plasma source ion implantation. Surf. Coat. Technol. 1998, *104*, 222–226.

140. Walter, K.C. Nitrogen plasma source ion implantation of aluminum. J. Vac. Sci. Technol. 1994, *B12*, 945–950.

141. Zhan, J.J.; Ma, X.X.; Feng, L.L.; Sun, Y.; Xia, L.F. Tribological behaviour of aluminum alloys surface layer implanted with nitrogen ions by plasma immersion ion implantation. Wear. 1998, *220*, 161–167.

142. Yamanishi, T.; Hara, Y.; Morita, R.; Azuma, K.; Fujiwara, E.; Yatsuzuka, M. Profile of implanted nitrogen ions in Al alloy for mold materials. Surf. Coat. Technol. 2001, *136*, 223–225.

143. Bolduc, M.; Popovici, D.; Terreault, B. Deep segregation and hardening in Al-Mg–Zn–Cu–Cr alloy treated by plasma source oxygen ion implantation. Surf. Coat. Technol. 2001, *138*, 125–134.

144. Bolduc, M.; Popovici, D.; Terreault, B. Giant segregation effect and surface mechanical modification of aluminum alloys by oxygen plasma source ion implantation. Nucl. Instrum. Methods 2001, *B175*, 458–462.

145. Popovici, D.; Bolduc, M.; Terreault, B.; Sarkissian, A.H.; Stansfield, B.L.; Paynter, R.W.; Bourgoin, D. Tribological modification of aluminum by electron-cyclotron resonance plasma source ion implantation. J. Vac. Sci. Technol. 1999, *A17*, 1996–2000.

146. Chen, A.; Blanchard, J.; Conrad, J.R. Wear improvement evaluation of non-homogeneous surface modification materials. Wear. 1993, *160*, 105–110.

147. Horswill, N.C.; Sridharan, K.; Conrad, J.R. A fretting wear study of a nitrogen implanted titanium alloy. J. Mater. Sci. Lett. 1995, *14*, 1349–1351.

148. Alonso, F.; Rinner, M.; Loinaz, A.; Onate, J.I.; Ensinger, W.; Rauschenbach, B. Characterization of Ti6Al4V modified by nitrogen plasma immersion ion implantation. Surf. Coat. Technol. 1997, *93*, 305–308.

149. Johns, S.M.; Bell, T.; Smandi, M.; Collins, G.A. Wear resistance of plasma immersion ion implanted Ti6Al4V. Surf. Coat. Technol. 1996, *85*, 7–14.

150. Loinaz, A.; Rinner, M.; Alonso, F.; Onate, J.I.; Ensinger, W. Effects of plasma immersion ion implantation of oxygen on mechanical properties and microstructure of Ti6Al4V. Surf. Coat. Technol. 198, *104*, 262–267.

151. Redsten, A.M.; Sridharan, K.; Worzala, F.J.; Conrad, J.R. Nitrogen plasma source ion implantation of AISI S1 tool steel. J. Mater. Process Technol 1992, *30*, 253–261.

152. Collins, G.A.; Hutchings, R.; Short, K.T.; Tendys, J.; Li, X.; Samandi, M. Nitriding of austenitic stainless-steel by plasma immersion ion implantation. Surf. Coat. Technol. 1995, *74/75*, 417–424.

153. Blawert, C.; Knoop, F.M.; Weisheit, A.; Mordike, B.L. Plasma immersion ion implantation of stainless steel: Austenitic stainless steel in comparison to austenitic-ferritic stainless steel. Surf. Coat. Technol. 1996, *85*, 15–27.

154. Mandl, S.; Gunzel, R.; Richter, R.; Moller, W. Nitriding of austenitic stainless steels using plasma immersion ion implantation. Surf. Coat. Technol. 1998, *100/101*, 372–376.

155. Moller, W.; Parascandola, S.; Telbizova, T.; Gunzel, R.; Richter, E. Surface process and

diffusion mechanisms of ion nitriding of stainless steel and aluminum. Surf. Coat. Technol. 2000, *136*, 73–79.

156. Hutchings, R.; Collins, G.A.; Tendys, J. Plasma immersion ion implantation-duplex layers from a single process. Surf. Coat. Technol. 1992, *51*, 489–494.

157. Samandi, M.; Shedden, B.A.; Smith, D.I.; Collins, G.A.; Hutchings, R.; Tendys, J. Microstructure, corrosion and tribological behavior of plasma immersion ion-implanted austenitic stainless-steel. Surf. Coat. Technol. 1993, *59*, 261–266.

158. Mandl, S.; Gunzel, R.; Richter, E.; Moller, W. Nitrogen and boron implantation into austenite stainless steel. J. Vac. Sci. Technol. B 1999, *17*, 832–835.

159. Collins, G.A.; Hutchings, R.; Short, K.T.; Tendys, J. Ion-assisted surface modification by plasma immersion ion implantation. Surf. Coat. Technol. 1998, *104*, 212–217.

160. Wei, R.; Vajo, J.J.; Matossian, J.N.; Wilbur, P.J.; Davis, J.A.; Williamson, D.L.; Collins, G.A. A comparative study of beam ion implantation, plasma ion implantation and nitriding of AISI 304 stainless steel. Surf. Coat. Technol. 1996, *83*, 235–242.

161. Leigh, S.; Samandi, M.; Collins, G.A.; Short, K.T.; Martin, P.; Wielunski, L. The influence of ion energy on the nitriding behavior of austenitic stainless steel. Surf. Coat. Technol. 1996, *85*, 37–43; 254.

162. Fewell, M.P.; Priesta, J.M.; Baldwin, M.J.; Collins, G.A.; Short, K.T. Nitriding at low temperature. Surf. Coat. Technol. 2000, *131*, 284–290.

163. Chen, J.; Conrad, J.R.; Dodd, R.A. Structure and properties of amorphous diamond-like carbon films produced by ion beam assisted plasma deposition. J. Mater. Eng. Perform. 1993, *2*, 839–842.

164. Hakovirta, M.; Lee, D.H.; He, X.M.; Nastasi, M. Synthesis of fluorinated diamond-like carbon films by the plasma immersion ion processing technique. J. Vac. Sci. Technol. 2001, *A19*, 782–784.

165. Walter, K.C.; Nastasi, M.; Munson, C. Adherent diamond-like carbon coatings on metals via plasma source ion implantation. Surf. Coat. Technol. 1997, *93*, 287–291.

166. Malaczynski, G.W.; Elmoursi, A.A.; Leung, C.H.; Hamdi, A.H.; Campbell, A.B. Surface enhancement by shallow carbon implantation for improved adhesion of diamond-like coatings. J. Vac. Sci. Technol. 1999, *B17*, 813–817.

167. Ji, H.B.; Xia, L.F.; Ma, X.X.; Sun, Y.; Sun, M.R. Tribological behaviour of duplex treated Ti6Al4V: combining nitrogen PSII with a DLC coating. Tribol. Int. 1999, *32*, 265.

168. Malaczynski, G.W.; Hamdi, A.H.; Elmoursi, A.A.; Qiu, X. Ion implantation and diamond-like coatings of aluminum alloys. J. Mater. Eng. Perform. 1997, *6*, 223–239.

169. Malaczynski, G.W.; Hamdi, A.H.; Elmoursi, A.A.; Qiu, X. Diamond-like carbon coating for aluminum 390 alloy-automotive applications. Surf. Coat. Technol. 1997, *93*, 280–286.

170. Hatada, R.; Baba, K. Preparation of hydrophobic diamond like carbon films by plasma source ion implantation. Nucl. Instrum. Methods B 1999, *148*, 655–658.

171. Miyagawa, S.; Nakao, S.; Saitoh, K.; Miyagawa, Y. Deposition of diamond like carbon films using plasma source ion implantation with pulsed plasmas. Surf. Coat. Technol. 2000, *128*, 260–264.

172. Anders, A.; Anders, S.; Brown, I.G.; Dickinson, M.R.; Macgill, R.A. Metal plasma immersion ion implantation and deposition using vacuum arc plasma sources. J. Vac. Sci. Technol. 1994, *B12*, 815–820.

173. Hamdi, A.H.; Qiu, X.; Malaczynski, G.W.; Elmoursi, A.A.; Simko, S.; Militello, M.C.; Balogh, M.P.; Wood, B.P.; Walter, K.C.; Nastasi, M.A. Microstructure analysis of plasma immersion ion implanted diamond-like carbon coatings. Surf. Coat. Technol. 1998, *104*, 395–400.

174. Walter, K.C.; Nastasi, M.; Baker, N.P.; Munson, C.P.; Scarborough, W.K.; Scheuer, J.T.; Wood, B.P.; Conrad, J.R.; Sridharan, K.; Malik, S.; Bruen, R.A. Advances in PSII techniques for surface modification. Surf. Coat. Technol. 1998, *103/104*, 205.

175. Tonosaki, M.; Okita, H.; Takei, Y.; Chayahara, A.; Horino, Y.; Tsubouchi, N. Nano-indentation testing for plasma-based ion implanted surface of plastics. Surf. Coat. Technol. 2001, *136*, 249–251.

176. Nie, X.; Wilson, A.; Leyland, A.; Matthews, A. Deposition of duplex Al_2O_3/DLC coatings on Al alloys for tribological applications using a combined micro-arc oxidation and plasma-immersion ion implantation technique. Surf. Coat. Technol. 2000, *121*, 506–513.

177. Lee, D.H.; He, X.M.; Walter, K.C.; Nastasi, M.; Tesmer, J.R.; Tuszewski, M.; Tallant, D.R. Diamond-like carbon deposition on silicon using radio-frequency inductive plasma of Ar and C_2H_2 gas mixture in plasma immersion ion deposition. Appl. Phys. Lett 1998, *73*, 2423.

178. Zeng, Z.M.; Tian, X.B.; Kwok, T.K.; Tang, B.Y.; Fung, M.K.; Chu, P.K. Effects of plasma excitation power, sample bias, and duty cycle on the structure and surface properties of amorphous carbon thin films fabricated on AISI440 steel by plasma immersion ion implantation. J. Vac. Sci. Technol. 2000, *A18*, 2164–2168.

179. Adler, R.J.; Picraux, S.T. Repetitively pulsed metal ion beams for ion implantation. Nucl. Instrum. Methods 1985, *B6*, 123–128.

180. Zeng, Z.M.; Zhang, T.; Tang, B.Y.; Tian, X.B.; Chu, P.K. Surface modification of steel by metal plasma immersion ion implantation using vacuum arc plasma source. Surf. Coat. Technol. 1999, *120–121*, 659–662.

181. Brown, I.G. Vacuum arc metal plasma production and the transition of processing mode from metal ion beam to dc metal plasma immersion. Surf. Coat. Technol. 2001, *136*, 16–22.

182. Anders, A. Metal plasma immersion ion implantation and deposition: a review. Surf. Coat. Technol. 1997, *93*, 158–167.

183. Tian, X.B.; Zeng, Z.M.; Tang, B.Y.; Fu, K.Y.; Kwok, D.T.K.; Chu, P.K. Properties of titanium nitride fabricated on stainless steel by plasma-based ion implantation/deposition. Mater. Sci. Eng. 2000, *A282*, 164–169.

184. Tian, X.B.; Zhang, T.; Zeng, Z.M.; Tang, B.Y.; Chu, P.K. Dynamic mixing deposition/ implantation in a plasma immersion configuration. J. Vac. Sci. Technol. 1999, *A17*, 3255–3259.

185. Wood, B.P.; Reass, W.A.; Henins, I. Plasma source ion implantation of metal ions: synchronization of cathodic arc plasma production and target bias pulses. Surf. Coat. Technol. 1996, *85*, 70–74.

186. Bunshah, R.F. Plasma-assisted vapor deposition processes: Overview. In *Handbook of Deposition Technologies for Film and Coatings*; Bunshah, R.F., Ed.; Noyes Publications: NJ, 1994; 485–505.

187. Schneider, J.M.; Rohde, S.; Sproul, W.D.; Matthews, A. Recent developments in plasma assisted physical vapor deposition. J. Phys. D: Appl. Phys. 2000, *33*, R173–R186.

188. Ahmed, N.A.G. *Ion Plating Technology–Developments and Applications*; John Wiley & Sons: NY, 1987.

189. Mattox, D.M. A concise history of vacuum coating technology—Part 2: 1940–1975. Plating Surf. Finish. 2000, *87*, 58–61.

190. Khandagle, M.J.; Gangal, S.A.; Karekar, R.N. Analysis of the properties of radio-frequency ion-plated Cu films based on the calculation of the ion energy available film atom. J Appl. Phys. 1993, *74*, 6150–6157.

191. Piot, O.; Machet, J. Study of the growth mechanisms of chromium nitride films deposited by ion plating. Mat Sci Forum 1998, *287*, 539–542.

192. Palmers, J.; VanStappen, M. Deposition of (Ti-Al)N coatings by means of electron beam ion plating with evaporation of Ti and Al from two separate crucibles. Surf. Coat. Technol. 1995, *76*, 363–366.

193. Lee, J.K.H.; Park, C.H.; Yoon, Y.S.; Lee, J.J. Structure and properties of (Ti1-xCrx)N coatings produced by the ion-plating method. Thin Solid Films 2001, *385*, 167–173.

194. Colligon, J.S. Physical vapor deposition. In *Non-Equilibrium Processing of Materials*; Suryanarayana, C., Ed.; Pergamon: Amsterdam, 1999; 225–226.

195. Anderson, C.A.; Hinthorne, J.R. Ion microprobe mass analyzer. Science 1972, *175*, 853–860.

196. Werner, H.W. The use of secondary ion mass spectrometry in surface analysis. Surf. Sci. 1975, *47*, 301–323.

197. Carter, G.; Colligon, J.S. *Ion Bombardment of Solids*; Heinemann: Oxford, 1968.

198. Sigmund, P. Theory sputtering I. Sputtering yield of amorphous and polycrystalline targets. Phys Rev 1969, *184*, 383–416.
199. Thornton, J.A.; Greene, J.E. Sputter deposition processes. In *Handbook of Deposition Technologies for Film and Coatings*; Bunshah, R.F., Ed.; Noyes Publications: NJ, 1994; 275–345.
200. Wehner, G.K.; Anderson, G.S. The nature of physical sputtering. In *Handbook of Thin Film Technology*; Maissel, L., Glang, R., Eds.; McGraw Hill: NY, 1970; 31-1–31-28.
201. Harrison, D.E.; Kelly, P.W.; Garrison, B.J.; Winograd, N. Low energy ion impact phenomena on single crystal surface. Surf. Sci. 1978, *76*, 311–322.
202. Mahan, J.E. *Physical Vapor Deposition of Thin Films*; John Wiley & Sons: New York, 2000.
203. Pauleau, Y. Physical vapor deposition techniques I: Evaporation and sputtering. In *Advanced techniques for surface engineering*; Gissler, W., Jehn, H.A., Eds.; Kluwer Academic Publishers, Brussels and Luxembourg: Netherlands, 1992; 135–179.
204. Yamamura, Y.; Itikawa, Y.; Itoh, N. Angular Dependence of Sputtering Yields of Monatomic Solids. Research Information Center, Institute of Plasma Physics, Nagoya University, Japan, 1983.
205. Murakami, Y.; Shingyoji, T.; Hijikata, K. Uniformity and composition of Tb_xFe_{1-x} films prepared by magnetron sputtering using different types of targets. J. Appl. Phys. 1990, *68*, 1866–1868.
206. Shah, S.I. Sputtering: Introduction and general discussion. In *Handbook of Thin Film Process Technology*; Institute of Physics Publishing: London, 1995; A2,0–A3.0:18.
207. Matthews, A.; Teer, D.G. Characteristics of thermionically assisted triode ion-plating system. Thin Solid Films 1981, *80*, 41–48.
208. Swarnalatha, M.; Sravani, C.; Gunasekhar, K.R.; Muralidhar, G.K.; Mohan, S. Estimation of density of charge species in a triode discharge system. Vacuum 1997, *48*, 845–848.
209. Hashimoto, M.; Onozaki, Y.; Uchida, H.; Matsumura, Y. MgO thin film formation by ion enhanced deposition processes. Rev. Sci. Instrum. 2000, *71*, 999–1001.
210. Thornton, J.A. Recent developments in sputtering—magnetron sputtering. Met. Finish. 1979, *77*, 45–49.
211. Penfold, A.S. Magnetron sputtering. In *Handbook of Thin Film Process Technology*; Institute of Physics Publishing: London, 1995; A3.2:1–A3.2:27.
212. Kukla, R.; Krug, T.; Ludwig, L.R.; Wilmes, K. Highest rate self-sputtering magnetron source. Vacuum 1990, *41*, 1968–1970.
213. Window, B.; Savvides, N. Charged-particle fluxes from planar magnetron sputtering sources. J. Vac. Sci. Technol. A 1986, *4*, 196–202.
214. Kadlec, S.; Musil, J.; Munz, W.D. Sputtering systems with magnetically enhanced ionization for ion plating of tin films. J. Vac. Sci. Technol. 1990, *48*, 1318–1324.
215. Teer, D.G. In Proceedings 7th Conference on Ion and Plasma-Assisted Techniques (IPAT). CEP Consultants Ltd.: Edinburg, 145 pp.
216. Richthofen, A.V.; Cremer, R.; Domnick, R.; Neuschutz, D. Use of Auger- and photoelectron lines in the identification of chemical states of novel ternary Ti–Al–O films prepared by reactive magnetron sputtering ion plating. Thin Solid Films 1998, *315*, 66–71.
217. Ulrich, S.; Theel, T.; Schwan, J.; Batori, V.; Scheib, M.; Ehrhardt, H. Low-temperature formation of beta-silicon carbide. Diamond Relat. Mater. 1997, *6*, 645–648.
218. Schwan, J.; Ulrich, S.; Roth, H.; Ehrhardt, H.; Silva, S.R.P.; Robertson, J.; Samlenski, R.; Brenn, R. Tetrahedral amorphous carbon films prepared by magnetron sputtering and dc ion plating. J. Appl. Phys. 1996, *79*, 1416–1422.
219. Ulrich, S.; Scherer, J.; Schwan, J.; Barzen, I.; Jung, K.; Scheib, M.; Ehrhardt, H. Preparation of cubic boron nitride films by radio frequency magnetron sputtering and radio frequency ion plating. Appl. Phys. Lett 1996, *68*, 909–911.
220. Ulrich, S.; Theel, T.; Schwan, J.; Ehrhardt, H. Magnetron-sputtered superhard materials. Surf. Coat. Technol. 1997, *97*, 45–59.
221. Martin, P.J. Cathodic arc deposition. *Handbook of Thin Film Process Technology*; Institute of Physics Publishing: London, 1995; A1.4:1–A1.4:16.

222. Bunshah, R.F. Critical issues in plasma-assisted vapor deposition processes. IEEE Trans. Plasma. Sci. 1990, *18*, 846–854.

223. Schuelke, T.; Witke, T.; Scheibe, H.J.; Siemroth, P.; Schultrich, B.; Zimmer, O.; Vetter, J. Comparison of DC and AC arc thin film deposition techniques. Surf. Coat. Technol. 1999, *121*, 226–232.

224. Sanders, D.M.; Anders, A. Review of cathodic arc deposition technology at the start of the new millennium. Surf. Coat. Technol. 2000, *133/134*, 78–90.

225. Pierson, H.O. *Handbook of Chemical Vapor Deposition (CVD): Principles, Technology, and Applications*; Noyes Publ: Park Ridge, NJ, 1992.

226. Vossen, J.L.; Kern, W. *Thin Film Processes II*; Academic Press: Boston, 1991.

227. Sherman, A. Growth and properties of LPCVD titanium nitride as a diffusion barrier for silicon device technology. J Electrochem Soc 1990, *137*, 1892–1897.

228. Sherman, A. Plasma-assisted chemical vapor deposition processes and their semiconductor applications. Thin Solid Films 1984, *113*, 135–149.

229. Dagostino, R.; Favia, P.; Fracassi, F.; Lamendola, R. Plasma-enhanced chemical vapor deposition. In *Advanced Techniques for Surface Engineering*; Gissler, W., Jehn, H.A., Eds.; Kluwer Academic Publishers, Brussels and Luxembourg: Netherlands, 1992; 105–133.

230. Prange, R.; Cremer, R.; Neuschutz, D. Plasma-enhanced CVD of (Ti, Al)N films from chloridic precursors in a DC glow discharge. Surf. Coat. Technol. 2000, *133*, 208–214.

231. Samokhvalov, N.V.; Strelnitskij, V.E.; Belous, V.A.; Zubar, V.P.; Timchuk, A.G.; Obraztsova, E.D.; Ralchenko, V.G. Production of polycrystalline diamond films by d.c. glow discharge CVD. Diamond Relat. Mater. 1995, *4*, 964–967.

232. Pai, M.P.; Musale, D.V.; Kshirsagar, S.T. Low-pressure chemical vapour deposition of diamond films in a radio-frequency-plasma-assisted hot-filament reactor. Diamond Relat. Mater. 1998, *7*, 1526–1533.

233. Kim, D.S.; Lee, Y.H. Room-temperature deposition of a-SiC:H thin films by ion-assisted plasma-enhanced CVD. Thin Solid Films 1996, *283*, 109–118.

234. Dalal, V.L.; Maxson, T.; Han, K.; Haroon, S. Improvements in stability of amorphous silicon solar cells by using ECR-CVD processing. J Non-Cryst Solids 1998, *30*, 1257–1261.

235. Conde, J.P.; Schotten, V.; Arekat, S.; Brogueira, P.; Sousa, R.; Chu, V. Amorphous and microcrystalline silicon deposited by low-power electron-cyclotron resonance plasma-enhanced chemical-vapor deposition. Jpn. J. Appl. Phys. 1997, *36* (1A), 38–49.

236. Rie, K.T.; Gebauer, A.; Woehle, J. Investigation of PA-VCD of tin-relations between process parameters, spectroscopic measurements and layer properties. Surf. Coat. Technol. 1993, *60*, 385–388.

237. Kawata, K.; Sugimura, H.; Takai, O. Characterization of (Ti–Al)N films deposited by pulsed d.c. plasma-enhanced chemical vapor deposition. Thin Solid Films 2001, *386*, 271–275.

238. Jarms, C.; Stock, H.R.; Mayr, P. Mechanical properties, structure and oxidation behaviour of $Ti_{1-x}Al_xN$-hard coatings deposited by pulsed dc plasma-assisted chemical vapour deposition (PACVD). Surf. Coat. Technol. 1998, *109*, 206–210.

239. Endler, I.; Wolf, E.; Leonhardt, A.; Beger, A.; Richter, V. Preparation, characterization and wear behaviour of TiN_x-coated cermets obtained by plasma-enhanced chemical-vapor-deposition. J. Mater. Sci. 1994, *29*, 6097–6103.

240. Laimer, J.; Störi, H.; Rödhammer, P. Titanium nitride deposited by plasma-assisted chemical vapor-deposition. Thin Solid Films 1990, *191*, 77–89.

241. Crummenauer, J.; Stock, H.R.; Mayr, P. Flow visualization studies for optimization of the growth of tin by PAVCD. Surf. Coat. Technol. 1995, *74/75*, 516–521.

242. Kugler, C.; Bauer, F.; Laimer, J.; Stori, H. Is there a way to improve the uniformity of TiN deposition conditions in large pulsed d.c. plasma CVD reactors? Vacuum 2001, *61*, 379–383.

243. Beer, T.A.; Laimer, J.; Stori, H. Study of the ignition behavior of a pulsed dc discharge used for plasma-assisted chemical-vapor deposition. J. Vac. Sci. Technol. 2000, *A18*, 423–434.

244. Peter, S.; Richter, F.; Tabersky, R.; Konig, U. Optical emission spectroscopy of a PCVD process used for the deposition of TiN on cemented carbides. Thin Solid Films. 2000, *377*, 430–435.

245. Archer, N.J. The plasma-assisted chemical vapor-deposition of TiCTiN and TiC_xN_{1-x}. Thin Solid Films 1981, *80*, 221–225.

246. Arai, T.; Fujita, H.; Oguri, K. Plasma-assisted chemical vapor-deposition of TiN and TiC on steel-properties of coatings. Thin Solid Films 1998, *165*, 139–148.

247. Jang, S.S.; Lee, W.J. Effect of gas composition on TiN thin-film fabrication in N-2/H-2/Ar/ TiCl4 inductively coupled plasma-enhanced chemical vapor deposition system. Jpn J Appl. Phys. 2001, *40*, 4819–4824.

248. Ma, S.L.; Li, Y.H.; Xu, K.W. The composite of nitrided steel of H13 and TiN coatings by plasma duplex treatment and the effect of pre-nitriding. Surf. Coat. Technol. 2001, *137*, 116–121.

249. Bentzon, M.D.; Barholmhansen, C.; Hansen, J.B. Interfacial shear-strength of diamond-like coatings deposited on metals. Diamond Relat. Mater. 1995, *4*, 787–790.

250. Bhushan, B.; Kellock, A.J.; Cho, N.H.; Ager, J.W. Characterization of chemical bonding and physical characteristics of diamond-like amorphous-carbon and diamond films. J. Mater. Res. 1992, *7*, 404–410.

251. Robertson, J. Ultrathin carbon coatings for magnetic storage technology. Thin Solid Films 2001, *383*, 81–88.

252. Korzec, D.; Fedosenko, G.; Georg, A.; Engemann, J. Hybrid plasma system for diamond-like carbon film deposition. Surf. Coat. Technol. 2000, *131*, 20–25.

253. Erdemir, A.; Nilufer, I.B.; Eryilmaz, O.L.; Beschliesser, M.; Fenske, G.R. Friction and wear performance of diamond-like carbon films grown in various source gas plasmas. Surf. Coat. Technol. 1999, *121*, 589–593.

254. Soltani, A.; Thevenin, P.; Bath, A. Formation and characterisation of c-BN thin films deposited by microwave PECVD. Diamond Relat. Mater. 2001, *10*, 1369–1374.

255. Vilcarromero, J.; Carreno, M.N.P.; Pereyra, I. Mechanical properties of boron nitride thin films obtained by RF-PECVD at low temperatures. Thin Solid Films 2000, *373*, 273–276.

256. Montero, I.; Galan, L.; Osorio, S.P.; Martinezduart, J.M.; Periere, J. Structural-properties of BN thin-films obtained by plasma enhanced chemical-vapor-deposition. Surf. Coat. Technol. 1994, *21*, 809–813.

257. Ianno, M.J.; Ahmed, A.U.; Englebert, D.E. J. Electrochem. Soc. 1991, *138*, 500.

258. Freller, H.; Lorenz, H.P. Hybrid processes. In *Advanced Techniques for Surface Engineering*; Gissler, W., Jehn, H.A., Eds.; Kluwer Academic Publishers, Brussels and Luxembourg: Netherlands, 1992; 253–273.

259. Hiramatsu, K.; Ohnishi, H.; Takahama, T.; Yamanishi, K. Formation of TiN films low Cl concentration by pulsed plasma chemical vapor deposition. J. Vac. Sci. Technol. 1996, *A14*, 1037–1040.

260. Shimozuma, M.; Date, H.; Iwasaki, T.; Tagashira, H.; Yoshino, M.; Yoshida, K. Three-dimensional deposition of TiN film using low frequency (50 Hz) plasma chemical vapor deposition. J. Vac. Sci. Technol. 1997, *A15*, 1897–1901.

261. Grill, A. Diamond-like carbon: state of the art. Diamond Relat. Mater. 1999, *9*, 428–434.

262. Zou, J.W.; Schmidt, K.; Reichelt, K.; Dischler, B. The properties of a-C:H films deposited by plasma decomposition of C_2H_2. J Appl. Phys. 1990, *67*, 487–494.

263. Grill, A.; Patel, V. Stresses in diamond-like carbon films. Diamond Relat. Mater. 1993, *2*, 1519–1524.

264. Donnet, C.; Fontaine, J.; Grill, A.; Le Mogne, T. The role of hydrogen on the friction mechanism of diamond-like carbon films. Tribol. Lett. 2000, *9*, 137–142.

265. Ronkainen, H.; Varjus, S.; Koskinen, J.; Holmberg, K. Differentiating the tribological performance of hydrogenated and hydrogen-free DLC coatings. Wear 2001, *249*, 260–266.

266. Sun, Z. Morphological features of diamond-like carbon films deposited by plasma-enhanced CVD. J. Non-Cryst. Solids 2000, *261*, 211–217.

267. Hirakuri, K.K.; Minorikawa, T.; Friedbacher, G.; Grasserbauer, M. Thin film characterization of diamond-like carbon films prepared by rf plasma chemical vapor deposition. Thin Solid Films 1997, *302*, 5–11.

268. Donnet, C. Recent progress on the tribology of doped diamond-like and carbon alloy coatings: a review. Surf. Coat. Technol. 1998, *100–101*, 180–186.

269. Ensinger, W. Ion bombardment effects during deposition of nitride and metal films. Surf. Coat. Technol. 1998, *99*, 1–13.

270. Cotell, C.M.; Hirvonen, J.K. Effect of ion energy on the mechanical properties of ion beam assisted deposition (IBAD) wear resistant coatings. Surf. Coat. Technol. 1996, *81*, 118–125.

271. Nag, A. Ion Plating: optimum surface performance and material conservation. Thin Sold Films. 1994, *241*, 179–187.

272. Chan, K.S.; He, M.Y.; Hutchinson, J.W. Cracking and stress redistribution in ceramic layered composites. Mater. Sci. Eng. 1993, *A167*, 57–64.

273. Sugimura, Y.; Lim, P.G.; Shih, C.F.; Suresh, S. Fracture normal to a biomaterial interface: effects of plasticity on crack-tip shielding and amplification. Acta. Metall. Mater. 1995, *43*, 1157–1169.

274. Huang, Y.; Zhang, H.W. The role of metal plasticity and interfacial strength in the cracking of metal-ceramic laminates. Acta. Metall. Mater. 1995, *43*, 1523–1530.

275. He, M.Y.; Heredia, F.E.; Wissuchek, D.J.; Shaw, M.C.; Evans, A.G. The mechanics of crack growth in layered materials. Acta. Metall. Mater. 1992, *43*, 1223–1228.

276. Wäppling, D.; Gunnars, J.; Ståhle, P. Crack growth across a strength mismatched biomaterial interface. Int. J. Fract. 1998, *89*, 223–243.

277. Wiklund, U.; Hedenqvist, P.; Hogmark, S. Multilayer cracking resistance in bending. Surf. Coat. Technol. 1997, *97*, 773–778.

278. Bell, T.; Dong, H.; Sun, Y. Realising the potential of duplex surface engineering. Tribol. Int. 1998, *31*, 127–137.

279. Kessler, O.H.; Hoffmann, F.T.; Mayr, P. Combinations of coating and heat treating processes: establishing a system for combined processes and examples. Surf. Coat. Technol. 1998, *109*, 211–216.

280. Freller, H.; Habler, H. Ti$_x$Al$_{1-x}$N Films deposited by ion plating with an arc evaporator. Thin Solid Films 1987, *153*, 67–74.

281. Coll, B.F.; Fontana, R.; Gates, A.; Sathrum, B. (Ti-Al)N advanced films prepared by are process. Mater. Sci. Eng. 1991, *A140*, 816–824.

282. Wang, H.W.; Stack, M.M.; Lyon, S.B.; Hovsepian, P.; Munz, W.D. Wear associated with growth defects in combined cathodic arc/unbalanced magnetron sputtered CrN/NbN superlattice coatings during erosion in alkaline slurry. Surf. Coat. Technol. 2000, *135* (1), 82–90.

283. Munz, W.D.; Donohue, L.A.; Hovsepian, P.E. Properties of various large-scale fabricated TiAlN- and CrN-based superlattice coatings grown by combined cathodic arc-unbalanced magnetron sputter deposition. Surf. Coat. Technol. 2000, *125*, 269–277.

284. Van Duyn, W.; Van Lochem, B. Chemical and mechanical characterisation of WC. H amorphous. Thin Solid Films 1989, *181*, 497–503.

285. Freller, H.; Hempel, A.; Lilge, J.; Lorenz, H.P. Influence of intermediate layers and base metals on adhesion of amorphous carbon and metal-carbon coatings. Diamond Relat. Mater. 1992, *1*, 563–569.

286. Navinsek, B.; Panjan, P.; Gorenjak, F. Improvement of hot forging manufacture with PCD and duplex coatings. Surf. Coat. Technol. 2001, *137*, 255–264.

287. Vetter, J.; Michler, T.; Steuernagel, H. Hard coatings on thermochemically pretreated soft steels: application potential for ball valves. Surf. Coat. Technol. 1999, *111*, 210–219.

288. Walkowicz; Smolik, J.; Miernik, K.; Bujak, J. Duplex surface treatment of moulds for pressure casting of aluminum. Surf. Coat. Technol. 1997, *97*, 453–464.

289. Dingremont, N.; Bergmanne, E.; Collignon, P.; Michel, H. Optimization of duplex coatings built from nitriding and ion plating with continuous and discontinuous operation for construction and hot-working steels. Surf. Coat. Technol. 1995, *72/73*, 163–168.

290. Fu, Y.Q.; Loh, N.L.; Wei, J.; Yan, B.B.; Hing, P. Friction and wear behavior of carbon nitride films deposited on plasma nitrided Ti-6Al-4V. Wear 2000, *237*, 12–19.

291. Fu, Y.Q.; Wei, J.; Yan, B.B.; Loh, N.L. Characterization and tribological evaluation of

duplex treatment by depositing carbon nitride films on plasma nitrided Ti6Al4V. J. Mater. Sci. 2000, *35*, 2215–2227.

292. Podgornik, B.; Vizintin, J.; wanstrand, O.; Larsson, M.; Hogmark, S.; Ronkainen, H.; Holmberg, K. Tribological properties of plasma nitrided and hard coated AISI 4140 steel. Wear. 2001, *249*, 254–259.

293. Podgornik, B.; Vizintin, J.; Wanstrand, O.; Larsson, M.; Hogmark, S. Wear and friction behaviour of duplex-treated AISI 4140 steel. Surf. Coat. Technol. 1999, *121*, 502–508.

294. Jeong, G.H.; Hwang, M.S.; Jeong, B.Y.; Kim, M.H.; Lee, C. Effects of the duty factor on the surface characteristics of the plasma nitrided and diamond-like carbon coated high-speed steels. Surf. Coat. Technol. 2000, *124*, 222–227.

295. Kaufmann, H. Industrial applications of plasma and ion surface engineering. Surf. Coat. Technol. 1995, *74/75*, 23–28.

296. Rie, K.T.; Broszeit, E. Plasma diffusion treatment and duplex treatment recent development and new applications. Surf. Coat. Technol. 1995, *77*, 425–436.

297. Park, J.R.; Song, Y.K.; Rie, K.T.; Gebauer, A. Hard coating by plasma-assisted CVD on plasma nitrided stellite. Surf. Coat. Technol. 1998, *98*, 1329–1335.

298. Nomoto, K.; Nishijima, S.; Katagiri, K.; Nunogaki, M.; Nishiura, T.; Okada, T.; Mori, H.; Iwamoto, K. Effects of ion-implantation and plasma treatment on tribological properties of aluminum and Al–Mg alloy. Surf. Coat. Technol. 1992, *51*, 157–161.

299. Kanno, I.; Nomotok, K.; Nishiura, S.; Okada, T.; Katagiri, K.; Mori, H.; Iwamoto, K. Tribological properties of aluminum modified with nitrogen ion-implantation and plasma treatment. Nucl. Instrum. Methods 1991, *B59*, 920–924.

300. Blawert, C.; Kalvelage, H.; Mordike, B.L.; Collins, G.A.; Short, K.T.; Jiraskova, Y.; Schneeweiss, O. Nitrogen and carbon expanded austenite produced by PI3. Surf. Coat. Technol. 2001, *136*, 181–187.

301. Collins, G.A.; Hutchings, R.; Short, K.T.; Tendys, J.; Li, X.; Samandi, M. Nitriding of austenitic stainless steel by plasma immersion ion implantation. Surf. Coat. Technol. 1995, *74–75*, 417–424.

302. Moller, W.; Parascandola, S.; Kruse, O.; Gunzel, R.; Richter, E. Plasma-immersion ion implantation for diffusive treatment. Surf. Coat. Technol. 1999, *119*, 1–10.

303. Tian, X.B.; Leng, Y.X.; Kwok, T.K.; Wang, L.P.; Tang, B.Y.; Chu, P.K. Hybrid elevated-temperature, low/high-voltage plasma immersion ion implantation of AISI 304 stainless steel. Surf. Coat. Technol. 2001, *135*, 178–183.

304. Matossian, J.N. Plasma ion implantation technology at Hughes Research laboratories. J. Vac. Sci. Technol. 1994, *B12*, 850–853.

305. Ensinger, W.; Usedom, K.J.; Rauschenbach, B. Characteristic features of an apparatus for plasma immersion ion implantation and physical vapour deposition. Surf. Coat. Technol. 1997, *93*, 175–180.

306. Ensinger, W.; Volz, K.; Enders, B. An apparatus for in-situ or sequential plasma immersion ion beam treatment in combination with RF sputter deposition or triode DC sputter deposition. Surf. Coat. Technol. 1999, *120–121*, 343–346.

307. Brown, I.G.; Anders, A.; Anders, S.; Dickinson, M.R.; Ivanov, I.C.; MacGill, R.A.; Yao, X.Y.; Yu, K.M. Plasma synthesis of metallic and composite thin films with automatically mixed substrate bonding. Nucl. Instrum. Methods 1993, *B80/81*, 1281–1287.

308. Brown, I.G.; Godechot, X.; Yu, K.M. Novel metal ion surface modification technique. Appl. Phys. Lett. 1991, *58*, 1392–1394.

309. Wood, B.P.; Rej, D.J.; Anders, A.; Brown, I.G.; Faehl, R.J.; Malik, S.M.; Munson, C.P. Fundamental of plasma immersion ion implantation and deposition. In *Handbook of Plasma Immersion Ion Implantation and Deposition*; Anders, A., Eds.; John Wiley & Sons: New York, 2000; 243–301.

310. Sano, M.; Teramoto, T.; Yukimura, K.; Maruyama, T. TiN coating to three-dimensional materials by PBII using vacuum titanium arc plasma. Surf. Coat. Technol. 2000, *128/129*, 245–248.

311. Schoser, S.; Forget, J.; Kohlhof, K. PIII-assisted thin film deposition. Surf. Coat. Technol. 1997, *93*, 339–342.

312. Tian, X.B.; Wang, L.P.; Zhang, Q.Y.; Chu, P.K. Dynamic nitrogen and titanium plasma ion implantation/deposition at different bias voltages. Thin Solid Films. 2001, *390*, 139–144.

313. Tian, X.B.; Zhang, T.; Zeng, Z.M.; Tang, B.Y.; Chu, P.K. Dynamic mixing deposition/implantation in a plasma immersion configuration. J. Vac. Sci. Technol. 1999, *A17*, 3255–3259.

314. Tian, X.B.; Zeng, Z.M.; Tang, B.Y.; Fu, K.Y.; Kwok, D.T.K.; Chu, P.K. Properties of titanium nitride fabricated on stainless steel by plasma-based ion implantation/deposition. Mater. Sci. Eng. A 2000, *282*, 164–169.

315. Wang, X.; Martin, P. Morphology modification by high energy ion beam bombardment with Tin film growth. J. Mater. Sci. Lett. 1994, *13*, 1317–1319.

316. Spies, H.J.; Hoeck, K.; Broszeit, E.; Matthes, B.; Herr, W. PVD hard coatings on prenitrided low alloy steel. Surf. Coat. Technol. 1993, *60*, 441–445.

317. Gredic, T.; Zlatanovic, M.; Popovic, N.; Bogdanov, Z. Effect of plasma nitriding on the properties of (Ti–Al)N coatings deposited onto hot work steel substrates. Thin Solid Films 1993, *228*, 261–266.

318. Zlatanovic, M. Deposition of (Ti–Al)N coatings on plasma nitrided steel. Surf. Coat. Technol. 1991, *48*, 19–24.

319. Subramanian, C.; Strafford, K.N.; Wilks, T.P.; Ward, L.P.; McMillan, W. Influence of substrate roughness on the scratch adhesion of titanium nitride coatings. Surf. Coat. Technol. 1993, *62*, 529–535.

320. Hock, K.; Spies, H.J.; Larisch, B.; Leonhardt, G.; Buecken, B. Wear resistance of prenitrided hardcoated steels for tools and machine components. Surf. Coat. Technol. 1997, *88*, 44–49.

321. Spies, H.J.; Larisch, B.; Hock, K.; Broszeit, E.; Schröder, H.J. Adhesion and wear resistance of nitrided and TiN coated low alloy steels. Surf. Coat. Technol. 1995, *74/75*, 178–182.

322. Damaschek, R.; Strydom, I.L.; Bergmann, H.W. Improved adhesion of TiN deposited on prenitrided steels. Surf Eng 1997, *13*, 128–132.

323. Haen, J.D.; Quaeyhaegens, C.; Stals, L.M.; Stappen, M.V. Interface study of physical vapor deposition Tin coatings on plasma nitrided steels. Surf. Coat. Technol. 1993, *61*, 194–200.

324. Baek, W.S.; Kwon, S.C.; Lee, S.R.; Rha, J.J.; Nam, K.S.; Lee, J.Y. A study of the interfacial structure between the TiN film and the iron nitride layer in a duplex plasma surface treatment. Surf. Coat. Technol. 1999, *114*, 94–100.

325. Podgornik, B.; Vizintin, J.; Ronkainen, H.; Holmberg, K. Friction and wear properties of DLC-coated plasma nitrided steel in unidirectional and reciprocating sliding. Thin Solid Films. 2000, *377*, 254–260.

326. Meletis, El; Erdemir, A.; Fenske, G.R. Tribological characteristics of DLC films and duplex plasma nitriding/DLC coating treatments. Surf. Coat. Technol. 1995, *73*, 39–45.

327. Harris, A.A.; Weiner, A.M.; Tung, S.C.; Simko, S.J.; Militello, M.C. A diamond-like carbon film for wear protection of steel. Surf. Coat. Technol. 1993, *62*, 550–557.

328. Buecken, B.; Leonhardt, G.; Wilberg, R.; Hoeck, K.; Spies, H.J. Direct combination of plasma nitriding and PVD hard coating by a continuous process. Surf. Coat. Technol. 1994, *68*, 244–248.

329. Walkowicz, J.; Smolik, J.; Tacikowski, J. Optimization of nitrided case structure in composite layers created by duplex treatment on the basis of PVD coating adhesion measurement. Surf. Coat. Technol. 1999, *119*, 370–379.

330. Baek, W.S.; Kwon, S.C.; Lee, J.Y.; Lee, S.R.; Rha, J.J.; Nam, K.S. The effect of Ti ion bombardment on the interfacial structure between TiN and iron nitride. Thin Solid Films 1998, *323*, 146–152.

331. Sun, Y.; Bell, T. Plasma surface engineering of low alloy steel. Mater. Sci. Eng. A 1991, *140*, 419–434.

332. Dingremont, N.; Pianelli, A.; Bergmann, E.; Michel, M. Analysis of the compatibility of plasma nitrided steels with ceramic coatings deposited by the ion-plating technique. Surf. Coat. Technol. 1993, 187–193.

16
Surface Engineering with Lasers

Jeff Th. M. De Hosson, Vašek Ocelík, and Yutao Pei
University of Groningen, Groningen, The Netherlands

I. INTRODUCTION

Interfaces between metals and ceramic materials have been the subject of extensive research in recent years because they control, to a great extent, the properties of metal–ceramic composites, protective coatings, thin metal/ceramic films in electronic devices, etc. For instance, low hardness and poor wear resistance are the principal limitations on the potential applications of aluminum alloys. Hardfacing on aluminum alloys, for example, by electro-deposition and anodizing, may significantly improve hardness and wear resistance. However, there are still major drawbacks in these conventional methods. The bonding between these coatings and aluminum alloys is usually weak and may cause failure during application. The amorphous anodizing layer is brittle and with a relatively low hardness ($HV_{0.1}$ 250 –500), neither of which is useful in resisting abrasive wear. For these reasons, it would be of great practical importance to have a basic fundamental understanding of the bonding between dissimilar materials such as metals and protective ceramic layers. Their obvious technological importance notwithstanding, our basic understanding of interfaces, even relatively simple interfaces as grain–boundaries, is still rudimentary, particularly in relation to materials properties. The importance of interfaces is primarily determined by their inherent inhomogeneity, that is, the fact that physical and chemical properties may dramatically change at or near the interface itself. An important property of a heterophase interface is its free energy per unit area, and the closely related work of adhesion. Thermodynamic and mechanical properties of the interface have been found to depend on these parameters. Experimental determination of the interface energy is an important step toward understanding heterophase interfaces. In principle, there are several ways in which information on the interface energy can be extracted from experiments, for example, either by measurement of wetting angles or by study of interface fracture behavior by four-point bending tests.

In general, metals and ceramic materials do not bond easily and some thermodynamic trick is necessary to manufacture ceramic–metal layers of a size that may be macroscopically tested on its mechanical behavior. For that reason, CO_2 and Nd–YAG laser systems are used within the framework of the laser melt particle injection technique. For a review on this topic, the reader is directed to Ref. 1. Here the interest is to obtain heterophase interfaces of ceramic particles that are firmly bonded to the surrounding metallic substrate. This chapter concentrates on a control of the microstructure of laser treated materials so as to achieve an enhancement in mechanical properties and wear resistance. The focus is on lightweight materials, such as Al- and Ti-based systems.

II. THERMAL CONTROL OF LASER TREATMENT

A. Optical Properties and Laser Radiation

A breakthrough in the application of lasers in the field of interface engineering by lasers is inconceivable without thorough knowledge of the thermal control during laser processing [2,3]. Processes such as phase transitions and chemical reactions depend significantly on the temperatures attained. In addition to the maximum temperature, the temperature gradients are of crucial importance. The cooling rate determines if the phase transformations that occur are conformable to the equilibrium phase diagram, and also influences the refinement of the microstructure [4,5]. Additionally, thermal stresses will be induced in the modified layer as a consequence of the thermal gradient [6]. Both microstructural composition and residual stress state will determine the final performance of the laser-modified layer. To gain insight into the thermal control, this chapter describes those phenomena relating to this. Therefore, the optical properties of a surface have to be examined in connection with the optical temperature measurements and the absorption of the laser radiation. An analytical model is presented that is used to calculate the temperature field under different processing conditions, by using Green's functions [7]. Special attention is paid to the influence of finite-sized specimen. The temperature of a body can be measured in several ways [8]. In principle, a distinction can be made as to which way detection occurs—that is, by either direct contact or without contact. The first method, where the thermocouple is a well-known tool, has the disadvantage of low response time. Further, the measurement is rather static. As a consequence, this method is not suitable for measuring locally high-temperature gradients. Therefore, the contactless way of measuring is preferable because it determines the temperature of a body by measuring its emitted radiation. Three different types of optical pyrometer systems are available at the moment, namely monochromatic, two-color, and multiwavelength pyrometers [9]. The basic principle of the optical pyrometer involves measuring infrared (IR) radiation, which is related to temperature as described by Planck's radiation law. However, as most surfaces do not behave like perfect blackbodies, intensity also depends on emissivity. The latter two mentioned pyrometers circumvent the emissivity measurement issue by measuring at two or more wavelengths. However, temperature calculations are based on the assumption that the emissivity at both wavelengths and their temperature dependence are equal, which may introduce error. At first glance, the monochromatic pyrometer only measures temperature accurately when the emissivity data are well known for the entire temperature range. However, when these data are not exactly known, it is still possible to measure the temperature accurately for metals by choosing the appropriate wavelength to detect. This is attributed to the fact that the IR radiation originating from a body depends more strongly on temperature than on emissivity. As a consequence, deviations in emissivity result only in relatively small errors in temperature.

To study the energy exchange at a surface, it is necessary to know the radiation properties, in particular, absorption, reflection, and emission characteristics [10]. In the case of blackbody radiation, these properties are well established. However, in practice, one often has to deal with surfaces that do not behave like perfect blackbodies.

One of the fundamental laws of thermal radiation theory is Kirchhoff's law. This law expresses the relation between the emission and absorption of energy for a nontransmitting material. If the incoming hemispherical radiation is uniformly polarized and uniformly distributed over all angles, absorbance is equal to emittance. In that case, only one radiation property is actually independent. The emittance is defined by the ratio between the emitted energy of a surface and the emitted energy of a blackbody at the same wavelength and temperature. For metals, the monochromatic emittance decreases as a function of wavelength, whereas for insulators there is a tendency to increase. The aforementioned radiation properties concern energy exchange with a complete hemispherical space above the surface. Radiation properties also exhibit an angular dependence and therefore the intensity is considered within a small solid angle $d\omega$.

Different surface preparation techniques are applied for laser processing to maximize the absorption of the laser beam. For metals, this is necessary because of the relatively poor absorption of IR radiation. The main concern in measuring temperature with an optical monochromatic pyrometer is to be familiar with the emissivity of the different surface conditions of the specimen. Therefore, emittance measurements are performed in advance of the final temperature measurements during laser processing. In practice, the surface preparation consists of increasing the roughness and/or painting the surface in order to approximate a blackbody. Coarsening is realized by using sandpaper or by sand blasting. Painting is carried out using a carbon spray.

To know the absorbed amount of energy, it is necessary to study the interaction of the laser beam with the specimen [11]. In case of an optically smooth surface, electromagnetic theory predicts the monochromatic reflection in the specular direction. The description of the optical phenomena is different for metals and ceramics [12]. As for experiments, titanium and aluminum are used as substrate materials, and so only the situation for metals will be considered. Whereas in vacuum, the refractive index $n = 1$, it becomes $\tilde{n} = n + ik$ for metals. The factor n represents the refraction index and k the extinction coefficient. When an incident wave arrives at the surface, it splits into a reflected and a transmitted wave. The specular reflection can be derived, by imposing the boundary conditions for the electromagnetic field at both sides of the interface. In case of normal incidence, the normal specular reflectance for an optically smooth surface is given by [13]

$$R_0 = \frac{(n-1)^2 + k^2}{(n+1)^2 + k^2} \tag{1}$$

As this degree of smoothness is rarely achieved, a modification is required to compensate for the surface roughness. To obtain some idea of the influence of roughness, the surface is assumed to consist of planes of random size and shape, all aligned parallel to the mean surface level and located at random levels relative to the surface. In addition, the height distribution is considered to be Gaussian. In that case, the normal incidence at an isotropic opaque medium and for which results in the total reflectance is given by [14,15]

$$R = R_0 \left[\exp -\left(\frac{4\pi\sigma}{\lambda}\right)^2 + \left(1 - \exp -\left(\frac{4\pi\sigma}{\lambda}\right)^2\right)\left(1 - \exp -\left(\frac{\pi^2\xi^2}{2\lambda}\right)^2\right) \right] \tag{2}$$

where σ is the root mean square (rms) roughness and ξ is the autocovariance length of the surface. The autocovariance is defined by the change of height as a function of the displacement in the plane parallel to the surface. The first term of Eq. (2) represents the specular reflection or the coherent reflection, whereby the second term represents the incoherent or diffuse reflection. As the wavelength of the CO_2 laser is a factor 10 higher than the wavelength of the YAG laser, the percentage of specular reflection will be much higher for the CO_2 laser.

To measure absorption as a function of the surface condition, three different surface preparation techniques have been performed on titanium substrates. Subsequently, the temperature is measured at a fixed distance from the laser track. The laser power amounts to 300 and 600 W, and the scan velocity is 15 mm/sec. Moreover, the specimen size may be assumed to be semi-infinite. To prevent oxidation, argon is used as the shielding gas at a rate of 6 l/min. The measured temperatures are shown in Fig. 1.

If we compare the ground and sandblasted specimen, it may be concluded that the increased surface roughness of the last mentioned specimen considerably increases the absorption. However, the highest absorption is attained for the black-like surface. In contrast to 600 W, the laser tracks do not show any trace of melting at 300 W. As a consequence, there are, in fact, two absorption conditions for the specimen processed at 600 W, because of the transformation to liquid titanium. If we consider only the ground specimen at 300 and 600 W, it can be seen that the absorption is higher for the liquid state

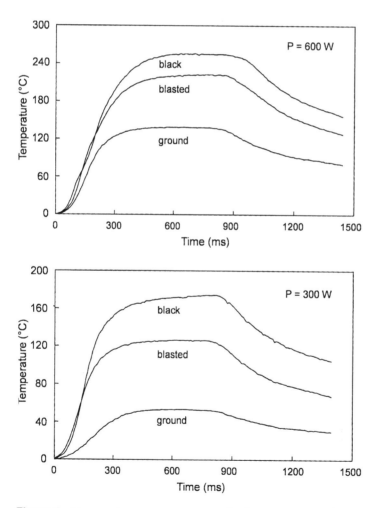

Figure 1 Temperature measurements during laser treatment of Ti for different surface conditions. The measurements are performed for CO_2 laser powers of $P = 300$ and 600 W. The pyrometer is fixed to the specimen coordinate system and is focused on the side of the specimen.

than for the solid state, as the maximum temperatures differ by a factor 2.5. The higher absorption can be explained by the disappearance of interband absorption upon melting [16]. In addition, it should be mentioned that the transition between solid and liquid states requires a certain amount of energy, represented by the latent heat. The sandblasted and black surfaces show a ratio of the maximum temperatures that is smaller than 2 when the laser power is doubled. Nevertheless, it is inaccurate to conclude that, for both surface conditions, the absorption is higher than in the liquid state.

B. Temperature Fields Induced by Heat Sources

In this section, those analytical solutions of the temperature field in a solid that are induced by an instantaneous heat source will be described. Subsequently, those influences of different energy distributions of the heat source will be examined as they correspond to the different types of lasers used in experiments. Because of the limited specimen size in practice, the influence of finite dimensions of the specimen will be incorporated into the solution by specific boundary conditions of zero heat flux. In addition, the effect of different types of cooling will be incorporated into the model. This will be carried out for different sets of boundary conditions. In addition to the absolute temperatures, the temperature gradient also influences the final performance of the modified layer. Therefore, the cooling rate will also be discussed. Phase transformations and temperature-dependent physical properties will not be taken into account. In that case, numerical methods have to be used.

Whenever a temperature gradient exists within a body, transport of energy takes place via heat transfer. There exist three distinct processes by which the transfer of heat occurs—conduction, convection, and radiation. For temperature differences in a solid, the transport of heat is mainly restricted to conduction. The flow of heat in a rigid isotropic medium can be described by the heat conduction equation

$$\nabla^2 T(\mathbf{r},t) = \frac{1}{\kappa}\frac{\partial T(\mathbf{r},t)}{\partial t} \tag{3}$$

where $T(\mathbf{r},t)$ is the temperature at position $\mathbf{r}$ at time t, and κ is the thermal diffusivity ($= K/\rho c_p$, where K is the thermal conductivity, ρ is the density, and c_p is the specific heat). To solve the heat conduction equation, Green's functions will be utilized [17,18]. When applied in the field of heat conduction, this type of function represents the temperature at a point (x,y,z) at time t because of an instantaneous point source of strength unity positioned at (x',y',z') at time t'. Furthermore, the solid is supposed to be at zero temperature and the surface is at zero temperature. Green's function, $G(|\mathbf{r}-\mathbf{r}'|, t-t')$, which satisfies the heat conduction equation, is given by

$$G(|\mathbf{r}-\mathbf{r}'|,t-t') = \frac{1}{(4\pi\kappa(t-t'))^{3/2}}\exp\left(-\frac{|\mathbf{r}-\mathbf{r}'|^2}{4\kappa(t-t')}\right) \tag{4}$$

where $t' < t$. Moreover, $\lim_{t\to t'} G = 0$ at every point with the exception of the position of heat generation. In the event of the production of heat in the solid, Eq. (3) can be replaced by a modified version

$$\frac{\partial T(\mathbf{r},t)}{\partial \tau} = \kappa < \nabla^2 T(\mathbf{r},t) + \frac{\kappa < Q}{K} \quad (\tau < t) \tag{5}$$

where Q describes the intensity distribution of the heat source. The solution of the temperature field in an infinite solid due to the continuous heat source Q can be obtained by the superposition in space of the fundamental solution, represented by Eq. (4). Additionally, the heat source can be made continuous by integrating over the time interval, where in between the heat is supplied. This results in the temperature field described by [17]

$$T(\mathbf{r},t) = \frac{\kappa}{K} \int_0^t \int_{-\infty}^{\infty} \int_{-\infty}^{\infty} \int_{-\infty}^{\infty} Q(\mathbf{r}',t')G(|\mathbf{r}-\mathbf{r}'|,t-t')dx'dy'dz'dt' \tag{6}$$

Thus $T(\mathbf{r},t)$ gives the temperature at a time t and position due to a continuous heat flux Q at time t' and position. Furthermore, the fundamental solution $G(|\mathbf{r}-\mathbf{r}'|, t-t')$ can be split up for the different coordinate directions. In Ref. 19, the complete solution of the heat conduction equation is derived, including initial surface temperature and an initial body temperature.

So far, we have summarized the fundamental solution of the heat conduction equation for a stationary heat source. The moving heat source can be introduced by formulating a moving medium relative to a fixed source, which is described in more detail by Rosenthal [20]. This method implies that the source distribution Q does not become time-dependent, but the observer position does. When the source is assumed to move along the x-direction in reality, the transformation toward a moving medium can be realized by defining the x-component as $x + v(t-t')$. As a consequence, the x-component of the temperature field is defined by

$$G(x) = \frac{1}{(4\pi\kappa(t-t'))^{1/2}} \exp\left(-\frac{(x + v(t_0 - t_0') - x)^2}{4\kappa(t-t')}\right) \tag{7}$$

The most convenient way to visualize the absorbed laser power is by an internal heat source. Furthermore, the heat is assumed to be generated within an infinitesimally thin surface layer. In this section, different heat sources will be examined for the case in which the medium is infinite. At first, the different sources are assumed to be time-independent, which corresponds to the continuous wave type of laser. Second, a method is described to encompass the time-dependent source term corresponding to the pulsed laser type.

1. Point Source

The most straightforward way to define an energy distribution is via a point source Q. At $r' = 0$, the distribution is given by $Q(\mathbf{r}') = P\delta(\mathbf{r}')$, whereby P is the generated power. Substitution of the source in Eq. (6) results in the temperature profile described by

$$T(r) = \frac{P}{2\pi K r} \exp\left(-\frac{v(r + x)}{2\kappa}\right) \tag{8}$$

where r is the radial distance from the point source. The temperature distribution is formulated in a coordinate system fixed to the source [21]. At $r = 0$, $T(\mathbf{r})$ becomes infinite, which is of course not the actual distribution of a temperature field in reality. From Eq. (8), it can be seen that the exponential term including velocity v causes a radially symmetrical exponential decrease of temperature with increasing v, but an asymmetrical temperature change in the direction of movement. Figure 2 depicts the temperature profile for a point source.

2. Gaussian Energy Distribution

The more realistic representation of the heat source is given by the Gaussian energy distribution. In contrast to the point source, there is no longer a singularity at $r' = 0$. In

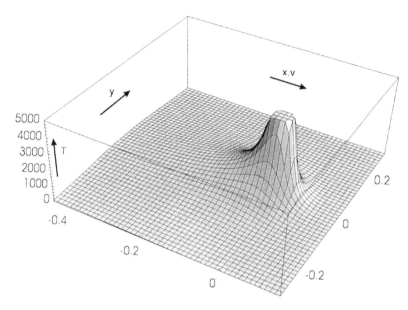

Figure 2 Profile of the calculated temperature rise at $z=0$ for a point source. The absorbed laser power and scan velocity are 200 W and 20 mm/sec, respectively.

practice, this distribution corresponds with the TEM_{00} mode for laser processing. If the total amount of absorbed power is P, then the energy distribution is described by

$$Q(|\mathbf{r}'|) = \frac{P}{2\pi\rho^2}\exp\left(-\frac{x'^2+y'^2}{2\rho^2}\right) \qquad -\infty < x', y' < \infty \qquad (9)$$

where ρ is the radius at which the Gaussian distribution has dropped to $1/e$ of its peak value. Substitution of the distribution function in Eq. (6) and integrating over x', y', z' from $-\infty$ top $+\infty$ yields

$$T(r,t) = \frac{\alpha P\kappa}{4\pi K}\int_0^t \frac{(2\kappa(t-t')+\rho^2)^{-1}}{\sqrt{4\pi\kappa(t-t')}}\exp\left(-\frac{(x+v(t-t'))^2+y^2}{4\kappa(t-t')+2\rho^2}-\frac{z^2}{4\kappa(t-t')}\right)dt' \qquad (10)$$

where α is the absorptivity. It can be seen that, at a large distance from the heat source, the temperature fields of point source and Gaussian beam will coincide. In the near field, whereby the point source idealization fails, the temperature field is expected to be modulated by the Gaussian distribution. Figure 3 depicts the temperature profile for a Gaussian intensity distribution of the heat source. The solution given in Eq. (10) is given for infinite beam size. If the size of the beam is finite, the solution will be modified as described in Ref. 19.

3. Homogeneous Energy Distribution

Another realistic representation of the heat source is the homogeneous energy distribution. In laser applications, this can be achieved by making use of beam integrators or by transport of the beam through optical fibers. If we assume the shape of the beam to be rectangular and the amount of absorbed power to be P, the intensity distribution is described by

$$Q(\mathbf{r}') = \frac{P}{4\rho_1\rho_2} \qquad \text{where} \qquad -\rho_1 < x' < \rho_1, -\rho_2 < y' < \rho_2 \qquad (11)$$

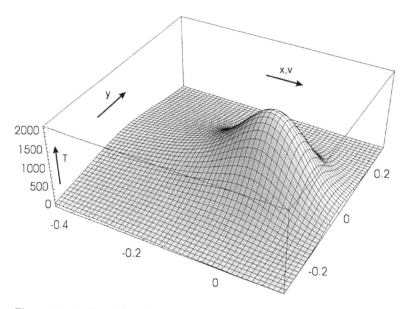

Figure 3 Profile of the calculated temperature rise at $z = 0$ for a gaussian intensity distribution of the laser beam. The applied laser power and scan velocity are 200 W and 20 mm/sec, respectively. For the substrate, the thermophysical properties of Ti are used.

$2\rho_1$ and $2\rho_2$ are the length and the width of the beam, respectively. The source term has to be integrated only over its finite dimensions. Substitution of the energy distribution in Eq. (6) results in the temperature field given by

$$T(\mathbf{r}, t) = \frac{\alpha P \kappa}{8K\rho_1\rho_2} \int_0^t \frac{dt'}{\sqrt{4\pi\kappa(t - t')}} F(x)F(y)\exp\left(-\frac{z^2}{4\kappa(t - t')}\right)dt' \qquad (12)$$

where

$$F(x) = \text{erf}\left(\frac{x + v(t - t') + \rho_1}{\sqrt{4\kappa(t - t')}}\right) - \text{erf}\left(\frac{x + v(t - t') - \rho_1}{\sqrt{4\kappa(t - t')}}\right) \qquad (13)$$

and

$$F(x) = \text{erf}\left(\frac{y + \rho_2}{\sqrt{4\kappa(t - t')}}\right) - \text{erf}\left(\frac{y - \rho_2}{\sqrt{4\kappa(t - t')}}\right) \qquad (14)$$

Figure 4 depicts the temperature profile for a heat source with a homogeneous intensity distribution.

4. Multiple Passes

In most practical applications, the laser treatment is not confined to single tracks. To modify a surface area, overlapping tracks are made. If the time between successive tracks is small, there already exists a temperature profile caused by the preceding pulsed laser track. Mathematically, this can be described by a second source at a distance Δx and Δy removed from the first one. This can be performed several times in the case of multiple tracks. Note

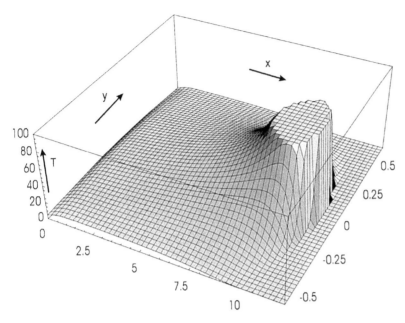

Figure 4 Profile of the calculated temperature rise at $z = 0$ for homogeneous intensity distribution of the laser beam. The applied laser power and scan velocity are 250 W and 60 mm/sec, respectively. For the substrate, the thermophysical properties of Ti are used.

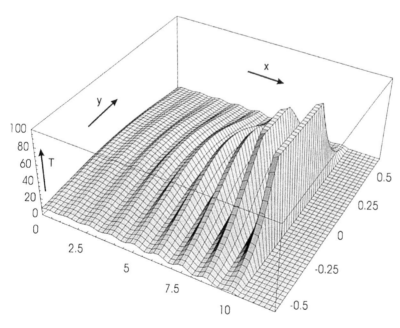

Figure 5 Profile of the calculated temperature rise at $z = 0$ for homogeneous intensity distribution of a pulsed laser beam. The applied laser power and scan velocity are 250 W and 60 mm/sec, respectively. The pulse frequency is 5 Hz, whereas the on/off time intervals are equal. For the substrate, the thermophysical properties of Ti are used.

that the time between successive tracks can be controlled by the distance Δx. This method can be incorporated into the model by changing the source distribution for the Gaussian intensity distribution by

$$Q(x',y') = \frac{P}{2\pi\rho^2} \sum_m \exp\left[-\frac{(x'-(m-1)\Delta x)^2+(y'-(m-1)\Delta y)^2}{2\rho^2}\right] \tag{15}$$

5. Finite Medium

This is the description given so far for the semi-infinite body. However, for industrial laser applications, this situation is not realistic in many cases. Therefore, it should be worthwhile to make the influence of finite dimensions of the specimen visible. This can be carried out by making use of image sources [17].

Imagine that a point source scans with a velocity v in the x-direction of the specimen. In the x-direction, the specimen is assumed to be infinite. The height of the sample amounts to d and, in y-direction, the size amounts to $R_1 + R_2$, representing the distance between the source and the boundaries in the positive and negative y-direction, respectively. Starting from the fundamental solution, the temperature field, represented by Green's functions, has to be determined for all directions. The temperature field in the x-direction is given by Eq. (7). To fulfill the boundary conditions in the y- and z-directions, multiple image sources are introduced. Figure 6 illustrates the positioning of the imaginary sources. To provide zero heat flux, the image sources are positioned at $-2R_1$ and $2R_2$. However, these sources provide a heat flux at the opposite boundary. Therefore, additional heat sources are required to cancel this out. When required, a sequence of image sources can be applied. As a result, the temperature field in the y-direction can be described by

$$\begin{aligned}
G(y) = \frac{1}{\sqrt{4\pi\kappa(t-t')}} &\left[\sum_{n=-\infty}^{\infty} \exp\left(-\frac{(y-y'-2n(R_1+R_2))^2}{4\kappa(t-t')}\right)\right. \\
&+ \sum_{n=1}^{\infty} \exp\left(-\frac{(y-y'-2nR_2-2(n-1)R_1)^2}{4\kappa(t-t')}\right) \\
&+ \left.\sum_{n=1}^{\infty} \exp\left(-\frac{(y-y'+2nR_1+2(n-1)R_2)^2}{4\kappa(t-t')}\right)\right]
\end{aligned} \tag{16}$$

In the z-direction, the image sources are used as in the aforementioned method, which results in a temperature field in the z-direction

$$G(z) = \frac{1}{\sqrt{4\pi\kappa(t-t')}} \sum_{n=-\infty}^{\infty} \exp\left(-\frac{(z-2nd)^2}{4\kappa(t-t')}\right) \tag{17}$$

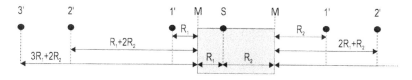

Figure 6 Construction of image sources in the y-direction to fulfill zero heat flux at the boundaries of the specimen. The heat source interacts with the specimen at point S. The edges of the specimen act as mirror planes.

In Ref. 19, the complete solution of the temperature field is given for a Gaussian intensity distribution of the heat source, including the use of image sources. Subsequently, the questions as to when it is necessary to use the image sources and how many are required will be asked. In principle, this is determined by the size of the specimen and the physical properties of the material. Therefore, a criterion should be formulated for the use of image sources that will be treated on the basis of two materials with completely different thermal conductivities.

When numerically integrating the 1-D Green function over a certain time interval, e.g., 1 sec, and over the width of the specimen, the outcome should be equal to unity in the case for which the specimen has infinite dimensions. However, for a small specimen, a value lower than unity is obtained. Which of the values taken is selected as a critical value depends on the required accuracy, that is, which image sources should be included in the model. When this is necessary, it is sufficient to include one single image source at every boundary for most cases. In addition to the specimen size, the outcome is also a function of the thermal diffusivity of the material. This will be illustrated on the basis of Ti and Mo, both of which exhibit a large difference in thermal diffusivity.

Two specimens, Mo and Ti, are assumed to be subjected to the same laser treatment. The width and the thickness of the specimen are 8 and 4 mm, respectively. The temperature profiles are calculated in the y-direction at a distance of 3 mm behind the laser beam for two different situations, i.e., with or without the use of image sources. Figure 7 shows the calculated temperature profiles. In the case of Mo, there is a significant temperature increase when making use of image sources. However, for Ti, the temperature profiles deviate only at the edge of the sample, which is hardly visible. To prove that the aforementioned criterion is capable of predicting if the use of an image source makes sense, evaluation is carried out on the basis of these different materials. Therefore, we consider Green's function in the y-direction, over a time interval of 1 sec. The integration yields values of 0.88 and 0.50 for Ti and Mo, respectively. Depending on the required accuracy of the calculations, a critical value should be stated. Nevertheless, the obtained values deviate significantly, which indicates the need for image sources in the case of Mo when calculating the appropriate temperature profile.

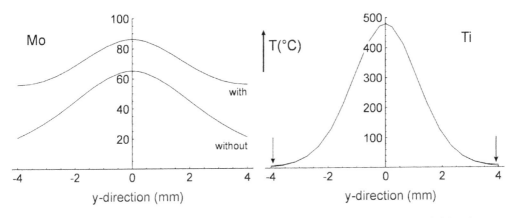

Figure 7 Calculated temperature profile in the y-direction for Mo (left) and Ti (right) substrate, with and without making use of image sources. The laser power and scan velocity are 400 W and 30 mm/sec, respectively. The temperature is calculated 3 mm behind the laser beam.

6. Heat Loss

In the previous model, the boundary conditions impose zero heat flux. In reality, energy transfer occurs at the surface due to temperature differences with surrounding media. For laser processing, the energy transfer takes place via different phenomena. If the specimen is in direct contact with a heat sink, conductive cooling takes place at $z = d$. As a consequence of the shielding gas, convective cooling (i.e., Newtonian cooling) takes place at $z = 0$. Furthermore, heat transfer also takes place by radiation at $z = 0$, in accordance with the Stefan–Boltzmann law. To take into account the different kinds of heat transfer, an adaptation is made to the model.

For our description, only cooling in the z-direction will be taken into account at $z = 0$. When starting from the temperature held in the z-direction, as given by Eq. (17), followed by the Laplacian transform, the transfer of heat can be incorporated by including an extra term in the following way:

$$\overline{G}_z = \frac{1}{2\sqrt{\kappa q}} \exp\left(-\sqrt{\frac{q z^2}{\kappa}}\right) + A \exp\left(-\sqrt{\frac{q z^2}{\kappa}}\right) \tag{18}$$

where A is a constant that can be determined by the boundary condition. If the transfer of heat is supposed to be linearly dependent on the temperature difference over the surface, the boundary condition can be represented by

$$\frac{\mathrm{d}G}{\mathrm{d}z} = \xi G \tag{19}$$

and

$$T_{\mathrm{corr}}(\mathbf{r}, t) = -\frac{\xi \kappa}{K} \int_0^\infty \exp[\xi z + \kappa \xi^2 (t - t')] \mathrm{erf}\left[\frac{z}{2\sqrt{\kappa(t - t')}} + \xi\sqrt{\kappa(t - t')}\right]$$
$$\times \int_0^\infty \int_0^\infty \frac{1}{4\pi\kappa(t - t')} \exp\left[\frac{(x + v(t - t') - x')^2 + (y - y')^2}{4\kappa(t - t')}\right] \tag{20}$$
$$\times Q(x', y') \, \mathrm{d}x' \, \mathrm{d}y' \, \mathrm{d}t'$$

where $x = h/K$, with h representing the heat transfer coefficient and K the thermal conductivity. After performing the Laplace transform on the boundary condition and substitution of Eq. (18), the solution of the modified temperature field can be obtained. In fact, the former solution of the temperature field is extended by a temperature correction term represented [22] as shown in Eq. (20).

In reality, the description of the physical processes, corresponding to the different ways heat transfer occurs, often exhibits a nonlinear temperature dependency. Where the transfer takes place by radiation, it depends on the fourth power of the temperature. However, for convective cooling at $z = 0$ caused by the shielding gas, the linear temperature dependency is a satisfactory approximation.

So far, the derivation of the temperature correction takes into account only the boundary condition at $z = 0$. The boundary condition at $z = d$ can be incorporated simultaneously, by introducing an extra term with prefactor B in Eq. (18). Furthermore, the use of image sources also can be included whereby the inverse Laplacian transformation becomes rather complex. A disadvantage of the aforementioned description method is that factors A and B do not represent any physical property. Therefore, a different approach is used in which a representative cooling factor arises that has a more physical meaning.

First, the so-called Biot number will be introduced [23]. In the general transient heat conduction formulations (as put forward in the previous section), the temperature field varies both with time t and position. However, in various engineering applications, temperature variations within the materials can be neglected and the temperature is considered to be only a function of time t. Such assumptions are made in the so-called lumped system formulation. To provide some criteria for the range of validity, the internal thermal resistance is considered vs. the external resistance. The former is described by $L/(K A)$, whereas the latter is equal to $1/(hA)$ (L is the characteristic length and h is the heat transfer coefficient). The ratio of the internal thermal resistance vs. the external resistance is called the Biot number

$$B = \frac{hL}{K} \tag{21}$$

If the internal thermal resistance is small, the temperature distribution remains sufficiently uniform during transients, that is, B is small and the lumped system formulation is valid. This formulation is applicable in most engineering systems if $B \leq 0.1$. Now, let us consider the solution of a homogeneous boundary value problem of heat conduction of a finite slab ($0 \leq z \leq L$) initially at a temperature $T(\mathbf{r})$ that dissipates heat for times $t > 0$ into an environment at zero temperature (we are interested only in the temperature change). The mathematical formulation is within Eqs. (3) and (5) with the boundary conditions

$$-K_1 \frac{\partial T}{\partial z} + h_1 T = 0 \qquad (z = 0, t \geq 0); \tag{22}$$

$$K_2 \frac{\partial T}{\partial z} + h_2 T = 0 \qquad (z = L, t \geq 0); \tag{23}$$

and

$$T = T(\mathbf{r}) \qquad (\text{for } t = 0 \text{ in } 0 \leq z \leq L) \tag{24}$$

If the system dissipates heat at the boundary $z = L$ by convection or conduction and the boundary at $z = 0$ is kept insulated, Green's function can be derived [24]

$$G(|z - z'|, t - t') = \frac{2}{L} \sum_{m=1}^{\infty} \exp\left(\frac{-\beta_m^2 \kappa(t - t')}{L^2}\right)$$
$$\times \frac{\beta_m^2 + B_2^2}{\beta_m^2 + B_2^2 + B_2} \cos\left(\beta_m \frac{z}{L}\right)\cos\left(\beta_m \frac{z'}{L}\right) \tag{25}$$

where z' is the position of the heat source that operates at $t' > 0$, κ is the thermal diffusivity, and the β_m values are the positive roots of

$$\beta_m \tan\beta_m = B_2 \tag{26}$$

with

$$B_2 = \frac{h_2 L}{K} \tag{27}$$

For $B_2 < 2$, the eigenvalues of Eq. (27) can be approximated by

$$\beta \approx \sqrt{3}\sqrt{\frac{B_2}{B_2 + 3}\left(1 - \frac{B_2^2}{5(B_2 + 3)^2}\right)} \tag{28}$$

If the system dissipates heat by convection/radiation both at $z = 0$ and $z = L$, Green's function can be derived by solving a special case of the Sturm–Liouville equation, as described in Ref. 19.

7. Cooling Rate

After the laser beam has passed, the cooling rate influences the final microstructure in several ways. First, the microstructure will change as a function of the rate. Near the solidification temperature, the stability of the solidification front is directly related to the cooling rate. For high cooling rates, the liquid may be undercooled, resulting in the formation of protrusions at the solidification front. These protrusions will develop as cellular or dendritic structures depending on the cooling rate. Besides, when during cooling down a phase transition takes place, the rate determines the degree of diffusion during this transition. For example, for titanium, a martensitic phase transformation takes place only from β- to α-phase for high cooling rates. Second, the cooling rates determine the thermal stresses, which are induced in the layer. These could result in crack formation when exceeding a critical stress state. Therefore, it is very important to obtain insight into cooling rates during laser processing.

In principle, the cooling rate is related to the temperature gradient in the direction of motion multiplied by the velocity of the solidification front. If the velocity of the solidification front is assumed to be equal to the laser scan velocity, the cooling rate is defined by

$$\frac{\partial T}{\partial t} = -v_\ell \frac{\partial T}{\partial x} \tag{29}$$

For most intensity distributions of the heat source, it is rather complicated or impossible to obtain an analytical expression for the temperature gradient. As a consequence, the gradient has to be determined numerically. However, for one specific situation, it is possible to derive an analytical expression. In case the observer is far removed from the center of the beam, an analytical expression for the derivative of the temperature field can be obtained for the point source. Differentiation of Eq. (8) yields a temperature gradient defined by

$$\frac{\partial T}{\partial t} = -v \left[\frac{x}{r^2} + \frac{v}{2\kappa} \left(1 + \frac{x}{r} \right) \right] T \tag{30}$$

III. EXAMPLES AT AN ATOMIC, MESO-, AND MACROSCOPIC SCALE

This section exemplifies the method of laser melt injection (LMI) by focusing on the creation of interfaces between dissimilar materials at a meso- and a macroscopic scale. Examples are presented of matrix composites of SiC particles into an Al substrate and SiC in a Ti matrix. An extremely small operational parameter window has been found for successful injection processing. It is shown that the final injection depth of the particles is mainly controlled by the temperature of the melt pool, rather than by the particle velocity. A theoretical model that takes into account wetting behavior and particle penetration processes has been developed on the basis of the observed particle velocity, the thickness, and the area fraction of the oxide skin that partially covers the surface of a heated Al melt pool. The model reveals the role of the oxide skin in such a way that it is relatively strong at a low temperature and acts as a severe barrier for the injection process. It has been found that a preheating of the Al substrate results in a higher temperature of the melt pool and in partial dissolution of the oxide skin, through which the injected particles are able to penetrate.

A. Laser Melt Particle Injection in Aluminum Alloys: On the Role of the Oxide Skin

Treatments of metal surfaces with high-power lasers are appropriate for improving the mechanical, tribological, and chemical properties of metallic surfaces [25]. The laser melt injection (LMI) method [26–30] is aimed at producing a metal matrix composite (MMC) layer on top of a substrate. The laser beam melts the substrate locally, while simultaneously particles of additional material (usually ceramics) are injected. These particles are trapped when the melt pool rapidly resolidifies after the laser beam has passed. In the commonly applied laser cladding process, a coating is created by fully melting the additional powder material and forming a new alloyed layer after mixture with the substrate materials. Unlike laser cladding [31], contact between the laser beam and the added material in LMI is limited to a level just necessary to form a strong bonding interface between the ceramic particles and the metallic matrix in the final MMC layer.

During laser treatments of Al alloys, one has to face certain problems [32]. For the wavelength of radiation produced by a Nd–YAG laser ($\lambda = 1.06\,\mu$m), as is used in this work, the reflectivity of solid Al is about 90% [33]. Therefore, a high-energy density of the laser beam is needed to create a melt pool in Al. Another problem is the oxide skin that is naturally formed on an Al melt. This oxide layer on molten Al has a substantial influence on the wetting behavior of ceramic particles with liquid Al [34]. Usually, the oxide skin forms an energy barrier for the particles to penetrate. However, the strength of this barrier depends on the temperature of the melt pool. For instance, at lower temperatures (640–850°C), the contact angle between the SiC particle and Al melt is about 130°, while at higher temperatures (>1100°C) the contact angle is decreased to about 50° [34].

In this work, the LMI method has been scrutinized on the SiC$_p$/Al system. Experimentally observed phenomena are discussed within the framework of the laser process and the microstructural features and they are supported by a theoretical analysis.

1. Experimental Procedures

The laser experiments have been carried out with a continuous wave 2 kW Rofin Sinar Nd–YAG laser. The laser beam is positioned at an 11° angle with respect to the substrate surface normal to avoid harmful reflections from the specimen back into the optical fiber. Argon was used as a shielding gas to protect the lens as well as to reduce oxidation of the specimen. The spot size of the laser beam on the work piece was 2.4 mm. A computer-controlled X–Y table was employed for specimen movement.

Aluminum plates with 99.6 wt.% purity and dimensions of $100 \times 40 \times 10$ mm were used as the substrate material. Before the laser treatment, the surface has been sandblasted to promote the absorption of the laser light. A powder feeding system Metco 9MP supplied 6H–SiC particles with mean size of about 80 μm, using Ar as a carrier gas with flow rates in the interval of 25–67 ml/sec. The particles were injected into the melt pool at an angle of 35° with respect to the surface normal. The injection velocity of the SiC particles was measured by recording their flight out of the powder nozzle with a KODAK 4540 Ektapro High Speed Camera.

The temperature of the substrate just before and during the laser processing was acquired via a monochromatic optical pyrometer (Sensys). The optical sensor was focused onto the center of the border of the substrate block.

The coatings were observed by the standard optical (Olympus Vanox-AHTM) microscope and a Philips XL-30 FEG scanning electron microscope (SEM). The oxidation layer was analyzed by a special small-spot ultrahigh vacuum (UHV) scanning electron–

scanning Auger microscope based on a JEOL JAMP 7800F, in which depth profiling has been performed in situ by 1 keV Ar^+ bombardment.

The experimental setup of the laser melt injection process is displayed in Fig. 8. A powder feeder apparatus connected to a cyclone makes a constant particle flow. The direction of the particle flow is important with respect to the moving direction of the substrate. In this case, the so-called "over-hill" direction is used where the particles are traveling over the coating toward the center of the melt pool. To achieve injection conditions for SiC particles into Al melt, the upper carrier gas outlet of the cyclone is partially or fully closed with the aim of increasing the speed of SiC particles. This also helps to reduce the interaction time between the laser beam and SiC_p, which is important because SiC has a high absorption of the laser radiation [35]. For the same reason, the focal point of the laser beam lies 6 mm below the substrate surface, which results in lower beam energy densities just above the surface where injected particles come into contact with the laser beam.

An appropriate combination of process parameters for the production of a single laser track, was found with a laser beam power density of 310 MW/m^2, scanning speed 8.3 mm/sec, powder feeding rate 8.3 mg/sec, glass nozzle diameter 1.7 mm and 60 ml/sec carrier gas flow. Before the LMI process, the substrate is preheated to 300°C by a plate heater under a constant temperature and is kept on the heater during injection. The temperature of the Al substrate increases to about 30°C after one track due to the absorbed beam energy, which provides the possibility of setting different preheating temperatures. The injection process is sensitive to small changes in power density, preheating temperature, powder flow, and particle velocity, which results in an extremely small operational window of laser and powder flow parameters.

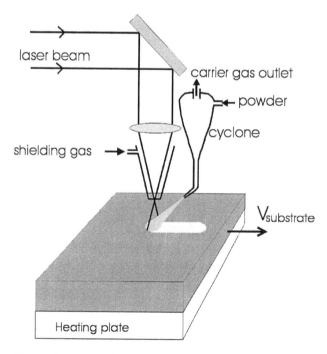

Figure 8 Schematic picture of the laser melt injection process. The particles are injected in the so-called "over-hill" direction.

As Fig. 9 demonstrates, the SiC volume fraction in the MMC track is about 35%, and the width of the track is about 1 mm. The penetration depth of the injected particles depends significantly on the temperature of the melt pool, as shown in Fig. 10. When the preheating temperature is lower than 300°C, the particles are unable to penetrate in the melt pool, whereas at higher temperatures the particles are successfully injected. Increasing the preheating temperature results in more, and also much deeper, injected particles. In all experiments, the melt pool is wider than the area where particles are injected.

The effect of the melt pool temperature on the injection depth is also shown in a longitudinal cross section (Fig. 11). During production of a laser track with a length of about 30 mm, the substrate temperature gradually increases and therefore the particles are injected deeper near the end of laser track. The increase in temperature during a single laser track depends on the length of the track and the size of the substrate, which acts as a heat sink. In Fig. 11, the substrate temperature increase is about 30°C, measured on the side rim of the substrate block. An increase to a maximum injected depth near the end of a track as well as local oscillations in the depth of injected particles are also observed.

2. Calculation of Particle Penetration

For a better understanding of the particle penetration process, a simple model is constructed. The problem can be divided into two parts: penetration through the melt surface and particles moving in a liquid. To simplify the problem, the particle is assumed to be of a spherical shape, with a radius R and vertical component of velocity v_0. To overcome the molten surface barrier and to propagate further into the melt, the particle kinetic energy is:

$$E_{kin} = \frac{2}{3} \pi R^3 \rho_{SiC} v_0^2 \qquad (31)$$

where ρ_{SiC} is the density of SiC (3217 kg/m^3).

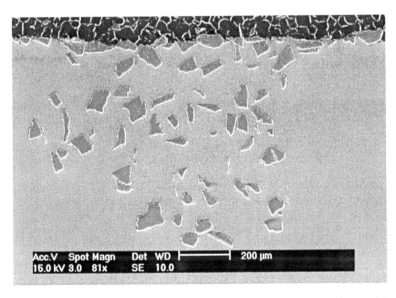

Figure 9 SEM of the cross-section of laser track with injected SiC particles.

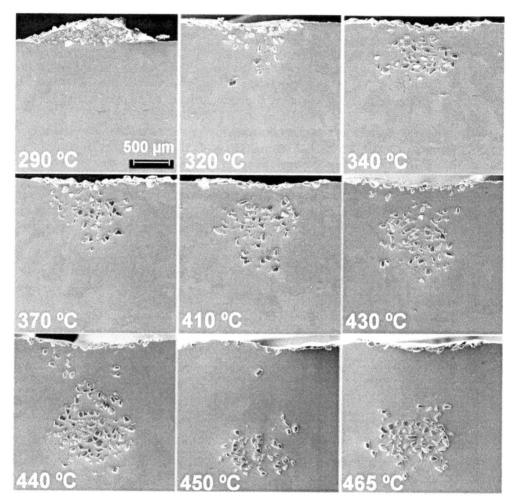

Figure 10 Cross-sections of single laser tracks that are produced by different preheating tempera-tures. Above 300°C, the SiC particles are injected, and the higher the preheating temperature, the deeper the particles are injected. The particles are not injected across the melt pool, i.e., at the edges no particles are injected.

The total interface energy for a particle partially immersed into a melt at depth x, $0 \leq x \leq 2R$ (see Fig. 12), is given by [36]:

$$G_{\text{interface}} = (R - x)^2 \pi \sigma_{\text{lv}} + 2R\pi x \sigma_{\text{lp}} + 2R\pi(2R - x)\sigma_{\text{pv}} \tag{32}$$

where σ is the interface energy (tension) between the phases indicated by the indices: l=liquid Al, p is the solid SiC particle, and v refers to vapor. The equilibrium depth x_{eq} and contact angle Θ can be calculated from Eq. (32) by minimizing the interface energy:

$$x_{\text{eq}} = R\frac{(\sigma_{\text{lv}} + \sigma_{\text{pv}} - \sigma_{\text{lp}})}{\sigma_{\text{lv}}}$$

$$\Theta = \pi\frac{-}{2} + \arcsin\left(\frac{R - x_{\text{eq}}}{R}\right) \tag{33}$$

Figure 11 Longitudinal cross-section of a laser track. The laser beam proceeded from the right to the left. The tendency to increase the injection depth as well as the local oscillations in the maximal injection depths is demonstrated.

This is basically the same as Young's equation for the equilibrium contact angle. The strength of the surface barrier can be estimated by the amount of change in interface energy when a single particle is moved from the depth where the interface energy of the system is minimal, that is, the wetting angle, to the depth where the particle is entirely inside the melt:

$$\Delta G_{\text{barrier}} = G_{(x=x_{\text{eq}})} - G_{(x=2R)} = \frac{\left(\sigma_{\text{lv}} + \sigma_{\text{lp}} - \sigma_{\text{pv}}\right)^2}{\sigma_{\text{lv}}} \pi R^2 \tag{34}$$

Combining Eqs. (31) and (34), the minimal vertical velocity of the particle v_{min}, required to overcome the melt surface barrier can be obtained:

$$v_{\text{min}} = \sqrt{\frac{3}{2\sigma_{\text{lv}} R \rho_{\text{SiC}}} \left(\sigma_{\text{lv}} + \sigma_{\text{lp}} - \sigma_{\text{pv}}\right)} \tag{35}$$

If the initial particle velocity is higher than v_{min}, the particle overcomes the surface barrier and may propagate farther into the melt with the reduced velocity v

$$v = \sqrt{v_0^2 - v_{\text{min}}^2} \tag{36}$$

where the Stoke's force

$$F = 6\pi\eta R v \tag{37}$$

and gravity corrected by Archimedes' principle, which work against each other, drive the particle penetration farther. The quantity η in Eq. (37) denotes the viscosity of the Al melt. The values of the interface energies for both oxidized and nonoxidized surfaces of liquid Al are given in Table 1 [37]. All this indicates that the velocity of particles and the particular state of the melt pool surface play important roles in the actual laser melt injection process.

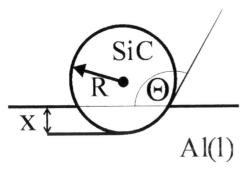

Figure 12 A spherical SiC particle immersed into an Al melt at depth x corresponding with a contact angle.

Table 1 Interface Energies of Oxidized and Nonoxidized Al in the Al(1), SiC$_p$ System

	Interface energy at 900 K (mJ/m^2)	
Physical quantity	Al, oxidized	Al, not oxidized
σ(SiC, solid–vapor)	1920	1920
σ(Al, liquid–vapor)	870	1100
σ(Liquid Al–solid SiC)	2480	1210

3. Velocity of Injected Particles

A high-speed camera was used to monitor the flight of the particles after they leave the nozzle, allowing their velocity to be measured and hence their kinetic energy to be calculated. Experiments with different glass nozzle diameters ($\phi = 1.7$ and 2.3 mm) and different carrier gas flows (25–70 ml/sec) were recorded with a camera speed of 27×10^3 frames/sec. Afterward, the particle velocity was measured on the video screen with the standard frame speed of 25 frames/sec by counting the number of frames that the particle needs to travel between two markers at a constant distance. A velocity of more than 50 particles was traced in each experiment.

The results of the high-speed camera observations are summarized in Fig. 13. The mean velocity of the particles shows a linear increase with carrier gas flow rate. A nozzle diameter of 1.7 mm increases the velocity by a factor of about 1.5, with respect to a nozzle diameter of 2.3 mm. The variation in particle speed is about 15–20%. Distinction is made between particles traveling in the upper, central, and lower parts of the powder stream. The particles have the highest velocity in the center of the stream; at the higher and lower parts of

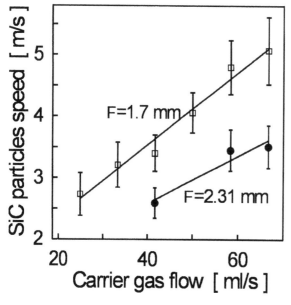

Figure 13 Particle velocity as a function of carrier gas flow for two diameters of glass nozzles, obtained by high-speed camera observations.

the stream the velocity is about 90% of the velocity in the center. In our LMI experiments, the particle velocities vary in the range of 2.5–5 m/sec.

4. Oxide Layer on Aluminum

To get an idea about the behavior of the oxide layer during the laser process, single laser tracks were produced with the same laser parameters as in the injection experiments but without particle flow. The appearance of the surface of a track displays two different areas. In the center, there is a shiny strip with a width of about 1.1 mm, while at both sides rather dull strips are present with an approximate width of 0.2 mm. The borders between these areas are quite sharp. The FEG-SEM micrographs taken from both the center and the side areas are shown in Fig. 14. In the central area of the laser track, about 45–50 % of the surface is not covered by a thick oxide layer. On the other hand, only 20% of the surface is not covered at the edge of the melt pool.

The thickness of the oxide layers of different areas is analyzed by depth profiling in the UHV SEM-SAM (scanning Auger microscopy). The surface layer is removed by Ar^+ bombardment while, after each etching step, the intensities of Al, O, and C element peaks in Auger electron spectra were analyzed. In this way, the thickness of the oxide layer can be estimated by recording the etching time needed to remove the oxygen from the spectra. The results depicted in Fig. 15(a,b,c) represent the depth profiles measured from the places defined in Fig. 14. It is clear that the oxide layer from the noncovered area at the center of laser track is removed after 100–150 sec of Ar^+ bombardment, as Fig. 14(a) shows. To remove the oxide layer from the noncovered edge region, an etching time of about 800 sec is necessary [Fig. 15(b)] and removal of the oxide layer from the covered edge region needs more than 5000 sec [Fig. 15(c)]. From these etching times, an estimate of the thickness of the oxide layer can be made [38] if one knows the principal etching parameters and the so-called sputtering yield [39]. Such estimates for all three characteristic places are marked in Fig. 15 by dashed lines.

5. Discussion

The first point of discussion concerns the laser melt injection process itself. The powder stream is suitable to produce tracks of laser melt particles and the particle velocity in the powder stream is quite constant and can be adjusted by the nozzle diameter and carrier gas flow. The extremely small operational window of laser and powder flow parameters can be explained by the substantial difference in the absorptivity of laser radiation between Al and SiC on one hand, and the oxide skin of Al on the other. A high laser energy input is needed to create a melt pool in Al while the same energy input will heat up the particles enormously when they travel through the laser beam, because of their high absorptivity for laser radiation. This will damage the particles and lead to undesired Al_4C_3 formation [40,41], which may subsequently have a negative effect on the mechanical properties of the coating [42]. Another effect of the high absorptivity is that the powder flow shields the surface. Therefore, the feeding rate should be low in comparison to the standard cladding parameters [43].

Preheating the substrate makes the laser process much more efficient because the coupling between the laser light and substrate increases with rising temperature. Furthermore, when the substrate is preheated to 300°C, the heating up to the melting temperature (660°C), where most of the laser light is reflected, is about halved. Although all these reasons explain that, by using the same laser energy, the melt pool temperature is higher when the substrate is preheated, it is hard to believe that this explains the reason why preheating is essential for an appropriate particle injection process. For example, the particles

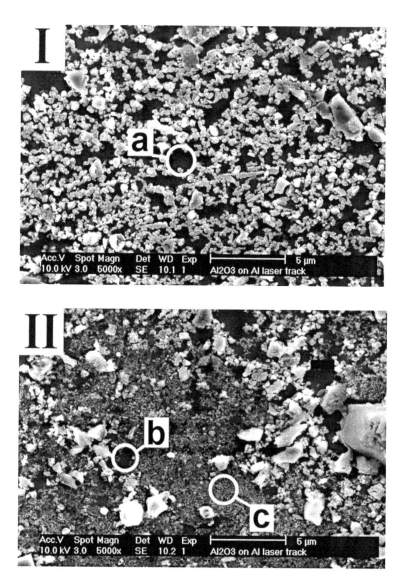

Figure 14 SEM micrographs of the surface of aluminum of two different areas after laser processing—I: in the center of the laser track where the open areas (a) represent approximately 50% of the whole surface; and II: at the edge of the melt pool where the open parts of the surface are covered with a "spongy" looking phase (c). Open areas (b) represent approximately 20% of the whole surface.

are injected to a depth of 500 μm when preheating to 340°C, whereas they are hardly injected at a preheating temperature of 300°C. To understand this, we will focus on the role of the oxide skin.

It is necessary to emphasize that the measured fractions and thickness of the oxide surfaces *after* the laser process are probably not the same as they have been *during* the laser process on the Al melt. To a first approximation, it is reasonable to assume that the areas where an oxide layer of 3 nm is detected by UHV SEM-SAM did not possess any oxide layer during the laser experiments. This is because a 3 nm-thick oxide layer is typical for an

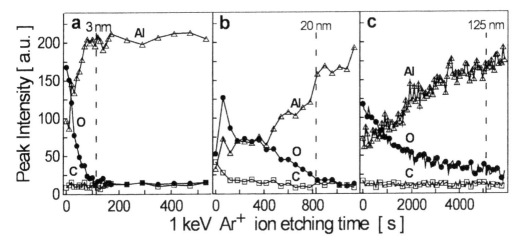

Figure 15 Aluminum, oxygen, and carbon element peak intensity in Auger electron spectrum as a function of Ar$^+$ bombardment time, giving an element depth profile. Panels a, b, and c show the depth profiles measured on the marked areas from Fig. 14.

amorphous oxide skin reported as a layer grown on a solid Al surface [44]. In addition, the fraction of "open" areas may be higher during laser treatments. Nevertheless, the ex-situ experiments show that the behavior of this oxide skin is a function of temperature [36]. Actually, the oxide skin present on the melt up to 850°C prevents direct contact between the particles and Al liquid and leads to a nonwetting behavior. Above this temperature, chemical interactions between Al and its oxide at the interface may occur, leading to the formation of gaseous suboxides. Consequently, the Al surface takes over at temperatures above 850°C, allowing the particles to have direct contact with the Al liquid. This results in chemical interactions between the particle and the Al melt. These interactions lower the energy of the particle/Al interface and result in an improved wetting behavior. At about 1100°C, the oxide is completely removed and consequently, a transition from nonwetting to wetting behavior takes place in a small temperature interval of 850 and 1100°C. In this interval, the surface is partially covered with oxide.

The energy barrier of partially covered surfaces can be obtained, realizing that the particle size ($R = 40$ µm) is much larger than the oxide free parts (<5 µm in Fig. 14), by adjusting Eq. (34):

$$\Delta G_{\text{barrier}} = \frac{\left(a\sigma_{\text{lv,ox}} + a\sigma_{\text{lp,ox}} + (1-a)\sigma_{\text{lv,al}} + (1-a)\sigma_{\text{lp,al}} - \sigma_{\text{pv}} \right)^2}{a\sigma_{\text{lv,ox}} + (1-a)\sigma_{\text{lv,al}}} \pi R^2 \qquad (38)$$

where a is the area fraction of the surface that is covered with an oxide skin and indices ox and al indicate the interface energies of the oxidized and not oxidized case, respectively. Eq. (38) can be modified in the same way to calculate the minimum velocity to overcome the surface energy barrier, that is, not only in the case of clean and fully covered surfaces, but also in the case of partially covered surface. Figure 16 shows that the minimum required velocity to overcome the barrier is a function of particle size for different oxide coverage percentages. Here, curves are depicted that characterize a surface that is fully covered, together with two experimentally observed, partially covered surfaces and a surface that is free of an oxide skin. In Fig. 16, the area corresponding to the particle velocities and the sizes used in the injection experiments also are marked.

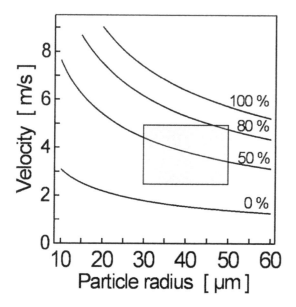

Figure 16 Minimal particle velocity, needed to overcome the surface energy barrier as a function of radius of the particle for different ratios of surface covering by an oxide skin. The area of particle sizes and velocities in the experimental setup is marked.

One may conclude that for 50% coverage, which is characteristic for near the center of laser track, a velocity of about 4 m/sec is sufficient to penetrate the surface layer. On the other hand, if the surface is covered 80% or more by an oxide layer, the particles cannot be injected with the present experimental arrangements. Indeed, this kind of behavior has been observed in our experiments where particles are injected only in the central part of the laser track and not near the sides. The distribution of the injected depth observed in the cross sections of the laser track (Fig. 9) is not controlled by the distribution of the particle speed across the particle stream, but is determined by the increase of coverage with oxide from the center of the laser track to its sides. Figure 16 also clearly indicates that laser injection is not a suitable technique for the preparation of SiC/Al MMCs with particles that are smaller than $R = 20$ μm. A preplacement technique is reported to be successful for particles with a size smaller than 45 μm [45,46].

The effects of the oxide skin behavior on the laser melt injection process will be discussed in more depth in the following. In the second part of the injection problem, that is, for a particle moving in a liquid, we have calculated the injection depth as a function of time. The result of such a calculation is displayed in Fig. 17 for a particle with a radius of 40 μm, an initial velocity of 4 m/sec, and having 50% of its surface covered by an oxide skin. The particle reaches its equilibrium velocity within 0.01 sec. This velocity is very small and, therefore, we may assume that the particle gains its final depth in this short time interval. The viscosity is taken to be a constant in the calculations, as that for Al liquid at $927°C$: 0.9 mPa sec, and this is only valid when the temperature in the melt pool is constant. At any rate, for such a short time interval, which is 40 times shorter than the estimated cooling time, the latter assumption seems to be quite reasonable. However, the temperature gradient with respect to the depth results in an increase in viscosity as a function of depth, which promotes the deceleration effect due to the Stokes' force. Therefore, the calculated final depths are somewhat overestimated. Both the temperature dependence of the viscosity

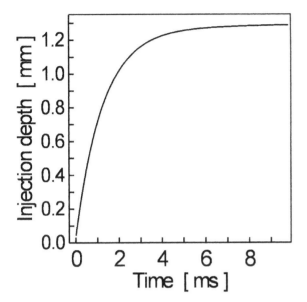

Figure 17 Injection depth of the particle as a function of time. The injection velocity is 4 m/sec and 50% of the surface is covered with an oxide skin and with particle radius of $R = 40$ μm.

and the convection are not taken into account. Looking at cross sections of the laser tracks, there are no indications that convection plays an important role.

In Fig. 18, the depth at $t = 0.01$ sec is presented as a function of the injection velocity for different oxide percentages. The steepness of the graphs confirms that if the injection velocity is sufficient to penetrate through the surface, a relatively great depth can be reached. This explains why small changes in laser process parameters may result in a substantial change in injection depth. This is shown in Fig. 10, where the preheating temperature only increases from 290 to 320°C, and this results in a very considerable difference in the actual injection depth. For preheating temperatures higher than 440°C, the particles with the lowest speed are able to reach relatively large depths, as Fig. 10 also shows.

Thus, without preheating, the temperature of the melt pool is probably close to 850°C, at which the oxide skin is still present and no particles can be injected. Preheating to 290°C increases the temperature, but not high enough to remove a sufficient part of the oxide. By increasing the temperature further to 320°C, the oxide layer is removed far enough to allow particle penetration. This substrate temperature increase of 30°C leads to a much higher temperature increase in the melt pool because of the better coupling factor of the laser beam at higher temperatures. In addition, the fact that we are in the temperature range of 850–1100°C, where small temperature changes result in a substantial change in wetting behavior, explains the impact of preheating the substrate.

Indeed, the idea that the temperature of the melt pool lies in the range of 850–1100°C is supported by the fact that the particles are injected only in the center of the melt pool and not near the edges. In the center, the temperature is high enough, while at the edges the temperature is too low to make any injection possible. Consequently, a transition temperature from wetting to nonwetting behavior is present in the melt pool, indicating that the temperature lies in the range of 850–1100°C. The sharp transition from injection up to several hundreds of microns to no injection at all at the edges of the laser track is explained once more by the rather steep slopes in the graphs shown in Fig. 18. The increase of injection

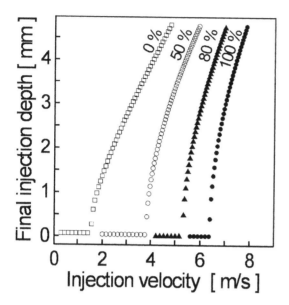

Figure 18 Final injection depth as a function of injection velocity for different percentages of covered surfaces by oxide skin.

depth during laser processing (Fig. 11) is explained by disappearance of the oxide layer as well. At higher temperatures, the decrease in kinetic energy loss during the penetration of the particles through the surface causes a deeper penetration of the particles. The depth oscillations, with a frequency of 5–10 Hz as observed in Fig. 11, can be explained by either the oscillatory kinetic behavior of the melt pool surface or the oscillatory changes in the powder stream.

Another apparent solution to create better injection conditions, i.e., other than preheating the substrate, involves increasing laser power to increase the melt pool temperature or to enhance the particle velocity. An increase in laser power is not advisable because it will lead to further damage of the SiC_p and the formation of undesired Al_4C_3 [30]. In addition, an increase in particle velocity may result in further damage or blowing away of the melt pool by the bombardment of the particles and the strong carrier gas flow. Consequently, preheating is an efficient and elegant method to be able to inject SiC in Al.

6. Conclusions

The small parameter window for laser processing of the SiC_p/Al MMC layer on an Al substrate is caused by a large difference between the laser light absorptivity of SiC and Al, as well as the presence of an oxide skin on Al melt. The final injection depth of the SiC particle in the Al melt is mainly controlled by the fraction of surface covered with oxide, i.e., by the temperature of the melt pool. A thick oxide skin on the Al melt at lower temperatures leads to loss in kinetic energy of the particles in the wetting process, and therefore they are not able to propagate into the melt.

Preheating the substrate to over 300 °C is an effective way to successfully inject SiC particles into the Al melt, by achieving the necessary high temperatures of the melt pool at which the oxide skin is fully or partially dissolved and avoiding the use of high laser power which leads to the overheating of SiC particles.

Because of the transversal gradient of the temperature in the melt pool, SiC particles are injected only in the center area of the melt pool. The two side areas are almost fully covered by a relatively thick oxide layer. It does not seem to be possible to inject SiC particles smaller than 40 μm into Al with the LMI process.

B. Laser Melt Particle Injection of Ceramic Particles in Titanium

Several experimental methods can be applied for the production of a protective coating with high power lasers. In principle, these are based on three different methodologies: 1) laser cladding—the addition of a layer by melting powder with modest mixing with the substrate material (e.g., Stellite coatings) [47]; 2) laser alloying—the addition of alloying elements in the melt pool to strengthen the substrate material (e.g., carburizing and nitriding) [48]; and 3) laser embedding—addition of particles in the melt pool which are frozen in the matrix [49]. In the case of laser embedding, two different processes can be used. The first method is to predeposit the powder on the substrate [50]. The disadvantage is that the laser beam does not initially interact with the substrate material, but with the powder particles. To melt the substrate material, the heat has to be transported through the powder layer, which might result in the melting of the powder. Depending on thermophysical properties, new phases will grow from the melt that do not necessary correspond with the original addition. The second method is the particle injection process during laser treatment [51–54]. Here, the laser beam primary interacts with the substrate and, to a lesser extent, with the particles. The resulting microstructure is determined by several parameters—the positioning of the powder flow, the particle size, the particle velocity, and the amount of powder. Therefore, the relevance of these different parameters will be examined first.

Based on the model presented in Section II, temperature profiles are calculated to obtain an estimate of the shape of the melt pool in the direction of the laser beam. The calculations are performed for a laser beam with a homogeneous temperature distribution, and Ti–6Al–4V has been taken as the substrate material. The model did not incorporate either melting or convection. Figure 19 shows the temperature profile as a function of depth and the relative position with respect to the laser beam. From these calculations, it appeared that: (1) the highest temperature is situated behind the middle of the laser beam; and (2) the melt pool is extended behind the laser beam. Actually, this means that the dissolution of the particles can be controlled by the positioning of the powder flow as depicted in Fig. 20. Situation A will result in much more dissolution of the powder than B, which is caused by the longer time spent in the melt pool. For the second case, some particles do not even cross the laser beam, which might result in a considerable rise in temperature.

In addition to the positioning, the interaction time of the SiC particles with the laser is also determined by the particle velocity. Although the interaction time is of the order of milliseconds, there is a considerable increase in temperature because of the high absorption of SiC particles (~90%) at a wavelength of 1.06 μm. Furthermore, the particle size also influences the temperature increase by its surface/volume ratio. The optimum injection velocity is found to be ~1.2 m/sec, resulting in a temperature increase of ~2000°C in the case of a SiC particle with a diameter of 80 μm, injected in the center of the laser beam. The temperature is assumed to increase homogeneously through the particle, whereas gas cooling is neglected. If relatively small dissolution with respect to the particle size is desired, large particles should be used. Another requirement is that the particle velocity should be high enough for the particles to reach the bottom of the melt pool (see Section III.A). Note that angle α should be large enough to prevent rebounding.

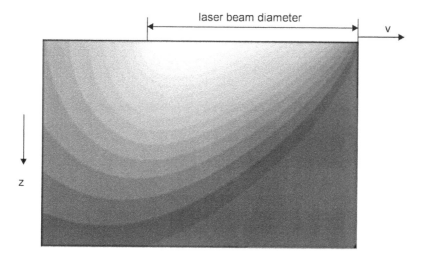

Figure 19 Temperature contours as a function of depth and position relative to the laser beam. The thermophysical constants of Ti–6Al–4V are used for the substrate ($P = 1$ kW, $v = 9$ mm/sec, and $d = 2$ mm).

In the case of SiC injection, an increase in the number of particles resulted in a higher total absorbed laser energy. As a consequence, the dimensions of the melt pool will increase. However, too much powder may lead to cladding of SiC particles onto the substrate, which yields a very rough laser track and shields the substrate from laser radiation.

To characterize materials, it is obvious that besides the microstructural observations with SEM, additional information is needed to gain insight into the behavior of materials during engineering use. Therefore, it is worthwhile to obtain and combine microstructural observations with chemical analysis and crystallographic information. This has been possible for many years with the transmission electron microscope (TEM). However, depending on the research goal, a drawback might be that this technique acquires very local information, without a good broader range of information. Furthermore, the area to be investigated is limited to the transparent region of a specimen. Another technique to obtain crystallographic data, X-ray diffraction analysis, has a relatively simple specimen preparation in comparison with TEM. However, it is impossible to obtain information on phase orientation for a single grain because this technique measures a distribution of orientations. The electron backscatter diffraction (EBSD) technique has now increased the potential of the

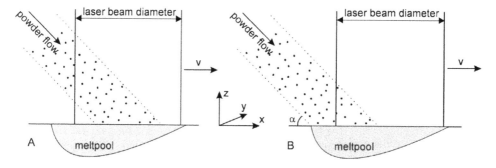

Figure 20 Illustration of the powder flow positioning during laser treatment: A) aligned at the center of the laser beam, B) aligned at the back of the laser beam.

SEM in the field of materials science [55,56]. The EBSD technique is able to obtain crystallographic information from single grains. Chemical analysis via energy-dispersive X-ray (EDX) has been already a conventional accessory for SEM, where it is anticipated that crystallographic analysis via EBSD will become a standard tool in time [57]. At this moment, it is possible to integrate EDX with EBSD in SEM, with the goal of simultaneously obtaining information on the chemical composition and the atomic arrangement as well within a single grain.

Electron backscattering diffraction involves the formation of diffraction patterns in a manner first described by Kikuchi [Jpn. J. Phys. 5,83 (1928)]. Kikuchi showed that the phenomenon occurred for electrons transmitting a thin crystal. The EBSD analysis makes use of a different geometry, where the electrons are backscattered instead of being transmitted. However, the same explanation is valid for the appearance of Kikuchi patterns. In 1973, Venables has been the first to introduce the technique in the SEM [58]. At the same time, Dingley has been recording backscattering Kossel X-ray diffraction patterns in SEM, which are formed in a similar way. However, this technique detects the characteristic X-ray fluorescence generated in a specimen alongside the backscattered electrons. Over time, the technique became better as a result of, among other things, the advances in low-light imaging systems. Recent developments in computer simulations may help to recognize the Kikuchi patterns in a routine manner and this will substantially contribute to the breakthrough of EBSD in the near future.

To understand the formation of an EBSD pattern, consider the spread of an incident electron beam beneath the specimen surface, which is tilted over 70°. As the electrons travel in all directions, they will also impinge on crystal planes at the Bragg angle. As a consequence, there will be a diffracted electron beam at angle θ with respect to the corresponding set of crystal planes. As the electrons impinge on a specific set of planes from all radial directions, a diffracted electron cone will be formed with its axis normal to the plane, as can be seen in Fig. 21(a). In addition, these cones can be present at both sides of a crystal plane. There will be one pair of cones for each set of crystal planes. Outside the specimen, the cones can be imaged by the projection on a phosphor screen. The pattern consists of pairs of parallel lines, known as Kikuchi bands, which intersect each other at various places. The same phenomenon can be observed in TEM when using the diffraction mode [59]. Here, the Kikuchi lines are formed by inelastically scattered electrons in the forward direction, as shown in Fig. 21(c). The spacing between the lines of one pair is identical to the spacing of the corresponding diffraction spots and, therefore, inversely proportional to the spacing between crystal planes. For EBSD, only high-energy (10–30 keV) electrons with an energy spread of 200 eV contribute to the backscatter pattern. Therefore, it is essential to tilt the specimen significantly with respect to the incident beam in order to have a significant amount of high-energy electrons escape from the surface. The minimum angle amounts to 60°, where 70° is commonly used in practice. As the high-energy electrons originate from a maximum depth of 100 nm from the specimen surface, it is necessary to make special demands on surface preparation. The spatial resolution of the EBSD technique depends on the interaction volume of the primary electron beam. The maximum lateral resolution in the SEM is achieved with a field emission gun and amounts to 0.2 μm. As a consequence, this technique is extremely suitable for obtaining crystallographic information from single crystals.

1. Experimental Setup EBSD

The specimen detector geometry of the Philips integrated EBSD system is depicted in Fig. 22. The electrons coming from above impinge on the specimen under an angle of 70°. The phosphor screen with a diameter of 45 mm is positioned on the right, with a minimum

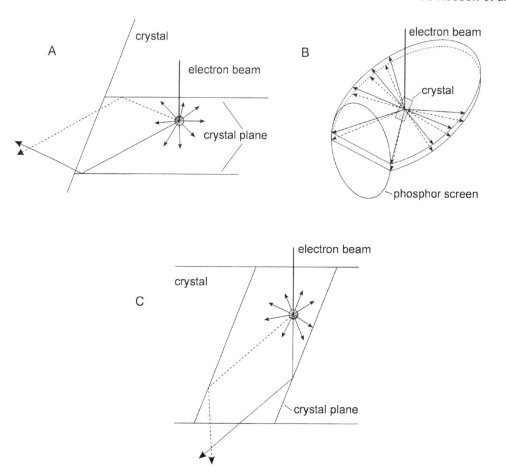

Figure 21 The EBSD pattern formation in the SEM. The electrons are backscattered outside the crystal. A) Formation of one pair of Kikuchi lines from diffraction of the inelastically scattered electrons with one family of crystal planes. B) Formation of one Kikuchi band at the phosphor screen. C) Kikuchi pattern formation in the TEM. The electrons are inelastically scattered in the forward direction.

distance of ~20 mm, ensuring a solid angle exceeding 90°. Furthermore, the distance from phosphor screen to specimen can be changed. Behind the phosphor screen, a camera is mounted to record the pattern. The patterns are imaged with the use of a computer monitor, where they have been subsequently adapted to increase image quality. In fact, the symmetry of the EBSD pattern is a 2-D projection of the symmetry in three dimensions of the atomic arrangement within a crystal. Therefore, the obtained backscatter pattern contains crystallographic information concerning the crystal structure and the orientation of the analyzed crystal. Using the crystal orientation software (COS), it is possible to simulate a pattern obtained. Figure 23 shows an example of a simulated pattern with the Miller indices superimposed on an experimental Kikuchi pattern. To translate the orientations obtained to the specimen, the COS software provides a transformation matrix that relates the crystallographic direction to the specimen directions. This can be depicted by pole figures projections. Figure 22(b) shows a 3-D projection of the lattice vectors in relation to the specimen directions, which will be used to present the experimental results. In addition, it is possible

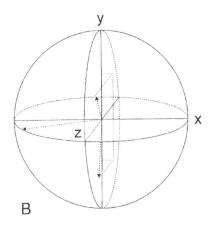

Figure 22   A) EBSD set-up in specimen chamber of the SEM. Specimen is rotated over 70°. At the right side, the phosphor screen is visible. B) Three-dimensional representation of the crystallographic directions projected relatively to the specimen axis.

to perform texture measurements by measuring the distribution of orientations. Therefore, the movement of the electron beam is controlled by the EBSD equipment. By imposing an orientation criterion between two successive measurements, it is possible to determine if a grain boundary is traversed. Furthermore, it is possible to detect plastic strain in the lattice, which manifests itself by the blurring of Kikuchi lines. However, at this moment, it is difficult to obtain quantitative information from this method as the quantitative relation between blurring and deformation is not accurately known. In addition, this method requires a well-defined surface condition, as contamination will introduce inaccuracy.

Figure 23   Recorded EBSD pattern of TiC with an overlay of the simulated pattern, including Miller indices.

2. Microstructural Characterization

During laser treatment of Ti–6Al–4V, particles of SiC, TiC, and TiN are consecutively injected in the melt pool. Because the particle sizes are almost the same, all tracks show approximately the same tendency. Therefore, the microstructural features of the SiC/Ti–6Al–4V system will be described most extensively. This system results in the formation of various reaction products [60], in contrast to the TiC/Ti–6Al–4V and TiN/Ti–6Al–4V system, where only two phases are found to be present after laser treatment. However, the description of wear behavior in a following paragraph concerns all three different coatings.

By changing the position of the particle flow from behind to beyond the center of the laser beam, there is a transition from modest dissolution to considerable dissolution of SiC particles. For the latter case, this results in a melt pool filled with reaction products of Ti with Si and C. The following microstructural description is given for the situation in which the particles are injected somewhat behind the middle of the laser beam. However, by changing the particle injection position, the presence and amount of the different micro-structures can be controlled. Figure 24 shows the cross section of a laser track with em-bedded SiC particles. The depth of the laser track is ~0.5 mm. The particles are homogeneously distributed over the melt pool, whereby the volume increase due to the addition of SiC amounts to 10–15%. First, the change of microstructural composition of the matrix will be described. Second, the appearance of the reaction layer around the SiC particles, which is present under all injection conditions, will be studied in more detail. Con-clusions concerning the phases present in the melt pool are mainly based on the combination of X-ray diffraction measurements and EDX analysis, whereas, in some specific cases, TEM is used.

3. Matrix

As the SiC particles partially dissolve, the matrix composition changes. The matrix between SiC particles consists of various phases where the percentage of each gradually changes as a function of depth. Figure 25 shows the structure near the bottom of the laser track. A

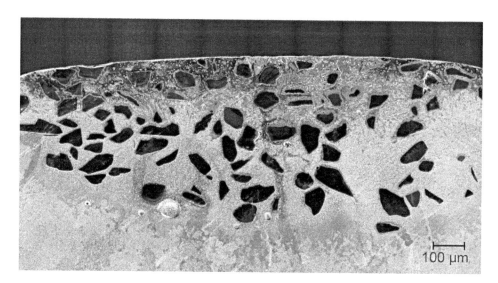

Figure 24 SEM micrograph of cross section showing SiC particles embedded in a Ti–6Al–4V matrix.

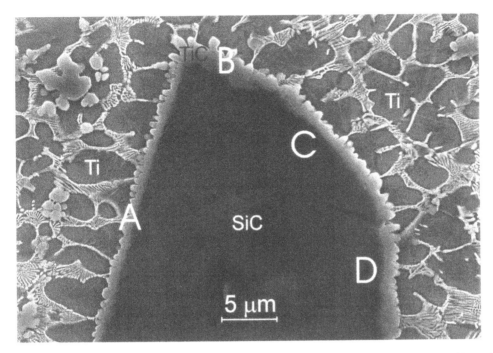

Figure 25 SEM micrograph of the structure at the bottom of the laser track showing SiC particles surrounded by a TiC reaction layer. The matrix consists of Ti–6Al–4V grains (dark) surrounded by a eutectic structure (white).

eutectic structure is observed surrounding grains of alloyed titanium. These Ti grains contain the alloying elements Al, V, and Si and exhibit a lenticular martensitic structure. The eutectic lamellae consist of alternating Ti and Ti_5Si_3 layers. This observation corresponds to the binary phase diagram of Ti and Si, depicted in Fig. 26, where a eutecticum is present at 13.7 at.% Si. Thus, solidification starts with the formation of Ti grains followed by eutectic solidification. In the middle of the laser track, TiC dendrites are randomly oriented and distributed over the laser track, as shown in Fig. 27. This indicates that a reasonable amount of carbon is dissolved in the melt for this region. Furthermore, the amount of eutectic structure is increased. Faceted Ti_5Si_3 grains situated near the top of the melt pool have an angle between the facets that points to a hexagonal phase of Ti_5Si_3, as shown in Fig. 28. In addition, this region also contains a small amount of eutectic structure and TiC dendrites.

With TEM, we have been able to see more details of the matrix in the upper part of the melt pool. Several Ti_5Si_3 grains with a size of approximately 0.6 µm are observed, as depicted in Fig. 29. The black particle in the image corresponds to a Ti_5Si_3 grain oriented along its [0001] pole with d_{100} of 0.645 nm. In comparison with SEM observations, these Ti_5Si_3 grains are differently shaped and have different dimensions. Therefore, these grains are expected to be part of the eutectic structure. In between these grains, Ti needles are detected. The Ti_5Si_3 particles seemed to be roughly spherical, but a closer look reveals many facets.

As already mentioned, the microstructure can be controlled by changing the injection position. The laser particle injection process is performed for six different injection positions, starting from the middle, with step sizes of 0.3 mm toward the back of the laser

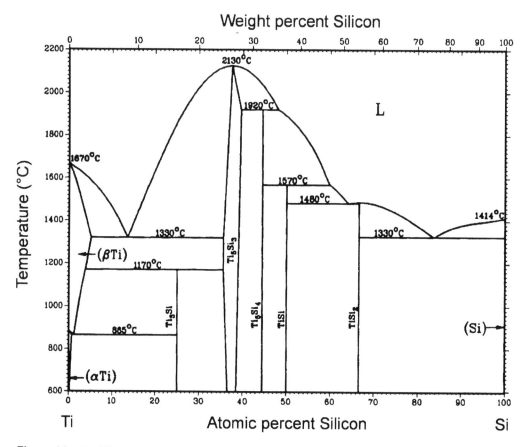

Figure 26 Equilibrium phase diagram of TiSi.

beam (negative x-direction). All other laser parameters have been kept constant. In the case of particle injection in the center of the laser beam, the microstructure contains regions at the top of the track that consist almost completely of Ti_5Si_3 and TiC cells. On the other hand, it is found that the last laser track consists mainly of Ti grains surrounded by the eutectic structure, as shown in Fig. 25. The intermediate laser injection position exhibits a smooth transition between these extremes. Similar effects of the different degree of dissolution can be obtained by increasing the scan velocity [61]. However, the laser power has to be increased for obtaining the same track depth, which will increase the absorbed power by the SiC particles.

4. Reaction Layer

The injected SiC particles are always surrounded by a reaction layer. At high temperatures, the SiC decomposes, where both components form a new phase with Ti. With increasing reaction time or temperature, the thickness of the layer will increase, simultaneously resulting in a decrease of the diameter of the remaining SiC particles. Both EBSD and EDX analysis indicate that the reaction layer consists mainly of TiC. In the case of TiN and TiC particle injection, there is no reaction layer present. The melting of these particles results only in the formation of TiC and TiN dendrites in the environment of the particle.

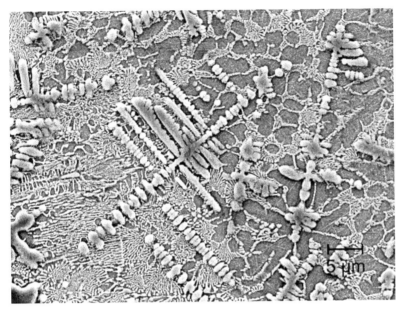

Figure 27 SEM micrograph of the structure at the middle of the laser track showing TiC dendrites. Released Si results in eutectic Ti_5Si_3/Ti on the grain boundaries of Ti. The reaction layer is of the cellular type.

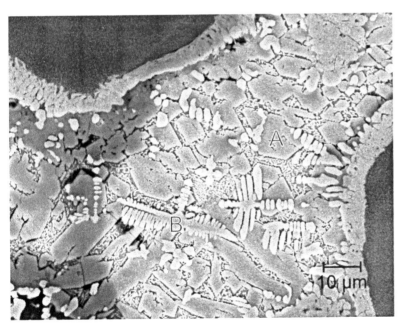

Figure 28 SEM micrograph of the structure at the top of the laser track showing SiC particles surrounded by a TiC reaction layer. The matrix consists of strongly faceted Ti_5Si_3 grains (A) and TiC dendrites (B).

Figure 29 Bright-field TEM micrograph of eutectic structure showing a strongly diffracting Ti$_5$Si$_3$ grain (A) and differently oriented Ti$_5$Si$_3$ grains (B). The structure in between consists of Ti needles.

The reaction layer of TiC around the SiC particles is of special interest because it will determine the mechanical properties of the metal–matrix composite (MMC), to be described in the following section. In principle, the reaction layer appears in two different ways—a cellular reaction layer and an irregular reaction layer, which can be observed in Figs. 25 and 29, respectively. The cellular reaction layer is relatively thin and very regularly shaped all around the particle. Furthermore, the layer is often well connected to the particle and exhibits minor porosity. To obtain crystallographic information, the reaction layer is examined with the use of EBSD. This is shown in Fig. 30, where the four EBSD patterns represent the orientation of the reaction layer at each side of a SiC particle. Along each side of a cellular reaction layer, the pattern remains unchanged. Furthermore, a 3-D representation is given of the $\langle 100 \rangle$ directions with respect to the specimen axis. Although it is unknown under which angle of inclination the surface of SiC is oriented, it is found that the cellular type of reaction layer always grows in the [100] direction, nearly perpendicular to the SiC surface. In addition, the remaining $\langle 100 \rangle$ directions might give an indication of the angle of inclination of the sides of the SiC particle. A specific relation is not found between the orientation of SiC and TiC reaction layers. Therefore, the preferred orientation of TiC is a growth phenomenon related to the steepest thermal gradient, as the thermal conductivity of TiC is maximum in the [100] direction. The 3-D representation exhibits only a slight change in orientation between the [100] direction perpendicular to the surface of grains B and C. Examination of the irregular reaction layer with EBSD indicates that the reaction layer consists of TiC grains, which are randomly oriented. In contrast to the cellular type, this type of layers is relatively thick. In addition, large pore volumes are observed within the reaction layer, which can be partially explained by the volume decrease of ~5% as a consequence of the formation of TiC. However, this becomes important only for solid-state reactions. Cracks sometimes arise in the thick reaction layers as a consequence of thermal stresses.

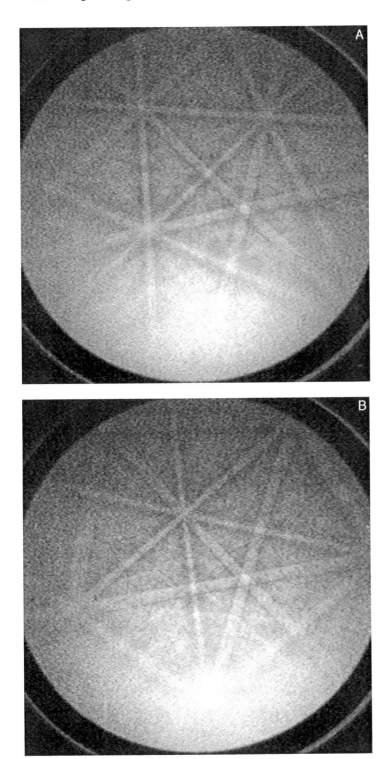

Figure 30 EBSD measurements are performed at positions A–D of the reaction layer indicated in Fig. 25. The corresponding backscatter Kikuchi patterns are shown for A, B, C, and D.

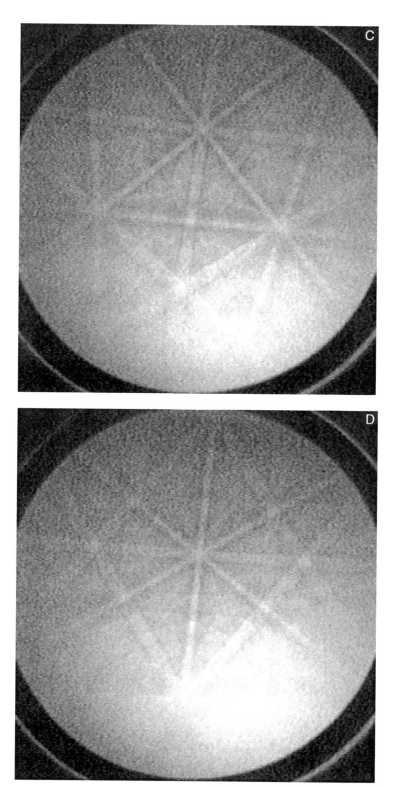

Figure 30 Continued.

To study this type of reaction layer in more detail, conventional transmission electron microscopy (CTEM) and high-resolution transmission electron microscopy (HRTEM) examinations are carried out. Electron diffraction analysis of the irregular reaction layer reveals the presence of another phase, namely Ti_3SiC_2. Although the symmetry in the diffraction pattern of Ti_3SiC_2 and 6H–SiC is equal, the c-lattice constant differs significantly (1.764 vs. 1.509 nm) [62]. Figure 31 depicts an atomic resolution image of Ti_3SiC_2 obtained via HRTEM, where the inset shows the simulated patterns which show good agreement. The atomic configuration of Ti_3SiC_2 exhibits the 8H-like structure, as depicted in Fig. 32. It is observed that the ternary phase is present as small plates around the round and randomly oriented TiC grains, as shown in Fig. 33. In the literature, similar structures are observed for TiC and Ti_3SiC_2, obtained with a specific sintering process [63]. In most cases, a specific orientation relationship between the TiC grains and plates of Ti_3SiC_2 is not observed. However, for one specific situation, the orientation relation (OR) appeared to be present between TiC and Ti_3SiC_2. This is shown in Fig. 34(a), where a Ti_3SiC_2 plate is connected to a relatively large, round TiC grain, when simultaneously viewed along a low-index zone axis. TiC and Ti_3SiC_2 are viewed along the $\langle 110 \rangle$ and $\langle 11\bar{2}0 \rangle$, respectively. Based on chemical and geometrical considerations, the lowest interfacial energy occurs if the basal plane of Ti_3SiC_2 is parallel to the (111) of TiC. Here, TiC (111) consists of alternating close-packed planes of either Ti or C atoms, which are identical to the fully Ti or C basal planes, as stacked in Ti_3SiC_2. It is remarkable, therefore, that the basal planes of Ti_3SiC_2 are found to be aligned parallel to the elongated direction of the plate, which implies that the latter mentioned stacking does not occur. The TiC{111} and Ti_3SiC_2{0001} are found to be rotated over $5\pm0.5°$, with respect to each other. Instead, two other planes are found to be more or less parallel, that is, the $\{111\}_{TiC}//\{10\bar{1}3\}_{Ti_3SiC_2}$ and the $\{002\}_{TiC}//\{10\bar{1}4\}_{Ti_3SiC_2}$. An obvious reason for the parallelism of these planes is not found, because the image does not reveal a clear and sharp interface. The reason for the orientation of the basal planes parallel to the surface of the plates and about perpendicular to the (macroscopic) surface of the TiC grain might be the lower surface free energy of the close-packed plane. However, it is anticipated

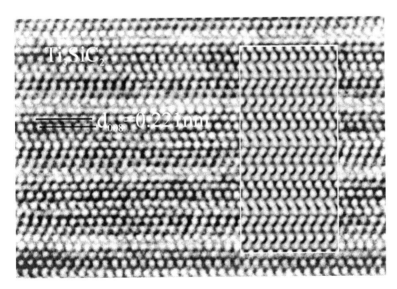

Figure 31 HRTEM image showing atomic structure of Ti_3SiC_2 viewed along $\langle 11\bar{2}0 \rangle$. The inset depicts the simulated structure.

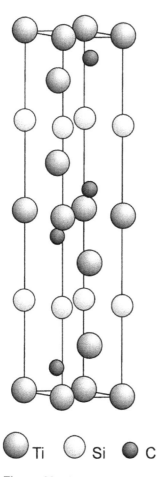

○ Ti   ○ Si   ● C

Figure 32 Atomic cph structure of Ti$_3$SiC$_2$, consisting of alternating hexagonal planes of Ti, Si, and C.

that a more probable, but not independent, reason is the growth direction during solidification. In that case, the grains grow away from the TiC surface along the basal planes because these planes are probably fastest growing. In Fig. 34(b), an adjacent Ti$_3$SiC$_2$ plate is tilted in a strong diffraction condition. It can be seen that the TiC grain contains stacking faults (SF), on two sets of slightly inclined {111} planes. Furthermore, the former Ti$_3$SiC$_2$ grain exhibits a stacking fault situated on the basal plane. It would be interesting to know how the stacking fault occurs with respect to the chemical composition of the basal planes in the crystal structure, in particular with regard to the possibility of Ti$_3$SiC$_2$ deforming plastically.

Although all different phases in the laser track are known, the solidification route is difficult to determine because of both the absence of the ternary phase diagram over the complete temperature range and the nonequilibrium conditions. Nevertheless, it is possible to relate a certain area in the ternary phase diagram [64] (Fig. 35) to a specific microstructural region in the melt pool. Additional information concerning the temperature can be derived from the binary phase diagrams, which are depicted in Figs. 36 and 37. The phases that consecutively solidify are schematically depicted in Fig. 38. In principle, we start from

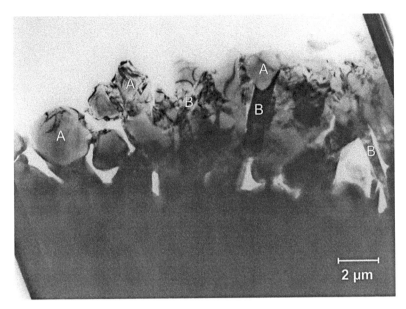

Figure 33 Bright-field TEM micrograph of reaction layer around SiC consisting of round TiC grains (A) and plate-like structures of Ti_3SiC_2 (B).

the condition where only SiC and Ti are present, if we neglect the alloying elements. After partial decomposition or melting of SiC, solidification starts with the formation of the reaction layer. This situation corresponds with the region within the triangle in the Ti–Si–C phase diagram formed by TiC, SiC, and Ti_3SiC_2. Based on the different solidification structure of these layers, it is concluded that, for the cellular layer, the solidification front starts at the SiC surface. During growth of this layer, Si is rejected, as EDX measurements do not indicate a significant amount of Si. In the case of the irregular reaction layer,

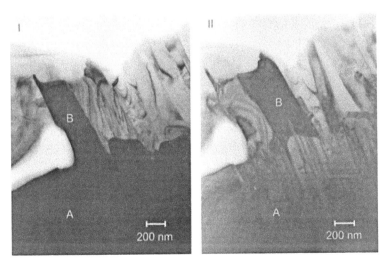

Figure 34 TEM micrographs of TiC grains (A) with adjacent plates of Ti_3SiC_2 (B), I) both strongly diffracting, II) Ti_3SiC_2 plate strongly diffracting. Stacking faults are visible in TiC.

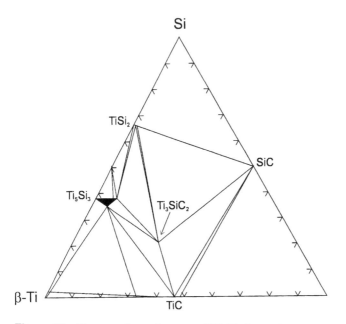

Figure 35   Ternary phase diagram of Ti–Si–C. Isothermal section at 1200°C.

solidification starts in the melt near the particle. (See the aforementioned results from the random orientation of the TiC grains.) In addition, this can be explained by the difference in melting temperature, which is higher for stoichiometric TiC (3067°C) than for SiC (2545°C). The presence of Ti_3SiC_2 might be explained by entrapped Si, which is in solution and cannot be rejected because of the solidified TiC grains. The solidification of the matrix corresponds to the triangle in the phase diagram formed by Ti, TiC_{1-x}, and $Ti_5Si_3C_y$. The solidification route of the matrix can be explained quite easily because of the different solidification temperatures. The lower solidification temperature of the TiC dendrites in comparison with the TiC reaction layer stems from the lower carbon content measured with EDX. Depending on the amount of Si in the liquid after solidification of TiC dendrites, solidification starts with nucleation of Ti grains or Ti_5Si_3. Finally, eutectic solidification takes place according to the Ti–Si phase diagram.

5. Existence of Ti_3SiC_2

There exists a controversy over the presence or absence of Ti_3SiC_2 in the reactive SiC/Ti systems, as explained in Ref. 60, that needs to be clarified. By adjusting the amount of material removed from the top or the bottom of the disks, TEM specimens corresponding to a certain chosen depth in the melt pool could be prepared. Ion milling was performed by using two beams of 4 kV Ar^+ ions having an incidence angle of either 13° or 6°, with the bottom and top surfaces of the disks (using a Gatan dual ion mill 600 or a Gatan PIPS 691). As the SiC has been preferentially thinned by the Ar^+ ions, holes in the sample started in SiC regions. In this way, electron-transparent regions in and around the SiC particles could be obtained rather nicely. For TEM, a JEOL 4000EX/II operating at 400 kV with a point-to-point resolution of 0.17 nm equipped with an on-axis Gatan 666 parallel electron energy loss spectrometer, located in Groningen, was used. For analytical TEM, a JEOL 200 CX operating at 200 kV equipped with a Gatan 666 parallel electron energy loss spectrometry

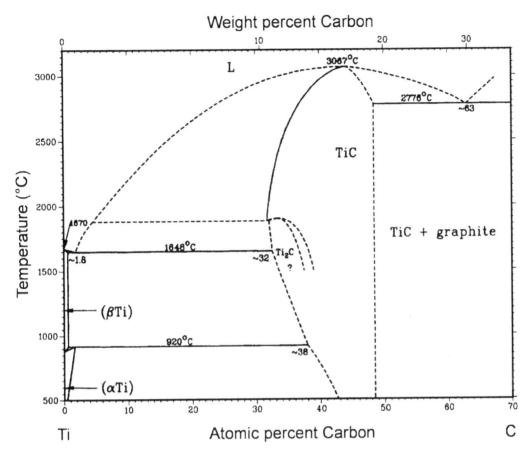

Figure 36   Equilibrium phase diagram of Ti–C.

(PEELS) spectrometer as well as two Kevex system 8000 energy-dispersive X-ray spectroscopy (EDS) detectors, located at the National Center for Electron Microscopy (NCEM) of the Lawrence Berkeley Laboratory, was used. The EDS detectors have a high-angle detector with a resolution of 155 eV for Mn-Kα radiation and an ultrathin window detector with a resolution of 109 eV for F-Kα radiation allowing quantitative analysis for elements with $Z \geq$ 6. Although the used quantification procedures (performed with software at the NCEM) do not have standards for both PEELS and EDS, the procedures for quantification have been internally calibrated on the embedded SiC particles. This allowed a relatively accurate quantification of the chemical composition of, for example, TiC. For instance, the ratios obtained of Si/C and Ti/C are sensitive to the foil thickness during EDS analysis. By varying the foil thickness, such that correct results are obtained for SiC, and by measuring EDS spectra, comparable count rates for both SiC and TiC, the composition of TiC (according to EDS in SEM and TEM, PEELS in TEM, and to selected area electron diffraction patterns of more than hundred grains) could be obtained. Finally, the results of PEELS and EDX could be directly compared in a number of cases to check for consistency.

Near the bottom of the melt pool, the dissolution of SiC is modest and the amount of Si rejected into the melt gives rise to a small amount of eutectic Ti_5Si_3/α-Ti on the grain boundaries of the α-Ti grains. In Fig. 25, the amount of eutectic is shown as having increased

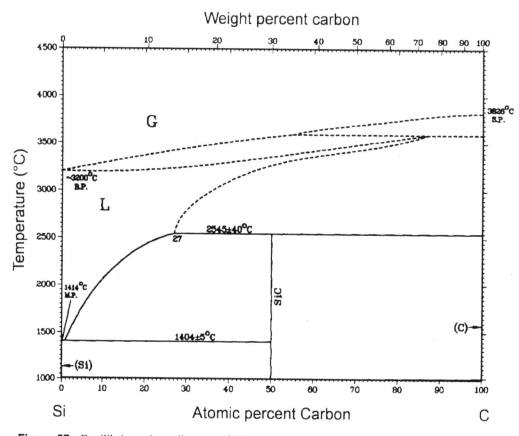

Figure 37 Equilibrium phase diagram of Si–C.

already as a result of the somewhat higher level in the melt pool, causing an increased Si dissolution. Approaching the top of the melt pool, the amount of Si rejected increases considerably and finally gives rise to large faceted pre-eutectic Ti$_5$Si$_3$ grains with eutectic Ti$_5$Si$_3$/α-Ti in between (Fig. 27). Carbon can be in solid solution in Ti$_5$Si$_3$ up to about 8 at.%. Furthermore, TiC dendrites can be observed within the matrix region in between the SiC particles from about the middle to the top of the melt pool. The C/Ti ratio in the TiC dendrites is expected to be lower than in the TiC reaction layer, which gives a lower solidification temperature for the dendrites.

Two types of TiC reaction layers around the SiC particles have been observed using SEM: (a) cellular layer in which the TiC grains give the appearance of having been heterogeneously nucleated at the solid SiC/melt interface with subsequent cellular growth of the TiC into the melt (Fig. 25); and (b) a cloud-like layer of spherical TiC grains, which points at "homogeneous" nucleation of the TiC within the melt very near to the solid SiC surface (more or less the case for the reaction layer observable in Fig. 27).

The TEM analysis of the cloud-like reaction layer revealed that besides a predominant fraction of spherical TiC grains, Ti$_3$SiC$_2$ plates also appeared to be present. Figure 33 shows a TEM overview of a cloud-like reaction layer, in which the two phases, TiC and Ti$_3$SiC$_2$, have been indicated for several grains. Much smaller Ti$_3$SiC$_2$ plates, not distinguishable in Fig. 33, are also present in the layer. The SiC particle to which this layer has been attached has been removed by ion milling and has been thus originally located in the region of the hole

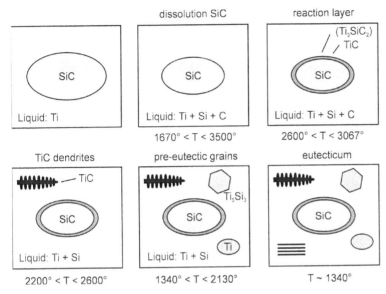

dissolution SiC   reaction layer

SiC

Liquid: Ti

SiC

Liquid: Ti + Si + C

1670° < T < 3500°

(Ti₃SiC₂)
TiC
SiC

Liquid: Ti + Si + C

2600° < T < 3067°

TiC dendrites   pre-eutectic grains   eutecticum

TiC

SiC

Liquid: Ti + Si

2200° < T < 2600°

Ti₅Si₃

SiC

Ti

Liquid: Ti + Si

1340° < T < 2130°

SiC

T ~ 1340°

Figure 38 Dissolution and consecutive solidification processes after the SiC particles are injected into liquid Ti during laser processing.

in the foil in Fig. 33. The Ti_3SiC_2 plates have dominant (0001) facets because they grow most easily along directions present within the basal plane. The identification of Ti_3SiC_2 is based on selected area electron diffraction (SAED) patterns. Because of similarities in the crystallography of 6H–SiC (with lattice parameters: $a = 0.308$ nm and $c = 1.509$ nm [65]) and of Ti_3SiC_2 (with $a = 0.306$ nm and $c = 1.764$ nm [66–69]), the diffraction patterns of the latter are at first glance, easily misinterpreted as being SiC. This is despite the fact that the c-lattice constants of the SiC poly-types are sufficiently different. An HRTEM image of Ti_3SiC_2, showing its 8H-like stacking, is shown in Fig. 31 with, as seen in the insert, the simulated image for a thickness of 5 nm, a defocus values of −6 nm and a beam tilt of 1.3 mrad.

In the Ti_3SiC_2, stacking faults on basal planes and dislocations with Burgers vector $1/3\langle11\bar{2}0\rangle$ have been observed. The dislocations showed a tendency to be confined within a basal plane. In Fig. 39, two-beam bright-field images (with other strong beam $g\langle11\bar{2}0\rangle$) show a Ti_3SiC_2 grain with a relatively large number of dislocations. To obtain strongly contrasting dislocations, the basal plane in Fig. 39 is tilted (with a tilt of about 12° out of the $\langle01\bar{1}0\rangle$ zone axis), but examples of dislocations confined within the basal plane are indicated by an A; their dislocation lines have predominantly a horizontal component in the image. The B dislocations, predominantly with a vertical line component in the image, are not confined in the basal plane. Only recently has the type of Burgers vector for dislocations in Ti_3SiC_2 been determined [70]. The $1/3\langle11\bar{2}0\rangle$ Burgers vector is common for hexagonal crystals (corresponding to stacking of hexagonal planes) and, in combination with the (0001) $\langle11\bar{2}0\rangle$ slip system, not striking. On the other hand, the preference of the dislocation lines for confinement within a basal plane of Ti_3SiC_2 appears more interesting. This confinement is commonly observed and seems independent from the way the Ti_3SiC_2 is produced [71]. The actual reason of interest is that the 8H-like stacking of Ti_3SiC_2 corresponds to the following repeat sequence of hexagonal planes: Ti–C–Ti–Si–Ti–C, where the C atoms can be regarded as located in interstitial sites. Hence the Si or C planes are in between two Ti planes and no adjacent Si and C planes occur. Now the confinement of the dislocations within a

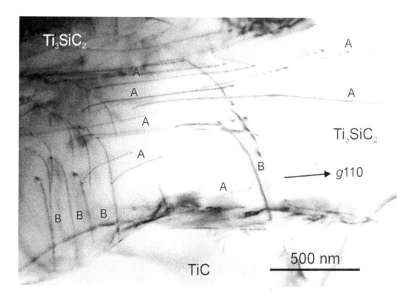

Figure 39 Two-beam bright-field image (with other strong beam $g = 11\overline{2}0$ as indicated and with the Ti$_3$SiC$_2$ crystal tilted about 12° out of the $\langle 01\overline{1}0 \rangle$ zone axis) showing dislocations in Ti$_3$SiC$_2$. The dislocations indicated by A are confined within a basal plane of Ti$_3$SiC$_2$.

basal plane of Ti$_3$SiC$_2$ may indicate that the dislocation core has a preference for a certain basal plane type. For instance, the core may have a preference to be linked directly to an Si basal plane, because in between the Si and Ti planes, the interstitial sites are free from C. In the same context, certain types of stacking faults in Ti$_3$SiC$_2$ will also be preferential over others. To provide more precise answers to these questions of the location of dislocation cores and stacking faults, HRTEM observations are required, and results will be published elsewhere.

The TEM analysis of the cellular reaction layer did not reveal the presence of Ti$_3$SiC$_2$. The different way in which the Si is rejected from the TiC grains in the two types of reaction layers can be held responsible for the presence or absence of Ti$_3$SiC$_2$. Where there is a cloud-like reaction layer, volumes in between the TiC grains can become enclosed, thereby preventing the rejection of excess Si into the melt pool. In the enclosed volumes, the Si concentration is subsequently raised in sufficient amount to nucleate at the interface between TiC and SiC the phase Ti$_3$SiC$_2$, in accordance with the ternary Ti–Si–C phase diagram (which is known only for the temperature range 1100–1250 °C [67,72,73]). This sequence of a primary TiC nucleation followed by a nucleation of Ti$_3$SiC$_2$ can be expected because the melting point of TiC is probably highest, that is to say: 3061 °C for TiC, 2545 °C for SiC, and a melting point for Ti$_3$SiC$_2$ probably between these two values. For the cellular reaction, areas in the material do not become trapped and, accordingly, Ti$_3$SiC$_2$ is not observed. The excess of Si rejected into the "open" melt results (in accordance with the ternary Ti–Si–C phase diagram) in the nucleation of Ti$_5$Si$_3$ at the interface between TiC and Ti. This occurs around both types of reaction layers. The Al and V present in the original Ti–6Al–4V are according to EDS in TEM present in Ti$_5$Si$_3$ and α-Ti and not detectably present in TiC.

The present sequence of primary TiC formation, followed by Ti$_3$SiC$_2$ at the "interface" between TiC and SiC (if sufficient supersaturation with Si occurs), is similar to the mechanism proposed for solid-state reactions at the Ti/SiC interface [60], but it is in contrast with the results of Ref. 83. For solid-state reactions, the phase TiC gives also the largest decrease in Gibbs free energy of the system and is expected to form first. However, for both solidification processes and solid-state reactions, the actual phases formed also naturally depend sensitively on the rate with which different elements are supplied locally and the total amount of different elements available. For instance, in the case of a thin Ti film on a SiC surface, the reaction products after annealing [74] are different from the case in which the amount of available Ti is more abundant.

In a considerable fraction of TiC grains in both types of reaction layers, but predominantly in the cellular layer, a high density of widely extended stacking faults (SFs) on TiC {111} planes have been observed with TEM; i.e., observations typical for (deformed) material with a very low SF energy. In Figs. 40 and 41, two-beam bright-field images are presented of TiC grains with SFs. The SF density in the TiC grains shown in Figs. 40 and 41 is still relatively low, but this gives a clearer view than the images full of overlapping contrast of SFs that are also often obtained.

Because the SF energy of TiC is very high and SFs are not observable after severe deformation [75–78], no directly observable SFs have been expected in TiC. Depending on the C/Ti ratio, the SF energy lies in the range between 130 and 300 mJ/m^2 [75]. This first would suggest that the high SF density grains have been composed of β-SiC instead of TiC, because the SF energy of SiC is much lower than of TiC. This is in accordance with the existence of many SiC polytypes that are based on different stacking sequences of basal hexagonal planes. The crystallography of TiC and β-SiC is almost identical (both NaCl-type structure with lattice parameters $a = 0.433$ nm and $a = 0.444$ nm, respectively) and, therefore,

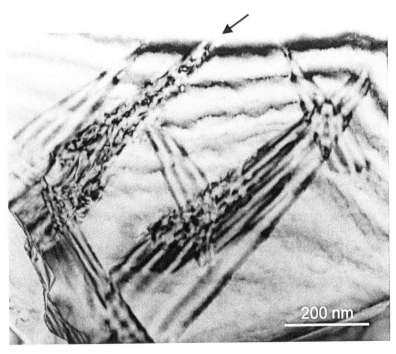

Figure 40 Two-beam bright field image showing (unexpected) stacking faults in TiC on {111}. The arrow points at an SF where the usual alternation of bright and dark fringes is distorted by the presence of small precipitates.

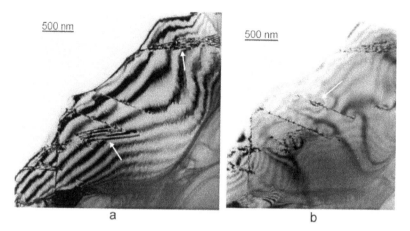

Figure 41 Two-beam bright field images showing the same TiC grain of the reaction layer containing (unexpected) stacking faults in TiC on {111}. (a) A two-beam condition with $g200$ near to a $\langle 110 \rangle$ zone axis of TiC. The lower white arrow indicates a widely extended SF and the upper white arrow indicates an SF where the usual alternation of bright and dark fringes is distorted by the presence of precipitates nucleated on the SF. (b) A two-beam condition with $g220$ near a $\langle 110 \rangle$ zone axis of TiC. The white arrow indicates a clear example of contrast due to long-range strain fields around precipitates nucleated on an SF. Three long, nearly edge-on SFs can be observed and the visibility of these SFs largely originates from their decoration with precipitates.

did not allow for precise discrimination between the two phases based on SAED patterns. However, subsequently performed EDS and PEELS analyses in TEM clearly indicated that these originally supposed β-SiC grains have been, in fact, TiC grains. A second possible explanation for the SFs in TiC is the presence of impurities. Planar defects with extrinsic stacking-fault character observed before in TiC have been ascribed to the effect of boron impurities [76–79]. Triangular planar defects nucleated at dislocation nodes of networks with $1/2\langle 110 \rangle$ Burgers vectors on {111} TiC have been identified as very thin TiB$_2$ platelets starting possibly as impurity-stabilized stacking faults. Dissociation of isolated dislocation lines resulting in (widely extended) stacking faults is less frequently observed than the triangular planar faults centered on the dislocation nodes. The impurity content of the present TiC grains is locally, probably substantially higher than in the aforementioned TiC material that contains additional traces of boron. Hence a strong reduction of the SF energy is thus conceivable and serves to explain the observed high density of widely extended SFs. In particular, Si, released during the reaction SiC + Ti→TiC + Si, may be enclosed in the TiC grains and may reduce the SF energy. The idea that TiC in reaction layers adjacent to SiC contains a high Si content has been earlier proposed in Ref. 60, but without any clear evidence though.

The observed faults are generally not of simple single-plane intrinsic or extrinsic stacking-fault type, as can be clearly observed in the HRTEM image depicted in Fig. 42. The faults are, in general, complexes of several closely spaced stacking faults, also often involving small twinned regions with typical thickness of 2–5 {111} planes. A high impurity content in these faults is probably responsible for their deviation from pure stacking-fault character. Detailed two-beam bright- and dark-field images and (many-beam) HRTEM images revealed the presence of small precipitates on many SFs in the TiC grains, with an extension perpendicular to the SF plane usually below 20 nm. The regularly spaced bright

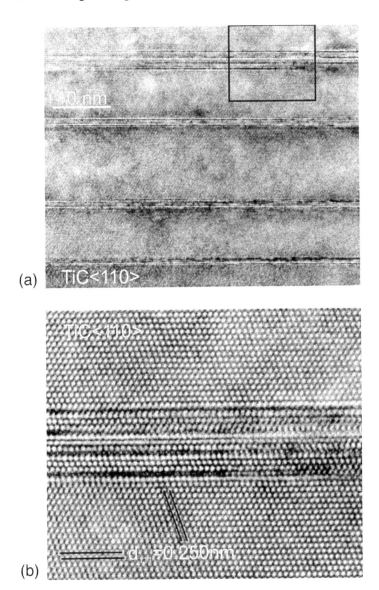

Figure 42 HRTEM image of TiC in a cellular reaction layer viewed along ⟨110⟩ containing a high density of stacking faults. In the image shown in (a) all SFs are parallel and edge-on observed. In (b) an enlarged part of (a) is shown. The faults are not of simple intrinsic or extrinsic type, but are complexes of closely spaced faults also often containing small twinned regions of 2–5 {111} planes. Within these faults, a strong Si enrichment is probably present.

(BF) and dark fringes (DF) associated with "clean" SFs imaged in two-beam BF and DF have been often observed to become largely distorted by the presence of these precipitates at the SFs. The arrow in Fig. 40 and the upper white arrow in Fig. 41a indicate that such SFs have been the alternating bright and dark fringes, and have been distorted by the presence of precipitates on the SFs. Furthermore, SFs with nearly edge-on orientation with respect to the viewing direction which, in the case of a clean SF, corresponds to a single, hardly observable line show strain-field contrast extending long distances away from the defect plane due to the presence of precipitates. A clear example of this phenomenon can be observed in Fig. 41b and is indicated by the white arrow. In fact, the visibility of all three, about 1-μm-long nearly edge-on SFs in Fig. 41b is strongly enhanced by their decoration with precipitates. These long-range strain fields indicate at least partial coherence of the precipitates with the TiC. Because of the limited size of the precipitates, the identification of their phase is difficult. One relatively large, lenticularly shaped precipitate (length 250 nm, maximum width of 25 nm) could be unambiguously identified on the basis of an SAED pattern as Ti_5Si_3 viewed along the $\langle 11\bar{2}3 \rangle$ direction.

Three examples of precipitates nucleated on SFs, as imaged by HRTEM, are depicted in Fig. 43. Using the known interplanar spacings of TiC as reference, the lattice spacings in the precipitates could be accurately determined in the HRTEM images. In Fig. 43a, the precipitate corresponds to Ti_5Si_3 viewed along the $\langle 11\bar{2}3 \rangle$ direction and appears to be nucleated on the crossing of two edge-on seen stacking faults in the TiC. The orientation relation (OR) between the precipitate and matrix is such that $\{1\bar{2}12\}$ $Ti_5Si_3//\{200\}$TiC and $\langle 1\bar{2}1\bar{3} \rangle$ $Ti_5Si_3//\langle 011 \rangle$TiC. The mismatch between the $\{1\bar{2}12\}$ Ti_5Si_3 and $\{200\}$TiC is only about 2.2%, and these planes give the impression of continuing (semicoherently) across the interface. In Fig. 43b, the precipitate corresponds to Ti_5Si_3 viewed along the $\langle 14\bar{5}0 \rangle$ direction and the OR and interface orientation is: $(0001)Ti_5Si_3//\{11\bar{1}\}$ TiC and $\langle 14\bar{5}0 \rangle$ $Ti_5Si_3//\langle 011 \rangle$TiC. The precipitate has a large aspect ratio with very large (0001) facets parallel to the original stacking fault plane of TiC on which it is nucleated. The spacing between the atomic columns along the interface as seen end-on in the HRTEM image is 0.244 nm for Ti_5Si_3 and 0.265 nm for TiC, which also corresponds to a relatively small mismatch.

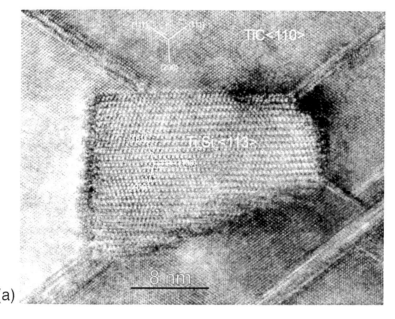

(a)

Figure 43 HRTEM images of Ti_5Si_3 precipitates nucleated on stacking faults in TiC present in the reaction layer. The TiC is viewed along the $\langle 110 \rangle$ direction. (a) Ti_5Si_3 precipitate viewed along $\langle 11\bar{2}3 \rangle$ on the crossing of two edge-on observed stacking faults in TiC. (b) Ti_5Si_3 precipitate viewed along $\langle 14\bar{5}0 \rangle$ having its (0001) plane parallel to the stacking faulted $\{111\}$ TiC plane. (c) Two merged Ti_5Si_3 precipitates. One of these precipitates is viewed along its $\langle 11\bar{2}3 \rangle$ axis and the interface is formed by the $\{10\bar{1}0\}$ Ti_5Si_3 plane parallel to the stacking faulted $\{111\}$ TiC plane.

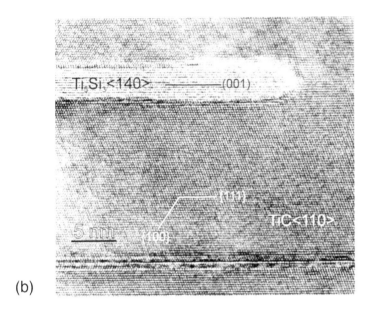

(b)

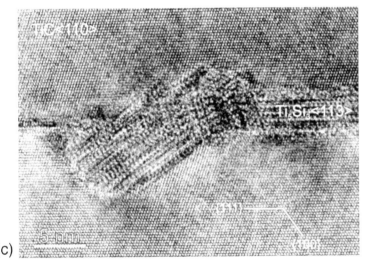

(c)

Figure 43 Continued.

Actually, in Fig. 43c, two Ti_5Si_3 precipitates are merged. One of these precipitates with its interface parallel to the {111}TiC on which the SF occurred, is viewed along its $\langle 11\bar{2}3 \rangle$ axis, i.e., the same as for the precipitate in Fig. 43a. However, this precipitate has the same shape as the one in Fig. 43b, but now the {$10\bar{1}0$} instead of the (0001) plane of Ti_5Si_3 is parallel to {111} of TiC. Examples in Fig. 43 show that the precipitates can nucleate and grow with a variety of orientations with respect to the TiC matrix and the SF present. The combined SAED and HRTEM information clearly indicates that the precipitates nucleated on SFs in TiC correspond to Ti_5Si_3. Precipitates shown in Fig. 43 are already well developed, and it should be realized that all the stages in between stacking faults and of these well-developed precipitates are actually present in the TiC.

The nucleation of the Ti_5Si_3 precipitates on SFs in TiC indicates that TiC is super-saturated with Si. Apparently, apart from the rejection of the excess Si in the molten matrix, Si also appears to become trapped in the TiC when it is formed at high temperatures (2500–3000°C). At these high temperatures, the solubility of Si in TiC is unknown, but at 1200°C, the solubility is very low according to the ternary Ti–Si–C phase diagram. Therefore, during cooling, the supersaturated Si likes to segregate out of the TiC. A possible mechanism for this tendency within the TiC grains is that Si induces SFs in the TiC and is enriched at these SFs. This mechanism explains the presence of the unusually wide extension and high density of SFs in the TiC grains. The SFs in face-centered cubic TiC correspond to locally close-packed hexagonal stacking and all currently relevant phases containing Si (Ti_5Si_3, Ti_3SiC_2, and all but one of the SiC polytypes) are based on this stacking as well. Thus, Si segregated to SFs is expected to be the most stable state for Si in TiC. If the local enrichment of Si to the SFs is sufficiently strong, new phases containing Si tend to nucleate on the SFs during cooling down, explaining the occurrence of the Ti_5Si_3 precipitates on the SFs in TiC as observed. According to the ternary phase diagram at 1200°C, excess Si in TiC for Ti/C ratios in between 0.33 and 0.46 will result in the formation of Ti_5Si_3. The experimental concentration of C in the TiC of the reaction layers, quantified using EDS and PEELS in the TEM, appeared to be 0.402 ± 0.022 and 0.386 ± 0.011, respectively, and is in accordance with the observed Ti_5Si_3 precipitation on SFs in TiC.

The Si gathered in the precipitates is drained from the SFs in the neighborhood and, subsequently, the stability of the remaining SF parts is dramatically changed. These SF parts will have a strong tendency to become unfaulted. Several clear impressions (both in HRTEM and in overviews in BF and DF) so far obtained support that, indeed, portions of originally long SFs become unfaulted in the neighborhood of precipitates. A clear example of an unfaulted part of an SF is already shown in Fig. 43b in front of the precipitate's nose, where, again at some further distance from the nose, the SF and more precipitates become present.

Although the laser treatment does not correspond at all to thermodynamic-equilibrium conditions, the ternary phase diagram is still useful in predicting the reaction products at "interfaces" on a local scale. On the tie-line of SiC–TiC, a relatively small increase of the Si content is sufficient to yield Ti_3SiC_2. The Si supersaturating TiC with a carbon content in between 33 and 46 at.% leads to Ti_5Si_3 formation. On the tie-line TiC–Ti, an increase of the Si content results in Ti_5Si_3. Rejection of Si in Ti with small Si concentration gives eutectic Ti/Ti_5Si_3 with pre-eutectic Ti, and rejection of more than 15 at.% Si results in pre-eutectic Ti_5Si_3 instead. All these phase formations taken from the ternary phase diagram, which are in fact the result of the release of Si during the exothermic reaction of SiC with Ti, are in agreement with the observed microstructure after laser embedding of SiC particles in Ti–6Al–4V. Applied with care, similar types of reasoning will, for example, hold for SiC fiber-reinforced Ti and Ti/SiC diffusion couples.

C. Mechanical Properties

The mechanical performance of the MMC strongly depends on the properties of the interfacial region between the reinforcement and the matrix [80,81]. A number of properties will be examined in more detail and, possibly, in correlation with the laser particle injection process. The SiC particle injection into liquid Ti results in chemical reactions that occur at the interface, which might have detrimental consequences for mechanical performance. Furthermore, high thermal stresses will be present in the modified layer, as the MMC is cooled down from high temperatures. These stress concentrations play an important role in damage initiation. Finally, hardness indentations are performed to initiate cracks in the matrix close to the reaction layer in order to study the crack initiation and propagation behavior at the interface region.

1. Reaction Layer Growth

The majority of metal matrix composites are nonequilibrium systems as a result of the presence of a chemical potential gradient across the interface. This means that, under certain conditions, diffusion and/or chemical reactions will take place at the interface. These specific conditions concern temperature and interaction time. Under controlled circumstances, the formation of a limited reaction layer might be desirable to obtain strong bonds. However, a thick reaction layer might have a detrimental affect on the composite properties. Therefore, it is extremely important to be aware of the influence of the process conditions in order to produce MMCs via a laser for practical applications. The most commonly used method to obtain a chemical bonding is a diffusion welding process, where the layer thickness shows a parabolic dependence in time and can be estimated by $l \propto \sqrt{Dt}$, see Refs. 82 and 83. The literature has reported on degrading of mechanical properties of a diffusionally bonded Ti–SiC composite to degrade for layer thickness exceeding 1 μm [84]. However, this is not comparable with laser processing as the formation of the reaction layer takes place mainly when the matrix is in the liquid state. In principle, this seems to be an uncontrollable situation due to the convective flow in the melt pool. In particular, the reaction of Ti with SiC exhibits a large thermodynamic driving force [85]. Nevertheless, it turned out to be feasible to minimize the reaction layer thickness by adjusting the particle injection position. If we compare the interaction time in diffusional processes with that of the laser embedding process, it differs by several orders of magnitude. For diffusional processes, the interaction time is given in hours, whereas in laser embedding it is given in seconds. If the length of the melt pool is 4 mm, the maximum time the SiC particles spend in the liquid Ti amounts to 0.44 sec. By changing the injection position, this interaction time will decrease.

It is found that the reaction layer thickness decreases by changing the particle injection position toward the back of the laser beam in the negative x-direction, which is depicted in Fig. 44. The values are obtained by taking the average of at least 10 reaction layer thicknesses in the middle of the track. Despite the undulation of the curve, there is a general tendency for the thickness to decrease. Thus, the different appearance of the reaction layer is related to the time the particles spend in the liquid titanium and the processing temperature. In principle, the thin reaction layers are of cellular type, whereas the irregular type is present for the thicker reaction layer. This transition takes place around a thickness of 2.5 μm, but it must be noted that this is certainly not a sharp transition.

The presence of Ti_3SiC_2 in the irregular reaction layer is interesting as the ternary compound Ti_3SiC_2 combines typical properties of ceramics with plastic-like behavior, which is rather exceptional for nonmetallic materials. This can be understood by taking into consideration the atomic arrangement in the crystal. As the hexagonal plane of Si is sandwiched between two hexagonal Ti planes, this configuration might exhibit ductile behavior. Because of this behavior, industrial interest in producing this peculiar ceramic is considerable, because the application of constructional ceramics can then be applied beyond the limit of brittle materials. In addition, it is suggested that Ti_3SiC_2 be used to obtain high joint strength between SiC grains. Therefore, a considerable amount of research is devoted to finding out the required process conditions to obtain this ternary phase [86]. In diffusion welding experiments, it has been reported that for 2-hr annealing time in the temperature range of 1200–1500°C, more than 90% of the reaction product consists of Ti_3SiC_2 [87]. For

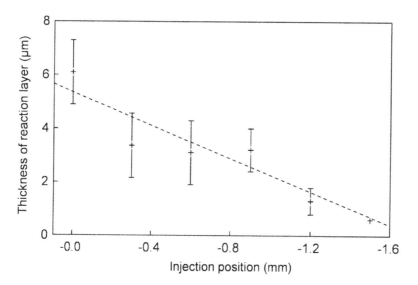

Figure 44 Thickness of the reaction layer as a function of the injection position relative to the center of the laser beam. (The dashed line only indicates the general decrease of the reaction layer thickness.)

laser embedding, this phase only arises in the irregular reaction layer, growing from the TiC surface.

2. Thermal Stresses

In general, metals have larger thermal expansion coefficients than ceramics. As the laser embedding process takes place at high temperatures, thermal stresses will arise during cooling down. The misfit strain at the interface is simply given by $\Delta\alpha\Delta T$, i.e., the difference in thermal expansion coefficient $\times$ the temperature route. In case of an isolated spherical particle with radius a, embedded in a infinite elastic continuum, the stress field in the matrix is given by

$$\sigma_{\mathrm{Mr}}(r) = -\frac{Pa^3}{r^3}; \qquad \sigma_{M\theta}(r) = \frac{Pa^3}{r^3} \tag{39}$$

where σ_{Mr} is the radial and $\sigma_{M\theta}$ represents the tangential stress field. Quantity P represents the internal pressure for the particle and is given by [88]

$$P = \frac{(\alpha_M - \alpha_P)\Delta T}{\dfrac{1 - v_M}{2E_M} + \dfrac{3v_M(1 - 2v_P)}{E_P(1 - 2v_M)}} \tag{40}$$

where v is Poisson's ratio and E is Young's modulus. In case of SiC particles in a Ti matrix, the hydrostatic compressive stress in the particle amounts to ~ 1.0 GPa. The stress distribution around the particle represented by Eq. (39) indicates a compressive stress in the radial direction and a tensile stress in the tangential direction. However, by local plastic flow, the stress state can be redistributed, which especially reduces the tangential stresses. As a consequence, the dislocation density in the vicinity of the particle will increase, resulting in a hardened zone [89]. This description is given for spherical particles, while the injected particles during laser processing show acuity. This acuity influences the stress field by locally inducing high stresses, which can be relieved by plastic flow. Furthermore, the presence of an intermediate TiC reaction layer influences the stress distribution by reducing the matrix stress [90]. This effect become more important with increasing reaction layer thickness.

3. Bond Strength

To study the bonding between the TiC reaction layer and the SiC particle, hardness indentations are made to initiate crack formation in the matrix in the vicinity of the reaction layer. In case of crack initiation, we examine the crack path through the reaction layer toward the SiC particle. The indents are obtained via a Vickers hardness tester with an applied load of 20 N. Although the applied force is constant, it must be mentioned that the crack propagation conditions all differ as far as the energy is concerned, because of the different matrix structures and the different distance between the corner of the pyramidal indent and the reaction layer. In addition to yielding the crack propagation information, these tests also reveal the crack initiation behavior of the matrix material, as the indents did not always result in crack formation.

Fracture of materials can be described from an internal energy point of view by the creation of free surfaces from bulk materials [91]. In case of brittle materials, without stress concentrators or plastic deformation, these amount to equating the release of strain energy to the energy required for creating free surfaces. The critical stress intensity factor correlates the fracture stress of a material to the flaw size in the material and is defined as

$$K_{\text{Ic}} = \sigma_{\text{c}}\sqrt{\pi a} \tag{41}$$

where σ_{c} is the critical stress for a flaw to propagate and a is the radius of the flaw. The values of the fracture toughness for SiC, TiC, and TiN are shown in Table 2. In plane strain conditions, the critical stress intensity factor is related to the strain energy release rate at fracture by

$$G_{\text{c}} = \frac{K_{\text{Ic}}^2(1 - v^2)}{E} \tag{42}$$

This is a measure of the amount of energy required for a crack to propagate through a monocrystalline brittle material.

The indents are made in the aforementioned set of laser tracks with six different particle injection positions, which resulted in two different types of reaction layers. In the case of the cellular reaction layer, the initiated cracks are found to propagate straight ahead through the TiC reaction layer and to advance in the SiC particle without deflection at the interface, as can be seen in Fig. 45. This is an indication of a good load transfer at the interface. Furthermore, there is no clear crack path observed in the matrix, which could have been only plastically deformed. The irregular reaction layer exhibits the crack as following the pores, as shown in Fig. 46. Two different phenomena are found to occur at the interface. In the first case, the crack traverses the interface without significant deflection, whereas the second case shows the crack to deflect first along the interface followed by penetration into the SiC particle. The latter situation is shown in Fig. 47. The deflection can be partially explained by the presence of pores at the interface. However, in between the pores, SiC and TiC are still detached from each other, which indicates that the interface exhibits locally poor bonding. This might be explained by the different growth conditions as this type of reaction layer grows from the melt toward the SiC particle, possibly resulting in a

Table 2 Material Properties

Physical properties (room temperature)	cp-Ti (grade 2)	Ti–6Al–4V	Mo	TiN	TiC	SiC
Density (Mg/m^3)	4.51	4.43	10.2	5.4	4.92	3.22
Thermal Melting point (°C)	1670	1650	2617	3290	3067	2545
Thermal conductivity (W/m K)	15.57	6.7	105	28	22	145
Thermal expansion (10^{-6}/K)	8.6	8.6	4.9	6.3	6.3	4.4
Specific heat (J/kg K)	518	564	250	630	557	690
Mechanical Young's modulus (GPa)	102	113	255	430	435	415
Poisson's ratio	0.36	0.36	0.32	0.23	0.24	0.15
Tensile strength (MPa)	172	827	480		295	355
Hardness (GPa)	0.6	1.2	8	20	28.5	29
Fracture toughness (MPa m$^{1/2}$)	60–120	55–123	20	2–3	2–3	3–5

weaker bond than the situation in which the cellular reaction layer grows from the surface of the SiC particle. If we consider the crack initiation in the matrix, this is found to occur depending on the matrix structure. In the case of high dissolution of SiC, which implies the presence of thick reaction layers, the indents often initiated cracks in the matrix. On the other hand, for limited dissolution of SiC, which implies thin reaction layers, the indents just occasionally initiated cracks. This is caused by the difference in yield stress (see Table 2) of

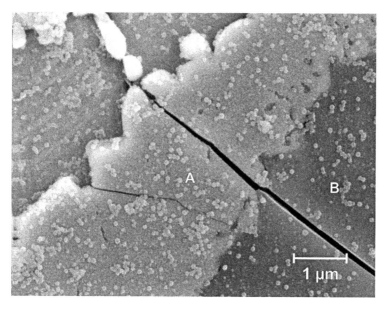

Figure 45 SEM micrograph of a cellular reaction layer (A) showing crack propagation straight through the reaction layer of TiC into the SiC particle (B) without deflection at the interface.

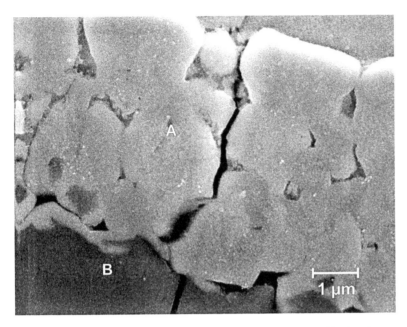

Figure 46 SEM micrograph of an irregular reaction layer (A) showing crack propagation along the pores of TiC. At the interface the crack slightly deflects but finally propagates into the SiC particle (B).

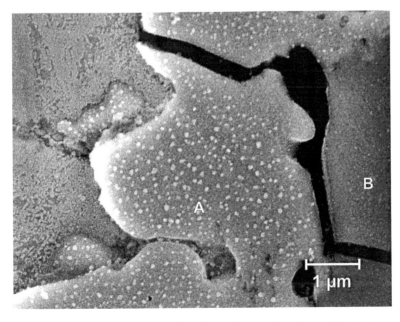

Figure 47 SEM micrograph of an irregular reaction layer (A) showing crack deflection at the interface between TiC and SiC. Finally, the cracks penetrate the SiC particle (B).

the microstructures, which are described in the previous section. For regions containing large amounts of Ti_5Si_3 and TiC, cracks are initiated more easily than for regions with Ti grains surrounded by eutectic structures. Therefore, the latter matrix structure is favorable, because of its better capacity to relieve stress by plastic deformation.

Hardness measurements have been performed to determine the hardness of the different phases present in the laser track with SiC particles. The hardness of the substrate material and the SiC particles amounts to 350 and ~2500 HV, respectively. The heat-affected zone below the laser track exhibits an increase in hardness of up to 400 HV. It must be noted that all other hardness indents are performed at composite structures. The hardness of the structure consisting of Ti grains surrounded by the eutectic structure, depicted in Fig. 25, is measured to be ~650 HV. This value increases with decreasing Ti grain size. The areas, which consist mainly of Ti_5Si_3, exhibit a hardness of 1100 HV. However, this value can be influenced by the presence of TiC cells and dendrites. As the composition of the matrix changes with the particle injection position, it is possible to control the average matrix hardness in the range between 650 and 1100 HV. It is anticipated that the lowest value is preferable with respect to its ductile behavior [92].

Abrasive wear is the displacement of material caused by the presence of hard particles and might involve both plastic flow and brittle fracture. The phenomena related hereto, concern microploughing, microcutting, microfatigue, and microcracking. The simple theory of abrasive wear by plastic flow is based on the concept of an abrasive particle that forms a groove whose depth depends upon applied load and is inversely proportional to the hardness of the material. Furthermore, it is assumed that all material from the wear groove is detached in a single cutting event. In that case, the volume loss of the material can be described by Archard's wear law.

$$\frac{W_v}{s} = k\frac{F_N}{H} \tag{43}$$

where W_v is the volume loss, s is the sliding distance, k is the wear coefficient, F_N is the applied load, and H represents the hardness of the wearing surface. Thus the wear rate is inversely proportional to the hardness of the material. However, for many materials, the hardness is not the only factor that determines the wear rate. In the case of brittle materials, the wear rate obtained by the aforementioned equation will be considerably surpassed in practice, because of a different wear mechanism. Therefore, this law can be considered only as an upper limit for ductile materials, such as pure metals and metallic alloys. In the case of brittle materials, the wear behavior strongly deviates. Above a critical loading, the abrasive wear will be controlled by the formation and propagation of cracks. In fact, it is not the hardness, but the fracture toughness that becomes the predominant factor to describe the wear behavior of brittle materials. However, the wear properties of MMCs differ from those of the single constituents. The wear resistance of composites depends on the hardness of the reinforcing phase and its volume fraction. Therefore, the wear resistance can be described by a rule of mixture, whereas the type of rule depends on the microstructural configuration. For this description, it is required that the reinforced phase is very well bonded to the matrix. Otherwise, it will not optimally contribute to reinforcement. Furthermore, the size of the reinforcement with respect to the abrasive particles plays an important role. It has been found that the response of the composite to deformation is more homogeneous in the case of small particles in relation to abrasive particles, which results in better wear resistance [93].

Abrasive wear tests are performed to examine the properties of the embedded SiC, TiC, and TiN particles. The particles are not easily pulled out during testing, from which it may be concluded that the bonding between particle and matrix is rather good. Figure 48

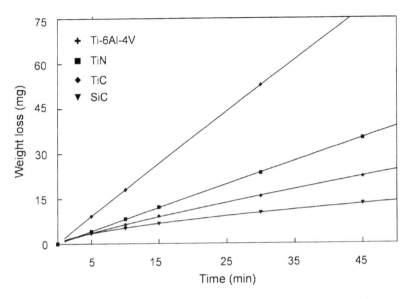

Figure 48   Abrasive wear as a function of time for different composites.

depicts the weight loss as a function of time. The improvement in wear resistance can be described by the relative wear resistance, which is defined as the ratio between the wear rate of the substrate material and the MMC under the same conditions. The measured relative wear resistance of SiC, TiC, and TiN with respect to Ti–6Al–4V amounts to 7, 4, and 2.5. It must be noted that the quality, especially the porosity, of the different powders has been not the same, which could have influenced the final result.

Sliding wear can be characterized as a relative motion between two smooth solid surfaces in contact under load. The surfaces may either be metallic or nonmetallic in nature, and may be lubricated or otherwise. The wear mechanisms, which accompany sliding wear, are adhesion, surface fatigue, tribochemical reaction, and abrasion. To what extent these mechanisms are present is determined by the type of deformation, plastic or elastic, the mechanical properties of both materials, the interfacial elements, and the loading conditions.

To minimize wear rate and friction coefficient, the appropriate material combination has to be chosen for the two interacting bodies. These might be both single-phase materials, but it could be favorable to combine two constituents in one body. However, the complexity of sliding wear increases when one material consists of composite materials, as the mechanical behavior of the two constituents is different. For composites, the most important properties concerning friction and sliding wear are the identity, shape, size, orientation, volume fraction of the reinforcing constituents, and the matrix properties. In addition, the interfacial bonding between the two phases is extremely important. In case of an MMC, the reinforcement affects wear by supporting the applied load with less deformation than the pure matrix, due to greater strength and elastic modulus. This load-bearing capacity increases with increasing particle size. Another influence of the reinforcement is that it obstructs the matrix from deforming plastically by reducing the mobility of dislocations. On the other hand, the strength and elastic modulus of the matrix must be sufficient to support the reinforcement in an appropriate way. However, if the matrix becomes brittle, this will result in higher wear rates.

The objective of this research is to determine the suitability of particle-strengthened MMCs for application in bearings that are sparsely lubricated. Therefore, three different metal matrix composites obtained by laser particle injection process are subjected to a sliding wear test under conditions of boundary lubrication. An indication for boundary lubrication conditions is the coefficient of friction, which should be ~0.10 or higher [94]. For each MMC, at least two different injection conditions are used during laser processing. This position is given with respect to the center of the laser beam. Furthermore, the laser-nitrided coatings are tested as well. The results will be compared to commercially obtained MMCs on the basis of Al substrate and qualified "tribo materials" [95]. The results of the wear tests are listed in Table 3. Overall, it may be concluded that the performance of the laser coatings is good. Some coatings even exhibit a specific wear rate below 0.01×10^{-6} mm^3/Nm, i.e. SiC (−2.0), TiC (−1.0), and TiN (gas), corresponding to the commercially obtained MMCs. Note that the wear rate of the obtained "tribo-materials" is substantially higher. The remarkable fact is that both TiN gas-alloyed coatings exhibit good wear resistance despite the presence of crack patterns. Although there is a slight decrease in the specific wear rate of SiC with increasing injection position, it is beyond the reach of this test to conclude that there is a relation between these factors. Nevertheless, it is beyond any doubt that there is a difference in the matrix composition and the reaction layer. Therefore, it may be concluded that the matrix and reaction layer appearance are not critical under these test conditions. A different test that might be critical is the fretting test, where one of the main concerns is the initiation of fatigue cracks. It is anticipated that this test will reveal

Table 3 Specific Wear Rate and Friction Coefficient of Sliding Wear Experiments Under Boundary Lubrication Conditions

No.	Material/injection position (mm)	f	k_{pin} (10^{-6} mm^3/Nm)	k_{ring} (10^{-6} mm^3/Nm)
1	Ti–6Al–4V(ref.)	0.15	274	0.02
2	SiC(−0.5)	0.15	0.03	0.47
3	SiC(−1.0)	0.14	0.02	0.24
4	SiC(−1.5)	0.15	0.02	0.35
5	SiC(−2.0)	0.16	<0.01	0.17
6	TiC (−0.5)	0.17	0.09	0.88
7	TiC (−1.0)	0.15	0.15	0.43
8	TiN (−0.5)	0.15	0.15	0.12
9	TiN (−1.0)	0.14	0.02	0.12
10	TiN (gas)	0.14	<0.01	0.04
11	TiN (gas)/cp-Ti	0.14	<0.01	<0.01
12	CuAl	0.15	3.6	<0.01
13	GG25	0.16	0.15	<0.01
14	CuSnPb	0.25	7.5	<0.01
15	AlMgSi/Al$_2$O$_3$	0.14	0.02	<0.01
16	AlCuMn/SiC	0.14	<0.01	<0.01
17	AlMgSi/SiC	0.14	<0.01	<0.01
18	AlMgSi/Al$_2$O$_3$	0.14	<0.01	0.08

Materials 1–11: experimentally obtained MMCs by laser processing.
Materials 12–14: commercially obtained "tribo-materials."
Materials 15–18: commercially obtained MMCs.

differences, because there is a difference in the ability to initiate cracks in the matrix for different injection positions, as already mentioned in the previous section. Another way to examine the metal–ceramic interface is under conditions of rolling contact, where, in particular, the properties of the interface are of crucial importance. Although the wear resistance of the pins is good, a problem is that the counter body exhibits severe wear. This is expected to be caused by surface roughness and the shape of the injected particles. Therefore, it should be favorable to inject spherical particles instead of acute particles exhibiting large attack angles. In our experiments, the type of wear observed is adhesion, which takes place especially between the reinforcement and the ring material, as EDX measurements indicate the presence of Fe upon the embedded particles.

D. Discussion and Conclusions

The laser particle injection process appeared to be very sensitive to the positioning of the powder injection, the interaction time of the particles with the laser, and the amount of powder used. Changing this in a controlled manner yields the possibility of varying the dissolution of the SiC particles and, at the same time, the composition of the matrix. The latter is very important because too much dissolution could result in a brittle matrix, which is undesirable with respect to the mechanical performance. In addition, the structure and properties of the reaction layer around the SiC particle changes for different injection positions.

Laser embedding of SiC particles in Ti–6Al–4V results in reaction layers of TiC around SiC with either a cellular or cloud-like structure. The Si released during the reaction $SiC + Ti \rightarrow TiC + Si$ is rejected into the melt pool and results in Ti_5Si_3 formation. Depending on the extent of local Si rejection, pre-eutectic Ti occurs at the bottom and pre-eutectic Ti_5Si_3 occurs at the top of the melt pool. If, in the cloud-like reaction layer, regions between the TiC grains become enclosed, the rejected Si content increases locally and Ti_3SiC_2 plates with dominant (0001) facets nucleate. In TiC grains, particularly those of the cellular reaction layer, a high density of widely extending stacking faults is observed and on these faults, in many instances, small Ti_5Si_3 precipitates are present. First, the Si supersaturating the TiC induces the SFs. Second, it causes nucleation of Ti_5Si_3 precipitates on the SFs. Third, parts of the SFs become unfaulted because the precipitates withdraw Si from the SFs and hence strongly increase the SF energy locally toward the high energy value pertaining to (pure) TiC. An alternative explanation for the presence of a high density of stacking faults in TiC is also suggested in literature (see Ref. 96). It is based on a reaction between Ti and SiC after laser treatment. However, this requires an additional step in the total sequence that cannot be verified by electron microscopy in a straightforward manner. What have been the principal observations? Besides the high density of SFs in TiC, we observed the presence of Ti_5Si_3 precipitates nucleated on the SFs and saw that parts of original SFs became unfaulted in the neighborhood of precipitated Ti_5Si_3. We suggest that the observed TiC is formed directly in an exothermic reaction between the Ti matrix and the SiC particles that have been injected during a process employing a high-power laser. The TiC phase has the highest melting point and is the first phase to form in the melt pool near or at the surface of the dissolving SiC particles. At the high formation temperature and during the subsequent high cooling rates associated with laser treatment, a relatively large amount of Si becomes trapped in the TiC. At lower temperatures, the Si supersaturates the TiC, induces SFs in the TiC, and becomes enriched at these SFs. If the local Si enrichment is sufficiently high, Ti_5Si_3 precipitates on the SFs in the TiC, in accordance with the Ti–Si–C ternary phase diagram for the Ti/C ratio we measured using PEELS and EDX. During the precipitation of Ti_5Si_3, Si is withdrawn from the SFs and, as a consequence, they become unstable, because in pure TiC the SF energy is high and SFs are not expected even after severe cold work. Thus, parts of the SFs in the neighborhood of the Ti_5Si_3 precipitates become unfaulted, as has been observed. In the alternative explanation [96], the hexagonal phase Ti_3SiC_2 is formed first in the exothermic reaction between the Ti matrix and the injected SiC particles. Because of its environment, this Ti_3SiC_2 is unstable and a topotactic reaction occurs, in which Si diffuses outwards and a cubic $Ti(C_{0.67},Si_{0.06})$ remains. Given the atomic mechanism of this topotactic reaction, the developing cubic phase will naturally contain a high density of SFs. The nice point about the explanation in Ref. 96 is that the SFs are a logical consequence of the reactions involved. However, a weak point of this explanation is that Barsoum assumes that an additional step occurs, in which a phase forms and disappears, and this step is not directly verifiable in the postmortem sample. Indeed, it can only be inferred from the traces it leaves behind in the newly developed phase, which is the only phase directly accessible for observations. Although the explanation contains a weak link, and as it needs an additional step and indirect evidence, it still might hold. However, we present several arguments that render this explanation improbable. The transformation of the hexagonal Ti_3SiC_2 to the cubic phase $Ti(C_{0.67},Si_{0.06})$, as proposed in Ref. 96, would generate, at least in domains, SFs parallel to one set of close-packed planes in the cubic phase that has been originally parallel to the basal plane in the hexagonal Ti_3SiC_2 phase. This is not in agreement with our observation of SFs on the *different* sets of $\{111\}$ in the same location within the TiC (see the crossing SFs in Figs. 40 and 41). Furthermore, we did observe Ti_3SiC_2. This phase has been always plate-like in shape, with the dominant facet parallel to the basal plane of Ti_3SiC_2. Our explanation for the occurrence of this phase is that, after the spherical TiC particles are formed in the melt near the surface of the SiC particles, volumes in between the TiC particles become enclosed. Silicon becomes trapped in these volumes and cannot be rejected into the melt pool. The increasing Si content in the enclosed volumes causes the nucleation of Ti_3SiC_2 in accordance with the Ti–Si–C phase diagram. This sequence of first forming TiC is logical because the melting point of TiC is higher than that of Ti_3SiC_2. The spherical shape of the TiC indicates that it is not likely as a result of transformed Ti_3SiC_2. In agreement with observations in the literature [68,70,71,97], hexagonal Ti_3SiC_2 would have formed in plate form. It is rather unlikely that, during transformation, these plates become reshaped into spheres. On the other hand, spherical particles are much more likely if the cubic TiC nucleates first in the melt near the SiC particles. Finally, during laser processing, high cooling rates are obtained. Because of the short time scale allowed during cooling, only very local bulk diffusion is possible. Our explanation already given here is possible with only local Si rearrangements in the TiC. On the other hand, the explanation presented in Ref. 96 would require bulk diffusion of Si out of the Ti_3SiC_2 on much larger length scales, of the order of the TiC grain size, which is near a micrometer. This process is rather unlikely in light of the high cooling rates. To summarize, in our opinion [98], there are sufficient arguments to state that the mechanism proposed in Ref. 96 for the explanation of the high density of SFs in TiC and the supersaturation with Si is not applicable to our process of laser embedding of SiC particles in a Ti matrix. It has to be emphasized, however, that in other lower cooling-rate processes, in which Ti reacts with SiC, the topotactic/peritectic transformation of Ti_3SiC_2 into $Ti(C_{0.67},Si_{0.06})$ still might be feasible.

Crack formation induced by hardness indents indicates that the cellular reaction layer exhibits a better bonding with the SiC particle than the irregular reaction layer. Experimental research presented in the literature indeed confirms that both materials might develop very strong bonds, which is important for mechanical performance [99]. The performed abrasive wear tests show a good relative wear resistance of SiC particles embedded in Ti–6Al–4V. On the other hand, the relative wear resistance of TiC and TiN particles embedded in Ti–6Al–4V is significantly lower, which could be partially caused by the higher porosity of these particles. The sliding wear tests under boundary lubrication conditions exhibit good results for all MMCs obtained via laser embedding. However, the wear of the counter body should first be diminished by embedding less acute particles. Nevertheless, in principle, these laser-modified layers are suitable for use in bearing applications.

IV. SUMMARY AND OUTLOOK

In this chapter, examples are presented of laser surface modification of lightweight materials, in particular aluminum and titanium, in order to enhance its wear resistance under conditions of high contact load. Other examples of enhanced wetting by laser treatment can be found in Ref. 27, e.g., firmly bonded coating of Cr_2O_3 on steel. In particular, interface layers, i.e., mullite and a high Si-content Al_2O_3, have been identified between the reaction coating on the Al6061 substrate. In this reaction coating, the wetting process is mainly governed by the contact of liquid Al with the reaction products of mullite and α-Al_2O_3. The formation of a mullite/Al interface may be essential to form a good bond by reaction coating on Al alloy, and a high Si-content Al_2O_3 interface layer may be wetted much better than pure α-Al_2O_3 by liquid Al. Despite the numerous advantages of Ti, such as its high strength-to-weight ratio and excellent corrosion resistance, it exhibits poor tribological properties. The research presented concentrates on processing control, thermal control, and an explanation of both microstructural features and mechanical performance. The microstructure is studied in great detail by scanning and transmission electron microscopy. In addition, the Kikuchi backscatter technique can be applied to obtain information concerning crystal structures in the SEM.

To explain the microstructural properties of the modified layer, it is necessary to know the temperature trajectory during laser processing. The maximum temperature attained is related to the occurrence of phase transitions and chemical reactions. Furthermore, high-temperature gradients are involved during cooling down, which determine the size and type of the microstructural features. As the thermal expansion of the applied coating constituents and the substrate material differ, high thermal stresses arise during cooling down, which finally may result in crack formation. In Section II, an analytical model is presented to predict the temperature profile depending on various process conditions. Therefore, the heat conduction equation is solved using Green's functions. Temperature fields are calculated for different intensity distribution of the laser beam, possibly including a pulsed laser mode. The model takes into account the finite dimensions of the specimen. This is achieved by making use of image sources, which provide zero heat flux at the boundary. The relevance of image sources is indicated by a case study of Ti and Mo, which exhibit a large difference in thermal conductivity. Finally, a description is presented to incorporate heat loss due to convective and conductive cooling at the boundary of the specimen.

This process of laser melt particle injection is found to be very sensitive for the positioning of the powder injection, the interaction time of the particles with the laser, and the amount of powder used. By changing the injection position in the negative x-direction relative to the laser beam, the time the particles spent in the liquid decreases. In addition, the temperature of the region where the particles penetrate into the melt pool is lower. Both effects result in a decrease of the degree in which the injected particles dissolve in the matrix. As a consequence of the dissolution of SiC, new phases in Ti will be formed in the matrix, such as the eutectic structure, strongly faceted Ti_5Si_3 grains, and TiC dendrites. Furthermore, a reaction layer of TiC is formed around the SiC particle. The thickness of this layer increases with increasing interaction time and temperature. A distinction can be made between the relatively thin cellular and thick irregular types of reaction layer, which distinction of the types of layers corresponds, respectively, to growth of TiC from the SiC surface and from the melt toward the surface of SiC. This dissimilarity is caused by the different temperature conditions during solidification. The latter situation results in the presence of Ti_3SiC_2 plates in the reaction layer, which represents a ceramic with plastic-like behavior.

The mechanical performance depends very much on the degree of dissolution and the type of reaction layer. Therefore, hardness indentations are performed to examine the crack initiation and propagation behavior. As the aim is to produce a hard phase for a ductile substrate, the dissolution of SiC should be minimized to avoid matrix embrittlement. This is confirmed by the ability to initiate cracks, which is found to be difficult for small amounts of dissolution and can be explained by the high tensile strength of Ti–6Al–4V relative to the ceramic phases. In the case of the cellular reaction layer, the cracks propagate straight through this layer into the SiC particle without deflection at the interface. For the irregular reaction layer, the cracks propagate along the pores and possibly deflect along the interface before penetrating into the SiC particle. The latter indicates a slightly inferior bonding between SiC and TiC, which is explained by the different growth mechanism of the irregular layer during solidification. The modified layers are subjected to abrasive and sliding wear tests. The sliding wear tests under boundary lubrication conditions exhibit a specific wear rate close to or below 0.01×10^{-6} mm^3/Nm for both the laser alloyed and embedded types of layers. Therefore, practical applications as bearing materials should be within reach. In particular, the SiC particles in Ti–6Al–4V show promise, provided control of the microstructure is at stake. Considerable progress has been made in recent years on functionally graded coating in lightweight materials [100–103].

In the future, the fields of nanometer-sized particles and laser treatment will meet, in particular for coating applications. Although wet-chemical processing of nanometer-sized particles, also called sol–gel processing, has been being explored in the 1930s, it has only recently become popular. Sol–gel coatings can be used not only for protection from an environment, but also for optical and electronic components. Technically, wet-chemical processing offers many advantages. The sol–gel method allows films to be made with almost any composition and degree of porosity. The problem of homogeneity, often encountered in the processing of powders, is absent in the sol–gel preparation technique because no comminuting is required. Moreover, processing temperatures can be significantly reduced, and combining different coating liquids (hybrid systems) is easy. The sol–gel concept can be combined with inkjet technology and laser treatment of surfaces into what is called a "stereostiction" (see Refs. 104 and 105 and references therein). However, behavior such as grain growth of nanoparticles in the sol–gel coating is a crucial aspect, and very little microscopic information is available until now [106–108]. Here, the message for the direction of the research is to have control over the microstructural features by exerting control over the temperature fields involved in laser surface engineering.

ACKNOWLEDGMENTS

The work described in this chapter is part of the research program of the Foundation for Fundamental Research on Matter (FOM, Utrecht) and has been made possible through financial support from the Netherlands Organization for Research (NWO, The Hague), the Foundation for Fundamental Research on Matter (Utrecht), the Foundation for Technical Sciences (STW, Utrecht), and the Netherlands Institute for Metals Research. Thanks are due to our collaborators over the years, Bart Kooi, George Palasantzas, Arjen Kloosterman, Edzo Zoestbergen, Arjan Vreeling, Dimitri van Agterveld, and the technicians Henk Bron, Klaas Post, Jan Harkema, and Uko Nieborg, for their stimulating discussions and contributions described in this chapter.

REFERENCES

1. De Hosson, J.Th.M. Chapter in book 'laser synthesis and properties of ceramic coatings'. In *Intermetallic and Ceramic Coatings*; Dahotre, B. Sudarshan, T.S., Eds.; Marcel Dekker: N.Y., 1999; 307–441.
2. Ignatiev, M.; Smurov, I.; Flamant, G. Meas. Sci. Technol. 1994, *5*, 563.
3. Ion, J.C.; Shercliff, H.R.; Ashby, M.F. Acta Metallurgica 1992, *40* (7), 1539.
4. Kurz, W.; Fisher, D.J. *Fundamentals of Solidification*; Trans Tech Publications: Switzerland, 1985.
5. Ashby, M.F.; Easterling, K.E. Acta Metallurgica 1984, *32* (11), 1935.
6. Johns, D.J. *Thermal Stress Analysis*; Pergamon Press Ltd.: Oxford, 1965.
7. Mathews, J.; Walker, R.L. *Mathematical Methods of Physics*; Addison-Wesley: Reading, MA, 1970.
8. Darling, C.R. *Pyrometry*; Spon & Chamberlain: New York, 1920.
9. Smurov, I.; Ignatiev, M. NATO series E Appl. Sci. 1996, *307*, 529.
10. Sparrow, E.M.; Cess, R.D. *Radiation Heat Transfer*; Wadsworth, California, 1966.
11. von Allmen, M. *Laser and Electron Beam Processing of Materials*; Academic Press: London, 1980.
12. De Hosson, J.Th.M.; Zhou, X.B.; van den Burg, M. Acta Metallurgica et Materialia 1992, *40*, S139–S142.
13. Sokolov, A.V. *Optical Properties of Metals*; Blackie & Son: London and Glasgow, 1967.
14. Benett, H.E.; Porteus, J.O. J. Opt. Soc. Am. 1961, *51*, 123.
15. Porteus, J.O. J. Opt. Soc. Am. 1963, *53* (2), 1394.
16. Faber, T.E. *Theory of Liquid Metals*; Cambridge Univ. Press, 1972; 309 pp.
17. Carslaw, H.S.; Jaeger, J.C. *Conduction of Heat in Solids*; In DGM (Deutsche Gemeinshaft füz Metallkunde); Informations Gesselschaft Verlag: Oberursel, Germany, 1959.
18. Geissler, E.; Bergmann, H.W. In *Laser Treatment of Materials*; Mordike, B.L., Ed.; Clarendon Verlag: Oxford , 1987.
19. De Hosson, J.Th.M.; Kooi, B.J. Structure and properties of heterophase interfaces, Chapter. In *Handbook of Surfaces and Interfaces in Materials*; Nalwa, H.S., Ed.; Chapter 1. Academic Press: San Diego, 2001; Vol. 1, 1–114 pp. *invited chapter.*
20. Rosenthal, D. Trans. A.S.M.E. 1946, *11*, 849.
21. Cline, H.E.; Anthony, T.R. J. Appl. Phys. 1977, *48*, 3895.
22. Porteus, J.O. J. Opt. Soc. Am. 1963, *53*, 1394.
23. Kreith, F. Principles of heat transfer. Harper & Row publishers: New York, 1973; Necati Ozisik, M. *Heat Conduction,* 2nd ed. ;Wiley: New York, 1993.
24. Morse, P.M.; Feshbach, H. *Methods of Theoretical Physics*; McGraw-Hill, 1953; In *Heat Conduction Using Green's Functions*; Beck, J.V., Cole, K.D., Haji-Sheikh, A., Litkouhi, B., Eds.; Himishere Pub. Comp. London, 1992.
25. Laser Processing: Surface Treatment and Film Deposition; Mazumder, J., Conde, O., Villar, R., Eds.; NATO ASI Series E. Kluwer Academic Publishers: Dordrecht, 1996; Vol. 307.
26. Abboud, J.H.; West, D.R.F. J. Mater Sci. Lett. 1991, *10*, 1149.
27. De Hosson, J.Th.M. In *Intermetallic and Ceramic Coatings*; Dahotre, B., Sudarshan, T.S., Eds.; Marcel Dekker: N.Y., 1999; 307–441.

28. Baker, T.N.; Xin, H.; Hu, C.; Mridha, S. Mater. Sci. Technol. 1994, *10*, 536.
29. De Mol van Otterloo, J.L; De Hosson, J.Th.M. Acta Mater. 1997, *45*, 1225.
30. Vreeling, J.A.; Ocelík, V.; Pei, Y.T.; De Hosson, J.Th.M. In *Surface treatment IV*; Brebbia, C.A., Kenny, J.M., Eds.; Computational Mechanics Publications: Southampton, 1999; 269 pp.
31. Zhou, X.B.; De Hosson, J.Th.M. Acta Metall. Materialia. 1994, *42*, 1155–1162.
32. Pilloz, M.; Pelletier, J.M.; Vannes, A.B.; Bignonnet, A. J. Phys. IV Coll. C7 1991, *1*, C7.
33. Mondolfo, L.F. *Aluminium Alloys: Structure and Properties*; Butterworths: London Boston, 1976; 108 pp.
34. Kaptay, G. Mater. Sci. Forum 1991, *77*, 315.
35. Kloosterman, A.B.; Kooi, B.J.; De Hosson, J.Th.M. Acta Mater. 1999, *46*, 6205; Kooi, B.J.; Kabel, M.; Kloosterman, A.B.; De Hosson, J.Th.M. Acta Mater. 1998, *47*, 3105.
36. Kaptay, G. Mater. Sci. Forum. 1996, *215–216*, 459.
37. Kaptay, G. Mater. Sci. Forum 1996, *215–216*, 475.
38. Hofmann, S. Dept Profiling in AES and XPS. In *Practical surface analysis, Vol. 1—Auger and X-ray photoelectron spectroscopy*; Briggs, D., Seah, M.P., Eds.; John Wiley & Sons: New York, 1990;143–199.
39. Kelly, R.; Lam, N.Q. Radiat. Eff. Radiation Effects 1973, *19*, 39–48.
40. Hu, C.; Xin, H.; Baker, T.N. J. Mater. Sci. 1995, *30*, 5985.
41. Pantelis, D.; Tissandier, A.; Manolatos, P.; Ponthiaux, P. Mater. Sci. Technol. 1995, *11*, 299.
42. Vreeling, J.A.; Ocelík, V.; Hamstra, G.A.; Pei, Y.T.; De Hosson, J.Th.M. Scripta Mater. 2000, *42*, 589–595.
43. Pelletier, J.M.; Sahour, M.C.; Pilloz, M.; Vannes, A.B. J. Mater. Sci. 1993, *28*, 5184.
44. Nylund, A.; Olefjord, I. Surface and Interface Analysis 1994, *21*, 283.
45. Hu, C.; Xin, H.; Baker, T.N. Mater. Sci. Technol. 1996, *12*, 227.
46. Hegge, H.J.; Boetje, J.; De Hosson, J.Th.M. J. Mater. Sci. 1990, *25*, 2335.
47. De Hosson, J.Th.M.; de Mol van Otterloo, J.L. Surface Engineering 1997, *13* (6), 471–481.
48. Katayama, S.; Matsunawa, A.; Morimoto, A.; Ishimoto, S.; Arata, Y. ICALEO '83 LIA, Los Angeles, 1983.
49. Ayers, J.D.; Schaefer, R.J.; Robey, W.P. J. Met. 1981, *8*, 19.
50. Mridha, S.; Baker, T.N. J. Mater. Proc. Techn. 1997, *63*, 432.
51. Abboud, J.H.; West, D.R.F. J. Mater. Sci. Lett. 1991, *10*, 1149.
52. Ayers, J.D.; Tucker, T.R. Thin Solid Films 1980, *73*, 201.
53. Abboud, J.H.; West, D.R.F. J. Mater. Sci. Lett. 1992, *11*, 1675.
54. Baker, T.N.; Xin, H.; Hu, C.; Mridha, S. Mater. Sci. Technol. 1994, *10*, 536.
55. Dingley, D.J.; Baba Kishi, K.Z.; Randle, V. *Atlas of Backscattering Kikuchi Diffraction Patterns*; IOP Publishing: London, 1995.
56. van der Wal, D.; Dingley, D.J. Philips Electron Optics Bull. 1996, *134*, 19.
57. Randle, V. *Microtext Determination and Its Applications*; Institute of Materials: London, 1992.
58. Venables, J.A.; Harland, C.J. Philos. Mag. 1973, *27*, 1193.
59. Loretto, M.H. *Electron Beam Analysis of Materials*; Chapman & Hall: London, 1994.
60. Choi, S.K.; Chandrasekaran, M.; Brabers, M.J. J. Mater. Sci. 1990, *25*, 1957.
61. Abboud, J.H.; West, D.R.F. Mater. Sci. Technol. 1989, *5*, 725.
62. Arunajatesan, S.; Carim, A.H. Mater. Lett. 1994, *20*, 319.
63. Morgiel, J.; Lis, J.; Pampuch, R. Mater. Lett. 1996, *27*, 85.
64. Brukl, C.E. In *Ternary Phase Equilibria in Transition Metal–Boron–Carbon–Silicon systems*, TMS, Springfield, II (VII), Illinois, USA.
65. JCPDS 29-1128 and 29-1131.
66. Jeitschko, W.; Nowotny, H. Monatsh. Chem. 1967, *98*, 329.
67. Nickl, J.J.; Schweitzer, K.K.; Luxenberg, P. J. Less-Common Met. 1972, *26*, 335–353.
68. Goto, T.; Hirai, T. Mater. Res. Bull. 1987, *22*, 1195–1201.
69. Arunajatesan, S.; Carim, A.H. Materials Letters 1994, *20*, 319–324.
70. Faber, L.; Barsoum, M.W.; Zavaliangos, A.; El-Raghy, T. J. Am. Ceram. Soc. 1998, *8*, 1677–1681.

71. Morgiel, J.; Lis, J.; Pampuch, R. Materials Letters 1996, *27*, 85–89.
72. Brukl, C.E. Techn. Rep. No. AF ML TR-65-2 part II. US Department of Commerce: Springfield, Virginia, 1965; Vol. VII, 425.
73. Wakelkamp, W.J.J.; van Loo, F.J.J.; Metselaar, M. J. Eur. Ceram. Soc. 1991, *8*, 135.
74. Morozumi, S.; Pailler, R.; Lahaye, M.; Naslain, R. J. Mat. Sci 1994, *19*, 2749.
75. Tsurekawa, S.; Yoshinaga, H. J. Jap. Inst. Metals 1994, *58*, 390–396.
76. Zhao, Q.H.; Wu, J.; Chadda, A.K.; Chen, H.S.; Parsons, J.D.; Downham, D. J. Mater. Res. 1994, *9*, 2096–2101.
77. Chien, F.R.; Nutt, S.R.; Cummings, D. Phil. Mag. A 1993, *68*, 325–348.
78. Hollox, G.E.; Smallman, R.E. J. Appl. Phys. 1968, *37*, 818–823.
79. Venables, J. Phys. Stat. Sol. 1970, *15*, 413–416; Phil. Mag., 873–890; Metall. Trans. 1967, *1*, 2471–2476.
80. Clyne, T.W.; Withers, P.J. *An Introduction to Metal Matrix Composites*; Cambridge University Press: Cambridge, 1993.
81. Chawla, K.K. In *Materials Science and Technology*; Cahn, R.W., Haasen, P., Kramer, E.J., Eds.; VCH: Weinheim, 1993; Vol. 13.
82. Warren, R.; Andersson, C.H. Composites 1984, *15* (2), 101.
83. Martineau, P.; Pailler, R.; Lahaye, M.; Naslain, R. J. of Mat. Sci. 1984, *19*, 2749.
84. Reeves, A.J.; Taylor, R.; Clyne, T.W. Mat. Sci and Eng. 1991, *A141*, 129.
85. Pan, Y.; Baptista, J.L. J. Am. Ceram. Soc. 1996, *79* (8), 2017.
86. Lis, J.; Pampuch, R.; Piekarczyk, J.; Stobierski, L. Ceramics Int. 1993, *19*, 219.
87. Gottselig, B.; Gyarmati, E.; Naoumidis, A.; Nickel, H. J. of Eur. Cer. Soc. 1990, *6*, 153.
88. Lee, J.K.; Earme, Y.Y.; Aaronson, H.I.; Russel, K.C. Met. Trans. 1980, *11A*, 1837.
89. Arsenault, R.J.; Fisher, R.M. Scr. Metall. 1983, *17*, 67.
90. Brooksbank, D.; Andrews, K.W. J. of Iron and Steel Inst. 1972, *4*, 246.
91. Broek, D. Elementary Engineering Fracture Mechanics. Noordhoff Int. Publ: Leyden, 1987, *257*.
92. Zum Gahr, K.H. *Microstructure and Wear of Materials*; Elsevier: Amsterdam, 1987.
93. Hutchings, I.M. Mat. Sci. and Techn. 1994, *10* (6), 513.
94. Habig, K.H.; Broszeit, E.; de Gee, A.W.J. Wear 1981, *69*, 43.
95. Mens, J.W.M.; Huis in't Veld, A.J.; de Gee, A.W.J. internal report TNO Apeldoorn.
96. Barsoum, M.W. Scripta Mater. 2000, *43*, 284.
97. Goto, T.; Hirai, T. Mater. Res. Bull. 1987, *22*, 1195–1201.
98. Kooi, B.J.; De Hosson, J.Th.M. Scripta Mater. 2000, *287*.
99. de Mestral, F.; Thevenot, F. In *The Physics and Chemistry of Carbides, Nitrides and Borides*; Freer, R., Ed.; Kluwer: Dordrecht, 1989.
100. Pei, Y.T.; Ocelík, V.; De Hosson, J.Th.M. Acta Materialia 2002, *50*, 2035–2051.
101. Vreeling, J.A.; Ocelík, V.; De Hosson, J.Th.M. Acta Materialia 2002, *50*, 4913–4924.
102. Pei, Y.T.; De Hosson, J.Th.M. Acta Materialia 2000, *48*, 2617–2625.
103. Pei, Y.T.; De Hosson, J.Th.M. Acta Materialia 2001, *49*, 561–571.
104. De Hosson, J.Th.M.; de Haas, M.; Teeuw, D.H.J. In *Advances in Electronmicroscopy*; Valdre, U., Ed.; Kluwer, 1999; 109–134.
105. De Hosson, J.Th.M.; Teeuw, D.H.J. Nano-ceramic coatings produced by laser treatment. Surface Engineering 1999, *15*, 235–241.
106. De Hosson, J.Th.M.; Teeuw, D.H.J. In *Lasers in Surface Science and Engineering*; Dahotre, N.B., Ed.; ASM Surface Engineering Series; ASM International: Materials Park, OH, 1998; Vol. 1, 205–255 chapter 6.
107. Vreeling, J.A.; Ocelík, V.; Pei, Y.T.; van Agterveld, D.T.L.; De Hosson, J.Th.M. Acta Materialia 2000, *48*, 4225–4233.
108. De Hosson, J.Th.M.; Popma, R.; Hooijmans, J. Surface Engineering 2000, *16*, 245–249.

17

Nanometer-Scale Surface Modification Using Scanning Force Microscopy in Chemically Active Environments

J. Thomas Dickinson and Steve C. Langford
Washington State University, Pullman, Washington, U.S.A.

I. INTRODUCTION

Simultaneous application of tribological and chemical stimuli are often much more effective in removing material than either stimulus alone and relate to the general areas of tribo- and mechanochemistry. These effects can be exploited to produce extremely flat surfaces, as in the planarization of surfaces by chemical–mechanical polishing (CMP). Conversely, corrosive wear, environmentally induced crack growth, and related phenomena can dramatically shorten useful device lifetimes. Scanning force microscopy (SFM) is a particularly valuable tool for the study of this synergism, being able to both localize the tribological stimulation and to image the resulting wear with nanometer-scale resolution [1–3]. In many respects, the SFM tip can simulate a single asperity or abrasive particle interacting with nearly ideal substrates.

Material removal is often a relatively inefficient means of obtaining a smooth surface. Energetically, filling in small pits can be much more effective. An SFM tip can also be used to nucleate, accelerate, and control deposition along step edges on brushite surfaces in supersaturated aqueous solutions. We present strong evidence that the tip sweeps adsorbed ions to nucleation sites by assisting them over an Ehrlich–Schwoebel-type barrier parallel to the steps [4,5]. The resulting deposition provides a unique means of generating nanometer-scale structures, growing atomically flat surfaces, and controlling biomineralization. The SFM allows us to stimulate the surface with controlled scanning parameters and use the same tip to image the changes in a quantitative fashion.

The removal of submicron particles poses a severe challenge in the production of optical components (mirrors and lenses) and high-density integrated circuits. Adhesion of metal particles to semiconductor substrates is a significant issue, for instance [6–8]. Whole technologies for particle removal have been developed, including laser-assisted particle removal [9,10]. In integrated circuit manufacture, chemical–mechanical polishing serves this function [11]. These processes typically employ both a liquid phase to reduce adhesion and a mechanical stimulus to actually remove the particle. However, the mechanisms of adhesive force reduction and the mechanical details of removal are not well understood. For particle removal, we employ a model particle–substrate system composed of single-crystal NaCl grown on soda lime glass substrates. Water vapor is a potent chemical stimulus on these hydrophilic surfaces and has a strong effect on crack growth in glass itself [12]. In this work, we apply stress to adhering particles with an SFM tip instrumented to monitor both normal and lateral forces.

In the above work, the SFM tip is a tool to modify various surfaces. Clearly one expects the tip to be modified in a similar fashion under appropriate conditions. In the role of a model asperity, the tip experiences "asperity wear." We show that the wear of silicon nitride SFM tips is dominated by chemical–mechanical effects involving not only the solution but also the chemical nature of the substrate. In the present work, the only substrates associated with significant tip wear are characterized by high densities of metal-hydroxide bonds. Wear also requires an aqueous environment. Solution pH has little influence on tip wear rates. The role of water is assumed to replenish hydroxide bonds on the substrate surface that are consumed in mechanically mediated reactions with the tip.

Combined chemical and mechanical attack is most effective for material and particle removal in a variety of contexts. Together, they provide an especially effective "one-two punch" to surfaces that can be exploited to produce desired structural features or, conversely, atomically flat surfaces. The formation of submicron structures for micro/nano-fluidic, sensor, and microelectromechanical system (MEMS) devices are areas of possible application.

II. DISSOLUTION ALONG MONOLAYER STEPS IN CALCITE

Calcite is readily imaged by SFM in aqueous solutions [13]. The dissolution and growth of calcite crystals in aqueous solution has been previously studied by SFM [14–21] and by other means [22–26], because of its importance in mineral formation, global CO_2 exchange, and strong surface interactions with heavy metals in the environment. We have shown that dramatic corrosive wear can be induced by scanning the SFM tip back and forth across the edge of a naturally occurring etch pit at high contact forces [2]. Indenting the surface with an SFM tip near the edge of an etch pit also locally enhances dissolution along steps near the tip. We attribute these effects to increased rates of double-kink nucleation in the strain field of the SFM tip [2].

Etch pits on calcite typically form parallelograms bounded by two pairs of crystallo-graphically distinct steps. During steady-state dissolution, one pair moves much more rapidly than the other [18]. We designate these steps "fast steps" and "slow steps," respectively. Physically, the difference between fast and slow steps is related to the inclination of the plane of the carbonate ions relative to the sample surface—a steric effect. The rate-limiting step in pit growth is believed to be double-kink nucleation along the pit edges [18]; the weakly bound ions at the resulting kinks are readily incorporated into solution, resulting in rapid kink motion to the corners of the pit. Fast steps are more vulnerable to dissolution than slow steps because the CO_3^{2-} ions along fast steps are more exposed to the surrounding water. This lowers the activation energy for double-kink nucleation (ion removal) on fast steps. A similar model was employed by Hirth and Pound to describe evaporation from crystal surfaces [27].

Drawing the SFM tip back and forth across the edge of a monolayer deep etch pit creates a wear pattern that can be directly interpreted in terms of double-kink nucleation. The geometry of the experiment and resulting features are shown in Fig. 1. The path of the SFM tip during wear is marked by the white line. The maximum step displacement in the

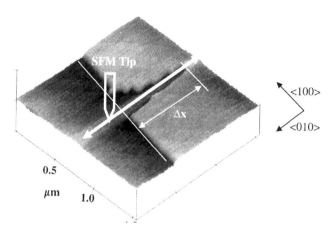

Figure 1 SFM image of a slow step on calcite after 3800 linear scans at a normal force F_N = 160 nN.

scanned region, Δx, was measured from the end of the "wear track" to the line defined by the unperturbed portions of the original step (as far as possible from the wear track). Similar treatments on flat terraces (away from steps) had no detectable effect on the surface. Likewise, repeated linear scans across cleavage steps on dry calcite surfaces (in ambient air) had no effect when imaged on this scale. This synergism between the corrosive environment *and* mechanical loading clearly marks this as an example of corrosive tribology.

The geometry of the wear track in Fig. 1 illustrates two important features of molecular-scale dissolution. The jogs along the edges of the wear track reflect the tendency of atomic-scale kinks to aggregate into larger structures. Furthermore, the two edges of the wear track show significantly different patterns of jogs, reflecting the different propagation behavior of crystallographically distinct kinks along these steps. These differences are also responsible for the marked contrast in dissolution along the two halves of the original step. Dissolution along the portion of the original step to the left of the wear track (accomplished by kinks moving in the $+\langle 100 \rangle$ direction) has been much more rapid than dissolution along the portion of the original step to the right of the wear track (accomplished by kinks moving in the $-\langle 100 \rangle$ direction). Analysis of material dissolution after wear track formation suggests that the jogs produced by wear are especially vulnerable to dissolution and play an important role in the planarization of stepped material.

This experimental geometry was chosen because the length of the wear track (Δx in Fig. 1) is quantitatively related to the rate of double-kink nucleation where the SFM tip crosses the step. Furthermore, the wear track growth rate ($\Delta x / \Delta t$) is a strong function of contact force. Plots of growth rate vs. contact force for two solution concentrations are shown in Fig. 2. These measurements employed 500×500 nm^2 scans acquired at a scan rate of 24 Hz (tip velocity of 12 μm/sec). The dark lines in Fig. 2 represent a least squares fit of the data to a model function described below (an exponential function of the surface stress). Growth rate measurements taken with different cantilevers show a high degree of consistency, as indicated by the data points represented by different symbols in Fig. 2(a). Each symbol represents measurements made on different days, with different calcite samples (from the same block) and cantilevers (from the same wafer).

Wear tracks across fast steps grow much faster than wear tracks across slow steps. At the low solution concentration employed in Fig. 2(a) (60 μM solution flowing at ~10 μL/

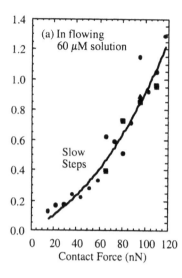
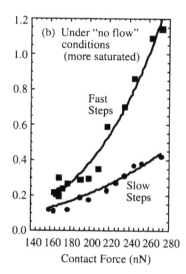

Figure 2 Wear track growth rate vs. contact force (a) under flowing, 60 µM solution and (b) in more saturated solution (obtained by turning off flow). The data points represented by different symbols were made on different calcite samples with different cantilevers. The dark line is a least squares fit of Eq. (2) (described below) to the data.

sec), wear tracks across fast steps grew too quickly for practical measurements. We therefore reduced the growth velocity by providing a more saturated solution, where a certain fraction of the nucleated double kinks are annihilated by redeposition of material from solution. Even at these higher solution concentrations, the SFM tip effectively mixes the nearby solution, preventing the development of concentration gradients that would complicate analysis.

The resulting wear track growth rates are shown in Fig. 2(b), where a high degree of saturation was provided by stopping the flow of solution for about 30 min. Under these conditions, a contact force of 270 nN is required to produce growth rates on slow steps comparable to 70 nN in flowing solution. Nevertheless, the strong dependence of growth rate on contact force confirms that reaction-limited conditions prevail—that is, wear track growth is not limited by concentration gradients. Over most of the range of contact forces probed in Fig. 2(b), wear tracks across fast steps grew at least twice as fast as wear tracks across slow steps. Thus fast steps are considerably more vulnerable to tribologically enhanced dissolution than slow steps.

We have observed similar, strong enhancements in dissolution during 2-D SFM imaging [2]. However, 2-D scanning nucleates kinks at many points along the step; mutual annihilation of kinks nucleated at different points along the step makes it difficult to infer the rate of kink nucleation from the step velocity. In contrast, kinks nucleated by linear scanning are formed along a narrow portion of the step and propagate away from this point in opposite directions. At the highest contact forces employed in this work, the average number of double kinks nucleated per scan is about 0.2, which corresponds to about 5 per second. This allows sufficient time for kinks nucleated along one row of ions to propagate away from the line scan before the next kink is nucleated, so that the next kink is nucleated along a new row. Then the total number of nucleated kinks can be estimated by dividing the length of the wear track, Δx, by the distance between ionic rows (3.2 Å).

Despite the high stresses applied by the SFM tip, we see no evidence for plastic deformation in this work or in SFM tip indentation experiments [2]. (We cannot rule out deformation-related features that do not survive long enough to be imaged—at least a few seconds.) The lowest contact forces employed in this work (15 nN) correspond to average compressive stresses of about 2 GPa; Vickers indentation (which employs millimeter-sized tips) at these stresses would produce an indent some tens of microns across [28]. We attribute the absence of SFM-induced deformation to the small size of the SFM tip (tip radius ~40 nm). Deformation is strongly hindered when the indentor is smaller than a typical slip band (usually ~1 μm) [29]. Deformation is also hindered at high strain rates [30], such as those associated with the motion of the tip across the surface. The absence of a threshold stress for the onset of enhanced dissolution also argues against dislocation emission and twinning as possible sources of double kinks. We conclude that plastic deformation does not contribute to mechanically enhanced dissolution under these conditions.

In terms of the available mechanical energy, the operation of mechanical effects is striking. At the highest contact forces employed (assuming elastic interactions only), the total work done by the tip on the substrate is less than 200 eV—and this mechanical energy is distributed over thousands of bonds. Furthermore, double-kink nucleation is not likely under the center of the tip, where the stresses are highest, because the compressive stresses there hinder the escape of solvated ions. The (albeit weaker) tensile stresses along the surface surrounding the SFM tip are much more likely to promote dissolution. Stress effects are strongly enhanced along steps, where surface ions are less constrained by surrounding ions. Ion displacement can play a key role in volume-activated processes, where the work done (force × displacement) can directly reduce the binding energy.

The magnitude of the stresses adjacent to the SFM tip can be estimated for the case of isotropic, elastic behavior. The maximum tensile stress involves the radial component, σ_r, along the circle where the SFM tip contacts the substrate. Double-kink nucleation will be enhanced over a modest range of distances from edge of tip contact (a few tip radii, from Saint Venant's principle), so that the relevant average stress will be somewhat lower. This suggests that mechanical enhancements are insensitive to variations in the geometry of the SFM tip itself but also introduces some uncertainty in the value of the (average) stress responsible for enhanced dissolution. The peak tensile stress given by the Hertz relation for an infinitely stiff, spherical tip is [31]:

$$\sigma_r = \frac{(1 - 2v)}{\pi} \left(\frac{2F_N E^2}{9(1 - v^2)^2 r^2} \right)^{1/3},\tag{1}$$

where E and v are Young's modulus and Poisson's ratio, respectively, for the substrate; F_N is normal force between the sphere and substrate; and r is the sphere radius. σ_r is relatively insensitive to errors in the contact force ($F_N^{1/3}$ dependence), and somewhat more sensitive to errors in the tip radius ($r^{-2/3}$). To avoid the complications of material anisotropy, we use technical moduli (directionally averaged values) appropriate for $CaCO_3$: $E = 81$ GPa, $v = 0.32$ [32]. For example, a tip radius of curvature $r = 40$ nm and an applied normal force of 100 nN yields a maximum radial stress of about 560 MPa.

The dependence of the wear track growth rate (expressed as a velocity, V) on contact force is readily modeled with a Zhurkov–Arrhenius expression [33].

$$V = V_o \exp\left[-\left(\frac{E_{act} - v^* \sigma}{kT} \right) \right] = V_o' \exp\left(\frac{v^* \sigma}{kT} \right),\tag{2}$$

where V_o is the appropriate preexponential, E_{act} is the zero stress activation energy for double-kink nucleation, and v^* is an activation volume. The best fit curve of Eq. (2) to the data of Fig. 2(a) (flowing solution), using the stresses given by Eq. (1), is shown by the dark line. The best fit values of the parameters correspond to $V_o' \approx 6 \pm 3$ pm/sec [where $V_o' = V_o \exp(-E_{act}/kT)$] and $v^* \approx 3.7 \pm 0.3 \times 10^{-29}$ m^3, respectively. This activation volume is slightly larger than the average volume per ion in the CaCO$_3$ lattice (3.3×10^{-29} m^3), making it reasonable to suppose that this activation volume corresponds to the displacement of one or perhaps two ions from a step site. (The CO$_3^{2-}$ ion is considerably larger than the Ca^{2+} ion.) We note that the displacement of ions in flat terrace sites is limited by the surrounding ions; this may explain why scratching does not nucleate new etch pits over the range of contact forces used here. The large size of the CO$_3^{2-}$ ion and the fact that a component C–O bond is directed into the solution would make it especially vulnerable to combined mechanical and chemical effects.

A similar analysis of the wear track growth rates at slow steps under "no flow" conditions [Fig. 2(b)] yields $v^* \approx 3.9 \pm 0.5 \times 10^{-29}$ m^3. The agreement with v^* under flowing solution is well within the numerical uncertainty of the curve fitting procedure. This agreement is consistent with the expectation that the degree of saturation controls the lifetime of nucleated kinks and not the (stress-enhanced) nucleation rate. The best fit value for v^* for dissolution along fast steps is somewhat larger, $v^* \approx 6.0 \pm 0.5 \times 10^{-29}$ m^3. A higher activation volume is consistent with the reduced steric constraint experienced by CO$_3^{2-}$ ions along fast steps, which would allow for larger displacements at a given stress and render them more vulnerable to water attack.

An estimate of the activation energy for double-kink nucleation at zero stress can be derived from Eq. (2), assuming that V_o corresponds to a typical "attempt frequency," f_o, multiplied by the step displacement per successful attempt (one lattice spacing: 3.2 Å). Setting f_o equal to typical vibrational frequencies (10^{13} sec^{-1}), the measured preexponential [V_o' from Fig. 2(a)] at $T = 293$ K corresponds to an activation energy of 0.8 ± 0.2 eV—a plausible value for reaction-limited nucleation. This analysis neglects the effects of possible redeposition reactions due to finite kink lifetimes. [Redeposition under "no flow" conditions is strongly affected by kink lifetimes. Therefore, this calculation uses the data taken in flowing solution from Fig. 2(a).] At the highest stresses employed in this work (corresponding to a normal force of 270 nN), this activation energy for double-kink nucleation is reduced by about 0.2 eV.

Thus, it is reasonable to interpret the observed step growth in terms of a thermally activated process. In addition, the Zhurkov–Arrhenius description of the observed stress dependence of the wear track growth also appears to be valid. We point out that crack velocities in environmental crack growth [34] often display a similar dependence on stress intensity, although in this case the relation between stress intensity and stress at the crack tip can be very difficult to deduce.

III. DISSOLUTION ALONG MONOLAYER STEPS IN BRUSHITE

The calcium phosphates are important biological minerals, occurring in both normal (e.g., enamel, dentine, bone) and pathological (e.g., dental cavities, kidney stones) calcifications. In weak acid solution (pH 4–5), equilibrium conditions favor the formation of brushite (monoclinic calcium hydrogen phosphate dihydrate—CaHPO$_4 \cdot$H$_2$O). Because urine often displays a pH in this range, kidney stones are often formed of this material. Brushite is also used as an abrasive in toothpaste and as an intermediate in phosphate fertilizer production. Stress-induced surface modification in brushite therefore has potentially important medical and commercial implications. It also provides another model inorganic material for the study of planarization mechanisms, where we can generate atomically smooth surfaces, of interest to the important technological area of CMP.

Single-crystal brushite forms crystalline plates with broad {010} surfaces. In undersaturated solutions, these surfaces develop triangular etch pits with three crystallographically distinct steps bounded by steps along the [101], [201], and ~[001] directions [35]. These directions correspond to rows of strong ionic bonds with contrasting bonding environments. This variability makes brushite a valuable system for testing models of corrosion and corrosive wear.

Step velocity measurements made at very low contact forces reflect the step stabilities against dissolution [36]. These measurements indicate that the [201] steps are most vulnerable to dissolution and that the [001] steps are significantly less vulnerable. No dissolution along [101] steps was observed on timescales of tens of minutes. The [101] steps are extremely resistant to dissolution at low contact forces.

SFM images obtained immediately after high- contact-force linear scans across two different monolayer etch pits on brushite are shown in Fig. 3. As in the case of calcite, crystallographically distinct steps show contrasting resistance to wear. From Fig. 3, wear across [101] steps is much faster than wear across [201] steps. Similar measurements on [001] steps show much less wear. In brushite, the steps that are most vulnerable to dissolution, the [201], are not the most vulnerable to wear. This contrasts with calcite, where fast steps are more vulnerable than slow steps in both dissolution and wear experiments.

Like calcite, the two edges of the wear tracks on brushite dissolve at different rates. The situation in brushite is especially severe, however. Wear tracks aligned along [001] directions (as in Fig. 3) show especially dramatic contrasts: one edge of the wear track (designated the +[001]) is quite stable, while the other literally falls apart into segments made up of the other two stable steps. Significant differences in dissolution are also observed along the original steps adjacent to the wear track. For instance, much more material has been removed from the [201] steps in Fig. 3 below the wear track than from this same step above the wear track. We attribute this to the different propagation velocities of crystallographically distinct kinks along these steps.

Wear track growth for brushite can be analyzed in much the same fashion as for calcite. The wear track growth rates for all three distinct steps are plotted as a function of

(a) $F_N = 80$ nN   (b) $F_N = 135$ nN

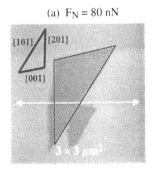
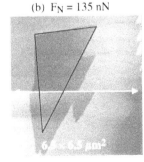

Figure 3 SFM images of triangular etch pits on brushite immediately after linear scanning along the white line to form wear tracks. The orientation of the steps along the edges of the pit are indicated in the inset of (a). The black outline shows approximate location of the original etch pit. Note the contrasting sizes of the images in (a) and (b).

contact force in Fig. 4(a). At all contact forces yielding measurable wear tracks, the wear track growth rates were highest along the [101] and lowest along the [001] steps. As noted above, [201] steps are more vulnerable to zero-force dissolution than [101] steps. Indeed, when the model curves in Fig. 4(a) are extrapolated to zero, the curve for [101] steps drops below the curve for [201] steps. In Eq. (2), this corresponds to a lower prefactor for [101] steps; as the contact force is raised, the higher v^* for the [101] steps ensures that the [101] growth rate overtakes the [201] growth rate.

The remarkable contrast between the stabilities of $+[001]$ and $-[001]$ steps may account for the presence of triangular etch pits in this material. Most ionic materials (including gypsum, $CaSO_4 \cdot 2H_2O$, which has an analogous structure yet lacks the HPO_4^{2-} hydrogen bonds), display four-sided etch pits, composed of the two most stable steps. The triangular steps in brushite would be accounted for if the step stabilities were ranked according to

$$+[001] \gg \pm[201], \ \pm[101] \gg -[001]. \tag{3}$$

We attribute the exceptional stability of $+[001]$ steps relative to $-[001]$ steps to the effect of the HPO_4^{2-} hydrogen bond. The [001] steps run parallel to the HPO_4^{2-} rows that make up the anion sublattice. Assuming that the HPO_4^{2-} ions lie along the step edges (which is physically reasonable), $+/-[001]$ steps would have the structure shown in Fig. 4(b). This orientation was confirmed by high-resolution SFM images, which exploit the relative positions of the uppermost HPO_4^{2-} oxygen ions to positively orient the lattice relative to the observed steps. The hydrogen bonds are localized on the $+[001]$ side of the HPO_4^{2-} ions, reducing the effective negative charge on the $+c$ side of these ions. This reduced charge would weaken the water-anion interaction. Because the $+c$ sides of the HPO_4^{2-} ions face the solvent on $+[001]$ steps, hydrogen bonding would stabilize these steps. Conversely, the $-[001]$ side of the HPO_4^{2-} ions lacks the hydrogen bond, so that $-[001]$ steps present greater negative charge to the solvent and thus interact more strongly with water. Neutron diffraction studies confirm the presence of long-range order in the hydrogen bonding in brushite [37], as required if hydrogen bonding were to account for a consistent difference in step stability.

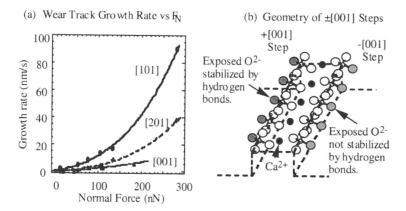

Figure 4 (a) Wear track growth rate vs. contact force in undersaturated solution for [101], [201], and [001] steps on single-crystal brushite. (b) Geometry of $\pm$ [001] steps, showing how the $+$[001] steps can be stabilized by hydrogen bonds.

A semiquantitative comparison between the expected and observed step stabilities is presented in Table 1. The effective bond strength was estimated by assuming that all ionic bonds in the lattice have the same strength and by counting the number of bonds per unit area along each of the observed step directions. This parameter correlates with the step dissolution rates on many ionic surfaces. For brushite, the correlation between bond strength and the zero-stress (low contact force) dissolution rates is poor. A much better correlation is obtained with the number of "missing bonds" per ion, i.e., the number of ionic bonds per ion that have been disrupted by creating the step. Missing bonds reflect uncompensated charge (similar to dangling bonds on covalent materials), which renders a step vulnerable to attack by water. This parameter accounts for the observed step stability if we also factor in the stabilizing effect of hydrogen bonds on the observed $+[001]$ steps. On more symmetric ionic crystals, each of the observed steps usually display the same number of missing bonds per ion. On such materials, second-order effects, such as effective bond strength, should dominate the stability ranking.

The effective bond strength also correlates poorly with the activation volumes determined by fitting the wear track growth rate data of Fig. 4(a) to Eq. (2). (Steps with high activation volumes are more sensitive to stress effects.) In the case of calcite, the difference in activation volume appears to be principally steric in nature: the CO_3^{2-} ions along one pair of ions are more free to move in response to an applied stress. In brushite, however, the most vulnerable steps, [101], are also the most constrained, with five strong ionic bonds holding each HPO_4^{2-} ion in place.

The most likely explanation for the variation of activation volume among the three brushite steps relates to the elastic anisotropy of the lattice. Because the SFM tip imposes a roughly uniform displacement on the surface ions, the mechanical energy delivered to the various bonds will vary in proportion to the bond stiffness. In this sense, the surface is analogous to a 2-D array of springs (bed springs), with stiff springs along one direction and soft springs along the other. Deforming the surface stretches both sets of springs, but the stiff springs resist the deformation more, resulting in higher stresses along the stiff direction. Although Eq. (2) assumes isotropic material behavior, this difference can be numerically accounted for by assigning higher activation volumes to steps lying along stiff directions.

The brushite lattice displays rows of strong, ionic, double bonds along the [101] direction, suggesting that the lattice stiffness is indeed maximum along the [101] direction ($v^* \sim 62$ Å^3). The activation volumes for the other two steps are lower in rough proportion to their inclination to the [101] direction. The [201] steps intersect the [101] steps at a 31° angle ($v^* \sim 48$ Å^3), whereas the [001] steps intersect the [101] steps at a 51° angle ($v^* \sim 41$ Å^3). Ignoring steric effects, a 35% variation in principal tensile stress as one moves from the [101] to the [001] direction would account for the observed variation in activation volume. This

Table 1 Step Stability in the CaHPO$_4$ Sheets of Brushite {010} Planes

	Predicted values		Measured values	
Step	Bond strength, bond/Å^2	Missing bonds per ion	V_{step} (zero stress), nm/sec	Activation volume (Å^3)
[101]	0.100	1.0	<0.1	62
[201]	0.121	1.5	1.5	48
[001]	0.067	1.5	0.5	41

effect would not apply to calcite, where the lattice stiffness along the two observed steps (fast and slow) are equal.

The case of brushite illuminates two factors that strongly affect the material response to combined chemical and mechanical attack: mechanical stiffness and the details of bonding, including possible hydrogen bonds. The importance of these factors is easily overlooked in more homogenous materials.

IV. DEPOSITION ALONG MONOLAYER STEPS IN BRUSHITE

One has to ask what happens to these surfaces under conditions of supersaturation? Are there any tip-induced effects? Indeed yes. We now show that atomic layer growth can be induced at step edges by simple scanning over the step edge [38].

Figure 5 shows several etch pits in supersaturated solution. The degree of supersaturation, σ, is in multiples of the concentration of nominally saturated solution; $\sigma = 5$ corresponds to Ca^{2+} and HPO_4^{2-} concentrations five times that of a nominally saturated solution. Even at these supersaturations, etch pits tend to grow (material locally dissolves) unless they are manipulated with sufficient force by the SFM tip. In more dilute solutions ($\sigma \sim 1$), dissolution produces etch pits with very sharp corners. However, at higher supersaturations, the corner joining [201] and [101] steps becomes rounded, as shown in Fig. 5(a) and (b) at $\sigma = 4$. This rounding reflects the influence of solution chemistry on the dynamics of ion removal from these two steps. Despite the fact that material is removed from the interior of the etch pits (material is dissolving), localized nucleation and growth of 3-D

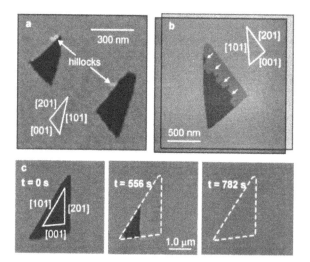

Figure 5 (a) SFM image of two spontaneously dissolving etch pits in a (0$\underline{1}$0) brushite surface exposed to supersaturated solution. 3-D hillock growth nucleates along the top edges of [201] steps. These hillocks are 0.8 nm high and are easily redissolved by local scanning with the SFM tip. (b) Fingering growth normal to the [201] step of a single atomic layer deep etch pit during 37 scans at a normal force of 5 nN and scan speed of 70 μm/sec. We have superimposed images acquired before and after growth, where the darker triangle is the original pit, and new single atomic layer growth is shown in a lighter shade. (c) A sequence of SFM images of a one atomic layer deep etch pit in a brushite (010) surface scanned a total of 23 times at a tip speed of 70 μm/sec and normal force of 5 nN.

hillocks is often observed along the top terrace edges of [201] steps, as shown in Fig. 5(a). Because the most favorable sites for ion attachment are actually on the lower terrace due to higher coordination, on-top nucleation indicates the presence of an Ehrlich–Schwoebel type barrier [4,5] that hinders ion diffusion from the top terrace down to the lower terrace.

Although spontaneous pit dissolution is generally observed at these supersaturations, localized, directional growth is readily induced by continuously scanning over an etch pit at low normal forces (F_N < 50 nN). Growth produced by 23 scans at F_N = 5 nN and σ = 5 is shown in the sequence of images in Fig. 5(c), eventually filling in the pit and producing an atomically flat surface. Growth is predominately along the [201] step and propagates normal to this step. Larger images taken before and after the images in Fig. 5(c) show that pits outside the continuously scanned area continue to dissolve even as the etch pit inside the continuously scanned area grows smaller. Thus we conclude that the observed deposition is "tip induced." This behavior (dissolution and tip-induced growth at the same time) is observed over a wide range of supersaturations (1 < σ < 6). Scanning at F_N > 50 nN immediately produces localized wear (dissolution) rather than growth as previously described [36].

At high saturations, growth along the [201] step induced by continuous scanning often shows a fingering instability. This is seen in the composite of two images in Fig. 5(b), one acquired before and one after 37 scans at F_N = 5 nN and a tip speed of 70 μm/sec. Despite this instability, continued scanning fills in the entire pit and leaves no visible defects in either the topographic (constant contact force) image or in the lateral force images, although point defects such as vacancies would not be detectable in this mode of imaging.

More localized growth can be induced by drawing the SFM tip back and forth along a line normal to the [201] step. Figure 6(a) shows a "single finger" growth feature generated by repeated linear scanning along the white line at a frequency of 2 Hz. The width of the resulting deposit is ~120 nm, several times wider than the region actually contacted by the tip. Assuming elastic contact, the width of the strip contacted by the moving tip is only

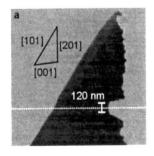

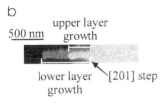

Figure 6 (a) A single atomic layer deposit ~120 nm wide growing normal to the [201] step produced by repeated linear scanning normal to the step. (b) Deposition in a two atomic layer deep pit showing that the lower layer grows faster than the upper layer.

~20 nm wide. Figure 6(b) shows "single finger" growth formed by scanning across the [201] edge of a two atomic layer deep pit. The finger on the lower atomic layer is slightly wider and grew at approximately twice the speed of the finger on the upper layer. Growth rates vary with pit size (smaller pits shrink faster) and the degree of supersaturation, with higher σ yielding higher growth rates. At $\sigma = 5$, we have measured growth rates during linear scanning as high as 7 nm/sec, corresponding to the deposition of ~3 rows of calcium phosphate ions per pass of the SFM tip over the step. We were unable to induce growth at [001] steps; [101] steps showed slow growth at rates <5% of the growth rate on [201] steps.

In the dissolution work described above [3,39], the material removal rate increases continuously with increasing stress, even at very low stresses. Therefore, stress applied at or near the step edge should hinder rather than favor deposition. However, the presence of an Ehrlich–Schwoebel barrier for motion of ions down the step suggests a possible mechanism for tip-induced deposition. In supersaturated solutions, one expects the nucleation and disappearance of transient, subcritical, 3-D clusters on the terraces [40]. If the tip can detach and sweep ions from these clusters over the step, this would increase the concentration of adsorbed ions on the bottom terrace at the step-edge and promote deposition. The growth pattern in Fig. 6(b) suggests that most of the swept material accumulates on the lower of the two terraces. Moving ions to the lowest terrace is of highest probability and would favor faster growth of the lowest finger. Diffusion along the bottom terrace would account for the broad (120 nm wide relative to the ~20-nm tip contact diameter) patches of deposited material in the pits near the linear scans in Fig. 6(a) and (b). The high growth rates along the [201] steps may be due to the zigzag rows of alternating Ca^{2+} and HPO_4^{2-} ions expected along these steps to maintain charge neutrality. This zigzag structure would provide high binding energy sites for the nucleation of new ion rows. New ion rows are much more difficult to nucleate along the other two steps.

To examine the possible role of adsorbed material swept from the upper terraces, we performed back and forth linear scans of exactly the same length (1.5 µm), normal force (10 nN), scan frequency (1.0 Hz), and number of scans (32) at the same supersaturation ($\sigma = 4.3$), but changed the fraction of the scan taking place on the top terrace. The length of the growing [201] deposit was recorded during each linear scan, and the growth rate was determined by differentiation. Figure 7 shows the measured growth rates for 10 experiments along a single [201] step, where we alternated between a 10:1 ratio of scan length on the top terrace to scan length on the bottom terrace (dark bars) and a 1:10 ratio (light bars). When the majority of the scan took place on the top terrace, the growth rates were 2–3 times higher than when the majority of the scan took place on the bottom terrace. Very little material is swept into the pit when most of the scan is confined to the bottom terrace, but scanning on the top terrace sweeps more adsorbed material over the Ehrlich–Schwoebel barrier into the pit, where it becomes available for deposition. Note that the growth rates for the 10:1 scans consistently decrease with time (the top of each bar slopes down), consistent with the depletion of subcritical, adsorbed clusters along the upper terrace. In contrast, the growth rates for the 1:10 scans consistently increase with time (the top of each bar slopes up). This suggests that on the bottom terrace near the step the adion concentration is initially low; as the 1:10 scans move ions into this depleted region the growth rate increases. (A suggested and plausible ion movement mechanism was suggested by a reviewer, namely, tip-enhanced diffusion of ions to and over the step edge.)

Although transient clusters of subcritical radii were not observed in any SFM images, the fact that at high σ and long exposure times we eventually observe 3-D nucleation on atomically flat terraces and away from step edges strongly supports their existence. We have acquired indirect evidence for their presence by examining small fluctuations ("noise") in the lateral twist of the SFM cantilever [41]. (The lateral twist of the cantilever is highly

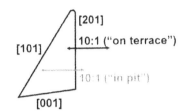

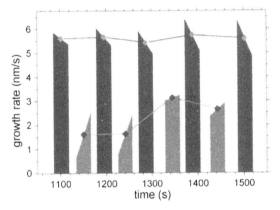

Figure 7 Comparison of growth rates along the [201] edge of a single etch pit during scanning primarily on the top terrace (10:1) (dark bars) vs. primarily on the bottom terrace (1:10) (lighter bars). The top of each bar shows the growth velocity vs. time during that particular set of scans. The points joined by lines join the average growth rates for each set of scans.

sensitive to surface roughness.) These fluctuations are much stronger during scanning in highly supersaturated solution ($\sigma = 5$) than during scanning in pure water (no cluster formation). Figure 8(a) shows power spectra of the fluctuations averaged over ten 1-μm linear scans along the crystal's c-axis at 1.25 μm/sec and $F_N = 5$ nN in supersaturated solution and in pure water. Both power spectra in Fig. 8(a) show peaks near ~2 and 4 kHz, but their amplitudes in supersaturated solution are higher by a factor of 2. The spatial frequency corresponding to the larger 2 kHz peak (given by the scan speed/frequency) is 0.63 nm. This corresponds well with the distance between the uppermost phosphate oxygen ions (which interact most strongly with tip asperities) along the c-axis of the unreconstructed (010) surface of 0.62 nm [37,42]. A schematic of an unrelaxed brushite cleavage plane (010) is shown in Fig. 8(b). We propose that this peak is due to a periodic rocking motion induced in the tip as asperities interact collectively with successive rows of oxygen ions. The second harmonic arises due to a nonlinearity in the cantilever response.

A related mechanism accounts for many images showing atomic-scale periodic structures (not true atomic resolution) with relatively large radii SFM tips acquired in contact mode on graphite [43], brushite [39], calcite (another crystal with protruding oxygen ions), [13,15,20] and other single crystals. The resonant frequency of the lateral twist of our cantilevers (estimated from the properties of Si_3N_4 and the dimensions of our cantilevers) is ~40 kHz, well above the observed spectral features. We propose that the increased amplitude of cantilever rocking at high saturations is caused by tip–cluster collisions and therefore serves as evidence of transient clusters on the terraces. These collisions may increase normal mode motion, which enhances the lateral deflection due to stronger interactions with the lattice. Numerous noise measurements in both pure water and

a

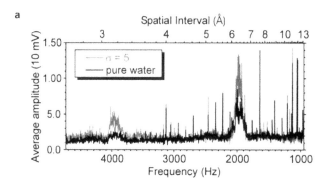

b

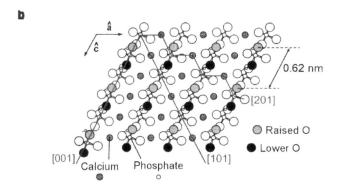

Figure 8 Characterization of fluctuations during scanning in pure water and in supersaturated solution on the (010) surface of brushite. (a) Power spectra of fluctuations in the lateral displacement of the SFM cantilever during scanning along the crystal c-axis. We attribute the higher amplitudes during scanning in supersaturated solution to increased periodic stimulation of the tip in the presence of small crystallites on the terrace. (b) Power spectra of fluctuations in the lateral displacement during scanning at $6°$ relative to the c-axis (perpendicular to the [201] steps. Although tip speeds were slightly different during acquisition of the data in (a) and (b) we can compare the spatial frequencies. (c) Schematic of the (010) surface of brushite.

saturated solutions at various scanning directions all show comparable increases in amplitude for the saturated solutions, supporting our hypothesis. Similarly, scanning atomically flat brushite terraces vs. regions with high densities of single atomic layer steps show precisely the same response: an increase in noise amplitude for the latter surfaces.

V. SALT PARTICLE REMOVAL FROM SODA LIME GLASS

When an SFM tip encounters a sufficiently large particle, the lateral force applied by the tip can produce significant shear stresses. If this stress is great enough, the particle can be literally fractured from the substrate. However, the required stress can be prohibitive in the absence of a chemically active ambient environment. In this work, we study a model particle substrate system involving salt particles bonded to soda lime glass. Modest partial pressures of water vapor dramatically lower the lateral force required to fracture the salt–glass bond as the SFM tip is drawn across the particle. Particle size also affects the interfacial shear strength, presumably due to variations in the size of interfacial flaws relative to the total interface area.

Submicron-sized NaCl crystals were deposited on soda lime glass substrates by dissolving a few grains (~1 mm^3) of commercial salt in a drop of deionized water on a clean microscope slide. The solution was spread across the slide with a cotton swab and allowed to evaporate to dryness. Both evaporation and sample storage were under ambient laboratory atmosphere conditions—typically 20–40% relative humidity (RH).

Particle observation and manipulation were performed on the stage of a scanning force microscope mounted in a closed chamber. The humidity was adjusted by introducing a controlled mixture of dry and humidified air. The RH in the chamber was continuously monitored with a BioForce Laboratory humidity sensor. Humidity variations in the course of an experiment were controlled to ± 1% absolute RH, with an estimated uncertainty of ± 2% absolute RH. This work employed triangular, 115-µm-long, "wide" Si$_3$N$_4$ cantilevers from Digital Instruments of Santa Barbara, CA.

Images taken before and after particle detachment are shown in Fig. 9 [44]. A suitable particle was located in a high scan rate (typically 42 µm/sec), low contact force image (5–20 nN). After zooming in on the chosen particle [particle A in Fig. 9(a)], the scan rate was reduced to 0.20 µm/sec. Scanning was continued until the tip was positioned to cross the center of the chosen particle [the white line in Fig. 9(b)]; then the contact force raised to a high value (≤320 nN).

After a single, high contact force sweep across the particle, the contact force was lowered and the scan rate increased to normal values for imaging. Finally, large-area scanning was resumed to search for the detached particle, marked A′ in Fig. 6(b). At higher relative humidities, locating the detached particle was often difficult. Circumstantial evidence suggests that the absorbed water film can bind the detached particle to the SFM tip, which then drags the particle across the surface. Sometimes, the detached particle was found adhering to another particle in the image.

The lateral force signal during a typical scan prior to particle detachment at 10% RH is shown in Fig. 10(a). This scan was acquired at a low normal force (10 nN) and high scan rate (42 µm/sec). The lateral force as the tip passes over the salt particle is significantly lower than the lateral force as the tip passes along the glass, reflecting salt's lower coefficient of friction. At 3% RH, the ratio of the lateral force measured along the salt to that measured along the glass ranges from 0.20 to 0.25. As the RH is increased to 68%, this ratio increases to 0.7. This increase is presumably related to the high affinity of water for NaCl; at high RH, capillary forces may contribute to the measured lateral force along the salt.

A typical lateral force signal during the linear scan that detaches a particle is shown in Fig. 10(b). Immediately prior to this scan, the scan rate was decreased from 42 to 0.20 µm/

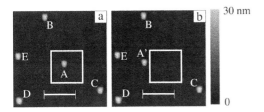

Figure 9 Low contact force images of NaCl particles (a) before particle detachment, and (b) after particle detachment. The line in (b) shows the orientation of the linear scan used to detach particle a, which was subsequently found at position A′.

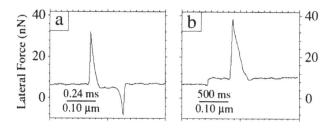

Figure 10 Lateral force signals during (a) a low contact force scan while aligning the tip on the particle, and (b) the slow, high contact force scan used to detach the particle from the substrate. In (b), the stepwise increase in lateral force coincides with the increase in normal force (not shown), and the sharp peak corresponds to the completion of crack growth along the particle–substrate interface.

sec. About a quarter of the way through the scan, the contact force was raised to 57 nN, producing a stepwise increase in lateral force. Detachment coincides with the peak in the lateral force signal. Tip height measurements during particle contact (not shown) indicate that the tip lifts slightly (~4 nm) up onto the edge of the particle. Because the tip is not in contact with the glass at the moment of detachment, the frictional force between the tip and the glass does not contribute to the lateral force at detachment. Thus the entire measured lateral force at detachment is applied to the particle, and it is this force that induces detachment.

An upper bound on the energy per unit area required to fracture the interface is provided by the area under the rising portion of the lateral force vs. displacement plot before failure. In Fig. 10(b), this amounts to about 0.05 J/m^2. In the fracture of more macroscopic, highly brittle samples (including soda lime glass and NaCl), it is not uncommon for two-thirds of this energy to be dissipated via plastic deformation and similar processes. Because plastic deformation is strongly hindered in nanometer-scale systems [29], its role in the removal of nanometer-scale particles is an open question.

The lateral force required for particle detachment can be reduced considerably by scanning even more slowly across the particle. Therefore, detachment is not merely due to the application of some critical stress greater than the intrinsic interfacial strength. We attribute failure under these conditions to chemically enhanced growth of interfacial cracks at stresses below the ultimate interfacial strength. Detachment occurs when the crack velocity is high enough to progress along the entire interface during the brief duration of tip-particle contact, typically ~35 msec. Relative to other fracture processes, chemically enhanced crack growth is typically quite slow (crack velocities $< 10^{-4}$ m/sec). The very small particle sizes in this work allow for interfacial fracture on timescales of tens of milliseconds, even at these low crack speeds.

Dividing the peak lateral force measured during detachment by the particle area, A, yields the nominal shear strength of the interface, σ_c. Nominal shear strengths determined for a large number of particles are displayed in Fig. 11(a) as a function of particle area for several relative humidities. We have scaled the failure stress for 3% RH (reduced by a factor of 10) for presentation purposes; the smallest particle at 3% RH failed at $\sigma_c = 55$ MPa! Raising the humidity from 3% to 11% dramatically lowers the interfacial shear strength. Increasing the particle size also tends to reduce σ_c. Both effects impose experimental limits on the size of particles amenable to study at the lowest humidities, where the SFM tip may break before a large particle will be detached. This is consistent with anecdotal reports of the difficulty of removing small particles under dry conditions. Importantly, increasing the

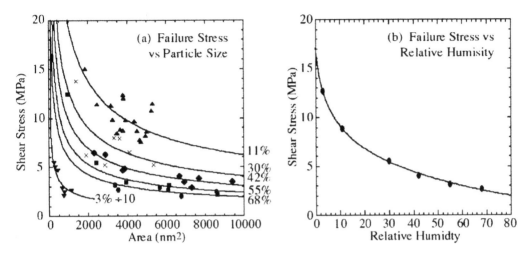

Figure 11   Nominal shear stress at failure as a function of (a) particle/glass contact area at relative humidities ranging from 3% to 68%, and (b) relative humidity for a fixed particle size of ~5000 nm^2. The dark lines in (a) represent one-parameter, least squares fits to the data of the form $\sigma_c \sim A^{-1/2}$. The point for 3% RH in (b) has been extrapolated from the data, due to the difficulty of detaching particles of this size at low humidities.

humidity beyond 50% has little additional effect on the shear stress at failure. Similarly, the particle-size dependence becomes weak for particles with contact areas larger than about 3000 nm^2.

Figure 11(b) shows the nominal shear strength as a function of humidity for a set of particles with contact areas of ~5000 nm^2. The failure stress drops rapidly with increasing RH at the lower humidities, and falls more gradually at higher humidities. The dark line represents a least squares fit of the data to a model based on the expected dependence of interfacial binding energy on humidity, described below.

At high relative humidities, particle detachment can be quite efficient. Figure 12 shows an attempt to sweep an 8×8 μm^2 area free of particles with a single scan. The resulting debris is piled around the edges. This image illustrates the effectiveness of combined chemical and mechanical stimuli in particle removal.

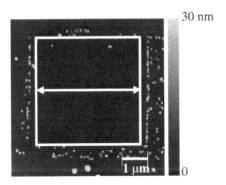

Figure 12   An SFM image of a region of the substrate cleared of adhering NaCl particles at 55% RH and a contact force of 51 nN.

To provide a framework for interpreting these results, we adopt several concepts from work on chemically enhanced crack growth. Although these chemical enhancements are not completely understood, studies by Wiederhorn, Michalske, Freiman, Bunker, Lawn, and others have advanced the field considerably. This work is reviewed by Lawn [12], and for the purposes of discussion, we adopt a similar chemical approach. Here, crack growth is viewed as a series of molecular-scale jumps of length a_o at frequencies given by the product of an "attempt frequency" f_o (a typical vibrational frequency) and a probability factor (the fraction of attempts yielding broken bonds). Ignoring the reverse process (crack healing), classical kinetic theory yields:

$$V = a_o f_o \exp(-\Delta F/kT),\tag{4}$$

where V is the crack speed, k is Boltzmann's constant, T is the temperature, and ΔF is the free energy change associated with fracture in a chemically active environment.

The effect of applied stress on ΔF is conveniently described in terms of the stress intensity factor, K, where ΔF is proportional to K. We note that this choice is somewhat controversial. ΔF is often taken to be proportional to the strain energy release rate, G, where $K = (GE)^{1/2}$ and E is Young's modulus. We adopt the stress intensity description because it provides a better description of our data. The difference between G and K is analogous to the difference between stress- and volume-activated processes during phase changes. The work on stress-enhanced dissolution described above required an analogous choice; there, volume-activated models described the data better than stress-activated models, consistent with the use of K here [3,36].

In general, K is a complex function of sample geometry, including crack length, and is generally treated numerically. Nevertheless, K is expected to scale as

$$K \sim d\sigma_{xy}c^{1/2}\tag{5}$$

where σ_{xy} is the nominal shear stress applied to the particle (lateral force divided by particle/substrate contact area), c is the crack length, and d is a constant of order 3 [45]. A complication in our loading scheme is the effect of the normal (compressive) stress exerted by the SFM tip on the adhering particles, σ_{yy}. This stress tends to close any existing crack and effectively reduces K. For the purpose of analysis, we assume this effect is independent of particle size.

Although we do not measure crack velocity per se, failure during the rather short tip–particle contact (typically 35 msec) requires a minimum crack velocity and thus a minimum K. The exponential dependence of crack velocity on ΔF (and thus σ_{xy}) in Eq. (4) ensures that the majority of crack growth will occur when σ_{xy} is close to the peak shear stress, σ_c. For the purposes of analysis, we replace σ_{xy} with σ_c. Furthermore, we assume that all particles of a given size (parameterized by A) have similar initial flaws, which we treat as initial "starter" cracks of length c in Eq. (5). Under these conditions, we expect σ_c to be proportional to $c^{-1/2}$.

Plots of σ_c as a function of the nominal particle–substrate contact area A are shown in Fig. 11(a) for several values of relative humidity. The data were initially modeled with a two-parameter curve fit ($\sigma_c \sim A^{-n}$) for each value of relative humidity. Each fit yielded $n = 0.5$ to within the uncertainty of the curve fitting procedure (typically ± 0.1). For simplicity, the curve fits in Fig. 8(a) represent one parameter least squares fits with $n \equiv 0.5$, i.e., $\sigma_c \sim A^{-1/2}$. Since we expect $\sigma_c \sim c^{-1/2}$, this dependence suggests that the initial flaw size is proportional to particle area ($c \sim A$). Interfacial flaws are expected to serve as starter cracks, and may be responsible for the particle-size dependence of the shear stress required for detachment.

To model the effect of humidity on failure, we assume that the energy required for crack growth is lowered by an amount equal to the net decrease in surface energy due to water sorption. Submonolayer coverages can often be described by the Langmuir isotherm, for which $\Delta F = 2\Gamma_m \ln(p/p_o)$, where Γ_m is the adsorption energy per unit area for full (monolayer) coverage, and p is the partial pressure of the adsorbate in the surrounding atmosphere (proportional to relative humidity). Assuming that interfacial fracture on experimental time scales occurs for cracks that reach a critical speed, V_c, and again replacing σ_{xy} with σ_c, the nominal shear stress at failure for particles of a given contact area will scale as:

$$\sigma_c = \gamma \ln[\delta/(1 + p/p_o)] \tag{6}$$

where γ, δ, and p_o are fit parameters. A fit of Eq. (6) to experimental data for cubes with contact areas of 5000 nm^2 appears in Fig. 11(b). The model describes the data quite adequately.

The application of the Langmuir isotherm at high relative humidities (several monolayers adsorbed) is somewhat strained. Nevertheless, it may be adequate for hydrophilic substances such as salt and glass. In these materials, the water/solid bonds are much stronger than water/water bonds, so that the total energy of adsorption is dominated by the adsorption energy of the first monolayer.

Several important avenues for future investigation present themselves. Experiments with better-characterized substrates (e.g., single-crystal silicon or noble metals) would allow for direct comparisons between experimental data and theoretical estimates of interfacial binding energies. Independent characterization of water adsorption isotherms would also facilitate detailed comparisons between theory and experiment. By changing the substrate, one can explore the effect of water vapor adsorption energy; this should strongly affect adhesion at low relative humidities, where substrate–vapor interactions are especially important. Equation (4) also suggests that particle removal can be a strong function of temperature. Measurements on variable temperature SFM stages would be of considerable interest.

VI. Si₃N₄ TIP WEAR DURING SCANNING ON REACTIVE SURFACES

Silicon nitride ceramics are hard, inert, and stable at high temperatures, making them attractive for use in extreme environments. However, conventional polishing and grinding operations often produce surface defects and cracks that lead to premature component failure. Tribochemical or chemical–mechanical polishing [46] can produce very smooth, defect-free, silicon nitride surfaces, but this process is not well understood. Hydrodynamic lubrication, which results in low friction and almost no wear, is observed with sufficiently smooth silicon nitride surfaces [47–49] and has potential applications in high-performance ceramic bearings. Wear of silicon nitride and silicon oxide is also of great interest in chemical–mechanical polishing of semiconductors for integrated circuits.

Geometrically, SFM wear experiments are analogous to the traditional pin-on-disk or ball-on-disk wear experiments [50–52]. Wear of silicon nitride SFM tips has been previously studied [53,54] and attributed to adhesive wear, plastic deformation, and low cycle fatigue fracture. Wear of silicon SFM tips can depend on the ambient solution [55,56], suggesting that chemical effects can be important.

In the work described here, tip wear was induced by scanning $3 \times 3\ \mu m^2$ patterns in a raster mode at a tip velocity of 16 $\mu m/sec$ on a variety of substrates. These scanning conditions correspond to a sliding distance of 1.5 mm per scan; the largest sliding distance reported here is 67.5 mm. Because of the important role of the substrate in tip wear, and because it is nearly impossible to reposition a tip on exactly the same portion of the surface after a wear measurement, each wear measurement required a new SFM tip and a previously unscanned portion of the substrate. The tip shape before and after each wear experiment was characterized by scanning sharp silicon spikes ~600 nm tall with a tip radius of curvature of <10 nm (MikroMasch, TGT01). Because the spikes are significantly sharper than the SFM tip, imaging the spikes produces an image of the SFM tip. Typical images before and after a wear experiment are shown in Fig. 13. Tip images were analyzed by first flattening the image and then counting the number of pixels at a series of height increments above the baseline, producing a plot of the cross-sectional area of the tip as a function of height above the baseline. After the tip was worn, this procedure was used to determine the area of the end of the worn tip. By comparing the area of the worn end of the tip to tip shape measurements made on the same tip before wear, the total change in tip length, H, due to wear was readily determined.

The effects of scan duration (T) (which determines the sliding distance) and contact or normal force (F_N) were explored using sodium trisilicate glass ($Na_2O \cdot 3SiO_2$) substrates and ammonium hydroxide solution (pH ~ 11). The progression of wear was observed by measuring the amount of material removed from a series of SFM tips, where each tip was treated with a different number of scans at the same contact force (~120 nN). The scan size and tip velocity were chosen so that each scan involved 1.5 mm of total tip travel along the substrate in 50 sec. The total change in tip length for each new tip is plotted as a function of the number of scans in Fig. 14(a). In terms of instantaneous depth of wear, tip wear is initially rapid and gradually slows. Because of the pyramidal tip shape, the area of the tip increases with time as wear occurs. The dark line shows a simultaneous least squares fit of a power law expression (described below).

Another set of measurements was undertaken as a function of contact force. Again using a fresh tip for each measurement, 15 scans were performed on a sodium trisilicate glass substrate at the contact force of interest, followed by tip wear characterization. The resulting change in tip height as a function of normal force is displayed in Fig. 14(b). The dark lines in Fig. 14(a) and (b) show the result of a least squares fit of the form ($F_N \times$ time)^M to both data sets simultaneously. The best fit yielded a functional dependence of ($F_N \times$ time)$^{0.51}$. No significant changes in tip wear with pH were observed.

Compared to macroscopic pin-on-disk wear of silicon nitride in similar solutions, the volume removal rates represented in Fig. 14 are high. For comparison with conventional

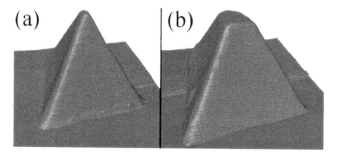

Figure 13 SFM images of (a) a fresh SFM tip and (b) the same tip after wear. The SFM tip was scanned 45 scan times across a sodium trisilicate glass substrate in ammonium hydroxide solution (pH ~ 11) at an applied force of 125 nN. Each image has been slightly cropped from the original 1500 × 1500 nm^2 image. The vertical (z) axis has the same scale as the x- and y-axes.

Height Change vs Number of Scans and Contact Force

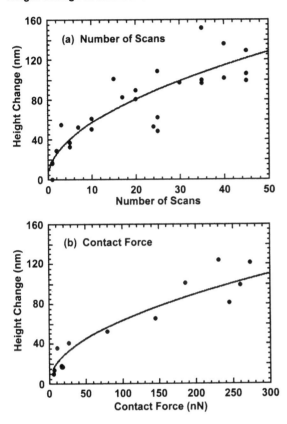

Figure 14   Height change in SFM tip (a) as a function of the number of scans on trisilicate glass in ammonium hydroxide (pH ~ 11) at a contact force of ~120 nN, and (b) as a function of contact force over 15 scans. Each scan was 3000 × 3000 nm² at 5.2 Hz. Each tip wear measurement was undertaken with a fresh SFM tip. Circles, experimental data. Line, least squares fit of $(F_N \times time)^m$ to both sets of data simultaneously.

wear measurements, we convert the height data of Fig. 14(a) to a volume removal rate per unit normal force per unit sliding distance. The slope of the wear rate data in Fig. 14(a) between 5 and 20 scans corresponds to a volume removal rate of 1×10^{-13} m³/N/m. This is an order of magnitude higher than the highest volume removal rates observed by Chen et al. [57], Jahanmir and Fischer [58], and Muratov et al. [59] This may in part be due to the exceptionally high stresses at the SFM tip contact (initially >1 GPa) and the single asperity nature of contact, as discussed below.

   The tip wear produced during 25 scans across a variety of substrates in distilled water (pH = 7) at a contact force of ~120 nN is shown in Fig. 15. Three tip wear measurements were made for each sample, starting with a fresh tip each time. Tip wear for five of the substrates was also measured in NH₄OH solution (pH ~ 11, data not shown); in all cases tip wear at pH = 11 was almost identical to tip wear at pH = 7.

   All of the substrates producing significant tip wear (i.e., quartz, fused silica, zirconia, titania, sodium trisilicate glass, polished soda lime glass, and silicon nitride) form modified hydroxide layers in aqueous solution. Twenty-five scans on these materials at a contact force

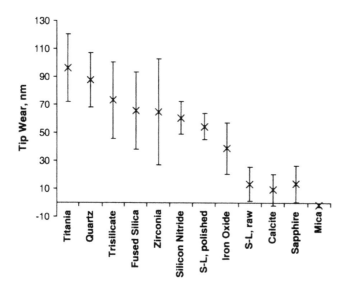

Figure 15   Amount of tip wear on various substrates in water produced by twenty-five 3000 × 3000 nm² scans at 5.2 Hz and a contact force of ~120 nN. A fresh tip was employed for each set of scans. Plot shows average and standard deviation for three measurements on each substrate material. "S-L" refers to soda-lime glass.

of 120 nN typically removed 50–100 nm of material from the tip. This wear is sufficient to produce a flat area of about 120 × 120 nm² at the end of the tip. Images of these tips showed that for all of these substrates this flat area was quite smooth, as in Fig. 13(b), with no detectable signs of roughening.

Conversely, substrates that do not form significant hydroxide layers in aqueous solution (calcite and mica) produced little tip wear. Typical wear during 25 scans of these substrates was 0–15 nm, with typical uncertainties less than ±2 nm. Although the mica structure possesses Si–OH bonds, these bonds are not exposed to the surface by cleavage. The Si–OH bonds on adjacent layers are linked by polyvalent cations, and cleavage occurs between surfaces lacking these bonds [60,61]. The lack of significant tip wear during scanning on these surfaces is evidence that hydroxide bonds are needed. Note that hardness of the substrate has no influence on tip wear.

Given the role of water in hydroxide formation, the effect of surface hydroxide formation was confirmed by performing wear measurements under ethyl acetate, which eliminates water. Neither the tip nor the sodium trisilicate glass substrates were given any special treatment to eliminate hydroxyl groups. The resulting tip wear was insignificant (10 ± 8 nm—not shown) at F_N = 120 nN; this wear is similar to the (lack of) tip wear produced by scanning calcite in water. Similarly, no tip wear is seen under ethyl acetate on a fused silica substrate. The exclusion of water prevents the replacement of hydroxide bonds consumed in initial wear-related reactions with the SFM tip. It is well known that glass substrate wear rates during polishing operations drop dramatically when water is excluded, presumably for similar reasons [62,63].

In this work, lateral force measurements (relative values only), such as those shown in Fig. 16, indicate a small reduction in friction (<20%) during the course of a typical wear experiment. Unlike some nanoscale systems [64,65] probed at a fixed normal force, friction here is definitely not proportional to the nominal tip-substrate contact area, i.e., friction

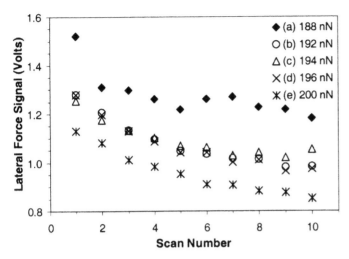

Figure 16   Lateral force signal during sequential scanning at five different locations on sodium trisilicate glass in ammonium hydroxide solution (pH ~ 11). Ten 750 × 750 nm² scans at 7.8 Hz were taken at the first location with a new tip [(a)], then the tip was moved to an unscanned area and another 10 scans were taken [(b), then (c), etc.].

decreases in Fig. 16 as the area *increases*. (The actual contact area may not be identical to the nominal contact area, but is expected to be close.) Furthermore, the small size of the decrease and the significant wear rates that prevail throughout the duration of scanning in this work argue strongly against any transition to hydrodynamic lubrication. The high stresses under the tip, even at our largest contact areas (~2.5 × 10⁻¹⁴ m²), prevent the formation of a continuous fluid film between the tip and the substrate.

The observed wear rate vs. time (as area increases) is higher than expected if the wear rate were simply proportional to the normal stress, i.e., the increasing tip area reduces the normal stress more than it reduces the wear rate. In contrast, the frictional force decreases even as the tip area increases. Our results are inconsistent with the hypothesis that the wear rate is proportional to friction. Friction and wear in this system appear to involve different chemical or mechanical processes.

To interpret these results, we wish to infer the instantaneous wear rate from measurements of the total wear after height change, $h(t)$, from measurements of total wear $[H(F_N, t_F)]$. The relationship between $h(t)$ and $H(F_N, t_F)$ is given by:

$$H(F_N, t_F) = \int_0^{t_F} \left. \frac{dh}{dt} \right|_{F_N} dt \tag{7}$$

where $h(t)$ is the instantaneous wear rate as a function of time.

The data of Fig. 14 clearly indicate that the wear rate is not simply proportional to the applied stress. For a square pyramid with a tip angle of 70°, the basal area $A = \alpha h^2$, where $\alpha = 1.99$. If the tip wear rate (dh/dt) were simply proportional to stress,

$$\frac{dh}{dt} = \frac{kF_N}{A} = \frac{kF_N}{\alpha h^2} \tag{8}$$

Integrating this expression yields $H(F_N, t_F) \sim (F_N t_F)^{1/3}$, i.e., $M = 0.33$, in contrast to the observed time and normal force dependence ($M = 0.51 \pm 0.05$). Equation (8) predicts that the rate of tip wear decreases much more slowly with time than is actually observed.

The data are better described by an instantaneous wear rate proportional to the product of applied stress times the length of the worn area, $l = \sqrt{\alpha h}$:

$$\frac{dh}{dt} = \frac{kF_N l}{A} = \frac{kF_N \sqrt{\alpha h}}{\alpha h^2} = \frac{k' F_N}{h} \tag{9}$$

where $k' = k/\sqrt{\alpha}$. When integrated, this yields $H(F_N, t_F) \sim (F_N t_F)^{0.5}$, in good agreement with the experimentally observed dependence $[(F_N t_F)^{0.51}]$.

Why the l dependence? The l dependence introduces an extra factor of h in the numerator of Eq. (9) and has the effect of increasing dh/dt at long times (large h) relative to Eq. (8). One possible source of this increase is a gradual, stress-activated production of a chemical precursor state on the substrate as the tip passes over. If the time constant for precursor formation is small relative to the duration of the applied stress, the precursor concentration will be approximately uniform and constant under the SFM tip. However, if the time constant for precursor formation is long relative to the duration of applied stress, the concentration of states under the tip will increase in a linear fashion from the leading edge of the tip to the trailing edge. Furthermore, the maximum concentration will be proportional to the time required for the tip to pass over any given point on the surface. This time is proportional to the length of the flat portion of the tip, $l = \sqrt{\alpha h}$. The stress at a given substrate location under tips with large flat areas is sustained for longer times than tips with small flat areas, producing higher concentrations of precursor states and yielding faster wear. Given the scan speeds and tip dimensions involved in this work, the relevant time scale for this effect is of the order of milliseconds.

Numerous studies have shown that on exposure to moist air or water, the silicon nitride surface rapidly oxidizes to form a thin layer of silicon oxides. Angle-resolved X-ray photoelectron spectroscopy (XPS) measurements by Hah et al. [66] showed 2-nm-thick oxide layers on silicon nitride surfaces formed by cleaving in air. Tribochemically polished surfaces showed a similar oxide layer, with approximate composition corresponding to 0.2–0.5 nm SiO_2 and 1.0–1.5 nm $SiO_x N_y$. Abrasively polished surfaces showed thicker oxide layers. Adhesion between silicon nitride SFM tips and a variety of substrates in water was inconsistent with predictions assuming that the tip was composed of Si_3N_4, but did match predictions for tips composed of SiO_2 [67]. In a separate study, XPS of Si_3N_4 SFM tips found considerable amounts of oxygen on the surface, which was attributed to an oxide layer [68]. Traces of surface oxide are also found in wear tracks on silicon nitride [58]. The ubiquity of oxide films on silicon nitride suggests that the wear of silicon nitride SFM tips is dominated by removal of the oxide, which is subsequently regenerated by further oxidation.

Significantly, SFM tip wear in this work is strongly affected by the chemical, as opposed to mechanical, properties of the substrate. Relatively soft substrates, such as the sodium trisilicate glass, often produced much more tip wear than much harder substrates, such as sapphire. All of the substrates yielding extensive tip wear form surface metal hydroxide species. Quartz, fused silica, zirconia, titania, sodium trisilicate glass, polished soda lime glass, and silicon nitride substrates in aqueous solution all have high densities of surface M–OH bonds and produced significant wear. Iron (II–III) oxide in aqueous solution is expected to display a high density of hydroxide bonds and produced intermediate levels of tip wear. Only one substrate with a strong potential for M–OH bonds (alumina) failed to produce significant tip wear, possibly due to kinetic effects. In contrast, all substrates expected to lack surface M–OH in aqueous solutions (calcite and mica) produced little if any tip wear. Therefore, we propose that tip wear results from reactions between hydroxide bonds on the tip with hydroxide bonds on the substrate.

On close contact, silicon hydroxyls on the SFM tip can form chemical bonds to hydroxyl groups on the substrate surface [55,62,69]. A schematic diagram of such tip–substrate bond formation is shown in Fig. 17(a) and (b). As the tip moves, these tip-surface bonds would be stretched [Fig. 17(c)], and eventually rebreak [Fig. 17(d)]. Simple bond breaking will not result in tip wear (requiring removal of Si). Breaking any connecting bond in Fig. 17(c) would leave the silicon atom attached to one of the surfaces, and breaking two bonds simultaneously is statistically unlikely. However, when a tip–surface bond breaks, the energy of the stretched bond may transfer kinetic energy to the atoms on each side of the bond [Fig. 17(d)] which weakens the other Si–O (or M–O) bonds toward reaction with water. We propose that this stretching and breaking process plays a critical role in tip wear.

Recoil energy as the stretched bonds break provides a mechanism to break the second bond after the first bond breaks, and it is the breaking of the second bond that leads to wear. The importance of this second step can also be inferred from the lack of wear in the absence of water. As long as surface hydroxyls are present, tip-substrate bonds can form in the absence of water, and if bond breaking [Fig. 17(c)–(d)] was all that was required for wear, then water would not be required for wear to occur. The fact that no wear occurs when ethyl acetate solution is used to exclude water strongly suggests that wear occurs not during bond breaking, but during subsequent reactions with water [Fig. 17(d)–(e)].

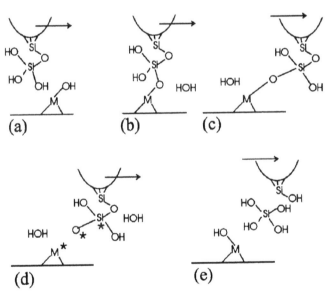

Figure 17   Chemical reaction model for tribochemical tip wear. In water, the silicon nitride SFM tip is coated with a layer of silicon oxide. (a) During scanning, a hydroxyl group on the tip (Si–OH) encounters a hydroxyl group on the surface (M–OH). (b) The two groups react to form a bond, releasing water. (c) As the tip continues its motion across the surface, the bridging bond is stretched. (d) The bond breaks. Energy that was formerly stored in the stretched bridging bond is deposited in the atoms on either side (asterisks). (e) These "activated" atoms react with water, which may also break additional bonds. In this case, a bond between the silicon atom and the SFM tip breaks, releasing the silicon atom from the tip and producing aqueous $Si(OH)_4$. Further tip motion yields additional reactions and more broken Si–O bonds on the tip surface, as well as M–O bonds on the substrate.

Tip–substrate bond formation plays a critical role in the proposed model of tip wear. One might expect that the formation of such bonds would depend on the ratio of protonated (M–OH) to deprotonated (M–O$^-$) hydroxyls on both surfaces, and thus vary with pH. Katsuki et al. [55] observed a weak dependence of silicon nitride SFM tip wear on pH, whereas Hah and Fischer [51] observed no significant pH dependence of the macroscopic wear of silicon nitride in the presence of various chemicals. No significant pH dependence was observed in our work. A weak or absent pH dependence may reflect the heterogeneity of the surface. With SiO$_x$N$_y$ and SiO$_2$ components, the tip surface may not have a well-defined isoelectric point. It is also possible that the local pH at the tip–substrate interface becomes nearly independent of the solution pH when the two surfaces are pressed tightly together and most of the solution is excluded.

In some models of chemical–mechanical polishing, material removal is attributed to the mechanical abrasion of a soft surface reaction layer. If, as we propose, this layer is formed between asperities, material removal will involve competition between the rate of the chemical reaction (to form the modified layer) and the rate of asperity wear (which limits the time for surface modification). In the context of commercial chemical–mechanical polishing operations, asperity contact is ideally confined to colloidal particles on the polishing pad, whereas chemical surface modification is largely confined to the substrate being polished. However, it is difficult to account for the effectiveness of certain abrasive materials, such as cerium oxide, in terms of mechanical effects alone. The effectiveness of some polishing materials has been attributed to a poorly defined property sometimes called "chemical tooth." In the present work, it is clear that the chemical reactivity of the substrate plays a key role in tip wear, which we feel is due to the formation of bridging oxide bonds between the tip and the substrate. The effectiveness of colloidal silica particles in chemical–mechanical polishing (although not as effective as some other oxides) suggests that bridging oxide bonds may play a similar role in many chemical–mechanical polishing systems.

VII. CONCLUSIONS

The tip of an SFM probe provides a useful surrogate for the study of nanoscale, single asperity wear. The synergism between mechanical (asperity) wear and chemical dissolution plays an important role in many commercial processes, including chemical–mechanical polishing. Much of this synergism can be attributed to the vulnerability of defect structures such as steps and kinks to combined mechanical–chemical attack. This vulnerability is readily observed on single-crystal surface materials such as calcite and brushite. These surfaces display remarkable anisotropies in wear track growth and evolution during and after linear scanning across molecular steps. These anisotropies reflect the important role of local geometry and bonding in determining step stabilities in aqueous environments, with and without applied stress.

In principle, knowledge of E_{act}, v^*, and the relevant prefactors for the key dissolution processes (e.g., pit nucleation, double-kink nucleation, and kink propagation) would form the basis for a complete predictive model for corrosive wear in these materials. Where deformation processes play an important role, dislocation emission and similar processes would also require characterization. Our understanding of how E_{act}, v^*, and other material parameters are affected by the local structure would also be enhanced by molecular dynamics simulations of water–surface interactions. The rather low rate of dissolution (one molecule removed per kink site in millions of vibrational periods) may complicate these calculations.

Low normal force scanning of the surface of an ionic crystal (brushite) with an SFM tip in supersaturated solution can induce and control atomic layer regrowth at step edges. This process can be exploited to produce atomically smooth surfaces by "filling" rather than "polishing." The chief role of the tip appears to be the controlled transport of sorbed ion clusters from upper terraces over the step edge, thereby raising concentration of adsorbed ions near the step inside the pit. Secondly, 3-D nucleation is totally suppressed by scanning, thereby maintaining flat surfaces. The use of fluctuations ("noise") in the lateral rocking of the cantilever to probe dynamic processes during deposition, such as the formation of transient sorbed clusters, is an exciting prospect.

Interaction between mechanical and chemical effects also plays an important role in particle removal. The dramatic effect of particle size and humidity on the removal of salt particles from a soda lime glass substrate is mirrored in practical CMP operations, where chemically active components of the solution lower the surface energy of the surfaces formed by removal. It is particularly significant that the humidity dependence in the salt–glass system can be quantitatively described in terms of the change in interfacial energy due to water absorption. This knowledge, in conjunction with the relevant particle size dependencies, would facilitate intelligent design and improvement of CMP processes in new materials systems. *Combined* chemical and mechanical attack is most effective for particle removal as well as for eliminating monolayer surface roughness.

Human-scale machining operations are often limited by wear of the tool, and it is no surprise that this is often true at the nanometer scale as well. Over a large range of contact forces and tip velocities, the wear of silicon nitride SFM tips is dominated by chemical–mechanical effects involving not only the solution but also the chemical nature of the substrate. In the present work, the only substrates producing significant tip wear are characterized by high densities of metal–hydroxide bonds. Wear also requires an aqueous environment. Solution pH has little influence on tip wear rates. Water is assumed to replenish hydroxide bonds on the substrate surface that are consumed in mechanically mediated reactions with the tip.

Measurements of SFM tip wear provide a powerful probe of single asperity wear— both in the tip itself and in the affected substrate. Ultimately, we hope to understand macroscopic wear dominated by asperity contacts in terms of the wear of individual asperities from a given initial size distribution as a function of applied stress and chemical environment. Together, stress and corrosion are much more effective than either alone and provide many variables for the control and localization of material wear or deposition.

ACKNOWLEDGMENTS

We thank Louis Scudiero, Forrest Stevens, and Rizal Hariadi of Washington State University, and Nam-Seok Park of Chungbuk University, South Korea, for performing portions of the SFM work discussed above. We thank John Hutchinson, Harvard University, Joe Simmons, Dan Bentz, and Ken Jackson, University of Arizona, for helpful discussions. We also thank Ryan Leach, Washington State University, for help with computer programming, and Larry Pederson, Pacific Northwest National Laboratories, for supplying sample materials.

This work was supported by the National Science Foundation under Grants CMS-98-00230, CMS-01-16196, CHE 02-34726, and a subcontract with the University of Florida on a KDI-NSF Collaboration under Grant No. 9980015.

REFERENCES

1. Nakahara, S.; Langford, S.C.; Dickinson, J.T. Surface force microscope observations of corrosive tribological wear on single crystal $NaNO_3$ exposed to moist air. Tribol. Lett. 1995, 1, 277–300.
2. Park, N.-S.; Kim, M.-W.; Langford, S.C.; Dickinson, J.T. Tribological enhancement of $CaCO_3$ dissolution during scanning force microscopy. Langmuir 1996, 12, 4599–4604.
3. Park, N.-S.; Kim, M.-W.; Langford, S.C.; Dickinson, J.T. Atomic layer wear of single-crystal calcite in aqueous solution using scanning force microscopy. J. Appl. Phys. 1996, 80, 2680–2686.
4. Ehrlich, G.; Hudda, F.G. Atomic view of self-diffusion: Tungsten on tungsten. J. Chem. Phys. 1966, 44, 1039–1049.
5. Schwoebel, R.L.; Shipsey, E.J. Step motion on crystal surfaces. J. Appl. Phys. 1966, 37, 3682–3686.
6. Meyer, E.; Luthi, R.; Howald, L.; Bammerlin, M. Instrumental aspects and contrast mechanism of force friction microscopy. In *Micro/Nanotribology and its Applications*; Bhusan, B., Ed.; Kluwer Academic: Dordrecht, 1997; 193–215.
7. Junno, T.; Deppert, K.; Montelius, L.; Samuelson, L. Controlled manipulation of nanoparticles with an atomic force microscope. Appl. Phys. Lett. 1995, 66, 3627–3629.
8. Lebreton, C.; Wang, Z.Z. Critical humidity for removal of atoms from the gold surface with scanning tunneling microscopy. J. Vac. Sci. Technol B 1996, 14, 1356–1359.
9. Tam, A.C.; Leung, W.P.; Zapka, W.; Ziemlich, W. Laser-cleaning techniques for removal of surface particulates. J. Appl. Phys. 1992, 71, 3515–3523.
10. Tam, A.C.; Park, H.K.; Grigoropoulos, C.P. Laser cleaning of surface contaminants. Appl. Surf. Sci. 1998, 127–129, 721–725.
11. Nanz, G.; Camilletti, L.E. Modeling of chemical-mechanical polishing: A review. IEEE Trans. Semicond. Manuf. 1995, 8, 382–389.
12. Lawn, B. *Fracture of Brittle Solids,* 2nd Ed.; Cambridge University Press: Cambridge, 1993.
13. Ohnesorge, F.; Binnig, G. True atomic resolution by atomic force microscopy through repulsive and attractive forces. Science 1993, 260, 1451–1456.
14. Hillner, P.E.; Gratz, A.J.; Manne, S.; Hansma, P.K. Atomic scale imaging of calcite growth and dissolution in real time. Geology 1992, 20, 359–362.
15. Hillner, P.E.; Manne, S.; Gratz, A.J.; Hansma, P.K. AFM images of dissolution and growth on a calcite crystal. Ultramicroscopy, 1992, 42–44, 1387–1393.
16. Stipp, S.L.S.; Eggleston, C.M.; Nielsen, B.S. Calcite surface structure observed at microtopographic and molecular scales with atomic force microscopy (AFM). Geochim. Cosmochim. Acta 1994, 58, 3023–3033.
17. Stipp, S.L.S.; Gutmannsbauer, W.; Lehmann, T. The dynamic nature of calcite surfaces in air. Am. Mineral. 1996, 81, 1–8.
18. Liang, Y.; Baer, D.R.; Lea, A.S. Dissolution of $CaCO_3(10\underline{1}4)$ surface. In *Evolution of Thin-Film and Surface Structure and Morphology*; Demczyk, B.G., Williams, E.D., Garfunkel, E., Clemens, B.M., Cuomo, J.J., Eds.; Materials Research Society: Pittsburgh, PA USA, 1995; Vol. 335, 409 pp.
19. Liang, Y.; Baer, D.R.; McCoy, J.M.; LaFemina, J.P. Interplay between step velocity and morphology during the dissolution of $CaCO_3$ surface. J. Vac. Sci. Technol. A 1996, 14, 1368–1375.
20. Liang, Y.; Lea, A.S.; Baer, D.R.; Engelhard, M.H. Structure of the cleaved $(10\underline{1}4)$ surface in an aqueous environment. Surf. Sci. 1996, 351, 172–182.
21. Liang, Y.; Baer, D.R. Anisotropic dissolution at the $CaCO_3(10\underline{1}4)$-water interface. Surf. Sci. 1997, 373, 275–287.
22. Schott, J.; Brantley, S.; Crerar, D.; Guy, C.; Borcsik, M.; Willaime, C. Dissolution kinetics of strained calcite. Geochim. Cosmochim. Acta 1989, 53, 373–382.
23. Compton, R.G.; Daly, P.J.; House, W.A. The dissolution of Iceland spar crystals: The effect of suface morphology. J. Colloid Interface Sci. 1986, 113, 12–20.

24. Paquette, J.; Reeder, R. New type of compositional zoning in calcite: Insights in to crystal-growth mechanisms. J. Geol. 1990, *18*, 1244–1247.
25. MacInnis, I.N.; Brantley, S.L. The role of dislocations and surface morphology in calcite dissolution. Geochim. Cosmochim. Acta 1992, *56*, 1113–1126.
26. MacInnis, I.N.; Brantley, S.L. Development of etch pit size distributions on dissolving minerals. Chem. Geol. 1993, *105*, 31–49.
27. Hirth, J.P.; Pound, G.M. Evaporation of metal crystals. J. Chem. Phys. 1957, *26*, 1216–1224.
28. Anuradha, P.; Raju, I.V.K.B. Study of plastic flow in $CaCO_3$ and $NaNO_3$ crystals under concentrated loads. Phys. Status Solidi A 1986, *95*, 113–119.
29. Bull, S.J.; Page, T.F.; Yoffe, E.H. An explanation of the indentation size effect in ceramics. Philos. Mag. Lett. 1989, *59*, 281–288.
30. Walker, W.W.; Demer, L.J. Effect of loading duration on indentation hardness. Trans. Metall. Soc. AIME 1964, *230*, 613–614.
31. Timoshenko, S.P.; Goodier, J.N. *Theory of Elasticity*, 3rd Ed.; McGraw-Hill: New York, 1970.
32. Chung, D.H.; Buessem, W.R. The elastic anisotropy of crystals. In *Anisotropy in Single-Crystal Refractory Compounds*; Vahldiek, F.W., Mersol, S.A., Eds.; Plenum: New York, 1968; 217–245.
33. Hillig, W.B.; Charles, R.J. Surfaces, stress-dependence surface reactions, and strength. In *High Strength Materials*; Zackay, V.F., Ed.; John Wiley: New York, 1965; 682–705.
34. Wiederhorn, S.M. Influence of water vapor on crack propagation in soda-lime glass. J. Am. Ceram. Soc. 1967, *50*, 407–414.
35. Ohta, M.; Tsutsumi, M.; Ueno, S. Observations of etch pits on as-grown faces of brushite crystals. J. Cryst. Growth 1979, *47*, 135–136.
36. Scudiero, L.; Langford, S.C.; Dickinson, J.T. Scanning force microscope observations of corrosive wear on single-crystal brushite in aqueous solution. Tribol. Lett. 1999, *6*, 41–55.
37. Curry, N.A.; Jones, S.W. Crystal structure of brushite, calcium hydrogen orthophospate dihydrate: A neutron diffraction investigation. J. Chem. Soc. London A 1971, *1971*, 3725–3729.
38. Hariadi, R. F.; Langford, S.C.; Dickinson, J.T. Controlling nanometer-scale crystal growth on a model biomaterial with a scanning force microscope. Langmuir 2002, *18*, 7773–7776.
39. Scudiero, L.; Langford, S.C.; Dickinson, J.T. Scanning force microscope observations of corrosive wear on single-crystal brushite ($CaHPO_4 \cdot 2H_2O$) in aqueous solution. Tribol. Lett. 1999, *6*, 41–55.
40. Toroczkai, Z.; Williams, E.D. Nanoscale fluctuations at solid surfaces. Phys. Today 1999, *52*, 24–28.
41. Hariadi, R.; Dickinson, J.T. *in preparation*.
42. Wyckoff, R.W.G. *Crystal Structures*, 2nd Ed.; Interscience: New York, 1965; Vol. 3.
43. Ruan, J.-A.; Bhushan, B.J. Atomic-scale and microscale friction studies of graphite and diamond using friction force microscopy. Appl. Phys. 1994, *76*, 5022–5035.
44. Hariadi, R.F.; Langford, S.C.; Dickinson, J.T. Atomic force microscope observations of particle detachment from substrates: the role of water vapor in tribological debonding. J. Appl. Phys. 1999, *86*, 4885–4891.
45. Hutchinson, J.W. Personal communication, 1998.
46. Muratov, V.A.; Fischer, T.E. Tribochemical polishing. Annu. Rev. Mater. Sci. 2000, *30*, 27–51.
47. Honda, F.; Saito, T. Tribochemical characterization of the lubrication film at the Si_3N_4/Si_3N_4 interface sliding in aqueous solutions. Appl. Surf. Sci. 1996, *92*, 651–955.
48. Saito, T.; Hoseo, T.; Honda, F. Chemical wear of sintered Si_3N_4, hBN and Si_3N_4-hBN composites by water lubrication. Wear 2001, *247*, 223–230.
49. Xu, J.; Kato, K. Formation of tribochemical layer of ceramics sliding in water and its role for low friction. Wear 2000, *245*, 61–75.
50. Xu, J.; Kato, K.; Hirayama, T. The transition of wear mode during the running-in process of silicon nitride sliding in water. Wear 1997, *205*, 55–63.
51. Hah, S.R.; Fischer, T.E. J. Tribochemical polishing of silicon nitride. Electrochem. Soc. 1998, *145*, 1708–1714.
52. Tomizawa, H.; Fischer, T.E. Friction and wear of silicon nitride and silicon carbide in water:

Hydrodynamic lubrication at low sliding speed obtained by tribochemical wear. ASLE Trans. 1986, *30*, 41–46.

53. Khurshudov, A.; Kato, K. Wear of the atomic force microscope tip under light load, studied by atomic force microscopy. Ultramicroscopy 1995, *60*, 11–16.

54. Bloo, M.L.; Haitjema, H.; Pril, W.O. Deformation and wear of pyramidal, silicon-nitride AFM tips scanning micrometer-size features in contact mode. Measurement 1999, *25*, 203–211.

55. Katsuki, F.; Kamei, K.; Saguchi, A.; Takahashi, W.; Watanabe, J. AFM studies on the difference in wear behaviour between Si and SiO_2 in KOH solution. J. Electrochem. Soc. 2000, *147*, 2328–2331.

56. Seta, S.; Nishioka, T.; Tateyama, Y.; Miyashita, N. Study of nano-scale wear of silicon oxide in CMP process. In *Chemical Mechanical Planarization IV*; Opila, R.L., Reidsema-Simpson, C., Sundaram, K.B., Seal S., Eds.; Electrochemical Society: Pennington, NJ, 2001; 28–33.

57. Chen, M.; Kato, K.; Adachi, K. Friction and wear of self-mated SiC and Si_3N_4 sliding in water. Wear 2001, *250*, 246–255.

58. Jahanmir, S.; Fischer, T.E. Friction and wear of silicon nitride lubricated by humid air, water, hexadecane and hexadecane + 0.5 percent stearic acid. STLE Trans. 1988, *31*, 32–43.

59. Muratov, V.A.; Luangvaranunt, T.; Fischer, T.E. The tribochemistry of silicon nitride: Effects of friction, temperature and sliding velocity. Tribol. Int. 1998, *31*, 601–611.

60. Bailey, S.W. Classification and structures of the micas. *Reviews in Mineralogy*; BookCrafters, Inc.: Chelsea, MI, 1984; Vol. 13.

61. Gaines, R.V.; Skinner, H.C.W.; Foord, E.E.; Mason, B.; Rosenzweig, A.; King, V.T. *Dana's New Mineralogy*; John Wiley & Sons Inc.: New York, 1997.

62. Kirk, N.B.; Wood, J.V. Glass polishing. Br. Ceram. Trans. 1994, *93*, 25–30.

63. Cook, L.M. Chemical processes in glass polishing. J. Non-Cryst. Solids 1990, *120*, 152–171.

64. Carpick, R.W.; Salmeron, M. Scratching the surface: Fundamental investigations of tribology with atomic force microscopy. Chem. Rev. 1997, *97*, 1163–1194.

65. Pietrement, O.; Troyon, M. Study of the interfacial shear strength pressure dependence by modulated lateral force microscopy. Langmuir 2001, *17*, 6540–6546.

66. Hah, S.R.; Burk, C.B.; Fischer, T.E. Surface quality of tribochemically polished silicon nitride. J. Electrochem. Soc. 1999, *146*, 1505–1509.

67. Jacquot, C.; Takadoum, J.J. A study of adhesion forces by atomic force microscopy. Adhes. Sci. Technol. 2001, *15*, 681–687.

68. Lee, G.U.; Chrisey, L.A.; O'Ferrall, C.E.; Pilloff, D.E.; Turner, N.H.; Colton, R.J. Chemically-specific probes for the atomic force microscope. Isr. J. Chem. 1996, *36*, 81–87.

69. Hoshino, T.; Kurata, Y.; Terasaki, Y.; Susa, K. Mechanism of polishing of SiO_2 films by CeO_2 particles. J. Non-Cryst. Solids 2001, *283*, 129–136.

18

Designing for Wear Life and Frictional Performance

O. O. Ajayi
Argonne National Laboratory, Argonne, Illinois, U.S.A.

Kenneth C Ludema
University of Michigan, Ann Arbor, Michigan, U.S.A.

I. INTRODUCTION

There are two major incentives for adopting rational methods of design for friction and wear performance. One is to reduce the great amount of time that is often lost during product development due to poorly understood issues in friction and wear. The other is to reduce the cost to the manufacturer of failed products: warranty costs due to unpredicted wear and undesirable frictional behavior exceed the cost for most other mechanical causes combined in several large industries.

The focus in this chapter is on the design of mechanical systems that contain sliding and rolling elements, assuming that such considerations in products as strength, weight, fatigue life, and others had already been covered. The major goals in such designs are to achieve targets in frictional behavior (level of friction and sliding vibrations) and wear life. Secondary goals might include the use of biodegradable lubricants and using materials that are readily recycled.

For friction and wear issues, designers have virtually no guidelines other than experience, certainly no directly applicable mathematical equations, and no charts or tables (other than those that apply to catalog components). The reason is that wear life and frictional stability are such complicated topics, involving so many variables, perhaps approaching the number of variables that affect human life. Sliding surfaces are not simple, but are usually composed of oxide, absorbed substances, and dirt on top of substrate material. Their behavior in sliding or rolling is also influenced by the environment in which they operate (or are stored), method of manufacture, as well as the conditions of operations. They suffer acute and/or progressive degeneration and they can often be partially rehabilitated by either a change in operating conditions or by some intrusive action such as by changing lubricant type after some wear has occurred.

Designing for specific wear life or frictional behavior of new products or upgraded products involves several steps which are described in the sections that follow. These steps, generally, are as follows.

1. Search for appropriate design methodologies (equations). As to equations, designers are faced with a choice: either they must use the few available equations that are likely not applicable to the present case or use methods that are not a part of their training. Life for a designer is much simpler in areas of product function, product strength, and fatigue limits of products.
2. Failing that, two other steps may be taken—tabulate the modes of wear and frictional behavior of products similar to the one under present consideration and seek solutions to these modes. Information on the modes of wear can be found in handbooks, textbooks, and published paper, *although* these are difficult to interpret.
3. If such solutions are not available, then testing of previous models under the expected new conditions of upgraded products (higher loads, higher speeds, and cheaper material) must be done. Where previous models are not available, that is, if the present design is for a new product, then prototypes must be made for testing.
4. If nothing else works, it may be time to engage outside expertise in designing prototypes, in choosing tests, and in interpreting results.

II. DESIGN PHILOSOPHY

Most sliding surfaces are redesigned rather than designed for the first time. Thus a designer working on an update of a product will usually have the benefit of access to people who have experience with previous versions of a product. Designing a product for the first time, however, requires very mature skill. The philosophy by which expected wear life or frictional behavior of a product is chosen may differ strongly within and between various segments of industry. Such considerations as acceptable modes of failure, product repair, controllability of environment, product cost, nature of product users, and the interaction between these factors receive different treatment for different products. For example, since automobile tires are easier to change than is an engine crank, the wear life of tires is not a factor in discussion of vehicle life. The opposite philosophy must apply to drilling bits used in the oil well industry. The cone teeth and the bearing upon which the cone rotates must be designed for equal life since both are equally inaccessible while wearing.

In some products or machines, function is far more important than manufacturing costs. One example is the sliding elements in nuclear reactors. The temperature environment of the nuclear reactor is moderate, but lubricants are not permitted and the result of wear is to compromise the function of the system. This is an example of a highly specified combination of materials and wearing conditions. Another complex example is that of artificial bone joints. A high cost is relatively acceptable, but durability may be strongly influenced by body chemistry and choice of life style, all beyond the range of influence by the designer.

There is no general rule whereby a designer can quickly proceed to selecting an acceptable sliding material for a product. One often-heard, but misleadingly simple recommendation for reducing wear is to increase the hardness of the material. A companion notion is that wear protection can be enhanced by applying a hard coating upon a substrate material that is inadequate by itself. Hardening is effective against abrasion and erosion if the hardness of the abrasive is greater than 1.3 times that of the wearing surface, but not in many other applications. One obvious exception is the case of bronzes, which are more successful as a gear material against a hardened steel pinion than is a hardened steel gear. The reason usually given for the success of bronze is that dirt particles are readily embedded into the bronze and therefore do not cut or wear the steel away. Another exception to the hardness rule is the cams in automotive engines. They are hardened in the range of 50 Rockwell "C" instead of the maximum available which may be as high as 67 Rc. A final example is that of buckets and chutes for handling some ores. Rubber is sometimes found to be superior to very hard white cast iron in these applications.

These examples indicate that properties other than hardness may control wear. The rubber offers resilience, whereas white iron is readily fragmented by impacting ore, and the cam material resists fatigue failure if it is not fully hardened. It is often argued that such as those cited above are rare or can be dealt with on a case-by-case basis. This view seems to imply that most sliding systems are "standard," thereby suggesting that generic wear resistance or friction of material are intrinsic properties of materials. Little real progress has been made in this effort, and very little is likely to be made in the near future. Wear resistance and frictional behavior are achieved by a balance of several very separate properties (hardness, toughness, wettability of lubricants, oxidation energy, etc.), not all of them intrinsic and different for each machine component or sliding surface. Selecting the best material for wear resistance is therefore a complex task and guidelines are needed in design. Guidelines will be more useful as our technology progresses and some guidelines are given below.

III. STEPS IN DESIGNING FOR WEAR LIFE OR FRICTIONAL BEHAVIOR WITHOUT SELECTING MATERIALS

A. The Search for Specialized Components

Designers make most of the decisions concerning material selection for products, *although* they may not be the most informed on materials (and coatings). Fortunately, for many cases, crucial sliding (or rolling) components are available that have fairly well specified performance capabilities. Examples are gearboxes, clutches, and bearings. Most such components have been well tested in the marketplace, having been designed and developed by very experienced designers. For designers of specialized components, very general rules for selecting materials are of little value. They must build devices with a predicted performance of $\pm 10\%$ accuracy or better. They know the range of capability of lubricants, they know the reasonable range of temperature in which their products will survive, and they know how to classify shock loads and other real operating conditions. Their specific expertise is not available to the general designer except in the form of the shapes and dimensions of hardware, the materials selected, and the recommended practices for use of their product. Some of these selections are based on tradition, and some are based on rational reasoning that is strongly tempered by experience. The makers of specialized components usually also have the facilities to test new designs and materials extensively before risking their product in real use. General designers, on the other hand, must often proceed without extensive testing.

The general designer must then decide whether to use specialized components or whether to risk designing every part, personally. Sometimes, the choice is based on economics, but sometimes, desired specialized components are not available. In such cases, components as well as other machine parts must be designed in-house.

B. In-House Design

Product improvement could be done by interpolating within or extrapolating beyond known experience, if any, using four sources.

1. Company Practice for Similar Items

If good information is available on similar items, a prediction of the wear life of a new product can be made within ±20% accuracy unless the operating conditions of the new design are very much beyond standard experience. Simple scaling of sizes and loads is often successful, but this technique usually fails after a few iterations. Careless comparison of a new design with "similar" existing items can produce very large errors for reasons discussed below.

When a new product must be designed that involves loads, stresses, or speeds beyond those previously experienced, it is often helpful to review the recent performance and examine the worn surface of a well-used previous model in detail. It is also helpful to examine unsuccessful prototypes or friction/wear test specimens as will be discussed below. An assessment should be made of the modes or mechanisms of surface change (different from wear) of each part of the product. For this purpose, it is also useful to examine old lubricants, the contents of the lubricant sump, and other accumulations of residue as well.

2. Vendors of Materials, Lubricants, and Components

Where a new product requires bearings or materials of higher capacity than now in use, it is frequently helpful to contact vendors of such products. Where a vendor simply suggests an existing item or material from old brochures, the wear life of a new product may not be predictable to an accuracy of better than ±50% of the desired. This accuracy is worse than the ±20% accuracy given above especially where there is inadequate communication between the designer and the vendor. Accuracy may be improved where an interested vendor carefully assesses the needs of a design, supplies a sample for testing, and follows the design activity to the end.

Contact with vendors, incidentally, often has a general beneficial effect. It encourages a designer to explore new ideas beyond the simple extrapolation of previous experience. Most designers need a steady flow of information from vendors to remain informed on both new products and on the changing capability of products.

3. Handbooks

There are very few handbooks that assist in selecting materials for wear resistance or frictional behavior. Materials and design handbooks usually provide lists of materials some of which are highlighted as having been successfully used by someone in sliding parts of various products. They usually provide little information on the rates of wear of products, the mode of wear failure, the variations of friction, the limit on operating conditions, or the method by which the sliding parts should be manufactured or run in (if necessary).

Many sources provide tables of coefficient of friction, and some sources will give wear coefficients which are purported to be figures of merit or rank placing of materials for wear resistance. A major limitation of friction and wear coefficients of materials as given in most literature is that there is seldom adequate information given on how the data were obtained. Usually, this information is taken from standard laboratory bench test, few of which simulate real systems and few of which rank (order) materials in the same way that production parts experience in the hands of the consumer. The final result of the use of handbook data is a design that will probably not perform to an accuracy of better than ±75%.

4. Equations

Wear is very complicated, involving up to seven basic mechanisms, operative in different balance or ratio in various conditions, and many of the mechanisms produce wear rates that are not linear in the simple parameters such as applied load, sliding speed, surface finish, etc. There is, at this time, no complete array of "first principles," equations or models available to use in selecting materials for wear resistance. This seems impossible in this era of science and unlikely when over 300 equations can be found in the literature on friction and wear. Often, product engineers feel compelled to include equations in their reports, but none should be used to guide design. Attempts to find good equations or to alter one in hand take more time than solving issues by the methods suggested in the following paragraphs.

IV. STEPS IN SELECTING MATERIALS, COATINGS, AND LUBRICANTS FOR FRICTION CONTROL AND WEAR RESISTANCE

When designing for wear resistance, it is necessary to ascertain that wear will proceed by the same combination of mechanisms throughout the substantial portion of the life of a product as found in a test or in a previous similar product: only then is reasonable prediction of life possible for the new design.

The following are those considerations that are vital in selecting materials and may be more important than selecting the most wear-resisting material.

A. Determine Whether There Are Restrictions on Material Use

In some industries, it is necessary for economic and other purposes to use, e.g., a gray cast iron, a material that is compatible with the human body, a material with no cobalt in it such as in a nuclear reactor, a material with high friction, or a selected surface treatment applied to a low-cost substrate. Furthermore, there may be a limitation on the surface finish available or the skill of the personnel to manufacture or assemble the product. Finally, there may be considerations of delivery, storage of the item before use, disposal after use, or several other events that may befall a wear surface.

B. Determine Whether the Sliding Surface Can Withstand the Expected Static Load Without Indentation or Excessive Distortion

Generally, this would involve a simple stress analysis.

C. Determine the Sliding Severity That the Materials Must Withstand in Service

Factors involved in determining sliding severity include the contact pressure or stress, the temperature due to both ambient heating and frictional temperature rise, the sliding speed, misalignment, duty cycle, and type of maintenance the designed item will receive. Each factor is explained below.

1. Contact Stress

Industrial standards for allowable contact pressure vary considerably. Some specifications in the gear and sleeve bearing industries limit the average contact pressures for bronzes to about 1.7 MPa, which is about 1% to 4% of the yield strength of bronze. Likewise, in pump parts and valves made of tool steel, the contact pressures are limited to about 140 MPa which is about 4% to 6% of the yield strength of the hardest state of tool steel.

One example of acceptable high contact pressure is the sleeve bearings in the landing gear of one commercial airplane, the DC9 and MD-80 series. These materials again are bronzes and have yield strengths up to 760 MPa. The design bearing stress is 415 MPa (55% of the yield strength) but with expectations of peak stressing up to 620 MPa. Another example is the use of tool steel in lubricated sheet metal drawing. Dies may be used for 500,000 parts with contact pressures of about 860 MPa, which is half the yield strength of the die steel.

2. Temperature Strongly Influences the Life of Some Sliding Systems

Handbooks often specify a material for "wear" conditions without stating a range of temperature within which the wear resistance behavior is satisfactory. The influence of temperature may be its effect on the mechanical properties of the sliding parts: high temperatures soften most materials and low temperature embrittles some. High temperature will degrade most lubricants, but low temperature will solidify a liquid lubricant.

Ambient temperature is often easy to measure, but the temperature rise due to sliding may have a larger influence. Thermal conductivity of material could be influential in controlling temperature rise in some cases, but a more important factor is μ, the coefficient of friction. If a temperature-sensitive wear mechanism is operative in a particular case, then high friction may contribute to a high wear rate, if not cause it. There is at least a quantitative connection between wear rate and μ when one compares dry sliding with adequately lubricated sliding, but there is no formal way to connect μ with temperature.

3. Sliding Speed and the PV Limits

Maximum allowable loads and sliding speeds for materials are often specified in catalogs in the form of PV limits, where P is the contact pressure and V is the sliding speed. A PV limit indicates nothing about the actual rate of wear of materials, only that above a given PV limit, a very severe form of wear may occur. It is often stated that the product of P and V can be taken as an indication of the temperature rise in sliding which is only partially true. The influence of sliding speed on temperature rise is not linear, and the temperature rise is a function of several factors beside P and V.

4. Misalignment

The difficulty with misalignment is that it is an undefined condition beyond that for which contact pressure between two surfaces is usually calculated. Where some misalignment may exist, it is best to use materials that can adjust or accommodate itself, that is, break in properly. Misalignment arises from manufacturing errors from a deflection of the system producing loading at one edge of the bearing, or it may arise from thermal distortion of the system, etc. *Thus* a designer must consider designing a system such that a load acts at the expected location in a bearing under all conditions. This may involve designing a flexible bearing mount, several bearings along the length of a shaft, a distribution of the applied loading, etc.

A designer must also consider the method of assembly of a device. A perfectly manufactured set of parts can be inappropriately or improperly assembled, producing misalignment or distortion. A simple tapping of a ball bearing with a hammer to seat the race may constitute more severe service than what occurs in bearing lifetime in a machine and often results in early failure.

Misalignment may result from wear. If abrasive species can enter a bearing, the fastest wear will occur at the point of entry of the dirt. In that region, the bearing will wear away and transfer the load to other locations. A successful design must account for such events.

5. Duty Cycle

Important factors in selecting materials for wear resistance are the extent of shock loading of sliding systems, stop–start operations, oscillatory operations, etc. It is often useful to determine also what materials surround the sliding system, such as chemical substances or abrasive particles.

6. Maintenance

A major consideration that may be classified under sliding severity is maintenance. One example is that whereas most phosphor bronze bushings are allowed a contact stress of about 1.4 to 7 MPa, aircraft wheel bushings made of beryllium bronze are allowed a maximum stress of 620 MPa as mentioned before. The beryllium bronze has a strength only twice that of the phosphor bronze, but the difference between industrial and aircraft use includes different treatment of bearings in maintenance. Industrial goals are to place an object into service and virtually ignore it or provide infrequently scheduled maintenance. Aircraft maintenance, on the other hand, is more rigorous and each operating part is under regular scrutiny by the flight crew and ground crew. There is scheduled maintenance, but there is also careful and continuous observation of the part. It is more likely that a poor lubricant will find its way into hardware in industry than in aircraft applications for example. Secondly, the aircraft wheel bearing operates in a much more standard or narrowly defined environment. Industrial machinery must operate in the dirtiest and hottest of places with the poorest care. These must be considered as severity conditions by the designer.

D. Determine Whether or Not a Break-in Procedure Is Necessary or Prohibited

It cannot be assumed that sliding surfaces made to a dimensional accuracy and specified surface finish are ready for service. Sliding alters surfaces. Frequently, sliding under controlled light loads can prepare a surface for a long life of high loading, whereas immediate operation at moderate loads may cause early failure.

It is useful here to distinguish between two surface-altering strategies. We refer to the first as "break-in," where a system is immediately loaded or operated to its design load. The incidence of failure of a population of such parts decreases with time of operation as the sliding surfaces change, and, frequently, the ability of the system to accommodate an overload or inadequate lubricant increases in the same time. The surfaces have changed in some way during running and this is "break-in." "Running-in," on the other hand, is the deliberate and planned action that is necessary to prepare surfaces for normal service.

The wear that occurs during run-in or break-in can be considered a final modification to the machined surface. This leads to the possibility that a more careful specification of manufacturing practice may obviate the need for run-in or break-in. This has been the case with automobile engines in particular, although part of a successful part finish specification often includes the exact technique for making the surface. Only 60 years ago, it was necessary to start and run an engine carefully for the first few thousand miles to insure a reasonable engine life. If run-in were necessary today, one would not see an engine survive the short trip from the assembly plant to the haul-away trucks!

It is difficult to determine whether the present conservative industrial design practices result from the impracticality of running in some products. For example, a gearbox on a production machine is expected to function immediately without run-in. If it was run in, its capacity might be greatly increased. But, for each expected severity of operation of a device, a different run-in procedure is necessary. A machine that has been operating at one level of severity may be no more prepared for a different state of severity than if it had never been run. A "safe" procedure therefore is to operate a device below the severity level at which run-in is necessary, but the device could already be overdesigned to avoid running-in.

E. Determine Acceptable Modes of Wear Failure, Surface Damage, or Debris Form

To specify a wear life in terms of a rate of loss of material is not sufficient. For example, when an automotive engine seizes up, there is virtually no loss of material, only a rearrangement such that function is severely compromised. In an engine, as well as on other precision slideways of machines, surface rearrangement or change in surface finish is less acceptable than attrition or loss of material from the system. Again, in metal working dies, loss of material from the system is less catastrophic than is scratching of the product. Finally, in some systems, particularly in artificial human joints and computer hard disks, the wear debris is a greater hazard than is a loss of dimension from the sliding members.

In truck brakes, some abrasiveness of brake linings is desirable, *although* it wears brake drums away but that wear removes microcracks and avoids complete thermal fatigue cracking. On the other hand, in cutting tools, ore crushing equipment, and amalgam filling in teeth, surface rearrangement is of little consequence, but material loss is to be avoided.

A final example of designing for an acceptable wear failure is the sleeve bearings in engines. Normally, they should be designed against surface fatigue. However, in some applications, corrosive conditions may seriously accelerate fatigue failure. This may require the selection of a material that is less resistant to dry fatigue than is the best bearing material, and this applies especially to the two-layer bearing materials. In all of these examples, a study of acceptable modes of wear may result in a different selection of material than if the goal is simply to minimize wear.

F. Decide Whether or Not to Begin Wear Testing

After some study of worn parts from a device or machine that most nearly approximates the new or improved product, one of several conclusions could be reached.

1. The same design and materials in the wearing parts of the example device will perform adequately in the redesign in terms of function, cost, and all other attributes.
2. A slight change in size, lubrication, or cooling of the example parts will be adequate for the design.
3. A significant change in size, lubrication, or cooling of the example parts will be necessary for the redesign.
4. A different material will be needed in the redesign.

The action to be taken after reaching one of the above conclusions will vary. The first conclusion above can reasonably be followed by production of a few copies of the redesign. These should be tested and minor adjustments should be made to ensure adequate product life.

The second conclusion should be followed by cautious action and the third conclusion should involve the building and exhaustive testing of a prototype of the redesign. The fourth conclusion may require tests in bench test devices, in conjunction with prototypes.

G. Testing and Simulation

Testing must be very well planned. It is costly and fruitless to purchase bench test machinery and launch into testing of materials or lubricants without experience and preparation. It is doubly futile for the novice to run accelerated wear tests, with either bench tests, prototypes, or production parts. Furthermore, an engineer learns very little by having wear tests done by a distant technician, who supplies the engineer with cleaned-up specimens and data on equilibrium wear rate at the end of the test.

The problem is that wear resistance is not a single property of any material. Hardness, Young's modulus, and density are single properties, which may be measured in standard tests. Wearing of material, by contrast, occurs by a succession of changing balances of mechanisms, controlled by the sliding situation. The life of tribological systems can only be determined in service where tribological conditions change, changing the wear mechanisms over time. Laboratory tests usually carry a product through an artificial lifetime, from which projections to lifetime in "real" environments can hardly be made. Experience with similar products helps both in establishing the "artificial lifetime" of products in laboratory tests and in projecting from test results to "real" product lifetimes.

Types of test, test parameters, and other details are discussed in the following paragraphs, followed by a suggested criterion for degree of correlation or simulation between wear test results and wearing of the product under study.

1. Standard Test and Test Devices

Standard test devices are described in several references. Some of them were developed as material wear testers, such as the dry-sand-rubber-wheel test. Many others were developed for testing lubricants, such as the four-ball tester. A few have been developed specifically to measure friction, such as the (tire) skid resistance machine. Most test devices simply slide two specimens together in a simple manner, whereas a few, such as a hip joint, simulate or attempt to emulate the complex motion of the "specimen" during walking or running.

Many, if not most, available test devices are named in the standards of the Society of Automotive Engineers (SAE), the American Society of Standards and Materials (ASTM), and standards making societies of other nations. Many "standards" are actually standard methods for operating the test devices so that everyone that uses identical devices will obtain very nearly identical results. This is a useful exercise where products, materials, and lubricants are to be compared or ranked for some particular quality. However, the measured or inferred quality may not be the quality connected with desired product life. In short, any particular standard test may not (probably does not) simulate the experience of material or lubricants in practical machinery. Methods for determining how well test devices and test procedures simulates practical machinery will be discussed in Section 4. Clearly, no standard test method assures simulation with any real product.

A clear indication of the problem with bench tests may be seen in some results with three test devices. These are:

1. Pin-V test in which a 1/4-in. diameter pin of 3135 steel rotates at 200 rpm with four-line contact provided by two V blocks made of 1137 steel.

2. Block on ring test where a rectangular block of a chosen steel slides on the outer (OD) surface of a ring of hard, case-carburized 4620 steel.
3. The four-ball test where a ball rotates in contact with three stationary balls, all of hardened 52100 steel.

The four-ball test and the ring on block test were run over a range of applied load and speed. The pin-V test was run over a range load only. All tests were run continuously, i.e., not an oscillating or stop–start sequence mode. All tests were run with several lubricants.

Results from the ring-block test were not sufficiently reproducible or consistent for reliable analysis. Results from the other two tests were adequate for the formulation of a wear equation from each, as follows:

Pin-V test: wear rate μ (Load)^2
Four-ball: wear rate μ (Load)$^{4.75}\times$(Speed)$^{2.5}$

These results may be compared with linear laws of wear discussed frequently in the literature, which would be of the form:

Linear law: wear rate μ (Load)$^{1.0}\times$(Speed)$^{1.0}$

No one can say in advance which equation fits the product under redesign. Likely none does. The linear law is the most widely used though, most likely because of its simplicity. Frequent use feeds on itself where accurate knowledge is lacking.

2. Necessary Variables to Consider in Wear Testing

The number of necessary variables that influence wear rate probably exceeds 30. Some variables are not usually explicitly mentioned but are embodied in designer's choice to use the same materials, prepared in the same way (surface processes, etc.) as that in the practical sliding pair under study.

Perhaps the single and most important error committed in wear testing is to assume that an adequate test is one in which the contact pressure (load) and sliding speed are the same as in the practical device under design. If this were the case, then the equations embodying the test results given above would have identical exponents. The fact that they do not most likely indicates that the results are sensitive to several of the unmentioned test parameters or material (including lubricants) variables. There are no methods for determining which parameters and variables are missing. There is no exhaustive list of available test variables, but for illustration, a few are mentioned briefly.

1. Contact Shape. A sphere-on-flat test produces different results from a cylinder-on-flat test probably because wear particles are recycled differently in the two tests. Particles are swept aside in the ball-on-flat test but are constrained to pass through the contact area in the cylinder-on-flat test. The result is a great difference in load-carrying capacity in boundary-lubricated sliding, for example.
2. System Vibration. The smallest effect of vibration is a time-varying contact stress, whereas the most significant is the alteration of wear particle movement.
3. Tracking variability in repeat pass sliding produces varying results.
4. Reciprocating sliding moves wear particles and fluid films differently than does circular repeat pass sliding.
5. Oxygen availability to the contact region influences the types and amounts of oxides that form.
6. Duty cycle and standing time influence temperature and surface chemistry.

3. Accelerated Tests

The most common way of conducting accelerated test is to increase either the load or sliding speed. The hazard of such procedures may be seen in two ways. The first is by considering the equations in "Standard Test and Test Devices" above. The influence of load and speed are different in each test and are certain to be different again in the device under design study. An accelerated test has no meaning if the influence of accelerated conditions is not known.

A second point is made by examining Fig. 1 [1]. If the design under study (using steel of 348 VPN hardness) is to operate with a load of 20 g, an accelerated test in which 200 g of load is applied would indicate an unacceptable wear rate. Note that the transition between 20 and 200 g load covers a range of about 2.5 orders of 10, i.e., about 130 times! Conversely, if the design uses a load of 200 g, a test where 2000 g is applied would produce deceiving results. It should be noted that the curve for the wear rate of 348 VPN steel can be characterized by a much more complicated equation than those given in "Standard Test and Test Devices" above. Figure 1 has been verified many times in student exercises. It shows a rather extreme variability but serves to illustrate the point that wear rates behave in a regular manner for very few materials over a wide range of variables. Most often, data are shown over rather narrow ranges of load, sliding speed, and other variables, and it may be suspected that usually, the simpler data are published.

4. Criterion for Adequate Simulation

Experience shows time after time that simple wear tests complicate the prediction of product life. The problem is correlation of assurance of simulation. For example, automotive company engineers have repeatedly found that engines on dynamometers must be run in a completely unpredictable manner to achieve the same type of wear as seen in engines of cars in suburban use. Engines turned by electric motors, although heated, wear very differently from fired engines. Separate components such as a valve train can be made to wear in a separate test rig nearly the same way as in a fired engine, with some effort, but cam materials rubbing against valve lifter materials in a bench test inevitably produce very different results from those in a valve train test rig.

Most machines and products are simpler than engines, but the principles of wear testing are the same, namely, the combination of wear mechanisms must be very similar in

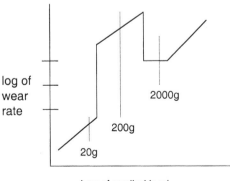

Figure 1 Wear rate versus load for 1050 steels of 348 VPN hardness.

each of the production designs, the prototype test, the subcomponent test, and the bench test. The wear rates of each test in the hierarchy should be similar, the worn surfaces must be nearly identical, and the transferred and loose wear debris must contain the same range of particle sizes, shapes, and composition. Thus it is seen that the prototype, subcomponent, and bench tests must be designed to correlate with the wear results of the final product. It would be best also if the measured coefficient of friction, electrical contact resistance, and approximate surface temperature were also similar. This requires considerable experience and confidence where the final product is not yet available. This is the reason for studying the worn parts of the product nearest to the redesign and a good reason for retaining resident wear expertise in every engineering group.

5. Measurements of Wear and Wear Coefficients and Test Duration

The measurements of material loss to be taken from a test could take any form that eventually can be used to predict the end of the wear life of a product. These include wear volume/unit distance of sliding, mass loss/unit of time, change in wear track width/unit distance of sliding, etc. (Conversion from one form to another may require knowledge of material density.)

The measurement of mass loss often requires longer time tests than other methods, which can cost more money than necessary. Generally, it is useful to devise ways to make precise measurements of volume loss than to wait for sufficient mass loss to measure. One precise method is to use a surface tracer system to measure the profile of the wear scar. With a properly programmed computer, several parallel traces can be converted into a volume of material loss. An important caveat in using precise methods is to assure first in a test series that wear is progressing in the same way early in a test as it does (would) late in a test.

H. Material Selection Criteria

Selecting materials for wear resistance requires a study of the details of wear in a wearing system (including the solids, the lubricant, and all of the wear debris) such as in an old product being redesigned or in a wear tester. This type of study need not be particularly sophisticated, and, in the examination of wearing surfaces, the materials engineer is no better prepared than is the mechanical engineer. [The best study is done with a variable-magnification (e.g., 3–40×) binocular microscope, with oblique or side lighting upon the object under study. It is not often useful to resort to a high-magnification metallurgical microscope and still less useful to use the scanning electron microscope early in an investigation.] Observe the wear debris, and note changes in surface texture and changes in the lubricant while wear is progressing. Keep a good record of observations, obtain photographs with and without magnification, and retain worn surfaces. Then attempt to name the type of wear that is seen, but do so in terms that indicate what mechanism of material failure has taken place.

There are very many terms in handbooks, technical papers, and in the language of engineers to describe types of wear. Most of the terms have meaning to people that use them, but they have little meaning to a person that had not observed the described wear processes. Two terms that are very widely used are "abrasive" wear and "adhesive" wear. Neither of these terms describes specifically how material is lost, and yet they do reflect two major types or mechanisms of wear.

1. Mechanisms of Wear

A wear mechanism is a basic cause or detailed explanation, usually on a microscopic scale, of how materials wear away. Two major causes or mechanisms of wear are

1.   Carving away material by sharp and hard particles (as in abrasion and erosion).
2.   Loosening of fragments of material by repeatedly imposing normal and friction stress on a surface, without abrasive materials present. The loosened fragments are eventually removed as wear debris. This mode of wear is often called adhesive wear, *although* there is usually a significant fatigue mode of materials failure involved.

The nature of the wear debris depends on the materials and the severity of abrasion or adhesion. As to materials, all metals in air except gold have oxides on them, some of which are hard and some are soft: many ceramic materials are covered with very thin films of soft reaction products (e.g., SiO_2 on Si_3N_4) and many polymers are covered with process fluids that bloom to the surface.

Very "light" abrasion or erosion will only remove the surface substances. Deeper abrasion will also remove substrate material. Abrasion of a ductile substrate will produce wear debris in the form of chips, whereas abrasion of a brittle substrate will produce very fine fragments or crystals.

Very "mild" nonabrasive sliding conditions will treat soft films on hard substrate material as lubricants. Soft oxides will rub off, and the expelled debris will contain metal ions, which constitutes wearing away of the metal. This is probably the predominant mode of wear of long-lasting products. When any of the soft films are rubbed off faster than they reform, the friction forces increase and, eventually, the sum of normal and friction forces will produce shearing of ductile substrates and fragmentation of brittle substrates. The ductile material will likely roughen more than wear away, rendering the sliding system unusable. This mode of surface failure is called galling (or perhaps scuffing or scoring). Metals with hard oxides such as that on aluminum and stainless steel usually have high friction and are very prone to galling: this occurs in dry sliding but is particularly likely in high-speed sliding in boundary lubrication conditions.

The above discussion relates to the single composition or monolithic materials from which many sliding/rolling components are made. With the exception of the oxides and other surface substances (nitrides, carburized layers, phosphate, etc.), monolithic materials have the same composition to some depth. But, increasingly, coatings are being specified for sliding surfaces. The general intent is to apply a wear-resisting coating or perhaps low friction coating on a low-cost substrate material. Much can be written about attempting to salvage a poor design by using coatings, but not here. The most important problem with wear-resisting coatings is that they are hard and brittle and when they chip off, they break up into abrasive particles.

There are very many types of coatings, but special attention will be given here to a thin hard coating called titanium nitride (TiN). A description of all coatings and the processes by which they are formed can be found in materials books. TiN will be cited here for two purposes. One is to include a hard coating in the table of wear mechanisms that follows later in the chapter, and the other is to emphasize a major problem in the specifying of materials for wear and friction control. The problem is the common assumption that a designated or generic material has only one set of properties. Designers are aware of the large range of wear-resisting properties of steels, aluminums, bronzes, and coatings, but new materials are often treated as "generic" materials with no differentiation across all available forms of the new product. The example of TiN is illustrated here, and some mention will also be made of diamond-like carbon (DLC), a modern and magic coating.

The coatings to be described are new compounds formed upon substrate materials which do not react to a great depth with or diffuse into substrate materials, but yet bond to the substrate. These processes differ from those processes that harden steel surfaces: this may be done by heating surfaces of steel to the temperature that produces the austenitic phase, followed by rapid cooling to form a hard phase known as martensite. Sometimes, a surface-hardening process of steel is combined with a prior chemical modification of the surface layer by diffusing carbon, nitrogen, or boron atoms into the near-surface layer of steel. Coating requires very different processes than those above.

2. Coatings and Coating Processes

Thin film coatings are usually of thickness between 1 and 10 μm. Some are of the same composition throughout, whereas some are multilayers of different chemical composition. Most thin coatings are produced by vacuum-based (or low pressure) deposition processes, and some are produced by an electrochemical deposition method. For most tribological applications, especially on new components, thin film coatings are preferred over "thick coatings" such as chrome plating, zinc plating, or polymer coating. Thin coatings can be easily incorporated into existing design without changing a part dimension because the coating thickness is well within the tolerance specifications for most tribological components. Thin coatings therefore need not be machined or ground to meet dimensional specifications [2,3].

Vacuum-based deposition is also referred to as vapor phase deposition. There are two broad groups of these processes, namely, physical vapor deposition (PVD) and chemical vapor deposition (CVD). There are numerous variations of these two methods in use.

The production of TiN films is most often done by physical vapor deposition (PVD). In a typical process, an intense electron beam is directed upon a "source" of titanium from which small groups of atoms are ejected or "sputtered." The stream of sputtered particles passes through and reacts with nitrogen: the product is directed towards the part to be coated by an electric field where it condenses. *In general,* the various forms of PVD processes are line-of-sight processes. Consequently, continuous rotation of the parts being coated is necessary during coating in order to obtain uniform coating coverage on the part.

Coating processes must satisfy several requirements. One is control of the uniformity of coating composition and thickness and another is good adhesion of the coating to the substrate. When a coating comes off in use or even by spontaneous separation due to residual stresses, it will usually fragment and circulate in the sliding system as abrasive particles. After all, a hard coating is often applied to a softer substrate and abrasion occurs by particles that are harder than the sliding members. Coating uniformity can be a major issue in complex-shaped parts such as gears. Lack of uniform coverage can result in nonuniform wear and all the associated problems.

a. Variability in Coating Composition, Microstructure, and Properties. One aspect of good design is being aware of the wide range of wear-resisting properties of materials. The wide range of properties of most metals, polymers, and ceramic materials is the subject of a very large literature, but not so for new materials and coatings. Actually, TiN comes in a wide range of composition. It is difficult to control the composition of residual gases and other constituents in a vapor deposition reactor, but, in addition, different vendors have different ideas on how best to make "TiN" coatings.

The two most important variables influencing the properties of TiN are grain morphology and the ratio of N to Ti. Grain morphology is strongly influenced by substrate temperature and nitrogen/argon partial pressure within the deposition chamber as shown in Fig. 2 [4]. The ratio of N to Ti atoms within commercial TiN coatings ranges between 0.5 and 1.2 and is spatially nonuniform in the coatings. This ratio has a large effect on hardness as shown in Fig. 3 [5]. The relative amount of N/Ti atomic ratio is shown in Fig. 3.

Once a coating of TiN is formed, it is generally not possible to alter its structure. By contrast, the structure of steel can be altered through heat treatment and other processes to get some desired properties. Thus a 1080 steel, for instance, could be made as soft as 20 Rc or as hard as 65 Rc by changing the microstructure from pearlite to martensite.

The variability highlighted for TiN is also true for other hard thin film coatings. Table 1 [6] shows hardness comparison of some bulk and thin film nitrides and carbide materials.

It is difficult to gain a perspective on new materials, particularly when vendors of new materials advertise them in glowing terms as miracle solutions to many problems. Consider the unwarranted claim that titanium is harder than steel and is therefore the better material for use in golf clubs. Consider also the general impression that diamond-like carbon (DLC) has tribological properties well beyond that of ordinary materials.

Diamond-like carbon, incidentally, is an amorphous form of carbon, with a structure and properties between those of graphite and diamond. It is usually made from methane and hydrogen in a vapor deposition process. The amount of hydrogen in the film governs the properties: 10% bonded hydrogen will produce a structure of about half (layered hexagonal

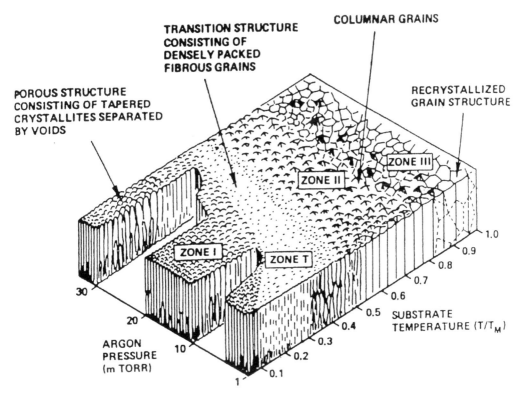

Figure 2 Coating grain morphology as a function of substrate temperature and partial pressure.

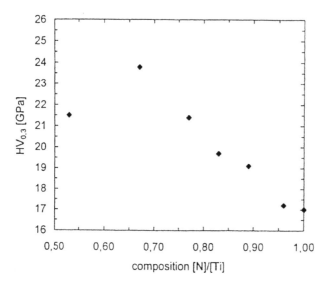

Figure 3 Variation of TiN coating hardness with N/Ti atomic ratio.

structure) graphitic and half (cubic, zinc blende structure) diamond with a hardness of about
30 GPa; 50% bonded hydrogen will produce about 30% diamond and 70% graphite with
a hardness of about 10 GPa. This latter form has very low friction in the range of 0.007 to
0.5 in a vacuum but in a more normal range of 0.05 to 1.0 in air. *Although* DLC produces
low friction in vacuum, it is no better in wear prevention than is any other solid lubricant.
One problem is that in the temperature range of 200–400°C, DLC changes to graphite-like
carbon.

    b. Bond Strength. The overall procedure in selecting materials, in particular,
coatings, must involve the testing of materials in conditions that simulate the expected
final production output. But, with coatings, an extra and vital consideration is the strength
of the bond between the coating and substrate. Rarely do vendors mention these properties.

Table 1 Hardness Values (VHN) for Bulk and Thin
Film Materials

Materials	Bulk	Films
TiN	2000	1500–4000
ZrN	1500	700–3600
HlN	1600	1850–2400
VN	1500	600–2000
NbN	1400	1100–3000
TaN	1000	—
Si_3N_4	1000–2000	2000–3000
TiC	3500	2770–4150
VC	2500	1900–2850
WC	2050	1800–2800
TaC	2200	1280–2200

Source: Ref. 6.

It is usually left for the user to try out a coating for their application. But when coatings flake off, there could be several reasons:

1. The interfacial bond may be weak.
2. A high residual stress (usually tensile) in the coating could cause cracking of the coating, with progressive peeling of islands of coating.
3. The contact pressures could plastically deform the substrate, thereby cracking the coating.

With these difficulties, it may be better to avoid coating. However, there are good general principles that can be followed for their successful use. From the failure mode analysis, the required coating attributes can be identified. For instance, if abrasive wear is identified as dominant in a component, then coating hardness is the most important property. If it is found that the temperature rise on sliding surfaces is detrimental, then low friction coatings and coatings with high thermal conductivity may alleviate the problems. In the case of corrosive or chemically dominated wear, corrosion-resistant coating is required. In general, most of the thin hard coatings and ceramic materials are more corrosion-resistant than are most metals and alloys. However, chemical reactivity may be necessary for good lubrication. Many commercial lubricants include chemically active constituents that react with iron and steel to form a "protective" film on the metal that survives severe contact conditions. TiN will not react with those chemical constituents and thus is a poor material for the prevention of scuffing and galling in lubricated high-speed contact.

I. Table of Wear Mechanisms

Table 2 is offered to provide a perspective on wear mechanisms. This table is based on the modes or mechanisms that remove material rather than on the common or usual subjective descriptions of wear. Designers really have no "starting point" in their search for materials, and they have no way to select materials on the basis of material properties since there is no useful connection between mechanical properties and wear resistance, except perhaps for simple abrasion processes which can be resisted by hardness in many cases. *Rather,* Table 2 is based on the idea that most activity of designers is redesigning of produces, for upgrading or for lowering cost, or to make a product slightly different from that of the competition. Materials selection then becomes a matter of observing what type of wear had taken place in previous products and either avoiding that type of wear by material replacement or reducing wear rates by altering the properties of presently used material. The exercise begins then with a description of surface damage and some connection with the property of material that may resist that mode of wear.

1. Seven Primary Mechanisms and One Auxiliary Mechanism of Material Loss from a Surface

a. Formation and Removal of Products of Corrosion/Oxidation/Chemisorption. New compounds form by chemical combination of an active constituent in a fluid lubricant or environment with a substrate material. The compounds, which contain ions from substrate materials, can be removed by very fine abrasives, by erosion, or by sliding without abrasives whether dry or lubricated. This is probably the mode of wear of the majority of long-lasting surfaces. Several terms have been suggested to describe this mode of wear, most often,

Table 2 Material Characteristics That Resist Several Mechanisms of Material Wear

Mechanisms of material loss	State of material to resist wear	Precautions to observe when selecting a material to resist mechanisms of material loss
Formation and removal of products of corrosion (including oxidation)	Reduce corrosiveness of surrounding region, increase corrosion resistance of metal by alloy addition, or select soft, homogeneous metal, ceramics, or polymers.	Total avoidance of new chemical species can result in high adhesion of contacting surfaces, and soft materials promote galling and seizure.
Cutting (may actually not be a unique mode of material failure, but rather a type of fracture.)	Achieve high hardness, either throughout, by surface treatments or by coatings. Add very hard particles or inclusions such as carbides, nitrides (ceramics), etc.	All methods of increasing cutting resistance cause brittleness and lower fatigue resistance.
Ductile fracture	High strength, achieved by any method other than by cold working or by heat treatments that produce internal cracks or large and poorly bonded intermetallic.	
Brittle fracture	Minimize tensile residual stress, for cold temperature, insure low-temperature transition, temper all martensite, use deoxidized metal, avoid carbides as in pearlite, *etc.* and assure a good bond between fillers and matrix in composites to deflect cracks.	Soft materials will not fail in a brittle manner, and will not resist cutting very well.
Low-cycle fatigue	Use homogeneous materials and high strength material that do not strain-soften, avoid overaged metals or other two-phase systems with poor adhesion between filler and matrix.	

High-cycle fatigue	For steel and titanium, apply stresses less than half the tensile strength (however achieved) and for other metals to be cycled fewer than 10^8 times, allow stresses less than 1/4 the tensile strength (however achieved), avoid retained austenite, select pearlite rather than plate structure, avoid poorly bonded second phases, avoid decarburization of surfaces, avoid platings with cracks, avoid tensile residual stress or form compressive residual stress by carburizing or nitriding.	Calculation of "contact stress" should include the influence of tractive stress.
Melting	Use material with high melting temperature and/or high thermal conductivity.	
Debonding	The strength of bond between the coating and substrate is determined by the parameters of the coating reactor, and the range of bond strength is very wide. There are tests for bond strength, some of which measure the shear strength of the bond, and some measure the tensile strength of the bond. Bond fracture occurs where stresses are imposed on interfaces by differential thermal expansion, by high friction on the coating, or by high differential straining due to contact pressures, and particularly, where the substrate plastically deforms. Since the bond is brittle in nature, its strength is much reduced when multiple cycles of strain are imposed.	Since hard coatings are available from a great number of vendors, careful comparison should be made among the products. None can be taken to be "generic."

General caveat: Metals of high hardness or strength usually have low corrosion resistance, all prone to early fatigue failure, polymers creep under constant load, and all materials with multiple phases and multiple desirable properties are expensive.

oxidative wear and mild wear. The terms corrosion, oxidation, and chemisorption are usually not considered to be synonymous.

Applied substances such as graphite, diamond-like carbon, molybdenum disulfide, phosphates, etc. all "wear off" in the same way as do the compounds formed by corrosion/oxidation/chemisorption. The first three are considered as solid lubricants, and their removal usually does not involve loss of substrate materials. Phosphates are usually considered a break-in, scuff avoiding coating, and *although* it functions as a solid lubricant, it is not called that because they do no lower friction very much.

    b. Cutting by Abrasive, Erosive, or Scratching Substances, which Removes Materials by Mechanisms Relating to Whether the Wearing Surface Is a Brittle Material or a Ductile Material. Materials of intermediate brittleness may produce mixed effects in cutting. Effective "cutting" tools or particles must be about 1.3 times as hard as the material being cut.

1. In cutting (or grinding) very brittle material, the surface fragments into very fine particles, as fine as the nanometer scale. Fragmentation occurs both by single encounters by brittle failure or by elastic fatigue, which in ceramic materials, for example, could occur in fewer than 10 cycles. Many fragmented particles will agglomerate by electrostatic attraction because of their very large surface area compared to their volume: many will fall out of the contact area immediately.

2. In cutting (grinding) ductile materials, there may be some material that is removed in the manner of a chip cut by a tool in a lathe. However, since few abrasive (erosive) substances are very sharp, most particles will form a groove in the substrate surface, plowing material aside. Subsequent particles will continue to plow surface material aside, repeatedly until the plowed mounds fracture by plastic fatigue.

    c. Nonabrasive Sliding and Rolling Imposes Stresses on Surfaces, Both Normal Stress and Shear Stress (Traction). Repeated stresses will eventually cause failure by either elastic (*high cycle*) fatigue, or plastic (*low cycle*) fatigue. Plastic fatigue requires only a few tens of cycles to fracture, whereas elastic systems are often expected, by selection of applied stresses, to survive millions of cycles. Asperities function in the plastic mode with little average applied load and will be the first to fracture. As materials wear away, elastic fatigue cracks may spread and penetrate deep into the substrate, or even through the body. This occurs in aluminum and some other metals where small amplitude cyclic shearing (fretting) or repeat pass rolling initiates a fracture into the substrate. Wearing in nonabrasive sliding is often referred to as "adhesive wear." This refers to the possibility that in some regions of contact, there may be nanoscale regions of primary (covalent, ionic, or metallic) bonding, which is likely to tear bits of material off by *plastic failure*. In most regions over a surface, however, high bonding forces will not occur because of the presence of soft oxides, soft products of chemisorption on ceramic materials, adsorbed liquids, and solid lubricants.

Nonabrasive sliding and rolling on thin hard coatings may shear off or debond segments of coating. High contact pressure may cause plastic flow of substrate material, thereby also causing debonding.

    d. Sliding Causes Temperatures to Rise on and in Surfaces. Temperature rise of hard coating can cause debonding or the more localized spalling where the temperature rise causes interface stresses to rise because of different thermal coefficient of expansion between the coating material and substrate material. At high sliding speed, there can be *melting*.

Table 2 also lists the material characteristics that resist several mechanisms of material loss or wear. Rarely is the wearing of any component due to a single mechanism, but rather it is due to some combination, which is likely to change during the wearing process. Surfaces should be observed and with experience, the various modes and types of wear will become apparent.

REFERENCES

1. Welsh, N.C. Philos. Trans. R. Soc. (Lond.) 1965, *257A*(part 2), 51.
2. Bunshah, R.F. *Handbook of Hard Coatings: Deposition Technologies, Properties and Applications*; Noyes, 2001.
3. Bhushan, B.; Gupta, B.K. *Handbook of Tribology: Materials, Coatings and Surface Treatments*; McGraw-Hill, 1991.
4. Thornton, J.A. Annu. Rev. Mater. Sci. 1977, *7*, 239.
5. Lengauer, W. In *Handbook of Ceramic Hard Materials*; Riedel, R., Ed.; Wiley-VCH, 2000; Vol. 1, 236 pp.
6. Strafford, K.N. Surf. Coat. Technol. 1996, *81*, 106.

19

Simulation Methods for Interfacial Friction in Solids

James E. Hammerberg and Brad Lee Holian
Los Alamos National Laboratory, Los Alamos, New Mexico, U.S.A.

I. INTRODUCTION

The microscopic atomistic understanding of dissipation at sliding material interfaces is a necessary ingredient in constructing physical models of the frictional interaction between materials at larger length scales. From an engineering point of view, one requires models of the interfacial stresses as functions of the relevant material variables for use in finite-element and finite-difference macroscopic computer codes. Relevant variables depend on materials, environment, and velocity regimes. Thus, when values of frictional coefficients are quoted for different material pairs, reproducibility requires, in most cases, specification of atmosphere, humidity, surface roughness, chemical composition, and microstructure at and away from the interface. Because a steady sliding state is ultimately achieved, the evolution of these variables toward the steady state is required as well. The character of the steady state depends on several length and time scales. Surface roughness in the form of the asperity distribution function clearly plays a role. The effects of microstructure, dislocation distributions, and their relationship to plastic deformation are important for both ductile metals and brittle materials. Defect production and mobility and crack propagation under strong shear loading become relevant. Surface chemistry (electronic states) is important when mechanical deformation plays a subsidiary role. Then, chemical kinetics, electron–electron, and electron–phonon interactions, as well as surface charge distributions are necessary elements to the description of dissipation. For lubricated interfaces, the rheology of complex fluids or hydrocarbon layers determines the local interfacial motion. If one crudely defines macro-, meso-, and microscale physics by length scales of mm, μm, and nm, and time scales of msec, μsec, and nsec, then the determination of the interfacial frictional force, measured at a distance from the interface, becomes quintessentially a multiscale problem in materials science.

What then, is the role of simulation in defining and modeling the processes involved in sliding friction? How do we use this information to predict the behavior of the average tangential force? What are the strategies for increasing or decreasing it in practical applications? Applications include macroscopic engineering, bioengineering, or nanoscale problems involving microelectromechanical systems (MEMS). There has been substantial progress in recent years as a result of an increase in computational capacity and speed, as well as a host of new experimental techniques, such as atomic force microscopy (AFM), quartz crystal microbalance (QCM), surface force apparatus (SFA), etc. These methods have been treated in recent reviews in great detail [1–7].

II. SOME HISTORY

A. Genesis of Molecular Dynamics

Molecular dynamics (MD) refers to the numerical solution of Newton's equations of motion for a large number of particles interacting with each other via pairwise interactions or by more complicated many-body interactions depending on the instantaneous particle configuration [8,9]. External forces representing experimental driving and constraints may also be included, as is the case for the frictional simulations discussed below. Once the interactions have been specified, the equations of motion, subject to initial and boundary conditions, specify the dynamical problem completely. The numerical difficulty of solving such a set of coupled nonlinear equations depends on the number of particles that it is feasible to follow for a given length of physical time. The interpretation of the measurable mechanical properties of the system, and their relation to statistical ensembles, and their approach to equilibrium or the nonequilibrium steady state is the principal task in the study of nonequilibrium irreversible processes in matter. The first studies of what may be termed "molecular dynamics" were conducted by Alder and Wainwright [10,11] in 1956 for two-dimensional (2-D) hard disks, which represent the simplest possible interaction potential.

The first MD simulation for atoms interacting via a continuous potential, other than 3-D hard spheres (or 2-D hard disks), and thus requiring a finite-difference integration scheme for the equations of motion, was carried out by George Vineyard and coworkers at the Brookhaven National Laboratory in the late 1950s. Their work on radiation damage cascades in copper, involving 500 atoms, was published in 1960 [12] in *Physical Review*. The tree-like cascade structure of interstitials and vacancies in a crystal is obtained by giving one atom—the so-called "primary knock-on atom" (PKA)—a large initial (nonthermal) velocity, as if it had been hit by an energetic neutron. As such, this is a relaxation simulation similar in spirit to Alder and Wainwright's early simulation of Boltzmann's H-Theorem. There, a highly nonequilibrium initial condition was imposed, namely, 100 hard spheres were lumped into one corner of the periodic boundary condition (PBC) computational box with initial random velocities. The system then equilibrated over time, arriving at a uniform low-density gas. The velocity distribution function, $f(v)$, was monitored as a histogram. The integral over velocity of $f \ln f$, $H(t)$, which is proportional to the negative of the entropy of an ideal gas, was seen to decay with time. In other words, the entropy gradually rose to the equilibrium value, as predicted a hundred years earlier by Boltzmann. Still, both of these calculations, radiation damage and equilibration of a gas, required the power of computing machines for their realization, and they gave far greater insight into the mechanisms at the atomic level in both solids and fluids than were possible by more approximate methods.

It is interesting to note that Vineyard's work on radiation damage was essentially ignored by the physics community, at least those concerned with the behavior of fluids. In his article discussing molecular dynamics simulations of the equilibrium properties [including the velocity autocorrelation function (VACF)] for the Lennard–Jones (LJ) fluid, Aneesur Rahman [13] of the Argonne National Laboratory gave only an oblique reference to the nonequilibrium molecular dynamics simulations of Vineyard's group, by way of referring to Joe Beeler's review article [14] in the same year (1964) as Rahman's. Three years later, Loup Verlet [15] "introduced" the central-difference equations that have come to bear his name, and it took a considerable number of years for the molecular dynamics community to notice that Vineyard had introduced them 7 years before Verlet.

The reasons for the historical disconnect between the fluid and solid communities are puzzling and not a little disturbing. If Vineyard and Rahman had been at national laboratories in the USSR and the United States, instead of both in the United States, one might have attributed the oversight to linguistic and political barriers to scientific exchange. One is tempted, reluctantly, to invoke egos and a cultural divide between academic physics vs. engineering materials science. A darker side to the issue is one of the legitimacy of equilibrium statistical mechanics, which was emphasized by the fluids community, vs. nonequilibrium statistical mechanics, which was represented by nonequilibrium MD (NEMD). The latter received its impetus in a complete body of work on shear and bulk viscosity, and thermal conductivity of fluids by Bill Ashurst and his Ph.D. advisor, Bill Hoover [16].

Nonequilibrium MD began its ascendancy in the early 1970s in response to Alder and Wainwright's discovery of the long-time tail of the VACF in dense fluids. To solve the evolution of the velocity distribution function, Boltzmann had to invoke the notion of "molecular chaos," or the exponential decay of the VACF, whereby a particle's velocity was quickly forgotten. Instead, using a hydrodynamic model, Alder showed that the particle set up a velocity field among its nearby neighbors that resembled a double vortex, whereby the particle eventually received a kick in the pants, and thereby a long-time enhancement of the memory of its earlier velocity history. To compare the autocorrelation function time integral (Kubo theory) for the transport coefficients, Hoover and Ashurst proposed a direct calculation of them, via setting up steady-state fluxes and gradients, whose ratios are the transport coefficients—long-time tails and all.

The success of NEMD and its comparison with equilibrium autocorrelation functions (ACFs) and their time integrals was a long time in coming, not because of errors in NEMD simulations, nor because of errors in the fundamental basis for transport coefficients in the limit of zero gradients and their relationship to ACFs. Rather, it turned out to be problematic because of errors in equilibrium calculations of the ACFs themselves. (It should be noted that any quantity that depends on measurements of fluctuations is inherently more difficult to evaluate than simple averages of experimental observables. NEMD, being a direct evaluation of the mechanical observables in nonequilibrium steady states, is simpler and more foolproof.)

B. NEMD Simulations in High Rate Deformation

The most difficult problem that has been studied by NEMD is the propagation of shock waves in dense media. Shock waves are caused by the collision of objects. Planar shock waves are studied via impacting plates, resulting in 1-D waves, which are naturally easier to interpret than fully 3-D shocks. Large gradients are inherent in the shock front, which separates unshocked and shocked material. Behind the propagating front, shocked material is at higher density, pressure, and temperature than the initial state. Fluid shock waves are, in general, simpler to interpret: Navier–Stokes transport coefficients are well understood, and the equilibrium equation state of a familiar material—at least computationally speaking—such as Lennard–Jonesium, is well known. As Alder and Hoover showed in their pioneering work, hydrodynamics persists down to the atomistic level—on distance scales approaching a few atomic spacings in dense fluids, and times as short as a few mean collision times (vibrational periods in solids). However, the process of shock propagation in crystalline solids takes on more serious difficulties by virtue of the larger-length scales involved in defects, particularly the carriers of plastic deformation: dislocations. One could say that defects are almost everywhere in fluids, but almost nowhere in solids, hence the larger scale needed for atomistic simulations of solids.

Shock waves in solids, even perfect crystals, can produce defects just by the nature of uniaxial compression. The pressure in the direction of shock propagation can greatly exceed that in the transverse directions, such that a shear stress is produced in the propagation of a 1-D planar shock. Materials, either liquid or solid, do not "prefer" to be in a state of mechanical disequilibrium, hence they tend to flow viscously in the case of fluids, and yield plastically in the case of solids. This transverse motion gives rise to the steadiness of true shock waves, distinguishing them from strong, elastic sound-wave pulses that do not leave the material in a state of higher entropy, density, pressure, or temperature. The calculation of shock propagation in solids by NEMD has intrinsic limitations: the plastic deformation that occurs when the solid yields has to be contained in the computational box. Transverse boundary conditions in NEMD shock simulations are typically periodic, mimicking the experimental shockwave conditions in the laboratory, where the central part of the impacting plates are diagnosed, taking care to eliminate the confusion caused by relief waves coming in from the edges.

From the point of impact in the NEMD shock simulations, the calculations must proceed long enough to establish a steadily propagating wave. Moreover, the transverse periodic dimensions must be sufficiently large to contain the natural length scale of plastic deformation structures—too small a dimension may suppress the mechanisms of plastic yielding. Beyond these considerations for homogeneous nucleation of plastic flow (or phase transformation), the problem of preexisting defects, which can be heterogeneous nucleation sites, imposes even more stringent limitations on feasible NEMD simulations. Because NEMD is expensive in proportion to the length of time needed for a simulation of a propagating wave, as well as to the number of particles, the window of opportunity can be small. Fortunately, shockwave phenomena are matched as ideally as possible for NEMD, because strong shocks can be quite thin (on the order of a few lattice spacings) and quite prompt (on the order of a few vibrational periods). Nevertheless, as we go down in shock strength (i.e., impact velocity), the limitations begin to loom dark: shock thicknesses increase and preexisting defects become increasingly important. Thus weak shocks strain the limits of modern-day computers, and as such, represent a challenge to "hand off" information that can be gleaned from NEMD to mesoscale models [17].

In addition, the long-time difficulties of achieving steady-state propagation when slow, thermally activated processes are occurring, such as annealing of defect structure, or "ripening" of microstructures, lead us to consider means of accelerating the time scale. The time scale in NEMD has always been tied to the atomistic vibrational period (subpicosecond), and computer speeds have not kept pace over the last few decades with computer memories. Massively parallel NEMD simulations have become possible, extending system sizes more rapidly than total simulated physical times.

Finally, all of these considerations come together in the problem of high-speed sliding friction between dry metal surfaces. As we will show, interesting plastic deformation produces marvelously complex microstructures, and the time and distance scale difficulties mentioned for solid shock waves apply equally well to the friction problem.

C. Thermostatting in NEMD Simulations

In this section, we discuss the problem of maintaining control of NEMD simulations under strong driving conditions, as well as the means of driving for sliding friction. The driving force for sliding friction can be determined via applying constant velocity boundary conditions to a pair of finite reservoir regions at the top and bottom sides of a computational box, the upper reservoir moving to the right at velocity $+u_p$ and the lower reservoir to the left at $-u_p$ (cf. Fig. 5). For a fluid sample, the long-time steady state would correspond to Couette shear-flow conditions, characterized by a linear x velocity profile as a function of y. If the sample is contained within $-L_y < y < L_y$, then the strain rate for the linear profile is $\dot{\varepsilon}_{xy} = \partial\dot{x}/\partial y = u_p/L_y$. On the other hand, for a solid sample sliding at the $y=0$ interface, the appropriate long-time conditions are a square-wave x-velocity profile. In either case, the reservoirs constrained to move at $\pm u_p$ will tend to gradually heat up the sample, so that some form of thermostatting, representing an infinite sample, needs to be applied to prevent overheating.

The simplest way to implement this driving and thermostatting is to rescale velocities in the reservoirs at each computational time step. If, for example, the upper reservoir is found to contain N_r particles (in d dimensions) at time step $t - \Delta t/2$ within $L_y < y < L_y + h$, where h is the thickness of the reservoir, then the mass, momentum, and temperature are given by

$$M = \sum_{i=1}^{N_r} m_i,$$

$$M\vec{u} = \sum_{i=1}^{N_r} m_i\vec{u}_i,$$

$$dN_r k_B T = \sum_{i=1}^{N_r} m_i|\vec{u}_i - \vec{u}|^2,$$

where $\vec{u} = \vec{u}_p \cdot \hat{e}_x$ is the imposed velocity in the x direction, d is the spatial dimensionality, and the temperature is set to $T \equiv T_0$. At the next time step $t+\Delta t/2$, the central difference approximation would give:

$$\vec{u}_i'\left(t + \frac{\Delta t}{2}\right) = \vec{u}_i\left(t - \frac{\Delta t}{2}\right) + \frac{\vec{F}_{ix}(t)}{m_i}\Delta t,$$

and the temperature would be $T(t + \Delta t/2)$. However, we then scale this intermediate update of the velocity by the factor f (near unity):

$$\vec{u}_i\left(t + \frac{\Delta t}{2}\right) = \vec{u} + f\left[\vec{u}_i'\left(t + \frac{\Delta t}{2}\right) - \vec{u}\right],$$

such that $T_0 = f^2 T$, or $f = \sqrt{(T_0/T)}$. In this way, the reservoirs are kept moving via an imposed external force (normal and transverse) and thermostatted at constant temperature. The sample is affected across the sample-reservoir boundary by the forces exerted mutually by reservoir atoms on sample atoms, and vice versa. In general, heat generated at the sliding interface is transported across the sample and into the reservoir, but otherwise, the equations of motion of the sample atoms are unaffected (i.e., Newton's, or more precisely, Hamilton's equations of motion). Other methods of thermostatting and driving can also be employed (e.g., Nosé–Hoover [18,19]), but the above differential feedback is the simplest.

III. LENGTH AND TIME SCALES FOR THE FRICTIONAL PROBLEM

The microscopic simulation of nonequilibrium phenomena requires the achievement of a steady state, unless one is only concerned with transient phenomena. Unreliable results can be obtained from simulations whose system sizes are too small or whose time scales are too short. This becomes particularly relevant in frictional studies, because these are, in many cases, multiscale steady states. For example, Fig. 1 shows a transmission electron micro-scopic (TEM) image taken by Rigney [20–22] of a Cu/Steel interface after sliding. It shows an initial grain boundary and subsequent development of subgrain structure, which is more refined as one approaches the interface. Hughes and Hansen [23] have observed similar structures in Cu sliding on stainless steel, shown in Fig. 2. In such strongly coupled inter-faces, structure from nanoscale to mesoscale is observed, consisting of grains, dislocations, twins, and rotational patterning. In more weakly coupled systems, e.g., in proximal probe experiments, a few dislocations or minimal structural change are the rule. When using atomistic simulations to understand the microscopic and mesoscopic phenomena involved, one must simulate for a time long enough that all relevant relaxation and transport times are encompassed, for system sizes large enough that all relevant structural features are resolved.

To obtain an idea of the computer time required for a friction NEMD simulation, without entering into an exhaustive discussion of parallel computing [24], let us suppose that we want to contain a reasonable amount of material in the computational space, namely, several defect structures approaching mesoscale size. As the atomic spacing is of order 0.3 nm, and supposing that the calculation is for a cube of material 0.1 μm (100 nm) on a side, or 300 atomic spacings, then the total computational size would be about 30 million atoms. For a sound wave to cross the metallic sample at a speed of 5 km/sec (or 5 nm/psec), the total computational time would be about 20 psec, or 60 atomic vibrational periods (typically about 0.3 psec). To integrate the equations of motion accurately, each vibrational period must be divided up into roughly 30 time steps (i.e., 0.01 psec or 10 fsec each), which means that a computation of 20 psec takes 2000 time steps.

If the 30 million particles are distributed onto 2000 parallel processors, then from Fig. 3, we see that each time step requires approximately 0.3 sec of wall-clock time; thus, the 2000 time-step simulation takes 10 min. But a sound traversal time is only a part of the story for NEMD simulations of sliding friction. We must move the interfaces far enough to achieve the steadily sliding state. At a relative sliding velocity of 1% of the sound speed (50 m/sec, all 300 atoms in a given sliding direction pass over each other at least once in 200,000 time steps (2 nsec), or less than a day of computing on the above hypothetical multiprocessor computer. Because all computations of any acceptable length are subject to the Principle of Exhaustion of the Observer (such that one has the hope of remembering just why the calculation was begun in the first place—say, 10 days of computing time, subject to available resources), there is a lower bound on relative sliding velocities for NEMD simulations of a relatively fast value of 5 m/sec.

The largest systems of particles (of the order 10^7 atoms) that have been simulated in NEMD for long periods of time (of the order 1 nsec) have been carried out by using the parallel MD computer code SPaSM (Scalable Parallel Short-ranged Molecular Dynamics) developed at Los Alamos National Laboratory by Beazley et al. [24,25]. Currently, cal-culations of this magnitude for realistic systems (including analysis and visualization) re-quire several days of dedicated time (if available) on a 2048-processor parallel architecture such as Los Alamos Nirvana. Changing aspect ratio and imposing periodic boundary con-ditions in certain directions can ameliorate the size restrictions somewhat, but straightfor-ward integration—even if an increase of a factor of a thousand in particle number is attained—is limited to 1-μm length scale, at least for the near future. This means that an unbiased atomistic simulation is severely limited in length and time scales. In general terms, if the system reaches a steady state before the size restrictions of the simulation cell are attained, the physics of the process can be trusted. If this is not the case, there are a number of ways to get around these restrictions, usually suggested by the excitations that arise in the

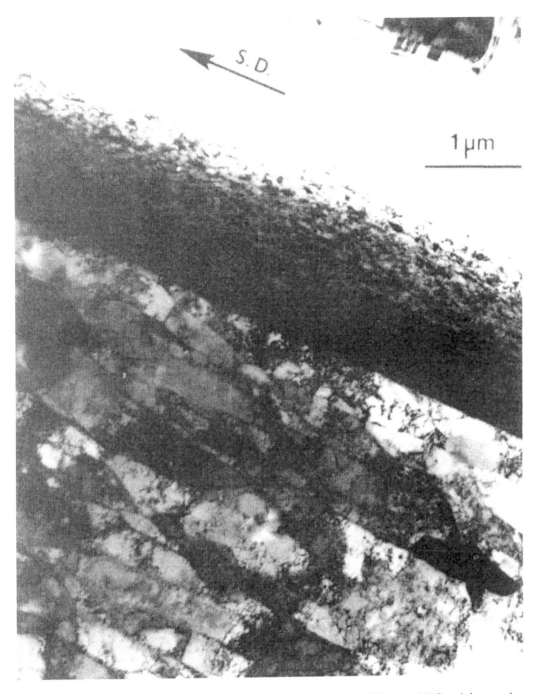

Figure 1 TEM micrograph (longitudinal section) for OFHC Cu sliding on 440C stainless steel, sliding at 1 cm/sec for 12 m at normal load 66.6 MPa. Arrow denotes sliding direction. (From Ref. 22.)

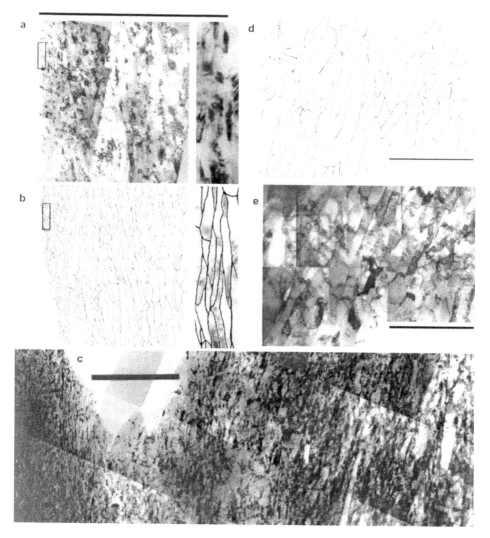

Figure 2 TEM micrographs for high purity Cu (99.99%) sliding on hardened steel. (a), (b): 12 MPa normal pressure. Surface at left. Width of enlarged rectangle: 70 nm. (c) 22 MPa normal pressure. Panels (d), (e) same as panels (a) and (b) but 20 μm below the surface. Scale markers in panels (c)–(e), 2 μm. (From Ref. 23.)

simulation and by the phase-space trajectory followed by the system. A generalized thermostat, which modifies the equations of motion, can be used to constrain certain average properties at the nonequilibrium steady state [9]. Examples are average global temperature [18,19], steady-state strain field [26], steady state strain rate [27], and in the field of shock wave physics, the Hugoniot relation, or energy conservation across a shock front [28]. One can also tailor the boundary conditions to the nonequilibrium flows. For example, if sound waves or mobile dislocations generated during a simulation reach the boundary, they can be absorbed by special reservoir regions. Such absorbing boundaries have been used in simulations of fracture [29].

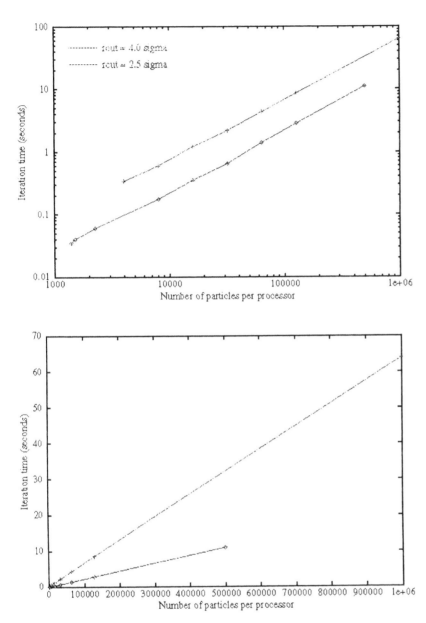

Figure 3 Typical iteration times vs. number of particles per processor for a 2048 processor array of SGI Origin 2000 processors (Los Alamos Nirvana), each with a clock speed of 250 MHz. Two curves are shown for a Lennard–Jones system with differing cutoff radii, r_{cut} ($\sigma = 2^{1/6} r_0$, cf. Section IV).

IV. SOME REMARKS ON POTENTIALS

The original MD simulation of Alder and Wainwright [10] used hard sphere interactions between atoms—a discontinuous potential. This was dictated mainly by the storage capabilities of the computers of that time. It was also a period of progress in the analytic treatment of the properties of hard-sphere fluids [30]. As mentioned in Section II, the first continuous potential, a local pair-wise interaction of the form

$$V(\vec{r}_i, \vec{r}_j) = \phi\left(|\vec{r}_i - \vec{r}_j|\right),$$

was the exponential repulsion between copper atoms that was used in NEMD simulations of radiation damage, and later, the Lennard–Jones (LJ) potential was used in equilibrium MD simulations of liquid argon [13]:

$$\phi(r) = \varepsilon\left[\left(\frac{r}{r_0}\right)^{-12} - 2\left(\frac{r}{r_0}\right)^{-6}\right],$$

with a minimum at r_0, for which $\phi(r_0) = -\varepsilon$ (the potential was truncated abruptly at radius $2.23r_0$). This is a particularly simple, short-ranged pairwise-additive potential. There are various ways to make this potential even shorter-ranged, such as truncating smoothly via a cubic spline (for the LJ potential, this is often referred to as LJ spline): $\phi(r) = \phi_{spl}(r)$ for $r > r_{spl}$, $\phi'(r_{spl}) = \phi'_{spl}(r_{spl})$, $\phi(r_{spl}) = \phi_{spl}(r_{spl})$, and $\phi(r) = 0$ for $r > r_{max} > r_{spl}$. Another common pair potential is the Morse:

$$\phi(r) = \varepsilon\left\{\left[\exp\left(-\alpha\left(\frac{r}{r_0} - 1\right)\right) - 1\right]^2 - 1\right\}.$$

In addition, volume-dependent pair potentials, derived from second-order perturbation theory (using pseudopotentials screened by a degenerate electron gas), and tight-binding potentials have been used [31].

A basic premise in most NEMD simulations is that the force term in the equations of motion may be evaluated in the adiabatic approximation. That is to say, in evaluating the potential energy of interaction $\Phi(\vec{R}_i, \vec{R}_j)$ between two atoms located at $\vec{R}_i(t)$ and $\vec{R}_j(t)$ at time t, the electrons respond rapidly enough that the electron density may be treated as a quantity depending only on the configuration of the atoms. Then the forces are spatial derivatives of Φ, calculated under this assumption. For ordered structures, this amounts to an electronic band-structure calculation. For a system with time-dependent disorder, the difficulties are extreme. However, approximate methods exist using density functional methods. The first such simulations were those of Car and Parrinello [32] and Martoňák et al. [33] for the dynamics of crystalline Si. Unfortunately, the system sizes that can be treated by this method are small, even when the simulation time is only 10 psec, namely, of the order 100–1000 atoms, mainly because of overhead in computing the electron density. This bottleneck can be avoided in many cases of interest, for example, close-packed metals, by using semiempirical methods [34,35]. The penalty that is paid for using the cheaper, easier semiempirical potentials is the quality of the representation of the true quantum mechanical system.

The most useful of the semiempirical potentials has allowed simulations on large scales. The Embedded Atom Method (EAM), developed by Daw and Baskes [36–39], incorporates some of the effects of local electronic densities, thereby becoming a many-body potential. EAM gives the potential energy Φ as a sum over atoms of the form:

$$\Phi = \sum_i \left[\frac{1}{2} \sum_{j \neq i} \phi(r_{ij}) + F(\rho_i) \right].$$

Here ϕ is a density-independent pair potential (such as LJ or Morse) and the second term in the sum over atoms is a density-dependent embedding energy, which depends on the local environment. The local density, ρ_i, is a sum of pairwise-additive terms over nearest neighbors,

$$\rho_i = \sum_{\langle j \rangle} w(r_{ij}),$$

The embedding energy F is a nonlinear function of ρ_i (if F were linear, then the interactions would be entirely pairwise-additive). EAM potentials are effectively short-ranged and have therefore enabled realistic simulations for dynamic properties of close-packed metals and alloys [40,41].

For NEMD simulations, there are several criteria to assess the quality of the potential energy function Φ for a given material. First, one would like to reproduce the zero-temperature and zero-pressure (ground-state) properties of the system, such as accurate values for the equilibrium density and cohesive energy E_{coh} for the experimentally observed structure. Next, it is essential that the ground-state bulk modulus B_0 be reproduced. Beyond this, one could, in principle, hope to fit the full phonon spectrum of the solid, or at least the elastic constants. Most importantly, however, for metals, defect energies (vacancy formation and stacking fault, or other surface energies), cannot be accurately obtained in the pairwise-additive approximation, but the EAM many-body form can be fitted to these, as well. Finally, during frictional sliding, subgrain structure can be generated, and material mixing may occur (both mechanical and diffusive), as well as alloying. Hence it may be important to consider a large region of the phase diagram, including melting, in judging the quality of the interatomic interactions.

To get some idea of how these properties are influenced by changes in the potentials, consider Fig. 4 where energy vs. density at $T = 0$ K is shown for a 2-D Cu system. The LJ pair potential is compared with Morse and a Cu EAM potential with particularly simple forms for the functions F and w [27]. The LJ and Morse potential parameters were chosen to reproduce ρ_0, E_{coh}, and B_0. Although all these interactions are similar near the ground state, the properties under compression differ considerably. Importantly as well for problems involving plastic deformation, the vacancy formation energy for pair potentials is $E_{vac} \approx 3\varepsilon$, whereas for many-body EAM, $E_{vac} \approx 1.2\varepsilon$. The temperature-dependent properties, which similarly depend on the nonlinear form of the potentials, are also significantly different. For example, the 2-D melting temperature for these potentials are $k_B T_{melt}/\varepsilon = 0.40$, 0.35, and 0.20 for the LJ, Morse, and EAM potentials, respectively. Thus, small differences in potentials can become important under strong loading of the system, i.e., when the local material pressure is large, the local temperature is high, the relative velocities and strain rates are high, or the local strain fields are large and inhomogeneous.

The LJ potential is often used in NEMD simulations because the computational overhead is minimal, so that it is very useful in investigating generic behavior for large systems (provided that the ground state is a close-packed crystal). However, for metals under steady-state frictional sliding, where significant damage can accumulate in the crys-

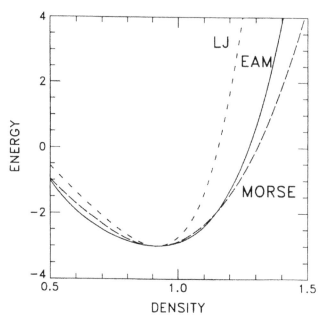

Figure 4   Energy vs. density for Lennard–Jones (LJ), Morse, and EAM Cu potentials at $T = 0K$. (From Ref. 27.)

talline state, the greater faithfulness of EAM many-body potentials to defect properties is an important advantage, far outweighing the factor of two computational penalty of EAM compared to a pair potential.

In systems where angular interactions may be important, the EAM potential has been modified (MEAM) by Baskes et al. [42–46]. The MEAM electron densities are written as sums of angular dependent terms corresponding to s, p, d, and f symmetries of the local electronic wavefunctions [47]:

$$\Phi = \sum_i \left[F_i(\rho_{h,i}) + \frac{1}{2} \sum_{j \neq i} \phi(r_{ij}) \right],$$

$$\rho_{h,i} = \rho^{(0)} G(\Gamma),$$

$$\Gamma = \sum_{l=1}^{3} t^{(l)} \left(\frac{\rho^{(l)}}{\rho^{(0)}} \right)^2,$$

where the $\rho^{(l)}$ are partial electron densities and the $t^{(l)}$ are orbital angular-momentum weighting factors. The EAM limit is obtained when $\Gamma \to 0$. MEAM has been applied to grain-boundary and surface structure simulations, but not yet to large-scale friction simulations.

V. PHYSICAL APPLICATIONS

Probably the first applications of direct integration of the classical equations of motion for atomistic systems in the context of friction and wear were those of McClelland and Glosli

[48,49]. They considered 3-D systems of size $N \sim 1000$ and were interested in the loading–unloading stick–slip phenomena observed in the AFM, SFA experiments carried out by themselves [48], Israelachvili [50], and Mate et al. [51]. Other simulations addressing tip–substrate interactions and transfer were carried out early on by Landman and collaborators [52–55] for a series of interfaces including Ni/Au. These more extensive simulations, using EAM potentials for system sizes of $\sim 10^4$ atoms, were the first attempts to model the full dynamics of a microscopic tip. Since these early attempts at microscopic simulations of sliding dissipation, a great deal of work has been carried out in three areas. The first of these is what may be called tribochemistry of lubricated interfaces. Harrison et al. [56,57] have recently reviewed this field. A second area is the lubrication at flat interfaces and the rheological effects of the lubricant, typically a hydrocarbon. Because of the complexity of the lubricant, the representation of the potential becomes an issue as well as the system size, as has been reviewed by Harrison et al. [56]. A third area, confined layers and ordering phenomena under shear, has been reviewed in Refs. 58–60. The emphasis of the present paper is the area of dry sliding, i.e., sliding in situations where contamination is not an issue and surfaces can be well characterized—notwithstanding the difficulty of achieving these conditions experimentally. There has been a divergence of effort in this field that has both a theoretical and an historical basis. The historical development of tribology has a very strong engineering and materials science emphasis, rooted in the need to understand macroscopic forces and coefficients of friction in practical engineering situations. Macroscopic experiments with subsequent analysis of the final microstructural state and wear debris have been traditional (cf. Tabor [61]). The development of local atomic probes, such as AFM, SFA, and QCM, has allowed a new class of precision measurements of time-dependent response to be made for weak loads and small velocities, with a degree of refinement unattainable with the traditional methods.

The above characterization highlights a similar distinction in theoretical efforts to understand interfacial sliding. In engineering applications using materials deformation computer codes, length scales of the order cm or mm are traditionally considered to be suitable for modeling macroscopic phenomena in unlubricated sliding. Moreover, the time scales are much longer than are characteristic of microscopic dynamics or thermal diffusion. Understanding macroscopic friction is then a problem of integrating multiple length and time scales, in order to arrive at such macroscopic models. The small-scale phenomena measured in local probes of interfacial phenomena are amenable to theories of linear response in condensed matter physics, and the weakly perturbed nature of many of these experiments has allowed atomistic simulations to achieve a qualitative—in some cases, quantitative—understanding of phenomena at flat, defect-free interfaces. The work of Robbins and Krim in the analysis of QCM experiments [62–66] is an example of this. The work of Hirano et al. [67,68] on the directional dependence of the frictional force also was an early application of atomistic simulations for smooth metal interfaces. As roughness of an interface increases, its statistical properties ultimately become important. Then the average interfacial stress, even in the limit of elastic distortions, is determined by the steady state distribution of micron-size contacting asperities. This asperity distribution, and the nonlinear elastic distortions associated with it, introduces mesoscale phenomena. Baumberger et al. [69,70] have made some progress in this direction. However, even for flat interfaces, as the interaction increases and local strain fields increase, other fundamentally nonlinear processes occur. At the local atomic scale, these may be associated with point defects and dislocation cores. At a somewhat larger scale, plastic deformation involving dislocation generation, motion, and pinning arises [71,72]. In brittle materials, fracture and delamination become important. At higher rates of sliding and deformation, the steady-state microstructural state of the material may change, resulting in grain formation [23,73] at a larger [10–100 μm] length scale. And finally, when elastic and inelastic processes are unable to transport energy away from an interface rapidly enough, local melting can be expected to occur [74].

We have emphasized some of the multiscale aspects of frictional dissipation in the above discussion, namely, the underlying steady-state configurations and excitations, as if the materials retained their Lagrangian identity up to some interface defining the line of separation between two materials. However, in many situations in dry sliding, there is either material transfer or material mixing [20,73] and the true material interface evolves with sliding. This complex state of affairs leads one to question whether a general theory of frictional interactions is possible [75]. However, for particular generic systems, one hopes the situation is not so bleak, although a general explanation of the "laws of friction" [76] may not be possible. The next section will review the role of atomistic simulation of friction and in particular, large-scale NEMD simulations.

VI. SIMULATIONS

Figure 5 shows the geometric layout of a generic friction experiment, as is used in NEMD simulations. This is a 2-D image, but the 3-D analogue is no different. There are boundary regions $B_1(t)$, $B'_1(t)$, and $B_2(t)$, $B'_2(t)$, where various time-dependent "external" perturbations may be applied or where time-dependent probes may be placed to analyze the phenomena which are taking place at and near the interface between two materials M_1 and M_2, which may be separated by a third body M_3. The above picture is a logically rectangular picture of the system that is mapped into physical space by specifying the shapes of the boundary interfaces. In the simulation of Glosli and McClelland [49], for example, B_2, B'_2, represented periodic boundary conditions, i.e., periodic images of the central regions. M_1 was an array (6×6) of alkane chains six monomers in length, M_2 was a similar array, B_1 and B'_1 were a combination of an isotropic harmonic pinning potential on a triangular lattice and an LJ interaction in the normal direction. Each film was independently thermostatted and the separation between B_1 and B'_1 was fixed. For more recent simulations

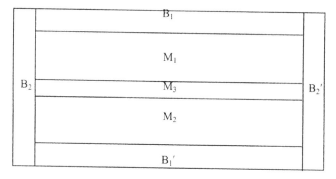

Figure 5   Logical representation of the geometrical layout in a generic NEMD friction simulation; boundaries (and boundary conditions) are denoted by B_i; material regions are denoted by M_i.

incorporating more sophisticated potentials and increased system sizes, the recent review by Harrison et al. [56] may be consulted.

The geometry of the AFM tip usually models M_1 by a pyramidal-shaped object coupled to a massive cantilever with boundary conditions at B_1, fixed normal load, and either fixed velocity or fixed force [77]. The flat upon which the tip moves is typically pinned by B'_1 and the boundary conditions B_2 for the tip M_1 are usually free and periodic for M_2. For example, the simulations by Landman et al. [53–55] had a pyramidal Si tip with varying contact area for M_1 modeled with Stillinger–Weber potentials and several layers of (111) Si for M_2 with lateral periodic boundary conditions for B_2, B'_2. B'_1 was a fixed boundary and B_1 a variety of boundary conditions including constant height and constant external normal load. These groundbreaking simulations (of size $N \sim 1000$) showed the utility of microscopic simulations for understanding tip–surface interactions. Through the use of LJ potentials, some of the qualitative influences of strong vs. weak materials in heterogeneous contact vis-à-vis adhesion were demonstrated, as well as material surface and subsurface deformation. A review of this generic tip–substrate problem may be found in Refs. 56 and 77.

Flat-on-flat geometry has received extensive attention beginning with the experiments of Israelachvili and collaborators, in which M_1 and M_2 were mica with atomically flat interfaces, and M_3 a wide variety of fluids and polymers [78]. The nano-rheology and surprising thickness dependence of the tangential force due to confined fluid lubrication has also been studied in detail by Granick and collaborators [58–60]. For simple fluids, where simple pair potentials are an adequate representation of the interatomic forces, much of this thickness dependence can be reproduced. Robbins and collaborators [66,79,80] have treated this problem in detail. For simple films several monolayers in thickness, an evident approximation is to replace M_1 and M_2 by a rigidly translating potential landscape with the 2-D space group of the appropriate crystalline face. For a single monolayer, this is the 2-D analogue of the Frenkel–Kontorova model [81]. Some of the language, techniques, and results of nonlinear dynamics can be used to describe pinning, depinning, and nonlinear excitations that one might expect, as well as the interplay between commensurability and incommensurability. Recent papers in this field have treated, for example, the multiple layers of Lennard–Jones atoms [82]. For moderate sliding velocities, it is possible to relate the velocity-dependent frictional force to the rate $\dot{\gamma} = \omega \rho(\omega) f_0^2 / 2m$ of intrinsic losses due to phonon–phonon scattering, where $\rho(\omega)$ is the imaginary part of a response function [82].

A simpler system, both experimentally and theoretically, is the one originally studied by Krim and coworkers using QCM, where rare gas atom layers were adsorbed onto metallic substrates (Au and Ag, typically), the substrate was oscillated, and the frequency-dependent response of the tangential force was measured. Robbins and coworkers analyzed these experiments in detail [5,62–66]. Relative sliding velocities v_r measured via QCM are typically less than 0.1 m/sec [83]. In this regime, whether the interface is incommensurate or fluid, the tangential force is linear in the velocity: $f_{tang}(v_r) = f_0 + \lambda v_r$. This result was attributable to the low-velocity dependence of the linear-response expression for dissipation due to phonon–phonon scattering. The velocity-dependent part of the force may be written as [64,65] $f_{tang} \sim M v_r / \tau$, where M is the total mass and τ is a velocity- and direction-dependent slip time, whose inverse is proportional to the square of the Fourier components of the interaction potential and the imaginary part of a response function—a constant in the limit of $v_r/c_t \ll 1$ (c_t is the transverse sound speed). Because in their analysis, the substrate could be approximated as a uniformly translating potential distribution, and because the interactions between rare gas atoms are reasonably well modeled by LJ potentials, they found good agreement between direct MD simulations, the linear response expression (with τ evaluated via MD), and experiment (assuming a realistic value of interfacial roughness) [65].

For weakly interacting interfaces, direct evaluation of the expression for the force between two lattices leads to expressions similar to those above, namely,

$$F_{12} = \frac{2\pi}{\Omega^2} \sum_{\vec{q}} \left(\vec{q} \cdot \frac{\vec{v}_r}{|\vec{v}_r|} \right) |\phi_{12}(\vec{q})|^2$$

$$\times \int d\omega \left[S_1(\vec{q}, -\vec{q}; \omega) \mathrm{Im} G_2^{\mathrm{ret}}(\vec{q}, -\vec{q}; \omega + \omega_0(\vec{q})) \right.$$

$$\left. + S_2(-\vec{q}, \vec{q}; \omega) \mathrm{Im} G_1^{\mathrm{ret}}(\vec{q}, -\vec{q}; \omega + \omega_0(\vec{q})) \right],$$

where $\phi_{12}(\vec{r})$ is the atom–atom interaction potential between M_1 and M_2 atoms, $\omega_0(\vec{q}) = \vec{q} \cdot \vec{v}_r$, S_i is the Fourier transform of the dynamic structure factor of medium i, G_i^{ret} is the density–density response function of medium i, and Ω is the system volume. Even for LJ systems, these expressions have not been investigated in detail. There is one point that is of some interest concerning expressions such as these. The expansion is in the interaction strength, that is, the potential of interaction between the two media M_1 and M_2, so that as long as the first term in the expansion is adequate, and the nonequilibrium state is only a weak perturbation from the equilibrium state, the full frequency dependence is, in principle, measurable. That is, there is a nonlinear dependence of the friction force on velocity, even in the linear response regime.

As the interaction between M_1 and M_2 increases, the steady-state description of the system changes from a uniform lattice, with a phonon bath of lattice oscillations, to a more complicated configuration, which may involve lattice defects, weakly interacting dislocations, stacking faults, strongly interacting dislocations, and nucleation of grain structures. Figure 2 from the paper of Hughes and Hansen [23] gives an indication of the complexity that can arise when large strains and strain rates are important. For such systems, NEMD approaches can provide some very useful information on the microscopic level, provided the system size is large enough to enable important structures to be resolved. The simulations of Hirano et al., Sørensen for Cu, Landman for Si and others [67,68,84,85] have been in the slow tip regime. Even here, Sørensen et al. find that, for Cu at velocities less than 20 m/sec, slip is characterized by mixed edge-screw dislocations for (111) surfaces. There have been recent theories of subsurface dislocation generation by Hurtado and Kim [66,67] for single-asperity friction, in which transitions from single-dislocation-assisted (SDA) slip to multiple-dislocation-cooperative slip are predicted (when the ratio of contact radius to Burger's vector is of order 10^5). The relevance of SDA slip for ductile asperities was demonstrated by Sørensen; however, the large tip size transition predicted by Hurtado and Kim [71,72] has not been investigated with MD, because it corresponds to planar sizes verging on 10^{10} atoms. It is in this regime that mesoscale approaches, such as dislocation dynamics, may become necessary.

Hammerberg et al. [86] performed large-scale NEMD simulations for 2-D EAM Cu at relative sliding velocities in the range $0.01 < v_r/c < 0.3$, where c is the longitudinal sound speed at the appropriate pressure, for pressures in the range of 1–30 GPa. The geometry (as in Fig. 5) for their simulations was blocks M_1, M_2 composed of identical triangular lattices of Cu sliding on each other (representing a 2-D version of Cu (111) sliding in the [101] direction). These simulations showed many of the aspects of dry sliding observed in the experiments of Rigney [73] and Hughes and Hansen [23]. The early-time behavior in these simulations showed an initial transition at the interface from phonon-dominated behavior and elastic distortions to dislocation-mediated slip, with dislocations confined near the interface. (The nonlinear dynamics of these early-time states were subsequently discussed in a model with fewer degrees of freedom by Röder et al. [87,88].) At longer times (of order 50 t_0, where t_0 is 1 or 2 phonon periods), the interfacial region transformed to one of a graded microstructure, namely, very fine-grained crystallinity very near the interface, and larger, highly strained grains farther away. In the temporal evolution of this state with localized dislocations along the interface to a steady state microstructure, there is a transient period in which dislocations generated near the interface move into the untransformed medium. The early- and late-time situations are shown in Figs. 6 and 7. Figure 6 shows a Cu/Cu interface sliding at $v_r = 0.12c$, under a pressure of 30 GPa. In this image, the colors denote the potential energy of the Cu atom. The edge dislocations are seen as the red regions moving in opposite directions on either side of the interface. Figure 7 shows a grain map in the steady state; the local orientation of the neighbors of an atom relative to an initial hexagonal symmetry axis are colored with a rainbow spectrum, where $-30°$ is red and $+30°$ is blue. The deformation observed in this steady state is to be compared to that seen by Chen and Rigney [89] and more recently by Hughes and Hansen [23] for Cu sliding on stainless steel.

Another aspect of dry sliding seen in these simulations is the mechanical mixing of material across the sliding boundary. Figure 8 shows the distribution of material from the two regions of Cu at late times ($t = 500t_0$) under the same conditions as Fig. 7. The small-scale mixing of material in these simulations occurred in structures generated close to the interface in the first 10–15 layers. The structures appeared to be growing diffusively in the steady state. These simulations demonstrate some of the important characteristics of

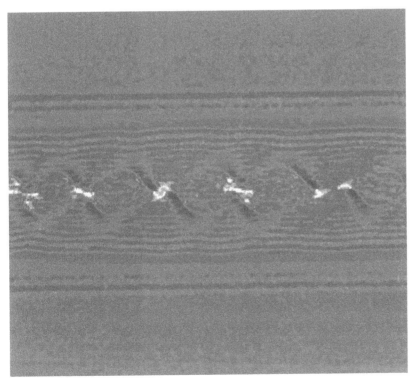

Figure 6   Early-time ($t = 10t_0$) behavior of a perfect Cu/Cu interface, $v_r = 0.12i$, $P = 30$ GPa. (From Ref. 85.)

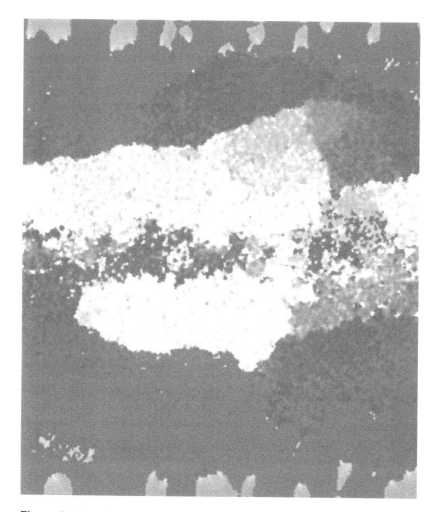

Figure 7 Late-time ($t = 500t_0$) behavior of a perfect Cu/Cu interface (same conditions as Fig. 6). (From Ref. 86.)

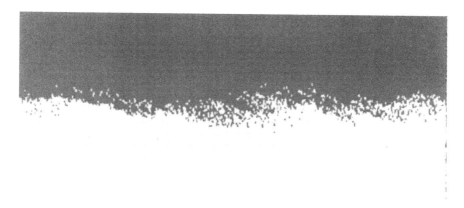

Figure 8 Cu/Cu interface at $t = 500t_0$. Same conditions as Fig. 7. (From Ref. 86.)

mechanical deformation in dry sliding when contact between ductile metals is strong: (1) The sliding process is multiscale in nature both spatially and temporally. (2) It is necessary to include both plastic deformation and subgrain growth. (3) It is necessary to include mechanical mixing. (4) The physics involved is strongly nonequilibrium and nonlinear.

By comparing the results of pair-potential vs. many-body EAM systems, it is possible to see whether there are regimes where frictional behavior is generic. The recent work of Germann et al. [90] for sliding of a 2-D system of LJ particles with an incommensurate interface (created by rotating a triangular lattice M_1 by $90°$ to give a rotated lower lattice M_2 with a $\sqrt{3}$ incommensurability) has shown a power-law velocity dependence of the frictional force at high velocities. This is shown in Fig. 9, where the tangential force is shown as a function of velocity for three different compressions. The curves have been scaled so that the maxima of the scaled frictional force coincide at the same-scaled velocity. For comparison, a $v_r^{-\beta}$ power law curve is shown with $\beta = 0.75$. In this high-velocity regime, the interfacial morphology consists of nanoscale grain structure for the 2-D LJ system, just as for the EAM Cu solid described above. The detailed nature of the interface is important at the lower velocities. A commensurate interface, for which M_1 and M_2 are identical (i.e., M_2 is unrotated), leads to locking (cold welding) of the interface at low velocities and a finite low-velocity asymptote, whereas the incommensurate interface results in a sliding frictional force approaching zero, consistent with the linear response predictions for nonlinear phonon dissipation [65]. Fu et al. [91,92] have pursued experiments and simulations of amorphous interfaces in 2-D using LJ pair potentials to model two component metallic glasses. The frictional force in their simulations shows similar power-law velocity dependence, although they have not analyzed the near-interface structural patterning in detail. We have proposed a model that assumes a critical plastic strain for grain formation, diffusive dislocation dynamics, and high strain-rate flow stress behavior with $\tau \sim \dot{\varepsilon}_{pl}^{\alpha}$, where ε_{pl} is the equivalent plastic strain. This model predicts that $\beta = 1 - \alpha \approx 0.75$.

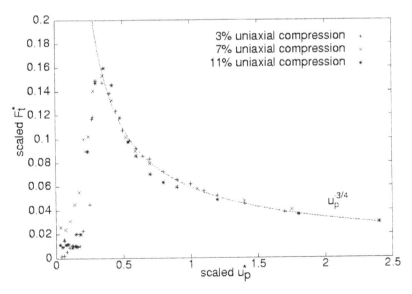

Figure 9   Scaled tangential force vs. scaled velocity for a 2-D LJ solid with an incommensurate interface, for three different compressions. (From Ref. 90.)

Three-dimensional atomistic simulations of sliding have concentrated on the physics of near-surface probes for the most part [55,64,84], or the rheology and chemistry of lubricating layers [56]. The 3-D analysis of some of the effects observed in dry sliding outlined above are becoming possible as computational power increases and becomes more widely available on local parallel arrays [93]. As indicated above, the plastic deformation of metal interfaces becomes the controlling factor in dry contact, when local strains become appreciable. For increased strains and strain rates, patterning and subgrain structure become important. Such phenomena are easier to resolve in 2-D simulations where system dimensions can be significantly larger than in 3-D. Mikulla et al. [94] have performed large-scale simulations for 3-D fcc [111] and [100] single crystal interfaces, for both LJ and EAM Cu. In particular, analogous near-interface, early-time dynamical structures (namely dislocation loops) were seen for 3-D commensurate interfaces, analogous to the edge dislocations observed in 2-D.

The model of Bowden and Tabor supposes that, at a heterogeneous interface, the deformation properties of the weaker material dominate the dissipation and hence the frictional force, as has recently been investigated by Germann et al. [95]. Figure 10 shows a 3-D simulation of a Cu/Ag sliding interface (5.5×10^6 atoms, interacting via EAM potentials). Local shear strains are represented by marker atoms that have been highlighted. An initial corrugation was imposed at the interface, which is still evident at 200ps. The near-surface deformation is a complicated tangle of stacking faults on multiple slip systems in the weaker Ag. At these fairly early times, on the other hand, the deformation in Cu consists mainly of a few single dislocations.

Ultimately, the length scales available to simulations are limited by computational technology if one considers time scales of order nsec. Currently, this limits 3-D length scales to 0.1 μm. (Much larger size simulations are possible, but currently system sizes of 10^9 atoms are limited to times of order 1.0 nsec.) At the continuum level, highly efficient 3-D Eulerian, Lagrangian, arbitrary Lagrangian Eulerian (ALE), and finite-element methods have been developed to treat engineering problems down to scales of 0.1 mm; below that scale, the physical models begin to depend on the time evolution of "subgrid" phenomena, such as dislocation distributions, grain structure, etc.

It is in this mesoscale region, from 0.1 to 100 μm, that much work remains to be carried out. From the microscale, one can attempt to measure relevant energies and interactions for incorporation into models of dislocation nucleation and dynamics. Such mesoscale dislocation models for plastic deformation can then form the basis for either a coarse-grained plasticity model to be used at the macroscale, or materials-dynamics continuum algorithms, which integrate the dynamics of subgrid densities [96–98]. Makarov and collaborators [98–100] incorporate plasticity into a Lagrangian finite-difference method, which treats subgrain rotation but not growth. The discrete element model (DEM) is similar in emphasis [102]. O'Connel and Thompson [103] and Broughton et al. [104] reformulate the boundary conditions, B_i (Fig. 5), into continuum ones that incorporating an appropriate model of the macro (or meso)-scale deformation. Smooth particle applied mechanics or smooth particle hydrodynamics (SPAM, SPH) [105] leads to a mathematical structure that is isomorphic to the atomistic equations of motion. SPAM has been applied mainly to elastic–plastic deformation.

These methods address primarily the length scale problem. The time scale problem, as was indicated above, involves the evolution to a steady state. When mesoscale or even nanoscale structures are involved in the final steady state, the length scale necessary to accommodate these structures requires large system sizes, and therefore, longer simulation times. One would hope that, by stochastic or statistical methods, the time between micro-

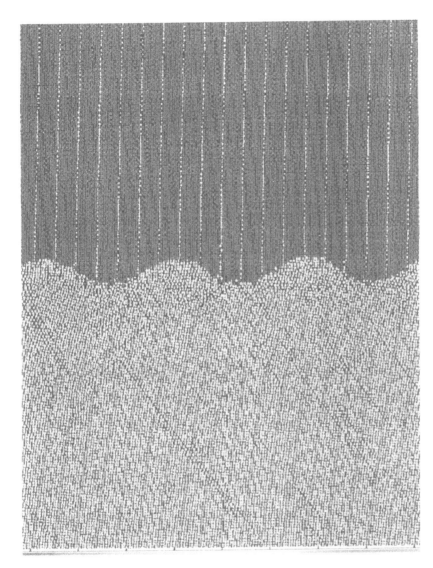

Figure 10   Sliding at a Cu/Ag interface, (100) faces sliding in the [010] direction, for $v_r = 379$ m/s at time $t = 200$ps, pressure $P = 5.1$ GPa. The figure shows a 2-D slice from a 3-D sample of 5.5×10^6 atoms; upper atoms are Cu, lower are Ag. Lines are marker atoms every 9th plane, initially vertical. (From Ref. 95.)

scopic configurations in an atomistic simulation could be accelerated while retaining physical fidelity, thereby boosting a large-scale simulation from a nsec- to a μsec-time scale. Such acceleration techniques have been developed for chemical kinetics, where there are isolated localized transitions in time, driven by stochastic equations [106–108]. This procedure runs into severe difficulties in situations such as interfacial slip and fracture problems, where long-range elastic fields, defect interactions, and defect evolution are important. There have been some attempts along these lines for tip–surface interactions, but not yet for the full sliding interface problem [109,110].

VII. OUTLOOK

We have attempted to address some of the multiscale issues in understanding sliding material interfaces, and we have indicated some of the microstructural issues which require that atomistic simulations be large scale, in order to adequately represent the underlying nonequilibrium steady-state physics at moving interfaces. The results of very large atomistic simulations can be used to build mesoscale dynamical models, whose results can, in turn, be incorporated into macroscopic engineering continuum codes, as were discussed in the preceding section. The dream of a coupled, integrated, atomistic–mesoscopic–macroscopic computational physics package still beckons and, as is implicit above, requires a solution of the problem of time-scale restrictions, more than the length-scale restrictions.

As is indicated in Ref. 88, the large-scale computational landscape is becoming much more democratic. In the next 5–10 years, the paradigm of the individual investigator at his workstation, possibly interacting with a more powerful computational engine, will be transformed into the individual investigator at his local parallel cluster. This situation will invigorate computational materials science across the research and applications enterprise—university, corporate, and governmental. Such developments will allow a much deeper fundamental understanding of many of the difficult issues in materials science involving dislocations, grain structure, structure nucleation, and structure evolution.

We close with a quotation from Hoover [9]:

> In problems involving homogeneous flows, without boundaries, the nonequilibrium algorithms have reduced number dependence in the results. But there are many important problems involving physical gradients, such as shock and detonation waves in which *all* the field variables cover a range of values. For such inhomogeneous systems larger and faster simulations are necessary. Because individual processors are reaching limiting speeds, improvements are being mainly obtained by combining processors in parallel. Simulations with *millions* of particles will soon be a reality. Likewise, the possibility of storing and processing *billions* of numbers makes it possible to characterize distribution functions in quantum problems and nonequilibrium problems, areas now made difficult simply by requirements in storage capacity.
>
> The flexibility in simulations will make it possible to follow the flow of relativity complicated molecules in channels, to solve problems involving friction and wear, potential energy surfaces with chemical reactions, and quantum mechanics. . . .

The "millions" and "billions" that Hoover referred to are now approaching the order of "billions" and "trillions."

ACKNOWLEDGMENTS

We would like to acknowledge our collaborators P.S. Lomdahl, T.C. Germann, J. Röder, S.J. Zhou, R.P. Mikulla, A.R. Bishop, and R. Ravelo for their contributions, insight, and encouragement of our work on interfacial dynamics over the last several years.

REFERENCES

1. Marti, O. Measurement of Adhesion and Pull-Off Forces with the AFM. In *Modern Tribology Handbook*; Bhushan, B., Ed.; CRC Press: New York, 2001; Vol. 1, 617–639.

2. Schwarz, U.D.; Hölscher, H. Atomic-Scale Friction Studies Using Scanning Force Microscopy. In *Modern Tribology Handbook*; Bhushan, B., Ed.; CRC Press: New York, 2001; Vol. 1. 641–665.

3. Bhushan, B. Friction, Scratching/Wear, Indentation, and Lubrication Using Scanning Probe Microscopy. In *Modern Tribology Handbook*; Bhushan, B., Ed.; CRC Press: New York, 2001; Vol. 1, 667–716.

4. Marti, O. AFM Instrumentation and Tips. In *Handbook of Micro/Nano Tribology*, 2nd ed.; CRC Press: New York, 1999; 81–144.

5. Krim, J. Atomic-scale origins of friction. Langmuir 1996, *12*, 4564–4566.

6. Mak, C.; Krim, J. Quartz-crystal microbalance studies of the velocity dependence of interfacial friction. Phys. Rev. B 1998, *58*, 5157–5159.

7. Borovsky, B.; Krim, J.; Syed, A.A.; Wahl, K.J. Measuring nanomechanical properties of a dynamic contact using an indenter probe and quartz crystal microbalance. J. Appl. Phys. 2001, *90*, 6391–6396.

8. Ciccotti, G., Frenkel, D., McDonald, I.R., Eds. In *Simulations of Liquids and Solids: Molecular Dynamics and Monte Carlo Methods in Statistical Mechanics*; North-Holland: New York, 1987.

9. Hoover, W.G. Molecular Dynamics. *Lecture Notes in Physics*; Springer-Verlag: Berlin, 1986; Vol. 258.

10. Alder, B.J.; Wainwright, T.E. Phase transition for a hard sphere system. J. Chem. Phys. 1957, *27*, 1207–1208.

11. Alder, B.J.; Wainwright, T.E. Studies in molecular dynamics I. J. Chem. Phys. 1959, *31*, 459–466.

12. Gibson, J.B.; Goland, A.N.; Milgram, M.; Vineyard, G.H. Dynamics of radiation damage. Phys. Rev. 1960, *120*, 1229–1253.

13. Rahman, A. Correlations in the motion of atoms in liquid argon. Phys. Rev. 1964, *136A*, 405–411.

14. Beeler, J.R., Jr. In *Physics of Many-Particle Systems*; Meeron, E., Ed.; Gordon and Breach Publishers, Inc.: New York, 1964.

15. Verlet, L. Computer "experiments" on classical fluids: I. Thermodynamical properties of Lennard–Jones molecules. Phys. Rev. 1967, *159*, 98–103.

16. Ashurst, W.T.; Hoover, W.G. Dense-fluid shear viscosity via nonequilibrium molecular dynamics. Phys. Rev. A 1975, *11*, 658–678.

17. Holian, B.L.; Hammerberg, J.E.; Lomdahl, P.S. The birth of dislocations in shock waves and high-speed friction. J. Comput.-Aided Mater. Des. 1998, *5*, 207–234.

18. Nosé, S. A unified formulation of the constant temperature molecular dynamics methods. J. Chem. Phys. 1984, *81*, 511–519.

19. Hoover, W.G. Canonical dynamics: equilibrium phase-space distributions. Phys. Rev. A 1985, *31*, 1695–1697.

20. Rigney, D.A. Sliding wear of metals. Annu. Rev. Mater. Sci. 1988, *18*, 141–163.

21. Heilmann, P.; Clark, W.A.T.; Rigney, D.A. Orientation determination of subsurface cells generated by sliding. Acta Metall. 1983, *31*, 1293–1305.

22. Rigney, D.A. Comments on the sliding wear of metals. Tribol. Int. 1997, *30*, 361–367.

23. Hughes, D.A.; Hansen, N. Graded nanostructures produced by sliding and exhibiting universal behavior. Phys. Rev. Lett. 2001, *87*, 1355031–1355034.

24. Beazley, D.M. Message-passing multi-cell molecular dynamics on the connection machine 5. Parallel Comput. 1994, *20*, 173–195.

25. Beazley, D.M.; Lomdahl, P.S.; Grønbech-Jensen, N.; Giles, R.; Tamayo, P. Parallel algorithms for short-range molecular dynamics. In *Annual Reviews of Computational Physics III*; World Scientific: Singapore, 1996; 119–175.

26. Wagner, N.J.; Holian, B.L.; Voter, A.F. Molecular-dynamics simulations of two-dimensional materials at high strain rates. Phys. Rev A 1992, *45*, 8457–8470.

27. Holian, B.L.; Voter, A.F.; Wagner, N.J.; Ravelo, R.J.; Chen, S.P.; Hoover, W.G.; Hoover,

C.G.; Hammerberg, J.E.; Dontje, T.D. Effects of pairwise vs. many-body forces on high-stress plastic deformation. Phys. Rev. A 1991, *43*, 2655–2661.

28. Maillet, J.B.; Mareschal, M.; Soulard, L.; Ravelo, R.; Lomdahl, P.S.; Germann, T.C.; Holian. B.L. Uniaxial Hugoniostat: a method for atomistic simulations of shocked materials. Phys. Rev. E 2000, *63*, 016121–016121–8.

29. Holian, B.L.; Lomdahl, P.S.; Zhou, S.J. Fracture simulations via large-scale nonequilibrium molecular dynamics. Physica A 1997, *240*, 340–348.

30. Frisch, H.L.; Lebowitz, J.L. *The Equilibrium Theory of Classical Fluids: A Lecture Note and Reprint Volume*; W.A. Benjamin: New York, 1964.

31. Sutton, A.P.; Finnis, M.W.; Pettifor, D.G.; Ohta, Y. The tight-binding bond model. J. Phys. C 1988, *21*, 35–66.

32. Car, R.; Parrinello, M. Unified approach for molecular dynamics and density-functional theory. Phys. Rev. Lett. 1985, *55*, 2471–2471.

33. Martonák, R.; Molteni, C.; Parinello, M. Ab initio molecular dynamics with a classical pressure reservoir: simulations of pressure-induced amorphization in a $Si_{35}H_{36}$ Cluster. Phys. Rev. Lett. 2000, *84*, 682–685.

34. Kress, J.D.; Goedecker, S.; Hosie, A.; Wasserman, H.; Lubeck, O.; Collins, L.A.; Holian, B.L. Parallel O(N) tight-binding molecular dynamics of polyethylene and compressed methane. J. Comput.-Aided Mater. Design 1998, *5*, 295–316.

35. Kress, J.D.; Bickham, S.R.; Collins, L.A.; Holian, B.L. Tight-binding molecular dynamics of shock waves in methane. Phys. Rev. Lett. 1999, *83*, 3896–3899.

36. Daw, M.S.; Baskes, M.I. Semiempirical, quantum mechanical calculation of hydrogen embrittlement in metals. Phys. Rev. Lett. 1983, *50*, 1285–1288.

37. Daw, M.S.; Baskes, M.I. Embedded-atom method: derivation and application to impurities, surfaces, and other defects in metals. Phys. Rev. B 1984, *29*, 6443–6453.

38. Daw, M.S.; Foiles, S.M.; Baskes, M.I. The embedded-atom method: a review of theory and applications. Mater. Sci. Rep. 1993, *9*, 251–310.

39. Foiles, S.M.; Baskes, M.I.; Daw, M.S. Embedded-atom-method functions for the face-centered-cubic metals, Cu, Ag, Au, Ni, Pd, Pt, and their alloys. Phys. Rev B 1986, *33*, 7983–7991.

40. Baskes, M.I. Modified embedded-atom potentials for cubic materials and impurities. Phys. Rev. B 1992, *46*, 2727–2742.

41. Kadau, K.; Entel, P.; Germann, T.C.; Lomdahl, P.S.; Holian, B.L. Large-scale molecular-dynamics study of the nucleation process of martensite in Fe–Ni alloys. J. Phys. IV 2001, *11*, 17–22.

42. Baskes, M.I.; Asta, M.; Srinivasan, S.G. Determining the range of forces in empirical many-body potentials using first-principles calculations. Philos. Mag. A 2001, *81*, 991–1008.

43. Lee, B.J.; Baskes, M.I.; Kim, H.; Cho, Y.K. Second nearest-neighbor modified embedded atom method potentials for bcc transition metals. Phys. Rev. B 2001, *64*, 4102–4112.

44. Lee, B.J.; Baskes, M.I. Second nearest-neighbor modified embedded-atom-method potential. Phys. Rev. B 2000, *62*, 8564–8567.

45. Cherne, F.J.; Baskes, M.I.; Deymier, P.A. Properties of liquid nickel: a critical comparison of EAM and MEAM calculations. Phys. Rev. B 2002, *65*, 4209–4217.

46. Aguilar, J.F.; Ravelo, R.; Baskes, M.I. Morphology and dynamics of 2D Sn–Cu alloys on (100) and (111) Cu surfaces. Model. Simul. Mater. Sci. Eng. 2000, *8*, 335–344.

47. Angelo, J.E.; Baskes, M.I. Interfacial studies using the EAM and MEAM. Interface Sci. 1996, *4*, 47–63.

48. McClelland, G.M.; Glosli, J.N. Friction on the atomic scale. In *Fundamentals of Friction: Macroscopic and Microscopic Processes*; Kluwer Academic Publishers: Dodrecht, 1992; 405–422.

49. Glosli, J.N.; McClelland, G.M. Molecular dynamics study of sliding friction of ordered organic monolayers. Phys. Rev. Lett. 1993, *70*, 1960–1963.

50. Israelachvili, J.N. Adhesion, Friction and lubrication of molecularly smooth surfaces. In *Fundamentals of Friction: Macroscopic and Microscopic Processes*; Kluwer Academic Publishers: Dordrecht, 1992; 351–381.

51. Mate, C.M.; McClelland, G.M.; Erlandsson, R.; Chiang, S. Atomic-scale friction of a tungsten tip on a graphite surface. Phys. Rev. Lett. 1987, *59*, 1942–1945.
52. Landman, U.; Luedtke, W.D. Nanomechanics and dynamics of tip–substrate interactions. J. Vac. Sci. Technol. B 1991, *9*, 414–423.
53. Landman, U.; Luedtke, W.D.; Burnham, N.A.; Colton, R.J. Atomistic mechanisms and dynamics of adhesion, nanoindentation, and fracture. Science 1990, *248*, 454–460.
54. Landman, U.; Luedtke, W.D.; Ribarsky, M.W. Structural and dynamical consequences of interactions in interfacial systems. J. Vac. Sci. Technol. A 1989, *7*, 2829–2839.
55. Landman, U. On nanotribological interactions: hard and soft interfacial junctions. Solid State Commun. 1998, *107*, 693–708.
56. Harrison, J.A.; Stuart, S.J.; Brenner, D.W. Atomic-scale simulations of tribological and related phenomena. In *Handbook of Micro/Nano Tribology*, 2nd Ed.; Bhushan, B., Eds.; CRC Press: New York, 1999; 525–594.
57. Mikulski, P.T.; Harrison, J.A. Packing-density effects on the friction of *n*-alkane monolayers. J. Am. Chem. Soc. 2001, *123*, 6873–6881.
58. Granick, S. Soft matter in a tight spot. Phys. Today 1999, *52*, 26–31.
59. Levent Demirel, A.; Granick, S. Origins of solidification when a simple molecular fluid is confined between two plates. J. Chem. Phys. 2001, *115*, 1498–1512.
60. Zhu, Y.; Granick, S. Rate-dependent slip of Newtonian liquid at smooth surfaces. Phys. Rev. Lett. 2001, *87*, 0961051–0961054.
61. Tabor, D. Friction as a dissipative process. In *Fundamentals of Friction: Macroscopic and Microscopic Processes*; Singer, I.L., Pollock, H.M., Eds.; Kluwer Academic Publishers: Dordrecht, 1992; 3–20.
62. Robbins, M.O.; Krim, J. Energy dissipation in interfacial friction. MRS Bull 1998, *23*, 23–26.
63. Krim, J.; Solina, D.H.; Chiarello, R. Nanotribology of a Kr monolayer: a quartz-crystal microbalance study of atomic-scale friction. Phys. Rev. Lett. 1991, *66*, 181–184.
64. Cieplak, M.; Smith, E.D.; Robbins, M.O. Molecular origins of friction: the force on adsorbed layers. Science 1994, *265*, 1209–1212.
65. Smith, E.D.; Robbins, M.O.; Cieplak, M. Friction on adsorbed monolayers. Phys. Rev. B 1996, *54*, 8252–8260.
66. Robbins, M.O.; Müser, M.H. Computer simulations of friction, lubrication, and wear. In *Modern Tribology Handbook*; Bhushan, B, Ed.; CRC Press: New York, 2001; Vol. 1, 717–765.
67. Hirano, M.; Shinjo, K. Atomistic locking and friction. Phys. Rev. B 1990, *41*, 11837–11851.
68. Hirano, M.; Shinjo, K.; Kaneko, R.; Murata, Y. Observation of superlubricity by scanning tunneling microscopy. Phys. Rev. Lett. 1997, *78*, 1448–1451.
69. Baumberger, T.; Caroli, C. Multicontact Solid Friction: A macroscopic probe of pinning and dissipation on the mesoscopic scale. MRS Bull. 1998, *23*, 41–46.
70. Bureau, L.; Baumberger, T.; Caroli, C. Jamming creep of a frictional interface. Phys. Rev. E. 2001, *64*, 0315021–0315024.
71. Hurtado, J.A.; Kim, K.S. Scale effects in friction of single-asperity contacts. I. From concurrent slip to single-dislocation-assisted slip. Proc. R. Soc. (Lond.) A 1999, *455*, 3363–3384.
72. Hurtado, J.A.; Kim, K.S. Scale effects in friction of single-asperity contacts. II. Multiple-dislocation-cooperated slip. Proc. Roy. Soc. (London) A 1999, *455*, 3385–3400.
73. Rigney, D.A.; Hammerberg, J.E. Unlubricated sliding behavior of metals. MRS Bull. 1998, *23*, 32–36.
74. Kennedy, F.A. Frictional heating and contact temperatures. In *Modern Tribology Handbook*; Bhushan, B., Ed.; CRC Press: New York, 2001; Vol. 1, 235–272.
75. Blau, P.J. Introduction to the Special Issue on Friction Test Methods for Research and Applications. Tribol. Int. 2001, *34*, 581–583.
76. Palmer, F. What about friction? Am. J. Phys. 1949, *17*, 181–187.
77. Gnecco, E.; Bennewitz, R.; Gyalog, T.; Meyer, E. Friction experiments on the nanometer scale. J. Phys: Condens. Matter 2001, *13*, 619–642.

78. Berman, A.; Israelachvili, J.N. Surface forces and microrheology of molecularly thin liquid films. In *Handbook of Micro/Nano Tribology*, 2nd ed.; Bhushan, B., Ed.; CRC Press: New York, 1999; 371–432 pp.

79. Denniston, C.; Robbins, M.O. Molecular and continuum conditions for a miscible binary fluid. Phys. Rev. Lett. 2001, 87, 1783021–1783024.

80. Thompson, P.A.; Robbins, M.O. Shear flow near solids: epitaxial order and flow boundary conditions. Phys. Rev. A 1990, 41, 6830–6837.

81. Braun, O.M.; Bishop, A.R.; Röder, J. Multistep locked-to-sliding transition in a thin lubricant film. Phys. Rev. Lett. 1999, 82, 3097–3100.

82. Braun, O.M.; Peyrard, M. Friction in a solid film. Phys. Rev. E 2001, 63, 0461101–04611019.

83. Mak, C.; Krim, J. Quartz-crystal microbalance studies of the velocity dependence of interfacial friction. Phys. Rev. B 1998, 58, 5157–5159.

84. Sørensen, M.R.; Jacobsen, K.W.; Stoltze, P. Simulations of atomic-scale sliding friction. Phys. Rev. B 1996, 53, 2101–2113.

85. Buldum, A.; Ciraci, S.; Batra, I.P. Contact, nanoindentation, and sliding friction. Phys. Rev. B 1998, 57, 2468–2476.

86. Hammerberg, J.E.; Holian, B.L.; Röder, J.; Bishop, A.R.; Zhou, S.J. Nonlinear dynamics and the problem of slip at material interfaces. Physica D 123, 330–340.

87. Röder, J.; Hammerberg, J.E.; Holian, B.L.; Bishop, A.R. Multichain Frenkel–Kontorova model for interfacial slip. Phys. Rev. B 1998, 57, 2759–2766.

88. Röder, J.; Bishop, A.R.; Holian, B.L.; Hammerberg, J.E.; Mikulla, R.P. Dry friction: modeling and energy flow. Physica D 2000, 142, 306–316.

89. Chen, L.H.; Rigney, D.A. Transfer during unlubricated sliding wear of selected metal systems. Wear 1985, 105, 47–61.

90. Germann, T.C.; Hammerberg, J.E.; Holian, B.L. *unpublished*.

91. Fu, X.Y.; Falk, M.I.; Rigney, D.A. Sliding behavior of metallic glass: Part I. Experimental investigations. Wear 2001, 250, 409–419.

92. Fu, X.Y.; Falk, M.I.; Rigney, D.A. Sliding behavior of metallic glass: Part II. Computer simulations. Wear 2001, 250, 420–430.

93. Germann, T.C.; Lomdahl, P.S. Recent advances in large-scale atomistic materials simulations. Comput. Sci. Eng. 1999, 1, 10–16.

94. Mikulla, R.P.; Hammerberg, J.E.; Lomdahl, P.S.; Holian, B.L. Dislocation nucleation and dynamics at sliding interfaces. Mater. Res. Soc. Symp. Proc. 1998, 522, 385–391.

95. Germann, T.C.; Hammerberg, J.E.; Holian, B.L. *unpublished*.

96. Ortiz, M.; Phillips, R. Nanomechanics of defects in solids. Adv. App. Mech. 1999, 36, 1–79.

97. Ortiz, M.; Cuitino, A.M.; Knap, J.; Koslowski, M. Mixed atomistic continuum models of material behavior: the art of transcending atomistics and informing continua. MRS Bull. 2001, 26, 216–221.

98. Shenoy, V.B.; Miller, R.; Tadmor, E.B.; Phillips, R.; Ortiz, M. Quasicontinuum models of interfacial structure and deformation. Phys Rev. Lett. 1998, 80, 742–745.

99. Makarov, P.V.; Schmauder, S.; Cherepanov, O.I.; Smolin, Yu.I.; Romanova, V.A.; Balokhonov, R.R.; Saraev, Yu.D.; Soppa, E.; Kizler, P.; Fischer, G.; Hu, S.; Ludwig, M. Simulation of elastic–plastic deformation and fracture of materials at micro-, meso- and macrolevels. Theor. Appl. Fract. Mech. 2001, 37, 183–244.

100. Makarov, P.V. Localized deformation and fracture of polycrystals at mesolevel. Theor. Appl. Fract. Mech. 2000, 33, 23–30.

101. Makarov, P.V. Microdynamic theory of plasticity and fracture of structurally heterogeneous materials. Russ. Phys. J. 1992, 35, 334–346.

102. Yano, K.; Horie, Y. Discrete-element modeling of shock compression of polycrystalline copper. Phys. Rev B 1999, 59, 13672–13680.

103. O'Connel, S.T.; Thompson, P.A. Molecular dynamics-continuum hybrid computations: a tool for studying complex fluid flows. Phys. Rev. E 1995, 57, R5792–R5795.

104. Broughton, J.Q.; Abraham, F.F.; Bernstein, N.; Kaxiras, E. Concurrent coupling of length scales: methodology and application. Phys. Rev. B 1999, 60, 2391–2404.

105. Hoover, W.G.; Posch, H.A.; Castillo, V.M.; Hoover, C.G. Computer simulation of irreversible expansions via molecular dynamics, smooth particle applied mechanics, Eulerian, and Lagrangian continuum mechanics. J. Stat, Phys. 2000, *100*, 313–326.

106. Sørensen, M.R.; Voter, A.F. Temperature-accelerated dynamics for simulations of infrequent events. J. Chem, Phys. 2000, *112*, 9506–9599.

107. Montalenti, F.; Sørensen, M.R.; Voter, A.F. Closing the gap between experiment and theory: crystal growth by temperature accelerated dynamics. Phys. Rev. Lett. 2001, *87*, 1261011–1261014.

108. Voter, A.F.; Montalenti, F.; Germann, T.C. Extending the time scale in atomistic simulations of materials. Annu. Rev. Mater. Res. 2002, *32*, 321–346.

109. Leng, Y.; Jiang, S. Atomic indentation and friction of self-assembled monolayers by hybrid molecular simulations. J. Chem. Phys. 2000, *113*, 8800–8806.

110. Leng, Y.; Jiang, S. Spanning time scales in dynamic simulations of atomic-scale friction. Tribol. Lett. 2001, *11*, 111–115.

Index